Introduction to the Physics of Lasers

This textbook provides an accessible introduction to laser physics by addressing three key questions: What is special about laser light? What is the physics underlying its generation? And how can laser light be manipulated and used?

The material is presented in a way that helps readers develop an understanding of the underlying physical principles and processes without becoming bogged down in mathematical detail, while still providing enough mathematics for that understanding to be quantitative.

The book begins by describing the special properties of laser light and then focuses on the physics of light-matter interaction to explain how these properties arise. It builds on Maxwell's equations for light and the classical electron oscillator model for atoms, augmented by a quantum-mechanical picture of atomic energy levels, to provide an accurate physical understanding of light-matter interaction. This approach avoids the need for a prior course in quantum theory.

Derivations of key equations are included not only to help readers follow the development of the subject, but also to illustrate that approximation, simplification, and the search for analytic results are integral to understanding complex physical phenomena. Topics such as nonlinear pulse propagation and nonlinear optics are discussed, but only at a level intended to help readers get started in these fields. The choice of topics is limited to atomic and molecular lasers so that the material can be covered in a one-semester course.

Problems, references, and illustrations are included throughout the book, making it suitable as a textbook for advanced undergraduate and graduate students in physics and engineering. Readers do not require prior knowledge of physics much beyond an undergraduate introduction to electromagnetic theory, including Maxwell's equations, polarization in atomic media, and electromagnetic-wave propagation.

Key features:

- Provides a qualitative and quantitative understanding of, and appreciation for, the special properties of laser light
- Explores the quantum statistics of laser light
- Presents an analytical approach to pulse formation

Introduction to the Physics of Lasers

Surendra Singh

CRC Press
Taylor & Francis Group
Boca Raton London New York

CRC Press is an imprint of the
Taylor & Francis Group, an **informa** business

First edition published 2027
by CRC Press
2385 NW Executive Center Drive, Suite 320, Boca Raton FL 33431

and by CRC Press
4 Park Square, Milton Park, Abingdon, Oxon, OX14 4RN

CRC Press is an imprint of Taylor & Francis Group, LLC

ISBN: 978-1-032-94277-3 (hbk)
ISBN: 978-1-032-95299-4 (pbk)
ISBN: 978-1-003-58417-9 (ebk)

DOI: 10.1201/9781003584179

Typeset in LM Roman
by KnowledgeWorks Global Ltd.

Publisher's note: This book has been prepared from camera-ready copy provided by the authors.

न तत्र सूर्यो भाति न चन्द्रतारकं
नेमा विद्युतो भान्ति कुतोऽयमग्निः।
तमेव भान्तमनुभाति सर्वं
तस्य भासा सर्वमिदं विभाति॥

na tatra sūryo bhāti na candratārakaṃ
nemā vidyuto bhānti kuto'yamagniḥ.
tameva bhāntamanubhāti sarvaṃ
tasya bhāsā sarvamidaṃ vibhāti
- Kathopanishad

There the sun shines not, nor the moon and stars,
Nor does this lightning, whence this radiance?
That alone shines, after whom shine all,
Its radiance makes all things shine.

Dedicated to Reeta,
my wife, constant companion, and best friend

Contents

Preface

Laser physics is a fascinating field in modern physics. Over the last six decades, laser physics and lasers have transformed many fields of physics and other sciences. Indeed, many fields came into existence because of lasers. From the vantage point of the mid-1970s, when I started my research in lasers, the performance of today's lasers in power output, frequency and phase stability, spectral range, pulse energy, pulse width, repetition rate, and the degree to which all of these can be controlled is amazing. From nonlinear optics and precision metrology to ultrafast spectroscopy in physics, chemistry, and biology, as well as gravitational-wave detection, lasers have opened new windows into physics at ultrashort spatial and temporal scales. They have also enabled the creation and study of exotic states of matter such as Bose–Einstein condensates and hot, dense matter for inertial fusion. Their use in consumer products, telecommunications, medicine, and surgery has been equally transformative. Lasers still hold the key to new capabilities, applications, and surprises in the future.

This textbook is based on a one-semester course in laser physics that I have taught since the 1990s to senior-level undergraduates and entry-level graduate students in physics, chemistry, mathematics, and engineering. The motivation behind writing this textbook was to provide a logically organized introduction to the physics of lasers using a uniform and accessible notation and formulation that does not require prior knowledge of physics much beyond an undergraduate-level introduction to electromagnetic theory, including Maxwell's equations, polarization in atomic media, and electromagnetic wave propagation. Knowledge of undergraduate optics and Fourier transforms, along with enough background in atomic physics to understand that atoms have discrete energy levels and that light can induce transitions between them, will be helpful. The choice of topics was intentionally limited to allow coverage within a single semester-long course.

The book begins by describing the special properties of laser light and focuses on the physics of light–matter interaction to explain how these properties arise. It builds upon Maxwell's equations for light and the classical electron-oscillator model for atoms, augmented with a quantum-mechanical picture of atomic energy levels, to provide an accurate physical understanding of light–matter interaction. This approach bypasses the need for a formal course in quantum theory. Derivations of key equations are included not only to help students follow the development of the subject, but also to illustrate that approximation, simplification, and the derivation of analytic results are integral to understanding complex phenomena.

Subjects such as nonlinear pulse propagation and nonlinear optics are discussed, but only at a level intended to help students get started in these fields. The choice of topics has been limited to atomic and molecular lasers to allow adequate coverage within a one-semester course. Certain topics, such as fiber lasers and semiconductor lasers, are left to other, more qualified colleagues. Problems, references, and illustrations are included throughout the book.

The first four chapters are devoted to the description of light using Maxwell's equations, electromagnetic waves and beams in homogeneous media, and the design, modes, and losses of laser resonators. In Chapter 5, the electron-oscillator model of atoms is introduced to describe energy exchange between light and atoms (light–matter interaction) and is used to calculate the atomic response to the electromagnetic field. Simple modifications of this response, motivated by physical arguments based on the quantized nature of atomic energy levels, lead to quantum-mechanically correct expressions for atomic susceptibility and gain. An optional Chapter 6 presents a fully quantum-mechanical treatment of atomic susceptibility, mainly to show that the approach based on the classical electron-oscillator model of light–matter interaction leads to the correct equations describing atom–field interaction. This chapter may, in fact, be skipped by readers not interested in this aspect of laser physics.

Chapter 7 uses the results of Chapter 5 to develop the basic equations of laser physics and to introduce the concepts of pumping, population inversion, amplification, gain, and saturation. A number of important and common laser systems are described, without attempting to provide a comprehensive handbook of laser systems.

Chapter 8 discusses laser threshold, power output, and the ways in which the nature of the atomic medium affects these characteristics. Chapter 9 is devoted to the quantum statistics of laser light and the quantum limit to laser linewidth, all in the context of a single-mode laser.

Chapter 10 discusses the complexities of laser dynamics, using spiking, Q-switching, and mode competition as examples. Chapter 11 may be considered a continuation of Chapter 10. It discusses inherently multimode phenomena such as mode locking, short-pulse generation, and pulse propagation. These topics require techniques more advanced than those used in previous chapters. They are introduced in the chapter itself, but only to a level that allows students to grasp the underlying physics.

Finally, Chapter 12 discusses nonlinear optical phenomena, limiting the discussion to processes that affect the operation of the laser itself or are relevant to extending the range of coherent radiation generated by lasers.

Writing a textbook is to rediscover and relearn the subject and, in the process, gain new insights. With those insights, obscure passages and comments in old papers and texts suddenly become meaningful, and one finally understands the authors' thoughts. Over the years, I have had the privilege of learning from great teachers and practitioners of quantum optics and laser physics. I have tried to emulate their style in my teaching. Alas, there can be only one Leonard Mandel and one Jan Hall. I learned a great deal from them. In discussions and conversations, if I asked them a question, the answer would be clear and insightful, but sometimes Jan Hall, especially, would answer the question I should have asked in the first place. Leonard Mandel, Emil Wolf, Robert Boyd, Joseph Eberly, and Carlos Stroud have left an indelible impression through their superb teaching. Many others have been teachers from afar.

I learned a great deal from *Lasers* by Anthony Siegman; *Laser Physics* by M. Sargent III, M. O. Scully, and W. E. Lamb, Jr.; *Principles of Lasers* by O. Svelto; and *Lasers* by Peter Milonni and Joseph Eberly. Discussions with Howard Carmichael, Reeta Vyas, and Greg Salamo over the years have been enlightening and very helpful. The teaching of these great mentors at Rochester and elsewhere, in one way or another, is integrated into this textbook. Is this the best I could do? Perfection is a moving target, and writing has to stop at some point. The errors are entirely mine. If you have suggestions or comments for improvement, I would enjoy hearing from you.

Starting and finishing a textbook is both a physical and an intellectual challenge. Many thanks to Rebecca Hodges-Davies at CRC for continued support and encouragement, to the production staff at KnowledgeWorks Global for help with typography, and to my department chairs, Lin Oliver and Julia Kennefick, for making it possible for me to finish the book.

Fayetteville, 2026 Surendra Singh

Author

Surendra Singh is University Professor of Physics at the University of Arkansas. He received B.Sc. (Honors) and M.Sc. degrees in Nuclear Physics from Banaras Hindu University, India, where he was awarded the Chancellor's Gold Medal, and a Ph.D. in Quantum Optics from the University of Rochester under the mentorship of Professor Leonard Mandel. His doctoral thesis received the Dexter Prize for Outstanding Physics Thesis.

His teaching and research span laser physics, quantum optics, nonlinear optics, electrodynamics, and quantum mechanics. He has contributed experimentally and theoretically to the study of noise in lasers and nonlinear optical systems and to the phase and polarization properties of laser beams. He has also developed laboratory courses in lasers and optics and co-authored a graduate-level textbook on quantum mechanics.

He has served as Chair of the Department of Physics at the University of Arkansas, was a Visiting Fellow at JILA, and has held honorary visiting professorships at IIT Madras and the University of Ulm. He is a Fellow of the American Physical Society and an APS Outstanding Referee.

1

Laser Light

Laser is an acronym for Light Amplification by Stimulated Emission of Radiation. In common usage, the term laser refers more often to a device based on this principle than to the principle itself. The term laser action is often used when referring to the process. The term light is used here in a broad sense to include electromagnetic radiation in the infrared, visible, and ultraviolet regions of the electromagnetic spectrum.

Light generated by lasers has certain special properties compared with light from ordinary sources such as an incandescent bulb or the sun. These properties include a high degree of spatial coherence and directionality, a well-defined spectrum, polarization, and high intensity, which make possible a myriad of scientific, medical, military, and commercial applications of laser light. The development of newer, smaller, and more efficient lasers at different wavelengths continues, as does the expansion of applications of laser light.

The physical principle of stimulated emission, which is responsible for laser action, was introduced by Einstein in 1916–1917. A device called the maser (microwave amplification by stimulated emission of radiation), based on this principle, was first operated in the microwave regime. The laser is an extension of this principle to the infrared, visible, and ultraviolet regions of the electromagnetic spectrum. A summary of the principal developments in the field of lasers follows.[1]

1.1 Major Developments in Laser Physics

1916: A. Einstein introduces stimulated emission as a fundamental process of light-matter interaction, in addition to the already known processes of absorption and spontaneous emission.

1924: Richard Tolman discusses "negative absorption" of light, in which the resulting radiation is coherent with the incident light and leads to amplification.

1928: Rudolf W. Ladenburg confirms the existence of stimulated emission and negative absorption.

1940: V. A. Fabrikant suggests a method for producing population inversion, a necessary condition for maser and laser action.

1950: Alfred Kastler proposes "optical pumping" as a method for orienting paramagnetic atoms or nuclei in the ground state. This work represents an important step toward the development of lasers. Kastler is awarded the 1966 Nobel Prize in Physics

[1]For a more detailed account of the development of our understanding of lasers, see M. Bertolotti, *Masers and Lasers: An Historical Approach*, Hilger (1983); C. H. Townes in Centennial Papers, I.E.E.E. Journal of Quantum Electronics **20**, 545–615 (1984); or J. Hecht, *Laser Pioneers*, Academic Press (1992).

DOI: 10.1201/9781003584179-1

for the discovery and development of optical methods for studying Hertzian resonances in atoms.

1951: Edward Purcell and Robert Pound observe an inverted population of nuclear spin states in a nuclear magnetic resonance experiment.

1952: Nikolay Basov and Alexander Prokhorov describe the principle of the maser (Microwave Amplification by Stimulated Emission of Radiation).

1954: C. H. Townes, J. P. Gordon, and H. J. Zeiger realize the first maser, utilizing a beam of excited ammonia molecules to amplify 24-gigahertz (GHz) microwaves by stimulated emission.

1958: Charles H. Townes and Arthur L. Schawlow introduce the concept of the laser.

1959: Gordon Gould introduces the term laser in a paper, "The LASER: Light Amplification by Stimulated Emission of Radiation."

1960: T. H. Maiman demonstrates laser action in ruby [Nature **187**, 493 (1960)]. The ruby laser is now known to be one of the more difficult laser systems to operate. Sorokin and Stevenson develop the first four-level solid-state laser at IBM. Ali Javan, William Bennett, and Donald Herriott at Bell Labs develop the first helium-neon (He-Ne) gas laser.

1961: Elias Snitzer reports the operation of a neodymium-glass laser. Charles Campbell and Charles Koester demonstrate an early medical use of the laser by destroying a retinal tumor with a ruby laser. P. A. Franken, A. E. Hill, C. W. Peters, and G. Weinreich demonstrate nonlinear optical generation of light by passing pulses from a ruby laser through a quartz crystal, converting red light into ultraviolet light at its second-harmonic frequency.

1962: Scientists at Bell Labs report the first yttrium aluminum garnet (YAG) laser, which continues to play an important role in materials-processing applications. Scientists at General Electric, IBM, and MIT Lincoln Laboratory develop gallium arsenide lasers that convert electrical energy directly into infrared light. F. J. McClung and R. W. Hellwarth develop the laser Q-switching technique to produce laser pulses of short duration and high peak power. Four groups in the United States (M. I. Nathan et al., R. N. Hall et al., T. M. Quist et al., and N. Holonyak and S. F. Bevacqua) nearly simultaneously demonstrate the first semiconductor diode lasers, which operate in pulsed mode at liquid-nitrogen temperatures. Semiconductor diode lasers represent an important step in the development of optical communications, optical storage, optical pumping of solid-state lasers, and many other applications.

1963: L. E. Hargrove, R. L. Fork, and M. A. Pollack report the first mode-locked operation of a laser, using a helium-neon laser with an acousto-optic modulator. Mode locking provides the basis for the generation of femtosecond laser pulses. Herbert Kroemer and, independently, Rudolf Kazarinov and Zhores Alferov propose concepts for semiconductor lasers based on heterostructures, work that contributes to the later development of practical semiconductor lasers. C. K. N. Patel develops the first carbon dioxide laser at Bell Labs.

1964: C. H. Townes, N. G. Basov, and A. M. Prokhorov are awarded the Nobel Prize in Physics for fundamental work in quantum electronics leading to the construction of oscillators and amplifiers based on the maser-laser principle. William B. Bridges develops the first noble-gas ion laser. J. E. Geusic, H. M. Marcos, and L. G. Van Uitert develop the neodymium-doped yttrium aluminum garnet (Nd) laser, which becomes one of the most widely used solid-state lasers, with applications ranging from cutting and welding to medicine and nonlinear optics. C. J. Koester and E. Snitzer develop a neodymium-doped fiber amplifier. Fiber amplifiers subsequently find important applications in optical communications and high-power lasers. Arno Penzias and Robert Wilson use a maser amplifier in their observation of the approximately 3-K cosmic microwave background radiation, providing important evidence for the Big Bang model. They are awarded the Nobel Prize in Physics in 1978.

1965: George C. Pimentel and Jerome V. V. Kasper demonstrate the first chemical laser. Chemical lasers derive their energy from chemical reactions and can reach very high output powers. A. J. DeMaria, D. A. Stetser, and H. A. Heynau report the generation of picosecond laser pulses using a neodymium-glass laser and a saturable absorber.

1966: P. Sorokin and J. R. Lankard demonstrate the first widely tunable organic dye laser, opening important new possibilities in spectroscopy and, subsequently, ultrafast science. C. K. Kao and G. Hockham of Standard Telecommunications Laboratories in England publish a landmark paper showing that optical fibers could transmit optical signals over long distances with sufficiently low loss if the glass were made pure enough. Alfred Kastler is awarded the Nobel Prize in Physics "for the discovery and development of optical methods for studying Hertzian resonances in atoms."

1968: NASA launches the first satellite equipped with a laser.

1969: Under the leadership of American physicist Carroll Alley, NASA's Lunar Laser Ranging experiments begin using retroreflectors placed on the Moon by Neil Armstrong and Buzz Aldrin. Scientists on Earth direct laser pulses toward these retroreflectors and measure the roundtrip travel time, enabling extraordinarily precise measurements of the Earth-Moon distance and tests of gravitational and relativistic theories. D. J. Spencer, T. A. Jacobs, H. Mirels, and R. W. F. Gross develop the first continuous-wave chemical laser. High-power chemical lasers subsequently reach megawatt power levels, leading, among other applications, to proposals for laser weapons.

1970: Nikolai Basov, V. A. Danilychev, and Yu. M. Popov of the Lebedev Physical Institute in Moscow develop an excimer laser. Excimer lasers subsequently become important in applications such as photolithography and laser eye surgery. Zhores Alferov's group at the Ioffe Physical Institute and, independently, Mort Panish and Izuo Hayashi at Bell Labs demonstrate continuous-wave (CW), room-temperature semiconductor lasers, paving the way for practical semiconductor lasers and fiber-optic communications. The world's first laser-driven lighthouse begins operation at Point Danger in Australia. J. Beaulieu develops the transversely excited atmospheric-pressure (TEA) CO_2 laser, which becomes useful in high-power pulsed applications. O. G. Peterson, S. A. Tuccio, and B. B. Snavely develop a CW dye laser, contributing to major advances in high-resolution spectroscopy. Arthur Ashkin demonstrates the acceleration and trapping of small particles by radiation pressure, laying the foundation for the later development of optical trapping and optical tweezers. Robert Maurer, Donald Keck, and Peter C. Schultz at Corning Glass Works produce optical fiber with losses of about 20 dB/km, a breakthrough that helps make optical-fiber telecommunications practical.

1971: Dennis Gabor is awarded the Nobel Prize in Physics for his invention and development of the holographic method.

1974: The first product carrying a Universal Product Code (UPC) is scanned at a supermarket checkout using a laser scanner. E. P. Ippen and C. V. Shank develop a subpicosecond mode-locked CW dye laser, helping to establish the field of ultrafast optical science.

1975: Laser Diode Labs develops a commercial continuous-wave semiconductor laser operating at room temperature. Continuous-wave operation is an important step toward the practical use of semiconductor lasers in optical communications.

1976: John Madey and his group at Stanford University demonstrate the first free-electron laser (FEL). Instead of a conventional gain medium, an FEL uses a beam of relativistic electrons passing through a periodic magnetic structure called an undulator. The resulting interaction generates coherent radiation whose wavelength can be tuned over a wide range by varying the electron energy and the parameters of the undulator.

1977: General Telephone and Electronics transmits live telephone traffic through an optical-fiber system at 6 Mbit/s in Long Beach, California.

1981: Nicolaas Bloembergen and Arthur L. Schawlow are awarded portions of the Nobel Prize in Physics for their contributions to laser spectroscopy; Bloembergen is recognized specifically for the development of nonlinear laser spectroscopy.

1982: Kanti Jain demonstrates excimer-laser lithography, a technology that later becomes important in the manufacture of semiconductor devices. P. F. Moulton develops the titanium-sapphire laser, which subsequently becomes one of the most important broadly tunable solid-state lasers and a principal source of ultrashort laser pulses.

1985: Steven Chu, Claude Cohen-Tannoudji, and William D. Phillips develop methods that contribute to the cooling and trapping of atoms with laser light. Gérard Mourou and Donna Strickland demonstrate chirped-pulse amplification (CPA), in which a short laser pulse is temporally stretched before amplification and subsequently recompressed. This permits much higher pulse energies to be amplified without damaging the gain medium and eventually enables the generation of ultrashort laser pulses with extremely high peak powers.

1987: Arthur Ashkin and coworkers demonstrate the trapping and manipulation of biological particles using optical tweezers, extending optical trapping to biological systems. Ashkin is awarded a share of the 2018 Nobel Prize in Physics for optical tweezers and their application to biological systems. Ophthalmologist Steven Trokel pioneers the use of the excimer laser for corneal surgery. Around this period, the development of the erbium-doped fiber amplifier by researchers including David Payne, Emmanuel Desurvire, and their coworkers provides a major advance toward long-distance fiber-optic communication.

1988: The first transatlantic fiber-optic telecommunications cable, TAT-8, enters service, marking an important milestone in global optical communications. The double-clad fiber laser is developed by E. Snitzer, H. Po, F. Hakimi, R. Tumminelli, and B. C. McCollum, providing an important approach to power scaling in fiber lasers.

1989: Norman F. Ramsey is awarded a share of the Nobel Prize in Physics for the invention of the separated oscillatory fields method and its use in the hydrogen maser and other atomic clocks. Hans G. Dehmelt and Wolfgang Paul share the other half of the prize for the development of the ion-trap technique.

1992: Eric Betzig and coworkers develop high-density magneto-optical data-storage techniques based on near-field optical methods.

1994: Following earlier proposals for intersubband semiconductor lasers, Jerome Faist, Federico Capasso, Deborah Sivco, Carlo Sirtori, Albert Hutchinson, and Alfred Cho at Bell Labs demonstrate the first quantum cascade laser.

1996: S. Nakamura and coworkers develop GaN- and InGaN-based semiconductor lasers, extending practical semiconductor laser operation into the blue and violet regions of the spectrum.

1997: Steven Chu, Claude Cohen-Tannoudji, and William D. Phillips are awarded the Nobel Prize in Physics for the development of methods to cool and trap atoms with laser light. Their research makes possible the study of fundamental phenomena and the measurement of important physical quantities with unprecedented precision. Researchers at MIT demonstrate the first atom laser.

1999: Ahmed Zewail is awarded the Nobel Prize in Chemistry for his studies of the transition states of chemical reactions using femtosecond spectroscopy. Using ultrashort laser pulses in pump-probe experiments, he follows the motion of atoms in molecules during chemical reactions on the femtosecond time scale.

2000: Zhores I. Alferov and Herbert Kroemer share one half of the Nobel Prize in Physics for developing semiconductor heterostructures used in high-speed electronics and optoelectronics. Around this period, John L. Hall and Theodor W. Hänsch make major advances in optical frequency-comb techniques for precision spectroscopy and frequency metrology. This work contributes to their receiving the 2005 Nobel Prize in Physics.

2001: Eric A. Cornell, Wolfgang Ketterle, and Carl E. Wieman are awarded the Nobel Prize in Physics for the achievement of Bose-Einstein condensation in dilute gases of alkali atoms and for early fundamental studies of the properties of the condensates.

2005: John L. Hall and Theodor W. Hänsch are awarded one half of the Nobel Prize in Physics for their contributions to the development of laser-based precision spectroscopy, including the optical frequency-comb technique. Roy J. Glauber is awarded the other half for his contribution to the quantum theory of optical coherence.

2009: Charles K. Kao is awarded one half of the Nobel Prize in Physics for groundbreaking achievements concerning the transmission of light in fibers for optical communication. Willard S. Boyle and George E. Smith share the other half for the invention of an imaging semiconductor circuit—the CCD sensor. The Linac Coherent Light Source (LCLS) at SLAC begins operation, producing intense, ultrashort coherent X-ray pulses and opening new possibilities for imaging matter on atomic length and time scales.

2018: Arthur Ashkin is awarded one half of the Nobel Prize in Physics for optical tweezers and their application to biological systems. Gérard Mourou and Donna Strickland share the other half for their method of generating high-intensity, ultrashort optical pulses.

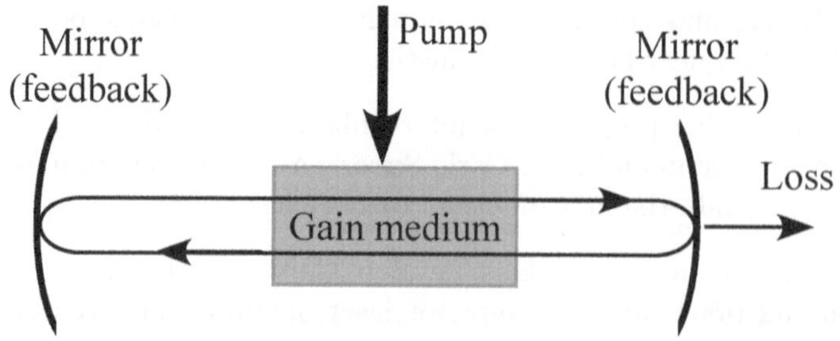

FIGURE 1.1
Basic elements of a laser.

1.2 What is in a Laser?

There are three essential elements of a laser [see Fig. 1.1]:

1. A gain or amplifying medium consisting of atoms, molecules, ions, or charge carriers, together with a mechanism (pump) for exciting these species to higher quantum-mechanical energy states. The energy stored in these excitations can be emitted as light, either spontaneously or by stimulated emission induced by pre-existing light, with the latter process leading to amplification.

2. A suitable arrangement of optical elements (lenses, mirrors, prisms, etc.) or some other mechanism that allows multiple passages of light through the gain medium, thereby providing optical feedback.

3. An output-coupling mechanism for extracting light energy from the device. From the viewpoint of the intracavity light, output coupling represents a useful or "desirable" loss. In addition, nonessential and generally unavoidable losses of light energy occur through absorption, diffraction, scattering, and transmission at mirrors and other optical elements.

These elements come in a great variety of forms and configurations, leading to lasers with a wide range of characteristics that can be used to classify them. For example, based on the physical state of the gain medium, lasers may be classified as solid-state, liquid, or gas lasers. Free-electron lasers constitute a special case in which the gain *medium* is a beam of relativistic free electrons passing through a spatially periodic magnetic field. Laser radiation extends over an enormous wavelength range, from the millimeter-wave and infrared regions to the X-ray region. On the basis of wavelength, lasers may therefore be classified as infrared, visible, ultraviolet (UV), and X-ray lasers.

Depending on their mode of operation, lasers may also be classified as continuous-wave (CW) or pulsed lasers. The output power of CW lasers can range from a few milliwatts to tens of kilowatts for lasers used in materials processing. For pulsed lasers, peak powers can reach the petawatt (PW; 10^{15} W) range and beyond. Pulse durations can also vary widely, from milliseconds to femtoseconds (1 fs = 10^{-15} s) and, with appropriate ultrafast sources, into the attosecond (1 as = 10^{-18} s) regime.

The physical dimensions of lasers also vary widely. In terms of cavity length, for example, laser cavities can range from micrometer dimensions to kilometer scales in specialized systems.

A quick survey of these elements follows. Its aim is not to provide an exhaustive catalogue of the types of lasers available, as any such account would soon become obsolete. Rather, it provides concrete examples of existing systems to illustrate important features and general principles.

1.2.1 Gain Media

Some important laser gain media and representative wavelengths or wavelength ranges include

Fiber lasers (glass fibers doped with rare-earth ions) (approximately 0.7–3 μm)

HCN (hydrogen cyanide) gas far-IR laser (311, 337, 545, 676, 744 μm)

H_2O (water-vapor) far-IR laser (28, 48, 120 μm)

CO_2 (carbon dioxide) gas laser (9.6–10.6 μm)

CO (carbon monoxide) gas laser (5.1–6.5 μm)

Excimer lasers (UV, approximately 190–350 nm)

Nd [neodymium-doped yttrium aluminum garnet (YAG)] laser (1.064 μm)

Titanium-sapphire solid-state laser (approximately 700 nm–1.0 μm)

He-Ne (helium-neon) gas laser (1.15 μm, 633 nm)

Ruby laser (694 nm)

Rhodamine 6G dye laser (approximately 560–640 nm)

Argon-ion laser (488–514 nm)

Metal-vapor lasers (visible and near-infrared)

Semiconductor lasers (near-infrared, visible, and near-ultraviolet)

GaAs semiconductor laser (approximately 870 nm)

Quantum cascade lasers (QCLs) (mid- to far infrared)

Pulsed N_2 discharge laser (337 nm)

Pulsed H_2 discharge laser (approximately 160 nm)

Free-electron lasers (widely tunable, from microwave/infrared to X-ray wavelengths)

1.2.2 Laser Pump

Depending on the gain medium, many different pumping mechanisms are used to supply energy to the gain medium. These include

Gas discharge, including dc, radio-frequency, and pulsed electrical discharges, involving direct electron excitation as well as multistage collisional pumping.

Optical pumping, using flash lamps, arc lamps, semiconductor light-emitting diodes (LEDs), other lasers, and even direct sunlight.

Chemical pumping, involving processes such as chemical reactions, photolysis, and combustion.

Direct electrical pumping, including high-voltage electron beams directed into high-pressure gas cells and direct current injection into semiconductor lasers.

Nuclear pumping, in which energy released by nuclear reactions is used to excite a laser medium.

Gas-dynamic pumping, in which a preheated gas, typically heated by a chemical reaction or electrical discharge, undergoes rapid expansion through a supersonic nozzle to produce the population conditions required for laser action.

Plasma pumping, in which hot, dense plasmas created by plasma pinches, focused high-power laser pulses, or intense electrical pulses provide the excitation required for laser action.

1.2.3 Optical Feedback

The principal mechanism for providing feedback in lasers involves optical cavities or resonators. Resonators store electromagnetic energy. At microwave frequencies, these resonators are often closed metallic cavities, whereas at optical frequencies they can be open. Open resonators are particularly useful at optical frequencies because they provide strong discrimination among the very large number of possible electromagnetic modes, allowing only a relatively small set of low-loss modes to build up through stimulated emission.[2]

The presence of boundaries in resonators, also referred to as cavities in laser physics, allows only certain field configurations, defined by their frequencies and spatial variations, to exist. These allowed field configurations are called modes of the resonator. In open resonators, modes corresponding to propagation away from the resonator axis experience very large losses because light emitted into such modes eventually escapes from the resonator. Only a group of *paraxial* modes, in which the electromagnetic energy remains localized near the resonator axis, can experience significant buildup through interaction with the gain medium.

Even these modes experience some losses due to diffraction, absorption, scattering, and transmission through the end mirrors. As a result, the definition of a mode as a strictly stationary field configuration cannot be applied to an open resonator. However, modes with quasi-stationary (long-lived) field patterns do exist. These modes experience very little loss of electromagnetic energy during one cavity roundtrip, so that energy emitted into them can remain in the cavity for many roundtrips.

Physically, the presence of a cavity increases the effective interaction length of the gain medium by allowing light to pass through it repeatedly. The cavity also determines the spatial characteristics of the laser beam, while its resonant modes determine the possible frequencies of oscillation. Examples of laser cavities include two or more high-reflectance mirrors arranged around a gain medium to form a closed optical path that allows light to pass repeatedly through it. The path may be unidirectional, as in a ring laser, or may retrace itself, as in a standing-wave laser. Edge-emitting semiconductor lasers receive feedback from reflections at the boundaries of the semiconductor medium. Cavities are not required for all lasers. Examples include so-called "mirrorless lasers" and random lasers.

[2] At optical frequencies, the density (number per unit frequency interval per unit volume of the resonator) of modes near a frequency ν is very large, $8\pi\nu^2/c^3$ [see Sec. 2.5]. If the atoms interact with modes in a frequency interval $\delta\nu$ (typically 10^9 Hz or more), the number of available modes per unit volume, $8\pi\nu^2\delta\nu/c^3$, is enormous. Without sufficient mode discrimination, the available energy would be distributed among many modes, making it difficult for any one mode to grow to a significant amplitude.

1.2.4 Losses

Mechanisms that lead to loss of light energy stored in the cavity include absorption in the gain medium, mirrors, and their coatings; diffraction and scattering by optical elements inside the resonator; and transmission through the mirrors. Although loss is generally detrimental to achieving laser action, not all loss is undesirable. In fact, the controlled loss of light energy through a partially transmitting mirror provides the useful output beam of a conventional laser. This process, known as *output coupling*, extracts a fraction of the intracavity light while allowing the remaining light to continue circulating in the resonator.

1.3 Basic Principles

The physical principles underlying laser operation may appear straightforward, but our progress in understanding them has been rather tortuous. The history of the laser usually begins in 1916–1917, when Einstein introduced stimulated emission as a fundamental process of light-matter interaction. Einstein considered thermodynamic equilibrium between the radiation field and a collection of atoms or oscillators with quantized energy levels at a temperature T and found that the Planck formula for the blackbody energy distribution could be derived only if, in addition to absorption and spontaneous emission, stimulated emission was included as a fundamental process of light-matter interaction.

From the perspective of laser physics, it is interesting to note that Einstein did not point out that stimulated emission could lead to coherent amplification of light. Experimental evidence for the existence of stimulated emission came in 1928. Thereafter, interest in stimulated emission waned because it was widely held that gains much larger than unity would be required for conceivable applications, and no methods were known for preparing a system sufficiently far from thermal equilibrium to generate significant optical gain. In addition, few physicists working in this area were familiar enough with electronic oscillators to appreciate the role of feedback in achieving oscillation in a system exhibiting gain.[3]

Things changed in 1954, when Townes, Gordon, and Zeiger demonstrated the ammonia maser ("microwave amplification by stimulated emission of radiation"), based on the amplification of microwaves by stimulated emission from excited ammonia molecules. The extension of the maser concept to the optical domain was not immediate, but in 1958 Schawlow and Townes proposed that two mirrors facing each other could form an open resonator and perform, at optical frequencies, the function of the closed resonant cavity used at microwave frequencies. The intense experimental activity that followed led to Maiman's demonstration of the first laser in 1960. Operation of the ruby laser was soon followed by demonstrations of laser action in a helium-neon gas mixture by Javan and coworkers and in carbon dioxide by Patel.

Among the striking advances that followed were the development of high-power lasers employing gain media such as neodymium-doped glass and doped optical fibers, and tunable systems employing dyes or solid-state materials such as titanium-doped sapphire. The semiconductor revolution has yielded lasers remarkable for their miniaturization, efficiency, and low cost, enabling laser technology to become widely used even in mass-produced consumer products such as CD and DVD players, barcode readers, and printers. The development of high-power fiber lasers has also revolutionized industrial applications of lasers, including cutting, welding, and marking.

[3]This historical fact lingers in our use of the term LASER to describe the device; the laser is not merely a light amplifier but a light oscillator—but by all accounts, the laser is a winner!

Lasers are examples of open, interacting light-matter systems. The term matter can refer to electrons, atoms, molecules, harmonic oscillators, and other entities that can exchange energy with light (or, more generally, with an electromagnetic field). We will use atoms as a generic term for these material entities. Interaction refers to the exchange of energy between light and atoms, while an open system is one that also exchanges energy with its environment. In describing the behavior of lasers, atomic energy must be considered quantized. Logically, therefore, light energy should also be quantized, and a fully rigorous treatment would require a quantum theory of light-matter interaction. However, such a comprehensive formulation of laser theory would be unnecessarily complex for much of what we wish to understand. The aim of this book is to introduce the fundamental concepts involved in laser operation and to develop an understanding of the physics that gives us remarkable control over the properties of laser light. Therefore, instead of becoming entangled in "beautiful but irrelevant" details, we will simplify and approximate the real world in order to understand the underlying principles and procedures.

We will use Maxwell's equations to describe light, and either the Schrödinger equation or Newton's second law together with the Lorentz force law to describe the behavior of atoms. A fully quantum-mechanical theory of light is needed only for a small subset of the remarkable properties of laser light [Chapters 6 and 9]. Modifications to the equations describing the behavior of light and matter that arise from a quantum-mechanical description will be motivated by physical arguments and by relating rates, parameters, and variables to experimentally measured quantities. This approach is well illustrated by Einstein's derivation of the spectral energy distribution of blackbody radiation, which we present in outline next.

1.3.1 Einstein's Theory of the Blackbody Spectrum

Einstein considered three basic processes involved in the exchange of energy between atoms and electromagnetic fields. These processes can be understood in terms of transitions between quantized energy levels of an atom or an oscillator [Fig. 1.2]:

- The absorption of a quantum (photon) of energy $h\nu_{21} = E_2 - E_1$ by an atom causes a transition from a lower energy level $|E_1\rangle$ to an upper energy level $|E_2\rangle$. Absorption is clearly a process induced (stimulated) by the presence of a field and results in attenuation of the energy (or amplitude) of the field.

- The spontaneous emission of a photon of energy $h\nu_{21}$ by an atom causes it to return from the upper level $|E_2\rangle$ to the lower level $|E_1\rangle$. The phase, polarization, and direction of propagation (the direction of the wave vector $\boldsymbol{k}$) of the emitted radiation are not determined by any field that may be present. Spontaneous emission thus produces radiation that is incoherent with the existing field. Fluorescence from excited atoms is an example of this type of radiation.

- Stimulated emission occurs when an atom in the upper level $|E_2\rangle$ interacts with an incident field whose frequency is near the resonance frequency $\nu_{21} = (E_2 - E_1)/h$, inducing a transition to the lower level $|E_1\rangle$. The radiation added by stimulated emission is coherent with the stimulating field and has the same frequency, polarization, phase, and direction of propagation. Stimulated emission can therefore produce coherent amplification of light and is fundamental to laser action.

Of the three processes listed above, absorption and spontaneous emission of light were known at the turn of the twentieth century. The third, stimulated emission, was introduced in 1916–1917 by Einstein to explain the experimentally observed spectral distribution of

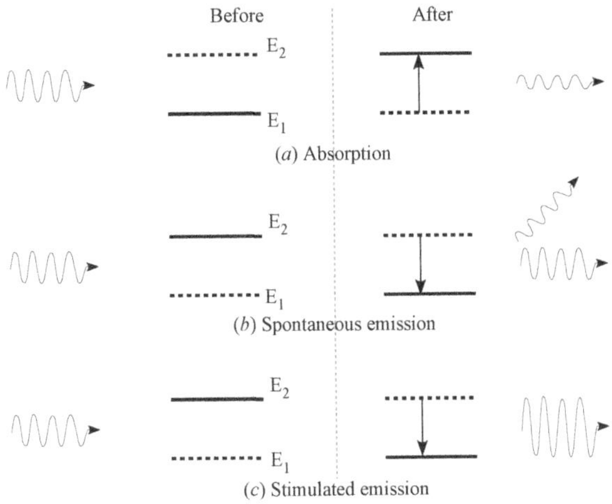

FIGURE 1.2
Three radiative energy-exchange processes in light-matter interaction.

electromagnetic energy in blackbody radiation,

$$u(\nu,T) = \frac{8\pi h\nu^3}{c^3}\frac{1}{e^{\beta h\nu}-1} = \left(\frac{8\pi\nu^2}{c^3}\right)\cdot h\nu\cdot n(\nu,T)\,. \tag{1.1}$$

Here $u(\nu,T)$ is the spectral energy density (J/m^3 Hz) of electromagnetic radiation, h is Planck's constant, T is the absolute temperature of the blackbody, and $\beta = 1/k_B T$, where k_B is Boltzmann's constant.

The second form of the spectral distribution makes its physical meaning particularly transparent. The term in parentheses is the spectral density of field modes (number of modes per unit volume per unit frequency interval) near frequency ν [see Sec. 2.5]; $h\nu$ is the energy of one quantum (photon) in a mode of frequency ν; and $n(\nu,T) = 1/(e^{\beta h\nu}-1)$ is the mean number of quanta in that mode. Thus, $u(\nu,T)d\nu$ is energy density contained in all modes with frequencies between ν and $\nu + d\nu$.

Einstein approached the derivation of the Planck distribution by considering the exchange of energy between light and matter in terms of three fundamental processes: absorption, spontaneous emission, and stimulated emission. The radiation field was represented by a collection of waves (modes of the cavity) of definite frequencies, while the atoms were described by a set of discrete (quantized) energy levels. The radiation and atoms in the blackbody cavity continually exchange energy and are in thermal equilibrium at temperature T.

Corresponding to a radiation frequency ν, consider two atomic levels, $|E_2\rangle$ and $|E_1\rangle$, with $E_2 - E_1 = h\nu$. Transitions between these levels are accompanied by the absorption or emission of energy $h\nu$. If the numbers of atoms in the two states are denoted by N_2 and N_1, respectively, absorption and emission continually change these populations. The absorption of photons by atoms in state E_1 increases N_2 and decreases N_1 at a rate given by

$$\left(\frac{dN_2}{dt}\right)_{\text{absorption}} = B_{12}u(\nu,T)N_1 = -\left(\frac{dN_1}{dt}\right)_{\text{absorption}}\,. \tag{1.2}$$

This rate is proportional to the population N_1 of atoms available to absorb radiation and to the energy density of the field at frequency ν, with a constant of proportionality $B12$.

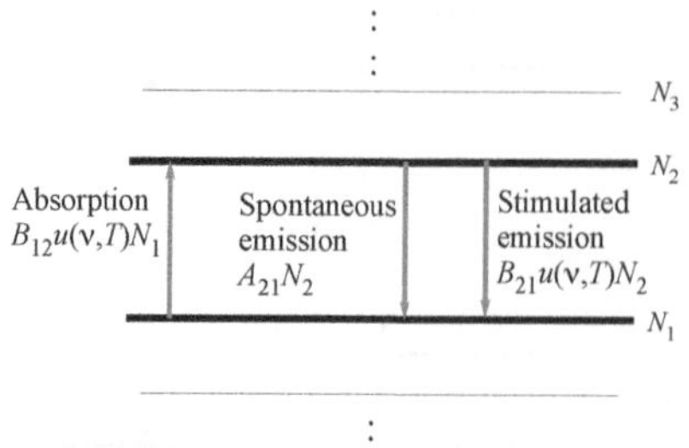

FIGURE 1.3
Radiative absorption and emission rates connecting populations N_1 and N_2 of levels E_1 and E_2, respectively.

Similarly, stimulated emission, which is the reverse of absorption, will decrease N_2 and increase N_1 at a rate given by

$$\left(\frac{dN_2}{dt}\right)_{\text{stimulated emission}} = -B_{21}u(\nu,T)N_2 = -\left(\frac{dN_1}{dt}\right)_{\text{stimulated emission}} . \tag{1.3}$$

This rate is likewise proportional to the energy density of the stimulating field and to the population N_2 of atoms available to undergo stimulated emission, with a constant of proportionality $B21$. Finally, the excited state population N_2 also decreases due to spontaneous emission at a rate

$$\left(\frac{dN_2}{dt}\right)_{\text{spontaneous emission}} = -A_{21}N_2 , \tag{1.4}$$

which is independent of the strength of the radiation field. Figure 1.3 shows these three radiative processes changing the populations of the two levels. Similar equations can be written for levels 2 and 3, or any other pair of atomic levels interacting with wave fields of appropriate frequencies.

Adding Eqs. (1.2)–(1.4) yields the overall rate of change of level populations due to radiative processes connecting the two levels

$$\frac{dN_2}{dt} = -A_{21}N_2 - B_{21}u(\nu,T)N_2 + B_{12}u(\nu,T)N_1 = -\frac{dN_1}{dt} . \tag{1.5}$$

In equilibrium, the net flow of population from level 1 to level 2 must be exactly balanced by the net population flow from level 2 to level 1. Such detailed balance for radiation-induced processes will hold between any two levels. Thus, in equilibrium, the net time rate of change of each level population must be zero. Setting $dN_2/dt = 0 = dN_1/dt$ then yields the ratio of the populations in states 2 and 1 in equilibrium:

$$\frac{N_2}{N_1} = \frac{B_{12}u(\nu,T)}{A_{21} + B_{21}u(\nu,T)} . \tag{1.6}$$

When the atoms and radiation field are in thermodynamic equilibrium, this ratio of populations is given by the Boltzmann factor

$$\frac{N_2}{N_1} = \frac{g_2}{g_1}e^{-\beta h\nu} , \tag{1.7}$$

where g_2 and g_1 are the degeneracies (numbers of states having the same energy) of levels E_2 and E_1, respectively. For example, in a hydrogen atom, there are $2n^2$ states having the same energy E_n, which depends only on the principal quantum number n. The concept of degeneracy appears in other contexts as well, such as in defining the density of states at energy E. Equating the right-hand sides of Eqs. (1.6) and (1.7), the spectral energy density is given by

$$u(\nu, T) = \frac{A_{21}}{B_{21}} \cdot \frac{1}{\frac{B_{12}g_1}{B_{21}g_2} e^{\beta h\nu} - 1} . \tag{1.8}$$

For this equation to take the form of the experimentally observed Planck distribution (1.1) for the spectral energy density, the parameters appearing in this formula must be related by

$$B_{12}g_1 = B_{21}g_2 , \tag{1.9a}$$

$$A_{21} = \frac{8\pi h\nu^3}{c^3} B_{21} . \tag{1.9b}$$

With these relations among the coefficients, Einstein's approach yields the same formula for the blackbody spectral energy density that Planck had derived. The success of Einstein's approach established stimulated emission as a fundamental process of light-matter interaction, which until then had not been recognized as such. It is this process that allows the amplification of light, a prerequisite for laser action. It is important to note that the relations between radiative transition rates depend on the quantized nature of material energy levels, not on whether the matter consists of atoms, molecules, or oscillators.

Equation (1.9) relates the three radiative energy-exchange processes in light-matter interaction. The first relation shows that the coefficients for stimulated emission and absorption connecting two levels of an atom are related by the degeneracies of the two levels. For levels having the same degeneracy, the stimulated emission and absorption coefficients are equal. The second relation connects stimulated emission to spontaneous emission. If we use the second form for the spectral energy density from Eq. (1.1), the total radiative emission rate from level 2 can be written as

$$A_{21} + B_{21}u(\nu, T) = A_{21}\left[1 + n(\nu, T)\right] . \tag{1.9c}$$

The form of this equation is the same as that obtained from a fully quantum-mechanical treatment of the radiation field. It shows that radiative emission consists of a spontaneous component and a component stimulated by the presence of quanta $n(\nu)$ in modes near the transition frequency ν. Einstein's approach, which assumed quantized atoms and introduced stimulated emission as a fundamental process of radiative interaction, thus anticipated an important result of the quantum theory of radiation.

Planck's earlier derivation of the blackbody spectral energy distribution introduced the idea that energy is exchanged between matter and radiation in discrete amounts, or quanta, each of magnitude $h\nu$. The subsequent development of quantum theory led to the description of the electromagnetic field itself in terms of quanta of energy $h\nu$, which we now call photons.

The coefficients A_{21} and B_{21} are known as the Einstein A and B coefficients. They are characteristic of the atomic transition and can be determined experimentally. They determine how an atom responds to radiation from any source, not just blackbody radiation. Einstein's approach does not give explicit expressions for these coefficients but establishes relations among the radiative transition rates involving them. These relations, given by Eqs. (1.9), are confirmed by detailed quantum-mechanical calculations, which also yield explicit expressions for these coefficients.

From the foregoing, we can gain some insight into what is required to make a laser work. For electromagnetic radiation to be amplified, the rate of stimulated emission must exceed that of absorption. For levels of equal degeneracy, this requires the population of the upper level to exceed that of the lower level, $N_2 > N_1$. According to Eqs. (1.6) and (1.7), such a population distribution is not possible in thermodynamic equilibrium. Amplification of radiation therefore requires an interacting light-matter system that is not in thermodynamic equilibrium. A laser is a prime example of such a system.

Example 1.1 *A He-Ne laser operating at 632.8 nm emits a 1.0-mW beam of 1.0-mm diameter. The laser linewidth is 150 MHz, and the transmission of the output mirror is* $T = 0.01$. *The beam inside the cavity is nearly uniform and has approximately the same diameter as the output beam. Assume that the power is uniform across the beam cross-section.*

(a) *Estimate the ratio of stimulated to spontaneous emission rates,* $B_{21}u(\nu)/A_{21}$. *[Recall that light intensity (power density) is related to energy density by* $I = (\text{energy density}) \times (\text{speed of light})$*, and the spectral power density is related to the spectral energy density by* $\frac{dI}{d\nu} \approx \frac{I}{\Delta\nu} = u(\nu)c$.

(b) *What would be the effective temperature of blackbody radiation having the same spectral energy density at this frequency?*

The ratio of the Einstein A and B coefficients at the transition frequency ν is

$$\begin{aligned}\frac{A_{21}}{B_{21}} &= \frac{8\pi\nu^2 h\nu}{c^3} = \frac{8\pi h}{\lambda^3} = \frac{16\pi^2\hbar c}{c\lambda^3}, \\ &= \frac{16\pi^2(197 \times 1.60 \times 10^{-19} \times 10^{-9})}{(3.00 \times 10^8)(0.6328 \times 10^{-6})^3} \approx 6.57 \times 10^{-14}\,\text{J/m}^3\text{·Hz},\end{aligned}$$

so that

$$\frac{B_{21}}{A_{21}} \approx 1.52 \times 10^{13}\,\text{m}^3\text{·Hz/J}.$$

The output power is related to the intracavity power by $P_{\text{out}} = TP_{\text{in}}$, which gives $P_{\text{in}} = \frac{P_{\text{out}}}{T} = 10^2 P_{\text{out}} = 0.10$ W. The spectral energy density in the laser cavity is therefore

$$\begin{aligned}u_L(\nu) &= \frac{\text{cavity power density}}{(\text{bandwidth}) \cdot c} = \frac{10^2(10^{-3})/\left[\pi(0.5 \times 10^{-3})^2\right]}{(1.5 \times 10^8)(3.00 \times 10^8)}, \\ &\approx 2.83 \times 10^{-12}\,\text{J/m}^3\text{·Hz}.\end{aligned}$$

Thus, the ratio of the stimulated to spontaneous emission rates is

$$\frac{B_{21}}{A_{21}}u_L(\nu) = \left(1.52 \times 10^{13}\,\text{m}^3\text{·Hz/J}\right)\left(2.83 \times 10^{-12}\,\text{J/m}^3\text{·Hz}\right) \approx 43.$$

Thus, under these conditions, stimulated emission occurs at a rate about 43 times the spontaneous emission rate.

Since $\frac{B_{21}u(\nu)}{A_{21}} = \bar{n}$, the same stimulated-to-spontaneous emission ratio requires $\bar{n}_{BB} \approx 43$. Therefore,

$$\begin{aligned}\bar{n}_{BB} &= \frac{1}{e^{h\nu/k_B T} - 1} \approx 43, \\ e^{h\nu/k_B T} - 1 &\approx \frac{1}{43} \approx 0.0233, \\ \frac{h\nu}{k_B T} &= \ln(1.0233) \approx 0.0230.\end{aligned}$$

At $\lambda = 632.8$ nm, $h\nu = \frac{1240}{632.8}$ eV ≈ 1.96 eV, and therefore[4]

$$T = \frac{h\nu}{k_B(0.0230)} = \frac{(2\pi\hbar c/\lambda)}{(k_B 300)/300} \frac{(1240/632.8)(300)}{(k_B 300)(0.0230)} =\approx 9.9 \times 10^5 \text{ K}.$$

Thus, blackbody radiation would require an effective temperature of approximately 10^6 K to produce the same spectral energy density at 632.8 nm as that estimated inside this modest He-Ne laser cavity.

Quantum-mechanical calculations of transitions between atomic levels under such conditions would require the atomic wave functions. Such calculations are complicated by the fact that an atom in a laser medium is not isolated but is subject to external and internal fields, interactions with neighboring atoms, collisions with other atoms in a gas or with lattice vibrations (phonons) in a solid, and other perturbations. Einstein's approach of using rate equations to describe transitions between quantum states bypasses many of these complicated quantum-mechanical calculations. It provides a key to analyzing nonequilibrium systems with rate equations: transitions between atomic levels induced by other processes can be treated in much the same manner as radiative transitions by relating the corresponding rates to experimentally measured quantities. This is not to minimize the importance of quantum mechanics in laser theory, but to point out that many features of light-matter interaction can be discussed and understood using a simpler phenomenological approach based on experimentally measured quantities. As will be seen in Chapter 5, this approach yields the laser gain equation in a relatively simple manner, and when laser gain is combined with optical feedback, laser oscillation emerges naturally. To be sure, not everything about lasers can be predicted and understood using a rate-equation approach. The statistical properties of laser light, for example, require a quantum-mechanical treatment of light-matter interaction. Even these properties, however, can often be motivated by physical arguments [Chapter 9].

1.4 Laser Light Characteristics

The light output from a laser is electromagnetic radiation and is not fundamentally different from the light emitted by other sources of electromagnetic radiation. Its characteristics, however, differ in several important ways from those of light emitted by thermal and other conventional sources. We will briefly review the characteristics of laser light that distinguish it from light produced by these other sources. The numbers stated below for lasers should not be taken as the final word; they continue to change as progress is made in improving laser performance.

1.4.1 Monochromaticity

The light emitted by a laser can have a high degree of spectral purity. This means that its frequency or wavelength is relatively well defined, so that the output field from an ideal single-frequency laser can be considered a nearly constant-amplitude sinusoidal wave. Two factors contribute to the spectral purity of laser light. First, light emission from atoms

[4] As discussed in Sec. 1.6, the temperature in this case can be evaluated by expressing photon energy as $h\nu = \frac{2\pi\hbar c}{\lambda} = \frac{1240}{\lambda(\text{in nm})}$ eV, the Boltzmann constant as $k_B = \frac{k_B(300\,\text{K})}{300\,\text{K}} = \frac{1}{39}\frac{1}{300\,\text{K}}$ eV, so that $T = \frac{h\nu}{k_B(0.0230)} = \frac{(1240/632.8)(300\,\text{K})}{(1/39)(0.0230)}$ K which yields the given value.

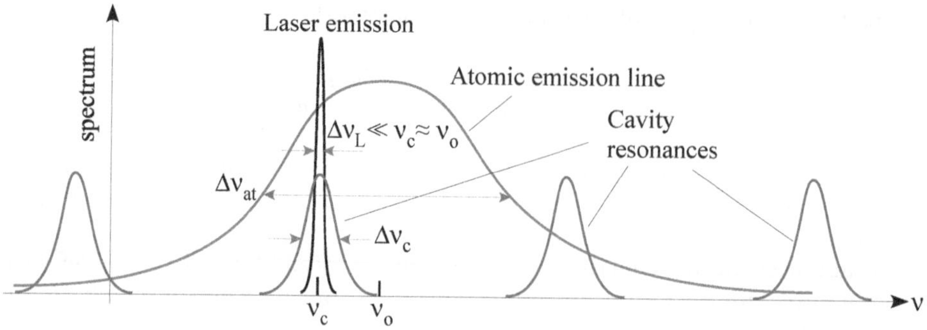

FIGURE 1.4
Relation between the spectral profiles of atomic emission, cavity resonance, and laser light, as labeled. The linewidth $\Delta\nu$, usually defined as the full width at half maximum (FWHM), is a measure of the width of the spectral distribution.

occurs over a narrow range of frequencies around an atomic transition frequency $\nu_o = (E_2 - E_1)/h$. This range of frequencies defines the atomic linewidth $\Delta\nu_{\text{at}}$ [see Fig. 1.4]. Consequently, only electromagnetic waves with frequencies sufficiently close to the atomic transition frequency can interact strongly with the atoms and be amplified. Second, the laser cavity forms a resonant structure. This means that strong optical feedback inside the cavity occurs only over narrow ranges of frequencies centered on certain discrete frequencies, called the resonance frequencies ν_c of the cavity. The narrow range of frequencies around each cavity resonance defines the cavity linewidth. Thus, amplification of light by stimulated emission can lead to laser oscillation at cavity resonance frequencies that lie within the atomic linewidth around the transition frequency. The resulting laser linewidth can be many orders of magnitude narrower than the atomic linewidth. The primary reason for the high spectral purity of laser light is the optical resonator (optical cavity), rather than the atomic transition alone. The laser cavity permits the buildup of field amplitude at certain discrete cavity resonance frequencies, while the atoms primarily provide gain at these frequencies.

The spectral width of a *single-frequency* laser can range from $\Delta\nu_{_L} = 10^6$ Hz for moderately stabilized lasers to $\Delta\nu_{_L} = 10$ Hz or less for well-stabilized lasers. Since visible light has frequencies of order $\nu_L = 5 \times 10^{14}$ Hz, a linewidth of 10 Hz corresponds to a fractional spectral width $\Delta\nu_{_L}/\nu_{_L} = 2 \times 10^{-14}$. Its inverse, the ratio $\nu_{_L}/\Delta\nu_{_L} = 5 \times 10^{13}$, provides a measure of the spectral purity of the laser output: the larger this ratio, the greater the spectral purity. Such large ratios are difficult to achieve with mechanical or electronic oscillators. In contrast, thermal sources such as the sun and incandescent solids generally emit broadband spectra. Some conventional sources, such as discharge lamps, emit only a few spectral lines or narrow bands of wavelengths, but the spectral widths of the light emitted by even the best of these sources are generally limited by the linewidths of the atomic transitions, which typically range from 10^8 to 10^{11} Hz. Table 1.1 compares the spectral purity of light emitted by several different sources [FWHM ≡ full width at half maximum].

The ultimate limit on the spectral purity of a free-running laser is set by quantum noise associated with spontaneous emission from atoms in the gain medium. Approaching this fundamental limit, however, requires careful control of numerous other sources of frequency noise and is possible only in highly stabilized laser systems.

TABLE 1.1
A comparison of spectral purity of different light sources

Light Source	Peak Wavelength	FWHM ($\Delta\lambda$)	FWHM ($\Delta\nu$)
He:Ne Laser	633 nm	10^{-8} nm	7.5×10^{3} Hz
Cadmium low pressure lamp	644 nm	10^{-3} nm	9.4×10^{8} Hz
Sodium discharge lamp	590 nm	0.1 nm	9×10^{10} Hz
Blackbody radiator at 5800 K	500 nm	600 nm	10^{14} Hz

1.4.2 Ultrashort Pulses

It is possible to produce light pulses of extremely short duration. Because of the Fourier-transform relationship between the temporal and spectral extents of physical signals, the ability to produce short pulses may be considered complementary to the monochromaticity of laser radiation. A laser can, in principle, produce radiation with a very narrow spectral width, but pulses of very short duration can be produced only by lasers having a sufficiently broad gain bandwidth. For gas lasers with relatively narrow gain bandwidths, laser pulse widths typically range from about 100 ps to 1 ns, whereas for some solid-state and liquid lasers with gain bandwidths 10^3–10^5 times larger, pulse widths can be as short as about 10 fs.

1.4.3 Coherence

If we picture the light wave from a laser as a sine wave, its amplitude and phase will, in fact, vary slowly but randomly in space and time.[5] Coherence refers to how well the values of these variables at different space-time points are correlated. While it is clear that a statistical description is required for a rigorous definition of coherence, such a treatment is not necessary here; a simple physical picture is sufficient to appreciate its meaning and significance.

Coherence is a measure of how well the values of, say, the phase (or amplitude) at two different space-time points maintain their relationship and how well we are able to predict their evolution in space and time. For example, if the difference between the phases of the field at two points in space remains constant in time, we say that the wave field has perfect spatial coherence between the two points. If this is true for any two points in the wave field, we say that the wave has perfect spatial coherence. In practice, a high degree of spatial coherence extends only between points separated by a finite distance. We will consider the temporal and spatial aspects of coherence separately.

Temporal coherence refers to strong correlations between the phase of a light wave at different times. If the difference in wave phase at two instants t and $t+\tau$ remains predictable as t varies, we say that the wave field has temporal coherence over a time interval τ. For real wave fields, this is true only for τ less than some characteristic interval τ_c, called the coherence time of the wave. We may picture τ_c as the time interval over which the wave phase remains predictable and $1/\tau_c$ as a measure of the rate at which phase coherence is lost. A high degree of temporal coherence in laser light means that the amplitude and phase of the nearly sinusoidal wave remain predictable over time intervals of order τ_c, which can encompass millions of optical cycles.

[5] If we write the electric field of a laser as the real part of a complex signal $\boldsymbol{\mathcal{E}}(\boldsymbol{r},t) = \mathcal{R}e[\boldsymbol{E}_o(\boldsymbol{r},t)e^{i(kz-\omega t)}]$, the phase here refers to the phase of the complex amplitude $\boldsymbol{E}_o(\boldsymbol{r},t)$ and the amplitude to its absolute value.

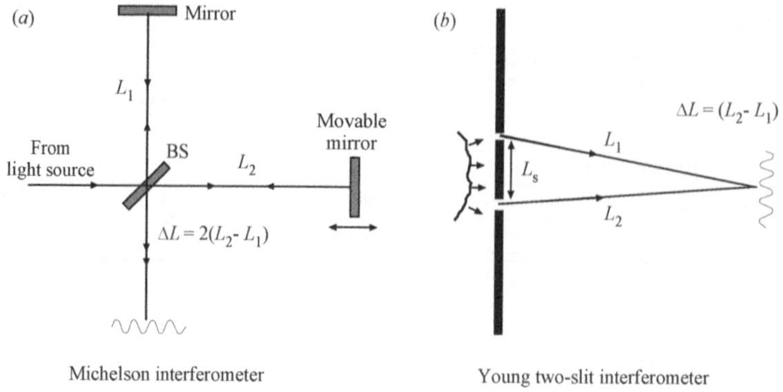

FIGURE 1.5
Coherence length can be measured using a Michelson interferometer (a), and transverse coherence length can be measured using Young's two-slit interferometer (b).

For stationary beams (whose statistical properties do not change with time), Fourier-transform relations show [see Sec. 4.3.4] that the coherence time τ_c is inversely related to the spectral width $\Delta\nu$. With the definitions used here, we write approximately

$$\tau_c \approx \frac{1}{2\pi\Delta\nu}. \tag{1.10}$$

The precise numerical factor depends on the definition of coherence time and on the spectral lineshape. For stationary periodic fields such as those from a mode-locked laser, the relationship between τ_c and the signal bandwidth is more subtle; the coherence time is not simply the inverse of the oscillation bandwidth and may, in fact, be much longer than $1/\Delta\nu_{\rm osc}$.

The distance traversed by light in one coherence time τ_c is called the (longitudinal) coherence length

$$\ell_c = c\tau_c. \tag{1.11}$$

The coherence time τ_c of laser light may be thought of as the characteristic time interval over which the laser light approximates a pure sine wave, and ℓ_c may similarly be thought of as the characteristic length over which the wave retains a predictable phase.

We can measure the coherence length of a light beam by splitting it and recombining the component beams after they have traveled different optical paths, as in Michelson or Mach-Zehnder interferometers [Fig. 1.5(a)]. The coherence length is of the order of the largest optical path difference over which high-contrast interference fringes can be observed. For lasers, ℓ_c can range from tens of meters to thousands of kilometers or more, depending on the laser linewidth, whereas for thermal sources, even those dominated by line emission, it is generally much smaller, as illustrated in the following table:

Source	τ_c (s)	ℓ_c (m)
He:Ne laser	$10^{-4} - 10^{2}$	$3 \times 10^{4} - 3 \times 10^{10}$
Sodium lamp	10^{-10}	3×10^{-2}

Spatial coherence refers to strong correlations between the light fields at different points in a plane transverse to the direction of propagation and is therefore sometimes referred to as transverse coherence. We can think of transverse coherence in terms of a two-slit interference experiment [Fig. 1.5(b)]. The transverse coherence length is the characteristic largest separation between two pinholes which, when illuminated by the same light field, produce high-contrast interference fringes. Another way to picture transverse spatial coherence is as the transverse spatial extent over which the field can be considered part of the same phasefront. At the output of a laser oscillator, the laser beam can have a very high degree of transverse spatial coherence. This property is also a consequence of the presence of the laser cavity. For thermal sources, transverse spatial coherence generally does not extend over distances much larger than a few wavelengths unless the source subtends a very small angle.

The concept of coherence outlined above is referred to as *first-order coherence*. Higher-order coherence, involving correlations of quantities such as light intensity, provides additional ways of distinguishing laser light from light emitted by ordinary thermal sources [Chapter 9].

1.4.4 Directionality or Collimation

Thermal sources emit light in many directions and generally over a broad range of wavelengths. We can capture a fraction of this light and collimate it with a lens or mirror to form a beam of light (a light wave with finite extent in directions transverse to the predominant direction of propagation), as in a searchlight or flashlight. Such a beam, because of its finite transverse extent, spreads as it propagates. The spreading may result from several effects, including finite source size, limited spatial coherence, and, ultimately, diffraction. The degree of collimation that can ordinarily be achieved with light from thermal sources is much lower than that obtainable with a laser.

Lasers, on the other hand, can emit beams that propagate over large distances with very little spreading. For an ideal diffraction-limited Gaussian laser beam, the angular divergence (half-angle) in radians in the far zone (far from the beam waist) is [Fig. 1.6]

$$\theta \approx \frac{\lambda}{\pi n w_0}, \tag{1.12}$$

where w_0 is the radius of the laser beam at its waist (the location where the beam has its smallest spot radius), and n is the refractive index of the medium of propagation. This is the minimum divergence for a Gaussian beam of waist radius w_0. Real laser beams may have a larger divergence.

The characteristic distance over which the beam remains approximately collimated is related to the Rayleigh range $z_R = \frac{\pi n w_0^2}{\lambda}$. The quantity

$$b \approx \frac{2\pi n w_0^2}{\lambda} \tag{1.13}$$

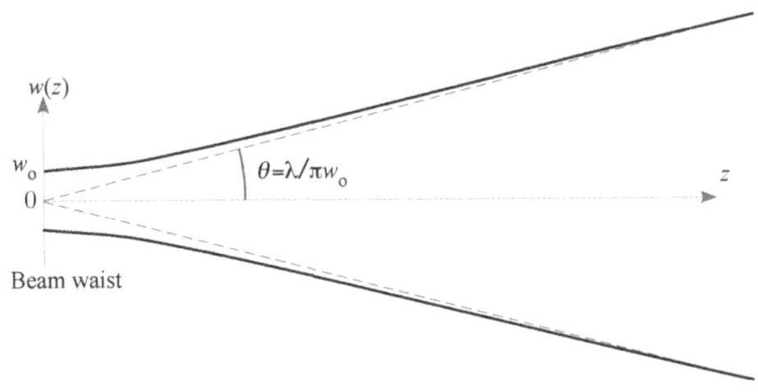

FIGURE 1.6
Divergence of a laser beam from its waist.

is called the confocal parameter and is equal to $2z_R$. Thus, the beam remains nearly collimated over a distance of order b centered on the beam waist.

The directionality of a laser beam is a consequence of the presence of the optical cavity, which favors light propagating close to the cavity axis. Such light can pass repeatedly through the gain medium and experience amplification, whereas light propagating at larger angles experiences much greater loss.

Consider a few concrete examples. A laser beam of wavelength $\lambda = 1.06 \times 10^{-6}$ m and waist radius $w_0 = 3$ mm in air ($n = 1$) has an angular divergence

$$\theta = \frac{\lambda}{\pi w_0} = \frac{1.06 \times 10^{-6}}{\pi(3 \times 10^{-3})} = 1.1 \times 10^{-4} \text{ rad} = 0.0064^o.$$

For a small helium-neon (He-Ne) laser emitting at wavelength $\lambda = 633$ nm, the beam waist might be $w_0 \approx 0.5$ mm. This corresponds to an angular divergence of

$$\theta \approx \frac{0.633 \times 10^{-6}}{\pi(0.5 \times 10^{-3})} = 4.0 \times 10^{-4} \text{ rad} = 0.023^o,$$

and a confocal parameter of approximately $b = 2.5$ m. For an argon-ion (Ar-ion) laser at 514 nm, the waist might be 5 mm, corresponding to an angular divergence of $\theta \approx 3.3 \times 10^{-5}$ rad and a confocal parameter of approximately $b = 306$ m. Compared with a typical flashlight, for which the divergence may be of order 25°, or a searchlight with a typical divergence of several degrees, the high directionality of laser light is evident.

1.4.5 Focusing

A laser beam can also be focused by a lens to a very small spot, with dimensions on the order of a few wavelengths. For an ideal Gaussian beam incident on a lens, the diameter d of the focused beam waist is approximately

$$d \approx \frac{2\lambda}{\pi w_0} f, \tag{1.14}$$

where f is the focal length of the lens and w_0 here denotes the radius of the beam incident on the lens. If the laser beam fills a lens aperture of diameter $D \approx 2w_0$, the conventional f-number of the lens is $f/\# = \frac{f}{D} \approx \frac{f}{2w_0}$. Consequently, $d \approx \frac{4\lambda}{\pi}(f/\#)$. For a lens with an f-number of order unity, the focused spot diameter is therefore of the order of a wavelength. The ability to focus laser light to such small spots is a consequence of its high spatial coherence and low beam divergence.

1.4.6 Brightness

Brightness B of a source, more precisely called *radiance* in radiometry, is defined as the power emitted per unit projected area of the emitting surface per unit solid angle. Its units are W/m^2·sr. The spectral brightness B_ν is the power emitted per unit projected area of the source per unit solid angle per unit frequency bandwidth. Its units are W/m^2·sr·Hz. The idea of brightness can be understood by considering a source that emits through a surface area S. Each surface element can emit light into the outward hemisphere, corresponding to a solid angle of 2π steradians. If a surface element δS emits power δP into a solid angle $\delta\Omega$ centered at an angle θ to the surface normal, the brightness of the source is

$$B = \frac{\delta^2 P}{\cos\theta, \delta S \delta\Omega}, \qquad [B] = \text{W/m}^2\text{·sr}.$$

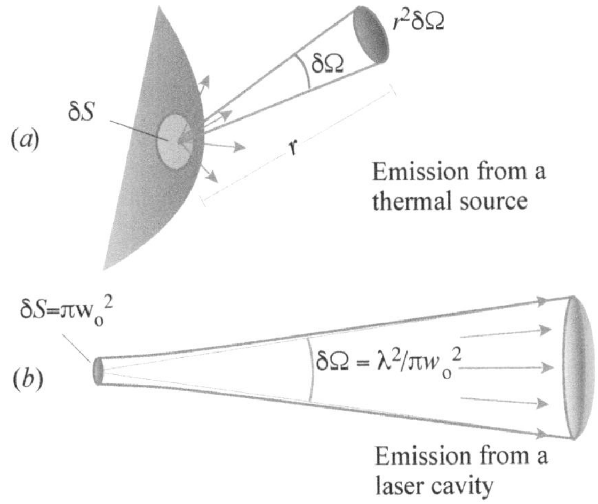

FIGURE 1.7
(a) Emission from the surface of a thermal source. (b) Emission from a laser cavity.

For radiation emitted close to the surface normal, $\cos\theta \approx 1$.

The power emitted per unit area by a blackbody at temperature T is given by the Stefan-Boltzmann law,

$$I = \sigma T^4, \qquad [I] = \mathrm{W/m^2},$$

where $\sigma = 5.6705 \times 10^{-8}$ W/m^2·K^4 is the Stefan-Boltzmann constant. For the Sun, approximated as a blackbody at $T = 6000$ K, this gives an efflux

$$I_{\text{sun}} = 5.6705 \times 10^{-8}(6 \times 10^3)^4 \approx 7.35 \times 10^7, \mathrm{W/m^2}.$$

A small He:Ne laser of modest power $P = 1$ mW and beam spot radius $w_0 = 0.5$ mm produces an efflux

$$I_{\text{L}} = \frac{P}{\pi w_0^2} = \frac{10^{-3}}{\pi(0.5 \times 10^{-3})^2} = 1.3 \times 10^3\, \mathrm{W/m^2}$$

The comparison becomes much more impressive when we take into account the directionality of the radiation. A blackbody surface is a Lambertian emitter, for which the radiance is independent of direction and the total hemispheric efflux is related to the brightness by $I = \pi B$. The brightness of the solar surface is therefore approximately

$$B_{\text{sun}} = \frac{I_{\text{sun}}}{\pi} \approx 2.3 \times 10^7, \mathrm{W/m^2{\cdot}sr}.$$

Because of the high directionality of the laser beam, its power is emitted into a small solid angle $\delta\Omega \approx \pi\theta^2 = \pi\left(\lambda/\pi w_0\right)^2 = \lambda^2/\pi w_0^2$. The brightness of the laser is therefore

$$B_{\text{L}} = \frac{P/\pi w_0^2}{\pi\theta^2} = \frac{P}{\lambda^2} = \frac{10^{-3}}{(0.6328 \times 10^{-6})^2} \approx 2.5 \times 10^9, \mathrm{W/m^2{\cdot}sr}.$$

Thus, even a 1-mW He:Ne laser can have a brightness substantially greater than that of the surface of the Sun.

The comparison becomes even more striking if we examine the spectral brightness. The Sun radiates approximately as a blackbody over a broad spectral bandwidth. The spectral energy density [energy per unit volume per unit bandwidth (J/m^3·Hz)] of blackbody radiation is given by

$$u(\nu) = \left(\frac{8\pi\nu^2}{c^3}\right)\frac{h\nu}{e^{\beta h\nu}-1} = \left(\frac{8\pi\nu^2}{c^3}\right)h\nu\bar{n}\,, \quad \beta = \frac{1}{k_B T}\,. \tag{1.15}$$

The quantity in the parentheses is the spectral mode density (#/m^3·Hz), $h\nu$ is the energy of a quantum, and $\bar{n} = \frac{1}{e^{\beta h\nu}-1}$ is the mean number of quanta in a mode of frequency ν.

$$B_\nu = \frac{1}{2}\frac{u(\nu)c}{\delta\Omega} = \frac{1}{2}\left[\frac{8\pi\nu^2}{c^3}\frac{h\nu}{e^{\beta h\nu}-1}\right]\frac{c}{2\pi} = \frac{2\nu^2}{c^2}\left[\frac{h\nu}{e^{\beta h\nu}-1}\right].$$

For the Sun, which radiates approximately as a blackbody at a temperature of $T \approx 6000$ K, we can estimate the spectral brightness at a wavelength in the green region [$h\nu \approx 2.5$ eV] as follows

$$\begin{aligned}
k_B T &= k_B \cdot 300\frac{T}{300} = \frac{1}{39} \times 20 \approx 0.51\,\text{eV} \\
\frac{h\nu}{k_B T} &\approx 4.9\,, \quad e^{\beta h\nu} = e^5 \approx 1.3 \times 10^2 \\
B_\nu &= \frac{2}{(633 \times 10^{-9})^2}\frac{2.5(1.6 \times 10^{-19})}{130-1} \approx 1.5 \times 10^{-8}\,\text{W/m}^2\text{-sr-Hz} \\
&= 1.5 \times 10^{-12}\,\text{W/cm}^2\cdot\text{sr}\cdot\text{Hz}\,.
\end{aligned}$$

Example 1.2 *Typical Spectral Brightness of Lasers*

For a 1-mW He:Ne laser ($\lambda = 633$ nm) of spectral width $\Delta\nu = 10^4$ Hz and spot radius 0.5 mm, the spectral brightness is

$$\begin{aligned}
B_\nu &= \frac{P}{\lambda^2\Delta\nu} = \frac{1 \times 10^{-3}}{(0.633 \times 10^{-6})^2 \times 10^4} \approx 2.5 \times 10^5\,\text{W/m}^2\cdot\text{sr}\cdot\text{Hz}\,, \\
&= 25\,\text{W/cm}^2\cdot\text{sr}\cdot\text{Hz}.
\end{aligned}$$

For a neodymium-glass (Nd-glass) laser with a peak power $P = 10^4$ MW, wavelength $\lambda = 1.06\,\mu$m, and a bandwidth limited by a pulse duration of 30 ps, we have

$$\begin{aligned}
\Delta\nu &= \frac{1}{2\pi\tau_p} = \frac{1}{2\pi \times 3 \times 10^{-11}} \approx 5.3 \times 10^9\,\text{Hz}\,, \\
B_\nu &= \frac{1 \times 10^4 \times 10^6}{(1.06 \times 10^{-6})^2 \times 5.3 \times 10^9} \approx 1.7 \times 10^{12}\,\text{W/m}^2\cdot\text{sr}\cdot\text{Hz}\,, \\
&\approx 1.7 \times 10^8\,\text{W/cm}^2\cdot\text{sr}\cdot\text{Hz}.
\end{aligned}$$

1.4.7 Photon Degeneracy

The mean number of photons per mode, independent of the volume and shape of the cavity, is referred to as the photon degeneracy parameter. Photons occupying the same mode are indistinguishable, having the same frequency, propagation direction, and polarization. For thermal radiation at 3000 K, the mean number of photons per mode $n(\nu, T)$ is a small number, of order 3×10^{-4} in the middle of the visible spectrum ($\lambda = 600$ nm or $\nu = 5 \times 10^{14}$

Hz). For solar radiation with $T = 5800$ K, $n(\nu, T)$ is of order 10^{-2} in the middle of the visible spectrum. In both cases, there are far fewer than one photon per mode on average.

Consider now a single-mode laser emitting a power P. If the output mirror has power transmission coefficient $\mathcal{T}$ and the laser cavity is a two-mirror cavity, as shown in Fig. 1.1, with the mirrors separated by a distance L, the power emitted can be related to the number of photons contained in the laser cavity as follows. Each photon traveling with speed c reaches the output mirror once every roundtrip time $2L/c$, or at a rate $c/2L$. At each encounter with the output mirror, a fraction $\mathcal{T}$ of its energy $h\nu$ is transmitted. If there are n_{cav} photons in the cavity mode, the output power P (the rate at which energy is emitted from the cavity) is therefore $P = \frac{c}{2L}\mathcal{T}\, h\nu\, n_{\text{cav}}$. Thus, the number of photons in the cavity mode is

$$n_{\text{cav}} = \frac{P}{h\nu} \frac{1}{c\mathcal{T}/2L}\,.$$

For a helium–neon laser with $L = 1$ m, 1% output-mirror transmittance ($\mathcal{T} = 0.01$), output power $P = 1$ mW, and wavelength $\lambda = 632.8$ nm, this gives $n_{\text{cav}} \approx 2 \times 10^9$ photons in the laser cavity mode. Such an enormous photon degeneracy is a characteristic feature of laser light that distinguishes it from ordinary thermal light. It reflects the concentration of a large number of photons into a very small number of electromagnetic modes and is closely associated with the high brightness and coherence of laser radiation. In contrast, the photon degeneracy parameter of visible thermal radiation is much less than unity at ordinarily accessible temperatures.

1.4.8 Laser Performance Records

The remarkable range of laser performance can be illustrated by the following representative examples. The numerical values should be regarded as indicative, since advances in laser technology continually extend these limits.

1. *Wide power range:* Continuous-wave (CW) powers extending to the kilowatt range and beyond are available from several types of lasers. In pulsed operation, peak powers in excess of 10^{15} W can be produced for extremely short durations.

2. *Extreme frequency stability:* The short-term fractional frequency stability of a highly stabilized laser can be extraordinarily high. A He:Ne laser operating at 3.39 μm and stabilized to a methane absorption line can achieve very high frequency reproducibility. Modern laser systems stabilized to narrow atomic or ionic transitions can attain fractional frequency stabilities approaching 10^{-18} under suitable conditions.

3. *Wide tunability:* Many common lasers operate only at sharply defined discrete frequencies. However, widely tunable sources of coherent radiation, including dye lasers, titanium-doped sapphire (Ti) lasers, and optical parametric oscillators (OPOs), can provide tuning over wavelength ranges of hundreds of nanometers, corresponding in some cases to frequency ranges of order 10^{13}–10^{14} Hz.

4. *Ultrashort pulses:* Mode-locked laser pulses shorter than 1 ps (picosecond) are routinely generated, and pulses only a few femtoseconds (fs) long, corresponding to only a few optical cycles, can be produced. Techniques based on high-harmonic generation can produce even shorter pulses, extending into the attosecond (as) regime.

5. *High efficiency:* Power-conversion efficiency is defined as the ratio of the optical power emitted by the laser to the power supplied to operate it. Laser efficiencies vary enormously with laser type. Some gas lasers have efficiencies well below 1%, whereas many

solid-state lasers have substantially higher efficiencies. Carbon dioxide (CO_2) lasers can achieve efficiencies of tens of percent, and semiconductor diode lasers can achieve electrical-to-optical conversion efficiencies exceeding 50%.

1.4.9 Laser Application Examples

The wide range of distinctive physical and operating characteristics of laser light—intensity, phase, directionality, coherence, monochromaticity, and spatial and temporal profiles—coupled with our ability to manipulate these characteristics, has given lasers enormous potential for applications in fundamental and applied sciences and in a very wide range of devices and systems. From the commonplace to the highly specialized, from barcode scanners in retail stores to LiDAR (Light Detection and Ranging) for remote sensing, distance measurement, mapping, and space exploration, lasers are everywhere. They touch our daily lives and are also integral to many cutting-edge scientific discoveries.

Industrial and manufacturing applications of lasers range from materials processing, including precision cutting, drilling, welding, and marking, to semiconductor manufacturing, where lasers are used in processes such as photolithography and precision material removal. Medical applications include high-resolution imaging and precision procedures such as LASIK for vision correction and laser lithotripsy for the treatment of kidney stones. Optical tweezers use highly focused laser beams to trap and manipulate microscopic particles, atoms, and biological objects.

Lasers are also the backbone of modern high-speed telecommunications, transmitting data through optical fibers. Information can be transmitted by modulating the output of continuous-wave (CW) lasers, and laser signals at several different wavelengths can be transmitted simultaneously through the same fiber using wavelength-division multiplexing (WDM). Lasers are widely used for optical data storage, where they read and write information on optical discs such as CDs and DVDs, and in retail systems for reading product barcodes.

Nonlinear optical effects produced by intense laser fields can be exploited to generate coherent light with a bandwidth exceeding one octave at optical frequencies. An octave bandwidth is a frequency range in which the upper frequency is twice the lower frequency. Under appropriate conditions, such broadband light can be converted into a spectrum consisting of uniformly spaced spectral lines—a laser frequency comb—which makes possible direct and extremely precise measurements of optical and molecular transition frequencies. Frequency combs have important applications in precision spectroscopy, astronomical measurements such as the detection and characterization of planets around other stars, and the development of optical clocks with accuracies surpassing those of traditional microwave frequency standards.

Another rapidly developing area is structured light, in which the spatial, temporal, spectral, and polarization properties of laser beams are tailored to produce beams with novel field distributions and propagation characteristics. Ultrashort laser pulses are used to study fast atomic and molecular processes on femtosecond and picosecond time scales, including molecular conformational changes, chemical reactions, energy transfer, and other dynamical processes in real time.

The nonlinear nature of light-matter interaction in lasers can also give rise to nontrivial dynamical behavior, including mode competition, bistability, hysteresis, self-pulsing, and transitions to chaos. Laser systems therefore provide a fertile setting for studying nonlinear dynamics and phenomena such as periodic and nonperiodic oscillations, bifurcations, phase transitions, chaotic behavior, and rogue waves. Similar nonlinear dynamical behavior appears in many other areas of science and engineering, including systems described by equations such as the Lotka-Volterra equations in population dynamics. These are only a few examples intended to convey the extraordinary versatility of lasers and the diversity of their applications.

1.5 The Organization of the Book

How laser light acquires its special properties is the fundamental question addressed in this book. Its aim is to introduce the fundamental concepts and processes involved in laser operation and to help the reader understand the physics that gives us remarkable control over the properties of laser light. The treatment is not intended to provide a comprehensive and rigorous formulation of laser theory, but it aims to describe the underlying physics accurately. The philosophy of this book is to be as pragmatic as possible and as rigorous as necessary, while pointing out important connections and their limitations. Exact results are often less useful in practice than physically transparent approximations. Therefore, we will avoid becoming entangled in beautiful but unnecessary details; instead, we will simplify and approximate real systems in order to understand the underlying principles, procedures, and limitations.

An understanding of lasers requires us to study an interacting light-matter system inside a resonator. It is an open system because energy can be added to or extracted from it. The atoms or molecules and the electromagnetic field interact and evolve simultaneously while also being coupled to the outside world. This situation is quite different from those normally encountered in electromagnetic theory, where the current and charge distributions are specified and the resulting fields are sought, or in mechanics, where the fields are specified and the resulting motion of charges is of interest. We will see that this seemingly formidable mathematical problem can nevertheless be treated quite effectively. The interacting light-matter system in a laser gives rise to light with special properties, and an understanding of the underlying physics gives us the ability to control these properties.

We will first study light and atoms separately by neglecting their dynamical interaction. We will then include the interaction between them, allowing the atoms and fields to evolve together and give rise to a wide range of dynamical behavior. In a quantum-mechanical treatment of light-matter interaction, the behavior of atoms is governed by the Schrödinger equation, whereas in a classical treatment it is described by Newton's second law together with the Lorentz force law for the electromagnetic force on charged particles. We will make use of both approaches, at different levels, to arrive at an accurate and physically transparent description of laser behavior.

Light, being an electromagnetic phenomenon, is described by Maxwell's equations. Its propagation in the form of rays or waves, as well as diffraction and interference, all follow from Maxwell's equations. In both the atomic and field descriptions, we will focus on behavior under conditions relevant to laser operation. In preparation for this development, we have begun by summarizing some of the special properties of laser light, which has also allowed us to introduce several concepts that will be useful later in the book.

1.6 Numerical Evaluation of Physical Quantities

In computing numerical values of physical quantities in laser physics, which typically involve fundamental constants, certain combinations of fundamental constants given in Table 1.2

TABLE 1.2
Some useful fundamental constants and their combinations.

Constant Name	Usual symbol and current value
Speed of light	$c = 2.997\,924\,58 \times 10^8$ m/s $\approx 3.00 \times 10^8$ m/s
Elementary charge	$e = 1.602\,177 \times 10^{-19}$ C
Electron-volt	1 eV $=1.602\,177\times10^{-19}$ J
Planck's constant	$h = 6.626\,070 \times 10^{-34}$ J s $\hbar \equiv h/2\pi = 1.054\,572 \times 10^{-34}$ J· s $\hbar c = 197.3270$ eV· nm (MeV· fm)
Free space permittivity Coulomb's law constant	$\epsilon_o = 8.8542 \times 10^{-12}$ F/m $1/(4\pi\epsilon_o) = 8.987\,552 \times 10^9$ N· m^2/C^2 $\approx 9.00 \times 10^9$ N· m^2/C^2
Free space permeability Ampere's law constant	$\mu_o = 1.256637 \times 10^{-6}$ H/m$\approx 4\pi \times 10^{-7}$ H/m $\mu_o/4\pi \approx 10^{-7}$ N/A^2 (T· m/A)
Fine structure constant	$\alpha = e^2/4\pi\epsilon_0\hbar c = 1/137.036 \approx 1/137$
Bohr radius	$a_0 = \hbar/m_e c\alpha = 5.2918 \times 10^{-11}$ m≈ 0.053 nm
Reduced Compton wavelength of electron	$\hbar/m_e c = 3.8616 \times 10^{-13}$ m
Photon energy	$\mathcal{E} = h\nu = \frac{2\pi\hbar c}{\lambda} = \frac{1240}{\lambda(\text{in nm})}$ eV
Boltzmann constant	$k_B = 1.380\,65 \times 10^{-23}$ J/K $\equiv 8.617 \times 10^{-5}$ eV/K $k_B T = 0.02525 \approx 1/40$ eV at 293 K; $0.02585 \approx 1/39$ eV at 300K
Stefan-Boltzmann constant	$\sigma = 5.6704 \times 10^{-8}$ W/m^2-K^4
Avogadro's constant	$\mathrm{N}_A = 6.022\,141\,79(30) \times 10^{23}$ /mol
Electron mass	$m_e = 9.109 \times 10^{-31}$ kg$=0.5110$ MeV/c^2
Proton mass	$m_p = 1.673 \times 10^{-27}$ kg$=938.2$ MeV/c^2
Proton-electron mass ratio	$m_p/m_e = 1836$
Neutron mass	$m_n = 1.675 \times 10^{-27}$ kg=940 MeV/c^2
Bohr magneton	$\mu_B = 9.274 \times 10^{-24}$ J/T $= 5.788 \times 10^{-5}$ eV/T
Nuclear magneton	$\mu_N = 5.051 \times 10^{-27}$ J/T$= 3.152 \times 10^{-8}$ eV/T
Pressure	1 atm = 760 Torr = 1.01325×10^5 Pa
Angle	1 degree = 1.745×10^{-2} rad

When accuracy is important, the accepted values of fundamental constants can be obtained from the NIST Fundamental Physical Constants database.

are often more useful than the values of the individual constants themselves. It is therefore a good practice to collect fundamental constants into one or more such combinations before substituting numerical values. For example, the energy of a photon may be calculated as

$$E \equiv h\nu = \frac{2\pi\hbar c}{\lambda} = \frac{2\pi(197\,\text{eV}\cdot\text{nm})}{\lambda} = \frac{1240}{\lambda(\text{in nm})}\,\text{eV}\,.$$

Similarly, an estimate of the Bohr radius of the hydrogen atom might proceed as follows

$$a_o = \frac{4\pi\epsilon_o\hbar^2}{m_e e^2} = \frac{4\pi\epsilon_o\hbar c}{e^2}\frac{\hbar c}{m_e c^2} = 137\frac{197\,\text{eV}\cdot\text{nm}}{0.511\times10^6\,\text{eV}} = 0.053\,\text{nm}$$

Energy and Frequency Units in Laser Physics
The multiplicative inverse (reciprocal) of wavelength, λ^{-1} with λ in cm is a frequently used unit of frequency in atomic and molecular spectroscopy. The reciprocal centimeter, cm^{-1}, is a commonly used unit of wavenumber.

A frequency of 1 Hz corresponds to $3.335640951 \times 10^{-11}\,\text{cm}^{-1}$ whereas a wavenumber of 1 cm^{-1} corresponds to $\nu = 2.99792458 \times 10^{10}$Hz

$$\frac{1}{\lambda} = \frac{\nu}{c} = \frac{1\,\text{s}^{-1}}{2.99792458\times10^{10}\,\text{cm}\cdot\text{s}^{-1}} = 3.335640951\times10^{-11}\,\text{cm}^{-1}$$

The energy of a photon is then

$$E = h\nu = \frac{2\pi\hbar c}{\lambda} = \frac{1240\,\text{eV}}{\lambda(\text{in nm})} = (1.240 \times 10^{-4}\,\text{eV})\nu(\text{in cm}^{-1})$$

This gives the following useful conversion factors:
$1\,\text{cm}^{-1} = 1.240 \times 10^{-4}\,\text{eV} = 1.439\,\text{K} = 1.986 \times 10^{-23}\,\text{J} = 2.998 \times 10^{10}\,\text{Hz}$
$1\,\text{eV} = 8.066 \times 10^{3}\,\text{cm}^{-1} = 1.160 \times 10^{4}\,\text{K} = 1.602 \times 10^{-19}\,\text{J} = 2.418 \times 10^{14}\,\text{Hz}$
$1\,\text{K} = 8.617 \times 10^{-5}\,\text{eV} = 0.6950\,\text{cm}^{-1} = 1.381 \times 10^{-23}\,\text{J} = 2.084 \times 10^{10}\,\text{Hz}$
$1\,\text{J} = 6.242 \times 10^{18}\,\text{eV} = 5.034 \times 10^{22}\,\text{cm}^{-1} = 7.244 \times 10^{22}\,\text{K} = 1.509 \times 10^{33}\,\text{Hz}$
$1\,\text{Hz} = 4.136 \times 10^{-15}\,\text{eV} = 3.336 \times 10^{-11}\,\text{cm}^{-1} = 4.799 \times 10^{-11}\,\text{K} = 6.626 \times 10^{-34}\,\text{J}$

Problems

1. The SI unit of energy, the joule (J), is inconveniently large for expressing the energy of a single photon, the separation between atomic energy levels, or the energy associated with the thermal motion of atoms or molecules. A more suitable unit of energy for these purposes is the electron-volt (1 eV $= 1.60 \times 10^{-19}$ J). Similarly, a convenient unit for atomic-scale lengths is the nanometer (1 nm $= 10^{-9}$ m). Use the table of constants [Table 1.2] to establish the following useful relations:

$$\hbar c = 197 \text{ eV-nm} \approx 200 \text{ eV-nm}\,, \quad E_\lambda = \frac{1240\,\text{eV}}{\lambda(\text{nm})}\,,$$

$$k_B T \approx 0.026 \text{ eV} \approx \frac{1}{40}\text{ eV} \quad \text{at room temperature } T = 300 \text{ K}\,.$$

   In writing the energy E_λ of a single photon of wavelength λ, we have used $\nu\lambda = c$, where ν, λ, and c are, respectively, the frequency, wavelength, and speed of light in free space. In estimating the magnitudes of physical quantities, it is often more convenient to remember these combinations of fundamental constants than the constants themselves. Use these relations to determine

   (i) the range of photon energies corresponding to visible wavelengths from 400 nm to 700 nm,

   (ii) the root-mean-square (rms) speed of sodium atoms in a sodium vapor at 127 °C ($T = 400$ K). Sodium has atomic mass number 23; take 1 atomic mass unit to be 931.5 MeV/c^2,

   (iii) the ratio of the populations of two nondegenerate atomic levels separated by an energy ΔE corresponding to a wavelength of 633 nm. Assume that the atoms are in thermal equilibrium at $T = 300$ K.

2. Using the ideal-gas law $PV = Nk_B T$, show that the number density of atoms in an ideal gas at pressure P (in Torr) and absolute temperature T (in K) is given by

$$\mathcal{N}(\text{atoms/cm}^3) = 9.65 \times 10^{18}\frac{P(\text{Torr})}{T(\text{K})}\,.$$

3. **Level population in thermal equilibrium**

   (a) An atom has two nondegenerate energy levels separated by an energy gap corresponding to a wavelength of 694.3 nm. Assume that all the atoms occupy one of these two levels and are in thermal equilibrium. Calculate the population ratio N_2/N_1 at 300 K and 500 K. What temperature is required for $N_2/N_1 = 0.1$?

   (b) Calculate N_2/N_1 for two nondegenerate levels separated by a frequency $\nu = 10$ GHz at (i) room temperature, $T = 300$ K, and (ii) liquid-helium temperature, $T = 4.2$ K.

4. **Spontaneous and stimulated emission rates compared**

   (a) Treating the radiation from a tungsten filament operating at 1500 K as thermal radiation, calculate the ratio of the spontaneous-emission rate to the stimulated-emission rate for an atomic transition of wavelength $\lambda = 600$ nm.

   (b) At what temperature are the spontaneous- and stimulated-emission rates equal for an atomic transition with $\lambda = 600$ nm?

5. The frequency spectrum of radiation emitted by a source has center frequency ν_o and spectral width $\Delta\nu$, with $\Delta\nu \ll \nu_o$. The corresponding spectrum expressed in terms of wavelength has central wavelength $\lambda_o = c/\nu_o$ and spectral width $\Delta\lambda$. Using $\lambda = c/\nu$ and the assumption $\Delta\nu \ll \nu_o$, show that $\Delta\lambda = (\lambda_o^2/c)\Delta\nu$, and, using the coherence-time convention $t_c = 1/(2\pi\Delta\nu)$, show that the coherence length is $\ell_c = \lambda_o^2/(2\pi\Delta\lambda)$. Compute ℓ_c for He:Ne light with $\lambda_o = 632.8$ nm and $\Delta\nu \approx 1.5$ GHz.

6. Lasers with frequency stability better than $\Delta\nu = 10^{-2}$ Hz have been demonstrated. For a laser of central wavelength λ_o, express the corresponding wavelength stability $\Delta\lambda$ in terms of λ_o. Estimate its coherence time and coherence length. Compare these quantities with those for light emitted by a sodium lamp with $\Delta\nu = 9 \times 10^9$ Hz.

7. Estimate the transverse spatial coherence length of sunlight at the surface of the Earth for visible light of wavelength $\lambda \approx 550$ nm. Using simple optical components, explain how the spatial coherence could be increased considerably, albeit at the expense of transmitted intensity.

8. **Laser beam vs. a searchlight beam:**

   (a) Consider a 1.0 mW He:Ne laser beam of wavelength 633 nm with waist diameter $2w_0 = 1.0$ mm. Assuming, for this estimate, that the beam power is distributed uniformly over its cross section and that its far-field divergence half-angle is $\lambda/\pi w_0$, express the power density in the beam at a large distance R from the waist.

   A fully dark-adapted human eye can detect a photon flux as small as approximately 100 photons/s entering a pupil of diameter about 8 mm. Ignoring atmospheric attenuation and other propagation losses, estimate the maximum distance at which the laser beam would be detectable by an observer located on the beam axis. This is an idealized sensitivity estimate and should not be interpreted as a safe viewing condition for a laser beam.

   (b) A 1.0 kW searchlight consists of an electric arc located at the focal point of a large mirror whose diameter and focal length are both 1.0 m. Assume that the arc radiates approximately isotropically, has an effective diameter of 10 mm, and emits 10% of its total radiated power in the visible region. What fraction of the total visible power is intercepted and collimated by the searchlight mirror? Estimate the divergence of the resulting beam and its power density as a function of distance. Ignoring atmospheric attenuation, estimate the maximum distance at which the searchlight could be detected by an observer located within the beam. Compare the advantages and limitations of the laser and searchlight beams for long-distance illumination or detection.

9. Sufficiently powerful laser pulses, when tightly focused, can trigger electrical breakdown of air. The breakdown electric-field strength of dry air at standard temperature and pressure is approximately 30 kV/cm. Using the plane-wave relation between electric-field amplitude and irradiance, estimate the corresponding laser power density.

10. Laser pulses of duration less than 150 fs and energy of the order of 10 J have been produced. When tightly focused, such pulses can reach peak irradiances of the order of 10^{21} W/cm^2. The irradiance I and electric-field amplitude E_0 of an electromagnetic wave in vacuum are related by $I = \frac{1}{2}\epsilon_0 c E_0^2$, where ϵ_0 is the permittivity of free space and c is the speed of light in vacuum. Estimate the corresponding electric-field amplitude and compare it with the Coulomb electric field experienced by an electron at the Bohr radius in a hydrogen atom.

2

Maxwell's Equations, Waves, and Rays

In the preceding chapter, we introduced the laser as an interacting light–matter system with suitable boundaries to provide optical feedback. This chapter provides an overview of light as an electromagnetic (EM) wave phenomenon, primarily to introduce important concepts and notation that will be used consistently throughout the rest of the book. Using Maxwell's equations, we show that EM wave fields obey a vector wave equation in which the atomic polarization appears in the source term. In a broad sense, this book is concerned with solving this equation under the varying conditions appropriate for laser operation, using suitable approximations. How polarization arises when charges in atoms and molecules interact with EM waves is considered in Chapter 5.

In this chapter, we consider transparent media in which the effect of medium polarization enters only through the (real) refractive index of the medium. Rather than working directly with the vector wave equation, we show that solutions of the vector wave equation—and, indeed, of Maxwell's equations—can be constructed from solutions of the scalar wave equation. This approach is simpler and is illustrated using the familiar plane- and spherical-wave solutions of Maxwell's equations in homogeneous media. These solutions are also used to develop a physical picture of wave propagation and energy transport in terms of phasefronts and rays. Intensity, the Poynting vector, and Poynting's theorem are introduced to describe power transport and power exchange between the EM wave and atoms in a medium.

The nature of EM wave fields in the presence of boundaries, which are inevitable in laser systems to provide optical feedback, is illustrated by considering a rectangular cavity. These solutions, referred to as modes, are used to show that open cavities are better suited for providing optical feedback in lasers than closed cavities.[1] Finding the EM wave modes of open resonators, however, is a difficult mixed-boundary-value problem. We avoid this difficulty by using the ray picture of energy transport in EM waves to find approximate solutions, which turn out to be excellent representations of laser beams.

In this chapter, we use Maxwell's equations to develop the laws governing the behavior of rays and introduce the so-called ABCD matrices to describe their propagation and transformation by optical elements such as mirrors and lenses, which are used to form open resonators. With the help of these matrices, we show that suitable open resonators can confine paraxial ray bundles and thus provide optical feedback in lasers. These matrices also prove useful later in describing the transformation of laser beams by optical elements. The insights gained from the study of paraxial ray confinement in this chapter are used in Chapter 3 to find wave solutions with phasefronts corresponding to paraxial rays. These waves are then used in Chapter 4 to find the EM wave modes of open resonators.

We begin this chapter by introducing the complex analytic-signal representation of wave fields, which is used throughout the book.

[1]The terms cavity and resonator are used interchangeably in the text. They emphasize different aspects of the same physical object, as will be discussed in Chapter 4.

DOI: 10.1201/9781003584179-2

2.1 Spectral Resolution: Analytic Signal

Electromagnetic waves are (space- and) time-dependent solutions of Maxwell's equations. In finding time-dependent solutions of Maxwell's equations, it is convenient to make use of the Fourier decomposition of a time-varying function. A time-dependent field can be expanded as a superposition of monochromatic waves of different frequencies. The nature of this expansion depends on the nature of the time dependence. In this discussion, we suppress the spatial dependence of the fields to simplify the equations. We consider a wave field $\mathcal{F}(\boldsymbol{r},t) \equiv \mathcal{F}(t)$ that can be represented by a Fourier integral. For this to be possible, $\mathcal{F}(t)$ must vanish sufficiently rapidly as $t \to \pm\infty$ such that $\int_{-\infty}^{\infty} |\mathcal{F}(t)|^2 dt$ is finite. Physically, this condition means that the energy of the wave field is finite. For fields produced by physical sources over a finite duration, this condition is satisfied. The function $\mathcal{F}(\boldsymbol{r},t)$ could, for example, represent a Cartesian component of the electric (or magnetic) field of a wave or the electric-field amplitude of a linearly polarized electromagnetic wave. The function $\mathcal{F}(t)$ can then be written as a Fourier integral,

$$\mathcal{F}(t) = \int_{-\infty}^{\infty} \tilde{\mathcal{F}}(\omega) e^{-i\omega t} \frac{d\omega}{2\pi}, \tag{2.1}$$

where the Fourier amplitude (or the spectral component) $\tilde{F}(\omega)$ is given by

$$\tilde{\mathcal{F}}(\omega) = \int_{-\infty}^{\infty} \mathcal{F}(t) e^{i\omega t} dt, \tag{2.2}$$

Since $\mathcal{F}(t)$ is a real function of time, it follows from Eq. (2.2) that the negative-frequency Fourier amplitude $\tilde{\mathcal{F}}(-\omega)$ is related to the positive-frequency amplitude $\tilde{\mathcal{F}}(\omega)$ by

$$\tilde{\mathcal{F}}(-\omega) = \tilde{\mathcal{F}}^*(\omega). \tag{2.3}$$

Thus the negative-frequency spectral components in the Fourier integral, Eq. (2.1), carry no additional information beyond that contained in the positive-frequency components. We can therefore write the Fourier integral in terms of positive frequencies alone

$$\begin{aligned}
\mathcal{F}(t) &= \int_0^{\infty} \tilde{\mathcal{F}}(\omega) e^{-i\omega t} \frac{d\omega}{2\pi} + \int_{-\infty}^{0} \tilde{\mathcal{F}}(\omega) e^{-i\omega t} \frac{d\omega}{2\pi} \\
&= \int_0^{\infty} \tilde{\mathcal{F}}(\omega) e^{-i\omega t} \frac{d\omega}{2\pi} + \int_0^{\infty} \tilde{\mathcal{F}}(-\omega) e^{i\omega t} \frac{d\omega}{2\pi} = \int_0^{\infty} \tilde{\mathcal{F}}(\omega) e^{-i\omega t} \frac{d\omega}{2\pi} + c.c. \\
&= \mathcal{R}e \int_0^{\infty} 2\tilde{\mathcal{F}}(\omega) e^{-i\omega t} \frac{d\omega}{2\pi}.
\end{aligned} \tag{2.4}$$

Thus, the function $\mathcal{F}(t)$ is the real part of the complex function

$$F(t) = \int_0^{\infty} \tilde{F}(\omega) e^{-i\omega t} \frac{d\omega}{2\pi}, \qquad \tilde{F}(\omega) = 2\tilde{\mathcal{F}}(\omega). \tag{2.5}$$

The function $F(t)$ contains only the positive-frequency spectral components and is known as the complex analytic signal associated with the real function $\mathcal{F}(t)$. In terms of the complex analytic signal $F(t)$, we can write the real function $\mathcal{F}(t)$ as

$$\mathcal{F}(t) = \frac{1}{2}\left[F(t) + c.c.\right] = \mathcal{R}e[F(t)]. \tag{2.6}$$

Using Parseval's theorem, the energy W of the wave can be written as

$$W = \int_{-\infty}^{\infty} dt\, [\mathcal{F}(t)]^2 = \int_{-\infty}^{\infty} |\tilde{\mathcal{F}}(\omega)|^2 \frac{d\omega}{2\pi} = \frac{1}{2}\int_{0}^{\infty} |\tilde{F}(\omega)|^2 \frac{d\omega}{2\pi} . \tag{2.7}$$

This equation shows that the condition of square integrability implies that the energy of the wave is finite. Equation (2.7) also allows us to identify the integrand as the spectral energy density associated with the wave:

$$\frac{dW}{d\omega} = \frac{|\tilde{\mathcal{F}}(\omega)|^2}{4\pi} . \tag{2.8}$$

The second type of Fourier expansion contains discrete frequencies. A periodic field will have an expansion of this type. Let $\omega_0 = 2\pi/T$ denote the fundamental frequency corresponding to period T, and let $\omega_n = n\omega_0$ denote its nth harmonic. Then a periodic function can be written as a Fourier series

$$\mathcal{F}(t) = \sum_{n=-\infty}^{\infty} \tilde{\mathcal{F}}_n e^{-i\omega_n t}, \qquad \omega_n = n\omega_0 \tag{2.9}$$

$$\tilde{\mathcal{F}}_n = \frac{1}{T}\int_{-T/2}^{T/2} dt\, \mathcal{F}(t) e^{i\omega_n t} \tag{2.10}$$

Again, since $\mathcal{F}(t)$ is a real function of time, the Fourier components at positive and negative harmonics are related by

$$\tilde{\mathcal{F}}_{-n} = \tilde{\mathcal{F}}_n^* . \tag{2.11}$$

The zero-frequency term is time-independent and has been omitted. Using Eq. (2.11), we can write the real field $\mathcal{F}(t)$ as

$$\mathcal{F}(t) = \sum_{n=1}^{\infty} \tilde{\mathcal{F}}_n e^{-i\omega_n t} + c.c. = \mathcal{R}e \sum_{n=1}^{\infty} 2\tilde{\mathcal{F}}_n e^{-i\omega_n t} . \tag{2.12}$$

We can introduce the complex analytic signal $F(t)$ associated with the real field $\mathcal{F}(t)$ by

$$F(t) = \sum_{n=1}^{\infty} \tilde{F}_n e^{-i\omega_n t} , \qquad \tilde{F}_n = 2\tilde{\mathcal{F}}_n . \tag{2.13}$$

In terms of $F(t)$, the real field $\mathcal{F}(t)$ is given by Eq. (2.6). The average power, averaged over one cycle of the fundamental frequency, is

$$\overline{\mathcal{F}^2}(t) \equiv \frac{1}{T}\int_{T/2}^{T/2} dt'\, \mathcal{F}^2(t+t') = 2\sum_{n=1}^{\infty} |\tilde{\mathcal{F}}_n|^2 = \sum_{n=1}^{\infty} \frac{1}{2} |\tilde{F}_n|^2 \tag{2.14}$$

From this, we identify the average power associated with the nth harmonic as

$$\overline{P}_n = \frac{1}{2} |\tilde{F}_n|^2 . \tag{2.15}$$

The preceding discussion allows us to associate a complex function $F(t)$ [Eqs. (2.5) and (2.13)] with a real $\mathcal{F}(t)$. The function $F(t)$ contains only positive frequencies. It is analytic in the lower half of the complex t-plane and is known as the complex analytic-signal representation of the real field $\mathcal{F}(t)$. Its real and imaginary parts form a Hilbert-transform pair. The function $\mathcal{F}(t)$ can represent any component of a vector field. The results derived here therefore generalize straightforwardly to vector fields.

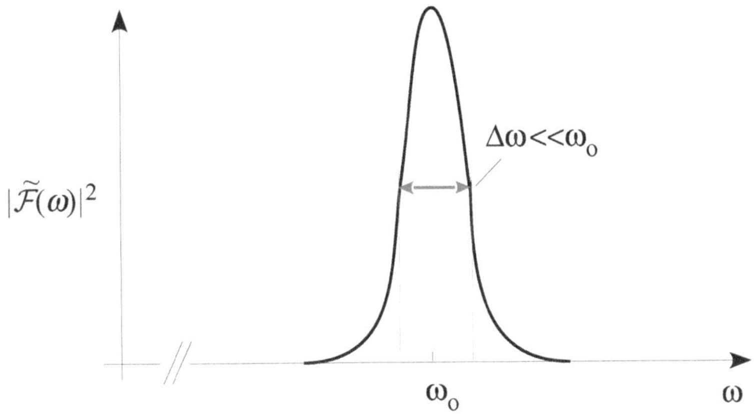

FIGURE 2.1
Spectrum of a quasi-monochromatic (narrowband) field. Carrier frequency ω_0 is large compared to the width of the spectrum $\Delta\omega$. The spectrum falls to zero well before ω reaches zero.

2.1.1 Monochromatic and Quasi-Monochromatic Fields

A field is said to be monochromatic if $\tilde{F}(\omega) = 2\pi\delta(\omega - \omega_0)F_0$. For such a field, we can write

$$F(t) = F_0 e^{-i\omega_0 t} \tag{2.16}$$

Strictly speaking, there are no perfectly monochromatic fields in practice. Nevertheless, the monochromatic approximation is quite adequate and useful in many applications. More realistic fields are the so-called quasi-monochromatic fields, which are characterized by a narrow spectrum such that the spectral amplitudes have appreciable values only within a small frequency interval $\Delta\omega$ centered around some mean frequency ω_0 [Fig. 2.1]. For such fields, the complex field $F(t)$ can be written as

$$F(t) = \int_0^\infty \tilde{F}(\omega)e^{-i\omega t}\frac{d\omega}{2\pi} = \left[\int_0^\infty \tilde{F}(\omega)e^{-i(\omega-\omega_0)t}\frac{d\omega}{2\pi}\right] e^{-i\omega_0 t} \equiv F_0(t)e^{-i\omega_0 t}, \tag{2.17}$$

$$\text{with} \quad F_0(t) = \int_0^\infty \tilde{F}(\omega)e^{-i(\omega-\omega_0)t}\frac{d\omega}{2\pi} = \int_{-\omega_0}^\infty \tilde{F}(\omega_0 + \Omega)e^{-i\Omega t}\frac{d\Omega}{2\pi}. \tag{2.18}$$

Since the spectral amplitude $\tilde{F}(\omega)$ is assumed to be appreciable only in a small neighborhood of width $\Delta\omega \ll \omega_0$ of the mean frequency ω_0, the amplitude $\tilde{F}(\omega_0 + \Omega)$ will be appreciable only for values of $|\Omega| \lesssim \Delta\omega$ centered around $\Omega = 0$. Hence, $F_0(t) = |F_0(t)|e^{i\phi(t)}$ is a superposition of harmonic components of low detuning frequencies. Since $\Delta\omega/\omega_0 \ll 1$, both the amplitude $|F_0(t)|$ and phase $\phi(t)$ vary slowly compared with $e^{-i\omega_0 t}$, such that

$$\left|\frac{dF_0(t)}{dt}\right| \sim \Delta\omega|F_0(t)| \ll \omega_0|F_0(t)|. \tag{2.19}$$

The function $F_0(t)$ is known as the slowly varying envelope function, and ω_0 is known as the carrier frequency. Equation (2.19) means that the fractional change in the envelope during one period, $2\pi/\omega_0$, of the carrier wave is negligible. We can think of $F_0(t)$ as modulating the sinusoidal variation due to the $e^{-i\omega_0 t}$ term. For continuous-wave (cw) optical fields,

$\omega_0/2\pi \approx 10^{14}$–$10^{15}$ Hz and typically $\Delta\omega/2\pi \lesssim 10^{10}$ Hz, so the inequality $\Delta\omega \ll \omega_0$ is clearly well satisfied. In such cases, the lower limit of integration in Eq. (2.20) can be extended to $-\infty$. We can then treat $\tilde{F}(\omega_0 + \Omega)$ and the slowly varying complex amplitude $F_0(t)$ as Fourier transform pairs:

$$F_0(t) \approx \int_{-\infty}^{\infty} \tilde{F}(\omega_0 + \Omega)e^{-i\Omega t}\frac{d\Omega}{2\pi}\,, \tag{2.20a}$$

$$\tilde{F}(\omega_0 + \Omega) = \int_{-\infty}^{\infty} F_0(t)e^{i\Omega t}dt\,. \tag{2.20b}$$

It should be kept in mind that $\tilde{F}(\omega_0 + \Omega)$ is the Fourier transform of a complex function $F_0(t)$. It is related to the Fourier transform $\tilde{\mathcal{F}}(\omega)$ [Eq. (2.2)] of the real function $\mathcal{F}(t)$ [Eq. (2.1)] by

$$\tilde{\mathcal{F}}(\omega) = \frac{1}{2}\left[\tilde{F}(\omega) + \tilde{F}^*(-\omega)\right]. \tag{2.21}$$

Spectral resolution of vector fields is similarly carried out by replacing the scalar functions $\mathcal{F}$, $\tilde{\mathcal{F}}$, F, and F_0 by vector functions $\boldsymbol{\mathcal{F}}$, $\tilde{\boldsymbol{\mathcal{F}}}$, $\boldsymbol{F}$, and $\boldsymbol{F}_0$.

In dealing with time-dependent fields, we often need time averages of quantities such as the energy density and the Poynting vector, which depend quadratically on the fields. A useful result can be established for time averages of such quantities when quasi-monochromatic fields are involved. Consider two quasi-monochromatic vector fields $\boldsymbol{\mathcal{F}}(t)$ and $\boldsymbol{\mathcal{G}}(t)$:

$$\boldsymbol{\mathcal{F}}(t) = \frac{1}{2}[\boldsymbol{F}_0(t)e^{-i\omega_0 t} + c.c.]\,, \qquad \left|\frac{d\boldsymbol{F}_0(t)}{dt}\right| \ll \omega_0|\boldsymbol{F}_0(t)|\,, \tag{2.22}$$

$$\boldsymbol{\mathcal{G}}(t) = \frac{1}{2}[\boldsymbol{G}_0(t)e^{-i\omega_0 t} + c.c.]\,, \qquad \left|\frac{d\boldsymbol{G}_0(t)}{dt}\right| \ll \omega_0|\boldsymbol{G}_0(t)|\,. \tag{2.23}$$

Then their dot or cross product (represented by $\hat{O}$) can be written as

$$[\boldsymbol{\mathcal{F}}(t)\hat{O}\boldsymbol{\mathcal{G}}(t)] = \frac{1}{4}[\boldsymbol{F}_0(t)\hat{O}\boldsymbol{G}_0(t)e^{-2i\omega_0 t} + \boldsymbol{F}_0(t)\hat{O}\boldsymbol{G}_0{}^*(t)] + c.c. \tag{2.24}$$

The first term inside the square brackets and its complex conjugate vary rapidly in time at a frequency $\sim 2\omega_0$. The second term and its complex conjugate vary slowly in time at frequencies of order $\Delta\omega \ll \omega_0$ and therefore change negligibly during one period (cycle), $T = 2\pi/\omega_0$, of the carrier frequency. On averaging the product over one cycle of the carrier frequency, the rapidly oscillating terms average to zero. We then obtain the cycle average of a quadratic quantity:

$$\begin{aligned}\overline{\boldsymbol{\mathcal{F}}(t)\hat{O}\,\boldsymbol{\mathcal{G}}(t)} &\equiv \frac{1}{T}\int_{-T/2}^{T/2} dt' \left(\boldsymbol{\mathcal{F}}(t+t')\hat{O}\boldsymbol{\mathcal{G}}(t+t')\right) = \frac{\omega_0}{2\pi}\int_0^{2\pi/\omega_0} dt' \left(\boldsymbol{\mathcal{F}}(t+t')\hat{O}\boldsymbol{\mathcal{G}}(t+t')\right), \\ &= \frac{1}{4}[\boldsymbol{F}_0(t)\hat{O}\boldsymbol{G}_0{}^*(t) + c.c.] = \frac{1}{2}\mathcal{R}e\,[\boldsymbol{F}_0(t)\hat{O}\boldsymbol{G}_0^*(t)]\,.\end{aligned} \tag{2.25a}$$

As special cases of this result, the cycle-averaged values of the scalar and vector products of two vectors are given by

$$\overline{\boldsymbol{\mathcal{F}}(t)\cdot\boldsymbol{\mathcal{G}}(t)} = \frac{1}{2}\,\mathcal{R}e\,[\boldsymbol{F}_0(t)\cdot\boldsymbol{G}_0^*(t)] = \frac{1}{2}\,\mathcal{R}e\,[\boldsymbol{F}_0^*(t)\cdot\boldsymbol{G}_0(t)]\,, \tag{2.25b}$$

$$\overline{\boldsymbol{\mathcal{F}}(t)\times\boldsymbol{\mathcal{G}}(t)} = \frac{1}{2}\,\mathcal{R}e\,[\boldsymbol{F}_0(t)\times\boldsymbol{G}_0^*(t)] = \frac{1}{2}\,\mathcal{R}e\,[\boldsymbol{F}_0^*(t)\times\boldsymbol{G}_0(t)]\,. \tag{2.25c}$$

For quasi-monochromatic fields, the cycle-averaged values can vary in time, but only slowly compared with the optical cycle. These results will be useful throughout the book when dealing with quasi-monochromatic fields.

2.2 Electromagnetic Waves in Homogeneous Media

We will use the complex analytic signal representation of the fields so that the real physical field $\mathcal{F}(\boldsymbol{r},t)$ is given by the real part of the complex field $F(\boldsymbol{r},t)$:

$$\mathcal{F}(\boldsymbol{r},t) = \mathcal{R}e\left[F(\boldsymbol{r},t)\right], \tag{2.26}$$

where $\mathcal{F}(\boldsymbol{r},t)$ represents any of the Cartesian components of the field. In a homogeneous, transparent, isotropic medium characterized by dielectric permittivity ϵ and magnetic permeability μ, the constitutive relations are simple proportionalities: $\boldsymbol{D}(\boldsymbol{r},t) = \epsilon\boldsymbol{E}(\boldsymbol{r},t)$ and $\boldsymbol{B}(\boldsymbol{r},t) = \mu\boldsymbol{H}(\boldsymbol{r},t)$. These relations hold for arbitrarily rapid variations of the fields, as long as the most significant part of the field spectrum lies within the transparency range of the medium. In this case, ϵ and μ may be taken to be real and approximately constant over the field spectrum. In the presence of strongly interacting atoms or molecules, the constitutive relations are modified to read

$$\boldsymbol{D}(\boldsymbol{r},t) = \epsilon\boldsymbol{E}(\boldsymbol{r},t) + \boldsymbol{P}_{\text{at}}(\boldsymbol{r},t), \tag{2.27a}$$

$$\boldsymbol{B}(\boldsymbol{r},t) = \mu\boldsymbol{H}(\boldsymbol{r},t), \tag{2.27b}$$

$$\boldsymbol{J}(\boldsymbol{r},t) = \sigma\boldsymbol{E}(\boldsymbol{r},t), \tag{2.27c}$$

where $\boldsymbol{J}$ represents the conduction current density, σ is the conductivity of the medium, and $\boldsymbol{P}_{\text{at}}$ is the atomic polarization induced by the interaction of light with the atoms. The conduction current accounts for dissipation of electromagnetic energy.

In writing the constitutive relation between $\boldsymbol{D}$ and $\boldsymbol{E}$, we have assumed that the strongly interacting atoms (active atoms) are embedded in a host medium. Let us denote the response of the host medium to the field by the polarization $\boldsymbol{P}_m$ and that of the active atoms by $\boldsymbol{P}$at. The interaction of the field with the host medium is weak, so we can take the medium response to be a linear function of the field: $\boldsymbol{P}_m = \epsilon_o\chi_m\boldsymbol{E}$, where χ_m is the linear dielectric susceptibility of the host medium. In contrast, the response $\boldsymbol{P}$at of the strongly interacting active atoms cannot, in general, be expected to be a linear function of the field. Indeed, we shall see that the nonlinear response of the active atoms is essential for the operation of a laser. Thus, we have written the overall polarization as the sum of two parts, $\boldsymbol{P} = \boldsymbol{P}_m + \boldsymbol{P}_{\text{at}}$, so that the constitutive relation for the electric field becomes

$$\boldsymbol{D}(\boldsymbol{r},t) = \epsilon_o\boldsymbol{E}(\boldsymbol{r},t) + \boldsymbol{P} = \epsilon_o(1+\chi_m)\boldsymbol{E}(\boldsymbol{r},t) + \boldsymbol{P}\text{at}(\boldsymbol{r},t). \tag{2.28}$$

This relation then defines the dielectric permittivity $\epsilon = \epsilon_o(1+\chi_m)$ in Eq. (2.27a). We still need to compute $\boldsymbol{P}$at using quantum mechanics (or Newton's second law) before proceeding further.

In writing the preceding relations, we have assumed that the magnetic response of the atoms is negligible compared to their electric response (magnetization $\boldsymbol{M}_{\text{at}} = 0$), which holds for most atomic media. We have also assumed that the fields are quasi-monochromatic.[2] Maxwell's equations then read

$$\boldsymbol{\nabla}\cdot\boldsymbol{D}(\boldsymbol{r},t) = 0, \tag{2.29a}$$

$$\boldsymbol{\nabla}\cdot\boldsymbol{B}(\boldsymbol{r},t) = 0, \tag{2.29b}$$

$$\boldsymbol{\nabla}\times\boldsymbol{E}(\boldsymbol{r},t) = -\frac{\partial\boldsymbol{B}(\boldsymbol{r},t)}{\partial t}, \tag{2.29c}$$

$$\boldsymbol{\nabla}\times\boldsymbol{B}(\boldsymbol{r},t) = \mu\boldsymbol{J}(\boldsymbol{r},t) + \mu\epsilon\frac{\partial\boldsymbol{E}(\boldsymbol{r},t)}{\partial t} + \mu\frac{\partial\boldsymbol{P}_{\text{at}}(\boldsymbol{r},t)}{\partial t}. \tag{2.29d}$$

[2]For quasi-monochromatic fields, we write the complex field as $\boldsymbol{F}(\boldsymbol{r},t) = \boldsymbol{F}_0(\boldsymbol{r},t)e^{-i\omega t}$, where the *envelope* $\boldsymbol{F}_0(\boldsymbol{r},t)$ changes negligibly over times of order $2\pi/\omega$. In terms of the Fourier decomposition of the fields, we say that the field contains frequencies in a narrow band $\Delta\nu$ centered around the carrier frequency ν ($\Delta\nu \ll \nu$). Parameters ϵ and μ are then taken to represent the values of the dielectric permittivity and magnetic permeability at the carrier frequency ν.

These are coupled first-order partial differential equations. By eliminating the magnetic or electric field, we can obtain a closed equation for the electric or magnetic field. For example, on taking the curl of Eq. (2.29c), we obtain

$$\nabla \times (\nabla \times \boldsymbol{E}) = -\frac{\partial}{\partial t}\nabla \times \boldsymbol{B}\,. \tag{2.29e}$$

Using the identity $\nabla \times (\nabla \times \boldsymbol{E}) = \nabla(\nabla \cdot \boldsymbol{E}) - \nabla^2 \boldsymbol{E}$, together with $\nabla \cdot \boldsymbol{E} = -\nabla \cdot \boldsymbol{P}_{at}/\epsilon \approx 0$ (negligible longitudinal polarization variation), and eliminating $\boldsymbol{B}$ with the help of Eq. (2.29d), we find that the electric field satisfies the equation

$$\nabla^2 \boldsymbol{E} - \mu\epsilon\frac{\partial^2 \boldsymbol{E}}{\partial t^2} - \mu\sigma\frac{\partial \boldsymbol{E}}{\partial t} = \mu\frac{\partial^2 \boldsymbol{P}_{at}}{\partial t^2}\,. \tag{2.30}$$

This is a driven, damped vector wave equation, with the $\boldsymbol{P}_{at}$ term providing the source and the finite conductivity σ of the medium providing the damping.

To make further progress with this equation, we need to know $\boldsymbol{P}_{\mathrm{at}}$. We approach this problem in several stages. The idea is to start with a simpler system without loss and source terms and solve for the fields. The effects of loss and source terms are then included by multiplying this solution by appropriate space- and/or time-dependent factors.

So we first consider a lossless ($\sigma = 0$) and source-free region ($\boldsymbol{P}_{at} = 0$), where the equation satisfied by the field becomes

$$\left[\nabla^2 - \mu\epsilon\frac{\partial^2}{\partial t^2}\right]\boldsymbol{E}(\boldsymbol{r},t) = 0. \tag{2.31}$$

It is easily checked that, in this case, the magnetic field $\boldsymbol{B}$ also satisfies the same equation. The quantity $1/\mu\epsilon$ has the dimensions of the square of a speed v given by

$$v = \sqrt{\frac{1}{\mu\epsilon}} = \sqrt{\frac{1}{\mu_0\epsilon_0}}\sqrt{\frac{\mu_0\epsilon_0}{\mu\epsilon}} = \frac{c}{n}\,, \tag{2.32}$$

where c is the speed of light in free space and n is the refractive index of the medium,

$$c = \sqrt{\frac{1}{\mu_0\epsilon_0}} = 2.99792458 \times 10^8 \text{ m/s} \approx 3.00 \times 10^8 \text{ m/s}. \tag{2.33}$$

$$n = \sqrt{\frac{\mu\epsilon}{\mu_0\epsilon_0}}\,. \tag{2.34}$$

Equation (2.31) is then readily identified as the homogeneous (source-free) wave equation, with v as the wave speed.

2.3 Solutions of the Wave Equation

To gain some insight into these wave-like solutions of Maxwell's equations, we consider a few special cases of the scalar wave equation for fields that depend only on a single spatial variable:

$$\left(\frac{\partial^2}{\partial z^2} - \frac{1}{v^2}\frac{\partial^2}{\partial t^2}\right)F(z,t) = 0\,, \tag{2.35}$$

where $F(z,t)$ stands for any of the three Cartesian components of the field. By means of the change of variables $\xi = t - z/v$ and $\eta = t + z/v$, the derivatives in the wave equation can be transformed to

$$\frac{\partial^2}{\partial z^2} = \left(-\frac{1}{v}\frac{\partial}{\partial \xi} + \frac{1}{v}\frac{\partial}{\partial \eta}\right)\left(-\frac{1}{v}\frac{\partial}{\partial \xi} + \frac{1}{v}\frac{\partial}{\partial \eta}\right), \tag{2.36}$$

$$\frac{1}{v^2}\frac{\partial^2}{\partial t^2} = \frac{1}{v^2}\left(\frac{\partial}{\partial \xi} + \frac{\partial}{\partial \eta}\right)\left(\frac{\partial}{\partial \xi} + \frac{\partial}{\partial \eta}\right). \tag{2.37}$$

In terms of ξ and η, the wave equation becomes

$$\frac{1}{v^2}\frac{\partial^2}{\partial \xi \partial \eta}G(\xi, \eta) = 0, \tag{2.38}$$

where $G(\xi, \eta)$ is the function $F(z,t)$ expressed in terms of ξ and η. The general solution to this equation is of the form

$$G(\xi, \eta) = K_1 f(\xi) + K_2 g(\eta), \tag{2.39}$$

where K_1 and K_2 are some constants, and $f(\xi)$ and $g(\eta)$ are arbitrary functions. Transforming back to z and t, we find that the general solution to the wave equation has the form

$$F(z,t) = C_1 f(t - z/v) + C_2 g(t + z/v), \tag{2.40}$$

where C_1 and C_2 are constants. We see that $F(z,t)$ is not an arbitrary function of z and t; rather, the spatial and temporal variables occur in the combination $t - z/v$ or $t + z/v$ in its argument. Examples of acceptable waveforms are periodic functions such as $e^{-i\omega(t \pm z/v)}$ or functions of finite duration (wave packets) such as $e^{-a(t \pm z/v)^2}$, where ω and a are constants having dimensions of frequency and frequency squared, respectively.

The function $f(t - z/v)$ represents a disturbance propagating in the $+z$ direction with speed v, while $g(t + z/v)$ represents a disturbance propagating in the $-z$ direction with speed v. To see this, consider a real-valued disturbance $F(z,t) = f(t - z/v)$ in the form of a pulse consisting only of the first term. A plot of $F(z,t)$ as a function of z when t is held fixed represents the pulse shape or profile. Figure 2.2 shows the profile of this pulse at time t. To see what happens to this pulse at a later time, consider a typical point, such as P_1, where the pulse amplitude has the value $f(t - z_1/v) = F_0$. At a later time $t + dt$, the pulse will have this same value F_0 at a point z_2 such that $F(z_2, t + dt) \equiv f(t + dt - z_2/v) = F_0 = f(t - z_1/v)$. For this to happen, the argument of f at time t and $t + dt$ must be the same. This means

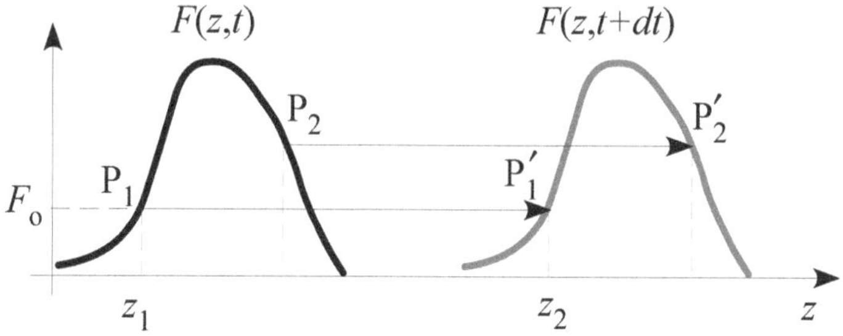

FIGURE 2.2
Pulse profile at times t and $t + dt$ for a real-valued waveform $F(z,t) \equiv f(t - z/v)$.

$t - z_1/v = t + dt - z_2/v$ or $z_2 = z_1 + vdt$. Hence, at time t, the disturbance has value F_0 at position z_1, and at time $t + dt$ it has the same value at point $z_2 = z_1 + vdt$. If dt is an infinitesimal interval, z_2 and z_1 will differ by an infinitesimal amount dz, so that $dz = vdt$. Similar considerations for other points on the pulse profile at time t show that, at time $t+dt$, the pulse has moved to the right, without changing its shape, by an amount $dz = vdt$. It is clear that the disturbance propagates with speed $v = dz/dt$ in the positive z-direction.

Similar considerations show that a real-valued function $F(z,t) = g(t+z/v)$ represents a wave propagating in the $-z$ direction with constant speed v. In general, a wave propagating in an arbitrary direction specified by the unit vector $\boldsymbol{\kappa} = \boldsymbol{k}/|\boldsymbol{k}|$ is given by

$$\mathcal{F}(\boldsymbol{r},t) = C_1 f(t - \boldsymbol{\kappa}\cdot\boldsymbol{r}/v)\,. \tag{2.41}$$

The term plane wave is used for such solutions because the surfaces of constant wave amplitude are planes perpendicular to the direction of propagation. For example, the wavefronts for $F(\boldsymbol{r},t) = Cf(t-\boldsymbol{\kappa}\cdot\boldsymbol{r}/v)$ are given by $t-\boldsymbol{\kappa}\cdot\boldsymbol{r}/v = const$. Different values of the constant define a family of planes perpendicular to $\boldsymbol{\kappa}$. These wavefronts move with constant velocity given by $d\boldsymbol{r}/dt = v\boldsymbol{\kappa}$. If an energy density is associated with the modulus squared of $F(\boldsymbol{r},t) = Cf(t-\boldsymbol{\kappa}\cdot\boldsymbol{r}/v)$, energy is transported along trajectories called rays, which, for the waves considered here, are straight lines parallel to $\boldsymbol{\kappa}$ and perpendicular to the wavefronts.

The most familiar examples of plane waves are the monochromatic (single-frequency) plane waves of angular frequency ω:

$$F(z,t) = Ae^{-i[\omega(t-z/v)-\phi_o]}\,, \tag{2.42}$$

and

$$F(\boldsymbol{r},t) = Ae^{-i[\omega(t-\boldsymbol{\kappa}\cdot\boldsymbol{r}/v)-\phi_o]}\,, \tag{2.43}$$

where A and ϕ_o are constants representing the wave amplitude and initial phase angle, respectively. Wave (2.42) has an amplitude A and phase $\phi(z,t) = \omega(t - z/v) - \phi_o$. The surfaces on which the wave has a definite amplitude and phase, at a fixed time t, are planes, $z = vt +$ const, perpendicular to the direction of propagation [Fig. 2.3(a)]. Similarly, for the wave (2.43), the wavefronts[3] are planes perpendicular to the direction of wave propagation $\boldsymbol{\kappa}$ [Fig. 2.3(b)].

It is possible to have other types of scalar solutions. For example, solutions of the form $F(\boldsymbol{r},t) = F(r,t)$ exist that satisfy the wave equation (2.31). Since F depends only on r, we can use the identity $\nabla^2 F(r,t) = \dfrac{1}{r}\dfrac{\partial^2}{\partial r^2} rF(r,t)$ and write the wave equation as

$$\left(\frac{\partial^2}{\partial r^2} - \frac{1}{v^2}\frac{\partial^2}{\partial t^2}\right) rF(r,t) = 0\,. \tag{2.44}$$

Noting the similarity of this equation to the one-dimensional wave equation (2.35), its solution can be written down at once as $rF(r,t) = C_1 f(t - r/v) + C_2 g(t + r/v)$, which can be rearranged to give

$$F(r,t) = C_1\frac{f(t-r/v)}{r} + C_2\frac{g(t+r/v)}{r}\,. \tag{2.45}$$

Here, the first term represents a spherically diverging wave, and the second term represents a spherically converging wave. Note that, for $F(r,t) = C_1 f(t - r/v)/r$, the surfaces of

[3]For a plane wave, the surfaces of constant phase are also the surfaces of constant wave amplitude. This is not the case in general. Strictly speaking, the term wavefront should be used only in the former case. Nevertheless, it is commonly used in situations where phasefront (surface of constant phase) is actually implied.

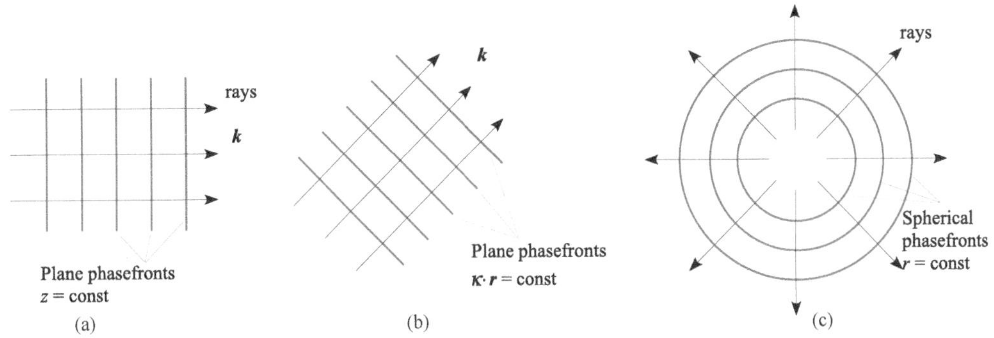

FIGURE 2.3
Wavefronts and rays for plane waves propagating (a) along the z axis and (b) in an arbitrary direction, and (c) for a spherical wave.

constant wave amplitude (wavefronts) are spheres centered at the origin, and energy is transported along radial lines diverging from the origin. The wave amplitude falls off as $1/r$, so that the total wave power remains constant as it propagates.

The simplest examples of spherical waves are the monochromatic (single-frequency) spherical waves of angular frequency ω corresponding to a point source:

$$F(r,t) = \frac{A}{r}e^{-i\omega(t-r/v)} = A\frac{e^{ikr}}{r}e^{-i\omega t}, \quad k = \omega/v = n\omega/c, \tag{2.46}$$

$$G(r,t) = \frac{A}{r}e^{-i\omega(t+r/v)} = A\frac{e^{-ikr}}{r}e^{-i\omega t}. \tag{2.47}$$

The first of these waves represents a spherical wave with phasefronts expanding radially outward from a point source at the origin [Fig. 2.3(c)], and the second represents a spherical wave with phasefronts collapsing toward a sink at the origin.

The spherical wave solution can be expressed in terms of plane waves as

$$\frac{e^{i(kr-\omega t)}}{kr} = \frac{ie^{-i\omega t}}{2\pi}\iint_{-\infty}^{\infty}\frac{1}{m}e^{ik(px+qy+m|z|)}dpdq\,, \quad m = \sqrt{1-p^2-q^2}\,.$$

This is the Weyl representation of a spherical wave. More complex scalar spherical wave solutions that behave as outgoing or incoming waves far from the origin are

$$F(\boldsymbol{r},t) = \begin{cases} \text{const} \times h_\ell^{(1)}(kr)Y_\ell^m(\theta,\varphi)\,e^{-i\omega t}, & \text{outgoing wave} \\ \text{const} \times h_\ell^{(2)}(kr)Y_\ell^m(\theta,\varphi)\,e^{-i\omega t}, & \text{incoming wave} \end{cases} \tag{2.48}$$

where $h_\ell^{(1)}(kr)$ and $h_\ell^{(2)}(kr)$ are spherical Hankel functions of the first and second kind and of (integer) order ℓ.

In cylindrical coordinates we have two-dimensional cylindrical waves

$$F(\rho,t) = C_1\frac{f(t-\rho/v)}{\sqrt{\rho}} + C_2\frac{g(t+\rho/v)}{\sqrt{\rho}}, \tag{2.49}$$

where the wavefronts are cylinders coaxial with the z-axis and the rays are radial lines diverging from the axis. Other solutions that involve Hankel functions and behave like outgoing or incoming cylindrical waves far from the z-axis also exist. We have mentioned only some of the simplest traveling wave solutions of the scalar wave equation. Many other solutions with more complex wavefronts representing standing or traveling waves are possible.

2.3.1 Monochromatic Plane Waves

Vector waves predicted by Maxwell's equations can be constructed from the solutions of the scalar wave equation because each Cartesian component of the field obeys a scalar equation. If the vector wave field has a fixed direction $\boldsymbol{\varepsilon}$ in space, then plane-wave solutions of the form $\boldsymbol{F}(\boldsymbol{r},t) = \boldsymbol{\varepsilon} f(t - \boldsymbol{\kappa}\cdot\boldsymbol{r}/v)$ exist. Thus, a monochromatic vector plane wave constructed from solutions of the scalar wave equation and propagating in the direction of the wave vector $\boldsymbol{k}$ has the form

$$\boldsymbol{F}(\boldsymbol{r},t) = \boldsymbol{F}_o e^{-i\omega(t-\boldsymbol{k}\cdot\boldsymbol{r}/v)+i\phi_o} = \boldsymbol{F}_o e^{i(\boldsymbol{k}\cdot\boldsymbol{r}-\omega t+\phi_o)}\,, \tag{2.50a}$$

where
$$\boldsymbol{k} = \frac{\omega}{v}\boldsymbol{\kappa} = \frac{n\omega}{c}\boldsymbol{\kappa}\,, \quad k^2 \equiv \boldsymbol{k}\cdot\boldsymbol{k} = \frac{\omega^2}{v^2} = \frac{n^2\omega^2}{c^2}\,, \tag{2.50b}$$

and $\boldsymbol{F}_o$ is the vector amplitude, assumed real for convenience. The wave vector magnitude k and angular frequency ω are related to the wavelength λ (in free space), frequency ν, and wave speed v (in the medium) by

$$\frac{\lambda}{n} = \frac{2\pi}{k}\,, \quad \nu = \frac{\omega}{2\pi}\,, \quad v = \frac{\lambda}{n}\nu = \frac{\omega}{k}\,. \tag{2.51}$$

Plane-wave electric and magnetic fields satisfying Maxwell's equations will also be of this form. Maxwell's equations place further restrictions on the amplitudes and relative orientations of the wave vector $\boldsymbol{k}$ and the electric and magnetic field vectors. Thus, a plane wave solution of Maxwell's equations is given by

$$\boldsymbol{E}(\boldsymbol{r},t) = \boldsymbol{E}_o e^{i(\boldsymbol{k}\cdot\boldsymbol{r}-\omega t+\phi_o)}\,, \qquad \boldsymbol{k}\cdot\boldsymbol{E} = 0\,, \tag{2.52a}$$

$$\boldsymbol{B}(\boldsymbol{r},t) = \frac{\boldsymbol{k}\times\boldsymbol{E}}{\omega} = \frac{\boldsymbol{k}\times\boldsymbol{E}_o}{\omega} e^{i(\boldsymbol{k}\cdot\boldsymbol{r}-\omega t+\phi_o)}\,, \qquad \boldsymbol{k}\cdot\boldsymbol{B} = 0\,. \tag{2.52b}$$

Note that the electric field vector can be expressed in terms of the magnetic field vector (using the Ampere-Maxwell equation and $\mu\epsilon = 1/v^2$) as $\boldsymbol{E} = -\boldsymbol{k}\times\boldsymbol{B}v^2/\omega = -\boldsymbol{\kappa}\times\boldsymbol{B}v$. These equations imply that, in a transparent medium, the electric, magnetic, and propagation vectors form a right-handed triad of mutually orthogonal vectors. This is expressed concisely by the Poynting vector $\boldsymbol{S}$ (energy flux density vector):

$$\boldsymbol{S} = \frac{\boldsymbol{\mathcal{E}}\times\boldsymbol{\mathcal{B}}}{\mu} = \frac{\mathcal{E}^2}{\mu\omega}\boldsymbol{k} = \epsilon\mathcal{E}^2 v\boldsymbol{\kappa} = \frac{\mathcal{B}^2}{\mu}v\boldsymbol{\kappa}\,. \tag{2.53}$$

Furthermore, using Eqs. (2.25), the cycle-averaged electric and magnetic energy densities $\bar{u}_e$ and $\bar{u}_m$ are equal and, therefore, contribute equally to the overall energy density $\bar{u}_{em}$ associated with the wave:

$$\overline{u}_e = \frac{1}{8}\left[\epsilon\,\boldsymbol{E}\cdot\boldsymbol{E}^* + c.c.\right] = \frac{1}{4}\mathcal{R}e\left[\epsilon\,\boldsymbol{E}_0\cdot\boldsymbol{E}_0^*\right] = \frac{1}{4}\epsilon\left|\boldsymbol{E}_o\right|^2\,, \tag{2.54a}$$

$$\overline{u}_m = \frac{1}{8}\left[\frac{1}{\mu}\boldsymbol{B}\cdot\boldsymbol{B}^* + c.c.\right] = \frac{1}{4}\mathcal{R}e\left[\frac{1}{\mu}\boldsymbol{B}_0\cdot\boldsymbol{B}_0^*\right] = \frac{1}{4\mu}\frac{k^2}{\omega^2}\left|\boldsymbol{E}_o\right|^2 = \frac{1}{4}\epsilon\left|\boldsymbol{E}_o\right|^2 = \overline{u}_e\,, \tag{2.54b}$$

$$\overline{u}_{em} \equiv \overline{u}_e + \overline{u}_m = \frac{1}{2}\epsilon\left|\boldsymbol{E}_o\right|^2\,. \tag{2.54c}$$

The cycle-averaged Poynting vector describing power flow in the wave is given by

$$\begin{aligned}\overline{\boldsymbol{S}} &= \frac{1}{2}\mathcal{R}e\left[\frac{\boldsymbol{E}_0\times\boldsymbol{B}_0^*}{\mu}\right] = \frac{1}{2}\mathcal{R}e\left[\frac{\boldsymbol{E}_0\times(\boldsymbol{k}\times\boldsymbol{E}_0^*)}{\omega\mu}\right]\,,\\ &= \left(\frac{1}{2}\epsilon|\boldsymbol{E}_o|^2 v\right)\boldsymbol{\kappa} = (\overline{u}_{em}v)\,\boldsymbol{\kappa} \equiv I\,\boldsymbol{\kappa}\,.\end{aligned} \tag{2.54d}$$

The Poynting vector thus points in the direction of wave propagation; that is, power flows in the direction of wave propagation. The direction of the Poynting vector defines the local ray direction. The magnitude of the Poynting vector, $I \equiv |\overline{\boldsymbol{S}}| = \frac{1}{2}\epsilon|\boldsymbol{E}_o|^2 v$, is referred to as the wave intensity in physics (irradiance in radiometry) and has units of W/m^2. For nonmagnetic media ($\mu \approx \mu_o$), the expression for wave intensity (irradiance in radiometry) reduces to

$$I = \frac{1}{2}\epsilon_o n|\boldsymbol{E}_o|^2 c, \qquad n = \sqrt{\frac{\epsilon}{\epsilon_o}}. \tag{2.55}$$

2.3.2 Power Flow in an Electromagnetic Wave: Poynting's Theorem

In this book we are interested in power exchange between light (an electromagnetic wave) and atoms. Energy flow in an electromagnetic wave is described by the Poynting vector $\boldsymbol{S}$, and the electromagnetic power flowing into a small volume element δV is $(-\boldsymbol{\nabla}\cdot\boldsymbol{S})\delta V$. Using Maxwell's equations (2.29c) and (2.29d), the electromagnetic power inflow per unit volume in the vicinity of any point in the wave field is given by

$$\begin{aligned} -\boldsymbol{\nabla}\cdot\boldsymbol{S} &= -\frac{1}{\mu}\boldsymbol{\nabla}\cdot(\boldsymbol{\mathcal{E}}\times\boldsymbol{\mathcal{B}}) = \frac{1}{\mu}\left[\boldsymbol{\mathcal{E}}\cdot\boldsymbol{\nabla}\times\boldsymbol{\mathcal{B}} - \boldsymbol{\mathcal{B}}\cdot\boldsymbol{\nabla}\times\boldsymbol{\mathcal{E}}\right], \\ &= \left[\epsilon\boldsymbol{\mathcal{E}}\cdot\frac{\partial\boldsymbol{\mathcal{E}}}{\partial t} + \frac{1}{\mu}\boldsymbol{\mathcal{B}}\cdot\frac{\partial\boldsymbol{\mathcal{B}}}{\partial t}\right] + \boldsymbol{\mathcal{J}}\cdot\boldsymbol{\mathcal{E}} + \boldsymbol{\mathcal{E}}\cdot\frac{\partial\boldsymbol{\mathcal{P}}_{at}}{\partial t}. \end{aligned} \tag{2.56}$$

The cycle-averaged electromagnetic power inflow per unit volume is then given by

$$\begin{aligned} -\boldsymbol{\nabla}\cdot\overline{\boldsymbol{S}} &= \frac{1}{2}\mathcal{R}e\left[\epsilon\boldsymbol{E}_)\cdot\frac{\partial\boldsymbol{E^*}_0}{\partial t} + \frac{1}{\mu}\boldsymbol{B}_0\cdot\frac{\partial\boldsymbol{B}_0^*}{\partial t}\right] + \frac{1}{2}\mathcal{R}e\left[\boldsymbol{J}_0\cdot\boldsymbol{E}_0^*\right] + \frac{1}{2}\mathcal{R}e\left[\boldsymbol{E}_0\cdot\frac{\partial\boldsymbol{P}_0^*}{\partial t}\right], \\ &= \frac{\partial}{\partial t}\left[\frac{1}{4}\epsilon|\boldsymbol{E}_0|^2 + \frac{1}{4}\frac{|\boldsymbol{B}_0|^2}{\mu}\right] + \frac{1}{2}\sigma|\boldsymbol{E}_0|^2 + \frac{1}{2}\mathcal{R}e\left[\boldsymbol{E}_0\cdot\frac{\partial\boldsymbol{P}_0^*}{\partial t}\right], \end{aligned} \tag{2.57}$$

where we have used the complex representation of narrowband fields and material responses (current and polarization) introduced in Sec. 2.1.1. The term on the left-hand side of this equation is the average electromagnetic power inflow per unit volume into the medium. The terms on the right-hand side describe how this power is distributed. The first term on the right-hand side of the second line is the rate at which the energy stored in the fields is increasing. (Energy associated with the polarization of the background medium is included in this term because we have used ϵ and μ in writing the energy stored in the fields.) The second term represents power delivered to the conduction current and dissipated as heat (Ohmic loss). The third and final term represents the average power per unit volume being delivered by the wave field to the atomic polarization $\mathcal{P}_{at}$. This term can be viewed as the rate of work done by the field in establishing the polarization or, equivalently, as the power transferred by the field to the atomic dipoles.

For the quasi-monochromatic (narrow-band) fields of interest in this book, if we write the field and polarization explicitly as $\boldsymbol{E} = \boldsymbol{E}_0(t)e^{-i\omega t}$ and $\boldsymbol{P}_{at}(t) = \boldsymbol{P}_0(t)e^{-i\omega t}$, respectively, then Poynting's theorem (2.57) for cycle-averaged quantities takes the form

$$-\boldsymbol{\nabla}\cdot\overline{\boldsymbol{S}}(t) = \frac{\partial}{\partial t}\bar{u}_{em}(t) + \frac{1}{2}\sigma|\boldsymbol{E}_0(t)|^2 + \frac{i\omega}{4}\left[\boldsymbol{E}_0(t)\cdot\boldsymbol{P}_0^*(t) - \boldsymbol{E}_0^*(t)\cdot\boldsymbol{P}_0(t)\right], \tag{2.58}$$

where
$$\bar{u}_{em}(t) = \frac{1}{2}\epsilon|\boldsymbol{E}_0(t)|^2. \tag{2.59}$$

The wave intensity can be written as

$$I(t) \equiv |\overline{\boldsymbol{S}}(t)| = \frac{1}{2}\epsilon_o nc|\boldsymbol{E}_0(t)|^2. \tag{2.60}$$

The time argument has been written explicitly here to emphasize that, for narrowband fields, cycle-averaged quantities inherit the slow variation of the field and polarization envelopes. For strictly monochromatic fields, $\bar{u}_{em}$ is constant, so that $\partial\bar{u}_{em}/\partial t = 0$. This means that there is no accumulation of electromagnetic energy in any volume element; energy entering a volume element is used in doing work on charges and dipoles within that volume. Similarly, the wave intensity is a constant for strictly monochromatic fields. Depending on the frequency spectrum of the slowly varying field envelope and on the variation of ϵ and μ with frequency (dispersion), further modifications to Eqs. (2.58) and (2.59) may be needed. Some of these modifications will be considered in later chapters.

Vector Waves

Vector spherical or cylindrical wave solutions are more complex even in the simplest case. For example, the simplest vector spherical wave allowed by Maxwell's equations has the form

$$\boldsymbol{F}(\boldsymbol{r},t) = \text{const} \times \left[\frac{e^{-i\omega(t-r/v)}}{kr} - i\frac{e^{-i\omega(t-r/v)}}{(kr)^2}\right] \sin\theta\, \boldsymbol{e}_\varphi\,, \quad k = \omega/v\,, \tag{2.61}$$

where ω is the frequency. We encounter this and other types of vector spherical waves in the context of scattering and radiation problems in electrodynamics. Even in these cases, however, the fields satisfy plane-wave-like relations, that is, at each point in the wave field, the vectors $\boldsymbol{\mathcal{E}}, \boldsymbol{\mathcal{B}}$, and the Poynting vector $\boldsymbol{S}$ form a right-handed triad of mutually orthogonal vectors, and wave intensity is given by

$$I(\boldsymbol{r}) = \frac{1}{2}\epsilon_o nc|E(\boldsymbol{r})|^2\,. \tag{2.62}$$

We conclude this section by observing that Maxwell's equations admit solutions that also satisfy the wave equation. However, not all solutions of the wave equation are admissible solutions of Maxwell's equations; only those that also satisfy the constraints imposed by Maxwell's equations describe electromagnetic waves. This distinction must be kept in mind, even though there are situations in which the vector character of the field can be ignored. This is discussed further in Secs. 2.6 and 3.3.

2.4 Monochromatic Wave Solutions with Boundaries

We have seen that waves of the type $F(\boldsymbol{r},t) = f(t - \boldsymbol{\kappa}\cdot\boldsymbol{r}/v)$ in free space retain their shape (profile) as they propagate. Strictly speaking, this is true only for monochromatic plane waves with a sinusoidal profile. For other waveforms, such as pulses, which can be considered, via Fourier decomposition, as superpositions of sinusoidal waves, this is possible only if the wave speed is independent of frequency. For light waves, free space provides the important case in which this condition is exactly satisfied. In all material media, including air, the speed of light $v = \sqrt{1/\mu\epsilon} = c/n$ [Eqs. (2.32) and (2.34)] depends on frequency because the material properties ϵ and μ and, therefore, the refractive index $n = \sqrt{\mu\epsilon/(\mu_0\epsilon_0)}$ depend on frequency, and pulse shapes evolve as they propagate. The dependence of the speed of light on frequency is referred to as dispersion. If the variation of the speed of light over the range of frequencies involved in the Fourier decomposition of the wave is small, the wave can still propagate without significant change in shape over limited distances. The conditions under which this is possible are of great practical interest in laser pulse propagation, for example, in optical fiber telecommunications, and will be considered in later chapters.

Dispersion of light is not solely due to material properties; the presence of boundaries in optical cavities and waveguides also contributes to dispersion. The presence of boundaries also modifies the spatial dependence of the fields. To appreciate these effects and set the stage for the discussion in the following chapters, we consider more general monochromatic wave solutions of Maxwell's equations. This allows us to see how dispersion arises and, by considering superpositions of these monochromatic solutions (inverse Fourier transforms), investigate how dispersion distorts a waveform as it propagates and gain some insight into how dispersion can be overcome.

Monochromatic plane waves (2.52) are the simplest solutions of the wave equation. To allow for more complex spatial dependence of the wave amplitude in the presence of boundaries and interactions with material media, we write the complex electric field as $\boldsymbol{E}(\boldsymbol{r},t) = \boldsymbol{E}(\boldsymbol{r})e^{-i\omega t}$, where the vector amplitude $\boldsymbol{E}(\boldsymbol{r})$ is now a function of position. Assuming that the atomic response follows the field, we can write the atomic polarization as $\boldsymbol{P}_{at}(\boldsymbol{r},t) = \epsilon_0\chi_{at}(\omega)\boldsymbol{E}(\boldsymbol{r})e^{-i\omega t}$. To avoid a proliferation of symbols, we have denoted the space-dependent part of the field by the same symbol as the total field. Using this expression in the wave equation (2.30), we find that the spatially dependent electric field amplitude $\boldsymbol{E}(\boldsymbol{r})$ satisfies

$$\left[\nabla^2 + \omega^2\mu\epsilon\left(1 + \epsilon_o\chi_{at}/\epsilon + i\sigma/\omega\epsilon\right)\right]\boldsymbol{E}(\boldsymbol{r}) = 0\,. \tag{2.63}$$

This is a fundamental equation of this course; given the atomic response characterized by $\chi_{at}(\omega)$ and the material properties $\mu(\omega)$ and $\epsilon(\omega)$, this equation governs the behavior of the field. A major part of our discussion will be devoted to solving this equation under different approximations, together with the equation that determines the atomic susceptibility $\chi_{at}(\omega)$ describing the atomic response to the field.

In a loss-free medium with no resonant interaction ($\sigma = 0 = \chi_{at}$), the field amplitude satisfies the homogeneous Helmholtz equation

$$\left[\nabla^2 + \omega^2\mu\epsilon\right]\boldsymbol{E}(\boldsymbol{r}) = 0\,. \tag{2.64}$$

The magnitude of the propagation constant (assuming μ and ϵ, both functions of frequency, real over the range of frequencies of interest) is given by

$$k = \sqrt{\mu\epsilon}\,\omega = \frac{\omega}{v} = \frac{n\omega}{c}\,, \tag{2.65}$$

where $v = c/n$ is the wave speed in terms of the speed of light in free space c and the refractive index n of the medium.

The plane-wave solution $\boldsymbol{E}(\boldsymbol{r}) = \boldsymbol{\varepsilon}E_o e^{i\boldsymbol{k}\cdot\boldsymbol{r}}$, where $\boldsymbol{\varepsilon}$ is a constant polarization vector (elliptical, in general) and E_0 is the field amplitude, is a special solution of this equation. Combining it with the time-dependence $e^{-i\omega t}$, we see that $\boldsymbol{E}(\boldsymbol{r},t) = \boldsymbol{\varepsilon}E_o e^{i(\boldsymbol{k}\cdot\boldsymbol{r}-\omega t)}$ with $\boldsymbol{k}\cdot\boldsymbol{\varepsilon} = 0$, represents a plane electromagnetic wave propagating in the direction $\boldsymbol{\kappa} = \boldsymbol{k}/|\boldsymbol{k}|$ in a loss-free medium [cf. Eq. (2.52)]. Plane-wave solutions like this exist for all frequencies in a homogeneous medium and are extremely useful in discussing propagation of electromagnetic waves in such media.

In the presence of boundaries, which will inevitably be present when we consider feedback in a laser, the form of the solution depends on the shape and material of the boundaries. There are restrictions on the frequencies and wave vectors that are allowed. We discuss the most important changes to the solutions of the wave equation and Maxwell's equations that arise in the presence of boundaries in the next section.

2.5 Solutions in a Cavity: Mode Density

Wave-like solutions of Maxwell's equations also exist in the presence of boundaries. Such solutions satisfy both the wave equation and Maxwell's equations, as well as the boundary conditions. This situation arises naturally in the context of lasers, which require resonators–arrangements of reflecting and refracting surfaces that provide for optical feedback.

Let us recall that the amplification of a propagating light signal requires its repeated passage (feedback) through a collection of excited atoms (gain medium). An excited population of atoms is necessary but not sufficient to produce sufficiently large amplification of light to overcome its propagation losses due to absorption and scattering and thereby increase its amplitude. Optical resonators, which provide optical feedback by causing multiple passages of light through the gain medium, are therefore an essential element of lasers. They are needed to

(i) store and build up light energy at the frequency of interest, since the rate of stimulated emission is proportional to the light energy density at that particular frequency;

(ii) act as spatial and frequency filters to select fields with specific spatial variation and frequencies. Spatial filtering is responsible for the collimation (directionality) of the output, while frequency filtering is responsible for the narrow spectral bandwidth of optical signals leaking out of the resonator.

A figure of merit quantifying how well a resonator performs these tasks is the so-called cavity-Q (quality) factor. Because of the presence of boundaries, only certain spatial field configurations and frequencies can exist in a resonator. They are referred to as the modes of the resonator. The specific form of the modes depends on the geometry (rectangular, cylindrical, spherical, etc.) and the nature of the boundaries (dielectric, metallic). Here we consider certain features common to all resonators by taking a rectangular cavity as an example. Although we consider a specific geometry, the conclusions are independent of the shape provided all dimensions are large compared to a wavelength in the medium inside the cavity.

For a rectangular cavity with perfectly conducting walls and each side of length L, the electric field satisfies the wave equation and the boundary condition that its tangential component (the component parallel to the wall) vanish at the walls $x = 0$, $x = L$, $y = 0$, $y = L$, $z = 0$ and $z = L$ [Fig. 2.4]. Only the normal component of the electric field can be

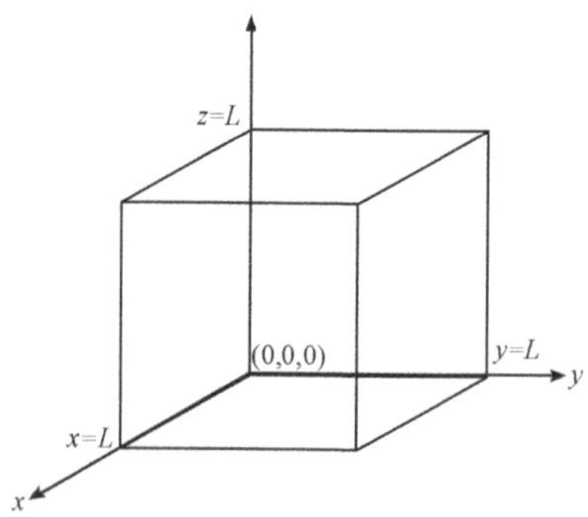

FIGURE 2.4
A rectangular cavity with conducting walls will support a discrete set of modes to allow the electric and magnetic field vectors to satisfy certain boundary conditions at the walls.

nonzero at the surface of a perfect conductor. For a monochromatic wave, the electric field has the form $\boldsymbol{E}(\boldsymbol{r},t) = \boldsymbol{u}(\boldsymbol{r})e^{-i\omega t}$, where the vector function $\boldsymbol{u}(\boldsymbol{r})$ satisfying these boundary conditions is given by[4]

$$u_x(\boldsymbol{r}) = A_{01}\cos\left(\frac{m_1\pi x}{L}\right)\sin\left(\frac{m_2\pi y}{L}\right)\sin\left(\frac{m_3\pi z}{L}\right), \tag{2.66}$$

$$u_y(\boldsymbol{r}) = A_{02}\sin\left(\frac{m_1\pi x}{L}\right)\cos\left(\frac{m_2\pi y}{L}\right)\sin\left(\frac{m_3\pi z}{L}\right), \tag{2.67}$$

$$u_z(\boldsymbol{r}) = A_{03}\sin\left(\frac{m_1\pi x}{L}\right)\sin\left(\frac{m_2\pi y}{L}\right)\cos\left(\frac{m_3\pi z}{L}\right), \tag{2.68}$$

where A_{01}, A_{02}, A_{03} are constants and m_1, m_2, m_3 are a set of nonnegative integers. Then monochromatic solutions of the form $\boldsymbol{u}(\boldsymbol{r})e^{-i\omega t}$ exist provided the mode frequency $\nu = \omega/2\pi$ is restricted to values

$$\nu \equiv \frac{\omega}{2\pi} = \frac{v}{2L}\sqrt{(m_1^2 + m_2^2 + m_3^2)}\,, \tag{2.69}$$

where v is the wave speed. Equation (2.69) for the mode frequency follows when Eqs. (2.66)–(2.68), together with Eq. (2.65), are used in Eq. (2.64). Note that all three functions u_x, u_y, and u_z, as well as the mode frequency ν, must be labeled by the same three integer indices (m_1, m_2, m_3) for the field to satisfy the wave equation. Also, no more than one of the indices may be zero if the field is to be nonzero. In addition, Maxwell's equations require that the divergence of the electric field vanish ($\boldsymbol{\nabla}\cdot\boldsymbol{E} = 0$) in the charge-free interior of the box, giving the additional constraint

$$m_1 A_{01} + m_2 A_{02} + m_3 A_{03} = 0\,. \tag{2.70}$$

The last equation implies that, for a given set of allowed integers (m_1, m_2, m_3), only two of the constants A_{0i} can be chosen independently. These correspond to two independent orientations (polarizations) of the electric vector. The set of integers (m_1, m_2, m_3) and the corresponding electric vector define a mode of the cavity. The magnetic field associated with this electric field can be calculated using Maxwell's equations.

If we introduce the wave vector $\boldsymbol{k}$ and the electric field amplitude $\boldsymbol{E}_o$ of a mode with indices (m_1, m_2, m_3) by

$$\boldsymbol{k} \equiv k_1\boldsymbol{e}_x + k_2\boldsymbol{e}_y + k_3\boldsymbol{e}_z = \frac{\pi}{L}\left[m_1\boldsymbol{e}_x + m_2\boldsymbol{e}_y + m_3\boldsymbol{e}_z\right], \tag{2.71}$$

$$\boldsymbol{E}_o \equiv E_{ox}\boldsymbol{e}_x + E_{oy}\boldsymbol{e}_y + E_{oz}\boldsymbol{e}_z = A_{o1}\boldsymbol{e}_x + A_{o2}\boldsymbol{e}_y + A_{o3}\boldsymbol{e}_z\,, \tag{2.72}$$

where $\hat{\mathbf{e}}_x, \hat{\mathbf{e}}_y$, and $\hat{\mathbf{e}}_z$ are unit vectors along the three Cartesian coordinate axes, we find that Eqs. (2.69) and (2.70) can be rewritten as

$$k = \sqrt{\boldsymbol{k}\cdot\boldsymbol{k}} = \frac{\pi}{L}\sqrt{(m_1^2 + m_2^2 + m_3^2)} = \frac{\omega}{v} = \frac{2\pi}{\lambda}\,, \tag{2.73}$$

$$\boldsymbol{k}\cdot\boldsymbol{E}_0 = 0\,. \tag{2.74}$$

Recalling that, in writing these equations, we have suppressed mode indices on wave vector, field amplitude, and frequency, these relations for each mode are analogous to Eqs. (2.50) and (2.52) for the solutions in unbounded space. The resulting mode functions satisfy the usual orthogonality condition

$$\int_V \boldsymbol{u}_{m_1m_2m_3}\cdot\boldsymbol{u}^*_{m_1m_2m_3}\,d^3r = \delta_{m_1m_1'}\delta_{m_2m_2'}\delta_{m_3m_3'}\,. \tag{2.75}$$

[4]For simplicity of writing, the labeling of mode functions u_x, u_y, u_z, mode amplitudes A_{oi}, frequencies, etc. by the integer indices m_1, m_2, m_3 has been suppressed.

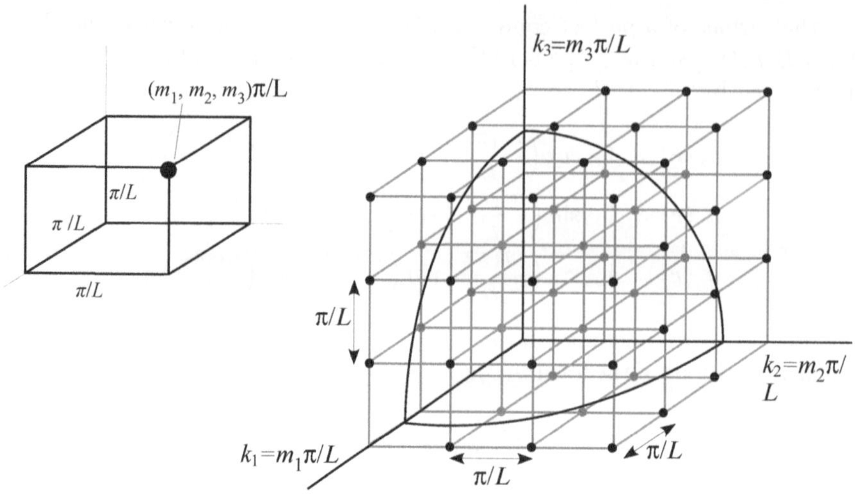

FIGURE 2.5
Each mode of a rectangular cavity, characterized by three nonnegative integers m_1, m_2 and m_3, can be represented as a point in the positive octant of (m_1, m_2, m_3) space. Note that the points on the axes do not represent allowed modes, since no more than one of the three mode indices may be zero. Each point is counted twice corresponding to two polarization states of the field associated with a given wave vector.

These modes of a rectangular cavity can be represented by discrete points in a three-dimensional space spanned by the (m_1, m_2, m_3) axes, as shown in Fig. 2.5. Each point is associated with a unique wave vector and represents two modes, which correspond to the two independent polarizations associated with that wave vector. With this mode structure, we can think of the mode frequency ν [Eq. (2.69)] as representing the distance of a point, $(m_1,\, m_2,\, m_3)v/2L$, from the origin. Then the number of modes in the frequency interval between ν and $\nu + d\nu$, which is the same as the number of modes whose $\boldsymbol{k}$-vectors have magnitudes between $k = 2\pi\nu/v$ and $k + dk = 2\pi(\nu + d\nu)/v$, is equal to the number of points inside a spherical shell of radius $k = 2\pi\nu/v$ and thickness $dk = 2\pi d\nu/v$. The points representing the modes are separated in k-space by π/L along each axis, so that the volume of k-space occupied by one mode is $(\pi/L)^3$. The volume of a spherical shell of radius k and thickness dk in k-space is $4\pi k^2 dk$. Therefore, provided the radius of the shell is large, the number of modes in the shell can be obtained by dividing the shell volume by the volume of one mode:

$$
\begin{aligned}
dN &= \frac{\frac{1}{8}(\text{volume of spherical shell})}{\text{volume occupied by one mode}} \times 2 = \frac{1}{4}\frac{4\pi k^2 dk}{(\pi/L)^3} = \frac{\pi\left(\frac{2\pi\nu}{v}\right)^2\left(\frac{2\pi d\nu}{v}\right)}{(\pi/L)^3}\,, \\
&= \frac{8\pi n^3 \nu^2 d\nu V}{c^3}\,, \quad v = c/n\,,
\end{aligned}
\tag{2.76}
$$

where $V = L^3$ is the cavity volume. The factor $\frac{1}{8}$ in the first line of this equation accounts for the fact that mode indices are positive integers; hence, only the points in the positive octant of the sphere in k-space count toward the number of modes. The factor of 2 takes into account the two polarizations of light allowed for each wave vector $k_{m_1 m_2 m_3}$. As the size L of the cavity increases, the spacing of the allowed modes decreases, and the number of modes in the interval from ν to $\nu + d\nu$ increases. However, the density of modes (the number of modes per unit volume per unit frequency interval),

$$g(\nu) \equiv \frac{1}{V}\frac{dN}{d\nu} = \frac{8\pi n^3 \nu^2}{c^3} \tag{2.77}$$

remains constant. In fact, for sufficiently large cavities, the mode density is independent of the size, shape, and material of the walls of the resonator. Free space may be considered as the $L \to \infty$ limit, in which the modes are distributed continuously in frequency (or wave vector) space, with the mode density given by Eq. (2.77).

If a cavity of the type we just considered is used as a laser resonator with a gain medium consisting of excited atoms with transition frequency ν_o inside the cavity, the number of cavity modes p falling within the laser transition will be

$$p = g(\nu_0)V\Delta\nu_{\rm at} = \frac{8\pi\nu_0^2}{c^3}V\Delta\nu_{\rm at} = 8\pi\left(\frac{V}{\lambda_0^3}\right)\frac{\Delta\nu_{\rm at}}{\nu_0}, \tag{2.78}$$

where $\Delta\nu_{at}$ is the width of the laser transition (approximately, the range of frequencies, centered at the transition frequency, that interact strongly with the atoms), and we have assumed the refractive index to be $n = 1$. For a frequency $\nu_0 = 5 \times 10^{14}$ Hz ($\lambda_0 = 600$ nm), $\Delta\nu_{at} = 1.5 \times 10^9$ Hz, and a cavity of volume V=1 cm^3, we find that the number of modes interacting strongly with the atoms is of order $p = 3.5 \times 10^8$. All of these modes will have access to the atomic excitation. They will have similar losses and feedback, so that the oscillation, if it occurs at all, would occur at a very large number of frequencies. Such a behavior would be highly undesirable because it would result in light from the laser being emitted in a wide spectral range and in all directions (and hence with no collimation).

How do we reduce the number of cavity modes strongly coupled to the atoms? One possibility (suggested by the number p being proportional to the cavity volume V) is to use a small cavity. Suppose we want $p = 1$ at λ_0=600 nm. This will require a cavity of volume

$$V = p\frac{\lambda_0^3}{8\pi}\frac{\nu_0}{\Delta\nu_{\rm at}} = 1 \times \frac{(0.6\,\mu{\rm m})^3}{8\pi}\frac{5 \times 10^{14}}{1 \cdot 5 \times 10^9} = (14\ \mu{\rm m})^3\ ! \tag{2.79}$$

Such a cavity, although not impossible nowadays, would not be practical because we need some room to place the amplifying medium. Moreover, even if we could fit the gain medium, the gain would be limited and perhaps not sufficient to overcome losses suffered by light due to absorption, scattering, diffraction, and other processes.

These problems can be overcome, at least partially, by employing open resonators, which were first suggested by Schawlow and Townes and Prokhorov.[5] In open resonators, there are no conducting side walls. The gain medium is placed between two carefully aligned reflecting (conducting) surfaces. To see how this might help, note that each standing-wave mode we have discussed can be considered a superposition of two equal-amplitude waves traveling in opposite directions (individually satisfying periodic boundary conditions over a roundtrip). In open resonators, only modes that correspond to a superposition of waves traveling very nearly parallel to the resonator axis will have low enough losses to store photons long enough to stimulate further emission by excited atoms. Confinement can be improved further by employing curved mirrors whose focusing action better confines waves traveling at small angles to the resonator axis. These modes, with wave vectors nearly parallel to the resonator axis, are called paraxial modes. If we choose the resonator axis

[5] A. L. Schawlow and C. H. Townes, Phys. Rev. 112, 1940 (1958); A. M. Prokhorov, Sov. Phys. JETP 1, 1140(1958).

to be the z-axis, the axial and transverse components of paraxial wave vectors satisfy the constraints

$$\frac{k_3}{k} = \frac{m_3\pi/L}{(\pi/L)\sqrt{m_1^2+m_2^2+m_3^2}} \approx 1\,, \tag{2.80a}$$

$$\frac{k_1}{k} = \frac{m_1\pi/L}{(\pi/L)\sqrt{m_1^2+m_2^2+m_3^2}} \approx \frac{m_1}{m_3} \ll 1\,, \tag{2.80b}$$

$$\frac{k_2}{k} = \frac{m_2\pi/L}{(\pi/L)\sqrt{m_1^2+m_2^2+m_3^2}} \approx \frac{m_2}{m_3} \ll 1\,. \tag{2.80c}$$

Thus, for each value of m_3, we have a small group of paraxial waves with frequencies close to $\nu = \frac{v}{2L}\sqrt{m_1^2+m_2^2+m_3^2} \approx m_3\frac{v}{2L}$, which have the potential to be amplified. We call $\nu_{m_3} = \frac{m_3 v}{2L}$ a resonance frequency of the open cavity and refer to the small group of paraxial waves associated with this frequency as a quasi-mode of the cavity, because true modes (stationary modes) do not exist for open cavities. A spatial superposition of paraxial waves associated with the frequency ν_{m_3} will create a spatial wave packet that spreads slowly as it propagates because of the presence of off-axis wave vectors in the spatial superposition. The electric field of such a quasi-mode will be of the form $\boldsymbol{E}(\boldsymbol{r},t) = E_o\boldsymbol{u}(\boldsymbol{r})e^{-i\omega t}$. However, even such a paraxial mode will suffer some energy loss due to diffraction every time the wave-packet hits the end mirrors; energy will also be lost due to the finite size of the end mirrors. There may be other losses due to scattering and absorption inside the medium filling the cavity or at the cavity walls. Because of the decay of field energy, the quasi-modes will have a space-time structure

$$\boldsymbol{E}(\boldsymbol{r},t) = E_o\boldsymbol{u}(\boldsymbol{r})e^{-i\omega t}e^{-t/2\tau_c}\,, \tag{2.81}$$

where E_o is the field amplitude at time $t = 0$, τ_c is a characteristic time scale over which the energy of the field decays, and the function $\boldsymbol{u}(\boldsymbol{r})$ represents a superposition of paraxial waves that determines the transverse extent of the quasi-mode. It is clear that, in the absence of amplification, the quasi-mode field amplitude decays with time and also decreases away from the resonator axis in the transverse directions. In the rest of this course, when we talk about modes of open resonators, we will be referring to these quasi-modes. These modes decay on a time scale millions of optical cycles long, and their spatial extent spans many wavelengths.

To determine the explicit form of the quasi-modes in open cavities, one must solve Maxwell's equations in the presence of mixed boundary conditions. This is a complex problem. Fortunately, a way around this formidable mathematical problem is to use the complementary description of wave propagation in terms of rays, which, as we have seen in the preceding section, are geometrical curves orthogonal to phasefronts, along which electromagnetic energy is transported. In essence, we will be looking for ray trajectories that stay confined within open resonator structures. Such rays are related to the phasefronts of long-lived quasi-modes of open resonators. Once we have identified this type of ray, we can look for waves with phasefronts matching these rays. This approach works very well for paraxial modes of open resonators at optical frequencies.

2.6 Ray Optics

In ray optics,[6] optical energy is regarded as being transported along certain curves called light rays. As we saw in Sec. 2.3.1, these are geometrical curves orthogonal to the phasefronts

[6]The branch of optics in which the wavelength λ is neglected in comparison with a characteristic length ℓ of the problem is known as geometrical optics. Examples of ℓ include the length scale over which the

along which light energy is transported. Since the Poynting vector describes the local flow of electromagnetic energy, rays indicate the local direction of the Poynting vector.

We will show that, in the short-wavelength limit, an electromagnetic wave locally has the same general character as a plane wave. This result holds quite generally when the characteristic length over which the properties of the medium or the field vary is much larger than the wavelength. Thus, the laws of reflection and refraction, which are usually established for plane waves incident on a plane boundary, remain valid under much more general conditions. For example, when a light ray encounters a sharp boundary, such as the surface of a lens, it is divided into a reflected ray and a transmitted ray according to the laws of reflection and refraction.

The preceding discussion implies that, when the wavelength is sufficiently small, optical phenomena can be understood in terms of geometrical considerations by determining the paths of light rays. In this limit, monochromatic wave solutions of Maxwell's equations can be written in the form

$$\boldsymbol{E}(\boldsymbol{r},t) = \boldsymbol{e}(\boldsymbol{r})\, e^{-i(\omega t - k_o \Psi(\boldsymbol{r}))}\,, \qquad \boldsymbol{B}(\boldsymbol{r},t) = \boldsymbol{b}(\boldsymbol{r})\, e^{-i(\omega t - k_o \Psi(\boldsymbol{r}))}\,. \tag{2.82}$$

Here $k_o = \omega/c = 2\pi/\lambda$, and the vector amplitudes $\boldsymbol{e}(\boldsymbol{r})$ and $\boldsymbol{b}(\boldsymbol{r})$ are slowly varying[7] functions of position, and $\Psi(\boldsymbol{r})$ is a real scalar function that remains to be determined. Substituting these into Maxwell's equations for transparent media (μ and ϵ real),

$$\begin{aligned} &\boldsymbol{\nabla}\cdot\epsilon\boldsymbol{E} = 0\,, && \boldsymbol{\nabla}\cdot\boldsymbol{B} = 0\,,\\ &\boldsymbol{\nabla}\times\boldsymbol{E} = i\omega\boldsymbol{B}\,, && \boldsymbol{\nabla}\times\tfrac{\boldsymbol{B}}{\mu} = -i\omega\epsilon\boldsymbol{E}\,, \end{aligned} \tag{2.83}$$

and cancelling the common factor $e^{-i[\omega t - k_o\Psi(\boldsymbol{r})]}$ from both sides of the resulting equations, we obtain

$$\epsilon\left[\boldsymbol{e}\cdot\boldsymbol{\nabla}\ln\epsilon + \boldsymbol{\nabla}\cdot\boldsymbol{e} + ik_0\boldsymbol{e}\cdot\boldsymbol{\nabla}\Psi(\boldsymbol{r})\right] = 0\,, \tag{2.84a}$$

$$\left[\boldsymbol{\nabla}\cdot\boldsymbol{b} + ik_0\boldsymbol{b}\cdot\boldsymbol{\nabla}\Psi(\boldsymbol{r})\right] = 0\,, \tag{2.84b}$$

$$\left[\boldsymbol{\nabla}\times\boldsymbol{e} + ik_0\boldsymbol{\nabla}\Psi(\boldsymbol{r})\times\boldsymbol{e}\right] = i\omega\boldsymbol{b}\,, \tag{2.84c}$$

$$\frac{1}{\mu}\left[-\boldsymbol{\nabla}\ln\mu\times\boldsymbol{b} + \boldsymbol{\nabla}\times\boldsymbol{b} + ik_0\boldsymbol{\nabla}\Psi(\boldsymbol{r})\times\boldsymbol{b}\right] = -i\omega\epsilon\boldsymbol{e}\,. \tag{2.84d}$$

Rearranging these equations, we find

$$\boldsymbol{e}\cdot\boldsymbol{\nabla}\Psi(\boldsymbol{r}) = \frac{1}{ik_0}\left[\boldsymbol{e}\cdot\boldsymbol{\nabla}\ln\epsilon + \boldsymbol{\nabla}\cdot\boldsymbol{e}\right] \tag{2.85a}$$

$$\boldsymbol{b}\cdot\boldsymbol{\nabla}\Psi(\boldsymbol{r}) = \frac{1}{ik_0}\left[\boldsymbol{\nabla}\cdot\boldsymbol{b}\right] \tag{2.85b}$$

$$\boldsymbol{\nabla}\Psi(\boldsymbol{r})\times\boldsymbol{e} - c\boldsymbol{b} = -\frac{1}{ik_0}\left[\boldsymbol{\nabla}\times\boldsymbol{e}\right] \tag{2.85c}$$

$$\boldsymbol{\nabla}\Psi(\boldsymbol{r})\times\boldsymbol{b} + c\mu\epsilon\boldsymbol{e} = -\frac{1}{ik_0}\left[-(\boldsymbol{\nabla}\ln\mu)\times\boldsymbol{b} + \boldsymbol{\nabla}\times\boldsymbol{b}\right] \tag{2.85d}$$

If the fractional changes in μ, ϵ, and the amplitudes $\boldsymbol{e}$ and $\boldsymbol{b}$ over distances of a few wavelengths are small, the terms on the right-hand side of these equations may be neglected. We

---

refractive index of a medium changes significantly and the radius of curvature of an interface between two media. In this approximation, the laws governing the transport of electromagnetic energy may be formulated in terms of geometrical considerations.

[7]That is, their fractional changes are negligible over distances of the order of a few wavelengths.

then obtain

$$\boldsymbol{e}(\boldsymbol{r}) \cdot \boldsymbol{\nabla}\Psi(\boldsymbol{r}) = 0\,, \tag{2.86a}$$
$$\boldsymbol{b}(\boldsymbol{r}) \cdot \boldsymbol{\nabla}\Psi(\boldsymbol{r}) = 0\,, \tag{2.86b}$$
$$\boldsymbol{\nabla}\Psi(\boldsymbol{r}) \times \boldsymbol{e}(\boldsymbol{r}) - c\boldsymbol{b}(\boldsymbol{r}) = 0\,, \tag{2.86c}$$
$$\boldsymbol{\nabla}\Psi(\boldsymbol{r}) \times \boldsymbol{b}(\boldsymbol{r}) + c\mu\epsilon\boldsymbol{e}(\boldsymbol{r}) = 0\,. \tag{2.86d}$$

Note that the first two equations follow from the last two by taking their scalar products with $\boldsymbol{\nabla}\Psi(\boldsymbol{r})$. It follows from Eq. (2.86) that $\boldsymbol{e}(\boldsymbol{r})$, $\boldsymbol{b}(\boldsymbol{r})$, and $\boldsymbol{\nabla}\Psi(\boldsymbol{r})$ form a right-handed triad of mutually orthogonal vectors at each point $\boldsymbol{r}$. Furthermore, since $\boldsymbol{\nabla}\Psi(\boldsymbol{r})$ is perpendicular to the surface $\Psi(\boldsymbol{r})$ = const, the vectors $\boldsymbol{e}(\boldsymbol{r})$ and $\boldsymbol{b}(\boldsymbol{r})$ are tangent to this surface. Since the phase is $k_o\Psi(\boldsymbol{r})$, this is a surface of constant phase, and may therefore be called the geometrical-optics wave surface or geometrical wavefront.

By eliminating $\boldsymbol{e}$ or $\boldsymbol{b}$ from Eqs. (2.86c) and (2.86d), we find that the condition for nontrivial solutions of Eq. (2.86) is

$$[\boldsymbol{\nabla}\Psi(\boldsymbol{r})]^2 = \mu\epsilon c^2 = n^2(\boldsymbol{r})\,, \tag{2.87}$$

where $n(\boldsymbol{r})$ is the refractive index of the medium given by

$$n(\boldsymbol{r}) = \sqrt{\frac{\mu(\boldsymbol{r})\epsilon(\boldsymbol{r})}{\mu_0\epsilon_0}}\,. \tag{2.88}$$

In general, $n(\boldsymbol{r})$ is a function of position because ϵ and μ are functions of position. Equation (2.87) is known as the eikonal equation and is the basic equation of geometrical optics.

By definition, $\boldsymbol{\nabla}\Psi(\boldsymbol{r})$ is perpendicular to the surface $\Psi(\boldsymbol{r})$ = const. According to the eikonal equation (2.87), its magnitude is $n(\boldsymbol{r})$. Thus the vector

$$\boldsymbol{s}(\boldsymbol{r}) = \frac{\boldsymbol{\nabla}\Psi(\boldsymbol{r})}{n(\boldsymbol{r})}\,, \tag{2.89}$$

has unit magnitude and is perpendicular to the surface $\Psi(\boldsymbol{r})$ =const. With the help of Eqs. (2.86c), (2.86d), and (2.89), we can express the magnetic and electric field vectors as

$$\boldsymbol{b}(\boldsymbol{r}) = \frac{\boldsymbol{s} \times \boldsymbol{e}(\boldsymbol{r})}{v}\,, \tag{2.90a}$$
$$\boldsymbol{e}(\boldsymbol{r}) = -v\boldsymbol{s} \times \boldsymbol{b}(\boldsymbol{r})\,, \tag{2.90b}$$

where $v = c/n$ is the wave speed in the medium. The time-averaged electric and magnetic energy densities are then

$$\overline{u}_e = \frac{1}{4}\epsilon|\boldsymbol{e}|^2\,, \tag{2.91a}$$
$$\overline{u}_m = \frac{1}{4}\frac{|\boldsymbol{b}|^2}{\mu} = \frac{1}{4}\frac{|\boldsymbol{e}|^2}{\mu v^2} = \frac{1}{4}\epsilon\frac{|\boldsymbol{e}|^2}{\mu\epsilon v^2} = \frac{1}{4}\epsilon|\boldsymbol{e}|^2 = \overline{u}_e\,. \tag{2.91b}$$

Thus, in the geometrical-optics limit, the time-averaged electric and magnetic energy densities associated with a monochromatic wave in a transparent medium are equal.

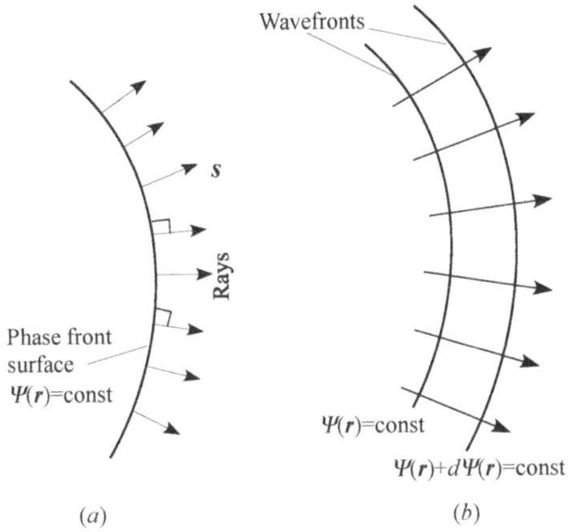

FIGURE 2.6
(a) A geometrical wavefront and the direction of the unit vector s. (b) Rays are directed trajectories perpendicular to phasefronts and pointing in the direction of energy flow.

The time-averaged Poynting vector (energy flux density vector) and the wave intensity are given by

$$\overline{\boldsymbol{S}} = \frac{1}{2}\mathcal{R}e\left[\frac{\boldsymbol{e}\times\boldsymbol{b}^*}{\mu}\right] = \frac{1}{2}\mathcal{R}e\left[\frac{\boldsymbol{e}\times(n\boldsymbol{s}\times\boldsymbol{e}^*)}{\mu c}\right] = \frac{1}{2\mu c}\mathcal{R}e\left[n\boldsymbol{s}(\boldsymbol{e}\cdot\boldsymbol{e}^*) - \boldsymbol{e}^*(\boldsymbol{e}\cdot n\boldsymbol{s})\right],$$
$$= \left(\frac{c}{n}\right)\left(\frac{1}{2}\epsilon\boldsymbol{e}\cdot\boldsymbol{e}^*\right)\boldsymbol{s} = v\overline{u}_{em}\boldsymbol{s}, \tag{2.92}$$
$$I(\boldsymbol{r}) = \frac{1}{2}\epsilon_o nc|\boldsymbol{e}(\boldsymbol{r})|^2. \tag{2.93}$$

Thus, the average Poynting vector is directed along the normal to the geometrical wavefront $\Psi(\boldsymbol{r}) = \text{const}$ [see Fig. 2.6]. Its magnitude is equal to the product of the average electromagnetic energy density and the wave speed $v = c/n$ in the medium. It follows from Eqs. (2.90)–(2.92) that, in the geometrical-optics limit, the fields locally have the character of a plane wave, even though both the fields and the ray direction may vary from point to point.

2.7 Ray Equation

We can now define geometrical light rays as the orthogonal trajectories of the geometrical wavefronts $\Psi(\boldsymbol{r}) = \text{const}$. We regard a light ray as a directed curve whose direction coincides everywhere with that of the average Poynting vector. We may therefore say that, in geometrical optics, light energy is transported along the light rays. The differential equation governing a ray is derived as follows. Let $\boldsymbol{r}(s)$ denote the position vector of a point P on a ray, considered as a function of the arc length s along the ray, measured from some fixed point on it [Fig. 2.7]. Then the unit vector $\boldsymbol{s} = d\boldsymbol{r}/ds$ is tangent to the ray and points in the direction of energy flow. Using the relation between $\boldsymbol{s}$ and $\boldsymbol{\nabla}\Psi(\boldsymbol{r})$ [Eq. (2.89)], the equation for a ray can be written as

$$\frac{d\boldsymbol{r}}{ds} \equiv \boldsymbol{s} = \frac{\boldsymbol{\nabla}\Psi(\boldsymbol{r})}{n} \qquad \Longrightarrow \qquad n\frac{d\boldsymbol{r}}{ds} = \boldsymbol{\nabla}\Psi(\boldsymbol{r}). \tag{2.94}$$

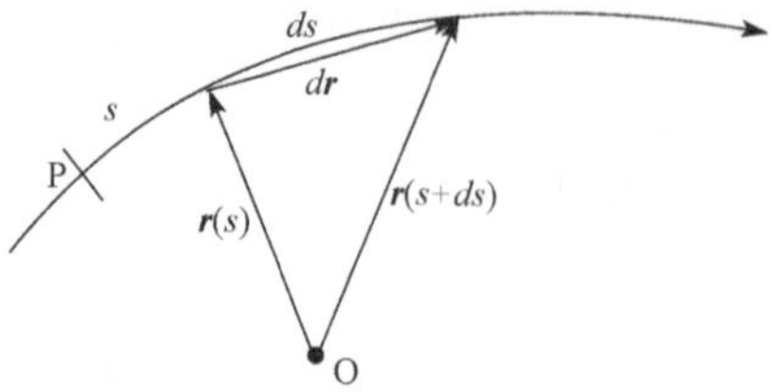

FIGURE 2.7
Rays are curves along which the energy of a wave is transported.

This equation is purely formal, since it specifies the rays in terms of $\Psi(\boldsymbol{r})$, which must first be determined from the eikonal equation. We can instead derive a more useful differential equation for the rays directly in terms of the refractive index $n(\boldsymbol{r})$. Differentiating Eq. (2.94) with respect to arc length gives

$$\frac{d}{ds}\left(n\frac{d\boldsymbol{r}}{ds}\right) = \frac{d}{ds}\left(\boldsymbol{\nabla}\Psi\right). \tag{2.95}$$

The right-hand side of Eq. (2.95) can now be expressed in terms of $n(\boldsymbol{r})$ as[8]

$$\begin{aligned}\frac{d}{ds}\left(\boldsymbol{\nabla}\Psi\right) &= (\boldsymbol{s}\cdot\boldsymbol{\nabla})\boldsymbol{\nabla}\Psi \\ &= \left(\frac{\boldsymbol{\nabla}\Psi}{n}\cdot\boldsymbol{\nabla}\right)(\boldsymbol{\nabla}\Psi) = \frac{1}{2n}\boldsymbol{\nabla}(\boldsymbol{\nabla}\Psi)^2 = \frac{1}{2n}\boldsymbol{\nabla}n^2 = \boldsymbol{\nabla}n\end{aligned} \tag{2.96}$$

Using this result in Eq. (2.95), we find the equation of a ray in terms of the variation of the refractive index

$$\frac{d}{ds}\left(n\frac{d\boldsymbol{r}}{ds}\right) = \boldsymbol{\nabla}n\,. \tag{2.97}$$

Further insight into this equation can be gained by writing it in another form. First, note that $\boldsymbol{s} = \boldsymbol{\nabla}\Psi/n$ [Eq. (2.89)] is a unit vector, so that $\boldsymbol{s}\cdot\boldsymbol{s} = 1$. Differentiating this relation with respect to s we find $\boldsymbol{s}\cdot(d\boldsymbol{s}/ds) = 0$ which means that $d\boldsymbol{s}/ds$ is a vector perpendicular to $\boldsymbol{s}$. In fact, from differential geometry

$$\frac{d\boldsymbol{s}}{ds} = \frac{\boldsymbol{\nu}}{R} = \frac{d^2\boldsymbol{r}}{ds^2}, \tag{2.98}$$

where R is the radius of curvature of the trajectory and $\boldsymbol{\nu}$ is a unit vector along the principal normal. Using these results, we find that the ray equation can be written as

$$\frac{dn}{ds}\boldsymbol{s} + n\frac{\boldsymbol{\nu}}{R} = \boldsymbol{\nabla}n\,. \tag{2.99}$$

Rearranging this equation gives

$$\frac{\boldsymbol{\nu}}{R} = \frac{1}{n}\left[\boldsymbol{\nabla}n - \boldsymbol{s}\frac{dn}{ds}\right]. \tag{2.100}$$

[8] The identity

$$\boldsymbol{\nabla}(\boldsymbol{A}\cdot\boldsymbol{B}) = \boldsymbol{A}\cdot\boldsymbol{\nabla}\boldsymbol{B} + \boldsymbol{B}\cdot\boldsymbol{\nabla}\boldsymbol{A} + \boldsymbol{A}\times(\boldsymbol{\nabla}\times\boldsymbol{B}) + \boldsymbol{B}\times(\boldsymbol{\nabla}\times\boldsymbol{A})$$

is useful here.

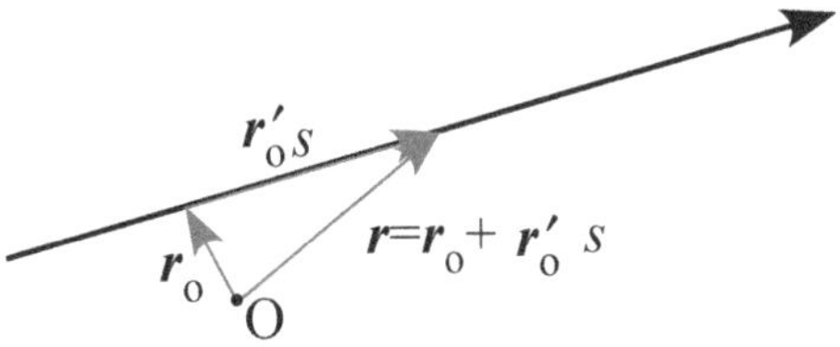

FIGURE 2.8
In a homogeneous medium, rays are straight lines.

Taking the scalar product with $\boldsymbol{\nu}$, we obtain

$$\frac{1}{R} = \frac{\boldsymbol{\nu}}{n} \cdot \boldsymbol{\nabla} n = \boldsymbol{\nu} \cdot \boldsymbol{\nabla} \ln n \,. \tag{2.101}$$

This equation shows that a ray bends toward the region of higher refractive index. A familiar special case is Snell's law: when a light ray enters a medium of higher refractive index from one of lower refractive index, it bends toward the normal.

2.7.1 Homogeneous Medium

As an application of the ray equation, we first consider a homogeneous medium for which $n = \text{const}$. In this case, Eq. (2.97) becomes

$$\frac{d^2\boldsymbol{r}}{ds^2} = 0 \,. \tag{2.102}$$

Integrating this equation, we obtain

$$\begin{aligned} \boldsymbol{r}'(s) &= \boldsymbol{r}'_o \,, \\ \boldsymbol{r}(s) &= \boldsymbol{r}'_o s + \boldsymbol{r}_o \,, \end{aligned} \tag{2.103}$$

where $\boldsymbol{r}'_0$ is the initial unit tangent vector to the ray at the point $\boldsymbol{r}_0$. This is the equation of a straight line.

2.7.2 Laws of Reflection and Refraction

Consider a region in which the refractive index depends only on y. In this case, the ray remains in a plane. Choosing this plane to be the y-z plane, we can write the ray as $\boldsymbol{r}(s) = y(s)\boldsymbol{e}_y + z(s)\boldsymbol{e}_z$. The equation for the z coordinate of the ray then becomes

$$\frac{d}{ds} n \frac{dz}{ds} = 0 \quad \Rightarrow \quad n \frac{dz}{ds} = \text{const} \,. \tag{2.104}$$

Using $\frac{dz}{ds} = \cos\alpha = \sin\theta$ [Fig. 2.9], we can write this result as $n\cos\alpha = \text{const}$. For the layered medium shown in Fig. 2.9, this leads to a ray trajectory satisfying $n_1 \cos\alpha_1 = n_2 \cos\alpha_2 = n_3 \cos\alpha_3 = \cdots$. Expressing the angle α in terms of the angle θ that the ray makes with the normal to the interface, we obtain

$$n_1 \sin\theta_1 = n_2 \sin\theta_2 = n_3 \sin\theta_3 = \cdots \,. \tag{2.105}$$

This is precisely Snell's law for a layered medium.

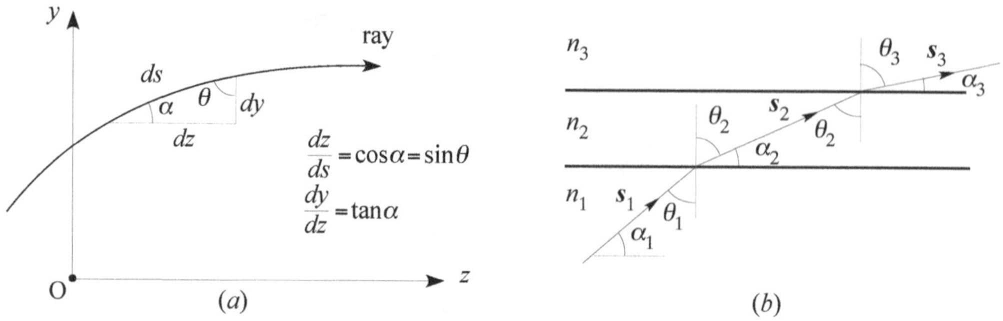

FIGURE 2.9
(a) A ray in a continuous inhomogeneous medium; (b) Ray path in a layered (piecewise homogeneous) medium. Note that $\alpha = \pi/2 - \theta$.

The laws of refraction and reflection at a surface across which the refractive index changes abruptly can be derived more rigorously from the relation $n\boldsymbol{s} = \boldsymbol{\nabla}\Psi$, which implies $\boldsymbol{\nabla} \times n\boldsymbol{s} = 0$. To see this, we replace the surface of discontinuity by a thin transition layer in which n changes rapidly but smoothly from its value n_1 on one side of the surface to its value n_2 on the other side [Fig. 2.10]. Consider a small rectangle $P_1Q_1Q_2P_2P_1$ straddling the interface between the two media [Fig. 2.10]. The integral of the normal component of $\boldsymbol{\nabla} \times (n\boldsymbol{s})$ over the surface of the rectangle can be converted, by Stokes' theorem, to a line integral around the periphery of the rectangle. Letting the thickness of the rectangle, δh, shrink to zero, the resulting line integral [Fig. 2.10] can be written as

$$[n_1\boldsymbol{s}_1 \cdot \boldsymbol{t}_1 + n_2\boldsymbol{s}_2 \cdot \boldsymbol{t}_2]\delta\ell = 0\,, \tag{2.106}$$

where $\delta\ell$ is the length of the line segment in which the rectangle intersects the interface. Expressing the unit vectors $\boldsymbol{t}_1$ and $\boldsymbol{t}_2$ in terms of the unit tangent vector $\boldsymbol{t}$ along the interface [Fig. 2.10], we obtain $\boldsymbol{a} \cdot [\boldsymbol{n}_{12} \times (n_2\boldsymbol{s}_2 - n_1\boldsymbol{s}_1)] = 0$, where the unit vector $\boldsymbol{a}$ is normal to the surface of the rectangle and $\boldsymbol{n}_{12}$ is a unit normal to the interface directed from medium 1 to 2 [Fig. 2.10]. Since the orientation of the rectangle, and therefore that of the unit vector $\boldsymbol{a}$, is arbitrary, we conclude that the tangential component of the ray vector is continuous across the surface of discontinuity.

$$\boldsymbol{n}_{12} \times (n_2\boldsymbol{s}_2 - n_1\boldsymbol{s}_1) = 0 \quad \Longrightarrow \quad \boldsymbol{n}_{12} \times n_2\boldsymbol{s}_2 = \boldsymbol{n}_{12} \times n_1\boldsymbol{s}_1\,. \tag{2.107}$$

This equation implies that the normal to the boundary, $\boldsymbol{n}_{12}$, and the ray vectors $\boldsymbol{s}_1$ and $\boldsymbol{s}_2$

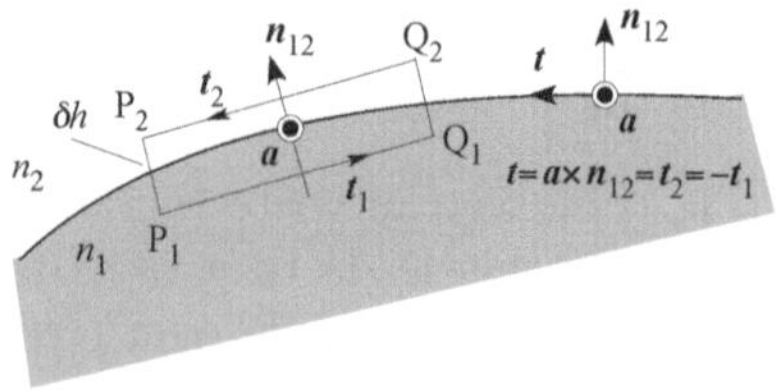

FIGURE 2.10
Ray transformation across a surface of discontinuity of refractive index.

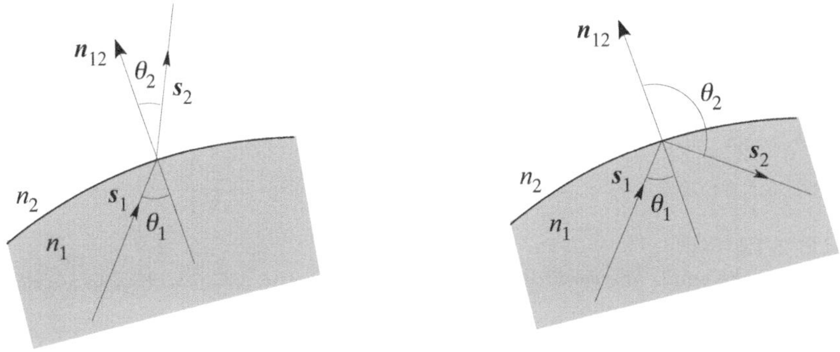

FIGURE 2.11
Laws of (a) refraction and (b) reflection. For reflection, $n_2 = n_1$, and Eq. (2.107) gives the familiar equality of the angles of incidence and reflection.

are coplanar, and that the angles the ray vectors make with the normal to the boundary are related by [Fig. 2.11]

$$n_1 \sin\theta_1 = n_2 \sin\theta_2 \,. \tag{2.108}$$

Here θ_1 and θ_2 are the angles that the rays in the two media make with $\boldsymbol{n}_{12}$. A similar procedure applied to the reflected ray leads to the law of reflection [Fig. 2.11(b)]. These laws are usually derived for plane waves incident on a plane refracting surface. Here we find that they are valid for more general waves and refracting surfaces, provided that the wavelength is sufficiently small compared with the characteristic dimensions of the wavefront and the interface. In particular, the radii of curvature of the incident wavefront and the interface must be large compared with the wavelength of the incident light.

Thus, in the short-wavelength limit, Maxwell's equations lead to a description of light propagation in terms of rays, which are geometrical curves along which light energy is transported. In this limit, the wave nature of light is largely masked, and phenomena such as diffraction can generally be neglected. This may seem like a drastic simplification, but it is extremely useful in instrument optics. As we will see, the laws of ray optics can also provide useful insight into diffraction effects. In what follows, we will limit our discussion to paraxial rays.

2.8 Paraxial Rays

In problems involving laser resonators, we are interested in rays that remain close to the optical axis, usually taken to be the z-axis of the system. Such rays are called paraxial rays. They make small angles with the optical axis, so that the sine and tangent of the angle can be approximated by the angle itself when the angle is expressed in radians:

$$\tan\alpha \approx \alpha \approx \sin\alpha \,. \tag{2.109}$$

This approximation is accurate to within about 3% for angles less than 18° ($\pi/10 \approx 0.314$ rad). For such rays, the arc length $ds \approx dz$, and the ray equation becomes

$$\frac{d}{dz} n \frac{d\boldsymbol{r}}{dz} = \boldsymbol{\nabla} n \tag{2.110}$$

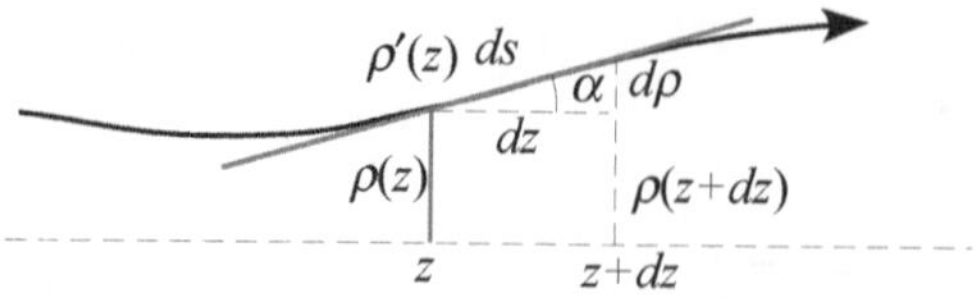

FIGURE 2.12
Paraxial rays make small angles with the optical axis, usually taken to be the z-axis, and remain close to it during propagation.

For a cylindrically symmetric medium, we can write $\boldsymbol{r} = \boldsymbol{e}_\rho \, \rho + \boldsymbol{e}_z \, z$, where ρ is the lateral displacement of the ray from the z-axis. The ray equation then becomes

$$\frac{d}{dz} n \frac{d\rho}{dz} = \boldsymbol{e}_\perp \cdot \boldsymbol{\nabla} n \,. \tag{2.111}$$

This equation determines the ray $\rho(z)$ from the lateral displacement $\rho_o = \rho(z_o)$ and the slope $\rho'_o = [d\rho/dz]_{z=z_o}$ at some fixed point z_o. To simplify the notation in what follows, we will use the symbol ρ to denote the lateral displacement of a paraxial ray from the axis and the abbreviation $\rho' = d\rho/dz$ to denote its slope. Both are functions of z.

Example 2.1 *In a homogeneous medium n =const, so that the equation of a ray becomes*

$$\frac{d}{dz} n \frac{d\rho}{dz} = 0 \,. \tag{2.112}$$

Integrating this equation from z_o to z, we find that the ray is a straight line given by

$$\begin{aligned} \rho'(z) &= \rho'_o \,, \\ \rho(z) &= \rho_o + \rho'_o (z - z_o) \,, \end{aligned} \tag{2.113}$$

where ρ_o and ρ'_o are, respectively, the ray displacement and slope at $z = z_o$.

Suppose now that n changes abruptly, as when a ray passes from one homogeneous medium of refractive index n_1 into another of refractive index n_2. If the ray has slope ρ'_1 and displacement ρ_1 just before reaching the interface at $z = 0$, integration of the ray equation across the interface gives

$$\begin{aligned} \rho'_2 &= \frac{n_1}{n_2} \, \rho'_1 \,, \\ \rho_2 &= \rho_1 + \frac{n_1}{n_2} \, \rho'_1 z \,. \end{aligned} \tag{2.114}$$

The first of these equations is Snell's law $n_1 \theta_1 = n_2 \theta_2$ in the paraxial approximation. The second equation gives the ray displacement in the second medium in terms of the displacement ρ_1 at the interface and the slope $\rho'_2 = (n_1/n_2) \, \rho'_1$ just inside the second medium.

2.8.1 Rays in a Duct

As another example, consider an axially symmetric medium with a quadratic index variation with distance ρ from the axis of symmetry, taken to be the z-axis:

$$n(\rho) = n_1 \left[1 - \frac{1}{2} \frac{\rho^2}{\Lambda^2} \right] , \tag{2.115}$$

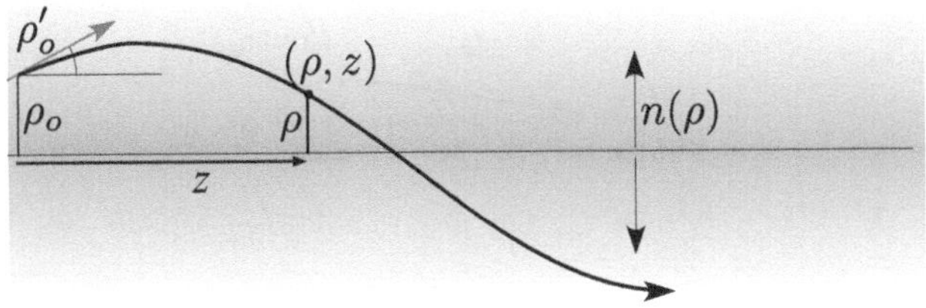

FIGURE 2.13
Rays in a duct with radially decreasing refractive index.

where the second term is assumed to be small, $0 < \rho^2/2\Lambda^2 \ll 1$, for all values of ρ of interest. The parameter Λ is a length scale that characterizes how rapidly the refractive index varies with ρ.

This is a useful model for a gradient-index fiber in which the refractive index varies from $n(0) = n_1$ on the axis to $n(a) = n_2 \equiv n_1 - \Delta n$ at the cladding, where $\rho = a$. For this quadratic index variation, we find $\Lambda = a\sqrt{n_1/2\Delta n}$. Typical values of $a = 50\,\mu\text{m}$, $n_1 = 1.47$, and $\Delta n = 0.05$ give $\Lambda = 192\,\mu\text{m}$. With this index profile, the ray equation (2.97), in the paraxial approximation $ds \approx dz$, becomes

$$\frac{d^2\rho}{dz^2} + \frac{1}{\Lambda^2}\rho = 0, \tag{2.116}$$

where, in obtaining the second term, we have used the approximation $n_1/n_1(1-\rho^2/2\Lambda^2) \approx 1$. This equation is easily solved to give the ray displacement as a function of z

$$\rho(z) = \rho_o \cos(z/\Lambda) + (\rho'_o\Lambda)\sin(z/\Lambda)\,, \tag{2.117}$$

where ρ_o is the initial ray displacement from the z-axis and ρ'_o is the initial slope of the ray at $z = 0$. Thus, the ray displacement oscillates about the z-axis.

We also note that a family of parallel rays, having the same slope but different displacements, periodically converges to a point in each of the planes $z_n = (4n+1)\pi\Lambda/2$, at a height $\rho_f = (\rho'_o\Lambda)$ from the z-axis. The rays emerge from each of these planes with their original slope. These planes are separated from each other by a distance $L_f = 2\pi\Lambda$.

Example 2.2 *Show that a section of the quadratic medium of length ℓ, surrounded by a medium of refractive index n_m, acts as a lens. Hint: Show that parallel rays entering at $z = 0$ with different displacements converge after emerging at $z = \ell$ to a common focus at a distance $f = \frac{\Lambda}{n_m}\cot(\ell/\Lambda)$.*

Example 2.3 *Find the equation for paraxial rays in a duct with refractive index profile $n = n_o\left[1 + \frac{1}{2}\frac{\rho^2}{\Lambda^2}\right]$. Answer: $\rho(z) = \rho_o\cosh(z/\Lambda) + (\Lambda\rho'_o)\sinh(z/\Lambda)$.*

2.8.2 Ray Matrices

The study of ray propagation is important in its own right in instrument optics. We will see shortly that, although describing light propagation in terms of ray bundles ignores diffraction, the laws of paraxial ray propagation turn out to be very useful for understanding the full diffractive propagation of light in optical resonators and laser beams. In paraxial optics, a ray is specified by its displacement ρ from the optical axis and its slope $\rho' = \frac{d\rho}{dz}$,

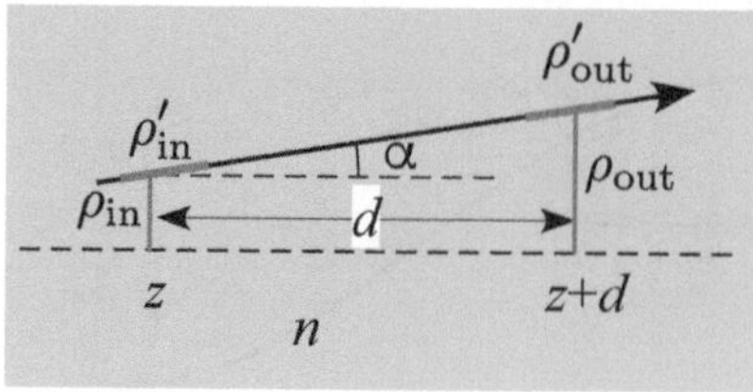

FIGURE 2.14
Propagation in a homogeneous medium of length d.

which, in the paraxial approximation, is equal to the angle the ray makes with the optical axis. Both quantities vary with z as the ray propagates through an optical system.

If we introduce a column matrix $\tilde{\rho}$, whose elements are ρ and ρ',

$$\tilde{\rho} = \begin{bmatrix} \rho \\ \rho' \end{bmatrix} , \tag{2.118}$$

we can describe the effect of an optical element on the ray parameters (ρ, ρ') by a 2×2 matrix of the form

$$M = \begin{bmatrix} A & B \\ C & D \end{bmatrix} . \tag{2.119}$$

The ray transformation matrices for most optical systems or elements of interest to us can be constructed from the transformation matrices of three basic elements:

(i) a homogeneous medium of length d and refractive index n;

(ii) reflection from a curved surface of radius of curvature R;

(iii) refraction at a curved interface of radius of curvature R between two media with refractive indices n_1 and n_2, for a ray incident from medium n_1.

2.8.2.1 Propagation in a Homogeneous Medium

Consider a ray propagating from plane z to plane $z + d$ in a homogeneous medium of refractive index n [Fig. 2.14]. The ray coordinates at the input and output faces are related by [Eq. (2.113)]

$$\begin{aligned} \rho_{\text{out}} &= \rho_{\text{in}} + d\,\rho'_{\text{in}} \,, \\ \rho'_{\text{out}} &= \rho'_{\text{in}} \,, \end{aligned} \quad \Rightarrow \begin{bmatrix} \rho_{out} \\ \rho'_{out} \end{bmatrix} = \begin{bmatrix} 1 & d \\ 0 & 1 \end{bmatrix} \begin{bmatrix} \rho_{in} \\ \rho'_{in} \end{bmatrix} \Rightarrow \quad \begin{bmatrix} A & B \\ C & D \end{bmatrix} = \begin{bmatrix} 1 & d \\ 0 & 1 \end{bmatrix} . \tag{2.120}$$

2.8.2.2 Reflection at a Curved Surface

From Fig. 2.15, we see that the ray displacement remains unchanged upon reflection

$$\rho_{\text{out}} = \rho_{\text{in}} \,. \tag{2.121}$$

To find the relation between the input and output ray slopes, we use the law of reflection, $\theta_2 = \theta_1$, to relate the outgoing and incoming ray angles in Fig. 2.15. We then find

$$\begin{aligned} \theta_2 &= \theta_1 \qquad \text{(law of reflection)} \,, \\ \text{or} \qquad \alpha_{\text{out}} - \phi &= \phi - \alpha_{\text{in}} \,, \\ \text{or} \qquad \alpha_{\text{out}} &= 2\phi - \alpha_{\text{in}} \,. \end{aligned} \tag{2.122}$$

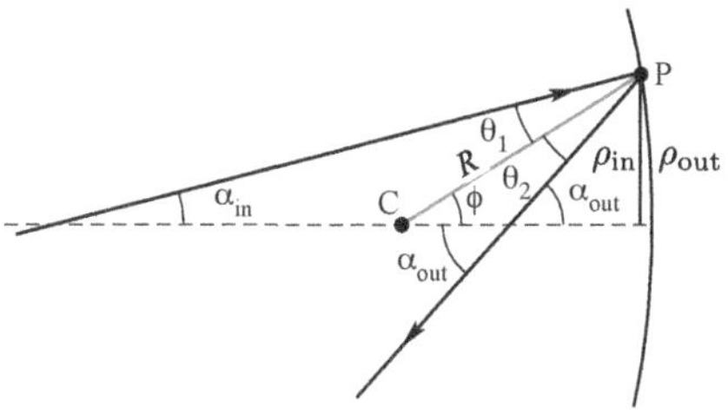

FIGURE 2.15
Reflection at a curved surface of radius of curvature R with center at C. From the laws of reflection, the angles of incidence and reflection are equal, $\theta_2 = \theta_1$.

We can now express these angles in terms of the ray coordinates [Fig. 2.15]. In the paraxial approximation,

$$\rho'_{\text{in}} = \alpha_{\text{in}}\,, \quad \rho'_{\text{out}} = -\alpha_{\text{out}}\,, \quad \phi = \frac{\rho_{\text{in}}}{R}\,, \tag{2.123}$$

where the sign convention is such that R is positive for a concave reflecting surface and negative for a convex reflecting surface. Using these relations, we find that the output ray slope is

$$\rho'_{\text{out}} = -\frac{2}{R}\,\rho_{\text{in}} + \rho'_{\text{in}}\,. \tag{2.124}$$

With the help of Eqs. (2.121) and (2.124), we obtain the ABCD matrix for a curved reflecting surface

$$\begin{bmatrix} A & B \\ C & D \end{bmatrix} = \begin{bmatrix} 1 & 0 \\ -\frac{2}{R} & 1 \end{bmatrix}. \tag{2.125}$$

2.8.2.3 Refraction at a Curved Interface

From Fig. 2.16, we see that the input and output ray displacements are equal

$$\rho_{\text{out}} = \rho_{\text{in}}\,. \tag{2.126}$$

To relate the input and output ray slopes, we first relate the outgoing and incoming ray

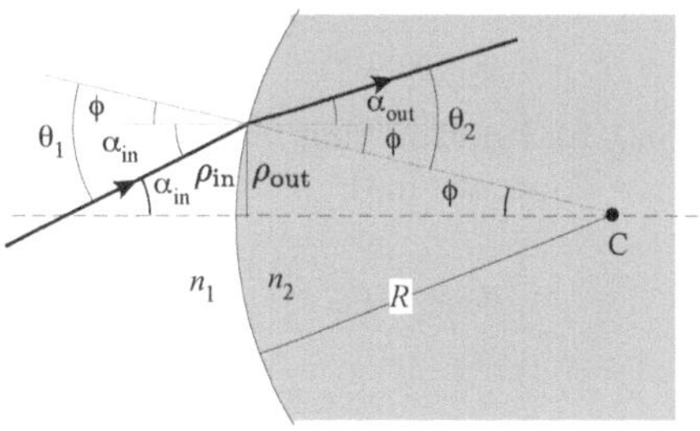

FIGURE 2.16
Refraction at a curved interface between two media with radius of curvature R and center at C. Note that in the figure ρ'_{out} is positive so that $\rho'_{\text{out}} = \alpha_{\text{out}}$.

angles using Snell's law and then express these angles in terms of the ray coordinates:

$$n_1\theta_1 = n_2\theta_2 \qquad \text{(Paraxial Snell's law)}\,,$$

$$\text{or} \qquad n_1(\alpha_{\text{in}} + \phi) = n_2(\phi + \alpha_{\text{out}})\,,$$

$$\text{or} \qquad \alpha_{\text{out}} = -\frac{(n_2 - n_1)}{n_2}\phi + \frac{n_1}{n_2}\alpha_{\text{in}}\,. \tag{2.127}$$

Now, noting that $\rho'_{\text{in}} = \alpha_{\text{in}}$ and $\rho'_{\text{out}} = \alpha_{\text{out}}$ and $\phi = \rho_{\text{in}}/R$, where R is taken to be positive for convex and negative for concave refracting interfaces, we obtain

$$\rho'_{\text{out}} = -\frac{(n_2 - n_1)}{n_2}\frac{\rho_{\text{in}}}{R} + \frac{n_1}{n_2}\rho'_{\text{in}} \tag{2.128}$$

From Eqs. (2.126) and (2.128), we find the ABCD matrix

$$\begin{bmatrix} A & B \\ C & D \end{bmatrix} = \begin{bmatrix} 1 & 0 \\ -\left(\dfrac{n_2 - n_1}{n_2}\right)\dfrac{1}{R} & \dfrac{n_1}{n_2} \end{bmatrix}. \tag{2.129}$$

The ABCD matrix of a thin lens can then be found by multiplying the ABCD matrices for its front and back refracting surfaces, while omitting the propagation matrix because the lens thickness is assumed to be negligible:

$$\begin{aligned} \begin{bmatrix} A & B \\ C & D \end{bmatrix} &= \begin{bmatrix} 1 & 0 \\ -\left(\dfrac{n_1 - n_2}{n_1}\right)\dfrac{1}{R_2} & \dfrac{n_2}{n_1} \end{bmatrix} \begin{bmatrix} 1 & 0 \\ -\left(\dfrac{n_2 - n_1}{n_2}\right)\dfrac{1}{R_1} & \dfrac{n_1}{n_2} \end{bmatrix} \\ &= \begin{bmatrix} 1 & 0 \\ -\left(\dfrac{n_2 - n_1}{n_1}\right)\left(\dfrac{1}{R_1} - \dfrac{1}{R_2}\right) & 1 \end{bmatrix} \equiv \begin{bmatrix} 1 & 0 \\ -\dfrac{1}{f} & 1 \end{bmatrix}. \end{aligned} \tag{2.130}$$

Here, the last step follows from the expression for the focal length of a lens in terms of the radii of curvature of its surfaces and the refractive indices of the lens material and the surrounding medium.

We can, of course, derive the matrix of a thin lens of focal length f directly with the help of Fig. 2.17 by recalling the thin-lens formula. Let ρ_{in} and ρ'in denote the displacement and slope of the incident ray just before the lens, and let ρout and ρ'out denote the corresponding quantities just after the lens. Then

$$\rho_{\text{out}} = \rho_{\text{in}}\,. \tag{2.131}$$

To determine the slope after the lens, we note that the incident ray may be regarded as originating from the axial point s_o, while the emergent ray may be regarded as proceeding toward the point s_i. These distances are related by the thin-lens formula

$$\frac{1}{s_o} + \frac{1}{s_i} = \frac{1}{f}. \tag{2.132}$$

Multiplying both sides by ρ_{in}, and noting that $\rho_{\text{in}}/s_o = \rho'_{\text{in}}$ and $\rho_{\text{in}}/s_i = -\rho'_{\text{out}}$, because the emergent ray has negative slope, we find

$$\rho'_{\text{out}} = \rho'_{\text{in}} - \frac{1}{f}\rho_{\text{in}}\,. \tag{2.133}$$

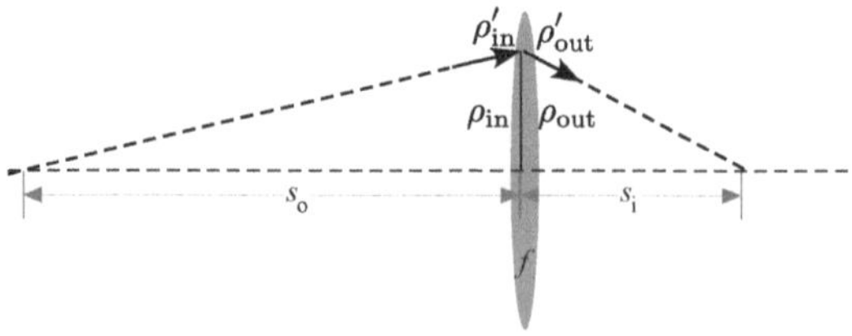

FIGURE 2.17
Input and output ray coordinates for a thin lens.

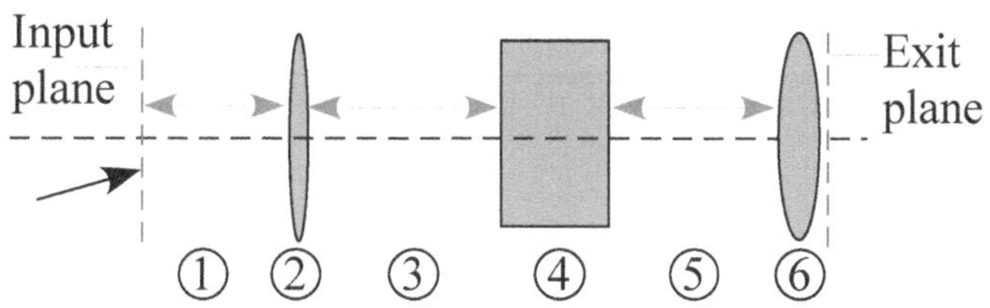

FIGURE 2.18
The overall transformation matrix for ray propagation through an optical system is obtained by multiplying the *ABCD* matrices for each optical element, including free-space sections, in the correct order [see the text].

Equations (2.131) and (2.133) then lead to the ray transfer matrix of a thin lens (2.130).

From the preceding examples, it is clear that we can write the relation between the input and output ray parameters in matrix form as

$$\begin{bmatrix} \rho_{\text{out}} \\ \rho'_{\text{out}} \end{bmatrix} = \begin{bmatrix} A & B \\ C & D \end{bmatrix} \begin{bmatrix} \rho_{\text{in}} \\ \rho'_{\text{in}} \end{bmatrix} . \tag{2.134}$$

This relation represents a transformation of the input ray parameters into the output ray parameters. The matrix representing this transformation is referred to as the ray transfer matrix, or the *ABCD* matrix. The ABCD matrices for the three basic optical elements considered above are given in Table 2.1.

Once these basic matrices are known, we can calculate the overall ABCD matrix for any system of optical elements. Consider, for example, the passage of a ray from left to right through the sequence of optical elements shown in Fig. 2.18. Labeling the various optical elements as $1, 2, 3, \ldots$ in the order in which the ray encounters them, we can write the ABCD matrix for the system as

$$\begin{bmatrix} A & B \\ C & D \end{bmatrix} = M_6 \cdot M_5 \cdots M_4 \cdot M_3 \cdot M_2 \cdot M_1 . \tag{2.135}$$

Note that the matrices are written from right to left in the order in which they are encountered by the ray.

2.8.3 Periodic Focusing Systems

An interesting and important application of ray matrices arises in the analysis of periodic focusing systems, in which a ray encounters the same sequence of optical elements repeatedly along a cascaded system. An optical resonator can be modeled as such an iterated periodic focusing system because the propagation of a ray through successive roundtrips in the resonator is mathematically equivalent to propagation through successive periods of a periodic lens guide.

As an example, consider the optical resonator shown in Fig. 2.19, formed by two spherical mirrors of radii of curvature R_1 and R_2 separated by a distance L. Imagine a ray propagating to the right, starting at the left end of the resonator. After one roundtrip, this ray will have propagated through a straight section of length L, reflected from a spherical mirror of radius R_2, propagated through another section of length L, and finally reflected from a spherical mirror of radius R_1. On each roundtrip, the ray encounters the same sequence of elements

TABLE 2.1
Paraxial ray matrices of some common optical elements.

Plate of thickness d, refractive index n, in a medium of refractive index n_m, near normal incidence 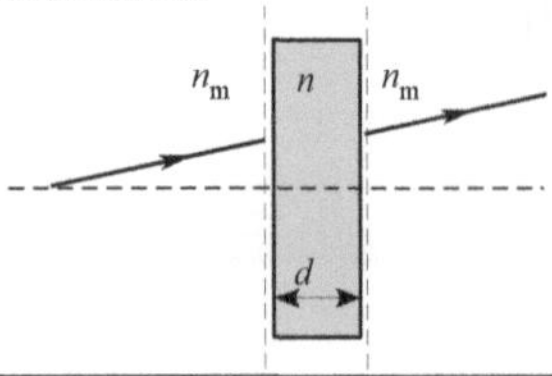	$\begin{bmatrix} 1 & \dfrac{d}{n} \\ 0 & 1 \end{bmatrix}$
Dielectric interface with radius of curvature R (+ for convex and − for concave interface) 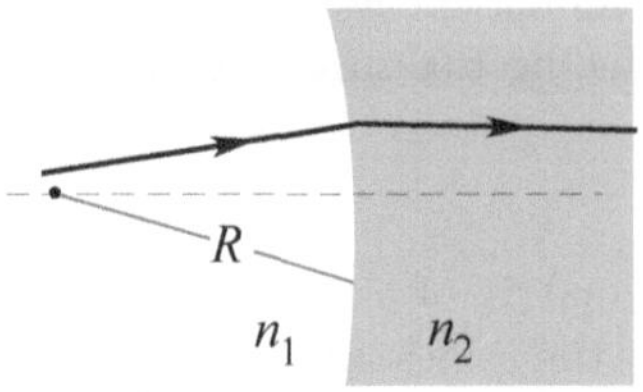	$\begin{bmatrix} 1 & 0 \\ -\dfrac{n_2 - n_1}{n_2}\dfrac{1}{R} & \dfrac{n_1}{n_2} \end{bmatrix}$
Spherical mirror of radius of curvature R, near normal incidence (+ for concave and − for convex mirror) 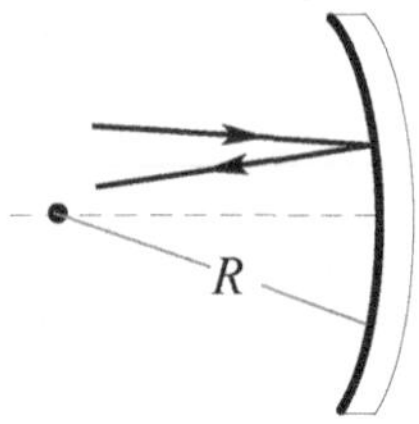	$\begin{bmatrix} 1 & 0 \\ -\dfrac{2}{R} & 1 \end{bmatrix}$
Thin lens (focal length f) in a medium of refractive index n_m near normal incidence 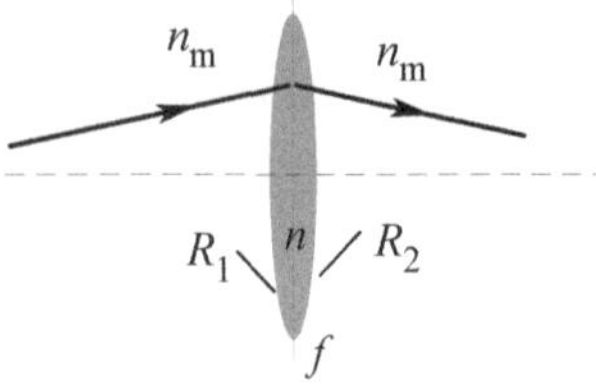	$\begin{bmatrix} 1 & 0 \\ -\dfrac{1}{f} & 1 \end{bmatrix}$ $\dfrac{1}{f} = \dfrac{n - n_m}{n_m}\left(\dfrac{1}{R_1} - \dfrac{1}{R_2}\right)$
Dielectric interface near Brewster's angle $\tan\theta_B = \dfrac{n_2}{n_1}$ 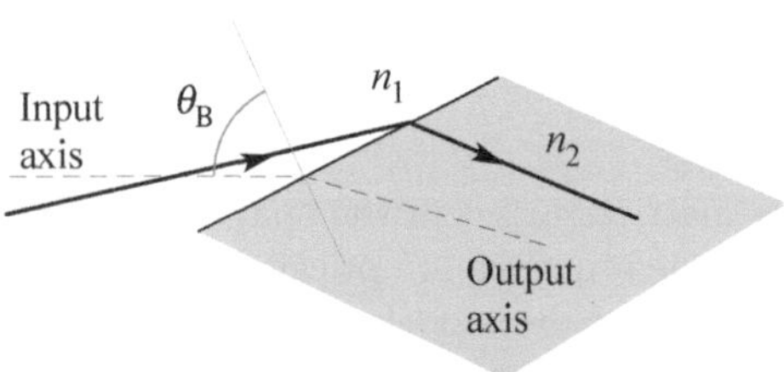	$\begin{bmatrix} \dfrac{n_2}{n_1} & 0 \\ 0 & \dfrac{n_1^2}{n_2^2} \end{bmatrix}$ plane of incidence (tangential plane) $\begin{bmatrix} 1 & 0 \\ 0 & \dfrac{n_1}{n_2} \end{bmatrix}$ perpendicular to the plane of incidence (sagittal plane)

and therefore undergoes the same ray transformation. This is equivalent to a lens waveguide in which lenses of focal lengths $f_1 = R_1/2$ and $f_2 = R_2/2$ are placed alternately, separated by a distance L.

The ABCD matrix M describing the ray transformation during one roundtrip through the resonator is given by $M = M_1 \cdot M_L \cdot M_2 \cdot M_L$, where M_1 and M_2 are the ray matrices

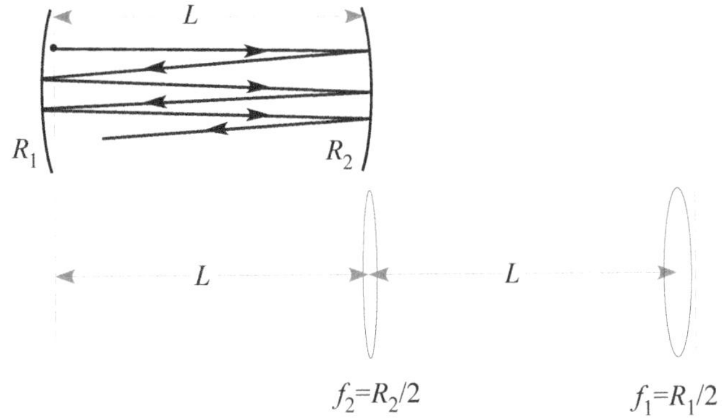

FIGURE 2.19
An optical cavity formed by two spherical mirrors of radii R_1 and R_2 separated by a distance L is equivalent to a periodic lens waveguide whose repeat unit is shown below the resonator.

of the mirrors with radii R_1 and R_2, respectively, and M_L is the ray matrix of a section of length L. Note that the matrices are written from right to left in the order in which the ray encounters the optical elements:

$$M \equiv \begin{bmatrix} A & B \\ C & D \end{bmatrix} = \begin{bmatrix} 1 & 0 \\ -\dfrac{2}{R_1} & 1 \end{bmatrix} \begin{bmatrix} 1 & L \\ 0 & 1 \end{bmatrix} \begin{bmatrix} 1 & 0 \\ -\dfrac{2}{R_2} & 1 \end{bmatrix} \begin{bmatrix} 1 & L \\ 0 & 1 \end{bmatrix}$$

$$= \begin{bmatrix} 1 - \dfrac{2L}{R_2} & 2L\left(1 - \dfrac{L}{R_2}\right) \\ -\dfrac{2}{R_1} - \dfrac{2}{R_2} + \dfrac{4L}{R_1 R_2} & 1 - \dfrac{4L}{R_1} - \dfrac{2L}{R_2} + \dfrac{4L^2}{R_1 R_2} \end{bmatrix}$$

$$\equiv \begin{bmatrix} 2g_2 - 1 & 2Lg_2 \\ \frac{2}{L}(2g_1g_2 - g_1 - g_2) & 4g_1g_2 - 2g_2 - 1 \end{bmatrix}, \quad g_i \equiv 1 - \frac{L}{R_i}. \tag{2.136}$$

To analyze the behavior of a ray after an arbitrary number of roundtrips, we introduce the ray vector $\tilde{\rho} \equiv \begin{bmatrix} \rho \\ \rho' \end{bmatrix}$. If $\tilde{\rho}_n$ denotes the ray vector after n roundtrips, then

$$\widetilde{\rho}_n = M \text{ Tr } \rho_{n-1} = M^2 \, \tilde{\rho}_{n-2} = M^n \, \tilde{\rho}_o \,. \tag{2.137}$$

Thus, the behavior of a ray for arbitrary n in a periodic system such as this is determined by the powers of the matrix M. This behavior is most conveniently analyzed by finding the eigenvalues and eigenvectors of M. In the present context, the eigenvectors correspond to special rays, which we call eigenrays, whose ray vectors are reproduced after each roundtrip apart from multiplication by a constant eigenvalue.

The eigenray $\tilde{r}$ and the corresponding eigenvalue Λ of matrix M satisfy the relation

$$M\widetilde{\rho} = \Lambda\widetilde{\rho} \qquad \text{or} \qquad \begin{bmatrix} A - \Lambda & B \\ C & D - \Lambda \end{bmatrix} \begin{bmatrix} \rho \\ \rho' \end{bmatrix} = 0 \,. \tag{2.138}$$

Nontrivial solutions of this equation require the determinant of $M - \Lambda I$ to vanish:

$$\det \begin{bmatrix} A - \Lambda & B \\ C & D - \Lambda \end{bmatrix} = \Lambda^2 - (A + D)\Lambda + (AD - BC) = 0 \,. \tag{2.139}$$

This equation can be simplified by noting that the determinant of the matrix M is unity.[9] Indeed, the determinant of a product of matrices is equal to the product of their determinants: $\det(M) = \det(M_1)\det(M_L)\det(M_2)\det(M_L)$ and each of the matrices in this product has unit determinant. Thus, $\det M \equiv AD - BC = 1$.

We also note that $A + D$ is the trace of the matrix M. It is therefore convenient to introduce the parameter

$$m = \frac{1}{2}\,\mathrm{Tr}\ M = \frac{1}{2}(A + D)\,. \tag{2.140}$$

Equation (2.139) for Λ then becomes $\Lambda^2 - 2m\Lambda + 1 = 0$, and its two eigenvalues are

$$\Lambda_a = m + \sqrt{m^2 - 1}\,, \qquad \Lambda_b = m - \sqrt{m^2 - 1}\,. \tag{2.141}$$

They satisfy

$$\Lambda_a + \Lambda_b = 2m \quad \text{and} \quad \Lambda_a\Lambda_b = 1\,. \tag{2.142}$$

There are two eigenrays $\widetilde{\rho}_a$ and $\widetilde{\rho}_b$ corresponding to Λ_a and Λ_b, respectively, such that

$$M\widetilde{\rho}_a = \Lambda_a\widetilde{\rho}_a, \qquad M\widetilde{\rho}_b = \Lambda_b\widetilde{\rho}_b\,. \tag{2.143}$$

These equations mean that after one roundtrip, propagation of an eigenray simply multiplies its ray vector by the corresponding eigenvalue.

Because the two eigenrays form a basis for the two-dimensional ray-vector space, any input ray $\widetilde{\rho}_o$ can be written as

$$\widetilde{\rho}_o = C_a\widetilde{\rho}_a + C_b\widetilde{\rho}_b\,, \tag{2.144}$$

where C_a and C_b are constants which are determined by the initial ray displacement ρ_o and slope ρ'_o. After n roundtrips, this ray becomes

$$\widetilde{\rho}_n = M^n\widetilde{\rho}_o = C_a\Lambda_a^n\,\widetilde{\rho}_a + C_b\Lambda_b^n\,\widetilde{\rho}_b\,. \tag{2.145}$$

Equation (2.145) shows that the long-term behavior of a ray is determined by the magnitudes of the eigenvalues. We can therefore classify periodic focusing systems according to their eigenvalues. When $m^2 < 1$, the eigenvalues are complex conjugates of unit magnitude, and the ray trajectories remain bounded. When $m^2 > 1$, the eigenvalues are real and reciprocal; except at the boundary cases, one has magnitude greater than unity and the other has magnitude less than unity. A generic ray then acquires a component that grows exponentially with the number of roundtrips, and the ray trajectory is unstable. We consider these two cases separately.

Case 1: $m^2 < 1$ or $-1 < m < 1$

Since m is real and has magnitude less than unity, we can parameterize m and the eigenvalues as

$$m \equiv \frac{1}{2}(A + D) \equiv \cos\phi\,, \quad 0 \le \phi < \pi, \tag{2.146}$$

$$\Lambda_{a,b} \equiv \cos\phi \pm i\sqrt{1 - \cos^2\phi} = \cos\phi \pm i\sin\phi = e^{\pm i\phi}\,, \tag{2.147}$$

where ϕ is uniquely determined by the first relation. The ray vector in a periodic system then takes the form

$$\widetilde{\rho}_n = C_a\widetilde{\rho}_a\,e^{in\phi} + C_b\,\widetilde{\rho}_b\,e^{-in\phi} \equiv \widetilde{c}_0\,\cos n\phi + \widetilde{s}_0\,\sin n\phi\,, \tag{2.148}$$

[9] This assumes that the optical system is surrounded by the same medium. If the refractive indices of the media on the input and output sides are $n_{\rm in}$ and $n_{\rm out}$, respectively, then $\det M = n_{\rm in}/n_{\rm out}$.

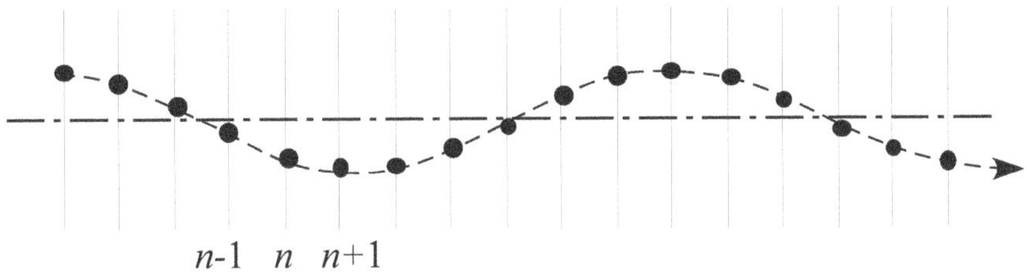

FIGURE 2.20
The dots indicate ray position in the reference planes after successive trips. The continuous curve joining the dots is simply a guide to the eye.

where

$$\widetilde{c}_0 = C_a \widetilde{\rho}_a + C_b \widetilde{r}_b \quad \text{and} \quad \widetilde{s}_0 = i(C_a \widetilde{\rho}_a - C_b \widetilde{\rho}_b)\,. \tag{2.149}$$

The ray displacement after n roundtrips therefore has the form

$$\rho_n = c_0 \cos n\phi + s_0 \sin n\phi\,, \tag{2.150a}$$

where c_0 and s_0 are the upper elements of the column matrices $\widetilde{c}_0$ and $\widetilde{s}_0$, respectively. By combining the sine and cosine terms, the expression for ray displacement ρ_n after n roundtrips can be written as

$$\rho_n = \rho_{\max} \sin(n\phi + \alpha)\,. \tag{2.150b}$$

The ray displacement thus remains bounded, since $|\rho_n| \leq |\rho_{\max}|$, independently of the number of trips n through the system, if $|m| < 1$. The constants ρ_{max} and α, which are related to c_o and s_o, are determined by the initial conditions:

$$\rho_o = \rho_{\max} \sin\alpha\,, \qquad \rho_1 \equiv \rho_{\max} \sin(\phi + \alpha) = A\rho_o + B\rho_o'\,. \tag{2.151a}$$

To find explicit expressions for $\rho_{\max}$ and α in terms of ρ_o and ρ_o', we expand the expression for ρ_1 and rewrite it as

$$\rho_{\max}(\sin\phi\cos\alpha + \cos\phi\sin\alpha) = \frac{\rho_o}{\tan\alpha}\sin\phi + \rho_o\cos\phi\,. \tag{2.151b}$$

Using $\cos\phi = \frac{1}{2}(A + D)$ in this expression and using the result in Eq. (2.151a), we find ρmax and α in terms of the initial ray displacement and slope:

$$\rho_{\max} = \frac{\rho_o}{\sin\alpha}\,, \qquad \tan\alpha = \frac{\rho_o\sqrt{1-(A+D)^2/4}}{\rho_o\left(\frac{A-D}{2}\right) + B\rho_o'}\,. \tag{2.152}$$

The ray displacement in the reference plane of a stable system thus oscillates back and forth about the axis, as shown in Fig. 2.20. Note that ρ_n gives the displacement of the ray in the reference plane after successive trips; it does not describe what happens to the ray displacement at other planes within the repeated section. In particular, although $\rho_{\max}$ represents the maximum possible excursion of the ray in the chosen reference plane, this maximum is not necessarily attained in that plane or in any other plane of the system.

It should also be remembered that, in the equation for ray displacement, the number of trips n through the repeated section is the variable, not the angle ϕ. For the ray trajectory to repeat after, say, N roundtrips, the condition

$$N\phi = 2\pi s \tag{2.153}$$

must be satisfied, where s is a positive integer. Since $0 \le \phi < \pi$, the ratio $s/N = \phi/2\pi < 1/2$. Thus, for a periodic ray trajectory, s is a positive integer less than $N/2$. The integer N can be interpreted as the number of trips through the repeated section after which the ray retraces its trajectory; that is, after N roundtrips, the ray vector returns to its initial value, so that both the ray displacement and slope have their initial values.

As an application of the foregoing results, consider a symmetric two-mirror cavity with mirror separation L and both mirrors having the same radius of curvature R. Suppose $L = R/2$, and we launch a ray at a height ρ_o above the optical axis with zero slope, $\rho'_o = 0$ (a ray parallel to the axis). Then, with $g_1 = g_2 = 1 - L/R = \frac{1}{2}$, the roundtrip matrix (2.136), with the input plane as in Fig. 2.19, takes the form

$$\begin{pmatrix} A & B \\ C & D \end{pmatrix} = \begin{pmatrix} 2g-1 & 2Lg \\ \frac{4}{L}(g^2-g) & 4g^2-2g-1 \end{pmatrix} = \begin{pmatrix} 0 & \frac{R}{2} \\ -\frac{2}{R} & -1 \end{pmatrix}. \tag{2.154}$$

With $\cos\phi = \frac{1}{2}(A+D) = -\frac{1}{2}$, $\frac{1}{2}(A-D) = \frac{1}{2}$, and zero initial slope $\rho'_0 = 0$, the auxiliary angle ϕ is given by

$$\phi = \cos^{-1}\frac{1}{2}(A+D) = \cos^{-1}\left(-\frac{1}{2}\right) = \frac{2\pi}{3}, \tag{2.155}$$

and using Eq. (2.152),

$$\tan\alpha = \frac{\rho_o\sqrt{1-(-1/2)^2}}{\rho_o(\frac{1}{2}) + B\times 0} = \sqrt{3},$$

which yields $\alpha = \pi/3$ and $\rho_{max} = \frac{2\rho_o}{\sqrt{3}}$. The ray displacement equation is then

$$\rho_n = \frac{2\rho_o}{\sqrt{3}}\sin(2\pi n/3 + \pi/3). \tag{2.156}$$

Let us use this equation to trace the ray displacement. At the end of the first roundtrip (and one unit cell later), $n = 1$, the ray position is given by

$$\rho_1 = \frac{2\rho_o}{\sqrt{3}}\sin(2\pi/3 + \pi/3) = 0.$$

After two roundtrips ($n = 2$), the ray position is

$$\rho_2 = \frac{2\rho_o}{\sqrt{3}}\sin(4\pi/3 + \pi/3) = -\rho_o,$$

and after three roundtrips ($n = 3$), we have

$$\rho_3 = \frac{2\rho_o}{\sqrt{3}}\sin(6\pi/3 + \pi/3) = \rho_o.$$

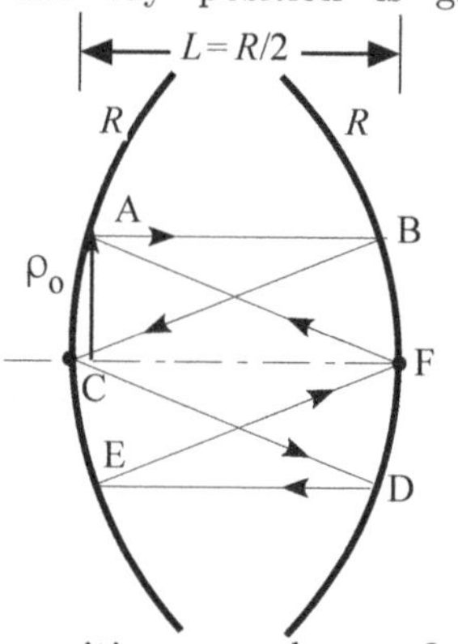

The sequence of ray positions then repeats. These ray positions can be confirmed by geometrical-optics ray tracing, as shown in the figure to the right. It should be emphasized that Eq. (2.156) says nothing about the ray position in any other plane. It is also important to note that $\rho_3 = \rho_0$ alone does not imply that the ray will retrace its path on its subsequent passage; for that to happen, both the ray displacement and slope must return to their initial values. In this case, it can be checked that after three roundtrips the ray slope also returns to its initial value of zero.

Let us consider a second example in which $R_1 \to \infty$ and $L = R/2$. In this case, $g_1 = 1$ and $g_2 = \frac{1}{2}$. The roundtrip matrix takes the form

$$\begin{pmatrix} A & B \\ C & D \end{pmatrix} = \begin{pmatrix} 0 & \frac{R}{2} \\ -\frac{2}{R} & 0 \end{pmatrix}.$$

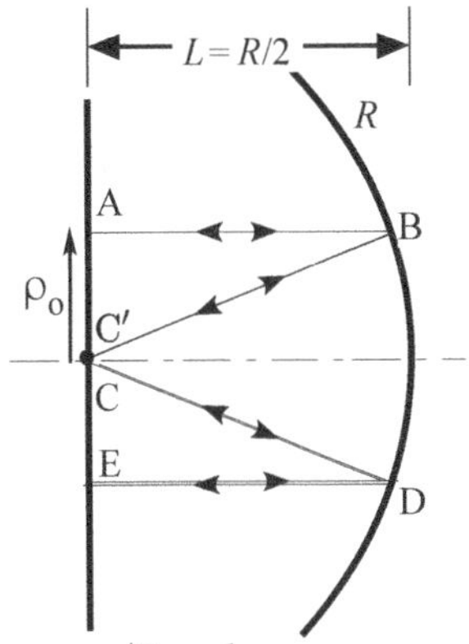

The auxiliary angle ϕ is now

$$\phi = \cos^{-1}\frac{1}{2}(A + D) = \cos^{-1}(0) = \frac{\pi}{2}.$$

For a ray launched at height ρ_o with slope $\rho_o' = 0$, we obtain $\alpha = \pi/2$ and $\rho_{\text{max}} = \rho_o/\sin\alpha = \rho_o$, yielding the ray position after n roundtrips,

$$\rho_n = \rho_o \sin(n\pi/2 + \pi/2). \tag{2.157}$$

The ray positions after 1, 2, 3, and 4 roundtrips are, respectively, 0, $-\rho_o$, 0, and ρ_o, and the cycle then repeats. Note that the ray visits $\rho = 0$ twice during each cycle, once with positive slope and once with negative slope. Geometrical-optics ray tracing, as shown in the figure, can be used to confirm the ray position after each roundtrip. Recall that a ray is defined by its position and slope at each point along its trajectory. Both must attain their initial values for the ray to retrace its path.

Case 2: $m^2 > 1$ ($|m| > 1$, or equivalently, $m < -1$ or $m > 1$).

In this case, the eigenvalues of M are real and are given by

$$\Lambda_a = m + \sqrt{m^2 - 1} \equiv \Lambda, \tag{2.158a}$$

$$\Lambda_b = m - \sqrt{m^2 - 1} \equiv \frac{1}{\Lambda}, \tag{2.158b}$$

with $\Lambda_a \Lambda_b = 1$. One of these eigenvalues has a magnitude greater than unity. Denoting this eigenvalue by Λ, and the other by $1/\Lambda$, the ray coordinates after n steps can be written as

$$\widetilde{r}_n = C_a \Lambda^n \widetilde{r}_a + C_b \Lambda^{-n} \widetilde{r}_b. \tag{2.159}$$

The ray displacement is therefore

$$\rho_n = C_a\, \Lambda^n\, \rho_a + C_b\, \Lambda^{-n}\, \rho_b. \tag{2.160}$$

Since $|\Lambda| > 1$, one term grows in magnitude with each step. For $\Lambda > 1$, the growing contribution has the same sign from one step to the next, whereas for $\Lambda < -1$ it alternates in sign and therefore grows in an oscillatory fashion, as seen in Fig. 2.21. Periodic systems with $|m| > 1$ are therefore unstable.

The ray solutions described here apply to both sets of transverse ray coordinates, (x_n, x_n') and (y_n, y_n'), with appropriate initial conditions for each transverse coordinate, provided the system has rotational symmetry about the z-axis. In fact, the same ray matrices apply to both sets of transverse ray coordinates. The ray in a stable periodic focusing system will thus oscillate about the z-axis with the same period in both x and y. In general, however, these oscillations will have different amplitudes and phases:

$$x_n = x_{\text{max}} \sin(n\phi + \alpha_x), \quad y_n = y_{\text{max}} \sin(n\phi + \alpha_y). \tag{2.161}$$

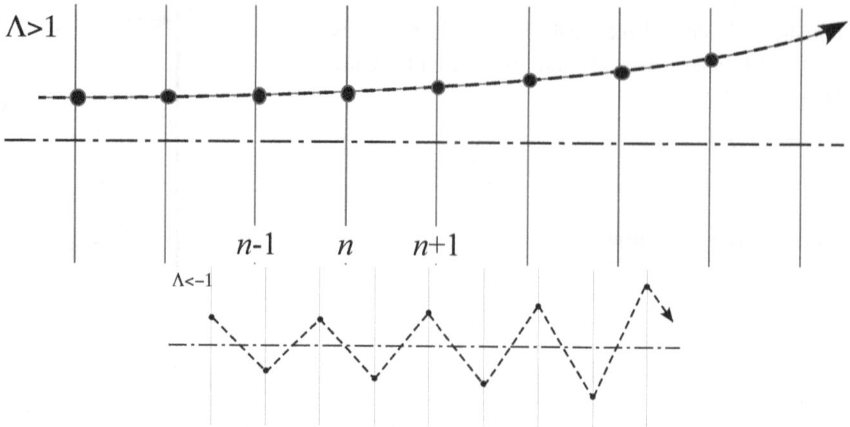

FIGURE 2.21
In an unstable periodic array, the ray displacement increases without limit, either monotonically or in an oscillatory fashion. Only the dots are meaningful; the continuous curve is simply a guide to the eye.

From these expressions for the transverse ray displacements, it is clear that successive ray positions (x_n, y_n) trace out Lissajous figures, except that the Lissajous figures here consist of discrete points rather than continuous curves. By adjusting the initial ray position and slope, the amplitudes and phases of the x and y oscillations can be varied, and circular, elliptical, or linear ray-spot patterns can be generated.

A simple periodic focusing system can be set up using a plane mirror and a curved mirror of focal length f, facing each other and separated by a distance $L < 2f$ to form a resonator. By injecting a beam (a good approximation to a ray) from a He-Ne laser and carefully adjusting the injection direction, mirror spacing, and mirror alignment, different types of periodic spot patterns can be generated on the end mirrors.

2.8.4 Two-Mirror Resonator Stability

For a two-mirror cavity, the roundtrip ABCD matrix is given by Eq. (2.136). The parameter $m = \frac{1}{2}(A + D)$ for a two-mirror cavity is then

$$m = 2g_1g_2 - 1\,. \tag{2.162}$$

The condition for stability, $-1 < m < 1$, therefore implies

$$-1 < 2g_1g_2 - 1 < 1\,. \tag{2.163}$$

Adding 1 throughout this inequality and dividing by 2, the stability condition for a two-mirror resonator can be written as

$$0 < g_1g_2 < 1\,. \tag{2.164}$$

The stability condition is a property of the resonator and not of a particular roundtrip matrix, which depends on the reference plane chosen and the direction in which the roundtrip is carried out. For example, if we choose a reference plane in the middle of the resonator and start toward mirror 2, the roundtrip matrix is found to be

$$\begin{pmatrix} A & B \\ C & D \end{pmatrix} = \begin{pmatrix} 2g_1g_2 - g_1 + g_2 - 1 & \frac{L}{2}(2g_1g_2 + g_1 + g_2) \\ \frac{2}{L}(2g_1g_2 - g_1 - g_2) & 2g_1g_2 + g_1 - g_2 - 1 \end{pmatrix}. \tag{2.165}$$

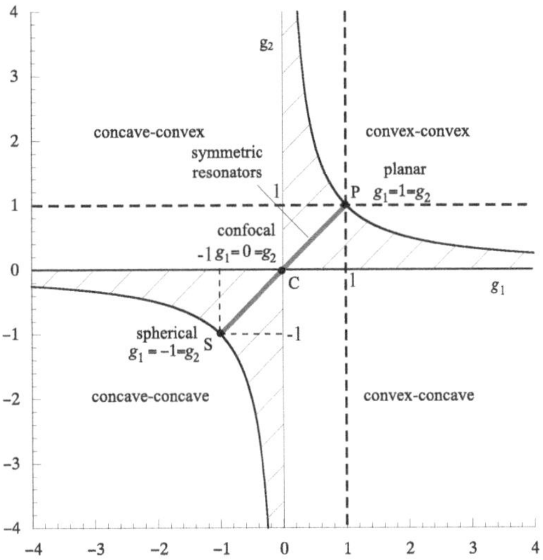

FIGURE 2.22
Stable resonators lie in the region bounded by the coordinate axes and the two branches of the hyperbola $g_1g_2 = 1$. Symmetric resonators lie along the line $g_2 = g_1$.

This matrix has unit determinant, as required for a valid roundtrip matrix, and its trace, $4g_1g_2 - 2$, again leads to the stability condition stated in Eq. (2.164). Recalling the definition $g_1 = 1 - \frac{L}{R_1}$ and $g_2 = 1 - \frac{L}{R_2}$, the stability condition (2.164) can be written explicitly in terms of mirror radii and separation:

$$0 < \left(1 - \frac{L}{R_1}\right)\left(1 - \frac{L}{R_2}\right) < 1 . \tag{2.166}$$

This relation specifies the region of stability of two-mirror optical resonators in terms of the mirror radii of curvature and their separation. An examination of Eq. (2.166) also shows that, for a two-mirror resonator to be stable, at least one mirror must be concave ($R > 0$).

Figure 2.22 shows a plot of the stability region for two-mirror resonators specified by Eq. (2.164) in the $g_1 - g_2$ plane. Stable resonators lie in the regions between the coordinate axes and the two branches of the hyperbola $g_1g_2 = 1$. We will return to the discussion of two-mirror resonators in Chapter 4.

Example 2.4 *As an application of Eq.* (2.166), *consider a two-mirror resonator formed by a convex mirror of radius* $R_1 = -1.0$ *m and a concave mirror of radius* $R_2 = 1.5$ *m [Fig. 2.23(a)]. What is the range of possible mirror separations for which this resonator is stable?*

For the given resonator ($R_1 = -1$ m and $R_2 = 3/2$ m), the stability condition becomes (with all lengths in meters)

$$0 < (1 + L)(1 - 2L/3) < 1 .$$

The lower stability limit, $0 < (1 + L)(1 - 2L/3)$, requires $L < \frac{3}{2}$, since $L > 0$. The upper stability limit requires $(1+L)(1-2L/3) = 1 + \frac{L}{3} - \frac{2L^2}{3} < 1$ or $\frac{L}{3}(1-2L) < 0 \quad \Rightarrow \quad L > \frac{1}{2}$. Combining these two results, we find that the cavity is stable when

$$\frac{1}{2} < L < \frac{3}{2} .$$

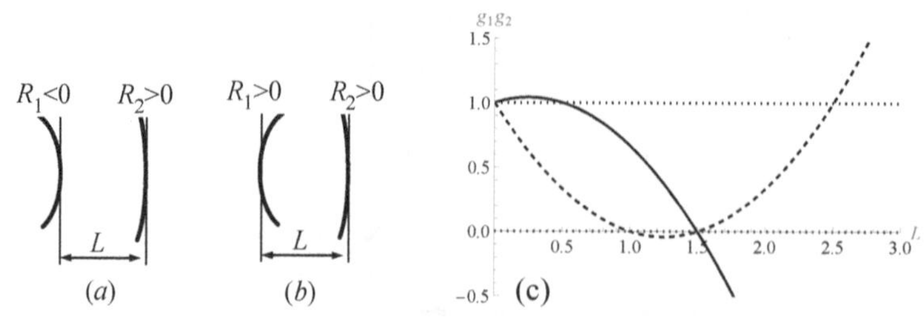

FIGURE 2.23
Two-mirror resonators formed by (a) a convex and a concave mirror and (b) two concave mirrors; (c) stability ranges for the resonators in (a) (solid curve) and (b) (dashed curve). See the examples in the text.

It is easier to arrive at this result by plotting $g_1g_2 = (1 + L)(1 - 2L/3)$ as a function of L. The range of stability corresponds to the range of mirror separations L for which g_1g_2 lies between 0 and 1 (dotted lines). As can be seen from the solid curve in Fig. 2.23(c), this occurs over the range 0.5 m$< L < 1.5$ m.

Example 2.5 *As another application, consider the resonator formed by two concave mirrors of radii 1 m and 3/2 m, respectively [Fig. 2.23(b)].*

In this case the stability ranges are $0 < L < 1$ and $3/2 < L < 5/2$, as shown by the dashed curve in Fig. 2.23(c).

2.8.5 Astigmatism

In optical systems with rotational symmetry, the same ray matrices apply to the (x, x') and (y, y') ray coordinates. However, many optical elements used in laser cavities introduce astigmatism. For example, a cylindrical lens with its cylinder axis aligned with the x-axis acts as a thin lens for the y transverse coordinate but has no focusing effect on the x coordinate of the ray. Similarly, the gain medium in gas lasers is usually enclosed in a discharge tube with Brewster windows at the ends, and many solid-state lasers use a gain medium in the form of a cylindrical rod with its ends polished at Brewster's angle [Fig. 2.24]. The purpose of terminating the gain region with Brewster windows is to minimize reflection losses as the beam enters or exits the gain medium.

In the presence of astigmatic elements, paraxial rays traverse different optical paths depending on whether they lie in the plane of incidence of the chief ray (the tangential plane) or in the plane perpendicular to it (the sagittal plane). As a result, the ray matrices are different for the ray coordinates in the two planes. For non-normal incidence at an

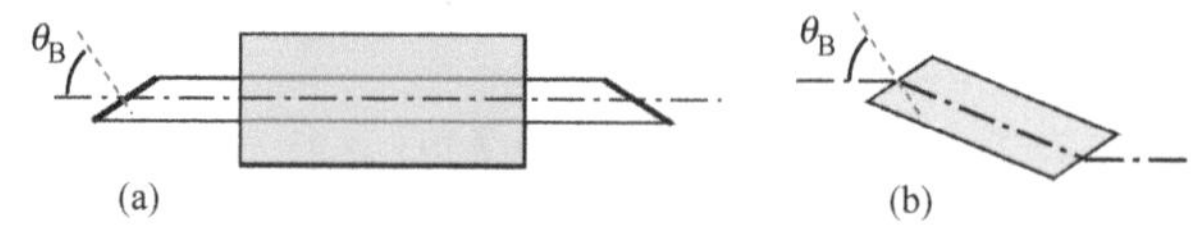

FIGURE 2.24
Gain media terminated by Brewster windows: (a) discharge tube in a gas laser; (b) gain rod in a solid-state laser.

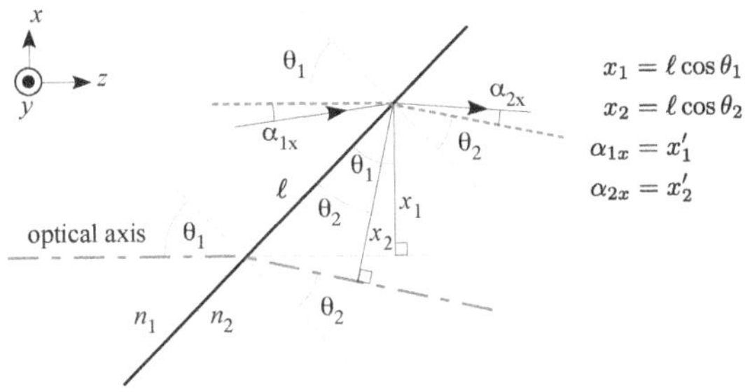

FIGURE 2.25
An interface at non-normal incidence.

interface between two media, the optical axis itself changes direction. Hence, the distance of a paraxial ray from the optical axis may change as the ray crosses the interface.

Figure 2.25 shows a paraxial ray in the tangential plane ($x - z$ plane) at an interface between two media with refractive indices n_1 and n_2, respectively, when the optical axis meets the interface at an oblique angle θ_1. The directions of the optical axis in the two media are then related by Snell's law: $n_1 \sin\theta_1 = n_2 \sin\theta_2$. For a ray incident at an angle close to θ_1 (i.e., paraxial to the optical axis), ray coordinates on the two sides of the interface between the media are given by $x_1 = \ell \cos\theta_1$, $x_2 = \ell \cos\theta_2$, and their slopes are related by Snell's law $n_1 \sin(\theta_1 + \alpha_{1x}) = n_2 \sin(\theta_2 + \alpha_{2x})$. With $\alpha_{1x}, \alpha_{2x} \ll 1$, these expressions lead to the following relation between ray coordinates

$$x_2 = \frac{\cos\theta_2}{\cos\theta_1} x_1 , \quad x_2' = \frac{n_1 \cos\theta_1}{n_2 \cos\theta_2} x_1' . \tag{2.167}$$

Thus, for oblique incidence, the interface ray matrix for ray coordinates in the tangential plane is

$$M_t = \begin{pmatrix} \frac{\cos\theta_2}{\cos\theta_1} & 0 \\ 0 & \frac{n_1 \cos\theta_1}{n_2 \cos\theta_2} \end{pmatrix} . \tag{2.168}$$

The interface ray matrix for ray coordinates in the sagittal plane (perpendicular to the tangential plane) is

$$M_s = \begin{pmatrix} 1 & 0 \\ 0 & \frac{n_1}{n_2} \end{pmatrix} . \tag{2.169}$$

For a plate of thickness t and refractive index $n_2 = n$ (surrounded by air, $n_1 = 1$), employed at an angle of incidence θ_1 in an optical system [Fig. 2.26], the effective length of the plate for rays in the x-z plane (the plane of incidence of the chief ray, or tangential plane) is

$$d_x = \frac{t \cos^2\theta_1}{n \cos^3\theta_2} \xrightarrow{\theta_1 = \theta_B} \frac{t\sqrt{n^2+1}}{n^4} , \tag{2.170}$$

whereas for rays in the y-z plane it is

$$d_y = \frac{t}{n \cos\theta_2} \xrightarrow{\theta_1 = \theta_B} \frac{t\sqrt{n^2+1}}{n^2} . \tag{2.171}$$

Sometimes a curved mirror is used at an angle of incidence that deviates significantly from normal incidence. Figure 2.27 shows a chief ray incident at an angle θ on a curved

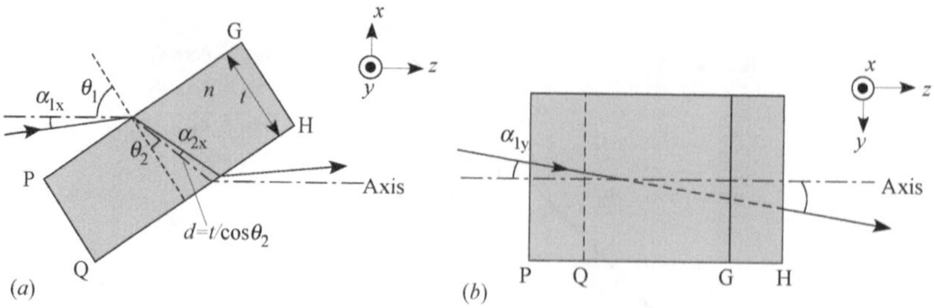

FIGURE 2.26
Ray coordinates for rays paraxial to a chief ray (dash-dot line) incident at an angle θ_1 on an interface are affected differently in (a) the tangential plane ($x - z$ plane) and (b) the sagittal plane ($y - z$ plane).

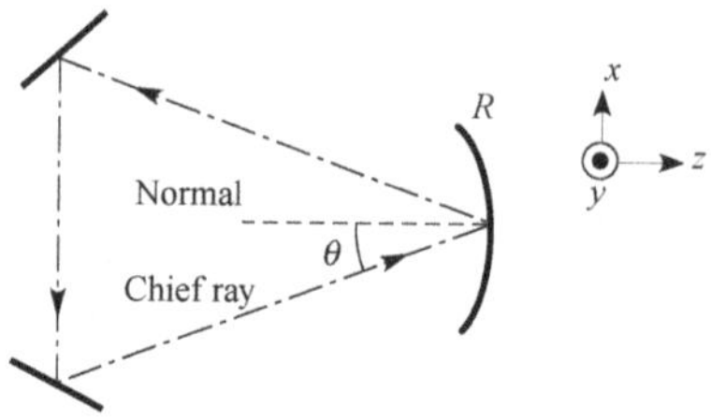

FIGURE 2.27
A curved mirror at non-normal incidence in an optical cavity.

mirror of radius of curvature R. Rays paraxial to the chief ray experience different effective focal lengths in the tangential and sagittal planes,

$$f_x = f\cos\theta\,, \quad f_y = f/\cos\theta\,. \tag{2.172}$$

Similarly, when a lens is used at an angle that deviates significantly from normal incidence, its effective focal lengths differ in the tangential and sagittal planes. The expressions for the two focal lengths are again given by Eq. (2.172). Astigmatism destroys cylindrical symmetry about the optical axis, and we must therefore treat ray trajectories in the $x - z$ and $y - z$ planes separately. This may lead to different conditions for ray confinement in the two planes. In laser cavities, astigmatism leads to elliptical beams.

If the optical system contains several astigmatic elements, but their principal axes are aligned along the same x and y axes, we can still analyze the ray behavior in each transverse coordinate independently, using separate ABCD matrices for the $x - z$ and $y - z$ planes. Astigmatic systems that can be described by separate ray matrices in two orthogonal planes (principal planes) are referred to as orthogonal systems. Systems that do not have this property are referred to as *nonorthogonal systems*. In general, nonorthogonal systems do not permit the ray matrices to be separated into two independent sets associated with orthogonal axes. Ray analysis of nonorthogonal systems generally involves 4 × 4 matrices and includes operations such as twist or image rotation; it is beyond the scope of this book.[10]

[10]See, for example, J. A. Arnaud, "Nonorthogonal optical waveguides and resonators," Bell Sys. Tech. J. **49**, 2311–2347 (1970) and "Mode coupling in first order optics," J. Opt. Soc. Am. **61**, 751–758 (1971).

Summary

This chapter began with an overview of light as an electromagnetic (EM) wave phenomenon and introduced the complex analytic representation of the electric field of an EM wave. Plane and spherical wave solutions were used to develop a physical picture of wave propagation and energy transport in terms of phasefronts and rays. Intensity, the Poynting vector, and Poynting's theorem were introduced to describe power transport and the exchange of power between an EM wave and atoms in a medium. The nature of EM wave fields in a rectangular cavity was examined to demonstrate how the presence of boundaries modifies the field and to point out that open cavities are better suited than closed cavities for providing optical feedback in lasers.

The ray picture of energy transport in EM waves was then developed as a prelude to finding the wave modes of open resonators. Maxwell's equations were used to derive the laws of ray optics, and the so-called ABCD ray matrices were introduced to describe the transformation of rays by optical elements such as mirrors and lenses used to form open resonators. With the help of these matrices, it was shown that suitably designed open resonators can confine paraxial ray bundles and thus provide optical feedback in lasers. This insight will be used in the next chapter to find paraxial-wave solutions of the wave equation.

Problems

1. **Monochromatic plane wave:** Consider the complex function $F(z,t) = Ae^{i(kz-\omega t+\phi_o)}$. What are the SI units of k and ω? Find the relation between k and ω required for $F(z,t)$ to satisfy the source-free wave equation for a wave speed v. This is an example of a waveform that is periodic in both space and time.

   (a) Express the spatial period λ (wavelength) and the temporal period τ of the wave in terms of the wave number k and angular frequency ω, respectively. Express the wave speed in terms of λ and τ.

   (b) Compute the numerical values of λ and τ for $v = 3.00 \times 10^8$ m/s and $\omega = 2\pi \times 5.00 \times 10^{14}$ rad/s. If phase diffusion gives the wave a spectral width $\delta\nu = 10$ MHz, what is the corresponding wavelength width $\delta\lambda$, expressed in nm?

2. Does the function $F(z,t) = A \sin kz\, e^{-i(\omega t-\phi_o)}$ represent a solution of the wave equation? If so, what relation must hold between k and ω? Explain the spatial and temporal phase dependence of this field and determine whether it represents a traveling wave or a standing wave. What can be said about its propagation speed and net energy flow? Write down electric and magnetic fields satisfying Maxwell's equations that have this standing-wave form, specifying their relative orientations and spatial dependence.

3. Consider the function $F(z,t) = A\, e^{-a^2(t-z/v_g)^2} e^{i(k_o z-\omega_o t)}$. Discuss the conditions under which this function can represent a propagating wave packet. Identify the carrier phase and determine the phase velocity of the carrier wave. What is the velocity of the pulse envelope? Under the usual wave-packet interpretation, what are the speeds of the phasefronts and of energy transport?

4. Cell phones contain low-power radio-frequency transmitters. For example, some transmitters operate with powers of the order of 1 W. Estimate the wavelength, average intensity I, and electric-field amplitude of the electromagnetic wave at a distance of 1 m from a cell phone radiating 1 W of power at a frequency of 1 GHz. Assume, for this estimate, that the power is radiated uniformly in all directions.

5. A radio station broadcasts a power of 100 kW at a frequency of 91.3 MHz. Assume that the power is radiated uniformly into a single hemisphere and that a radio receiver is located 40 km from the transmitter.

   (i) What are the average intensity and the electric-field amplitude at the location of the receiver?

   (ii) Assuming that the incident radiation is completely absorbed by the receiver, what radiation pressure does it exert on the receiver?

   (iii) What is the maximum magnitude of the electric force exerted by the wave on an electron in the receiver?

6. Instead of using the standing-wave modes appropriate to a closed metallic box to expand the electromagnetic field in a cubical region, we may choose a different complete set of functions. For example, consider the traveling-wave functions

$$\boldsymbol{u}_{\boldsymbol{k}}(\boldsymbol{r}) = \frac{\boldsymbol{A}_{\boldsymbol{k}}}{\sqrt{L^3}}\, e^{i(k_x x + k_y y + k_z z)}\,,$$

where $k_x = \dfrac{2\pi m_x}{L}, \quad k_y = \dfrac{2\pi m_y}{L}, \quad k_z = \dfrac{2\pi m_z}{L}, \qquad m_x, m_y, m_z = 0, \pm 1, \pm 2, \pm 3, \cdots$

and $k^2 \equiv \left(\dfrac{2\pi}{L}\right)^2 (m_x^2 + m_y^2 + m_z^2) = \dfrac{\omega^2}{v^2}\,, \qquad \boldsymbol{k}\cdot\boldsymbol{A}_{\boldsymbol{k}} = 0\,.$

These functions are periodic in x, y, and z with period L. Each vector function in this set is characterized by three integers m_i, or equivalently by the wave vector $\boldsymbol{k} = (2\pi/L)(m_x, m_y, m_z)$.

For a source-free electromagnetic field in a homogeneous medium, the transversality condition $\boldsymbol{k}\cdot\boldsymbol{A}_{\boldsymbol{k}} = 0$ implies that only two independent components of $\boldsymbol{A}_{\boldsymbol{k}}$ can be chosen. These correspond to the two independent polarizations associated with a given wave vector.

The modes may be represented by points (k_x, k_y, k_z) in three-dimensional k-space. Note that both positive and negative integer values, as well as zero, occur in the definitions of k_x, k_y, and k_z. By counting the number of modes in a spherical shell between k and $k+dk$, and using the relation $k = \omega/v = 2\pi\nu/v$, show that the number of modes per unit frequency interval per unit volume is the same as the mode-density expression derived in the text.

7. In deriving the mode density (number of modes per unit volume per unit frequency interval) in Eq. (2.76), we neglected the frequency dependence of the refractive index. Derive the corresponding expression for the mode density when material dispersion, $n = n(\nu)$, is included.

8. **Square-law medium:** The refractive-index profile of a medium is given by $n^2(x, y) = n_0^2 - n_2(x^2 + y^2)$, where n_0 and n_2 are constants and $n_2(x^2 + y^2) \ll n_0^2$ for all transverse positions of interest.

(a) Show that the ray equation is

$$\frac{d}{ds}\left(n\frac{d\mathbf{r}}{ds}\right) = \boldsymbol{\nabla} n = -\frac{n_2}{n(x,y)}\left(x\mathbf{e}_x + y\mathbf{e}_y\right) . \tag{2.173}$$

(b) Consider a ray launched at $z = 0$ in the x-z plane from $x = x_o$, with launch angle α measured with respect to the z-axis. Show that its transverse coordinate has the form

$$x(z) = A\sin\left[\frac{\sqrt{n_2}}{n(x_o)\cos\alpha}z + \phi_o\right] ,$$

where A and ϕ_o are determined by the initial position and slope of the ray.

(c) Consider now a ray launched from $x = x_o$, $y = 0$, and $z = 0$, with its initial transverse slope entirely in the y-direction and with launch angle α measured with respect to the z-axis. Show that the ray trajectory is

$$x(z) = x_o\cos\left(\frac{\sqrt{n_2}}{\beta}z\right) ,$$

$$y(z) = \left(\frac{\beta\tan\alpha}{\sqrt{n_2}}\right)\sin\left(\frac{\sqrt{n_2}}{\beta}z\right) ,$$

where $\beta = n(x,y)\cos\theta(x,y) = n(x_o,0)\cos\alpha$.

9. The paraxial ray-transfer matrix for the gradient-index fiber considered in Sec. 2.8.1, with refractive-index profile $n(\rho) = n_1[1-\rho^2/(2\Lambda^2)]$, where ρ is the radial distance from the axis (z-axis) of the fiber, n_1 is the refractive index on the axis, and $\Lambda \gg \rho$ is a length scale characterizing the transverse variation of the refractive index, is given by $\begin{pmatrix} \cos(L/\Lambda) & \Lambda\sin(L/\Lambda) \\ -\frac{1}{\Lambda}\sin(L/\Lambda) & \cos(L/\Lambda) \end{pmatrix}$, where L is the propagation distance in the fiber.

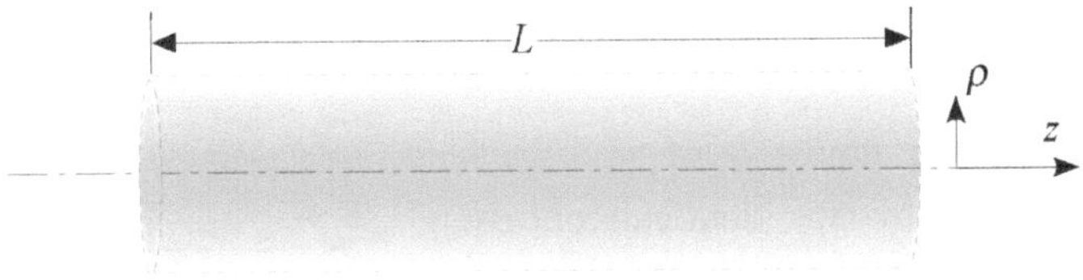

A length $L = \pi\Lambda/2$ of this fiber is placed in air and used as a lens. Find the ABCD matrix of this optical element by choosing the entrance and exit reference planes just outside the fiber, and hence determine its effective focal length in air.

10. **Brewster-angle elements:** In many solid-state lasers, the gain medium is in the form of a cylindrical rod or rectangular prism whose end faces are oriented at Brewster's angle to minimize reflection losses for one linear polarization. Consider paraxial rays lying in the plane of incidence. The beam travels a distance d inside the rod, parallel to its axis [Fig. (a)]. Find the ray-transfer matrix of the element in terms of d and the refractive index n of the rod, assuming that the surrounding medium is air. If the gain medium instead consists of a slab or sheet of thickness t used at Brewster's angle, express the ray-path length d inside the medium in terms of t and n [Fig. (b)].

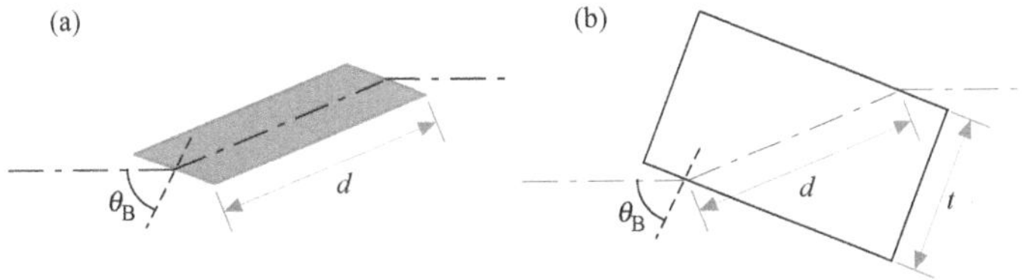

11. **Brewster plate tuner:** A Brewster plate may be used for several purposes in an optical cavity. Discuss its use (i) as a variable-loss element, (ii) for frequency tuning, and (iii) for polarization selection.

12. **Imaging condition:** Given the ABCD matrix of an optical system between two reference planes, show that the condition for the exit plane to be the image plane of an object in the input plane is $B = 0$. What, then, is the transverse magnification of the image relative to the object, expressed in terms of the elements of the system ray matrix?

13. **Focal length of a thin lens:** Use the ray-matrix technique and the imaging condition of the preceding problem to derive an expression for the focal length of a thin lens made of a material of refractive index n. The radii of curvature of the first and second surfaces are R_1 and R_2, respectively, and the lens thickness is negligible. Assume that the medium surrounding the lens has refractive index n_m. Compute the focal length of a biconvex lens with $R_1 = -R_2 = 50$ cm and $n = 1.5$ for (i) $n_m = 1.0$ and (ii) $n_m = 1.8$.

14. **Mirage mirror:** A toy optical system consists of two spherical mirrors of the same focal length facing each other. The right mirror has a hole at its center, and the separation between the mirrors is equal to their common focal length. If an object is placed at the center of the left mirror, a real image is formed at the hole in the right mirror. When you see it, you want to touch it! The ray path from the object to the image consists of propagation to the right mirror, reflection from the right mirror, propagation back to the left mirror, reflection from the left mirror, and finally propagation to the image plane.

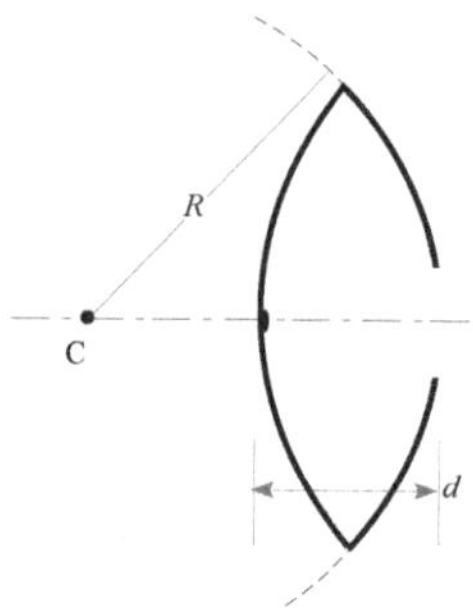

Find the overall ABCD matrix for this ray path, starting at the object plane and treating the mirror separation d as a variable. Determine the value of d for which the image is formed at the hole. Is the image inverted or upright relative to the object?

15. **Symmetric two-mirror cavity**

(a) A cavity is formed by two identical concave mirrors ($R_1 = R_2 = R$) facing each other and separated by a distance L. Starting at the reference plane in the middle of the cavity and traveling to the right, as shown in the figure, find the roundtrip matrix for the cavity and determine the range of values of L for which the cavity is stable.

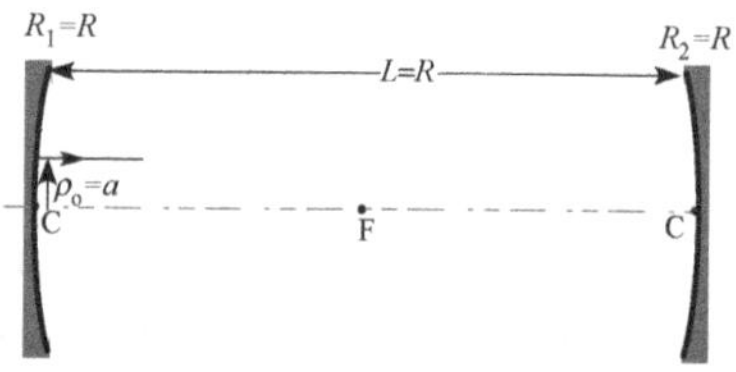

(b) A ray starts just in front of one mirror at a height $\rho_o = a$ and travels toward the other mirror with slope $\rho'_o = 0$ in a symmetric two-mirror confocal cavity ($L = R$). Find the ray displacement at the reference plane as a function of the number of roundtrips.

16. Examine the stability of the cavity shown here by constructing an equivalent periodic lens waveguide. Identify a unit cell by choosing a reference plane immediately after reflection from mirror M_1, and find the ray transfer matrix for the unit cell. Assume that the radius of curvature of mirror M_1 is $R_1 = -R$ and that of mirror M_2 is $R_2 = +R$ ($R > 0$). Apply the stability criterion to determine the relationship between L and R required for a stable cavity.

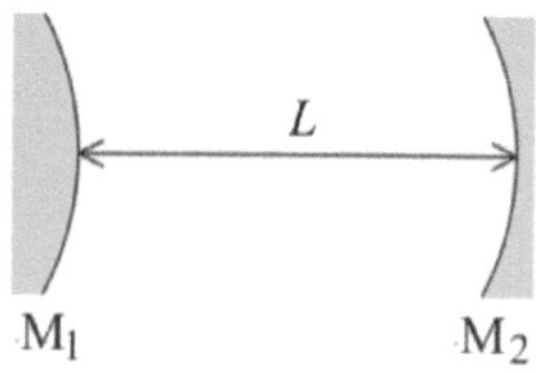

17. **Concave-convex mirror cavity:** Consider a resonator consisting of a convex mirror ($R_1 < 0$) and a concave mirror ($R_2 > 0$), separated by a distance L. Determine the range of values of L for which the resonator is stable for each of the two cases $|R_1| > R_2$ and $|R_1| < R_2$.

    (a) A two-mirror resonator is formed by a convex mirror of radius $R_1 = -45.0$ cm and a concave mirror of radius $R_2 = 60.0$ cm. What range of mirror separations results in a stable resonator?

    (b) A two-mirror resonator is formed by a convex mirror of radius $R_1 = -1.0$ m and a concave mirror of radius $R_2 = \frac{3}{2}$ m. What range of mirror separations results in a stable resonator?

18. **Cavity stability:**

    (a) Can a stable resonator be formed using a **convex** mirror with a radius of curvature of 2.0 m and a **concave** mirror with a radius of curvature of 4.0 m, separated by a distance L? Determine the range of L for which the resonator is stable.

    (b) A resonator consists of two spherical concave mirrors with radii of curvature 4.0 m and 2.0 m, respectively. The mirrors are separated by a distance L in air. Determine the range of L for which the mirrors form a stable resonator.

19. Consider a two-mirror cavity formed by a plane mirror and a concave mirror of radius of curvature R, separated by a distance L, with their reflecting surfaces facing each other.

    (a) Starting with a reference plane just in front of the plane mirror and traveling toward the concave mirror, write the roundtrip matrix M as a product of component matrices. Express the elements of M in terms of L and R, and, for a given R, determine the range of L for which the cavity is stable.

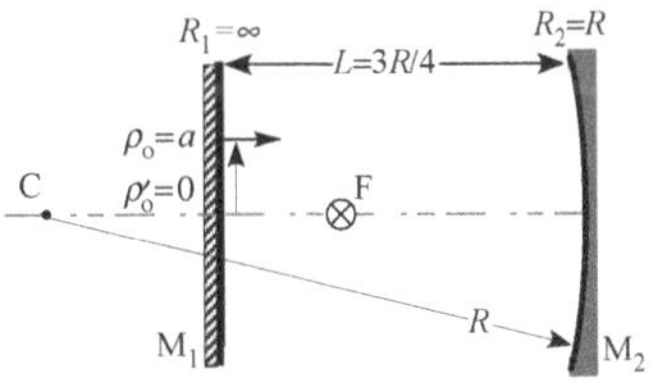

    (b) Now consider the special case $L = \frac{3}{4}R$. Is this configuration stable? Suppose a ray starts at a height a above the axis and perpendicular to the reference plane in part (a), so that $\rho_o = a$ and $\rho_o' = 0$. If the ray displacement in the reference plane after n roundtrips is given by $\rho_n = \rho_{\max} \sin(n\theta + \alpha)$, determine $\rho_{\max}$, θ, and α. After how many roundtrips does the ray trajectory repeat? Check your result by tracing the ray path using geometrical optics.

20. A White cell is an optical system consisting of two mirrors, one of which has a small hole through which light can enter and leave the cell. Usually, a cell containing an absorbing or fluorescing medium is placed between the mirrors. A laser beam is introduced through the hole so that it travels back and forth between the mirrors many times before exiting. Such a system is often used in spectroscopy to increase the interaction length between the light and the medium.

    The condition for obtaining multiple passes involves considerations similar to those used in the analysis of a stable optical cavity. Note that, in the paraxial limit, ray displacements in the two transverse directions can be treated independently. If a White cell of length $L = 20$ cm is to be constructed using identical mirrors of radius of curvature R, find the possible values of R for which the laser beam strikes each mirror ten times before exiting. How many passes does the beam make through the cell?

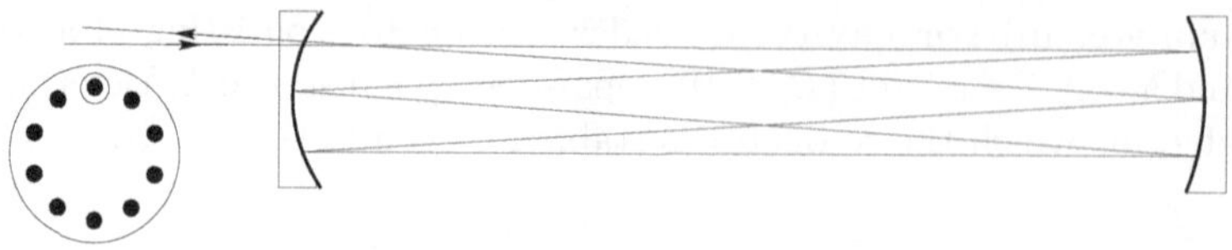

In a stable mirror configuration, the laser spots on a mirror will, in general, lie on an ellipse.

21. **Ring cavity:** A triangular ring cavity of perimeter L is formed by three plane mirrors with a positive lens (focal length f) inserted between two of the mirrors. Find a relation between L and f for the cavity to be stable.

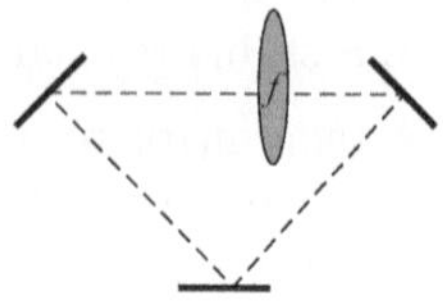

3

Paraxial Beam Solutions of the Wave Equation

In Chapter 2, we established that suitably chosen open resonators can support paraxial rays that remain spatially confined within the cavity boundaries, rendering them suitable for providing optical feedback in laser oscillators. A bundle of paraxial rays escaping through the partially transmitting end mirrors of such a resonator into free space will propagate along a relatively well-defined direction, converging or diverging slowly as it travels. Such a ray bundle defines an optical beam. In this chapter, we seek wave-like solutions whose phasefronts correspond to such paraxial ray bundles.

As discussed in the preceding chapter, wave-like solutions of Maxwell's equations in free space or in the presence of boundaries can be constructed from solutions to the scalar Helmholtz equation. Indeed, each Cartesian component of the monochromatic electric field can be written as $E(\boldsymbol{r},t) = E(\boldsymbol{r})e^{-i\omega t}$, where the space-dependent complex amplitude $E(\boldsymbol{r})$ satisfies the scalar Helmholtz equation derived from Eq. (2.30):

$$\left[\nabla^2 + \omega^2\mu\epsilon\left(1 + \frac{\epsilon_0\chi_{\rm at}}{\epsilon} + i\frac{\sigma}{\omega\epsilon}\right)\right] E(\boldsymbol{r}) = 0\,. \tag{3.1}$$

Here, the space-dependent amplitude is denoted by the same symbol as the total field to avoid a proliferation of symbols. Likewise, the Cartesian components of the resonant atomic polarization have been written as $P_{\rm at}(\boldsymbol{r}) = \epsilon_0\chi_{\rm at}(\omega)E(\boldsymbol{r})e^{-i\omega t}$. Alternatively, we can specify an electric field with uniform linear polarization by writing $\boldsymbol{E}(\boldsymbol{r},t) = \hat{\varepsilon}E(\boldsymbol{r})e^{-i\omega t}$, where ε is the polarization unit vector and $E(\boldsymbol{r})$ is the scalar field amplitude satisfying Eq. (3.1). Vector solutions to Maxwell's equations can then be constructed from these scalar Helmholtz solutions, as discussed later in Section 3.3.

A monochromatic plane wave propagating in the direction $\hat{\kappa} = \boldsymbol{k}/|\boldsymbol{k}|$ in a loss-free, non-interacting medium ($\chi_{\rm at} = 0$ and $\sigma = 0$) is represented by $E(\boldsymbol{r}) = E_0e^{i\boldsymbol{k}\cdot\boldsymbol{r}}$. For a non-magnetic medium where μ and ϵ are real, the magnitude of the propagation vector is given by:

$$k = \sqrt{\mu\epsilon}\,\omega = \frac{n}{c}\,\omega\,, \tag{3.2}$$

where n is the refractive index. We are now equipped to explore beam-like solutions to the wave equation by factoring the total scalar field into a slowly varying envelope and a fast carrier wave: $E(\boldsymbol{r}) = E_0(\boldsymbol{r})e^{i\boldsymbol{k}\cdot\boldsymbol{r}}$. Substituting this into Eq. (3.1) yields the equation specified by the spatial envelope $E_0(\boldsymbol{r})$:

$$\left[\nabla^2 + 2i\boldsymbol{k}\cdot\nabla + k^2\left(\frac{\chi_{\rm at}}{n^2} + i\frac{\sigma}{\omega\epsilon_0 n^2}\right)\right] E_0(\boldsymbol{r}) = 0\,, \tag{3.3}$$

where we have used the identity $\epsilon/\epsilon_0 = n^2$, which is valid for materials with negligible magnetic response ($\mu \approx \mu_0$).

DOI: 10.1201/9781003584179-3

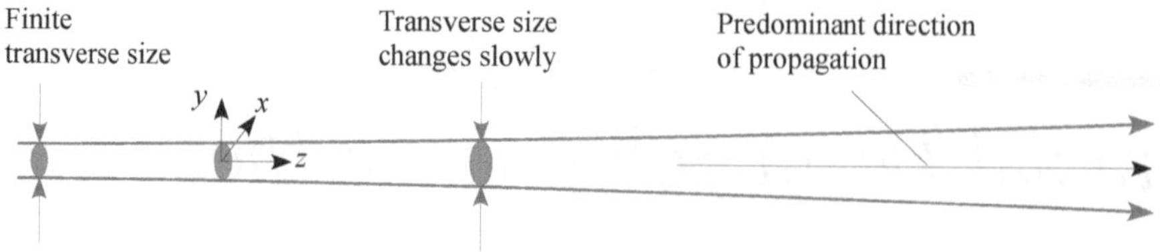

FIGURE 3.1
Geometry of an optical beam, illustrating a dominant direction of propagation and a finite spatial extent perpendicular to the direction of propagation.

3.1 Paraxial Beam Solutions

Paraxial, beam-like solutions to the Helmholtz equation are characterized by a predominant direction of propagation and a finite transverse cross-section An optical beam with a finite transverse profile inevitably experiences diffractive spreading, which modifies its cross-sectional area during propagation. Nevertheless, the field maintains a well-defined, predominant direction of propagation provided this diffractive divergence remains small over large distances. To construct a solution with these characteristics, we use a monochromatic plane wave propagating along the z-axis as a guide:

$$E(\boldsymbol{r}) = E_0 e^{ikz}\,, \tag{3.4}$$

where the longitudinal propagation constant is $k = n\omega/c = 2\pi n/\lambda$. Using this plane-wave solution as a guide, we look for solutions that correspond to a superposition of plane waves propagating in directions making small angles with the z-axis. For such plane waves paraxial to the z-axis, we can write the propagation vector as $\boldsymbol{k} = \boldsymbol{e}_x k_x + \boldsymbol{e}_y k_y + \boldsymbol{e}_z k_z$, with

$$k^2 = k_x^2 + k_y^2 + k_z^2 \equiv n^2\omega^2/c^2\,, \quad k_x, k_y \ll k, k_z\,, \tag{3.5a}$$

and $$k_z = \sqrt{k^2 - k_x^2 - k_y^2} \approx k - \frac{k_x^2 + k_y^2}{2k}\,. \tag{3.5b}$$

The space-dependent part of the field can now be synthesized as a two-dimensional Fourier integral

$$\begin{aligned} E(\boldsymbol{r}) &= \frac{1}{2\pi}\iint dk_x, dk_y, \tilde{E}_0(k_x, k_y) e^{i(k_x x + k_y y + k_z z)}\,, \\ &= \frac{e^{ikz}}{2\pi}\iint dk_x, dk_y, \tilde{E}_0(k_x, k_y) e^{i(k_x x + k_y y) - i(k_x^2 + k_y^2)z/2k}\,, \\ &\equiv e^{ikz} E_0(\boldsymbol{r})\,. \end{aligned} \tag{3.6}$$

To ensure that the predominant direction of propagation of the synthesized wave packet remains along the z-axis, the angular spectral amplitude $|\tilde{E}_0(k_x, k_y)|$ must decrease rapidly for large transverse wavevector components ($|k_x|, |k_y| \to \infty$).

Comparing Eq. (3.6) with the expression for a monochromatic plane wave (3.4), we observe that for paraxial beams, the field amplitude $E_0(\boldsymbol{r})$ depends on the spatial coordinates, whereas for an idealized plane wave it is a constant. Because $E_0(\boldsymbol{r})$ is a superposition of plane waves dominated by propagation vectors with small transverse components ($k_x, k_y \ll k$), its spatial profile will evolve slowly in the transverse directions (x–y plane). Substituting

Eq. (3.6) into the scalar Helmholtz equation (3.3) allows the spatial variation of $E_0(\boldsymbol{r})$ during propagation to be expressed as

$$\frac{1}{k^2}\frac{\partial^2 E_0(\boldsymbol{r})}{\partial z^2} + \frac{2i}{k}\frac{\partial E_0(\boldsymbol{r})}{\partial z} = -\frac{1}{k^2}\nabla_\perp^2 E_0(\boldsymbol{r}) - \left(\frac{\chi_{at}}{n^2} + i\frac{\sigma}{\omega\epsilon_o n^2}\right) E_0(\boldsymbol{r})\,, \tag{3.7a}$$

$$\text{where}\quad \nabla_\perp^2 = \frac{\partial^2}{\partial x^2} + \frac{\partial^2}{\partial y^2}\,. \tag{3.7b}$$

According to this equation, two effects contribute to the variation of the field amplitude with propagation. The first term on the right-hand side represents the change in the beam profile due to the presence of paraxial plane waves with $|k_x|, |k_y| \ll k$. This is diffraction at work, which arises from the variation of the field amplitude in the transverse directions. The second term represents the combined effects of interaction with the atoms (via the atomic susceptibility χ_{at}) and losses (via the medium conductivity σ). Let w denote the length scale over which the wave amplitude varies significantly in the transverse directions. If w is large compared with the reduced wavelength $\lambda/2\pi$, that is, $w \gg \lambda/2\pi$ or $kw \gg 1$, the diffraction-induced fractional change in the wave amplitude over a propagation distance of order $1/k = \lambda/2\pi$ will be small, that is, $\left|\frac{1}{E_0(\boldsymbol{r})}\frac{\partial E_0(\boldsymbol{r})}{\partial z}\right|\frac{\lambda}{2\pi} \ll 1$. We expect the wave amplitude for paraxial waves to obey this inequality because the spatial Fourier transform of the amplitude $E_0(\boldsymbol{r})$ is dominated by values $k_x, k_y \sim 1/w$ that are small compared with $k_z \approx k = 2\pi/\lambda$ ($k_x, k_y \ll k$). The paraxial wave amplitude is therefore expected to satisfy the following inequalities:

$$\frac{1}{k}\left|\frac{\partial E_0(\boldsymbol{r})}{\partial z}\right| = \frac{\lambda}{2\pi}\left|\frac{\partial E_0(\boldsymbol{r})}{\partial z}\right| \ll |E_0(\boldsymbol{r})|\,, \tag{3.8a}$$

$$\frac{1}{k}\left|\frac{\partial^2 E_0(\boldsymbol{r})}{\partial z^2}\right| = \frac{\lambda}{2\pi}\left|\frac{\partial}{\partial z}\left(\frac{\partial E_0(\boldsymbol{r})}{\partial z}\right)\right| \ll \left|\frac{\partial E_0(\boldsymbol{r})}{\partial z}\right|\,. \tag{3.8b}$$

The first inequality dictates that the change in the beam profile during propagation over a distance of order $\lambda/2\pi$ is a small fraction of $|E_0(\boldsymbol{r})|$. The second inequality states that the rate of change of the beam profile does not vary significantly over the same propagation distance. In view of the inequalities (3.8), the dominant variation of $E_0(\boldsymbol{r})$ with z in Eq. (3.7) is described by the first derivative term. Accordingly, neglecting the second derivative term $\partial^2 E_0(\boldsymbol{r})/\partial z^2$ in Eq. (3.7), the equation governing the spatial evolution of the paraxial beam envelope reduces to

$$\frac{\partial E_0(\boldsymbol{r})}{\partial z} = \frac{i}{2k}\nabla_\perp^2 E_0(\boldsymbol{r}) + i\frac{k\chi'_{at}}{2n^2}E_0(\boldsymbol{r}) - \frac{k}{2}\left(\frac{\chi''_{at}}{n^2} + \frac{\sigma}{\omega\epsilon_o n^2}\right) E_0(\boldsymbol{r})\,. \tag{3.9}$$

Here we have written the atomic susceptibility in terms of its real and imaginary parts as $\chi_{at} = \chi'_{at} + i\chi''_{at}$. Equation (3.9) is one of the basic equations of laser physics. It demonstrates that the field amplitude evolves slowly along the propagation axis z due to three distinct physical mechanisms: (i) diffraction, represented by the $\nabla_\perp^2$ term; (ii) refractive index and phase shifts due to interaction with the atomic medium, represented by the χ'_{at} term; and (iii) dissipation due to Joule heating, represented by the conductivity term, and absorption or amplification due to interaction with the atoms, represented by the χ''_{at} term.

Neglecting the loss and atomic interaction terms in Eq. (3.9), we find that the beam profile changes with propagation solely because of diffraction (due to the variation of the amplitude in the transverse directions). The resulting equation governing the evolution of the beam profile is

$$\left[\frac{\partial^2}{\partial x^2} + \frac{\partial^2}{\partial y^2} + 2ik\frac{\partial}{\partial z}\right] E_0(\boldsymbol{r}) = 0\,. \tag{3.10}$$

In this form, it is known as the scalar paraxial wave equation. We can now give a quantitative interpretation of the paraxial approximation. If w is the characteristic transverse length scale over which the field amplitude varies, then $\nabla_\perp^2 E_0 \sim \frac{E_0}{w^2}$. It follows from Eq. (3.10) that $\frac{\partial E_0}{\partial z} \sim \frac{E_0}{kw^2}$, and hence $\frac{1}{k}\left|\frac{1}{E_0}\frac{\partial E_0}{\partial z}\right| \sim \frac{1}{k^2w^2} \ll 1$ when $kw \gg 1$. Thus, a field that varies over a transverse length scale much larger than $1/k = \lambda/2\pi$ also varies slowly along the direction of propagation, as required by the paraxial approximation.

Equation (3.10) is written in Cartesian coordinates. It can also be expressed in other coordinate systems. For analyzing resonators with circular symmetry, we recast this equation into cylindrical polar coordinates (ρ, φ, z), defined in the following figure. The Cartesian coordinates (x, y) of a point P in a plane transverse to the z-axis (corresponding to a fixed value of z) are related to its cylindrical polar coordinates (ρ, φ) by

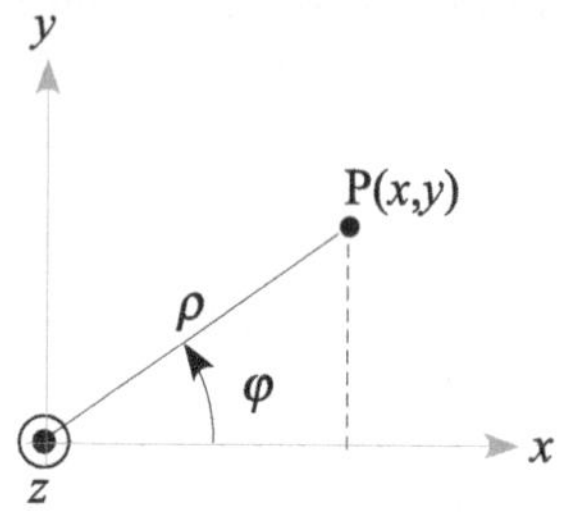

$$x = \rho\cos\varphi\,, \qquad y = \rho\sin\varphi\,, \tag{3.11a}$$

or
$$\rho = \sqrt{x^2+y^2}\,, \qquad \varphi = \tan^{-1}\left(\frac{y}{x}\right). \tag{3.11b}$$

Corresponding to the range $-\infty < x, y < \infty$ of Cartesian coordinates, the cylindrical coordinates vary over the range $0 \le \rho < \infty$ and $0 \le \varphi < 2\pi$. The angle φ is considered positive when measured in the counterclockwise direction. The radial variable ρ is the distance of point P from the axis of the beam (the z-axis), and φ is the angle (also called the azimuthal angle) that the radial line from the z-axis to point P makes with the x-axis. We also note that an infinitesimal area element (in a plane transverse to the direction of propagation) is $dx\,dy$ in Cartesian coordinates and $\rho\,d\rho\,d\varphi$ in cylindrical polar coordinates. In cylindrical polar coordinates, the paraxial wave equation becomes

$$\left[\frac{\partial^2}{\partial\rho^2} + \frac{1}{\rho}\frac{\partial}{\partial\rho} + \frac{1}{\rho^2}\frac{\partial^2}{\partial\varphi^2} + 2ik\frac{\partial}{\partial z}\right] E_0(\boldsymbol{r}) = 0\,. \tag{3.12}$$

We now derive and examine specific fundamental solutions to this differential equation

3.2 Fundamental Gaussian Beam Solution

The paraxial wave equation has many known solutions. These will be introduced in Sec. 3.4. The simplest of these has circular cylindrical symmetry about the direction of propagation and is given by ($\rho = \sqrt{x^2+y^2}$)

$$E_0(\rho, z) = A\left[\frac{w_0}{w(z)}\right] e^{i\frac{k\rho^2}{2q(z)} - i\psi(z)}\,, \tag{3.13a}$$

Rayleigh range
$$z_R = \frac{1}{2}kw_0^2 = \frac{\pi w_0^2 n}{\lambda}\,, \tag{3.13b}$$

Complex beam parameter
$$q(z) = z - iz_R\,, \tag{3.13c}$$

$$\frac{1}{q(z)} = \frac{1}{R(z)} + i\frac{2}{kw^2(z)}\,, \tag{3.13d}$$

Beam spot radius
$$w(z) = w_0\sqrt{1 + (z/z_R)^2}\,, \tag{3.13e}$$

Phasefront radius of curvature
$$R(z) = z + \frac{z_R^2}{z} = z\left(1 + \frac{z_R^2}{z^2}\right), \tag{3.13f}$$

Gouy phase
$$\psi(z) = \tan^{-1}\left(\frac{z}{z_R}.\right) \tag{3.13g}$$

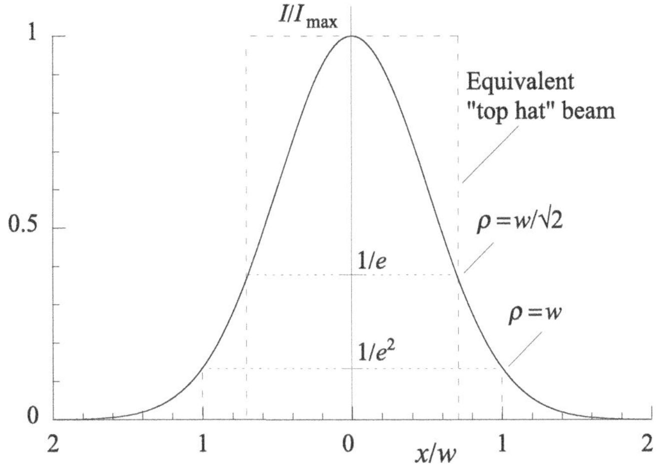

FIGURE 3.2
Intensity profile of a circularly symmetric Gaussian beam along the x (or$y$) axis ($\rho = x$).

To appreciate the physical meaning of the various terms in these equations, let us first consider the time-averaged intensity $I(\rho, z) \equiv \frac{1}{2}\epsilon_0 nc|E|^2$ given by

$$I(\rho, z) = I_0 \left[\frac{w_0}{w(z)}\right]^2 \exp\left[-\frac{2\rho^2}{w^2(z)}\right], \tag{3.14}$$

where $I_0 = \frac{1}{2}\epsilon_0 nc|A|^2$ represents the peak intensity at the beam waist. This has a Gaussian dependence on the radial coordinate ρ. For this reason, the beam is called a Gaussian beam. The total optical power P carried by the beam is obtained via integration over the transverse plane[1]

$$P = \int_0^{2\pi} d\varphi \int_0^{\infty} \rho d\rho I(\rho, z) = I_0 w_0^2 2\pi \times \frac{1}{4} = I_0 \frac{\pi w_0^2}{2}. \tag{3.15}$$

By expressing the peak intensity as $I_0 = 2P/(\pi w_0^2)$, the spatial intensity profile can be written in terms of the total power:

$$I(\rho, z) = \frac{2P}{\pi w^2(z)} e^{-2\rho^2/w^2(z)}. \tag{3.16}$$

The resulting profile is plotted as a function of the radial coordinate ρ in Fig. 3.2. The intensity reaches its maximum value, $I_{\max}(z) = 2P/[\pi w^2(z)]$, along the axis ($\rho = 0$) and decays to $1/e^2 \approx 14\%$ of this value when the radial coordinate reaches $\rho = w(z)$. The parameter $w(z)$ is defined as the $1/e^2$ intensity radius, or simply the beam spot radius. The intensity of a Gaussian beam falls off rapidly as ρ increases beyond the spot radius $w(z)$.

[1]Here and elsewhere in this book, the following integrals will be found to be useful:

$$\int_{-\infty}^{\infty} dx \exp\left[-ax^2 + ibx\right] = \sqrt{\frac{\pi}{a}} \exp(-b^2/4a)$$

$$\int_{-\infty}^{\infty} dx\, x \exp\left[-ax^2 + ibx\right] = ib\frac{\sqrt{\pi}}{2a^{3/2}} \exp(-b^2/4a)$$

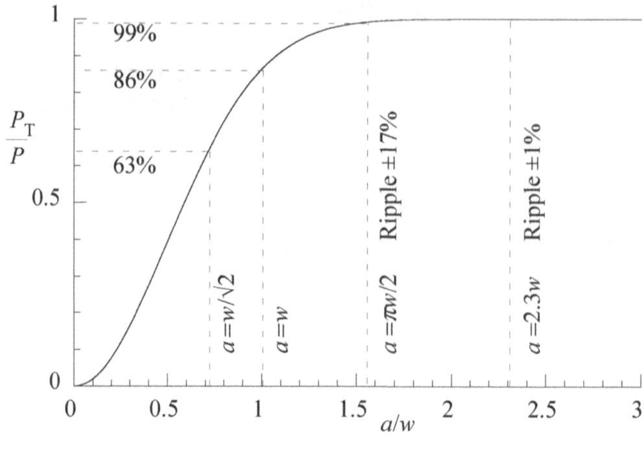

FIGURE 3.3
Fractional power of a Gaussian beam transmitted by a circular aperture of radius a centered on the beam axis.

The form of the peak intensity $2P/\pi w^2(z)$ suggests another measure of beam size. If we imagine a circular cylindrical beam of uniform intensity having the same total power and peak intensity as the Gaussian beam, the radius of such a beam would be

$$w_{\mathrm{TH}} = w/\sqrt{2}\,. \tag{3.17}$$

Such a beam, with uniform intensity over its cross-sectional area, is referred to as a "top-hat beam" because its intensity distribution has the shape of a top hat [see Fig. 3.2]. We may also refer to $w_{\mathrm{TH}} = w/\sqrt{2}$ as the $1/e \approx 0.37$ intensity radius of the Gaussian beam.

There are other measures to characterize the spatial extent of a beam. A common criterion is based on the fractional power transmitted through a centered circular aperture. A sharp-edged circular aperture of radius a placed symmetrically within a fundamental Gaussian beam will transmit a power fraction given by:

$$\frac{P_{\mathrm{T}}}{P} = \frac{2}{\pi w^2}\int_0^{2\pi} d\varphi \int_0^a \rho d\rho e^{-(2\rho^2/w^2)} = 1 - e^{-(2a^2/w^2)}\,. \tag{3.18}$$

This fraction as a function of the normalized aperture radius is plotted in Fig. 3.3. According to this relation, an aperture of diameter $2a = w_{\mathrm{FWHM}}$ transmits only 21% of the total power. An aperture of radius $a = w_{\mathrm{TH}}$ transmits 63.2% of the power, whereas an aperture with a radius matching the beam spot radius ($a = w$) transmits 86.5%. To transmit 99% of the total integrated power, the aperture radius must be $a = \pi w/2$. Consequently, physical apertures must be significantly larger than the spot radius $w(z)$ to transmit nearly all the power of a Gaussian beam.

Beyond optimizing power transmission of a Gaussian beam, one may also want to minimize diffraction ripples which severely distort the intensity profile of the transmitted beam. Such ripples occur whenever a sharp-edged circular aperture is used even if the aperture passes nearly all of the total incident power. For instance, a sharp circular aperture of radius $a = \pi w/2$, which transmits 99% of the beam power, still induces diffraction ripples that create intensity variations of up to 17% in the near field, alongside an identical reduction in the far-field peak intensity. To keep diffraction ripples below 1% in the beam transmitted, a clear aperture of radius $a \geq 2.3w(z)$ must be employed.

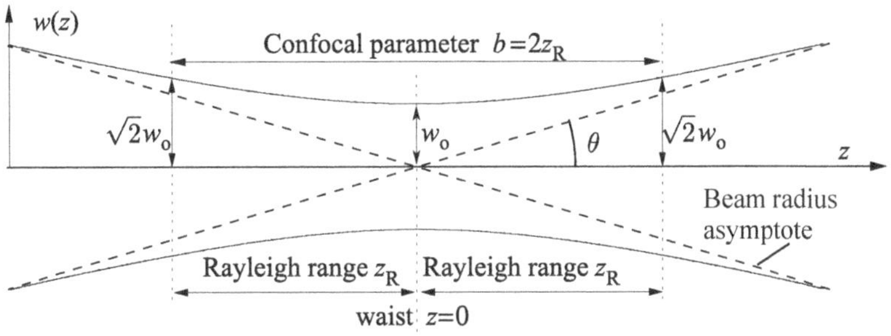

FIGURE 3.4
Variation of spot radius with propagation near beam waist.

The preceding discussion shows that the spot radius $w(z)$ provides a rigorous measure of the beam's transverse size, which expands and contracts as the beam propagates. Figure 3.4 shows the variation of the beam spot radius as a function of the propagation distance z measured from the beam waist. The beam waist is defined as the plane where the spot radius reaches its minimum value w_0. In writing the expression for the beam spot radius (3.13e), we have chosen the waist plane to be the $z = 0$-plane. An examination of Eqs. (3.13a)–(3.13g) shows that a Gaussian beam is uniquely determined by the location of its waist and the spot radius w_0 at the waist.

The Gaussian beam spot radius $w(z)$ varies nonuniformly as the beam propagates away from its waist [Fig. 3.4]. In the near zone ($z \lesssim z_R$), the increase is slow (quadratic for $z \ll z_R$),

$$w(z) = w_0\sqrt{1 + (z/z_R)^2} \approx w_0 \left[1 + \tfrac{1}{2}\left(\frac{z}{z_R}\right)^2\right]. \tag{3.19}$$

In this regime, the beam cross-section remains essentially unchanged, and the beam is said to be collimated. The characteristic distance the beam travels from the waist before the spot radius increases to $\sqrt{2}w_0$ (or the beam spot area doubles) is z_R given by Eq. (3.13b). This distance is called the Rayleigh range of the beam. There are two planes located on opposite sides of the beam waist where the spot radius reaches the value $\sqrt{2}w_0$. The distance between these two planes defines the confocal parameter of the beam, given by

$$b = 2z_R = \frac{2n\pi w_0^2}{\lambda}. \tag{3.20}$$

The confocal parameter b is a measure of the distance (centered on beam waist) over which a beam may be considered to have uniform cross-section (collimated) and is sometimes also referred to as the measure of the total depth of focus. It plays an important role in the theory of laser resonators and many applications of laser light.

In the far zone $z \gg z_R$, the spot radius grows linearly with z [Fig. 3.4],

$$w(z) \approx w_0 \frac{z}{z_R}. \tag{3.21}$$

The far-field divergence angle of the beam may be defined by the ratio of the far-field spot radius to the propagation distance,

$$\theta = \frac{w(z)}{z} = \frac{w_0}{z_R} = \frac{\lambda}{n\pi w_0}. \tag{3.22}$$

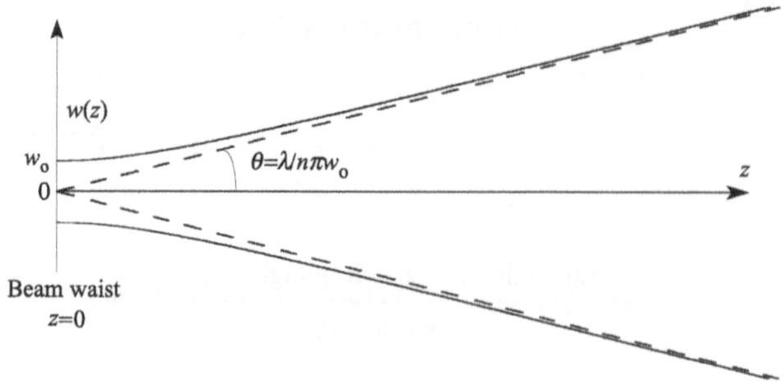

FIGURE 3.5
Long-range linear divergence profile and asymptotic angular boundaries of a fundamental Gaussian beam in the far-field zone ($z \gg z_R$).

As a practical criterion for the paraxial approximation, we take $\theta < 1/\pi$, which translates into a minimum spot radius $w_0 > \lambda/n$. From the dependence of the confocal parameter (3.20) and the beam divergence angle (3.22) on the spot radius, we see that as w_0 decreases, the beam stays collimated over shorter distances and spreads more rapidly.

The physical origin of Gaussian beam divergence is diffraction, an intrinsic wave property that manifests whenever an electromagnetic field is spatially confined to a finite transverse dimension. In fact, the far-field beam divergence angle is of the same order as the angle associated with Fraunhofer (far-field) diffraction of a plane wave by a circular aperture of radius $a \sim w_0$,

$$\theta_D = 0.61\frac{\lambda}{na}\,. \tag{3.23}$$

Thus, the fundamental Gaussian beam is a diffraction-limited beam: its divergence is of the order of the minimum imposed by diffraction for a beam of finite transverse size. Since diffractive phenomena cannot be described by ray optics, wave optics must be used when dealing with Gaussian beams.

The parameter $R(z)$ [Eq. (3.13f)] is the radius of curvature of the nearly spherical phasefronts at z. This can be seen by writing the full electric field of the Gaussian beam, $E_0(\rho, z)e^{i(kz-\omega t)}$ [Eq. (3.13a)], in the limit $z \gg z_R$:

$$\begin{aligned} E(\boldsymbol{r}, t) &= \frac{Ae^{-i\psi(z)}}{\sqrt{1 + z^2/z_R^2}}\, e^{-\rho^2/w^2(z)}\, e^{ik\left(z + \frac{\rho^2}{2R(z)}\right)}\, e^{-i\omega t} \\ &\approx -iAz_R\, e^{-\rho^2/w^2(z)} \left[\frac{e^{ik\rho^2/2z}}{z}\right] e^{i(kz-\omega t)}\,, \end{aligned} \tag{3.24}$$

where we have used the approximations

$$R(z) = z\left(1 + z_R^2/z^2\right) \approx z\,, \quad z \gg z_R\,, \tag{3.25}$$

and $\psi(z) \approx \pi/2$, so that $e^{-i\psi(z)} \approx -i$.

Let us compare this result with a monochromatic spherical wave [Eq. (2.46)] emitted by a point source located on the z-axis at $z = 0$. Close to the z-axis such that $\rho^2 \ll z^2$, we can use the approximation $kr = k\sqrt{\rho^2 + z^2} \approx kz(1 + \rho^2/2z^2)$ in the exponent and $r \approx z$ in the denominator. Then the spherical wave close to the z-axis can be approximated as

$$E_s(\boldsymbol{r}, t) = C\frac{e^{i(kr-\omega t)}}{r} \approx C\left[\frac{e^{ik\rho^2/2z}}{z}\right] e^{i(kz-\omega t)} . \tag{3.26}$$

A comparison of Eqs. (3.24) and (3.26) shows that far from the waist, a Gaussian beam has spherical phasefronts with radius of curvature $R(z) \approx z$ and center of curvature at $z = 0$, in the waist plane. Closer to the waist, $R(z)$ is still the radius of curvature of the phasefront of a Gaussian beam, although the center of curvature no longer lies in the waist plane.

It is interesting to note that the quantity inside square brackets in Eq. (3.26) is an exact solution of the paraxial wave equation (3.10), except at the source plane $z = 0$. This is not surprising, since the spherical wave [Eq. (2.46)] is an exact solution of the scalar Helmholtz equation, and its paraxial form satisfies the corresponding paraxial wave equation. If we replace z by $z - iz_R$, where z_R is a constant with dimensions of length, $\frac{e^{ik\rho^2/2(z-iz_R)}}{(z-iz_R)}$ still satisfies the paraxial wave equation. To see the effect of this replacement, we write

$$\begin{aligned}\frac{ik\rho^2}{2(z-iz_R)} &= \frac{ik\rho^2(z+iz_R)}{2(z^2+z_R^2)} ,\\ &= -\frac{k\rho^2 z_R}{2(z^2+z_R^2)} + i\frac{k\rho^2 z}{2(z^2+z_R^2)} ,\\ &= -\frac{\rho^2}{w^2(z)} + i\frac{k\rho^2}{2R(z)} ,\end{aligned} \tag{3.27}$$

where, on writing $z_R = kw_0^2/2$, we have $w^2(z) = w_0^2\left[1 + \left(\frac{z}{z_R}\right)^2\right]$ and $R(z) = z\left(1 + \frac{z^2_{*R}}{z^2}\right)$. The real part of the exponent produces the Gaussian transverse amplitude $e^{-\rho^2/w^2(z)}$, while the imaginary part produces the quadratic phase associated with a phasefront of radius of curvature $R(z)$. Indeed, if we write z_R in terms of a new length w_0 as $z_R = \pi n w_0^2/\lambda = kw_0^2/2$, we recover the fundamental Gaussian solution of the paraxial wave equation. Thus the fundamental Gaussian solution can be thought of as a paraxial spherical wave produced by a point source located on the imaginary z-axis at $-iz_R$. Sometimes this is the route followed to arrive at the fundamental Gaussian solution.

Returning to our fundamental Gaussian solution, we recall that $R(z)$ is negative when the center of curvature lies in front of the phasefront (characterizing a converging wave front) and positive when the center of curvature lies behind the phasefront (characterizing a diverging wave front). The radius of curvature $R(z)$ of the phasefront of a Gaussian beam has the following limiting forms

$$R(z) = \begin{cases} \pm\infty & |z| \ll z_R \\ 2z_R \equiv b & z = z_R \\ z & |z| \gg z_R \end{cases} . \tag{3.28}$$

ts variation with propagation distance z, measured from the waist, is shown in Figs. 3.6 and 3.7. The phasefront is flat or planar, corresponding to an infinite radius of curvature, in the waist plane. As the beam propagates away from the waist, the phasefront becomes curved, and the magnitude of its radius of curvature decreases, reaching a minimum at $|z| = z_R$. Beyond $|z| = z_R$, the magnitude of the radius of curvature increases, with $R(z) \approx z$ in the far field. The minimum magnitude of the radius of curvature is $|R|_{\min} = 2z_R = b$.

The center of curvature of the phasefront at $z = z_R$ is located at $z = -z_R$, and the center of curvature of the phasefront at $z = -z_R$ is located at $z = z_R$, as seen in Fig. 3.7.

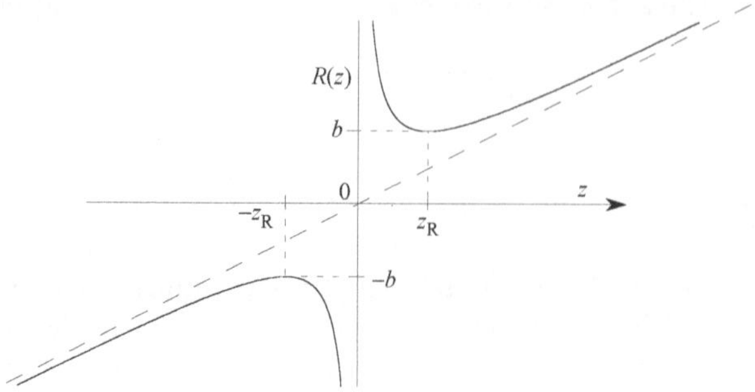

FIGURE 3.6
Variation of the radius of curvature $R(z)$ of a Gaussian beam with distance z from the waist.

The curved phasefronts at $z = \pm z_R$ have special significance in stable resonator theory. If these phasefronts are replaced by two mirrors with matching radii of curvature, they form a stable resonator. This resonator has mirrors with radius of curvature $R = b = 2z_R$ and spacing $L = 2z_R$, so that $R = L = b$. Since the focal length of a mirror of radius R is $f = R/2$, the focal points of the two mirrors coincide at the center of the resonator. The two mirrors therefore form a symmetric confocal resonator, thus giving rise to the *confocal parameter* $b = 2z_R$.

3.2.1 Ray Equation and Energy Flux for a Gaussian Beam

From the expression for the fundamental Gaussian beam, the phasefronts [Eq. (3.24)] are given by

$$k\Psi(\boldsymbol{r}) \equiv kz + \frac{k\rho^2}{2R} = \text{constant}\,, \tag{3.29}$$

where we have neglected the Gouy phase term, which gives a small z-dependent correction to the phasefronts near the waist. The phasefronts are shown in Fig. 3.7. The equation of paraxial rays in a Gaussian beam follows from the definition $\frac{d\boldsymbol{r}}{dz} = \boldsymbol{\nabla}\Psi(\boldsymbol{r})$ [Eq. (2.94)]. In

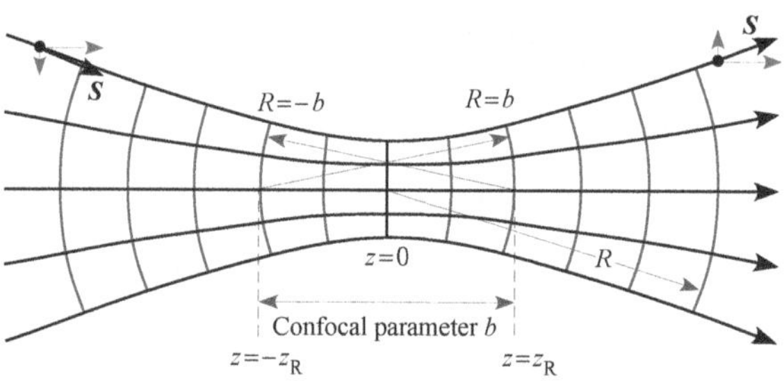

FIGURE 3.7
Phasefronts and ray trajectories in the waist region for a Gaussian beam traveling from left to right. Far from the beam waist, the phasefronts are parts of spheres with centers of curvature located at the center of the waist. The directed hyperbolas represent ray trajectories, perpendicular to the phasefronts, along which light energy flows in a Gaussian beam.

the radial direction,

$$\frac{d\rho(z)}{dz} = \frac{\partial \Psi}{\partial \rho} = \frac{\rho}{R(z)} = \frac{\rho z}{z_R^2[1+(z/z_R)^2]}, \tag{3.30}$$

which is easily integrated to give

$$\rho(z) = \rho_o\sqrt{1+(z/z_R)^2}, \tag{3.31}$$

where ρ_o is the ray coordinate at the beam waist $z = 0$. The rays are thus a family of hyperbolas, shown by the directed curves in Fig. 3.7 for a Gaussian beam traveling from left to right.

The Poynting vector $\boldsymbol{S}$, which describes the flow of energy (energy flux density, $\mathrm{W/m^2}$), is tangential to the rays. To the left of the waist, the Poynting vector has a small radial component pointing toward the axis, corresponding to a converging (focusing) beam. To the right of the waist, it has a small radial component pointing away from the axis, corresponding to a diverging (defocusing) beam. Energy flow across the waist is from left to right, as in a plane wave propagating in the z-direction.

Finally, we note that in passing through the waist, the phase angle $\psi(z)$ [Eq. (3.13g)] changes from $-\pi/2$ (far to the left of the waist) to $\pi/2$ (far to the right). Thus, the fundamental Gaussian beam acquires an extra phase of π in passing through its focal region. This phase shift, called the Gouy phase shift, is a special case of the Gouy effect, which states that a wave with a reasonably simple transverse profile acquires an extra phase shift of π in passing through a focal region. Higher-order Gaussian beams acquire larger Gouy phase shifts on account of their more complicated transverse profiles.

3.3 Gaussian Beam Solutions of Maxwell's Equations

We began this chapter by noting that each Cartesian component of the field vector satisfies the scalar wave equation. Each Cartesian component may satisfy the scalar paraxial wave equation individually, but an arbitrary collection of such scalar solutions does not necessarily constitute a solution of Maxwell's equations. We illustrate this by considering the field structure of a linearly polarized paraxial electromagnetic beam in free space.[2] We write the spatial part of the complex electric field of a paraxial beam traveling in the z-direction as

$$\boldsymbol{E}(\boldsymbol{r}) = [\boldsymbol{e}_1 E_x(\boldsymbol{r}) + \boldsymbol{e}_2 E_y(\boldsymbol{r}) + \boldsymbol{e}_3 E_z(\boldsymbol{r})]\, e^{ikz}, \tag{3.32}$$

where $\hat{e}_1$, $\hat{e}_2$, and $\hat{e}_3$ are unit vectors along the x, y, and z axes, respectively. This field must satisfy Gauss's law for the electric field in free space, $\boldsymbol{\nabla}\cdot\boldsymbol{E} = 0$, which leads to

$$\frac{\partial E_x(\boldsymbol{r})}{\partial x} + \frac{\partial E_y(\boldsymbol{r})}{\partial y} + \frac{\partial E_z(\boldsymbol{r})}{\partial z} + ikE_z = 0\,. \tag{3.33}$$

In the paraxial approximation, $\left|\frac{\partial E_z}{\partial z}\right| \ll |kE_z|$, we get

$$E_z(\boldsymbol{r}) \approx \frac{i}{k}\left[\frac{\partial E_x}{\partial x} + \frac{\partial E_y}{\partial y}\right]. \tag{3.34}$$

[2] W. L. Erikson and S. Singh, "Polarization properties of Maxwell-Gaussian beams," Phys. Rev. E **49**, 5778-5786 (1994); J. Peatross, M. Berrondo, D. Smith, and M. Ware, "Vector fields in a tight laser focus: comparison of models," Opt. Express **25**, 13990-14007 (2017).

From this equation, we see that a linearly polarized paraxial electromagnetic wave cannot, in general, be written simply as a constant unit polarization vector multiplied by a solution of the scalar paraxial wave equation. However, we can have a paraxial wave with one dominant Cartesian component. Let this dominant component be E_x, proportional to the fundamental Gaussian solution (3.13):

$$E_x(\boldsymbol{r}) = A\frac{w_0}{w}e^{ik(x^2+y^2)/2q-i\psi} \equiv g(\boldsymbol{r}), \tag{3.35a}$$

Here $w(z)$, $q(z)$, and $\psi(z)$ have their usual meanings given in Eq. (3.13). Then, if we take $E_y = 0$, we find that the longitudinal component of the electric field is

$$E_z(\boldsymbol{r}) = \frac{i}{k}\frac{\partial E_x(\boldsymbol{r})}{\partial x} = -\frac{x}{q}g(\boldsymbol{r}). \tag{3.35b}$$

The magnitude of the longitudinal field can be estimated by noting that, over the characteristic transverse extent of the beam, $x \sim w$. Since $|q|$ is of order kw^2, we have $\left|\frac{E_z}{E_x}\right| \sim \frac{w}{|q|} \sim \frac{1}{kw} \ll 1$. Thus, the longitudinal component is smaller than the dominant transverse component by a factor of the order of the paraxial parameter $1/kw$, consistent with our assumption of a predominantly x-polarized field. Similarly, using Faraday's law, we find that the magnetic field of the wave has nonzero y and z components

$$B_y(\boldsymbol{r}) = \frac{g(\boldsymbol{r})}{c}, \tag{3.35c}$$

$$B_z(\boldsymbol{r}) = -\frac{y}{cq}g(\boldsymbol{r}). \tag{3.35d}$$

Note that this magnetic field satisfies $\boldsymbol{\nabla}\cdot\boldsymbol{B} = 0$. We also note that E_z and B_z are proportional to the higher-order Hermite-Gauss modes HG_{10} and HG_{01} [see Sec. 3.5], respectively, and thus satisfy the scalar paraxial wave equation.

Using the electric and magnetic fields given in Eqs. (3.35), the time-averaged Poynting vector $\overline{\boldsymbol{S}} = \frac{1}{2\mu_0}\mathcal{R}e\,[\boldsymbol{E}^* \times \boldsymbol{B}]$ is

$$\begin{aligned} \boldsymbol{S}_{\text{av}} &= \frac{1}{2\mu_0}\mathcal{R}e\left[(\boldsymbol{e}_1E_x^* + \boldsymbol{e}_3E_z^*) \times (\boldsymbol{e}_2B_y + \boldsymbol{e}_3B_z)\right], \\ &= \frac{1}{2}\epsilon_o c\,|g(r)|^2\left[\boldsymbol{e}_3 + \frac{x}{R}\boldsymbol{e}_1 + \frac{y}{R}\boldsymbol{e}_2\right]. \end{aligned} \tag{3.36}$$

In cylindrical coordinates, the transverse component of the Poynting vector is therefore $\overline{S}_\rho = \frac{1}{2}\epsilon_0 c|g(\boldsymbol{r})|^2\frac{\rho}{R}$, while $\overline{S}_z = \frac{1}{2}\epsilon_0 c\,|g(\boldsymbol{r})|^2$. Consequently, $\frac{\overline{S}_\rho}{\overline{S}_z} = \frac{\rho}{R}$. But from the ray equation derived in Sec. 3.2.1, $\frac{d\rho}{dz} = \frac{\rho}{R}$. Thus, $\frac{d\rho}{dz} = \frac{\overline{S}_\rho}{\overline{S}_z}$, showing explicitly that the ray trajectories are tangent to the Poynting vector and therefore describe the trajectories of power flow in the Gaussian beam.

The quantity outside the square brackets is simply the intensity of the wave [W/m^2]. The vector inside the square brackets gives the direction of power flow. Recalling that the phasefront radius of curvature $R(z)$ is negative to the left of the beam waist and positive to the right of the waist (for a wave propagating to the right), we see that the Poynting vector has small transverse components $-\frac{x}{|R|}\boldsymbol{e}_1 - \frac{y}{|R|}\boldsymbol{e}_2$ directed toward the beam axis as the beam approaches the waist, and components $\frac{x}{R}\boldsymbol{e}_1 + \frac{y}{R}\boldsymbol{e}_2$ directed away from the axis as it recedes from the waist [Fig. 3.7]. Thus, the power-flow trajectories described by the Poynting vector in a Gaussian beam are precisely the ray trajectories described in Sec. 3.2.1, as they should be. Without the longitudinal field components, power flow would occur only

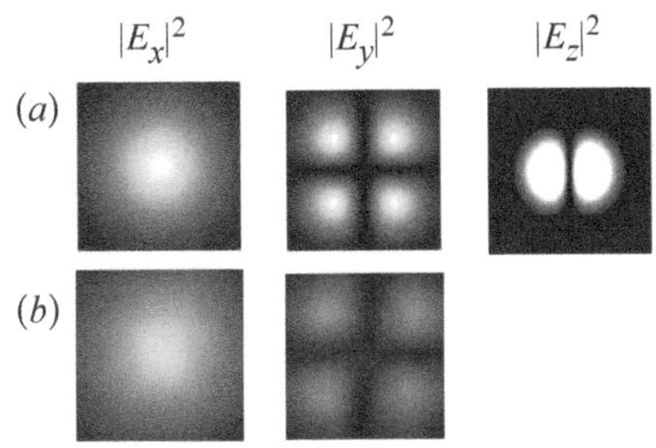

FIGURE 3.8
(a) Theoretical transverse intensity profiles for E_x (dominant polarization component), E_y (cross-polarization component), and E_z (longitudinal component) for the fundamental Maxwell-Gauss beam; (b) measured intensity profiles for the dominant and cross-polarization components. The latter was recorded after the beam had passed through a pair of crossed polarizers. [Adapted from J. Opt. Soc. Am. **A**29, 579-584 (2012).]

in the z-direction. Thus, the longitudinal field components are required for the Poynting vector to correctly describe the focusing and divergence of laser beams.

With some extra effort, we can show that for a predominantly linearly polarized wave, cross-polarization components $E_y(\boldsymbol{r})$ and $B_x(\boldsymbol{r})$ are also nonzero. The electric and magnetic fields of a predominantly x-polarized Maxwell-Gauss beam can then be written as

$$E_x(\boldsymbol{r}) = g(\boldsymbol{r})\,, \tag{3.37a}$$

$$E_y(\boldsymbol{r}) = \frac{1}{2k^2}\frac{\partial^2 g(\boldsymbol{r})}{\partial x \partial y} = \frac{1}{(kw_0)^2}\frac{2xy}{w^2}e^{-2i\psi}g(\boldsymbol{r})\,, \tag{3.37b}$$

$$E_z(\boldsymbol{r}) = \frac{i}{k}\frac{\partial g(\boldsymbol{r})}{\partial x} = -\frac{2i}{kw_0}\frac{x}{w}e^{-i\psi}g(\boldsymbol{r})\,, \tag{3.37c}$$

$$cB_x(\boldsymbol{r}) = \frac{1}{2k^2}\frac{\partial^2 g(\boldsymbol{r})}{\partial x \partial y} = \frac{1}{(kw_0)^2}\frac{2xy}{w^2}e^{-2i\psi}g(\boldsymbol{r})\,, \tag{3.37d}$$

$$cB_y(\boldsymbol{r}) = g(\boldsymbol{r})\,, \tag{3.37e}$$

$$cB_z(\boldsymbol{r}) = \frac{i}{k}\frac{\partial g(\boldsymbol{r})}{\partial y} = -\frac{2i}{kw_0}\frac{y}{w}e^{-i\psi}g(\boldsymbol{r})\,. \tag{3.37f}$$

From these expressions for the field components, we see that the longitudinal and cross-polarization components involve, respectively, first- and second-order transverse derivatives of $g(\boldsymbol{r})$ and are of order $g/(kw_0)$ and $g/(kw_0)^2$. For $kw_0 \gg 1$ (or $w_0 \gg \lambda/2\pi$ in free space), these terms are small. The solution therefore has the form of a perturbative series in powers of the small parameter $(kw_0)^{-1} = \lambda/(2\pi w_0)$. Consistent with this perturbative expansion, only the leading-order terms for each field component have been retained.

We have seen that, although the longitudinal field component is smaller than the dominant transverse component by a factor of order $1/(kw_0)$, it is required for a correct description of power flow in the beam. The cross-polarization component $E_y(\boldsymbol{r})$ is smaller still, being of order $1/(kw_0)^2$ relative to the dominant component. If the polarization properties of the wave are not of interest, the cross-polarization component may be ignored. However, for a correct description of the polarization properties, this component must be retained. For example, if the beam passes through a linear polarizer whose transmission axis is crossed with respect to the dominant direction of polarization, ideally, only the cross-polarization component will be transmitted. This has, in fact, been verified in a number of experiments.[3] Figure 3.8 shows the theoretically expected transverse intensity profiles for the fundamental Maxwell-Gauss beam polarized predominantly in the x-direction and propagating in the z-direction, together with the experimentally recorded profiles.

3.4 ABCD Law for Gaussian Beams

During the propagation of a Gaussian beam in a homogeneous medium, the beam spot radius $w(z)$ and the phasefront radius of curvature $R(z)$ evolve according to Eqs. (3.13e) and (3.13f) while the beam profile remains Gaussian. The evolution of $R(z)$ and $w(z)$ during propagation can be combined into a single transformation law for the complex beam parameter $q(z)$ [Eq. (3.13d)]. Indeed, when a Gaussian beam traverses a linear optical system with ray matrix $\begin{pmatrix} A & B \\ C & D \end{pmatrix}$, it is transformed into another Gaussian beam. The output complex beam parameter q_o immediately following the element is related to the incident parameter q_i by the bilinear transformation:

$$q_o = \frac{Aq_i + B}{Cq_i + D} . \tag{3.38}$$

This relation is the *ABCD* law for Gaussian beams and follows from paraxial diffraction theory (the paraxial wave equation). In using this relation, it is often convenient to express the complex beam parameter as

$$q(z) = z - iz_R , \tag{3.39}$$

where the distance from the waist z is positive when measured along the direction of propagation and negative when measured in the counter-propagating direction, and $z_R = \pi n w_0^2/\lambda_0$ is the Rayleigh range.

Example 3.1 *Gaussian beam propagation in a homogeneous medium (refractive index n) starting at its waist.*

Let us take the beam waist to be the $z = 0$ plane. Then the complex beam parameter at the waist is pure imaginary and can be expressed in terms of its Rayleigh range z_R as

$$q(0) \equiv q_0 = -iz_R = -i\frac{\pi n w_0^2}{\lambda} . \tag{3.40}$$

The matrix for free propagation over a distance z in a homogeneous medium is

$$\begin{pmatrix} A & B \\ C & D \end{pmatrix} = \begin{pmatrix} 1 & z \\ 0 & 1 \end{pmatrix} . \tag{3.41}$$

Note that this matrix is independent of the medium index. Applying the *ABCD* law, Eq. (3.38), the complex beam parameter at an axial displacement z from the waist evaluates to

$$q(z) = \frac{Aq_0 + B}{Cq_0 + D} = \frac{1 \cdot (-iz_R) + z}{0 \cdot (-iz_R) + 1} = z - iz_R , \tag{3.42a}$$

[3] J. Conry, R. Vyas, and S. Singh, "Cross-polarization of linearly polarized Hermite-Gauss laser beams," J. Opt. Soc. Am. A **29**, 579–584 (2012); J. Conry, R. Vyas, and S. Singh, "Cross-polarization of orbital angular momentum carrying laser beams," J. Opt. Soc. Am. A **30**, 821–824 (2013).

which yields the reciprocal form:

$$\frac{1}{q(z)} \equiv \frac{1}{R(z)} + i\frac{2}{kw^2(z)} = \frac{1}{z - iz_R} = \frac{z + iz_R}{z^2 + z_R^2} = \frac{z}{z^2 + z_R^2} + i\frac{z_R}{z^2 + z_R^2} \,. \tag{3.42b}$$

Equating the real and imaginary parts on the two sides of Eq. (3.42b) and substituting the identity $\frac{2}{kw^2(z)} = \frac{2}{(2\pi n/\lambda_0)w^2(z)} = \frac{1}{z_R}\left[\frac{w_0}{w(z)}\right]^2$ we obtain the familiar expressions for the wavefront radius of curvature and beam spot radius

$$R(z) = z + \frac{z_R^2}{z} \,, \tag{(3.13f*)}$$

$$w(z) = w_0\sqrt{1 + \frac{z^2}{z_R^2}} \,. \tag{(3.13e*)}$$

Example 3.2 *Transformation of a Gaussian beam passing through a thin lens.*

The ray transfer matrix for a thin lens of focal length f is written as:

$$\begin{pmatrix} A & B \\ C & D \end{pmatrix} = \begin{pmatrix} 1 & 0 \\ -\frac{1}{f} & 1 \end{pmatrix} . \tag{3.43}$$

Let q_1 denote the complex beam parameter just before the thin lens. According to the *ABCD* law, the complex beam parameter q_2 immediately after the lens is

$$q_2 = \frac{Aq_1 + B}{Cq_1 + D} = \frac{1 \cdot q_1 + 0}{-\frac{1}{f}q_1 + 1} = \frac{q_1}{1 - \frac{q_1}{f}} \,, \tag{3.44a}$$

or equivalently,

$$\frac{1}{q_2} = \frac{1}{q_1} - \frac{1}{f} \,. \tag{3.44b}$$

Decomposing the reciprocal beam parameters in Eq. (3.44b) into their respective phasefront curvature and spot radius components yields:

$$\frac{1}{R_2} = \frac{1}{R_1} - \frac{1}{f} \,, \tag{3.45a}$$

$$w_2 = w_1 \,. \tag{3.45b}$$

Equations (3.45) demonstrate that a thin lens changes the phasefront radius of curvature while leaving the beam spot radius unaffected. A related problem is the focusing of a Gaussian beam by a mirror of focal length $f = R/2$.

3.4.1 Waist-to-Waist Gaussian Beam Focusing

Consider a Gaussian beam incident from the left on a thin lens of focal length f. The incident beam waist is located at a distance d_1 before the lens, with an associated minimum spot radius denoted by w_{01}. After passing through the lens, the beam has a new waist at a distance d_2 after the lens with a minimum spot radius w_{02}.

The ray transfer matrix for beam passage from the first waist to the second waist is

$$\begin{pmatrix} A & B \\ C & D \end{pmatrix} = \begin{pmatrix} 1 & d_2 \\ 0 & 1 \end{pmatrix} \begin{pmatrix} 1 & 0 \\ -\frac{1}{f} & 1 \end{pmatrix} \begin{pmatrix} 1 & d_1 \\ 0 & 1 \end{pmatrix} = \begin{pmatrix} 1 - \frac{d_2}{f} & d_1 + d_2 - \frac{d_1 d_2}{f} \\ -\frac{1}{f} & 1 - \frac{d_1}{f} \end{pmatrix} . \tag{3.46}$$

Applying the *ABCD* law, Eq. (3.38), the complex beam parameter q_2 at the second waist is related to the initial waist parameter q_1 by

$$q_2 = \frac{Aq_1 + B}{Cq_1 + D} = \frac{(1 - d_2/f)q_1 + (d_1 + d_2 - d_1 d_2/f)}{-q_1/f + (1 - d_1/f)} \,. \tag{3.47}$$

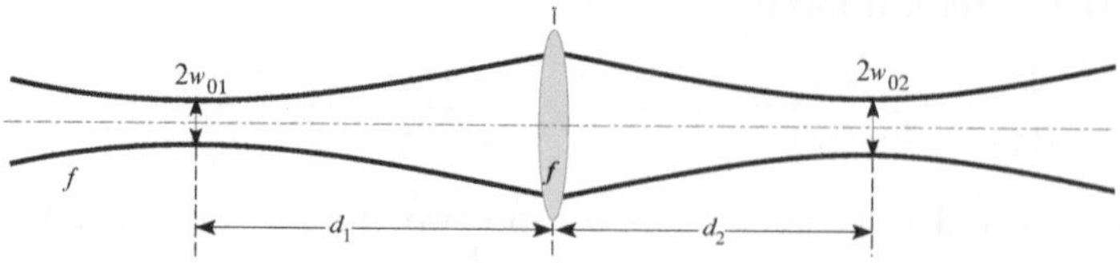

FIGURE 3.9
Geometry of the imaging of a Gaussian beam by a positive thin lens (f > 0).

Because both the input and output planes coincide with beam waists, the complex beam parameters are pure imaginary,

$$q_1 = -iz_{R1} \equiv -i\pi n w_{01}^2/\lambda, \quad q_2 = -iz_{R2} \equiv -i\pi n w_{02}^2/\lambda. \tag{3.48}$$

Using these in the transformation equation (3.47), we obtain

$$\begin{aligned}-iz_{R2} &= \frac{-iz_{R1}A + B}{-iz_{R1}C + D} = \frac{(-iz_{R1}A + B)(iz_{R1}A + B)}{(z_{R1}C)^2 + D^2} \\ &= \frac{(z_{R1}^2 AC + BD) - iz_{R1}(AD - BC)}{(z_{R1}C)^2 + D^2}\end{aligned} \tag{3.49}$$

Equating the real and imaginary parts from the two sides of this equation yields

$$0 = \frac{z_{R1}^2 AC + BD}{(z_{R1}C)^2 + D^2}, \tag{3.50a}$$

$$z_{R2} = \frac{z_{R1}(AD - BC)}{(z_{R1}C)^2 + D^2} = \frac{z_{R1}}{(z_{R1}C)^2 + D^2}, \tag{3.50b}$$

where in the last step, we used the determinant identity $AD - BC = 1$ for the ray transfer matrix. Because the denominator in Eq. (3.50a) remains finite, the numerator must vanish identically ($z_{R1}^2 AC + BD = 0$). Substituting the explicit matrix elements from Eq. (3.46) into this yields

$$z_{R1}^2\left(1 - \frac{d_2}{f}\right)\left(-\frac{1}{f}\right) + \left(d_1 + d_2 - \frac{d_1 d_2}{f}\right)\left(1 - \frac{d_1}{f}\right) = 0. \tag{3.51}$$

Solving this equation for d_2, we obtain

$$\frac{d_2}{f} = \frac{(z_{R1}/f)^2 + (d_1/f - 1)(d_1/f)}{(z_{R1}/f)^2 + (d_1/f - 1)^2} = 1 + \frac{(d_1/f - 1)}{(z_{R1}/f)^2 + (d_1/f - 1)^2}. \tag{3.52}$$

Recalling that $z_R = \pi n w_0^2/\lambda$, we find from Eq. (3.50b) that the new minimum spot radius is

$$w_{02} = \frac{w_{01}}{\sqrt{(z_{R1}/f)^2 + (1 - d_1/f)^2}} = \left(\frac{\lambda f}{\pi n w_{01}}\right)\frac{1}{\sqrt{1 + [(f - d_1)/z_{R1}]^2}}. \tag{3.53}$$

3.4.2 Gaussian Beams and the Geometrical Optics Limit

It is instructive to compare these wave-optics results with those of geometrical optics, where the exit beam waist is treated as the image (of size w_{02}) of the incident beam waist object

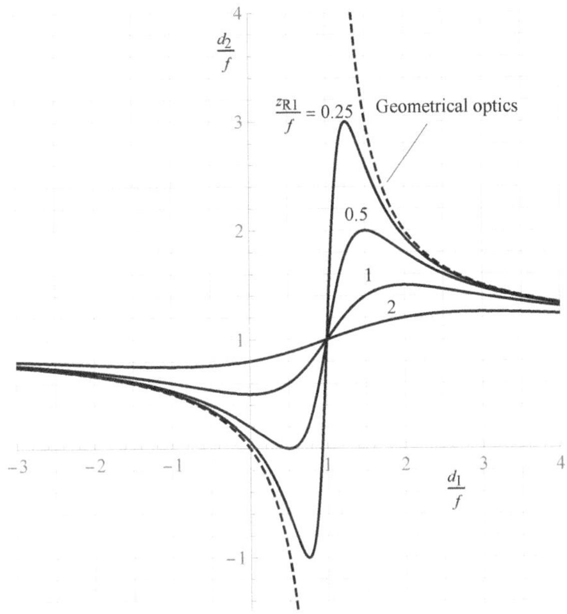

FIGURE 3.10
Normalized secondary beam waist position d_2/f (following the passage of a Gaussian beam through a positive thin lens ($f > 0$), plotted as a function of the normalized incident waist position d_1/f for different values of the ratio z_{R1}/f.

(of size w_{01}). The image distance d_2 and its size w_{02} are then given by

$$\frac{d_2}{f} = \frac{d_1/f}{d_1/f - 1} = 1 + \frac{1}{d_1/f - 1}, \tag{3.54}$$

$$w_{02} = w_{01} \times \left| \frac{d_2}{d_1} \right| . \tag{3.55}$$

By comparing the Gaussian beam results [Eqs. (3.52) and (3.53)] with the geometrical optics predictions [Eqs. (3.54) and (3.55)], we see that the two agree when $z_{R1} \ll |d_1 - f|$, that is, when the lens is located in the far zone of the incident beam. In particular, when d_1 and f are of the same order but not close to each other, this condition corresponds to $z_{R1}/f \ll 1$.

A plot of the exit waist position d_2/f as a function of the incident waist position d_1/f is shown in Fig. 3.10, with z_{R1}/f as a parameter. The prediction of geometrical optics for d_2/f is shown by the dashed curve in Fig. 3.10. The graph confirms that the Gaussian beam and geometrical optics predictions agree when $z_{R1} \ll |d_1 - f|$, that is, when the lens is located in the far zone of the incident beam.

Note also that, for a Gaussian beam, the exit waist location d_2/f is described by a single continuous curve passing through the point $d_1/f = d_2/f = 1$. Thus, when the input waist lies in the front focal plane of the lens, the exit waist is located in the back focal plane. The discontinuity in the image location predicted by geometrical optics at $d_1/f = 1$ is absent for a Gaussian beam. Indeed, the disagreement between the two predictions is particularly striking in this case. For $d_1 = f$, geometrical optics predicts $d_2 \to \infty$ and $w_{02} \to \infty$, whereas the Gaussian beam results are

$$d_2 = f \qquad \text{and} \qquad w_{02} = w_{01}\frac{f}{z_{R1}} = \frac{\lambda f}{\pi n w_{01}} = f\theta_1 \,. \tag{3.56}$$

Another noteworthy feature of the Gaussian beam result, which can be seen from Fig. 3.10, is that the distance d_2 of the second waist from the lens has a maximum and a minimum.[4] By differentiating d_2/f as a function of d_1/f, we find the maximum for the usual case of a positive lens and real object and image is

$$d_{2\max} = f + \frac{f^2}{2z_{R1}} \quad \text{for} \quad d_1 = f + z_{R1} \,. \tag{3.57}$$

Similarly, the minimum exit waist distance is

$$d_{2\min} = f - \frac{f^2}{2z_{R1}} \quad \text{for} \quad d_1 = f - z_{R1} \,. \tag{3.58}$$

Unlike the case of geometrical optics, where for a real object and image there is a minimum object-to-image distance $d_1 + d_2 = 4f$ when $d_1 = d_2 = 2f$, there is no corresponding universal minimum separation between the input and output beam waists for a Gaussian beam when both waists are real.

The variation of the exit waist spot radius w_{02} with d_1/f is discussed most conveniently in terms of the magnification, defined as the ratio of the exit waist spot radius to the input waist spot radius:

$$M \equiv \frac{w_{02}}{w_{01}} = \frac{1}{\sqrt{(z_{R1}/f)^2 + (1 - d_1/f)^2}} \,, \tag{3.59}$$

which reduces to the geometrical optics result when $z_{R1} \ll |d_1 - f|$. Figure 3.11 shows the variation of M as a function of d_1/f for different values of z_{R1}/f. The spot radius of a Gaussian beam after passing through a lens is always finite. The presence of the term $(z_{R1}/f)^2$ removes the singularity in the magnification of geometrical optics for $d_1 = f$. When the incident waist is in the front focal plane, the magnification has the finite maximum value $M_{\max} = f/z_{R1}$. Thus, this maximum magnification is unity when $f = z_{R1}$.

A finite exit spot radius under all circumstances implies that it is not possible to produce a truly collimated Gaussian beam with zero divergence. However, an approximately collimated beam can be obtained by minimizing its divergence, or equivalently by maximizing its waist spot radius. Since the output divergence is inversely proportional to w_{02} and hence to the magnification M, the output divergence reaches its minimum value when $d_1 = f$. The lens equation for a Gaussian beam, giving the position d_2 of the image waist formed by a thin lens of focal length f, can be written as

$$\frac{1}{d_1 + z_{R1}^2/(d_1 - f)} + \frac{1}{d_2} = \frac{1}{f} \,. \tag{3.60}$$

The radius of the image waist is $w_{02} = M w_{01}$, where $M = |f|/\sqrt{(d_1 - f)^2 + z_{R1}^2}$ is the magnification of the beam waist. We can also express the input beam parameters in terms of output beam parameters

$$\frac{1}{d_1} + \frac{1}{d_2 + z_{R2}^2/(d_2 - f)} = \frac{1}{f} \,, \tag{3.61}$$

which is useful in tracing the properties of the input beam when those of the output beam are known.

[4] Sidney A. Self, "Focusing of spherical Gaussian beams," Appl.Opt. 22, 658-661 (1983).

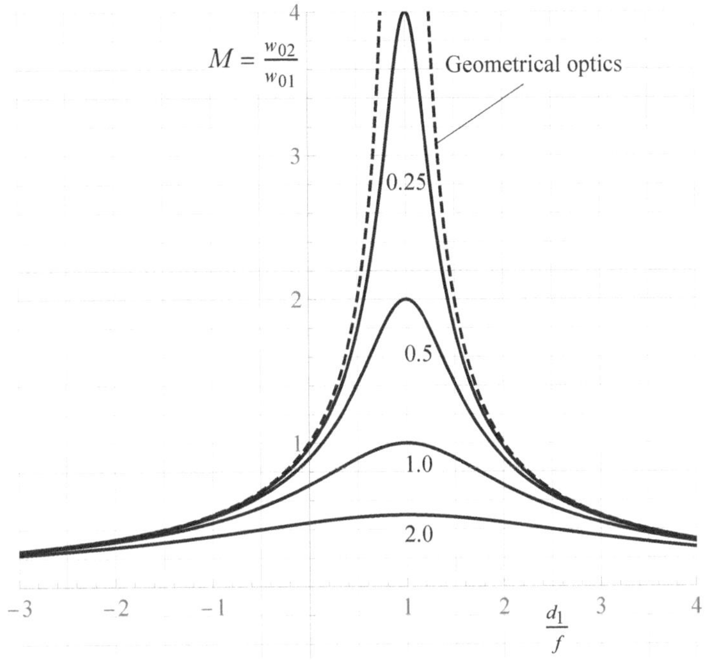

FIGURE 3.11
Gaussian beam spot radius magnification in focusing by a lens of focal length f for $z_{R1}/f = 0.25, 0.5, 1, 2$. The dashed curve shows the geometrical optics result.

Example 3.3 *A Gaussian beam (wavelength $\lambda = 1.0, \mu m$) with a planar phasefront and a spot radius of 2.0 cm is incident on a lens of focal length $f = 4.0$ cm. (a) If $z = 0$ is the location of the lens, where does the beam reach its minimum spot radius after the lens? (b) What is the new far-field divergence angle?*

Note that the incident beam has a waist at the lens (planar wavefront) and spot radius $w_{01} = 2.0$ cm. This means $d_1 = 0$ with

$$z_{R1} = \frac{\pi w_{01}^2}{\lambda} = \pi(0.02)^2/(1.0 \times 10^{-6}) = 1.257 \times 10^3\,\mathrm{m} \approx 1.26 \times 10^3\,\mathrm{m}\,,$$
$$\frac{f}{z_{R1}} = \frac{0.04}{1.257 \times 10^3} = 3.18 \times 10^{-5} \ll 1\,.$$

Using these results, we find that the minimum spot radius and its location after the lens are

$$w_{02} = \frac{w_{01}}{\sqrt{1 + (z_{R1}/f)^2}} \approx \frac{w_{01} f}{z_{R1}} = \frac{f\lambda}{\pi w_{01}}\,,$$
$$d_2 = \frac{f}{1 + (f/z_{R1})^2} \approx f[1 - (f/z_{R1})^2] \approx f\,.$$

Numerically,

$$w02 \approx \frac{(0.04)(1.0 \times 10^{-6})}{\pi(0.02)} = 6.37 \times 10^{-7}, \mathrm{m} = 0.637, \mu\mathrm{m}\,.$$

The Gaussian-beam divergence parameter is therefore

$$\theta = \frac{\lambda}{\pi w_{02}} = \frac{w_{01}}{f} = \frac{2}{4} = 0.50\,.$$

If this quantity is interpreted as the tangent of the divergence angle, then

$$\theta_{\text{angle}} = \tan^{-1}(0.50) = 0.464, \text{rad} \approx 27^\circ\,.$$

This angle is clearly not small, so the paraxial approximation used to obtain the Gaussian-beam formulas is no longer quantitatively reliable. Thus, this example illustrates both the strong focusing predicted by the paraxial Gaussian model and the breakdown of that model when the focused spot becomes comparable to the wavelength.

Example 3.4 *A Gaussian beam with* $\lambda = 633$ *nm has a focus with spot radius* $w_0 = 75, \mu m$*. A lens with focal length* 25 *mm is placed a distance* 50 *mm after the focus. At what distance is the light refocused after the lens, and what beam spot radius is obtained?*

The incident beam has $z_1 = 50$ mm and $z_{R1} = \frac{\pi w_{01}^2}{\lambda} = 27.93 \approx 28$ mm, giving

$$q_1 = 50 - i(27.93)\,\text{mm}\,.$$

Immediately after the lens, the complex beam parameter is

$$\begin{aligned} q_2 &= \frac{q_1}{-(q_1/f)+1} = \frac{50-28i}{-(50-28i)/25+1} = \frac{50-28i}{-1+i(1.117)} = -\frac{(50-28i)\big(1+i(1.117)\big)}{2.248} \\ &= -36.12 - i(12.42)\,\text{mm} \equiv z_2 - iz_{R2}\,. \end{aligned}$$

The negative value $z_2 = -36.12$ mm means that the new waist lies 36.12 mm downstream from the lens. Therefore, the transmitted beam is refocused approximately 36 mm after the lens, and its Rayleigh range is $z_{R2} = 12.42$ mm.
The spot radius at the new waist is thus

$$w_{02} = \sqrt{\lambda z_{R2}/\pi} = \sqrt{(0.633\times10^{-3})(12.42)/\pi} = 0.050\,\text{mm} = 50\,\mu\text{m}$$

3.4.3 Gaussian Beam Focusing

A question of practical importance in applications such as laser traps, cutting, drilling, and laser fusion is how small a focal spot can be obtained in order to increase the power density. For a given focal length f, we can reduce w_{02} by making z_{R1}/f large. Since $z_{R1} = \pi n w_{01}^2/\lambda$, this means that we need to make w_{01} as large as possible. However, w_{01} cannot be made arbitrarily large because it is limited by the lens aperture if significant beam power loss is to be avoided. It may need to be even smaller to allow for beam spreading from the first waist to the lens.

One way to address this is to use a nearly collimated incident beam with a confocal parameter many focal lengths long and place the lens in the near zone of the incident beam, so that the input beam spot radius does not change significantly between the input beam waist and the lens. Under these conditions, w_{01} is limited by the lens aperture. If we want the lens to transmit 99% of the incident power, then, from our earlier discussion of the power transmitted through a circular aperture, the lens radius must be approximately $\pi w_{01}/2$. Thus,

$$\frac{1}{2}\pi w_{01} = \frac{1}{2}D\,, \tag{3.62}$$

where D is the lens aperture (diameter). Under these conditions ($z_{R1}/f \gg 1$ and the constraint $\pi w_{01} = D$), Eq. (3.53) leads to the following expression for the spot radius at the second waist

$$w_{02} = \frac{\lambda f}{n\pi w_{01}} = \frac{\lambda}{n}\frac{f}{D} = \frac{\lambda}{n}(FN)\,, \tag{3.63}$$

where $FN = f/D$ is the f-number of the lens. A small FN corresponds to a fast lens (high light-gathering capability), whereas a large FN corresponds to a slow lens (low light-gathering capability). The best lenses have $FN \approx 1$, while most have $FN > 1$. Thus, the smallest focal spot radius that can be obtained with conventional focusing is of the order of a wavelength.

The location d_2 of the focal spot (second waist), from Eq. (3.52), under the same conditions is given by

$$\frac{d_2}{f} \approx 1 + \frac{f^2}{z_{R1}^2}\left(\frac{d_1}{f} - 1\right) = 1 - \frac{z_{R2}}{z_{R1}}\left(1 - \frac{d_1}{f}\right) \approx 1\,, \tag{3.64}$$

where the last step follows because $z_{R1} \gg f$ and the correction term is small. Thus, the second waist is located very nearly in the focal plane of the lens.

The peak power density at the second waist is

$$I_{02} = \frac{2P}{\pi w_{02}^2} = \frac{2P}{(\lambda/n)^2}\,\Omega_2\,, \tag{3.65}$$

where $\Omega_2 = \pi\theta_2^2$ is the solid angle into which the second beam waist radiates [Fig. 1.7], and θ_2 is the divergence angle of the focused beam. Similarly, the peak intensity at the first waist is

$$I_{01} = \frac{2P}{\pi w_{01}^2} = \frac{2P}{(\lambda/n)^2}\Omega_1\,, \tag{3.66}$$

where $\Omega_1 = \pi\theta_1^2$ is the solid angle into which the first waist radiates. The quantity $B = I_{01}/\Omega_1 = 2P/(\lambda/n)^2 = I_{02}/\Omega_2$ may be regarded as the on-axis brightness of the Gaussian beam, that is, the power per unit area per unit solid angle (W/m^2·sr). For an ideal Gaussian beam propagating through lossless linear optical elements such as mirrors and lenses, this quantity is invariant.

By expressing w_{02} in terms of w_{01}, the result for the peak intensity at the second waist can also be written as

$$I_{02} = \frac{2P}{f^2\Omega_1}\,. \tag{3.67}$$

From this expression, we see that the power density that can be obtained by focusing a beam of given power is inversely proportional to the solid-angle divergence of the beam being focused. Thus, the high degree of directionality (small Ω) of laser beams is of crucial importance for obtaining high power densities in the focal spot.

Example 3.5 *Focusing of a 1 mW He-Ne laser beam of wavelength λ=633 nm and spot radius w_0=1 mm.*

The divergence angle ($n = 1$), solid angle of emission, and intensity of the beam are

$$\begin{aligned}
\theta &= \frac{\lambda}{\pi w_0} = \frac{0.633\times 10^{-6}}{\pi\times 10^{-3}} \approx 0.2\times 10^{-3}\,\text{rad}\\
\Omega_1 &= \pi\theta^2 = \pi(0.2\times 10^{-3})^2 \approx 1.3\times 10^{-7}\,\text{sr}\\
I_{01} &= \frac{2P}{\pi w_0^2} = \frac{2\times 10^{-3}}{\pi\times(0.1)^2}\,\text{W/cm}^2 = 64\,\text{mW/cm}^2
\end{aligned}$$

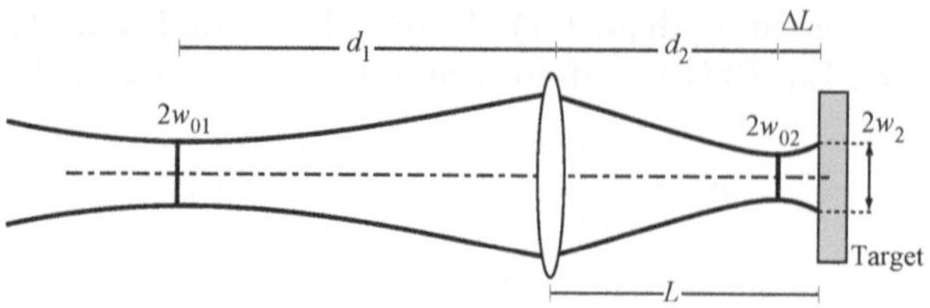

FIGURE 3.12
The minimum beam radius at a target occurs when the waist of the focused beam is located at a specific position before the target, rather than at the target itself.

If this laser beam is focused by a lens of focal length $f = 2.5$ cm (approximately the focal length of the human eye), the peak intensity will be

$$I_{02} = \frac{2P}{f^2 \Omega_1} = \frac{2 \times 10^{-3}}{(2.5)^2 (1.3 \times 10^{-7})} \approx 2.5 \times 10^3 \,\mathrm{W/cm^2}\,. \tag{3.68}$$

Thus, direct viewing of even a low-power laser beam can result in severe retinal damage. Thermal sources would require vastly greater power to produce comparable focused intensities. The large intensities achievable by focusing laser beams are a direct consequence of their low divergence. While care must be exercised when working with laser beams, their ability to deliver optical energy to a small region has also made them valuable in retinal photocoagulation and other medical procedures.

Example 3.6 *Optimization of beam spot radius w_2 at a sample.*

Another application of the preceding formulas is the optimization of the beam spot radius w_2 (not necessarily a waist) at a sample located a distance $L = d_2 + \Delta L$ from the focusing lens, where d_2 is the distance from the lens to the output beam waist and ΔL is the distance from the output beam waist to the sample (referred to as the focus shift). This problem can be handled by expressing the beam spot radius as [Problem 26]

$$w_2 = \frac{w_{01}}{z_{R1}} \left[\left(\frac{L - f}{M} \right)^2 - 2(d_1 - f)(L - f) + f^2 \right]^{1/2} \tag{3.69}$$

where M is the magnification of the beam waist spot radius given by Eq. (3.59). This relation expresses the beam spot radius at the sample as a function of f, d_1, L, w_{01}, and λ, and can be used to optimize the spot radius at the sample by varying one or several of these parameters, including w_{01} if it is possible to adjust the input waist size.

3.5 Hermite-Gauss Beams

So far we have discussed only the fundamental Gaussian beam solution. There are other solutions of the paraxial wave equation (3.10) that have more complex transverse spatial structure. With the help of spatial light modulators (SLMs), laser beams with a wide variety of transverse spatial profiles can be generated. Here we focus on two families of beams, namely, the Hermite-Gauss (HG) and Laguerre-Gauss (LG) families. They are arguably the most important from both fundamental and applied points of view. Most lasers naturally produce the HG family of beams, which can be converted to the LG family using a simple optical setup in the laboratory.[5]

The Hermite-Gauss family of beams consists of solutions of the paraxial scalar wave equation in Cartesian coordinates. We will use the fundamental Gaussian beam solution as a guide in searching for these higher-order beam solutions of the paraxial wave equation. The fundamental Gaussian beam solution can be written as

$$\begin{aligned} E_0(x,y,z) &= A\left[\frac{w_0}{w(z)}\right] e^{ik\rho^2/2q(z)-i\psi(z)}\,, \\ &= A\frac{e^{-i\psi(z)}}{\sqrt{1+(z/z_R)^2}} e^{-\left(x/w(z)\right)^2} e^{-\left(y/w(z)\right)^2} e^{ik(x^2+y^2)/2R(z)}\,. \end{aligned} \tag{3.13a*}$$

Here the expressions for $\psi(z)$, $w(z)$, and $R(z)$ [or the complex beam parameter $q(z)$] are defined by Eqs. (3.13). In the second line, we have rewritten the fundamental solution explicitly in terms of the transverse Cartesian coordinates x and y. This shows that, for fixed z, its transverse dependence is a product of a function of x and a function of y. We also note that it depends on the transverse coordinates scaled by the beam spot radius $w(z)$, through the combinations $x/w(z)$ and $y/w(z)$. The change of the phasefront from planar to spherical, represented by the $k(x^2+y^2)/2R(z)$ term, as well as the Gouy phase, represented by the $\psi(z)$ term, is a direct result of the finite transverse extent of the wave amplitude. This amplitude-phase coupling is a diffraction effect inherent to all wave fields of finite transverse extent. The dependence on scaled coordinates means that the beam shape remains unchanged even as its size increases, while the factor $1/\sqrt{1+(z/z_R)^2}$ in front ensures that the total power carried by the beam is conserved during propagation along the z-direction. Both are desirable features of a model for a laser beam.

We are now ready to look for higher-order solutions of the paraxial wave equation in Cartesian coordinates that can serve as good models of laser beams. Keeping in view all that has been said in the preceding paragraph, we take them to be of the form

$$E_0(x,y,z) = AX\left(\frac{x}{w(z)}\right)Y\left(\frac{y}{w(z)}\right)e^{i\frac{k(x^2+y^2)}{2q(z)}+iP(z)}\,, \tag{3.70}$$

where we take $w(z)$ and $q(z)$ to be the same as those for the fundamental Gaussian beam, X and Y are real functions of the indicated variables, and $P(z)$ is a function of z alone, possibly complex. These functions remain to be determined, but it is clear that beam solutions of this form will preserve their shape as the beam propagates in the z-direction. We note that for $X=1=Y$ and $P(z)=i\ln\sqrt{1+(z/z_R)^2}-\tan^{-1}(z/z_R)$, we recover the fundamental Gaussian beam solution.

By substituting Eq. (3.70) into the paraxial wave equation (3.10), we obtain the differential equations satisfied by X, Y, and P. The algebra is straightforward but a bit tedious. To simplify the notation, the functions are written without their arguments in the following. In addition, the first and second derivatives of a function $f(u)$ with respect to its argument u are denoted by f' and f'', respectively. We should keep in mind that, for example, X is in fact a function of x and z, since $w(z)$ is a function of z, and therefore X' also remains a function of x and z. On the other hand, $q(z)$, $w(z)$, and $P(z)$ are functions only of z, so

[5]Experiments to generate Hermite-Gauss and Laguerre-Gauss beams and study their intensity and phase profiles are described in H. Kogelnik and T. Li, "Laser beams and resonators," Appl. Opt. 5, 1550-1567 (1966); M. Padgett, J. Arlt, N. Simpson, and L. Allen,"An experiment to observe the intensity and phase structure of Laguerre–Gaussian laser modes," Am. J. Phys. **64**, 77-82 (1996); J. Vickers, M. Burch, R. Vyas, and S. Singh, "Phase and interference properties of optical vortex beams," J. Opt. Soc. Am. A**25**, 823-827 (2008); A. Bisht, N. Murillo, R. Vyas, and S. Singh "Experimental confirmation of phase profile of Hermite–Gauss beams," J. Opt. Soc. Am. A**41**, 2023-2029 (2024).

that their derivatives are written simply as $q' = dq/dz$, $w' = dw/dz$, and $P' = dP/dz$. With these remarks on notation, we can proceed to find X, Y, and P.

Noting that X and Y are functions of the scaled variables

$$\xi = \frac{x}{w(z)} \quad \text{and} \quad \eta = \frac{y}{w(z)}, \tag{3.71}$$

we first use the chain rule to express derivatives with respect to x and y in terms of derivatives with respect to ξ and η:

$$\frac{\partial X}{\partial x} = \frac{dX}{d\xi}\frac{\partial \xi}{\partial x} = \frac{dX}{d\xi}\frac{1}{w(z)} \equiv \frac{1}{w(z)}X', \tag{3.72a}$$

$$\frac{\partial^2 X}{\partial x^2} = \frac{1}{w^2(z)}X'', \tag{3.72b}$$

$$\frac{\partial X}{\partial z} = \frac{dX}{d\xi}\frac{\partial \xi}{\partial z} = \frac{dX}{d\xi}\left(-\frac{x}{w^2(z)}\frac{dw}{dz}\right) \equiv -\frac{x}{w^2(z)}\frac{dw}{dz}X'. \tag{3.72c}$$

We can write similar equations for Y. Using these we have

$$\frac{\partial E_0}{\partial z} = \left[XY\left(iP' - \frac{ik}{2q^2}(x^2+y^2)\right) + \left(-X'Y\frac{x}{w^2}w' - XY'\frac{y}{w^2}w'\right)e^{iP+ik\rho^2/2q}, \tag{3.73a}\right.$$

$$\frac{\partial E_0}{\partial x} = \left(\frac{ikx}{q}XY + \frac{1}{w}X'Y\right)e^{iP+ik\rho^2/2q}, \tag{3.73b}$$

$$\frac{\partial E_0}{\partial y} = \left(\frac{iky}{q}XY + \frac{1}{w}XY'\right)e^{iP+ik\rho^2/2q}, \tag{3.73c}$$

$$\frac{\partial^2 E_0}{\partial x^2} = \left(\frac{ik}{q}XY - \frac{k^2x^2}{q^2}XY + \frac{2ikx}{qw}X'Y + \frac{1}{w^2}X''Y\right)e^{iP+ik\rho^2/2q}, \tag{3.73d}$$

$$\frac{\partial^2 E_0}{\partial y^2} = \left(\frac{ik}{q}XY - \frac{k^2y^2}{q^2}XY + \frac{2iky}{qw}XY' + \frac{1}{w^2}XY''\right)e^{iP+ik\rho^2/2q}. \tag{3.73e}$$

Substituting Eqs. (3.72) and (3.73) into the paraxial wave equation (3.10) and dividing throughout by XY, we obtain

$$\frac{2ik}{q} - 2kP' + \frac{1}{w^2}\frac{X''}{X} - 2\frac{ikx}{w^2}\frac{X'}{X}\left(w' - \frac{w}{q}\right) + \frac{1}{w^2}\frac{Y''}{Y} - 2\frac{iky}{w^2}\frac{Y'}{Y}\left(w' - \frac{w}{q}\right) = 0. \tag{3.74}$$

With the help of Eqs. (3.13d)–(3.13f), we find that $w' - w/q = -2i/kw$. Using this result in the preceding equation and multiplying throughout by w^2, we find that X, Y, and P must satisfy

$$\left(\frac{2ik}{q} - 2kP'\right)w^2 + \frac{1}{X}(X'' - 4\xi X') + \frac{1}{Y}(Y'' - 4\eta Y') = 0, \tag{3.75}$$

for the paraxial wave equation to have solutions of the form (3.70).

We now note that the first term on the left depends only on z, whereas the second term depends only on the independent variable ξ and the third term depends only on the independent variable η. For this equation to be satisfied for all values of ξ, η, and z, each of these terms must separately be equal to a constant. Therefore, we write

$$\frac{1}{X}(X'' - 4\xi X') = -a, \tag{3.76a}$$

$$\frac{1}{Y}(Y'' - 4\eta Y') = -b, \tag{3.76b}$$

$$\left(\frac{2ik}{q} - 2kP'\right)w^2 = a + b. \tag{3.76c}$$

Here a and b are constants. This procedure is an example of the method of separation of variables.

Equations (3.76) can be brought to a more recognizable form by introducing the new variables

$$u = \sqrt{2}\xi \quad \text{and} \quad v = \sqrt{2}\eta \,. \tag{3.77a}$$

Then, since $\frac{dX}{d\xi} = \frac{dX}{du}\frac{du}{d\xi} = \sqrt{2}\frac{dX}{du}$ and $\frac{d^2X}{d\xi^2} = 2\frac{d^2X}{du^2}$, Eq. (3.76a) for X becomes

$$\frac{d^2X}{du^2} - 2u\frac{dX}{du} + \frac{a}{2}X = 0 \,. \tag{3.77b}$$

This has the form of the Hermite equation, which also arises in the quantum mechanics of the harmonic oscillator. Polynomial solutions of this equation, which combine with the Gaussian factor already present in Eq. (3.70) to produce physically acceptable beams of finite power, occur only when

$$a = 4m, \quad m = 0, 1, 2, 3 \cdots \tag{3.77c}$$

The corresponding solutions are the Hermite polynomials of degree m and argument $u = \sqrt{2}x/w$

$$X_m(x/w) = H_m(\sqrt{2}x/w) \,, \quad m = 0, 1, 2, 3, \cdots \,. \tag{3.78}$$

Similarly, the physically acceptable solutions for the function Y, obtained for $b = 4n$, are

$$Y_n(y/w) = H_n(\sqrt{2}y/w) \,, \quad n = 0, 1, 2, 3, \cdots \,. \tag{3.79}$$

With these solutions for X and Y and the expression (3.76c) for $q(z)$, Eq. (3.76c) for $P(z)$ becomes

$$\frac{dP}{dz} = \frac{iz}{z^2 + z_R^2} - \frac{(m+n+1)z_R}{z^2 + z_R^2} \,. \tag{3.80}$$

This can be integrated to give

$$P(z) = i \ln \sqrt{1 + \left(\frac{z}{z_R}\right)^2} - (m+n+1)\psi(z) \,,$$

or

$$e^{iP(z)} = \frac{e^{-i(m+n+1)\psi(z)}}{\sqrt{1 + (z/z_R)^2}} = \frac{w_0}{w(z)} e^{-i(m+n+1)\psi(z)} \,, \tag{3.81}$$

where the phase angle $\psi(z) = \tan^{-1}(z/z_R)$ is the same as in Eq. (3.13g).

Using the results of (3.78), (3.79), and (3.81) in Eq. (3.70) and denoting the resulting solution with indices m, n by HG_{mn}, we find that paraxial beam-like solutions separable in the Cartesian coordinates are of the form

$$HG_{mn}(\boldsymbol{r}) = A_{mn}\left(\frac{w_0}{w}\right) H_m(\sqrt{2}x/w) H_n(\sqrt{2}y/w) e^{-i(m+n+1)\psi + ik\rho^2/2q} . \tag{3.82}$$

Here we have suppressed the z-dependence of $w(z)$, the complex beam parameter $q(z)$, and the phase $\psi(z)$ for simplicity of writing. These quantities are independent of the beam indices and are given by Eqs. (3.13). Some low-order Hermite polynomials and recursion relations for computing the higher-order ones are listed below

$$\begin{aligned}
&H_0(u) = 1 \,, && H_1(u) = 2u \,, \\
&H_2(u) = 4u^2 - 2 \,, && H_3(u) = 8u^3 - 12u \,, \\
&H_4(u) = 16u^4 - 48u^2 + 12 \,, && \\
&H_m(u) = (-1)^m e^{u^2} \frac{d^m}{du^m} e^{-u^2} \,, && \\
&H_{m+1}(u) = 2uH_m(u) - 2mH_{m-1}(u) \,, && \frac{dH_m(u)}{du} = 2mH_{m-1}(u) \,.
\end{aligned} \tag{3.83}$$

The intensity distribution for the beam with indices m, n is given by

$$I_{mn}(x,y) = I_{0mn}\frac{w_0^2}{w^2}H_m^2(\sqrt{2}x/w)H_n^2(\sqrt{2}y/w)e^{-2(x^2+y^2)/w^2}\,, \tag{3.84a}$$

where the intensity coefficient I_{0mn} is given by

$$I_{0mn} = \frac{1}{2}\epsilon_0 n|A_{mn}|^2 c\,. \tag{3.84b}$$

The n appearing with ϵ_0 on the right-hand side of this equation is the refractive index of the medium, not the beam index. The total power of an HG_{mn}-beam is given by

$$\begin{aligned} P &= \iint_{-\infty}^{\infty} dx dy I_{mn}(x,y,z) = \frac{1}{2}I_{0mn}w_0^2\left[\int_{-\infty}^{\infty} dX H_m^2(X)e^{-X^2}\right]\cdot\left[\int_{-\infty}^{\infty} dY H_n^2(Y)e^{-Y^2}\right], \\ &= \frac{1}{2}I_{0mn}w_0^2\left[2^m\sqrt{\pi}m!\right]\cdot\left[2^n\sqrt{\pi}n!\right] \end{aligned} \tag{3.84c}$$

This gives the intensity coefficient, or equivalently the field amplitude A_{mn}, in terms of the total beam power:

$$I_{0mn} \equiv \frac{1}{2}\epsilon_0 n|A_{mn}|^2 c = \frac{2P}{\pi w_0^2}\frac{1}{2^{m+n}\,m!n!}\,. \tag{3.84d}$$

The intensity distribution can then be written directly in terms of the total beam power as

$$I_{mn} = \frac{2P}{\pi w^2}\frac{1}{2^{m+n}m!n!}H_m^2\left(\frac{\sqrt{2}x}{w}\right)H_n^2\left(\frac{\sqrt{2}y}{w}\right)e^{-2(x^2+y^2)/w^2}\,. \tag{3.85}$$

Figure 3.13 shows the intensity distributions for some low-order Hermite-Gauss beams. Note that I_{0mn} does not, in general, correspond to either the maximum intensity or the on-axis

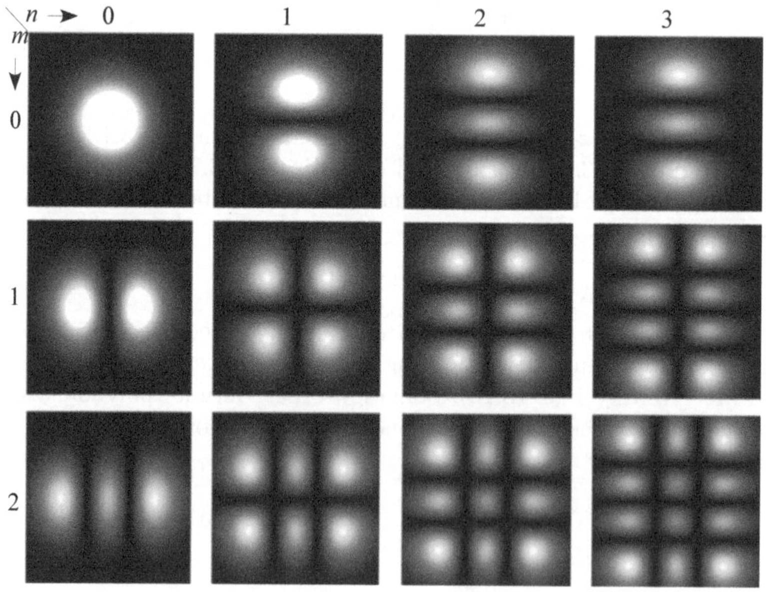

FIGURE 3.13
Intensity distribution in a plane transverse to the direction of propagation for some low-order HG_{mn} beams.

intensity, except for the fundamental Gaussian beam. For example, for the HG_{10} beam, the on-axis intensity is zero, and the maximum intensity is attained at $x = \pm w/\sqrt{2}\,, y = 0$. At the beam waist, its value is $I_{\max} = \frac{4}{e} I_{010} \approx 1.47\, I_{010}$. For an HG_{10} beam and a fundamental Gaussian beam carrying the same total power, this maximum is $2/e \approx 0.736$ times the peak intensity of the fundamental Gaussian beam.

It is easy to check that for $m = 0 = n$ we recover the intensity distribution of the fundamental Gaussian mode. The intensity distributions of some low-order Hermite-Gauss beams are explicitly

$$\begin{aligned}
I_{00}(x,y,z) &= \frac{2P}{\pi w^2} e^{-2(x^2+y^2)/w^2}\,, \\
I_{10}(x,y,z) &= \frac{2P}{\pi w^2}\frac{1}{2}\frac{8x^2}{w^2} e^{-2(x^2+y^2)/w^2}\,, \\
I_{01}(x,y,z) &= \frac{2P}{\pi w^2}\frac{1}{2}\frac{8y^2}{w^2} e^{-2(x^2+y^2)/w^2}\,, \\
I_{11}(x,y,z) &= \frac{2P}{\pi w^2}\frac{1}{4}\frac{8x^2}{w^2}\frac{8y^2}{w^2} e^{-2(x^2+y^2)/w^2}\,, \\
I_{20}(x,y,z) &= \frac{2P}{\pi w^2}\frac{1}{8} H_2^2\left(\frac{\sqrt{2}x}{w}\right) e^{-2(x^2+y^2)/w^2}\,.
\end{aligned} \tag{3.86}$$

3.5.1 Astigmatic Hermite-Gauss Beams

It is possible for Gaussian beams to have different spot radii in the two transverse dimensions. The fundamental beam will then have an elliptical cross-section and is therefore called an elliptical beam. Such beams provide good models for the light emitted by semiconductor lasers. Elliptical beams generally converge or diverge at different rates along the two principal axes of the ellipse and are therefore astigmatic. If an elliptical beam is passed through a lens, the beam waists after the lens do not, in general, lie in the same plane. The fundamental elliptical beam has the form

$$\begin{aligned}
&E_0(\boldsymbol{r}) = A\sqrt{\frac{w_{0x}w_{0y}}{w_x(z)w_y(z)}}\, e^{ik\left(\frac{x^2}{2q_x(z)}+\frac{y^2}{2q_y(z)}\right)-i\frac{1}{2}[\psi_x(z)+\psi_y(z)]}\,, \\
&\frac{1}{q_x(z)} = \frac{1}{R_x(z)} + i\frac{2}{kw_x^2(z)}, \quad \frac{1}{q_y(z)} = \frac{1}{R_y(z)} + i\frac{2}{kw_y^2(z)}\,, \\
&w_x^2 = w_{ox}^2\left[1+\left(\frac{2(z-z_{ox})}{kw_{0x}^2}\right)^2\right], \quad w_y^2 = w_{oy}^2\left[1+\left(\frac{2(z-z_{oy})}{kw_{0y}^2}\right)^2\right], \\
&R_x(z) = (z-z_{ox})\left[1+\left(\frac{kw_{0x}^2}{2(z-z_{ox})}\right)^2\right], \quad R_y(z) = (z-z_{oy})\left[1+\left(\frac{kw_{0y}^2}{2(z-z_{oy})}\right)^2\right], \\
&\psi_x(z) = \tan^{-1}\left[\frac{2(z-z_{ox})}{kw_{0x}^2}\right], \quad \psi_y(z) = \tan^{-1}\left[\frac{2(z-z_{oy})}{kw_{0y}^2}\right].
\end{aligned} \tag{3.87}$$

The beam reaches its minimum extent in the x direction at $z = z_{0x}$ and in the y direction at $z = z_{0y}$. These two waist planes do not, in general, coincide. The beam generally has an elliptical transverse profile whose widths and aspect ratio change during propagation. Since the principal axes in the present representation are fixed along the x and y directions, the ellipse does not rotate continuously; however, the direction of its major axis may interchange between the x and y axes as $w_x(z)$ and $w_y(z)$ change.

We can also write down the general astigmatic HG_{mn} beam

$$HG_{mn}(\boldsymbol{r}) = A_{mn}\sqrt{\frac{w_{0x}w_{0y}}{w_x(z)w_y(z)}}\, H_m\left(\frac{\sqrt{2}x}{w_x(z)}\right) H_n\left(\frac{\sqrt{2}y}{w_y(z)}\right) e^{ik\left(\frac{x^2}{2q_x(z)}+\frac{y^2}{2q_y(z)}\right)-i\psi_{mn}(z)} \tag{3.88}$$

$$\psi_{mn}(z) = \left(m+\frac{1}{2}\right)\tan^{-1}\left[\frac{2(z-z_{ox})}{kw_{0x}^2}\right] + \left(n+\frac{1}{2}\right)\tan^{-1}\left[\frac{2(z-z_{oy})}{kw_{0y}^2}\right] \tag{3.89}$$

It is easy to see that for $w_{0x} = w_{0y} = w_0$ and $z_{0x} = z_{0y} = z_o$, we recover the nonastigmatic HG beams with a common waist at $z = z_o$. Moreover, for $m = 0 = n$, the fundamental circularly symmetric Gaussian beam is recovered. Semiconductor lasers commonly emit astigmatic elliptical beams of the type described above. Such beams can be converted into nonastigmatic Hermite-Gauss beams by using prisms or cylindrical lenses.

3.6 Laguerre-Gauss Beams

The paraxial wave equation (3.10) admits beam-like solutions that reflect the symmetry of the circular cylindrical coordinate system. The procedure for finding such solutions is similar to that followed for the Cartesian coordinate system. For circular cylindrical symmetry about the z-axis, we look for solutions separable in the transverse radial and azimuthal coordinates introduced via

$$\rho = \sqrt{x^2+y^2}\,, \qquad \varphi = \tan^{-1}\frac{y}{x}\,; \qquad 0 \le \rho < \infty\,, \qquad 0 \le \varphi < 2\pi\,. \tag{3.90}$$

In terms of these coordinates, the paraxial wave equation takes the form

$$\left[\frac{\partial^2}{\partial\rho^2} + \frac{1}{\rho}\frac{\partial}{\partial\rho} + \frac{1}{\rho^2}\frac{\partial^2}{\partial\varphi^2} + 2ik\frac{\partial}{\partial z}\right] E_0(\rho,\varphi,z) = 0\,. \tag{3.12*}$$

In this case, we look for solutions separable in the ρ and φ coordinates that retain their shape as they propagate, apart from a uniform expansion in the transverse dimensions. Inspired by the success of our approach in obtaining the HG beam solutions, we try a solution of the form

$$E_0(\rho,\varphi,z) = A\,g\left(\frac{\rho}{w(z)}\right)\Phi(\varphi)e^{iP(z)+\frac{ik\rho^2}{2q(z)}}\,. \tag{3.91}$$

Here g is a function of the radial coordinate scaled by a size parameter $w(z)$, which may change with propagation as in the case of Hermite-Gauss beams. We shall see shortly that $w(z)$ turns out to be the spot radius of the fundamental Laguerre-Gauss beam, which coincides with the fundamental Gaussian beam (3.13).

The dependence of the field amplitude on the azimuthal angle φ, described by the function $\Phi(\varphi)$, is fundamentally different from its dependence on the radial coordinate. The azimuthal coordinate φ has a finite range, and a physical point in space is specified by ρ, z, and an azimuthal angle in the range $0 \le \varphi < 2\pi$. Increasing φ by 2π returns us to exactly the same physical point. Therefore, the φ-dependence of the field amplitude must satisfy

$E_0(\rho, \varphi + 2\pi, z) = E_0(\rho, \varphi, z)$. Thus the field amplitude must be periodic in φ with period 2π. The choice

$$\Phi(\varphi) = e^{i\ell\varphi}\,, \qquad \ell = 0, \pm 1, \pm 2, \pm 3, \ldots \tag{3.92}$$

ensures the single-valuedness of the beam amplitude as a function of the azimuthal angle φ.

To obtain the differential equation satisfied by the radial function g, we substitute Eq. (3.91), with the azimuthal dependence given by Eq. (3.92), into the paraxial wave equation (3.12*) and express the derivatives of $E_0(\rho, \varphi, z)$ in terms of derivatives of $g(u)$, $\Phi(\varphi)$, and $P(z)$. In doing so, we keep in mind that g is a function of the independent variable $u = \rho/w(z)$. Using the notation

$$u = \frac{\rho}{w(z)}\,, \qquad g' = \frac{dg(u)}{du}\,, \qquad g'' = \frac{d^2 g(u)}{du^2}\,, \tag{3.93}$$

and the chain rule, we obtain

$$\begin{aligned}
\frac{\partial g}{\partial \rho} &= \frac{\partial u}{\partial \rho}\frac{dg}{du} = \frac{1}{w}\frac{dg}{du} = \frac{g'}{w}\,, \\
\frac{\partial^2 g}{\partial \rho^2} &= \frac{1}{w^2}\frac{d^2 g}{du^2} = \frac{g''}{w^2}\,, \\
\frac{\partial g}{\partial z} &= \frac{\partial u}{\partial z}\frac{dg}{du} = -\frac{\rho w'}{w^2}\frac{dg}{du} = -\frac{\rho w' g'}{w^2}\,.
\end{aligned} \tag{3.94}$$

We can then express the derivatives of $E_0(\rho, \varphi, z)$ with respect to ρ, φ, and z in terms of the new independent variables u and z as

$$\begin{aligned}
\frac{\partial E_0}{\partial z} &= \left[-\frac{\rho w'}{w^2} g' + \left(iP' - \frac{ikq'}{2q^2}\rho^2\right) g\right] e^{i\ell\varphi} e^{iP(z) + ik\rho^2/2q} \\
&= \left[-\frac{w'}{w} u g' + \left(iP' - \frac{ikw^2 q'}{2q^2} u^2\right) g\right] e^{i\ell\varphi} e^{iP(z) + ik\rho^2/2q}\,, \\
\frac{\partial E_0}{\partial \rho} &= \left[\frac{1}{w} g' + \frac{ik\rho}{q} g\right] e^{i\ell\varphi} e^{iP(z) + ik\rho^2/2q} \\
&= \left[\frac{1}{w} g' + \frac{ikw}{q} u g\right] e^{i\ell\varphi} e^{iP(z) + ik\rho^2/2q}\,, \\
\frac{\partial^2 E_0}{\partial \rho^2} &= \left[\frac{1}{w^2} g'' + \frac{2ik\rho}{qw} g' - \frac{k^2\rho^2}{q^2} g + \frac{ik}{q} g\right] e^{i\ell\varphi} e^{iP(z) + ik\rho^2/2q} \\
&= \left[\frac{1}{w^2} g'' + \frac{2ik}{q} u g' - \frac{k^2 w^2}{q^2} u^2 g + \frac{ik}{q} g\right] e^{i\ell\varphi} e^{iP(z) + ik\rho^2/2q}\,, \\
\frac{\partial^2 E_0}{\partial \varphi^2} &= -\ell^2 g\, e^{i\ell\varphi} e^{iP(z) + ik\rho^2/2q}\,.
\end{aligned} \tag{3.95}$$

Using these relations in the paraxial wave equation (3.12*), dividing throughout by $ge^{i\ell\varphi}$ and the remaining common exponential factor, and multiplying by w^2, we obtain

$$\frac{g''}{g} + \left(\frac{1}{u} - 2ikww'u + \frac{2ikw^2}{q} u\right)\frac{g'}{g} - \frac{\ell^2}{u^2} + \left(\frac{2ik}{q} - 2kP'\right) w^2 + \frac{k^2 w^4 u^2}{q^2}(q' - 1) = 0\,. \tag{3.96}$$

This equation must be satisfied for all values of $u = \rho/w$. To obtain a radial equation independent of z, we choose $q(z)$ such that the last term, proportional to u^2, vanishes. This requires

$$\frac{dq}{dz} = 1 \qquad \Rightarrow \qquad q(z) = z - iz_R\,, \tag{3.97}$$

where the origin of z has been chosen at the beam waist. This is the same expression for $q(z)$ as that given in Eq. (3.13). Alternatively, we could have started by assuming that $q(z)$ has the form given by Eq. (3.13). With this choice for $q(z)$, the last term proportional to u^2 drops out. In addition, the following relations are satisfied:

$$\frac{w'}{w} = \frac{1}{R}\,, \qquad \frac{1}{q} = \frac{1}{R} + i\frac{2}{kw^2}\,. \tag{3.98}$$

With the help of these relations, the coefficient of $\frac{g'}{g}$ in Eq. (3.96) simplifies to

$$\begin{aligned}\frac{1}{u} - 2ikww'u + \frac{2ikw^2}{q}u &= \frac{1}{u} - \frac{2ikw^2}{R}u + 2ikw^2\left(\frac{1}{R} + \frac{2i}{kw^2}\right)u \\ &= \frac{1}{u} - 4u\,.\end{aligned} \tag{3.99}$$

Using Eqs. (3.97) and (3.99) in Eq. (3.96), we find that the functions $g(u)$ and $P(z)$ must satisfy the differential equation

$$\left[\frac{g''}{g} + \left(\frac{1}{u} - 4u\right)\frac{g'}{g} - \frac{\ell^2}{u^2}\right] + \left(\frac{2ik}{q} - 2kP'\right)w^2 = 0\,, \tag{3.100}$$

for solutions of the form (3.91) to satisfy the paraxial wave equation.

Equation (3.100) must be satisfied for all values of u and z. The first term in this equation, inside the square brackets, depends only on the independent variable $u = \rho/w$, whereas the second term depends only on the independent variable z. Therefore, for Eq. (3.100) to hold for all values of the independent variables u and z, each of these two terms must separately be equal to a constant. We therefore write

$$\left[\frac{g''}{g} + \left(\frac{1}{u} - 4u\right)\frac{g'}{g} - \frac{\ell^2}{u^2}\right] = -\alpha\,, \tag{3.101}$$

$$\left(\frac{2ik}{q} - 2kP'\right)w^2 = \alpha\,, \tag{3.102}$$

where α is a constant to be determined. Note that Eqs. (3.101) and (3.102) together satisfy Eq. (3.100).

Equation (3.101) for the radial function is a second-order differential equation with a regular singular point at $u = 0$. Therefore, a series solution for $g(u)$ must begin with a term $g(u) \sim a_0 u^c$. Substituting this form into Eq. (3.101), and requiring $a_0 \neq 0$, we find that the index c must satisfy $c(c-1) + c - \ell^2 = 0$, or $c = \pm|\ell|$. For $|\ell| > 0$, the solution proportional to $u^{-|\ell|}$ diverges on the beam axis and must be discarded. The regular solution therefore behaves as $u^{|\ell|}$. For $\ell = 0$, the regular solution approaches a constant. Thus, in all cases, the regular radial solution can be written in the form $g(u) = u^{|\ell|}h(u)$, where $h(u)$ is a function of u.

The equation for $h(u)$ can be brought to a more recognizable form by introducing the new variable $\xi = 2\rho^2/w^2 = 2u^2$. To proceed directly to the equation for $L(\xi)$, we write

$$\left.\begin{aligned} u = \sqrt{\xi/2}\,, \qquad \xi &= 2u^2\ (= 2\rho^2/w^2)\,, \\ g(u) = u^{|\ell|}h(u) &= \xi^{|\ell|/2}L(\xi)\,. \end{aligned}\right\} \tag{3.103}$$

where an unimportant constant factor $2^{-|\ell|/2}$ has been absorbed into $L(\xi)$.

We substitute this expression into Eq. (3.101) and express the derivatives in terms of the new variable ξ using

$$\frac{d\xi}{du} = 4u = 4\sqrt{\xi/2} = \sqrt{8\xi}\,, \tag{3.104a}$$

$$\begin{aligned}\frac{dg}{du} &= \frac{d\xi}{du}\frac{d}{d\xi}\left[\xi^{|\ell|/2}L(\xi)\right] \\ &= \sqrt{8\xi}\left[\frac{|\ell|}{2}\xi^{|\ell|/2-1}L + \xi^{|\ell|/2}L'\right],\end{aligned} \tag{3.104b}$$

$$\begin{aligned}\frac{d^2g}{du^2} &= \sqrt{8\xi}\frac{d}{d\xi}\left\{\sqrt{8\xi}\left[\frac{|\ell|}{2}\xi^{|\ell|/2-1}L + \xi^{|\ell|/2}L'\right]\right\} \\ &= 2|\ell|(|\ell|-1)\xi^{|\ell|/2-1}L + 4(2|\ell|+1)\xi^{|\ell|/2}L' + 8\xi^{|\ell|/2+1}L''\,,\end{aligned} \tag{3.104c}$$

$$\left(\frac{1}{u} - 4u\right)\frac{dg}{du} = (4-8\xi)\left[\frac{|\ell|}{2}\xi^{|\ell|/2-1}L + \xi^{|\ell|/2}L'\right]. \tag{3.104d}$$

With the help of these equations, Eq. (3.101) reduces to

$$8\xi L'' + 8(|\ell|+1-\xi)L' + (\alpha - 4|\ell|)\,L = 0\,,$$

$$\text{or}\quad \xi L'' + (|\ell|+1-\xi)L' + \left(\frac{\alpha}{8} - \frac{|\ell|}{2}\right)L = 0\,. \tag{3.105}$$

This equation has the form of the associated Laguerre equation, which also arises in the quantum mechanics of a two-dimensional harmonic oscillator. A polynomial solution of Eq. (3.105), which ensures that the complete beam carries finite power, is possible only if $\left(\frac{\alpha}{8} - \frac{|\ell|}{2}\right)$ is a nonnegative integer. Thus,

$$\left(\frac{\alpha}{8} - \frac{|\ell|}{2}\right) = p\,, \qquad p = 0,1,2,3,\ldots\,,$$

$$\text{or, equivalently,}\quad \alpha = 8p + 4|\ell|\,, \qquad p = 0,1,2,3,\ldots\,. \tag{3.106}$$

The corresponding solutions of Eq. (3.105) are the associated Laguerre polynomials,

$$L_p^{|\ell|}(\xi) = L_p^{|\ell|}\left(\frac{2\rho^2}{w^2(z)}\right), \qquad p = 0,1,2,3,\ldots\,; \qquad \ell = 0,\pm1,\pm2,\pm3,\ldots\,. \tag{3.107}$$

where we have recalled the change of variable $\xi = 2u^2 = 2\rho^2/w^2(z)$ from Eq. (3.103).

We still need to determine $P(z)$ to complete our derivation of the solution (3.91) in circular cylindrical coordinates. Using the value of α from Eq. (3.106) in Eq. (3.102), the equation for $P(z)$ becomes

$$\begin{aligned}\frac{dP}{dz} &= -\frac{8p+4|\ell|}{2kw^2} + \frac{i}{q} = -\frac{(2p+|\ell|)z_R}{z^2+z_R^2} + \frac{i}{z - iz_R}\,, \\ &= -\frac{(2p+|\ell|+1)z_R}{z^2+z_R^2} + i\frac{z}{z^2+z_R^2}\,.\end{aligned} \tag{3.108a}$$

This can be integrated to give

$$P(z) = -(2p+|\ell|+1)\psi(z) + i\ln\sqrt{1+\left(\frac{z}{z_R}\right)^2}\,, \tag{3.108b}$$

$$\text{or}\quad e^{iP(z)} = \frac{e^{-i(2p+|\ell|+1)\psi(z)}}{\sqrt{1+(z/z_R)^2}} = \frac{w_0}{w(z)}e^{-i(2p+|\ell|+1)\psi(z)}\,. \tag{3.108c}$$

Using the results of Eqs. (3.92), (3.107), and (3.108c) in Eq. (3.91), and recalling the change of variable in Eq. (3.103), we can finally write the beam-like solutions of the paraxial wave equation in cylindrical coordinates as

$$LG_{p\ell}(\boldsymbol{r}) \equiv [E_0]_{p\ell}(\boldsymbol{r}) = A_{p\ell}\frac{w_0}{w}\left(\frac{\sqrt{2}\rho}{w}\right)^{|\ell|} e^{i\ell\varphi}\, L_p^{|\ell|}\left(\frac{2\rho^2}{w^2}\right) e^{-i(2p+|\ell|+1)\psi+i\frac{k\rho^2}{2q}} . \tag{3.109}$$

Here $L_p^{|\ell|}(x)$ is an associated Laguerre polynomial, while $w(z)$, $q(z)$, and $\psi(z)$ have the same expressions and physical significance as for the fundamental Gaussian beam (3.13). These quantities are independent of the mode indices.

The radial index $p = 0, 1, 2, 3, \ldots$ gives the number of radial nodes at finite nonzero ρ, while the azimuthal index $\ell = 0, \pm1, \pm2, \pm3, \ldots$ determines the phase winding around the beam axis. On making one complete circuit around the beam axis, the phase changes by $2\pi\ell$; thus $|\ell|$ gives the magnitude of the phase winding, while the sign of ℓ determines its sense. For $\ell \neq 0$, the factor $\rho^{|\ell|}$ causes the field amplitude to vanish on the beam axis ($\rho = 0$). Beams represented by Eq. (3.109) are called Laguerre-Gauss (LG) beams.

Some low-order associated Laguerre polynomials and recurrence relations for computing higher-order polynomials are

$$\begin{aligned}
L_0^{|\ell|}(u) &= 1\,,\\
L_1^{|\ell|}(u) &= -u + |\ell| + 1\,,\\
L_2^{|\ell|}(u) &= \frac{1}{2}\left[u^2 - 2(|\ell|+2)u + (|\ell|+1)(|\ell|+2)\right],\\
L_p^{|\ell|}(u) &= \frac{1}{p!}\left(\frac{e^u}{u^{|\ell|}}\right)\frac{d^p}{du^p}\left[u^{p+|\ell|}e^{-u}\right],\\
(p+1)L_{p+1}^{|\ell|}(u) &= (2p+|\ell|+1-u)L_p^{|\ell|}(u) - (p+|\ell|)L_{p-1}^{|\ell|}(u)\,,\\
u\frac{dL_p^{|\ell|}(u)}{du} &= pL_p^{|\ell|}(u) - (p+|\ell|)L_{p-1}^{|\ell|}(u)\,.
\end{aligned} \tag{3.110}$$

From the expression for the field amplitude (3.109), the intensity distribution $I_{p\ell} = \frac{1}{2}\epsilon_0 cn|E_{p\ell}(\boldsymbol{r})|^2$ for the LG beams can be written as

$$I_{p\ell} = I_{0p\ell}\left(\frac{w_0}{w}\right)^2\left(\frac{\sqrt{2}\rho}{w}\right)^{2|\ell|}\left[L_p^{|\ell|}(2\rho^2/w^2)\right]^2 e^{-2\rho^2/w^2}, \tag{3.111}$$

where the constant $I_{0p\ell} = \frac{1}{2}\epsilon_0 cn|A_{p\ell}|^2$ is related to the total power carried by the beam by

$$\begin{aligned}
P &= \int_0^\infty \rho\, d\rho \int_0^{2\pi} d\varphi\, I_{p\ell}(\rho, z)\\
&= I_{0p\ell}\left(\frac{w_0}{w}\right)^2 2\pi\int_0^\infty \rho\, d\rho \left(\frac{\sqrt{2}\rho}{w}\right)^{2|\ell|}\left[L_p^{|\ell|}(2\rho^2/w^2)\right]^2 e^{-2\rho^2/w^2}\\
&= I_{0p\ell}\left(\frac{w_0}{w}\right)^2\frac{\pi w^2}{2}\int_0^\infty du\, u^{|\ell|}\left[L_p^{|\ell|}(u)\right]^2 e^{-u}\\
&= I_{0p\ell}\frac{\pi w_0^2}{2}\frac{(|\ell|+p)!}{p!}\,.
\end{aligned} \tag{3.112}$$

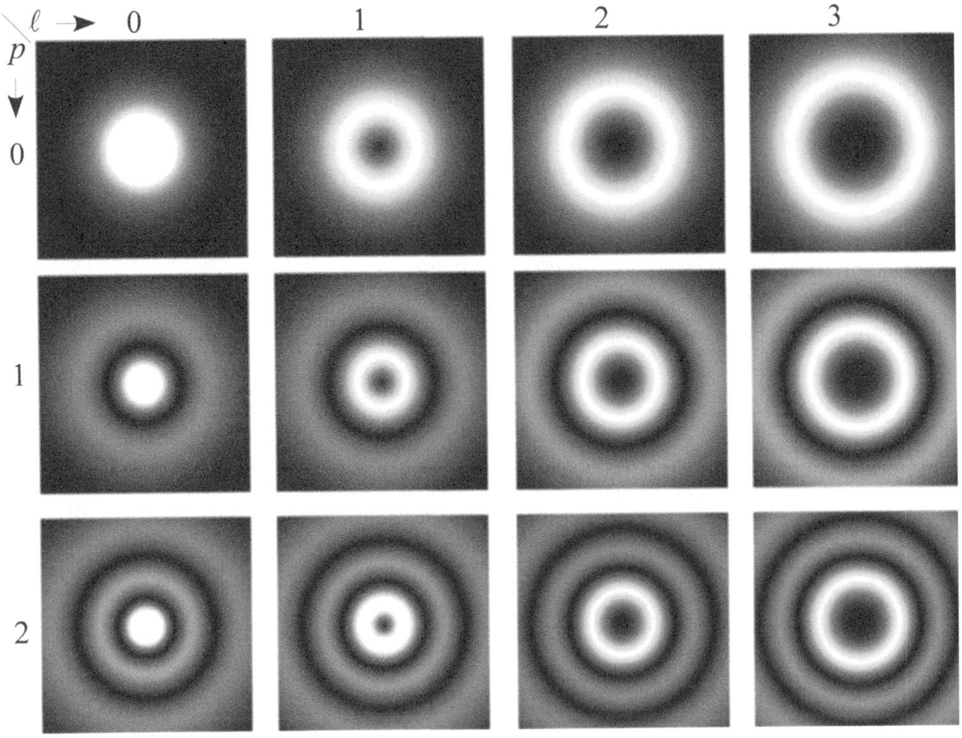

FIGURE 3.14
Intensity distribution in a plane perpendicular to the direction of propagation for several low-order $\mathrm{LG}_{p\ell}$ beams.

Using this relation to express $I_{0p\ell}$ in terms of the total beam power P, we can write the intensity of an $\mathrm{LG}_{p\ell}$ beam as

$$I_{p\ell} = \frac{2P}{\pi w^2} \frac{p!}{(|\ell| + p)!} \left(\frac{\sqrt{2}\rho}{w} \right)^{2|\ell|} \left[L_p^{|\ell|}(2\rho^2/w^2) \right]^2 e^{-2\rho^2/w^2} . \tag{3.113}$$

The lowest-order Laguerre-Gauss solution ($p = 0 = \ell$) coincides with the fundamental Gaussian beam solution, which is also the lowest-order member of the HG family. Figure 3.14 shows the intensity profiles of several low-order LG beams.

The field and intensity distributions for the Laguerre-Gauss beams with $p = 0$ take especially simple forms:

$$E_\ell(\boldsymbol{r}) = A_\ell \frac{w_0}{w} \left(\frac{\sqrt{2}\rho}{w} \right)^{|\ell|} e^{i\ell\varphi} \, e^{-i(|\ell|+1)\psi + i\frac{k\rho^2}{2q}} , \tag{3.114}$$

$$\begin{aligned} I_\ell(\rho) &= I_{0\ell} \frac{w_0^2}{w^2} \left(\frac{2\rho^2}{w^2} \right)^{|\ell|} e^{-2\rho^2/w^2} \\ &\equiv \frac{2P}{\pi w^2} \frac{1}{|\ell|!} \left(\frac{\sqrt{2}\rho}{w} \right)^{2|\ell|} e^{-2\rho^2/w^2} . \end{aligned} \tag{3.115}$$

These beams have helicoidal phasefronts (surfaces of constant phase). If we make a circuit around the beam axis in a transverse plane (fixed z), the continuously unwrapped phase changes by $2\pi\ell$ because of the presence of the $e^{i\ell\varphi}$ term. Nevertheless, the complex field is single-valued because $e^{i2\pi\ell} = 1$ for integer ℓ. On the beam axis, the azimuthal angle φ, and hence the phase associated with $e^{i\ell\varphi}$, is undefined. For $\ell \neq 0$, however, the field amplitude is proportional to $\rho^{|\ell|}$ and vanishes on the axis. The beam therefore has an on-axis phase singularity, or optical vortex. For this reason, these beams are referred to as Laguerre-Gauss vortex (LGV) beams. The index ℓ is sometimes referred to as the order, or topological charge, of the vortex. Note that for LG beams with nonzero radial index p, the Laguerre polynomial is real and therefore introduces only phase changes of π across its zeros rather than a continuous azimuthal phase winding.

The intensity distribution of LGV beams consists of a dark core surrounded by a bright ring, resembling a doughnut. For $p = 0$ and $\ell \neq 0$, the radius of maximum intensity is

$$\rho_{\text{max}} = w(z)\sqrt{\frac{|\ell|}{2}}, \tag{3.116}$$

and therefore grows as $\sqrt{|\ell|}$. For this reason, these beams are sometimes referred to as doughnut beams. It should be mentioned that a "doughnut"-shaped intensity distribution sometimes seen in lasers can also result from an incoherent mixture of HG_{01} and HG_{10} modes. LGV modes are distinguished from such a mixture by their characteristic azimuthal phase structure.

From the expression for the LG beams, the phasefronts are given by

$$k\Psi \equiv kz + \frac{k\rho^2}{2R} + \ell\varphi = \text{constant}, \tag{3.117}$$

where we have neglected the Gouy phase term, which gives a z-dependent phase correction near the waist but does not affect the azimuthal phase winding. The phasefronts are shown in Fig. 3.15. These phasefronts can be visualized by interfering the beam with a co-propagating plane wave. The resulting interference fringes have a spiral structure.

Laguerre-Gauss beams carry energy and linear momentum, just as Hermite-Gauss and other families of beams do. However, LGV beams—and, in fact, all Laguerre-Gauss beams with nonzero ℓ and azimuthal dependence $e^{i\ell\varphi}$—carry intrinsic (origin-independent) orbital angular momentum associated with their azimuthal phase structure. By contrast, the HG modes considered here do not carry intrinsic orbital angular momentum. For a pure LG mode, the orbital angular momentum along the propagation direction is $\ell\hbar$ per photon. Transfer of this orbital angular momentum to a microscopic particle can exert an azimuthal torque and cause orbital motion about the beam axis under suitable conditions. It may be recalled that electromagnetic beams can also carry spin angular momentum, which is associated with the state of polarization of the beam. In contrast, the orbital angular momentum carried by Laguerre-Gauss beams is associated with the spatial structure of the optical field and its energy flow.

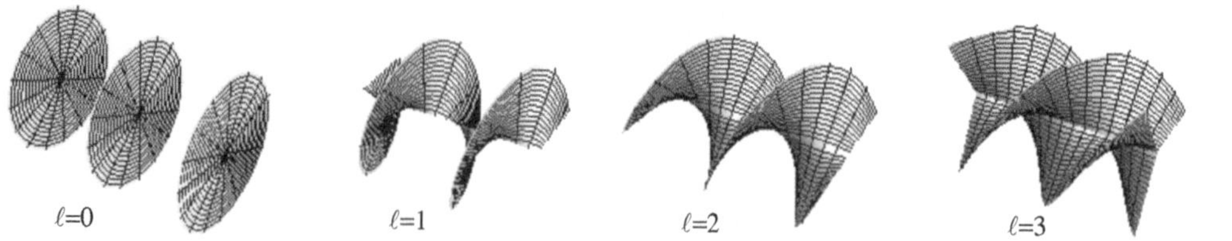

FIGURE 3.15
Phasefronts for LG_{00} [same as HG_{00}] and $LG_{0\ell}$ for $\ell = 1$, 2, and 3.

3.6.1 Ray Equation for LG Beams

The ray equations follow from Eq. (2.94) with refractive index $n = 1$, $d\boldsymbol{r} = dz\,\boldsymbol{e}_z + d\rho\,\boldsymbol{e}_\rho + \rho\,d\varphi\,\boldsymbol{e}_\varphi$, and the phase $\Psi(\boldsymbol{r}) = z + \frac{\rho^2}{2R(z)} + \frac{\ell\varphi}{k}$. The resulting ray trajectory in circular cylindrical coordinates is determined by

$$\frac{d\rho(z)}{dz} = \frac{\rho z}{z_R^2[1+(z/z_R)^2]}, \qquad \frac{d\varphi(z)}{dz} = \frac{\ell}{k\rho^2}. \tag{3.118}$$

Here ρ is the radial distance and φ is the azimuthal angle of a point on the ray. These equations are readily integrated to give ρ and φ as functions of z:

$$\rho(z) = \rho_o\sqrt{1+(z/z_R)^2}, \qquad \varphi(z) = \varphi_o + \frac{w_0^2\ell}{2\rho_o^2}\tan^{-1}\left(\frac{z}{z_R}\right), \tag{3.119}$$

where ρ_o and φ_o are the ray coordinates at the beam waist $z = 0$. The resulting ray is a curve that winds around a hyperboloid, with the winding rate decreasing as $|z/z_R|$ increases. The total angle through which the ray turns as it propagates from $z = -\infty$ to $z = +\infty$ is

$$\Delta\varphi = \frac{\pi w_0^2}{2\rho_o^2}\ell. \tag{3.120}$$

For a ray passing through the radius of maximum intensity at the waist, $\rho_o = w_0\sqrt{|\ell|/2}$, the total rotation becomes

$$\Delta\varphi = \pi\,\frac{\ell}{|\ell|}, \tag{3.121}$$

and is therefore independent of $|\ell|$; its sense is determined by the sign of ℓ.

3.6.2 LGs Beams

In our derivation of LG beams, we assumed the azimuthal dependence of the beam amplitude to be $e^{i\ell\varphi}$. Since modes with $+\ell$ and $-\ell$ have the same radial dependence, we can also form equal-amplitude linear superpositions having azimuthal dependence $\cos(\ell\varphi)$ or $\sin(\ell\varphi)$. The mathematical expressions for such beams can therefore be obtained by replacing $e^{i\ell\varphi}$ by either $\cos(\ell\varphi)$ or $\sin(\ell\varphi)$. These beams have zero net intrinsic orbital angular momentum because they contain equal contributions from modes with opposite orbital angular momenta. Their intensity profiles have azimuthal zeros in addition to radial zeros. For $p = 0$, they exhibit a flower-petal pattern, as shown in Fig. 3.16.

3.6.3 Relation Between HG and LG Families

Hermite-Gauss and Laguerre-Gauss modes are solutions of the same paraxial wave equation. It is therefore not surprising that a beam-like solution belonging to one family can be expressed as a linear superposition of solutions belonging to the other family.[6] There are two interesting aspects of this transformation:

(i) The transformation involves $\mathrm{LG}_{p\ell}$ and HG_{mn} modes of the same order, where the order is defined by $N = 2p + |\ell|$ for the $\mathrm{LG}_{p\ell}$ modes and by $N = m + n$ for the HG_{mn} modes. Thus, the $N = 1$ LG mode set $\{\mathrm{LG}_{01}, \mathrm{LG}_{0(-1)}\}$ can be expressed entirely in terms of the $N = 1$ HG mode set $\{\mathrm{HG}_{10}, \mathrm{HG}_{01}\}$.

[6] M. W. Beijersbergen, L. Allen, H. E. L. O. van der Veen, and J. P. Woerdman, "Astigmatic laser mode converters and transfer of orbital angular momentum," Opt. Commun. **96**, 123–132 (1993).

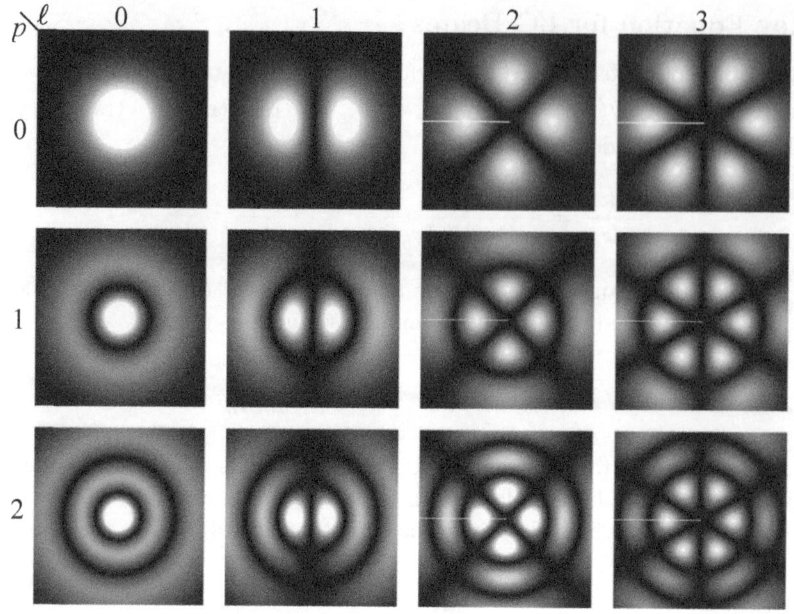

FIGURE 3.16
Intensity distribution in a transverse plane for a few low-order $\mathrm{LG}_{p\ell}$ beams with cosine azimuthal dependence.

(ii) An $\mathrm{HG}_{mn}(x', y')$ mode whose principal axes $[x' = (x+y)/\sqrt{2},\, y' = (x-y)/\sqrt{2}]$ are inclined at an angle of 45° to the (x, y) axes can be decomposed into $\mathrm{HG}_{m'n'}(x, y)$ modes of the same order.

We illustrate these relations for a few low-order modes. The lowest-order modes, HG_{00} and LG_{00}, belonging to the two families are, in fact, the same function expressed in Cartesian and cylindrical coordinates. The first-order ($N = 1$) modes provide an especially simple example. The first-order ($p = 0$, $\ell = \pm 1$) LG modes are given by

$$LG_{0(\pm 1)}(\rho, \varphi) = \frac{2}{\sqrt{\pi}\, w^2} \rho e^{\pm i\varphi} GW\,, \tag{3.122}$$

where $G = e^{-\rho^2/w^2}$ and $W = e^{ik\rho^2/2R - 2i\psi}$, and the two first-order Hermite-Gauss modes are given by

$$HG_{10}(x, y) = \frac{2\sqrt{2}}{\sqrt{\pi}\, w^2} xGW, \quad HG_{01}(x, y) = \frac{2\sqrt{2}}{\sqrt{\pi}\, w^2} yGW\,, \tag{3.123}$$

where the z-dependence of w, R, and ψ has been suppressed. Using $\rho e^{\pm i\varphi} = x \pm iy$, we can express the two first-order LG modes in terms of the two first-order HG modes as

$$LG_{0(\pm 1)} = \frac{2}{\sqrt{\pi}\, w^2}(x \pm iy) GW = \frac{1}{\sqrt{2}} \left[HG_{10} \pm iHG_{01}\right]. \tag{3.124}$$

The inverse relations expressing the HG modes in terms of the LG modes are

$$HG_{10} = \frac{1}{\sqrt{2}} \left[LG_{0(+1)} + LG_{0(-1)}\right], \tag{3.125a}$$

$$HG_{01} = \frac{-i}{\sqrt{2}} \left[LG_{0(+1)} - LG_{0(-1)}\right]. \tag{3.125b}$$

With a little more effort, we can show that for modes of order $N = 2$,

$$LG_{0(+2)} = \frac{1}{2}\left[HG_{20} + i\sqrt{2}HG_{11} - HG_{02}\right] , \tag{3.126a}$$

$$LG_{10} = \frac{1}{\sqrt{2}}\left[HG_{20} + HG_{02}\right] , \tag{3.126b}$$

$$LG_{0(-2)} = \frac{1}{2}\left[HG_{20} - i\sqrt{2}HG_{11} - HG_{02}\right] . \tag{3.126c}$$

This direct method becomes impractical for higher-order modes, for which a more systematic mathematical approach is useful.[7]

It is also interesting to note that an $\mathrm{HG}_{mn}(x', y')$ mode with $x' = (x + y)/\sqrt{2}$ and $y' = (x - y)/\sqrt{2}$ can be expressed as a superposition of HG modes referred to the (x, y) axes and having the same order. For example, for $N = 1$ modes, we have

$$HG_{10}(x', y') = \frac{2\sqrt{2}}{\sqrt{\pi}w^2}\frac{x+y}{\sqrt{2}}GW = \frac{1}{\sqrt{2}}\left[HG_{10}(x, y) + HG_{01}(x, y)\right] , \tag{3.127a}$$

$$HG_{01}(x', y') = \frac{2\sqrt{2}}{\sqrt{\pi}w^2}\frac{x-y}{\sqrt{2}}GW = \frac{1}{\sqrt{2}}\left[HG_{10}(x, y) - HG_{01}(x, y)\right] . \tag{3.127b}$$

Similarly, for $N = 2$ modes, we have

$$HG_{20}(x', y') = \frac{1}{2}\left[HG_{20}(x, y) + \sqrt{2}HG_{11}(x, y) + HG_{02}(x, y)\right] , \tag{3.128a}$$

$$HG_{11}(x', y') = \frac{1}{\sqrt{2}}\left[HG_{20}(x, y) - HG_{02}(x, y)\right] , \tag{3.128b}$$

$$HG_{02}(x', y') = \frac{1}{2}\left[HG_{20}(x, y) - \sqrt{2}HG_{11}(x, y) + HG_{02}(x, y)\right] . \tag{3.128c}$$

A comparison of Eqs. (3.124) and (3.126) with Eqs. (3.127) and (3.128) shows that the rotated HG modes and LG modes can be written in the forms

$$HG_{mn}(x', y') = \sum_{k=0}^{N} b(m, n, k) HG_{N-k,k}(x, y) , \tag{3.129}$$

$$LG_{p\ell} = \sum_{k=0}^{N} i^k b(m, n, k) HG_{N-k,k}(x, y) , \tag{3.130}$$

where the same coefficients $b(m, n, k)$ appear in both expansions, while the LG expansion contains the additional phase factor i^k. The coefficients $b(m, n, k)$ are given by

$$b(m, n, k) = \left(\frac{(N-k)!k!}{2^N m! n!}\right)^{1/2} \frac{1}{k!}\frac{d^k}{dt^k}\left[(1+t)^m(1-t)^n\right]_{t=0} , \tag{3.131}$$

where $N = m + n = 2p + |\ell|$.

The importance of this connection is that a rotated $\mathrm{HG}_{mn}(x', y')$ mode passed through an appropriate astigmatic mode converter can be transformed into an $\mathrm{LG}_{p\ell}$ mode. In the commonly used $\pi/2$ converter, consisting of a pair of parallel cylindrical lenses with the appropriate separation and orientation, the relative Gouy phase between the two transverse components is changed by $\pi/2$. The resulting LG mode has radial index $p = \min(m, n)$ and azimuthal index $\ell = m - n$, apart from conventions that may reverse the sign of ℓ.

[7]L. Allen, M. W. Beijersbergen, R. J. C. Spreeuw, and J. P. Woerdman, *Orbital angular momentum of light and the transformation of Laguerre-Gaussian laser modes*, Phys. Rev. A **45**, 8185–8189 (1992); I. Kimel and L. R. Elias, *Relations between Hermite and Laguerre Gaussian modes*, IEEE J. Quantum Electron. **29**, 2562–2567 (1993).

Beyond HG and LG Beam Families

Paraxial laser beams appropriate to several other coordinate systems have also been discussed and generated. For example, separation of the paraxial wave equation in elliptic coordinates leads to the Ince-Gauss family of paraxial beams. There are also Bessel-beam solutions, which are often referred to as nondiffracting beams. Ideal Bessel beams have infinite transverse extent and therefore carry infinite power; experimentally realizable Bessel-like beams are necessarily of finite extent and remain approximately nondiffracting only over a finite propagation distance.

In conventional laser resonators, Cartesian symmetry is particularly common because Brewster surfaces and other asymmetric optical elements tend to distinguish two orthogonal transverse directions. Consequently, Hermite-Gauss modes arise naturally in many laser systems, although Laguerre-Gauss and other structured modes can also be generated directly or by mode conversion. More generally, the generation and control of the spatial structure—and, increasingly, the spatiotemporal and spectral structure—of laser beams continue to be active areas of research, broadly referred to as *structured light*.[8]

3.7 Gaussian Beams in Graded Index Fibers

In a graded-index fiber, the refractive index varies across the cross-section of the fiber. A common configuration consists of a circular cylindrical fiber core whose refractive index is highest on the axis and decreases approximately quadratically with distance from the axis toward the cladding. Such fibers also support Gaussian beam modes. Consider a fiber of core radius a in which the refractive index varies quadratically from n_1 at the center to n_2 at the cladding. We can parametrize this variation as

$$n^2(\rho) = \frac{\mu\epsilon}{\mu_0\epsilon_0} = n_1^2\left(1 - \frac{\rho^2}{\Lambda^2}\right), \tag{3.132a}$$

with
$$\Lambda = a\sqrt{\frac{n_1^2}{n_1^2 - n_2^2}} \approx a\sqrt{\frac{n_1}{2\Delta n}} \quad \text{for } \Delta n \equiv |n_1 - n_2| \ll n_1 . \tag{3.132b}$$

The parameter Λ is a measure of the grading length scale of the fiber. Typically, Λ is several times the core radius, so that the fractional variation of the refractive index over distances of a few wavelengths is very small. Starting from Maxwell's equations in a source-free region, we then obtain the following equation for the electric field:

$$\nabla^2 \boldsymbol{E} - \mu\epsilon\frac{\partial^2 \boldsymbol{E}}{\partial t^2} = -\boldsymbol{\nabla}\left(\boldsymbol{E}\cdot\boldsymbol{\nabla}\ln\epsilon\right) \approx 0 . \tag{3.133}$$

The neglect of the term on the right is justified when the fractional change in ϵ over distances of a few wavelengths is small. As seen in Sec. 2.8.1, this is an excellent approximation for graded-index fibers encountered in practice. For example, for $n_1 = 1.47$, $\Delta n \approx 0.05$, and $a = 50\,\mu\text{m}$, we find $\Lambda \approx 192\,\mu\text{m}$, with still larger values for smaller Δn, as is typically the case.

[8] U. T. Schwarz, M. A. Bandres, and J. C. Gutierrez-Vega, "Observation of Ince-Gaussian modes in stable resonators," Opt. Lett. **29**, 1870–1872 (2004); J. C. Gutierrez-Vega, M. D. Iturbe-Castillo, G. A. Ramirez, E. Tepichin, R. M. Rodriguez-Dagnino, S. Chavez-Cerda, and G. H. C. New, "Experimental demonstration of optical Mathieu beams," Opt. Commun. **195**, 35–40 (2001); A. Forbes, "Structured Light from Lasers," Laser Photonics Rev. **13**, 1900140 (2019); A. Zdagkas, C. McDonnell, J. Deng, Y. Shen, G. Li, T. Ellenbogen, N. Papasimakis, and N. I. Zheludev, "Observation of Toroidal Pulses of Light," Nat. Photonics **16**, 523 (2022).

With this index profile, the wave equation for a linearly polarized monochromatic wave with time dependence $\boldsymbol{E} = \hat{\boldsymbol{\varepsilon}} E e^{-i\omega t}$ reduces to the scalar wave equation

$$\nabla^2 E + \frac{\omega^2 n_1^2}{c^2}\left(1 - \frac{\rho^2}{\Lambda^2}\right) E = \nabla^2 E + k^2\left(1 - \frac{\rho^2}{\Lambda^2}\right) E = 0\,, \tag{3.134a}$$

with $$k = \frac{\omega n_1}{c}\,. \tag{3.134b}$$

Here E represents the electric-field amplitude for a wave linearly polarized along the unit vector $\hat{\boldsymbol{\varepsilon}}$. Alternatively, E may be regarded as any one of the Cartesian components of the vector field within the scalar approximation.

We now look for solutions that propagate along the axis of the fiber while maintaining their transverse shape. Choosing the z-axis along the fiber axis, and recalling that a monochromatic wave propagating in the z-direction has a spatial dependence $e^{i\beta z}$, we seek solutions of the form $E(\boldsymbol{r}) = E(x,y)e^{i\beta z}$, where the transverse field amplitude $E(x,y)$ and the propagation constant β are to be determined. Substituting this form into Eq. (3.134a), we find that the transverse amplitude must satisfy

$$\left[\frac{\partial^2}{\partial x^2} + \frac{\partial^2}{\partial y^2} + \left(k^2 - \beta^2 - \frac{k^2\rho^2}{\Lambda^2}\right)\right] E(x,y) = 0\,. \tag{3.135}$$

This equation has the same mathematical form as the stationary-state Schrödinger equation for a two-dimensional harmonic oscillator. Its solutions can therefore be written down immediately.

For a circular fiber, the natural coordinates for describing the transverse field variation are ρ and φ. However, because the quadratic index profile is also separable in Cartesian coordinates, we can equally well use x and y. Writing the transverse field amplitude as a product, $E(x,y) = X(x)Y(y)$, and applying separation of variables, we find

$$X_m(x) = H_m\left(\sqrt{\frac{k}{\Lambda}}\,x\right)\exp\left(-\frac{kx^2}{2\Lambda}\right), \quad m = 0,1,2,\cdots\,, \tag{3.136a}$$

$$Y_n(y) = H_n\left(\sqrt{\frac{k}{\Lambda}}\,y\right)\exp\left(-\frac{ky^2}{2\Lambda}\right), \quad n = 0,1,2,3,\cdots\,, \tag{3.136b}$$

with propagation constant

$$\beta_{mn} = k\left[1 - \frac{2}{k\Lambda}(m+n+1)\right]^{1/2}. \tag{3.137}$$

Introducing a spot radius w through

$$w^2 = \frac{2\Lambda}{k} = \frac{\lambda\Lambda}{\pi n_1}\,, \tag{3.138}$$

we can write the field amplitude as

$$E_{mn}(\boldsymbol{r}) = \mathcal{A}_{mn} H_m\left(\frac{\sqrt{2}x}{w}\right) H_n\left(\frac{\sqrt{2}y}{w}\right) e^{-\rho^2/w^2} e^{i\beta_{mn}z}\,, \tag{3.139a}$$

$$\beta_{mn} = k\left[1 - \frac{4}{k^2w^2}(m+n+1)\right]^{1/2}. \tag{3.139b}$$

These modes have the same transverse form as the Hermite-Gauss modes in a homogeneous medium, with one important difference: the spot radius w is independent of z. In a homogeneous medium, diffraction causes an HG beam to expand away from its waist. In a graded-index fiber, however, the refractive index decreases with distance from the axis, producing a continuous focusing action toward the axis. For the modes in Eq. (3.139), this focusing action exactly balances diffraction, resulting in a transverse intensity distribution that remains unchanged as the beam propagates. Thus, the graded-index profile acts as a continuously distributed focusing system, converting the freely expanding Gaussian modes of a homogeneous medium into stationary transverse modes of the fiber.

Modes of fibers with a parabolic refractive-index profile can also be described by Laguerre-Gauss (LG) functions. This can be seen by expressing Eq. (3.135) in circular cylindrical coordinates, as was done for free-space propagation. The resulting equation has the form of the Schrödinger equation for a two-dimensional harmonic oscillator. Assuming that the field decays sufficiently rapidly away from the fiber axis, as we did for the HG-mode solutions, the solutions in terms of LG functions can be obtained. We will not repeat this calculation.

As an estimate of the accuracy of ignoring the boundary condition at $\rho = a$, consider a fiber of diameter 50 μm, refractive index $n_1 = 1.52$, $\Delta n = 0.01$, and wavelength $\lambda = 1.06\,\mu$m. Using these values, we find $\Lambda = \sqrt{n_1 a^2/(2\Delta n)} = 218\,\mu$m and $w = \sqrt{\lambda\Lambda/(\pi n_1)} = 6.96\,\mu$m. The exponential factor in the field amplitude at the boundary is then

$$\exp[-(a/w)^2] = \exp[-(25\,\mu\text{m}/6.96\,\mu\text{m})^2] = 2.46 \times 10^{-6}\,. \tag{3.140}$$

Thus, the field is already negligible at the core boundary. For higher-order modes, the field distribution extends to larger values of ρ, or of x and y in the HG representation, so that the boundary eventually becomes important. Nevertheless, for the low-order modes considered here, the exponentially decreasing field makes the neglect of the boundary an excellent approximation.

Note that even with a relatively small index change in going from the fiber axis to the outer radius, the beam can be tightly confined. The index change over a transverse distance comparable to the spot radius w is only about 1.3×10^{-3}. Thus, only a small transverse variation of the refractive index is required to produce appreciable guiding over many wavelengths. Conversely, a similarly weak index gradient of the opposite sign can produce defocusing rather than guiding.

For index profiles other than the parabolic profile, the fiber modes do not, in general, have the same transverse form as the free-space HG or LG solutions. Consequently, a mode emerging from a fiber does not necessarily correspond exactly to one of these free-space modes. Conversely, efficient excitation of a particular fiber mode requires that the incident field have a large spatial overlap with that mode, and beam shaping may therefore be required. For the important case of a weakly guiding step-index fiber with cylindrical symmetry, however, the fundamental fiber mode can be approximated quite accurately by a Gaussian beam.

The modes of fiber-optic waveguides are of great importance in telecommunication applications of laser light, but a fuller discussion requires considerably more waveguide theory than is necessary for our purposes. The same is true of other waveguides, such as slab and rectangular waveguides. Detailed discussions of the modes of different waveguides can be found in books and monographs devoted to this subject. We close this section with a few brief remarks about fiber-waveguide modes, sufficient to appreciate their similarities to and differences from the more familiar HG and LG modes of free space and open optical resonators.

The modes of a weakly guiding step-index fiber can be described approximately by linearly polarized LP modes. For a cylindrically symmetric fiber with core radius a, the transverse dependence of an $\mathrm{LP}_{\ell p}$ mode is obtained from the scalar Helmholtz equation.

Its radial dependence is oscillatory in the core and evanescent in the cladding and can be written in the form

$$LP_{\ell p}(\rho,\varphi) = A_{\ell p} e^{i\ell\varphi} \begin{cases} \dfrac{J_\ell\left(\nu_{\ell p}\rho/a\right)}{J_\ell\left(\nu_{\ell p}\right)} & \rho < a\,, \\ \dfrac{K_\ell\left(\mu_{\ell p}\rho/a\right)}{K_\ell\left(\mu_{\ell p}\right)} & \rho > a\,, \end{cases} \tag{3.141}$$

where $A_{\ell p}$ is a normalization constant, J_ℓ is the Bessel function of the first kind of order ℓ, and K_ℓ is the modified Bessel function of the second kind of order ℓ. The dimensionless quantities $\nu_{\ell p}$ and $\mu_{\ell p}$ characterize the transverse variation of the field in the core and cladding, respectively, and their allowed values are determined by the boundary conditions at $\rho = a$.

An important dimensionless fiber parameter is the normalized frequency, or V-number, defined by

$$V = \sqrt{\nu^2 + \mu^2} = \frac{2\pi a}{\lambda}\sqrt{n_{\text{core}}^2 - n_{\text{clad}}^2}\,. \tag{3.142}$$

Many fiber properties, including the number of guided modes, mode cutoff conditions, and propagation constants, can be expressed in terms of V. For example, for a step-index multimode fiber with $V \gg 1$, the approximate number of guided modes, including polarization degeneracy, is $V^2/2$. A weakly guiding step-index fiber is single-mode when $V < V_c = 2.405$.

The corresponding cutoff wavelength is

$$\lambda_c = \frac{2\pi a}{V_c}\sqrt{n_{\text{core}}^2 - n_{\text{clad}}^2}\,. \tag{3.143}$$

For wavelengths $\lambda > \lambda_c$, $V < V_c$, and only the fundamental LP_{01} mode is guided; the fiber is then single-mode. For wavelengths $\lambda < \lambda_c$, $V > V_c$, and higher-order modes can also become guided, so that the fiber becomes multimode.

The fundamental LP_{01} mode of a weakly guiding single-mode fiber is nearly linearly polarized, and its transverse spatial distribution, assuming polarization predominantly in the x-direction, can be approximated by a Gaussian,

$$E_x(x,y,z,\omega) = A_0 \exp[-(x^2+y^2)/w^2]e^{i\beta(\omega)z}\,, \tag{3.144}$$

with the spot radius approximately given by

$$\frac{w}{a} \approx 0.65 + 1.62V^{-3/2} + 2.879V^{-6}\,, \tag{3.145}$$

which is accurate to within about 1% over the range $1.2 < V < 2.4$ commonly encountered in single-mode fibers.

Summary

In this chapter we have considered the problem of finding paraxial beam solutions of the scalar wave equation in an unbounded region of space. We focused primarily on the Hermite-Gauss and Laguerre-Gauss families of beams because they maintain their transverse shape, apart from a scale change, as they propagate through space, and their phasefronts correspond to paraxial ray trajectories that can remain confined in open resonators formed by spherical mirrors. In the next chapter we will see that these beams are particularly well suited to solving the mixed boundary-value problem associated with the wave equation in open resonators with spherical mirrors and, thus, to determining the "modes" of open resonators that provide optical feedback in a laser.

In free space, Hermite-Gauss and Laguerre-Gauss beams provide important and widely used descriptions of laser beams. Their evolution during propagation is completely specified by the complex beam parameter $q(z)$, and their transformation by linear optical elements—interfaces, lenses, mirrors, etc.—can be described through the transformation of $q(z)$ by the corresponding ray matrices. We also discussed how these solutions of the scalar wave equation can be used to construct vector solutions of Maxwell's equations, as well as the conditions under which the vector character of the electromagnetic field becomes important.

Finally, we briefly considered guided-wave structures, including media with quadratic refractive-index profiles and step-index optical fibers. Such structures can provide optical confinement and feedback and are, in fact, used in fiber lasers. A great variety of specialized waveguide structures are employed in other lasers as well. Semiconductor lasers, for example, use ridge, channel, or slab waveguides to provide transverse optical confinement. A detailed discussion of these specialized waveguides lies outside the scope of this book and can be found in books and monographs devoted to guided-wave optics and semiconductor lasers.

Problems

1. Compare the irradiances at the retina when looking directly at the following sources:

   (a) The Sun. The Sun subtends an angle of 0.5 degrees. At the Earth's surface, the solar irradiance is 1 kW/m^2. Assume that the pupil of the bright-adapted eye is 2 mm in diameter and that the focal length of the eye is 22.5 mm.

   (b) A 1 mW He-Ne laser beam. Assume that the minimum spot radius of the laser beam is 1 mm and that the beam waist is located 1 m from the eye.

   (c) Which source presents the greater risk of eye damage? Eye-damaging irradiances are of the order of 10 μW/cm^2 or greater.

2. **Gaussian beam waist shifts**

   (a) A Gaussian beam traveling to the right has a waist at $z = 0$ with minimum spot radius w_0. A glass plate of thickness d and refractive index n is inserted in its path, as shown. Find the location of the new waist and the minimum spot radius there.

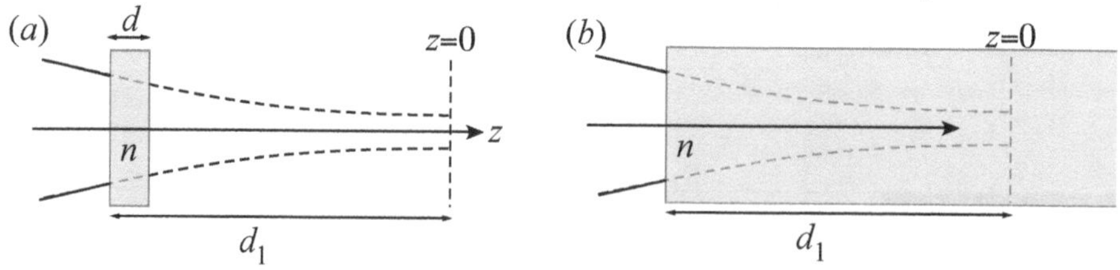

   (b) If the glass plate is now replaced by a thick slab (assumed to be semi-infinite) of refractive index n, find the location of the new waist and the minimum spot radius there.

3. **He-Ne laser beam:** A He-Ne laser operating in the fundamental Gaussian mode at $\lambda = 632.8$ nm generates a beam with an output power of $P = 5.0$ mW and a far-field

divergence half-angle of 1 mrad. Calculate the minimum spot radius, peak irradiance, and peak electric field amplitude at the waist.

4. **Argon-ion laser beam:** An argon-ion laser emits a fundamental Gaussian beam of power 1 W, wavelength $\lambda = 514.5$ nm, and minimum spot radius $w_0 = 2$ mm at its waist, $z = 0$.

   (a) How far from the waist will the beam propagate before its spot radius reaches $w = 1$ cm?

   (b) What is the radius of curvature of the phasefront at this distance?

   (c) What is the amplitude of the electric field at $\rho = 0$ and $z = 0$?

5. **Gaussian beam propagation:** In one of the Apollo space missions, astronauts placed laser retroreflectors on the Moon. A laser beam with spot radius w_0 is launched from the Earth toward the Moon. Use the propagation law for Gaussian beams to compute the diameter of the laser beam when it reaches the Moon. Take $\lambda = 694.3$ nm and the Earth-Moon distance to be 3.85×10^8 m. Consider two cases:

   (i) The laser beam has an initial diameter of 2.0 cm.

   (ii) The same laser beam is expanded by a telescope so that its initial diameter is 2.00 m.

   What are the far-field divergence half-angles in cases (i) and (ii)?

   Eye-damaging irradiances are of the order of $10\,\mu\text{W/cm}^2$ or greater. If the laser on Earth produces a pulse with a peak power of 10 MW, estimate the irradiance at the Moon in case (ii). Would this irradiance present a potential eye hazard to the astronauts?

6. **Projecting laser beams:** Suppose you have a laser that emits a Gaussian beam whose waist radius w_0 can be adjusted. You wish to use this beam to illuminate a target a distance L away with the minimum possible spot radius (or maximum possible irradiance) at the target, while keeping the waist at your location ($z = 0$). Given that the wavelength of the laser is λ, what value of w_0 would you choose? For this choice, what will be the spot radius at the target?

   As an application, consider a surveyor using a 633 nm laser to establish a baseline parallel to level ground and wishing to maximize the resolution at a distance of 100 m. How large should the beam radius at the surveyor's location be?

7. **Lasers to the Moon:** Suppose you want to measure the distance to the Moon by sending a laser pulse from the Earth and measuring the time required for the pulse to make a roundtrip to the Moon and back. Assume that the laser beam is Gaussian, has a wavelength of 500 nm and a waist radius of 1 m at the surface of the Earth. Suppose further that there is a flat mirror on the surface of the Moon large enough to reflect the entire beam back toward the Earth and oriented so that the reflected beam returns to a telescope on Earth whose aperture has a radius of 1.5 m.

   (a) What is the diameter of the beam when it reaches the Moon?

   (b) What fraction of the initially transmitted laser power is intercepted by the telescope aperture upon its return to Earth?

   (c) What beam diameter would be required on Earth so that the beam diameter increases by only a factor of $\sqrt{2}$ when it reaches the Moon? Note that telescopes with apertures of this size now exist on Earth.

Historical note: The Apollo astronauts placed arrays of retroreflecting cubes on the Moon's surface to enable such ranging experiments.

8. **Measuring the spot radius:** One method for measuring the spot radius of a Gaussian beam is to advance a knife edge into the beam with a micrometer and measure the power in the unblocked portion of the beam using a digital power meter placed downstream from the knife edge.

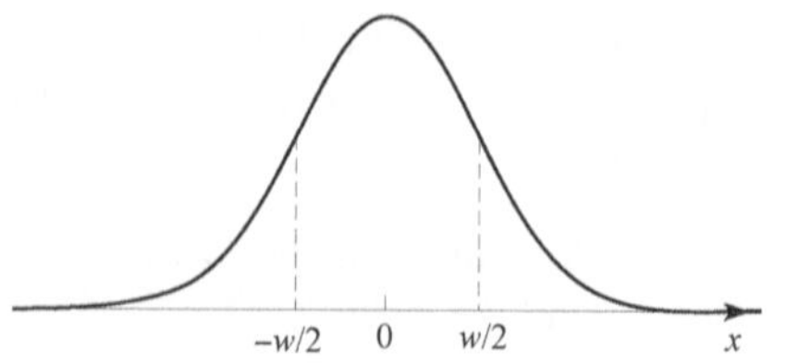

As the knife edge advances into (or retreats out of) the beam, the power meter records a decreasing (increasing) fraction of the total power. Suppose the knife edge advances through a distance equal to the spot radius w, with this displacement symmetric about the center of the beam. What fractions of the total power are recorded at the two knife-edge positions $x = -w/2$ and $x = w/2$?

9. **Optical data storage:** A common use of lasers is to read and write data on CDs and DVDs, where data are stored as changes in the surface reflectivity of the disk. These changes can be written and read using a laser beam focused onto the disk. Assume that the marks on a CD are $1.0\,\mu$m in diameter and that the Gaussian laser beam ($\lambda = 780$ nm) used to read them is focused so that its $1/e^2$ irradiance diameter equals the mark diameter.

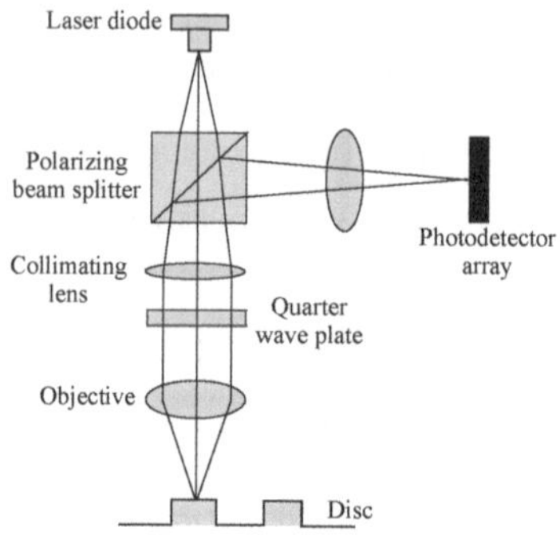

(a) Ideally, the beam waist lies on the surface of the disk. If the focusing lens moves away from its ideal position because, for example, the car in which the CD player is located encounters a bump, the spot radius on the disk will increase, degrading the ability to resolve the marks. If the acceptable tolerance is a factor of $\sqrt{2}$ increase in the spot radius, how far can the beam waist move from the disk surface? Maintaining such tight control over the lens position is one of the challenges in designing a robust CD player.

(b) If the lens is located 1.0 mm above the disk and its diameter is chosen so that it transmits 99% of the incident beam power, what must the lens diameter be?

10. **Gaussian beam waist location 1:** A Gaussian beam of wavelength $\lambda = 10.6\,\mu$m (emitted by a CO_2 laser) is found to have spot radii $w_1 = 1.699$ mm and $w_2 = 3.38$ mm at two locations P_1 and P_2, respectively, where P_2 is located a distance $d = 10$ cm downstream from P_1. Determine the possible locations of the beam waist relative to P_1 and P_2 and the corresponding minimum spot radii. Interpret the two solutions and illustrate them with a diagram. [**Hint:** You may find it convenient to work with the variable $u = \pi w^2/\lambda$. Thus, $u_o = \pi w_0^2/\lambda$, $u_1 = \pi w_1^2/\lambda$, etc.]

11. **Gaussian beam waist location 2:** The phasefront radii of curvature of a Gaussian beam at two points P_1 and P_2 on the same side of its waist are measured to be 1.0 m and 1.5 m, respectively. Point P_2 lies $d = 1.0$ m farther along the direction of beam propagation than P_1.

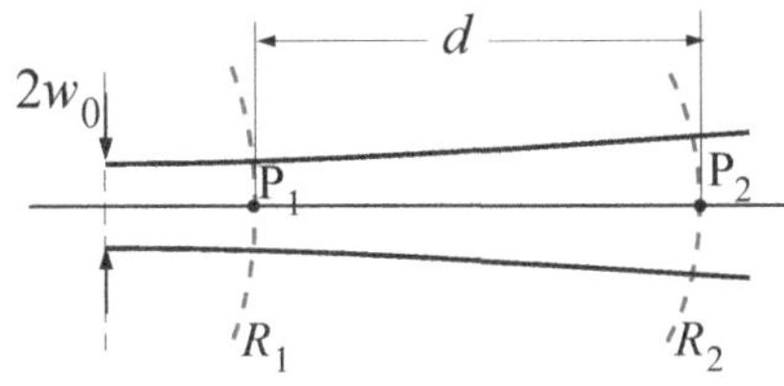

(a) Find the location of the beam waist relative to P_1 and P_2.

(b) What is the minimum spot radius of the beam if its wavelength is $1.0\,\mu$m?

(c) A 2.0 mm diameter circular aperture, centered on the beam axis, is placed at P_2. What fraction of the incident beam power is lost at the aperture?

12. A Gaussian beam with $w_0 = 1.0$ mm and $\lambda = 1.55\,\mu$m has its waist in the input plane of an optical system whose ray matrix is $\begin{pmatrix} 1 & 1 \\ -40 & 41 \end{pmatrix}$. Determine the location of the new waist of the beam emerging from the system relative to the output plane, and find the minimum spot radius there. Take $n = 1$ everywhere in this problem and all lengths to be in meters (m). Draw a neat diagram showing the location of the new waist relative to the optical system.

13. **Ar-ion beam focusing:** An argon-ion laser produces a Gaussian beam of wavelength $\lambda = 488$ nm and waist radius $w_0 = 0.50$ mm. Design a single-lens optical system that focuses the beam to a waist diameter of $100\,\mu$m. Specify a suitable lens focal length and its position relative to the initial laser beam waist. What is the shortest focal-length lens that can be used?

14. **Imaging a Gaussian beam:** A Gaussian laser beam of wavelength $\lambda = 532$ nm has a waist with minimum spot radius $50\,\mu$m in the plane $z = 0$. A thin lens of focal length 25 mm is placed 35 mm from the waist. Find the location of the new waist and its minimum spot radius. Compare the waist location with that predicted by geometrical optics for (i) a collimated input beam and (ii) an input beam diverging from a point focus at $z = 0$.

15. A lens of focal length $f > 0$ is placed at a distance d from the waist of a fundamental Gaussian beam of waist radius w_0. Find an expression for f in terms of d and w_0 such that the beam immediately after leaving the lens has a planar phasefront.

16. A thin positive lens is placed at the waist of a Gaussian beam of waist radius w_0. Show that the new waist formed by the lens cannot be located farther than $z_R/2$ from the lens, where $z_R = \pi w_0^2/\lambda$. What lens focal length corresponds to this maximum focusing distance?

17. **Laser beam expander:** A fundamental Gaussian laser beam with far-field divergence θ_1 is incident on a combination of two positive thin lenses of focal lengths f_1 and f_2, separated by a distance equal to the sum of their focal lengths, as shown in the figure.

(a) Starting with a reference plane just in front of the first lens f_1 and ending with a reference plane just after the second lens f_2, compute the ray-transfer matrix of the system by expressing it as the product of the ray-transfer matrices of the individual optical elements.

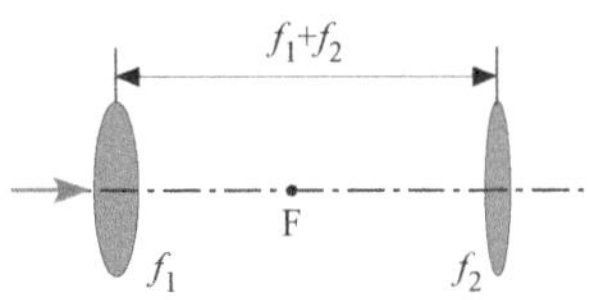

(b) Find the far-field divergence θ_2 of the beam emerging from the device and express it in terms of θ_1 and the two focal lengths. If $f_1 = 15$ mm and the desired outgoing beam divergence is one-third that of the incident beam, what value of f_2 should be chosen? Also, calculate the output waist radius w_{02} for $w_{01} = 2.00\,\text{mm}$, $f_1 = 50$ mm, and $f_2 = 20$ mm.

18. A Gaussian beam with far-field divergence θ_1 is incident on a telescope consisting of a pair of positive thin lenses separated by a distance equal to the sum of their focal lengths, as in the preceding problem. Find the far-field divergence of the beam emerging from the telescope. If you wish to expand a Gaussian beam with waist radius 1.0 mm to one with waist radius 5.0 mm, what ratio of lens focal lengths is required? Give one suitable pair of focal lengths and explain your choice.

19. A Gaussian beam is incident normally on a transparent plate of refractive index n and thickness d, surrounded by a medium of refractive index unity. If w_1 and w_2 denote the waist radii of the incident and transmitted beams, respectively, show that the transmitted beam has the same waist radius as the incident beam,

$$w_2 = w_1\,,$$

but that its waist is shifted in the direction of propagation by

$$\Delta z = \left(1 - \frac{1}{n}\right) d\,.$$

20. **Graded media:** Show that a medium with a quadratically decreasing refractive-index profile $n^2(\rho) = n_1^2\left(1 - \frac{\rho^2}{\Lambda^2}\right)$ supports monochromatic guided-wave solutions in cylindrical coordinates of the form

$$\mathcal{E}_{p\ell}(\boldsymbol{r}) = \mathcal{A}_{p\ell}\left(\frac{\sqrt{2}\,\rho}{w}\right)^{|\ell|} L_p^{|\ell|}\left(\frac{2\rho^2}{w^2}\right) e^{i\ell\varphi} e^{-\rho^2/w^2} e^{i\beta_{p\ell} z}\,,$$

$$\beta_{p\ell} = k\left[1 - \frac{4}{k^2 w^2}(2p + |\ell| + 1)\right]^{1/2}, \quad w = \sqrt{2\Lambda/k}, \quad k = \frac{\omega n_1}{c}\,.$$

21. **Paraxial beams in graded media:** Suppose we wish to find paraxial-beam solutions in a medium whose refractive index varies with radial distance from an axis according to $n^2(\rho) = n_1^2\left(1 - \frac{\rho^2}{\Lambda^2}\right)$, where $\rho^2/\Lambda^2 \ll 1$ over the region of interest. The refractive-index variation can then be written approximately as $n(\rho) \approx n_1\left(1 - \frac{\rho^2}{2\Lambda^2}\right)$.

Taking a cue from free-space propagation, suppose that this medium supports Hermite-Gauss modes characterized by a complex beam parameter q. The parameter q then transforms according to the ABCD law

$$q(z) = \frac{Aq(0) + B}{Cq(0) + D}\,,$$

where the transfer matrix for propagation through a distance z is

$$\begin{pmatrix} A & B \\ C & D \end{pmatrix} = \begin{pmatrix} \cos(z/\Lambda) & \Lambda\sin(z/\Lambda) \\ -\frac{1}{\Lambda}\sin(z/\Lambda) & \cos(z/\Lambda) \end{pmatrix}.$$

For the postulated beams to be normal modes of the medium, this relation must hold for an arbitrary choice of the origin $z = 0$. Consequently, q must be independent of z and therefore satisfy $q(z) = q(0) = q_0$. Find q_0 in terms of the parameters characterizing the refractive-index profile and hence determine the form of the fundamental beam. Compare your result with the fundamental beam ($m = 0 = n$) obtained from Eq. (3.139) or from the preceding problem.

22. **Retro-reflection of a Gaussian beam:** A Gaussian laser beam of wavelength $\lambda = 670$ nm has a waist radius of 200 μm. A spherical mirror is placed 10 cm from the waist. What radius of curvature must the mirror have so that, after reflection, the Gaussian beam retraces its propagation and refocuses at the original waist location? Compare the required mirror radius of curvature with the radius of curvature of the incident Gaussian-beam phasefront at the mirror.

23. **Determination of beam divergence:** An arrangement for measuring the divergence of a Gaussian beam uses a beam splitter in the beam path, as shown in the sketch. The reflected beam is focused by a thin lens of focal length f. Show that, independent of the position of the lens relative to the incident beam waist, the far-field divergence of the incident beam is given by

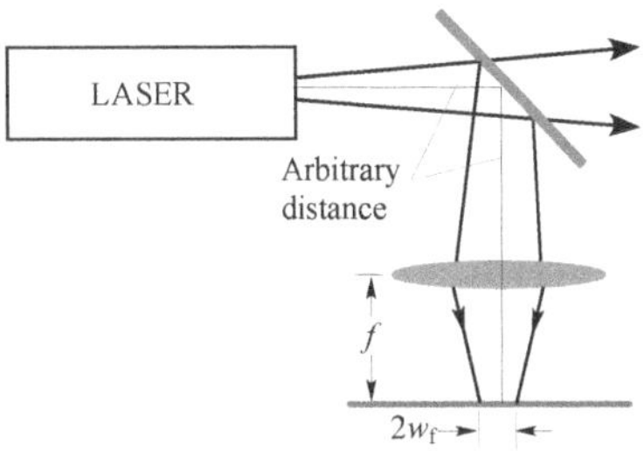

$$\theta = \frac{w_f}{f},$$

where w_f is the beam spot radius in the back focal plane of the lens.

24. A Gaussian beam is incident on a thin lens of focal length f with its waist located in the lens plane. Find the beam spot radius in the back focal plane of the lens. Note that the focal plane is not, in general, the waist plane of the beam emerging from the lens. Without any further calculation, how can you determine whether the new waist lies to the left or to the right of the focal plane?

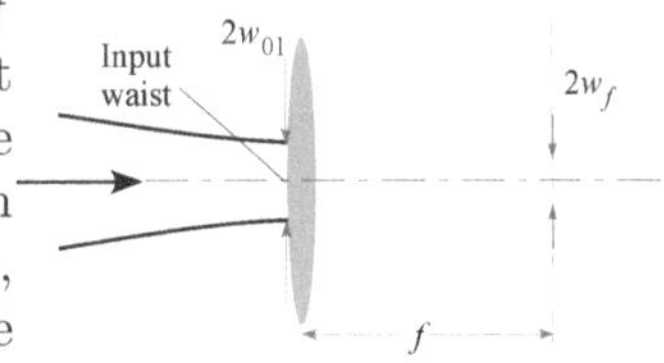

25. A diode-pumped solid-state laser produces a fundamental Gaussian beam of wavelength $\lambda = 530$ nm and waist radius 225 μm. The beam is incident on a thin lens of focal length $f = 15.0$ cm with its waist located in the lens plane, as in the figure in the preceding problem.

    (a) Write the ray-transfer matrix for propagation from the lens plane to its back focal plane as the product of the appropriate component matrices, and evaluate the product.

    (b) What is the complex beam parameter q_1 in the lens plane? Find the complex beam parameter q_2 in the back focal plane of the lens.

    (c) Find the location of the new beam waist after the lens and the minimum spot radius there. Demonstrate your understanding by indicating the location of the new waist on a diagram.

26. **Optimum focusing of Gaussian laser beams:** A laser-marking system uses an Ar-ion laser operating at 488.0 nm to write serial numbers on electronic components. The focused laser beam modifies the surface of the component, producing lines whose widths are approximately equal to the beam spot diameter $2w$ at the target. Design an optical system like that shown in the figure by choosing d_1, d_2, and the focal length f of the lens so that the line width is less than 0.03 cm. Assume that the laser produces a Gaussian

beam with a waist at the exit mirror M_2 and a waist radius of $430\,\mu$m. The solution is not unique; find at least one combination of d_1, d_2, and f that satisfies the design requirement.

[Ref: Y. Li and J. Katz, "Optimum focusing of Gaussian laser beams: beam waist shift in spot size minimization," Optical Engineering **33**, 1152–1154 (1994).]

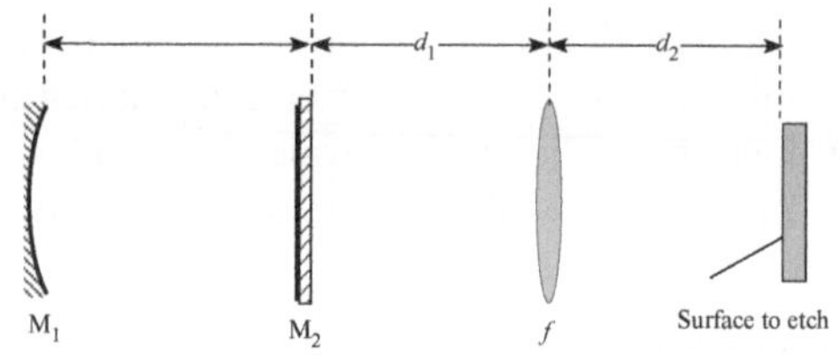

27. The spatially coherent beam from a diode laser has an elliptical spot at the exit face, with waist radii parallel and perpendicular to the junction given by $w_{o\|} = 2.5\,\mu$m and $w_{o\perp} = 0.5\,\mu$m, respectively. Assuming that the exit face is a waist plane in both transverse directions, derive an expression for the distance from the exit face at which the beam spot becomes circular. Evaluate this distance for $\lambda = 850$ nm.

4

Laser Resonators

In Chapter 3, we considered several examples of paraxial beams, most of them Gaussian. We now examine how Gaussian beams with suitably chosen parameters can be used to determine the modes of stable open resonators. Indeed, lasers incorporating such resonators naturally generate Gaussian beam modes. To determine these modes, we require that, after propagation through the optical elements during one cavity roundtrip, the beam recover its original Gaussian form. We first consider a two-mirror standing-wave resonator and then generalize the analysis to an arbitrary stable resonator. By further requiring that the beam phase remain unchanged modulo 2π after a roundtrip—a condition that can be met only for certain values of the wave number, or equivalently the frequency, since $k = 2\pi n\nu/c$—we obtain the Gaussian beam modes of stable resonators.

This analysis yields the location of the beam waist, the minimum spot radius, and the spectrum of allowed frequencies in terms of cavity parameters such as the mirror radii of curvature, the mirror separation, and the angles of incidence. These are the long-lived, preferred field modes of open cavities that provide the optical feedback required for laser operation, as discussed in Sec. 2.5.

These modes do not yet include the effects of cavity losses arising from absorption, diffraction, scattering, and finite mirror transmission. Such losses are an integral part of laser cavities. We first discuss how cavity losses are characterized and then examine their effects on the field-mode spectrum. In doing so, we introduce the concepts of cavity finesse, lifetime–the characteristic time over which a cavity stores electromagnetic energy–and linewidth, and show how these quantities are determined by loss. Cavity losses are not entirely detrimental. In fact, small losses, arising mainly from finite mirror transmission, are essential to many useful applications of optical cavities. Just as electromagnetic energy can leak out of a cavity through a partially transmitting mirror, it can also be coupled into the cavity through that mirror. We conclude the chapter by discussing applications of low-loss optical cavities for spectral analysis and for building up large electromagnetic fields inside the cavity.

4.1 Gaussian Beam Modes of Open Optical Cavities

Consider the reflection of a Gaussian beam of complex beam parameter q_i (just before the mirror) incident from the left on a mirror of radius of curvature R as in Fig. 4.1. The complex beam parameter q_o of the outgoing (reflected) beam is given by

$$q_o = \frac{q_i\,A + B}{q_i\,C + D}\,, \tag{4.1}$$

where $ABCD$ are the elements of the ray transfer matrix of the mirror

$$\begin{pmatrix} A & B \\ C & D \end{pmatrix} = \begin{pmatrix} 1 & 0 \\ -\frac{2}{R} & 1 \end{pmatrix}, \tag{4.2}$$

DOI: 10.1201/9781003584179-4

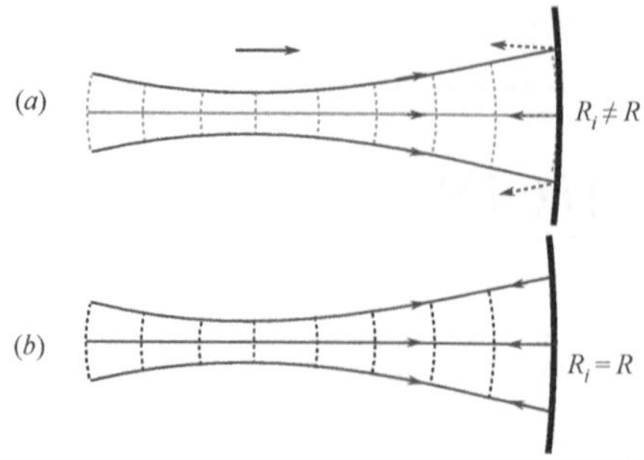

FIGURE 4.1
Reflection of a Gaussian beam by a mirror whose radius of curvature (a) differs from the radius of curvature of the incident phasefront and (b) matches the radius of curvature of the incident phasefront. In (b), the reflected beam retraces its path since, by definition, the rays are perpendicular to the phasefront, and normally incident rays are reflected back on themselves.

and R is the radius of curvature of the mirror (+ for concave and − for convex mirrors). Using this ray transfer matrix in Eq. (4.1), we find that the inverse of the complex beam parameter for the outgoing beam is

$$\frac{1}{q_o} = \frac{Cq_i + D}{Aq_i + B} = \frac{C}{A} + \frac{1}{q_i}\frac{D}{A} = -\frac{2}{R} + \frac{1}{q_i} . \tag{4.3}$$

Separating the real and imaginary parts of $1/q$ in the form $\frac{1}{q(z)} \equiv \frac{1}{R(z)} + i\frac{2}{kw^2}$, we find that the spot radius and radius of curvature of the outgoing beam are given by

$$w_0 = w_i \tag{4.4}$$

$$\frac{1}{R_o} = \frac{1}{R_i} - \frac{2}{R} . \tag{4.5}$$

Thus, the spot radius remains unchanged, but the radius of curvature of the phasefront changes according to Eq. (4.5). For a flat mirror ($R \to \infty$), the radius of curvature of the reflected wave also remains unchanged in magnitude, with $R_o = R_i$ according to the ray-matrix convention used here. The reflected wave, now propagating to the left, has the corresponding phasefront curvature appropriate to its reversed direction of propagation.

Now suppose that the mirror radius of curvature R matches the radius of curvature R_i of the incident phasefront, so that $R_i = R$ and the mirror surface coincides with the incident phasefront. Then the radius of curvature of the reflected wave is

$$R_o = -R_i . \tag{4.6}$$

Thus, the reflected phasefront has the same magnitude of curvature as the incident phasefront but the opposite sign. This means that if the incident wave traveling to the right is diverging, the reflected wave traveling to the left is converging and retraces the path of the incident wave. If a second mirror is now placed in the path of the reflected wave traveling to the left, such that the radius of curvature of the mirror matches the radius of curvature of the phasefront at the location of the second mirror, the beam is reflected to the right and again retraces its path. Thus, when two mirrors are positioned such that their reflecting surfaces coincide with phasefronts of the Gaussian beam, the beam is trapped between the mirrors, bouncing back and forth between them.

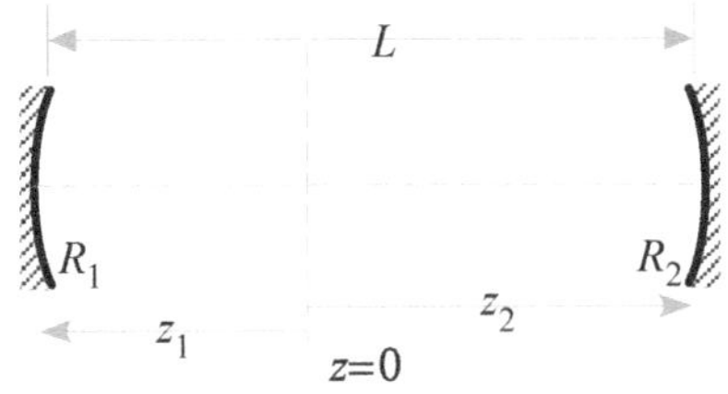

FIGURE 4.2
Relative to the beam waist ($z = 0$), the mirrors in a two-mirror cavity are located at $z = z_1$ and $z = z_2$.

The conditions for Gaussian beam trapping are easily written down. Suppose one mirror, with radius of curvature R_1, is placed at position z_1 and the other, with radius of curvature R_2, is placed at z_2, relative to the beam waist [see Fig. 4.2]. Trapping of a Gaussian beam then requires that the wavefront curvature at each mirror match the curvature of the corresponding mirror:

$$R(z_1) \equiv z_1 + \frac{z_R^2}{z_1} = -R_1, \tag{4.7a}$$

$$R(z_2) \equiv z_2 + \frac{z_R^2}{z_2} = R_2, \tag{4.7b}$$

where $z_R = \frac{1}{2}kw_0^2 = \pi n w_0^2/\lambda$ is the Rayleigh range of the beam. Note that when these conditions hold at the end mirrors, the reflected beam retraces the path of the incident beam in the opposite direction. The negative sign in the first equation arises because the wavefront radius of curvature is taken to be positive for a diverging beam and negative for a converging beam traveling to the right, whereas the radius of curvature of a mirror is taken to be positive for a concave mirror and negative for a convex mirror. For a Gaussian beam with minimum spot radius w_0, Eq. (4.7) determines the radii of curvature and the positions, relative to the waist, of the mirrors that will trap the beam.

In practice, the inverse of the problem we have just considered is of greater interest. Namely, given the radii of curvature R_1 and R_2 of the mirrors and their separation L, we wish to determine the minimum spot radius w_0 of the beam trapped between the mirrors and the location of the beam waist relative to M_1 (or M_2) in terms of R_1, R_2, and L. Whether or not a trapped Gaussian beam exists for given R_1, R_2, and L should, of course, also emerge from this analysis. Assuming that a trapped Gaussian beam exists, let us denote the locations of the mirrors M_1 and M_2 relative to the beam waist by z_1 and z_2, respectively. Then, solving Eqs. (4.7) subject to the constraint

$$z_2 - z_1 = L\,, \tag{4.8}$$

we find z_1, z_2, and z_R in terms of R_1, R_2, and L.

The resulting expressions for z_1, z_2, z_R, and w_0 can be written in a compact form by introducing the "g-factors"

$$g_1 = 1 - \frac{L}{R_1} \quad \text{and} \quad g_2 = 1 - \frac{L}{R_2}\,. \tag{4.9}$$

The resulting equations for z_1, z_2, z_R, and w_0 are then

$$z_1 = -\frac{g_2(1-g_1)}{g_1+g_2-2g_1g_2}L\,, \tag{4.10a}$$

$$z_2 = \frac{g_1(1-g_2)}{g_1+g_2-2g_1g_2}L\,, \tag{4.10b}$$

$$z_R = \left[\frac{g_1g_2(1-g_1g_2)}{(g_1+g_2-2g_1g_2)^2}\right]^{1/2} L\,, \tag{4.10c}$$

$$w_0 = \sqrt{\frac{L\lambda}{n\pi}}\left[\frac{g_1g_2(1-g_1g_2)}{(g_1+g_2-2g_1g_2)^2}\right]^{1/4} \tag{4.10d}$$

Using the propagation law for the spot radius of a Gaussian beam, we find that the spot radii at the end mirrors are given by

$$w_1 = w_0\sqrt{1+\frac{z_1^2}{z_R^2}} = \sqrt{\frac{L\lambda}{n\pi}}\left[\frac{g_2}{g_1(1-g_1g_2)}\right]^{1/4}, \tag{4.10e}$$

$$w_2 = w_0\sqrt{1+\frac{z_2^2}{z_R^2}} = \sqrt{\frac{L\lambda}{n\pi}}\left[\frac{g_1}{g_2(1-g_1g_2)}\right]^{1/4}. \tag{4.10f}$$

Inspection of Eqs. (4.10c)–(4.10f) shows that, for real and positive Gaussian beam parameters, g_1 and g_2 must satisfy the condition

$$0 < g_1g_2 < 1\,. \tag{2.164*}$$

We obtained this condition in Chapter 2 from the requirement of paraxial ray confinement. Resonators with R_1, R_2, and L satisfying Eq. (2.164) thus admit Gaussian beam solutions that satisfy the mirror boundary conditions and correspond to confined paraxial rays. Such resonators are said to be stable. Two-mirror resonators that do not satisfy this condition (2.164) are said to be unstable as they do not have stationary Gaussian beam solutions.

We have found that stable Gaussian beams can bounce back and forth between two spherical mirrors facing each other, provided the mirror radii of curvature and their separation satisfy Eq. (2.164). Resonators with three or more mirrors, such as those shown in Fig. 4.3, can also support stable Gaussian beams that travel along closed trajectories without necessarily retracing their paths. A general method for determining the Gaussian beam parameters for an arbitrary resonator is considered next.

4.1.1 Gaussian Beam in an Arbitrary Cavity

Consider an arbitrary cavity with a roundtrip matrix $M = \begin{pmatrix} A & B \\ C & D \end{pmatrix}$. Recall that the matrix elements A, B, C, and D depend on the choice of reference plane and are functions of the cavity parameters. It is important first to establish that the cavity is stable using the criterion $-1 < (A+D)/2 < 1$. If the cavity is stable, the complex beam parameter q of the confined Gaussian beam must be reproduced after one roundtrip; that is, the complex beam parameter must transform into itself:

$$q = \frac{Aq+B}{Cq+D} \Rightarrow q^2C - q(A-D) - B = 0\,. \tag{4.11}$$

This quadratic equation in q can be solved to yield

$$q = \frac{A-D}{2C} \pm \frac{1}{2C}\sqrt{(A-D)^2+4BC}\,. \tag{4.12}$$

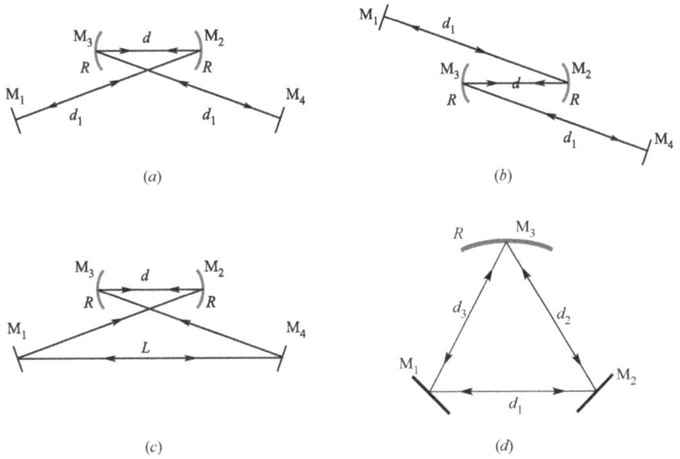

FIGURE 4.3
A few standing-wave [(a), (b)] and traveling-wave [(c), (d)] resonator configurations used in lasers. Unlike the resonators shown here, actual resonators need not be symmetric.

Using the result $AD - BC \equiv \det M = 1$, which holds for the roundtrip matrix of a cavity, we can write the radicand as $(A - D)^2 + 4(AD - 1) = (A + D)^2 - 4$. The result for q then becomes

$$q = \frac{A-D}{2C} \pm \frac{i}{2C}\sqrt{4-(A+D)^2}\,. \tag{4.13}$$

Recalling the definition of the complex beam parameter q,

$$q(z) = z - iz_R\,, \tag{4.14}$$

where z is the location of the reference plane relative to the waist (assumed to be at $z = 0$) and the Rayleigh range z_R, as usual, is related to the minimum spot radius w_0 by $z_R = \pi n w_0^2/\lambda$. The coordinate z is positive if the reference plane is downstream of the waist and negative if it is upstream. By equating the real and imaginary parts of the two sides of Eq. (4.13), we find that the distance z of the reference plane from the waist and the minimum spot radius w_0 are given in terms of the roundtrip matrix elements by

$$z = \frac{A-D}{2C}\,, \tag{4.15a}$$

$$w_0^2 = \frac{\lambda}{n\pi}\frac{\sqrt{4-(A+D)^2}}{2|C|}\,. \tag{4.15b}$$

For a physical confined Gaussian beam, the quantity w_0^2 in Eq. (4.15b) must be real and positive, requiring the stability condition $-1 < \frac{1}{2}(A + D) < 1$.

For a two-mirror cavity (two mirrors with radii of curvature R_1 and R_2 separated by a distance L in Fig. 2.19), with the reference plane just in front of mirror M_1 and the initial direction of propagation toward M_2, the roundtrip matrix is

$$\begin{bmatrix} A & B \\ C & D \end{bmatrix} = \begin{bmatrix} 2g_2 - 1 & 2Lg_2 \\ -\frac{2g_1}{R_2} - \frac{2g_2}{R_1} & 4g_1g_2 - 2g_2 - 1 \end{bmatrix}, \quad g_i \equiv 1 - \frac{L}{R_i}\,. \tag{2.136*}$$

From this matrix we find

$$\frac{1}{2}(A-D) = 2g_2(1-g_1)\,, \quad \frac{1}{2}(A+D) = 2g_1g_2 - 1\,. \tag{4.16}$$

Using these results in Eqs. (4.15), we find that the distance z_1 of the reference plane (which coincides with the location of mirror M_1) from the waist and the minimum spot radius w_0 are given by

$$z_1 = -\frac{g_2(1-g_1)}{g_1+g_2-2g_1g_2}L\,, \tag{4.10a*}$$

$$w_0 = \sqrt{\frac{L\lambda}{n\pi}}\left[\frac{g_1g_2(1-g_1g_2)}{(g_1+g_2-2g_1g_2)^2}\right]^{1/4}. \tag{4.10d*}$$

These expressions for z_1 and w_0 are the same as those found earlier [Eqs. (4.10a) and (4.10d)]. From the reality of beam spot radius w_0, we again recover the two-mirror stability condition $0 < g_1g_2 < 1$ derived earlier.

If the reference plane happens to coincide with the beam waist, its distance from the waist is $z = 0$. The expression for z in Eq. (4.15) then shows that the roundtrip matrix must satisfy $A = D$. At the beam waist, the phasefront is planar, $R \to \infty$, consistent with this same condition. With $A = D$, the spot radius at the waist is given by

$$w_0^2 = \frac{\lambda}{n\pi}\frac{\sqrt{1-A^2}}{|C|} = \frac{\lambda}{n\pi}\frac{\sqrt{1-D^2}}{|C|}\,, \qquad -1 < A(=D) < 1\,. \tag{4.17}$$

Another way of thinking about this problem is in terms of $\frac{1}{q} \equiv \frac{1}{R} + i\frac{2}{kw^2}$, where $R(z)$ and $w(z)$ are, respectively, the phasefront radius of curvature and beam spot radius at z. Beam waist corresponds to planar phasefronts, $R \to \infty$. In this case, we solve Eq. (4.13) for $1/q$ and write its real and imaginary parts in terms of the matrix elements A, B, C, and D.

Example 4.1 *A pair of mirrors with a nominal radius of 200 mm have actual radii of 203 mm and 197 mm. A problem will arise if these mirrors are used in a "confocal" cavity with a mirror separation of 200 mm. What is the problem, and what are the possible mirror spacings that can address it?*

Let us denote the error (3 mm) by ΔR. Then, for $L = R$, $g_1 = 1 - \frac{R}{R+\Delta R} > 0$, $g_2 = 1 - \frac{R}{R-\Delta R} < 0$, so that $g_1g_2 < 0$ and the cavity is unstable. For stability, we require

$$g_1g_2 = \left(1 - \frac{L}{R+\Delta R}\right)\left(1 - \frac{L}{R-\Delta R}\right) > 0$$
$$\Longrightarrow L^2 - 2RL + (R+\Delta R)(R-\Delta R) = (L-R)^2 - (\Delta R)^2 > 0\,,$$

which is satisfied for $L < R - \Delta R$ or $L > R + \Delta R$. The additional stability requirement $g_1g_2 < 1$ gives $L < 2R$. Hence, for $R = 200$ mm and $\Delta R = 3$ mm, the stable ranges are $L < 197$ mm or $203 \text{ mm} < L < 400$ mm. Therefore, for a small adjustment about the nominal 200-mm spacing, the mirrors must be moved more than 3 mm closer together or more than 3 mm farther apart. The spacings $L = 197$ mm and $L = 203$ mm themselves correspond to $g_1g_2 = 0$ and lie on the boundary of the stability region.

4.2 Mode Frequency Spectrum

So far, we have focused on the transverse spatial structure of paraxial beams and its dependence on resonator characteristics such as the geometric layout, mirror radii of curvature, and mirror separation. We have seen that a stable resonator can support a Gaussian beam that remains confined as it propagates around the resonator. For a stable standing-wave resonator, this means that the radii of curvature of the Gaussian beam phasefronts at the end mirrors coincide in magnitude with the radii of curvature of the mirrors. Thus, a right-going Gaussian beam reflected from the right end mirror M_2 remains a Gaussian beam as it travels to the left, and a left-going Gaussian beam reflected from the left end mirror M_1 remains a Gaussian beam as it travels to the right.

Between the mirrors, therefore, we have both right- and left-going waves, and the field at any point is a superposition of the two. For these Gaussian beams to form a stationary standing-wave pattern, the phase change of the right-going (or left-going) Gaussian beam in one roundtrip must equal an integer multiple of 2π, or, equivalently, the one-way phase change must equal an integer multiple of π. For Hermite-Gauss beams,

$$E_{\ell m}(\boldsymbol{r},t) = A_{\ell m}\frac{w_0}{w}H_\ell\left(\frac{\sqrt{2}x}{w}\right)H_m\left(\frac{\sqrt{2}y}{w}\right)e^{-\rho^2/w^2-i\omega t}e^{i\Phi_{\ell m}(z)}\,, \tag{4.18a}$$

where

$$w(z) = w_0\sqrt{1+z^2/z_R^2}, \tag{4.18b}$$

$$R(z) = z + z_R^2/z, \tag{4.18c}$$

$$z_R = \pi n w_0^2/\lambda, \tag{4.18d}$$

$$\Phi_{\ell m}(z) = kz + k\rho^2/2R(z) - \psi_{\ell m}(z), \tag{4.18e}$$

$$\psi_{\ell m}(z) = (\ell+m+1)\tan^{-1}(z/z_R)\,. \tag{4.18f}$$

This condition becomes

$$k(z_2-z_1) - (\ell+m+1)\left(\tan^{-1}\frac{z_2}{z_R} - \tan^{-1}\frac{z_1}{z_R}\right) + k\rho^2\left(\frac{1}{R_2}+\frac{1}{R_1}\right) = p\pi\,, \tag{4.19}$$

where we have used $R(z_1) = -R_1$ and $R(z_2) = R_2$ in the last term. In writing this equation, we have ignored phase shifts that might arise upon reflection. These can easily be included, if desired, and do not change the main results of this section. The term involving $k\rho^2$ is negligible compared with $k(z_2-z_1) = kL$, since the range of ρ over which the field is appreciably different from zero extends over only a few spot radii, and the beam spot radius is small compared with the characteristic dimensions R and L:

$$\frac{1}{kL}\left[\frac{k\rho^2(R_1+R_2)}{R_1R_2}\right] \approx \frac{w^2}{RL} \ll 1\,. \tag{4.20}$$

Neglecting the ρ^2 term, Eq. (4.19) reduces to

$$k(z_2-z_1) - (\ell+m+1)\left(\tan^{-1}\frac{z_2}{z_R} - \tan^{-1}\frac{z_1}{z_R}\right) = p\pi\,, \quad p \text{ an integer.} \tag{4.21}$$

When this condition is satisfied, the right- and left-going waves form a stationary standing-wave pattern whose transverse field distribution is that of the corresponding traveling Gaussian beam. This standing-wave pattern is referred to as a mode of the resonator, and Eq. (4.21) is referred to as the mode condition or resonance condition.

The first term, $k(z_2-z_1)$, in Eq. (4.21) represents the phase change that a plane wave accumulates in propagating from the plane z_1 to the plane z_2. The second term represents the additional phase change arising from the Gouy phase and its dependence on the transverse mode indices. For example, a Gaussian beam with its waist at $z = 0$, propagating from $z = -z_R$ to $z = z_R$, accumulates a phase

$$\begin{aligned}\Delta\Phi_{\ell m} &= k[z_R - (-z_R)] - (\ell + m + 1)\left[\tan^{-1}\left(\frac{z_R}{z_R}\right) - \tan^{-1}\left(\frac{-z_R}{z_R}\right)\right] \\ &= 2kz_R - (\ell + m + 1)\frac{\pi}{2}\,. \end{aligned} \tag{4.22}$$

In the general case, the $\tan^{-1}$ terms in Eq. (4.21) can be combined to give

$$\tan^{-1}\frac{z_2}{z_R} - \tan^{-1}\frac{z_1}{z_R} = \tan^{-1}\frac{(z_2 - z_1)z_R}{z_R^2 + z_1 z_2} = \tan^{-1}\frac{Lz_R}{z_R^2 + z_1 z_2}\,. \tag{4.23}$$

For a two-mirror cavity, this can be written in a more compact form by using the expressions for z_1, z_2, and zR given in Eqs. (4.10a)-(4.10d). This leads to

$$\tan^{-1}\frac{z_2}{z_R} - \tan^{-1}\frac{z_1}{z_R} = \tan^{-1}\sqrt{\frac{1 - g_1 g_2}{g_1 g_2}} = \cos^{-1}\pm\sqrt{g_1 g_2}\,, \tag{4.24}$$

where the + sign applies when both $g_1, g_2 > 0$ and the − sign applies when both $g_1, g_2 < 0$. Hence, for a two-mirror resonator, the resonance condition

$$kL - (\ell + m + 1)\cos^{-1}\pm\sqrt{g_1 g_2} = p\pi\,. \tag{4.25}$$

This equation determines k, which now carries three indices

$$k_{\ell mp} = \frac{\pi}{L}\left[p + (\ell + m + 1)\frac{\cos^{-1}\pm\sqrt{g_1 g_2}}{\pi}\right]\,. \tag{4.26}$$

It is clear from this equation that, unlike propagation in free space, only certain values of k are compatible with the cavity mode condition. Since the wavenumber and frequency of a wave are related by $k = n\omega/c$, it follows that only certain frequencies are compatible with the mode condition (4.21). These frequencies, referred to as the cavity mode frequencies or eigenfrequencies, are given by

$$\nu_{\ell mp} \equiv \frac{k_{\ell mp}c}{2\pi n} = \frac{c}{2nL}\left[p + (\ell + m + 1)\frac{\cos^{-1}\pm\sqrt{g_1 g_2}}{\pi}\right]\,. \tag{4.27}$$

The mode indices ℓ and m characterize the transverse spatial dependence of the field through the Hermite polynomials and are referred to as the transverse mode indices. They give the numbers of intensity zeros along the two transverse directions. The index p, which is related to the number of half wavelengths that fit along the cavity axis, is referred to as the axial or longitudinal mode index. From the expression for the mode frequencies, we see that associated with each axial mode are transverse modes with the same value of p and different values of ℓ and m. Consecutive axial and transverse modes are separated in frequency by, respectively,

$$\Delta\nu_{ax} \equiv \nu_{00p+1} - \nu_{00p} = \frac{c}{2nL}, \tag{4.28}$$

$$\Delta\nu_{tr} \equiv \nu_{(\ell+1)mp} - \nu_{\ell mp} = \nu_{\ell(m+1)p} - \nu_{\ell mp} = \frac{c}{2nL}\left[\frac{\cos^{-1}(\pm\sqrt{g_1 g_2})}{\pi}\right]. \tag{4.29}$$

The resonator mode functions can then be written as ($\omega_{\ell mp} = 2\pi\nu_{\ell mp}$)

$$E_{\ell mp} = \mathcal{A}_{\ell mp}\frac{w_0}{w(z)}H_\ell\left(\frac{\sqrt{2}x}{w(z)}\right)H_m\left(\frac{\sqrt{2}y}{w(z)}\right)e^{-\frac{\rho^2}{w^2(z)} - i\omega_{\ell mp}t}\sin[\Phi_{\ell mp}(z)], \tag{4.30}$$

where $$\Phi_{\ell mp}(z) = \left[k_{\ell mp}\left(z + \rho^2/2R(z)\right) - (\ell + m + 1)\psi(z)\right]\,. \tag{4.31}$$

Depending on the boundary conditions and the phase shifts upon reflection, a cosine function instead of the sine function may be appropriate in this expression.

Example 4.2 *A laser operating at 1.06 μm uses a resonator formed by two mirrors with radii of curvature $R_1 = 20.0$ cm and $R_2 = -40.0$ cm, separated by $L = 10.0$ cm. Find the location of the beam waist, the minimum spot radius, and the spot radii at the mirrors. What are the axial and transverse mode frequency separations?*

The inequality $0 < g_1 g_2 = \frac{5}{8} < 1$ is satisfied, so the resonator is stable. Matching the phasefront radii of curvature to those of the mirrors and using $z_2 - z_1 = 10$, or $z_1 = z_2 - 10$ (all distances in cm), we obtain

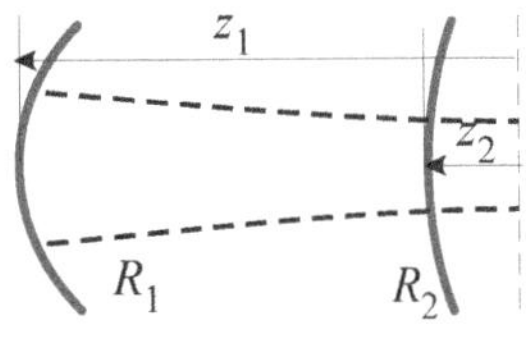

$$\text{Mirror 1:} \quad -20 = z_1 + \frac{z_R^2}{z_1}, \qquad \text{Mirror 2:} \quad -40 = z_2 + \frac{z_R^2}{z_2},$$

$$\Rightarrow \quad -20(z_2 - 10) = (z_2 - 10)^2 + z_R^2, \qquad -40 z_2 = z_2^2 + z_R^2,$$

$$\text{or} \quad 100 = z_2^2 + z_R^2,$$

Comparing the final equations for the two mirrors gives

$$100 = -40 z_2 \Rightarrow z_2 = -2.5\,\text{cm} \text{ and } z_1 = z_2 - 10 = -12.5\,\text{cm}.$$

Thus, the beam waist lies 12.5 cm to the right of mirror M_1, or 2.5 cm to the right of mirror M_2. Using $z_2 = -2.5$ cm in the equation for mirror 2, we find the Rayleigh range of the Gaussian beam to be

$$z_R = \sqrt{(-40 - z_2) z_2} = 9.682 \text{ cm}.$$

Recalling that $z_R = \frac{\pi w_0^2}{\lambda}$ (with $n = 1$), the beam spot radius at the waist is

$$w_0 = \sqrt{\frac{\lambda z_R}{\pi}} = \sqrt{\frac{(1.06 \times 10^{-4})(9.682)}{\pi}} = 1.807 \times 10^{-2}\text{cm} \approx 181\,\mu\text{m}.$$

To find the beam spot radii at the mirrors, we use the Gaussian beam propagation formulas:

$$w_1 \equiv w_0 \sqrt{1 + (z_1/z_R)^2} = 180.7 \times 1.63 = 295\,\mu\text{m}$$
$$w_2 \equiv w_0 \sqrt{1 + (z_2/z_R)^2} = 180.7 \times 1.033 = 187\,\mu\text{m}$$

The axial mode separation, assuming the refractive index $n = 1$, is

$$\Delta\nu_{ax} = \frac{c}{2nL} = \frac{3.00 \times 10^8}{2 \times 0.10} = 1.50 \times 10^9\,\text{Hz}.$$

To find the transverse mode separation, we note that

$$\frac{\cos^{-1}\sqrt{g_1 g_2}}{\pi} = \frac{1}{\pi}\cos^{-1}\left[\sqrt{0.50 \times 1.25}\right] = \frac{0.659}{\pi} = 0.210.$$

The transverse mode separation is therefore

$$\Delta\nu_{tr} = \Delta\nu_{ax} \times \frac{\cos^{-1}\sqrt{g_1 g_2}}{\pi} = (1.50 \times 10^9) \times [0.210] \approx 315\,\text{MHz}.$$

4.2.1 Symmetric Resonators

Symmetric resonators with $R_1 = R_2 \equiv R$ and $-z_1 = z_2 = L/2$ constitute an important class of stable resonators for which $g_1 = g_2 \equiv g = 1 - L/R$. These resonators lie along the line $g_1 = g_2$ in the g_1-g_2 plane [see Fig. 2.22]. By symmetry, the beam waist occurs at the center of the resonator, $z = 0$, and the minimum spot radius there is given by

$$w_0 = \sqrt{\frac{L\lambda}{2n\pi}}\left[\frac{1+g}{1-g}\right]^{1/4} = \sqrt{\frac{\lambda}{2n\pi}}\,[L(2R-L)]^{1/4}\,. \tag{4.32}$$

The spot radii at the two mirrors are equal and are given by

$$w_1 = w_2 = \sqrt{\frac{L\lambda}{n\pi}}\left[\frac{1}{1-g^2}\right]^{1/4} = \sqrt{\frac{\lambda}{n\pi}}\left[\frac{LR^2}{2R-L}\right]^{1/4}. \tag{4.33}$$

We now discuss three special cases of symmetric stable resonators.

Long-Radius Near-planar Resonators

Near-planar resonators consist of mirrors characterized by $R_1 \approx R_2 \equiv R \gg L$, so that $g_1 \approx g_2 = 1 - L/R \approx 1$ [Fig. 4.4]. The beam spot radii at the waist and the mirrors are then

$$w_0 = \sqrt{\frac{\lambda}{n\pi}}\left[\frac{RL}{2}\right]^{1/4}\left(1-\frac{L}{2R}\right)^{1/4}, \tag{4.34}$$

$$w_1 \approx w_2 = \sqrt{\frac{\lambda}{n\pi}}\left[\frac{RL}{2}\right]^{1/4}\left(1-\frac{L}{2R}\right)^{-1/4} \approx w_0\left[1+\frac{L}{4R}\right] \tag{4.35}$$

Thus, in a near-planar resonator, the spot radius is large at the waist and varies little over the length of the resonator ($w_1 = w_2 \approx w_0$). This is a useful feature for extracting energy efficiently from the gain medium of a laser. As an example, for an Ar-ion laser with $R = 8.0$ m and $L = 1.0$ m, operating at a wavelength of 514 nm, and with $n \approx 1$, the beam spot radius, neglecting the small correction of order L/R, is

$$w_0 \approx \sqrt{\frac{514 \times 10^{-9}}{\pi}} \times \left(\frac{8 \times 1}{2}\right)^{1/4} = 0.057\,\text{cm} = 570\,\mu\text{m}\,.$$

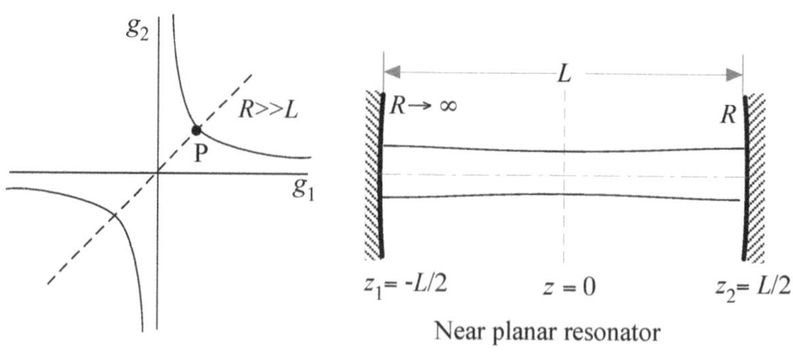

FIGURE 4.4
Long-radius near-planar two-mirror resonator.

A large spot radius at the waist corresponds to a small beam divergence. This advantage, however, is accompanied by increased sensitivity to mirror alignment. Since near-planar resonators operate close to the boundary of the stability region, small perturbations in mirror orientation can readily misalign the resonator axis. In addition, mirrors with large radii of curvature are difficult to manufacture accurately. For example, a mirror of diameter $D = 2.50$ cm and radius of curvature $R = 20$ m has a sag of only $D^2/8R \approx 4\,\mu$m. Nevertheless, resonators utilizing mirrors with radii of curvature $R \sim 10$ m are commonly used in gas lasers.

The frequency spectrum of near-planar resonators can also be examined. Since $g_1 = g_2 = 1 - L/R \approx 1$, the angle $\psi = \cos^{-1}\sqrt{g_1 g_2} = \cos^{-1}(1 - L/R)$ is small. Using the small-angle approximation $\cos\psi \approx 1 - \psi^2/2$, we have $1 - \frac{\psi^2}{2} \approx 1 - \frac{L}{R}$, so that $\psi = \sqrt{2L/R}$. The mode frequency spectrum for a near-planar resonator then becomes

$$\nu_{\ell mp} = \frac{c}{2nL}\left[p + \frac{(\ell + m + 1)}{\pi}\sqrt{\frac{2L}{R}}\right]. \tag{4.36}$$

The axial and transverse mode separations are therefore given by

$$\Delta\nu_{ax} = \frac{c}{2nL} \tag{4.37}$$

$$\Delta\nu_{tr} = \frac{c}{2nL}\left[\frac{1}{\pi}\sqrt{\frac{2L}{R}}\right] \ll \Delta\nu_{ax} \tag{4.38}$$

Thus, clustered on the high-frequency side of each axial mode is a group of closely spaced transverse modes, as shown in Fig. 4.5. Note that, for a fixed value of the axial index p, all transverse modes with $\ell + m =$ const. are degenerate. For example, the $01p$ and $10p$ modes are degenerate. Similarly, the $02p$, $11p$, and $20p$ modes are degenerate.

Symmetric Confocal Resonator

Another important symmetric resonator that lies, literally, at the heart of the stability region is the symmetric confocal resonator [Fig. 4.6]. This resonator is characterized by $g_1 = 0 = g_2$ (point C), corresponding to $R_1 = R_2 \equiv R = L$. Since the focal point of a concave spherical mirror lies a distance $R/2$ in front of it, it is clear that for $L = R$ the focal points of the two mirrors coincide. Hence the name confocal. Note that there are other confocal resonators defined by the relation $L = \frac{1}{2}(R_1 + R_2)$ or, equivalently, by the hyperbola $g_1 + g_2 = 2g_1g_2$. All such confocal resonators, except the symmetric one, are unstable. For the symmetric confocal resonator, the beam waist lies at the center of the resonator. The spot radii at the waist and mirrors are given by

$$w_0 = \sqrt{\frac{L\lambda}{2n\pi}}, \tag{4.39}$$

$$w_1 = \sqrt{\frac{L\lambda}{n\pi}} = \sqrt{2}\,w_0\,. \tag{4.40}$$

Thus, in propagating from the waist to an end mirror, the spot radius increases by a factor of $\sqrt{2}$. The Rayleigh range is $z_R = L/2$, so that the mirror separation in a confocal resonator is exactly equal to the confocal parameter $2z_R$ of the beam. Among symmetric stable resonators of fixed length, the confocal resonator has the minimum spot radius at the mirrors. This feature is useful when a high power density is desired over the resonator length. A small mode volume associated with a small beam radius, however, is not the most efficient design for energy extraction in a laser if the gain volume does not match the mode

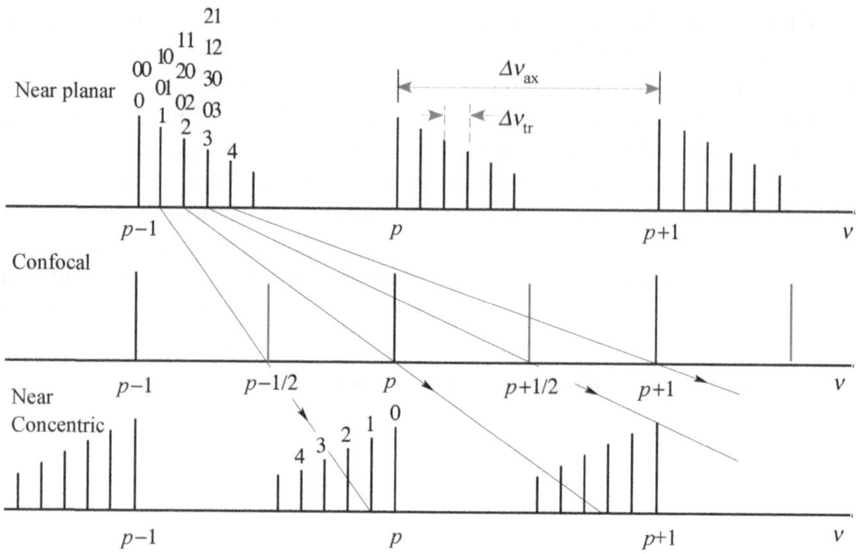

FIGURE 4.5
Frequency spectrum of symmetric two-mirror resonators with Hermite-Gauss modes. The dashed lines show the trajectories of mode frequencies as the resonator evolves from a plane-parallel to a concentric configuration.

volume. Furthermore, a small fundamental-mode volume may allow higher-order transverse modes to oscillate, thereby degrading the spatial profile of the laser beam. Matching the gain profile to the mode volume is therefore a fundamental consideration in the design of laser resonators.

The symmetric confocal resonator is relatively insensitive to small perturbations in mirror position and orientation. A small tilt of either mirror primarily displaces the resonator axis rather than strongly altering the mode geometry. This relative insensitivity, combined with the high degree of degeneracy of its mode spectrum, makes the symmetric confocal resonator particularly useful in scanning spectrum-analyzer applications.

The frequency spectrum of the confocal resonator has a high degree of degeneracy. Since $g = 0$, it follows that $\psi = \cos^{-1}(0) = \pi/2$, and the mode spectrum becomes

$$\nu_{\ell mp} = \frac{c}{4nL}(2p + \ell + m + 1). \tag{4.41}$$

This spectrum gives axial and transverse modes separated by

$$\Delta\nu_{ax} = \frac{c}{2nL}, \tag{4.42}$$

$$\Delta\nu_{tr} = \frac{c}{4nL}. \tag{4.43}$$

Thus, the spectrum has a particularly simple structure, with successive resonances in the combined axial-transverse spectrum separated by

$$\Delta\nu_c = \frac{c}{4nL}. \tag{4.44}$$

The spectrum has a high degree of degeneracy. For example, the modes $00p$, $20(p-1)$, $11(p-1)$, $02(p-1)$, $40(p-2)$, $31(p-2)$, $22(p-2)$, $13(p-2)$, $04(p-2)$, etc., are degenerate.

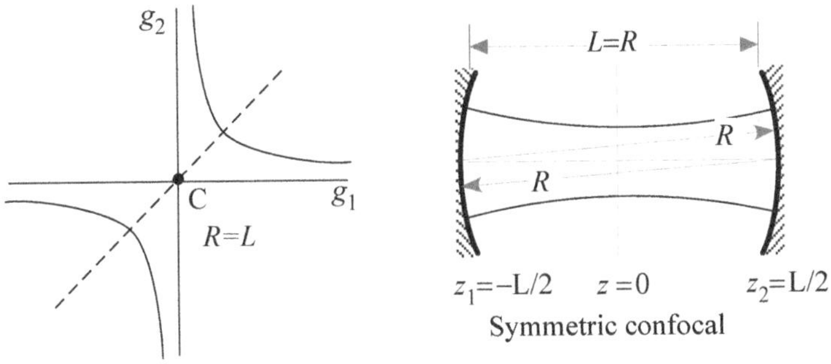

FIGURE 4.6
Symmetric confocal two-mirror resonator ($R_1 = R = R_2 = L$).

Near-concentric or Near-spherical Resonators

These resonators [Fig. 4.7] lie near the boundary of the stability region in the third quadrant of the g_1-g_2 plane. They are characterized by $L \approx 2R$. Writing $L = 2(R - \delta)$, where $0 < \delta \ll L, R$, we find

$$g_1 = g_2 = 1 - \frac{L}{R} = 1 - \frac{2(R-\delta)}{R} = -1 + \frac{2\delta}{R}. \tag{4.45}$$

Using this result, we obtain the beam spot radii

$$w_0 = \sqrt{\frac{L\lambda}{2\pi n}}\left[\frac{1+g}{1-g}\right]^{1/4} \approx \sqrt{\frac{L\lambda}{2\pi n}}\left[\frac{2\delta/R}{2}\right]^{1/4} = \sqrt{\frac{L\lambda}{2\pi n}}\left[\frac{\delta}{R}\right]^{1/4}, \tag{4.46}$$

$$w_1 = \sqrt{\frac{L\lambda}{\pi n}}\left[\frac{1}{1-g^2}\right]^{1/4} \approx \sqrt{\frac{L\lambda}{\pi n}}\left[\frac{1}{2\times(2\delta/R)}\right]^{1/4} = \sqrt{\frac{L\lambda}{2\pi n}}\left[\frac{R}{\delta}\right]^{1/4}. \tag{4.47}$$

In this case, the minimum spot radius w_0 at the center of the resonator becomes very small, whereas the spot radii at the mirrors, located at $z_2 = -z_1 \approx R$, become very large, $w_1 = w_2 \gg w_0$. The small central spot produces a very high power density and may be useful in some nonlinear optical applications. From the point of view of efficient energy extraction from the gain medium of a laser, however, this mode geometry is generally less

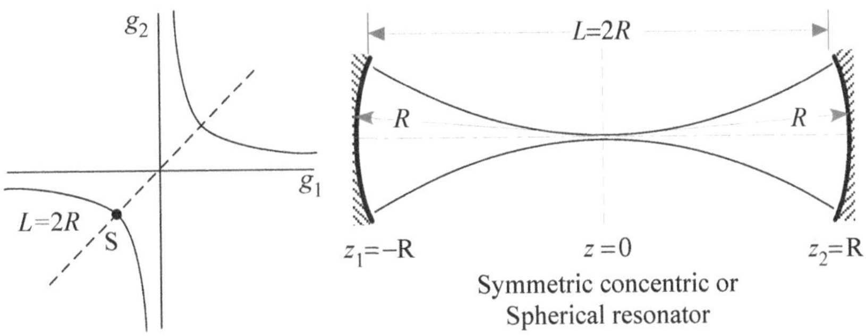

FIGURE 4.7
Symmetric nearly concentric resonator $R_1 = R_2 \approx L/2$.

favorable. The beam divergence is also very large. Near-concentric resonators are also highly sensitive to small mirror misalignments, which can produce large changes in the resonator axis.

The frequency spectrum of a near-concentric resonator is also readily obtained. From Eq. (4.45), we have $g_1 = g_2 = -1+2\delta/R \approx -1$. For a symmetric resonator on this branch of the stability diagram, the appropriate Gouy-phase angle is $\psi = \cos^{-1} g = \cos^{-1}[-1+2\delta/R] \approx \cos^{-1}(-1)$, which approaches π as the concentric limit is approached. Writing this angle as $\psi = \pi - \phi$, with $\phi \ll 1$, we obtain

$$\cos(\pi - \phi) = -\cos\phi \approx -\left(1 - \frac{\phi^2}{2}\right) = \left[-1 + \frac{2\delta}{R}\right] . \tag{4.48}$$

This gives $\phi \approx 2\sqrt{\delta/R}$ and $\psi = \pi - 2\sqrt{\delta/R}$. The mode spectrum for a near-concentric resonator can then be written

$$\begin{aligned} \nu_{\ell mp} &= \frac{c}{2nL}\left[p + \frac{(\ell+m+1)}{\pi}\left(\pi - 2\sqrt{\frac{\delta}{R}}\right)\right] \\ &= \frac{c}{2nL}\left[p + \ell + m + 1 - \frac{2(\ell+m+1)}{\pi}\sqrt{\frac{\delta}{R}}\right] . \end{aligned} \tag{4.49}$$

Mode frequency separations are then given by

$$\Delta\nu_{ax} = \frac{c}{2nL} , \tag{4.50}$$

$$\Delta\nu_{tr} = \frac{c}{2nL}\left(1 - \frac{2}{\pi}\sqrt{\frac{\delta}{R}}\right) = \Delta\nu_{ax}\left(1 - \frac{2}{\pi}\sqrt{\frac{\delta}{R}}\right) . \tag{4.51}$$

Thus, the transverse mode spacing is only slightly smaller than the axial mode separation. Consequently, the modes $\nu_{01p} = \nu_{10p}$ approach $\nu_{00(p+1)}$ from the low-frequency side, the modes $\nu_{02p} = \nu_{11p} = \nu_{20p}$ approach $\nu_{00(p+2)}$ from the low-frequency side, and so on. Hence, each fundamental axial mode in a near-concentric resonator has a cluster of transverse modes on its low-frequency side. The trajectories of the various transverse modes as a resonator changes from near-planar through confocal to near-concentric are shown in Fig. 4.5.

Example 4.3 *A He-Ne laser ($\lambda = 632.8$ nm) resonator consists of two spherical mirrors, each with a radius of curvature of* 50.0 *cm, separated by* 25.0 *cm. Determine the waist and mirror spot radii and the axial and transverse mode frequency separations.*

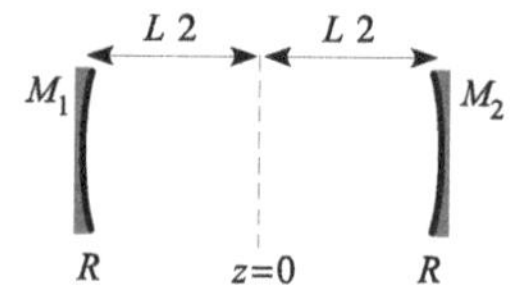

We have a symmetric resonator with $R = 0.500$ m, $L = 0.250$ m, $g = 1 - \frac{L}{R} = \frac{1}{2}$, and $0 < g^2 = \frac{1}{4} < 1$. Thus, the resonator is stable. The spot radii are then

$$w_0 = \sqrt{\frac{L\lambda}{2\pi}}\left[\frac{1+g}{1-g}\right]^{1/4} = \sqrt{\frac{0.25 \times 0.6328 \times 10^{-6}}{2\pi}}[3]^{1/4}\,\text{m} = 209\,\mu\text{m} ,$$

$$w_1 = w_2 = \sqrt{\frac{L\lambda}{\pi}}\left[\frac{1}{1-g^2}\right]^{1/4} = \sqrt{\frac{0.25 \times 0.6328 \times 10^{-6}}{\pi}}\left[\frac{4}{3}\right]^{1/4}\,\text{m} = 241\,\mu\text{m} .$$

We can find the spot radius at the waist directly by first noting that, by symmetry, the waist is located at the center of the cavity. The radius of curvature of the beam at the location of either curved mirror, $z = L/2$, must then match the radius of curvature of the mirror:

$$R = \frac{L}{2} + \frac{2z_R^2}{L} \quad \Rightarrow \quad z_R^2 = \frac{1}{2}\sqrt{(2R-L)L} \quad \Rightarrow w_0 = \sqrt{\frac{L\lambda}{2\pi}}\left[\frac{2R}{L} - 1\right]^{1/4} .$$

The mirror spot radii are then $w_1 = w_2 = w_0\sqrt{1 + (L/2z_R)^2}$. With $\cos^{-1}\sqrt{g_1 g_2} = \cos^{-1}(1/2) = \pi/3$, the mode frequency separations are

$$\Delta\nu_{\rm ax} = \frac{c}{2L} = \frac{3.00 \times 10^8}{2 \times 0.25}\,\text{Hz} = 600\,\text{MHz}$$

$$\Delta\nu_{\rm tr} = \frac{c}{2L} \times \frac{\cos^{-1}\sqrt{g_1 g_2}}{\pi} = (600\,\text{MHz}) \times \frac{1}{3} = 200\,\text{MHz}$$

Half-symmetric Resonators

Half-symmetric resonators consist of one flat and one curved mirror ($R_1 \to \infty$, $R_2 = R$). The flat mirror acts as a plane of reflection symmetry, so that the cavity is optically equivalent to one half of a symmetric resonator of twice the length. Half-symmetric resonators are among the most common resonator designs for medium- to low-power gas lasers. They are characterized by a beam waist located at the plane mirror, which requires the plane mirror to be perpendicular to the resonator axis, and a relatively large spot radius at the curved mirror. An important advantage of the half-symmetric design is its relative ease of alignment.

A near-hemispherical resonator (one half of a near-concentric resonator) with $L = R - \delta$ has $g_1 = 1$, $g_2 = 1 - \frac{L}{R} = \frac{\delta}{R} \approx \frac{\delta}{L} \ll 1$. It has a very small spot radius at the plane mirror and a very large spot radius at the curved mirror:

$$w_0 = w_1 \approx \sqrt{\frac{L\lambda}{\pi}}\left(\frac{\delta}{L}\right)^{1/4} , \tag{4.52}$$

$$w_2 \approx \sqrt{\frac{L\lambda}{\pi}}\left(\frac{L}{\delta}\right)^{1/4} . \tag{4.53}$$

Thus, by varying $\delta = R - L$, we can obtain the desired spot radius at the plane mirror, which can conveniently be used as the output mirror.

4.2.2 More Complex Resonators

A few examples of more complex resonators are presented here [see also Problem 14]. These resonators typically have more than one beam waist. Their mode conditions can be obtained by choosing a reference plane and unfolding the resonator. The phase change in one roundtrip can then be calculated as the phase accumulated by the wave in traversing one unfolded unit cell. Equating this phase change to an integer multiple of 2π yields the mode frequency spectrum, as illustrated by the examples below. The following three examples illustrate how the propagation and Gouy-phase contributions are accumulated for different resonator geometries.

$$\begin{aligned} L &= d_1 + d_2, \\ 2\pi p &= 2kd_1 - 2(\ell + m + 1)\tan^{-1}\frac{d_1}{z_{R1}}, \\ &\quad + 2kd_2 - 2(\ell + m + 1)\tan^{-1}\frac{d_2}{z_{R2}}, \\ &= 2k_{\ell mp}L - 2(\ell + m + 1)\left[\tan^{-1}\frac{d_1}{z_{R1}} + \tan^{-1}\frac{d_2}{z_{R2}}\right], \\ \nu_{\ell mp} &= \frac{c}{2nL}\left[p + \frac{(\ell + m + 1)}{\pi}\left(\tan^{-1}\frac{d_1}{z_{R1}} + \tan^{-1}\frac{d_2}{z_{R2}}\right)\right]. \end{aligned}$$

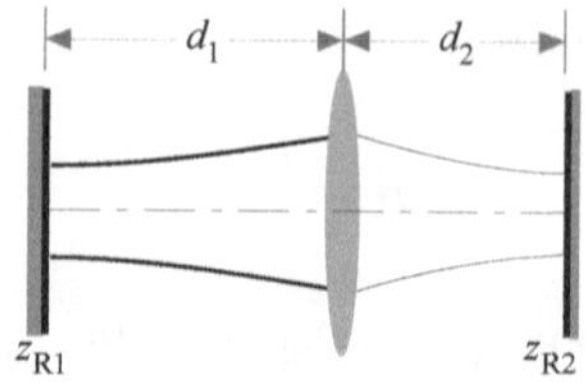

$$\begin{aligned} L &= d_1 + d_2 + d_3, \\ 2\pi p &= k_{\ell mp}L - 2(\ell + m + 1)\left[\tan^{-1}\frac{d_1}{z_{R1}} + \tan^{-1}\frac{d_2}{z_{R2}}\right], \\ \nu_{\ell mp} &= \frac{c}{nL}\left[p + \frac{(\ell + m + 1)}{\pi}\left(\tan^{-1}\frac{d_1}{z_{R1}} + \tan^{-1}\frac{d_2}{z_{R2}}\right)\right]. \end{aligned}$$

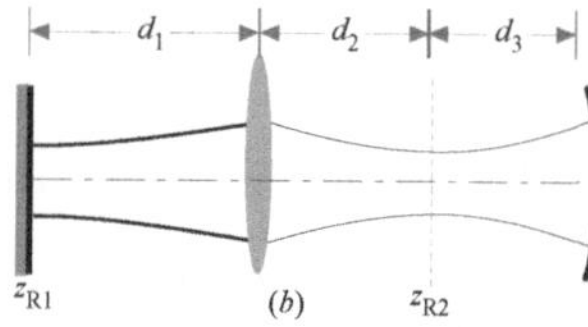

$$\begin{aligned} L &= 2d_1 + 2d_2, \\ 2\pi p &= k_{\ell mp}L - (\ell + m + 1)\left[2\tan^{-1}\frac{d_1}{z_{R1}} + 2\tan^{-1}\frac{d_2}{z_{R2}}\right], \\ \nu_{\ell mp} &= \frac{c}{nL}\left[p + \frac{(\ell + m + 1)}{\pi}\left(\tan^{-1}\frac{d_1}{z_{R1}} + \tan^{-1}\frac{d_2}{z_{R2}}\right)\right]. \end{aligned}$$

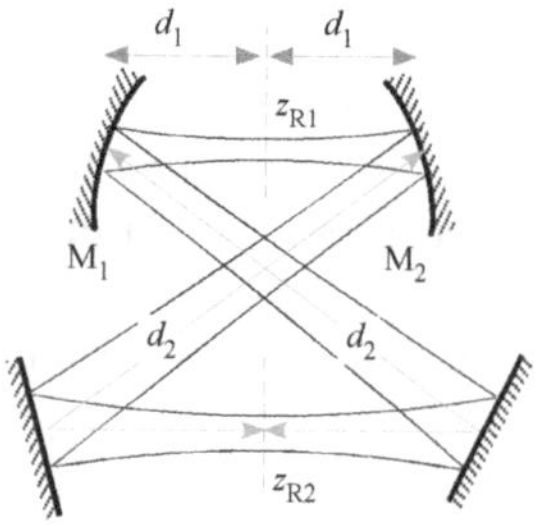

4.2.3 Effective Mode Volume

In many problems involving the interaction of light with a collection of active atoms in a cavity, we need to estimate the number of atoms interacting with a given cavity mode. If the atoms are uniformly distributed throughout the spatial extent of the mode, an estimate of this number can be obtained by calculating the effective volume occupied by the mode. Since the atoms respond to the field intensity, we can define an effective mode volume through the intensity integral

$$V_{\ell m} = \frac{1}{I_{0\ell m}}\int_0^L dz \int_{-\infty}^{\infty} dx \int_{-\infty}^{\infty} dy I_{\ell m} = \frac{\pi w_0^2}{2} L\left(\ell! m! 2^{\ell+m}\right). \tag{4.54}$$

In evaluating the integral along the cavity length L, we have assumed a traveling-wave Hermite-Gauss mode with transverse indices ℓ, m. For a standing wave, the expression for the mode volume must be multiplied by a factor of $\frac{1}{2}$, corresponding to the average value of $\sin^2 k_{\ell mp} z$ over the cavity length.

Example 4.4 *As an example of a mode-volume calculation, consider a half-symmetric two-mirror cavity with $R_1 = \infty, R_2 = 20.0$ m, and $L = 1.0$ m for a He-Ne laser.*

With the minimum spot radius $w_0 = \sqrt{\frac{\lambda}{\pi}}\left[LR_2(1 - L/R_2)\right]^{1/4} = 0.94$ mm, we find that the mode volume for the fundamental mode is $V_{00} = 1.38\,\text{cm}^3$. Applying the standing-wave correction gives $1.38 \times 0.5 = 0.69\,\text{cm}^3$. If a gain medium containing Ne at a pressure of 0.1 Torr fills the entire cavity, the number of atoms interacting with the mode is

$$N = \frac{6.0221 \times 10^{23}\,\text{atoms/mol}}{(760\,\text{Torr})(22.4 \times 10^3\,\text{cm}^3/\text{mol})} p(\text{Torr})V(\text{cm}^3) = (3.54 \times 10^{16})0.1(0.69) = 2.44 \times 10^{15}\,.$$

If each of these atoms is excited at a rate of 10 Hz and each excited atom emits a photon at 633 nm, the corresponding ideal maximum optical power is

$$P_{\text{max}} = (2.44 \times 10^{15}) \left(\frac{1240\,\text{eV}}{633}\right) (1.60 \times 10^{-19}\text{J/eV})(10\,\text{Hz}) \approx 7.6\,\text{mW}\,.$$

This idealized estimate shows that the available power can be increased by increasing the excitation rate, atomic number density (pressure), or mode volume, within limits determined by the particular laser system.

4.3 Resonator Losses

In our discussion of resonators so far, we have ignored losses. Losses are both necessary and unavoidable. For example, to couple energy into a cavity or extract energy from it, the mirrors must be partially transmitting; that is, their reflectance must be less than unity. In addition, resonator modes suffer losses due to diffraction, scattering, and absorption in the medium filling the cavity. These losses are usually specified for one roundtrip of the circulating field inside the cavity. These loss mechanisms remove energy from the wave, causing its amplitude to decrease as it propagates. We now consider different types of loss mechanisms and how they are taken into account in laser physics.

4.3.1 Absorption and Scattering Losses

Energy loss due to scattering and absorption in the medium filling the cavity occurs continuously as the wave propagates through the cavity. Such losses are referred to as distributed or internal losses. They are specified in terms of an absorption coefficient α, with units $[\alpha]$=m^{-1} or, more practically, cm^{-1}. Distributed losses can be modeled by an effective conductivity σ of the medium filling the resonator in the paraxial wave equation (3.9):

$$\frac{\partial E_o(\boldsymbol{r})}{\partial z} = \frac{i}{2k}\nabla_\perp^2 E_o(\boldsymbol{r}) + i\frac{k\chi'_{at}}{2n^2}E_o(\boldsymbol{r}) - \frac{k}{2}\left(\frac{\chi''_{at}}{n^2} + \frac{\sigma}{\omega\epsilon_0 n^2}\right)E_o(\boldsymbol{r})\,. \tag{3.9*}$$

For nonzero conductivity σ and $\chi at = 0$, this equation becomes

$$\frac{\partial E_o(\boldsymbol{r})}{\partial z} = \frac{i}{2k}\nabla_\perp^2 E_o(\boldsymbol{r}) - \frac{k\sigma}{2\omega\epsilon_0 n^2}E_o(\boldsymbol{r})\,. \tag{4.55}$$

This equation shows that, during propagation, the wave amplitude evolves both because of diffraction, represented by the transverse Laplacian on the right-hand side, and because of absorption and scattering losses, represented by the effective-conductivity term. We have seen that Gaussian beams already incorporate the effect of diffraction in open resonators. To account for the effect of distributed loss, we write the spatial part of the field as

$$E_o(\boldsymbol{r}) = A(z)u_{\ell m}(\boldsymbol{r})\,, \tag{4.56}$$

where $u_{\ell m}(\boldsymbol{r})$ represents the spatial dependence appropriate to a Hermite-Gauss mode or, more generally, to any other family of paraxial beams. This has the same form as the field of a Hermite-Gauss beam [Eq. (3.82)], except that the wave amplitude $A(z)$ now depends on the propagation distance z, allowing for attenuation due to absorption and scattering. For simplicity of notation, we have dropped the mode indices from the amplitude.

Substituting Eq. (4.56) into Eq. (4.55), and recalling that both A and $u_{\ell m}$ depend on z, whereas only $u_{\ell m}(\boldsymbol{r})$ depends on transverse coordinates, we obtain

$$A(z)\frac{\partial u_{\ell m}(\boldsymbol{r})}{\partial z} + u_{\ell m}\frac{\partial A(z)}{\partial z} = \frac{i}{2k}A(z)\nabla_{\perp}^2 u_{\ell m}(\boldsymbol{r}) - \frac{k\sigma}{2\omega\epsilon_0 n^2}A(z)u_{\ell m}(\boldsymbol{r})\,. \tag{4.57}$$

Now, using the fact that $u_{\ell m}(\boldsymbol{r})$ satisfies the paraxial wave equation $\frac{\partial}{\partial z}u_{\ell m}(\boldsymbol{r}) = \frac{i}{2k}\nabla_{\perp}^2 u_{\ell m}(\boldsymbol{r})$, we find that the equation determining the amplitude $A(z)$ is

$$\frac{\partial A(z)}{\partial z} = -\frac{\sigma}{2\epsilon_0 nc}A(z)\,. \tag{4.58}$$

Introducing the absorption coefficient

$$\alpha = \frac{\sigma}{\epsilon_o nc}\,, \tag{4.59}$$

the solution of this equation can be written as

$$A(z) = A(0)e^{-\alpha z/2}\,, \tag{4.60}$$

where $A(0)$ is the wave amplitude at $z = 0$. Since the absorption coefficient defined by Eq. (4.59) is a positive quantity, this equation shows that the field amplitude decays exponentially as a result of distributed losses, which have been represented collectively by an effective conductivity of the medium filling the cavity. Since the wave intensity (or power density) is proportional to the square of the absolute value of the wave amplitude, it follows that

$$I(z) = I(0)e^{-\alpha z}\,. \tag{4.61}$$

This relation is known as Beer's law for intensity. The absorption coefficient α has dimensions of inverse length and is related to the penetration depth d, defined as the distance over which the wave intensity decreases to $1/e$ of its initial value.

4.3.2 Cavity Losses

Distributed losses

For a wave circulating inside a cavity containing an absorbing medium, a fraction $e^{-\alpha 2L_a}$ of its initial power survives each roundtrip (back and forth in a standing-wave cavity), and a fraction $1 - e^{-\alpha 2L_a}$ is lost to absorption. Here $2L_a$ is the total distance traveled through the absorbing region in one roundtrip. The quantity $f_s \equiv e^{-\alpha 2L_a}$ is the survival factor, representing the fraction of the initial wave power that survives one cavity roundtrip.

Other cavity losses can be characterized similarly in terms of a survival factor. If f_s is the fraction of the initial wave power that survives one roundtrip as a result of a particular loss mechanism, we can regard f_s as the reflectance of a fictitious mirror and $1 - f_s$ as its transmittance, or the fraction of wave power removed by that loss mechanism. Thus, if η represents the fractional power loss per roundtrip due to a particular loss mechanism, the survival factor is $f_s = 1 - \eta$.

In laser physics, loss is conveniently specified in terms of a loss factor (sometimes also referred to as a logarithmic loss factor), defined as the negative of the natural logarithm of the fraction of the initial wave power that survives one roundtrip:

$$\mathcal{L}_i \equiv -\ln f_s = -\ln(1-\eta) = -\ln[e^{-\alpha 2L_a}] = \alpha 2L_a\,. \tag{4.62}$$

For small fractional loss, $\eta \ll 1$, the logarithmic loss factor reduces to $\mathcal{L}_i \approx \eta$. An important advantage of defining the loss in logarithmic form is that loss factors from independent mechanisms are additive because their survival factors multiply.

Localized losses

Losses incurred by the circulating wave due to finite transmittance of a mirror or absorption in mirror coatings are obviously localized at the mirror. When a wave circulating inside the cavity encounters a mirror of reflectance $\mathcal{R}$, a fraction $\mathcal{R}$ of incident power is reflected back into the cavity, and a fraction $1-\mathcal{R} = \mathcal{T} + \mathcal{A}$ is lost due to mirror transmission (a fraction $\mathcal{T}$) and absorption (a fraction $\mathcal{A}$) in mirror coatings. Mirror losses are usually dominated by transmission. For this reason, mirror losses are often referred to simply as transmission losses. They are characterized by the loss factor

$$\mathcal{L} \equiv -\ln \mathcal{R} = -\ln(1-\mathcal{T}-\mathcal{A})\,. \tag{4.63}$$

For small mirror transmission and absorption, the loss factor reduces approximately to the fractional power loss suffered by the wave at the mirror:

$$\mathcal{L} = -\ln(1-\mathcal{T}-A) \approx \mathcal{T}+\mathcal{A}\,. \tag{4.64}$$

For example, for a mirror with 5% transmittance and $\mathcal{A} = 0$, the loss factor is $-\ln(1-0.05) = 0.0513$, which is close to the fractional power loss of 0.05. Even for a transmittance as large as 20%, the loss factor $-\ln(1-0.20) \approx 0.223$ is still reasonably close to the fractional power loss. For large mirror losses, on the other hand, $\mathcal{L}$ can exceed unity. For example, for a mirror with 70% transmittance, the loss factor is $-\ln(1-0.70) = 1.20$.

In a two-mirror cavity, the wave is reflected once from each mirror during one roundtrip, so that a fraction $\mathcal{R}_1\mathcal{R}_2$ of its initial power remains inside the cavity and a fraction $1-\mathcal{R}_1\mathcal{R}_2$ is lost through mirror transmission and absorption. The corresponding loss factor is $\mathcal{L} = -\ln(\mathcal{R}_1\mathcal{R}2)$. Generalizing this result to an N-mirror cavity, we find that the loss factor due to mirror transmission and absorption is

$$\mathcal{L}_{_T} = -\ln[\mathcal{R}_1\mathcal{R}_2\cdots\mathcal{R}_N] = \sum_{j=1}^{N} -\ln\mathcal{R}_j = \sum_{j=1}^{N}\mathcal{L}_j \tag{4.65}$$

Diffraction losses

Diffraction caused by optical elements of finite aperture in a laser cavity also results in loss of energy from the cavity mode. Finite-diameter mirrors, the finite transverse extent of the gain medium, and apertures inserted into the cavity to suppress higher-order transverse modes can all cause diffraction losses. Accurate calculations of these losses require solving the finite-aperture diffraction problem, for example by using the Huygens–Fresnel diffraction integral. Such calculations determine not only the diffraction losses but also the actual mode distributions and additional phase shifts introduced by the finite apertures. In the limit in which the Gaussian-beam approximation is valid, these phase shifts approach those associated with the Gouy phase $\psi(z) = \tan^{-1}(z/z_{_R})$ and determine the corresponding corrections to the transverse-mode frequencies.

An important parameter for discussing diffraction losses in finite-diameter optical resonators is the resonator Fresnel number N_F, defined by

$$N_F = \frac{na^2}{L\lambda}, \tag{4.66}$$

where $2a$ is the transverse width of the resonator end mirrors, L is the length of the resonator, λ is the wavelength of light, and n is the refractive index of the medium filling the resonator. The Fresnel number is approximately the number of Fresnel zones extending from the center to the edge of one end mirror as viewed from the center of the opposite mirror.

Recall that the spot radius at an end mirror of a symmetric confocal resonator of length L is $w_1^2 = L\lambda/\pi n$, and other stable resonators have spot radii of the same general order at the mirrors. The expression for N_F can therefore be written as

$$N_F = \frac{\pi a^2}{\pi}\frac{\pi n}{\pi L\lambda} = \frac{1}{\pi}\frac{\pi a^2}{\pi w_1^2}. \tag{4.67}$$

Thus, apart from the factor π, N_F measures the ratio of the mirror aperture area πa^2 to the characteristic mode area πw_1^2.

If we further note that the transverse extent of a high-order Hermite-Gauss mode increases approximately as the square root of its transverse order, $\rho_n \sim \sqrt{n}\, w_1$, an approximate estimate of the largest transverse order that can fit within the mirror aperture is

$$n_{\text{max}} = \frac{a^2}{w_1^2} = \pi N_F. \tag{4.68}$$

In practice, for $N_F \geq 10$, the Gaussian approximation to the mode shape is excellent for the lowest few modes. Examples of empirical formulas for the single-pass diffraction loss factor $\mathcal{L}_{\text{diff}}$ for several simple resonator geometries are

Confocal circular mirrors:
$$\mathcal{L}_{\text{diff}} = \begin{cases} 16\pi^2 N_F\, e^{-4\pi N_F}, & N_F \geq 1, \\ 1 - (\pi N_F)^2, & N_F \to 0, \end{cases} \tag{4.69a}$$

Planar circular mirrors:
$$\mathcal{L}_{\text{diff}} = 0.33\, N_F^{-3/2} \qquad N_F \geq 1. \tag{4.69b}$$

The corresponding single-pass survival factor is $e^{-\mathcal{L}_{\text{diff}}}$ and the fractional power loss is $1 - e^{-\mathcal{L}_{\text{diff}}}$. The total cavity loss factor is the sum of the internal loss factor (distributed absorption and scattering) and the external loss factor (mirror transmission and absorption, together with diffraction and other localized losses). Let $\mathcal{R}_i$ denote the roundtrip survival factor associated with internal losses other than those represented by the absorption coefficient α. The total cavity loss factor can then be written as

$$\mathcal{L}_c = \mathcal{L}_i + \mathcal{L}_e = 2\alpha L_a - \ln \mathcal{R}_i - \ln \mathcal{R}_1\mathcal{R}_2 \cdots \mathcal{R}_N. \tag{4.70}$$

Diffraction losses can enter this equation simply as another term $\mathcal{L}_{\text{diff rt}}$ if it is not already included in $\mathcal{R}_i$ or another survival factor.

Diffraction losses can also be deliberately introduced by apertures placed in the cavity to limit the transverse modes that the cavity can support. An estimate of the diffraction loss at an aperture can be obtained by calculating the fraction of the incident mode power transmitted through the aperture. This fraction can be treated as an effective survival factor or reflectance. For example, for the first few low-order Hermite-Gauss modes characterized by spot radius w, the fractions transmitted through a circular aperture of radius a are

$$\mathcal{R}_{00} = 1 - \exp\left[-2\left(\frac{a}{w}\right)^2\right], \tag{4.71}$$

$$\mathcal{R}_{01} = \mathcal{R}_{10} = 1 - \left[1 + 2\left(\frac{a}{w}\right)^2\right]\exp\left[-2\left(\frac{a}{w}\right)^2\right]. \tag{4.72}$$

Similarly, when a Gaussian beam is incident on a finite-diameter mirror, the fraction of the incident power that falls outside the mirror aperture is lost. For a circular mirror of aperture diameter D, the fractions of power lost by spillover are

$$\mathcal{T}_{00} = \exp\left[-\frac{1}{2}\left(\frac{D}{w}\right)^2\right], \tag{4.73}$$

$$\mathcal{T}_{01} = \mathcal{T}_{10} = \left[1 + \frac{1}{2}\left(\frac{D}{w}\right)^2\right]\exp\left[-\frac{1}{2}\left(\frac{D}{w}\right)^2\right]. \tag{4.74}$$

These estimates, although quite small for typical mirrors used in lasers, generally overestimate the actual diffraction loss. The Gaussian mode description is usually excellent over most of the mirror aperture, but finite-aperture eigenmode calculations show that the field near the mirror edges can be smaller than predicted by simply truncating an unperturbed Gaussian mode.

4.3.3 Cavity Lifetime

As a consequence of absorption, transmission, and other losses, any field inside a cavity loses energy as it propagates. If there are no sources to replenish the lost energy, the field amplitude and, therefore, the intensity decay with time constants determined by the cavity losses. This means that the field amplitude A introduced in Eq. (4.56) must now be allowed to vary with time as well. Thus, the space and time dependence of the field is written as

$$E(\boldsymbol{r}, t) = A(z, t)u(\boldsymbol{r})e^{i(kz-\omega t)}, \tag{4.75}$$

where $A(z,t)$ is a slowly varying function of z and t. By slowly varying, we mean that the fractional change in the field amplitude due to absorption and cavity losses over a distance of a few wavelengths is small. Similarly, the fractional variation of the field amplitude over a few optical cycles is small. These conditions can be expressed by the inequalities

$$\left|\frac{\partial A}{\partial z}\right| \ll k|A|, \quad \left|\frac{\partial^2 A}{\partial z^2}\right| \ll k\left|\frac{\partial A}{\partial z}\right|, \tag{4.76a}$$

$$\left|\frac{\partial A}{\partial t}\right| \ll \omega|A|, \quad \left|\frac{\partial^2 A}{\partial t^2}\right| \ll \omega\left|\frac{\partial A}{\partial t}\right|. \tag{4.76b}$$

Using Eq. (4.75) for the field in Eq. (2.30) with $P_{at} = 0$, and using the fact that $u(\boldsymbol{r})$ is a solution of the paraxial wave equation (3.10), we find that the field amplitude $A(z,t)$ satisfies

$$\frac{\partial A}{\partial t} + \frac{c}{n}\frac{\partial A}{\partial z} + \frac{c\alpha}{2n}A = 0. \tag{4.77}$$

This equation differs from Eq. (4.58), but it is just as easy to solve by introducing the variables

$$\tau = t - z/v, \quad \zeta = z. \tag{4.78}$$

Then Eq. (4.77) becomes

$$\frac{\partial A(\tau,\zeta)}{\partial\zeta} + \frac{\alpha}{2}A(\tau,\zeta) = 0\,, \tag{4.79}$$

with the solution $A(\tau,\zeta) = A(\tau,0)e^{-\alpha\zeta/2}$.

Let us now apply this result to a two-mirror cavity of length L filled with a medium of refractive index n. The roundtrip distance is $2L$, and the corresponding roundtrip time is

$$\tau_R = 2nL/c\,. \tag{4.80}$$

During one roundtrip, the field experiences distributed and localized losses as well as reflections from both mirrors. Thus, one roundtrip time τ_R later, the field amplitude at a given point in the cavity is

$$A(\tau+\tau_R,\zeta) = \sqrt{\mathcal{R}_1\mathcal{R}_2(1-\eta)}\,e^{-\alpha L}A(\tau,\zeta) = e^{-\alpha L+\ln\sqrt{\mathcal{R}_1\mathcal{R}_2(1-\eta)}}A(\tau,\zeta)\,. \tag{4.81}$$

Here $\mathcal{R}_i$ are the power reflection coefficients of the mirrors, and η denotes the combined roundtrip fractional power loss due to diffraction, scattering from intracavity optical elements, and other losses not included explicitly. Thus, after each cavity roundtrip, the survival factor for the light intensity ($\propto |A|^2$) at any point in the cavity is $e^{-2\alpha L}\mathcal{R}_1\mathcal{R}_2(1-\eta) = e^{-\mathcal{L}_c}$. At a cavity resonance, the total roundtrip phase—including any Gouy and reflection-phase contributions—is an integer multiple of 2π. For the simplified axial phase factor shown here [See Eq. (4.75)] is $2kL = 2\pi p$, where p is an integer.

For small losses in one roundtrip, we can write the amplitude $A(\tau+\tau_R,\zeta)$ and the survival factor multiplying $A(\tau,\zeta)$ as

$$A(\tau+\tau_R,\zeta) \approx A(\tau,\zeta) + \tau_R\frac{dA}{d\tau}\,, \tag{4.82a}$$

$$e^{-\alpha L+\ln\sqrt{\mathcal{R}_1\mathcal{R}_2(1-\eta)}} \approx 1 - \frac{1}{2}\big(2\alpha L - \ln\mathcal{R}_1\mathcal{R}_2(1-\eta)\big) \tag{4.82b}$$

Using these expressions in Eq. (4.81), we find that the dependence of A on τ is given by

$$\frac{dA}{d\tau} \approx -\frac{1}{2}\left(\frac{2\alpha L - \ln\mathcal{R}_1\mathcal{R}_2(1-\eta)}{\tau_R}\right)A = -\frac{1}{2}\left(\frac{\mathcal{L}_c}{\tau_R}\right)A\,, \tag{4.83}$$

Here the second step follows by recognizing that the numerator in parentheses is the cavity loss factor $\mathcal{L}_c$ defined in Eq. (4.70). For fixed z, the derivative with respect to τ can be replaced by an ordinary time derivative. Then the time evolution of the field amplitude A and intensity $I = \frac{1}{2}\epsilon_0 nc|A|^2$, at any point in the cavity, is given by

$$A(z,t) = A(z,0)e^{-t/2\tau_c}\,, \tag{4.84a}$$

$$I(z,t) = I(z,0)e^{-t/\tau_c}\,, \tag{4.84b}$$

where

$$\tau_c = \frac{\tau_R}{\mathcal{L}_c}\,, \tag{4.84c}$$

with

$$\mathcal{L}_c = 2\alpha L - \ln\mathcal{R}_1\mathcal{R}_2(1-\eta)\,. \tag{4.84d}$$

Equation (4.84b) shows that the intracavity intensity, and hence the stored energy, decays with the time constant τ_c defined by Eq. (4.84c); it is the time required for the cavity intensity (or stored energy) to decrease to $1/e \approx 37\%$ of its initial value [see Fig. 4.9]. The time τ_c is referred to as the cavity lifetime or photon lifetime; it is a measure of the time for which the cavity can store electromagnetic energy deposited in it. Its inverse, $\gamma_c = 1/\tau_c$,

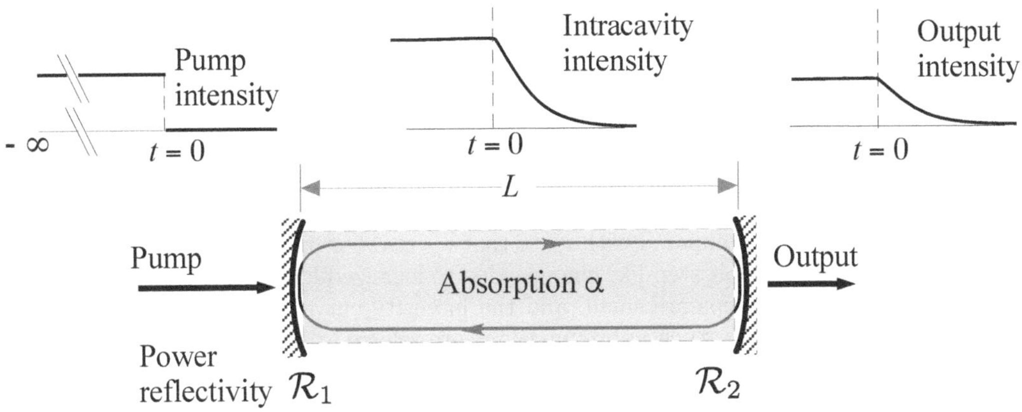

FIGURE 4.8
Some loss mechanisms (e.g., transmission) allow an external pump field to build up steady intracavity and transmitted field amplitudes. If the pump is removed, cavity losses cause the intracavity and transmitted field amplitudes to decay. For simplicity, the field transmitted by only one mirror is shown.

defines the cavity decay rate. Both τ_c and γ_c are naturally related to the cavity loss factor $\mathcal{L}_c$ [Eqs. (4.84c) and (4.84d)].

Because $\mathcal{L}_c$ is defined as a logarithmic loss factor, the exact survival factor for the intracavity intensity after one roundtrip is $e^{-\mathcal{L}c}$, and the exact fractional power loss in one roundtrip is $\left|\frac{\Delta I}{I}\right|_{\text{rt}} = 1 - e^{-\mathcal{L}_c}$. For small losses, $\mathcal{L}c \ll 1$, this becomes

$$\left|\frac{\Delta I}{I}\right|_{\text{rt}} = 1 - e^{-\mathcal{L}_c} \approx \mathcal{L}_c = \frac{\tau_R}{\tau_c} \equiv \gamma_c \tau_R \,, \tag{4.85a}$$

$$\gamma_c = \frac{\mathcal{L}_c}{\tau_R} = \frac{1}{\tau_c} \,. \tag{4.85b}$$

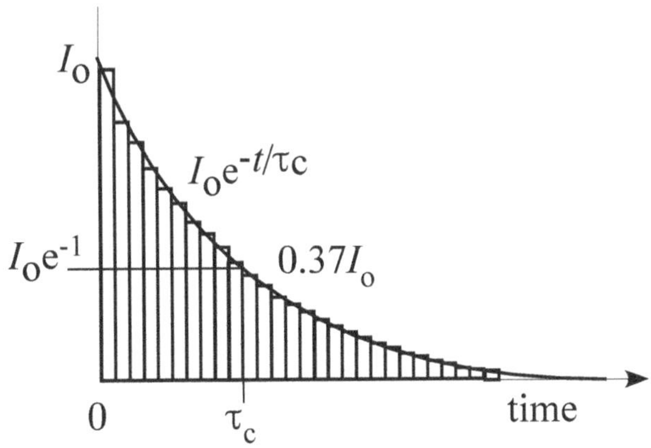

FIGURE 4.9
In the absence of gain, the initial intracavity field intensity in a lossy cavity decays exponentially in time with a time constant τ_c in the coarse-grained time approximation.

Thus, the cavity loss factor $\mathcal{L}_c$ is the logarithmic power loss per roundtrip, while $1 - e^{-\mathcal{L}_c}$ is the exact fractional power loss per roundtrip. For small cavity losses, the two are approximately equal. The cavity decay rate γ_c gives the fractional energy decay rate per unit time.

The smooth temporal evolution of the field amplitude given by Eq. (4.84a) and of intensity given by Eq. (4.84b) provides a good approximation to the actual evolution when the cavity loss per roundtrip is small. The coarse-graining time interval Δt should be long compared with τ_R but short compared with τ_c, so that $\tau_R \ll \Delta t \ll \tau_c$. The actual time evolution of the intensity may exhibit step-like changes due to localized losses in the cavity [Fig. 4.9]. For small losses, these steps are small, and the intensity curve is well approximated by a smooth exponential decay. Such approximations are referred to as coarse-graining approximations; they are common in describing light-matter interaction processes in both space and time domains because, at the fundamental level, matter consists of discrete atoms, and emission and absorption of radiation are discrete events involving discrete amounts of energy. We will encounter more examples of this in later chapters.

For high-reflectivity mirrors ($\mathcal{R}_1, \mathcal{R}_2 \approx 1$), negligible absorption in the mirror coatings, and no other cavity losses, we can write $\mathcal{R}_1 = 1 - \mathcal{T}_1$ and $\mathcal{R}_2 = 1 - \mathcal{T}_2$, where the mirror transmittances $\mathcal{T}_1, \mathcal{T}_2 \ll 1$. In this case, the expressions for the cavity loss factor and cavity lifetime simplify to

$$\mathcal{L}_c \approx 2\alpha L + \mathcal{T}_1 + \mathcal{T}_2\,, \tag{4.86a}$$

$$\tau_c = \frac{2nL}{c(2\alpha L + \mathcal{T}_1 + \mathcal{T}_2)} \equiv \frac{2nL}{c\mathcal{L}_c}\,. \tag{4.86b}$$

The cavity loss factor $\mathcal{L}_c = 2\alpha L + \mathcal{T}_1 + \mathcal{T}_2$ is thus, in the small-loss approximation, simply the sum of the distributed absorption loss and the mirror transmission losses per roundtrip. These equations are easily generalized to cavities containing additional optical elements by including the losses introduced by those elements in the calculation of the total loss.

For example, for a standing-wave cavity with two mirrors and a lens having a scattering loss $S = 2\%$ per surface per pass, and with the absorbing medium confined to a section of length L_a, the cavity loss factor is

$$\mathcal{L}_c = 2\alpha L_a - \ln[\mathcal{R}_1\mathcal{R}_2(1-S)^4] \approx 2\alpha L_a + \mathcal{T}_1 + \mathcal{T}_2 + 4S\,, \tag{4.87}$$

where the second expression holds if each loss is small. Note that in one roundtrip, each of the two lens surfaces is encountered twice, so the survival factor associated with scattering from the lens is $(1-S)^4$. As always, we must be careful in making the low-loss approximation because a large number of individually small losses can add up to become significant.

Example 4.5 *An estimate of cavity lifetime: Let us estimate τ_c for a cavity consisting of two mirrors with reflectance $\mathcal{R}_1 = 0.98 = \mathcal{R}_2$, separated by $L = 90$ cm in air with negligible absorption.*

Since the mirror reflectances are large (and therefore the transmission losses are small), we can use Eqs. (4.86) to obtain

$$\mathcal{L}_c = 2\alpha L + \mathcal{T}_1 + \mathcal{T}_2 = 0.04\,,$$

$$\tau_c = \frac{2nL}{c\mathcal{L}_c} = \frac{2 \times 0.90\,\text{m}}{(3.00 \times 10^8\,\text{m/s})(0.04)} = 150\,\text{ns}\,.$$

This differs by less than 2% from the result $\tau_c = 148$ ns obtained using Eqs. (4.84c) and (4.84d).

4.3.4 Cavity Linewidth

From the decay of the amplitude $A(z,t)$ [Eq. (4.84a)], it follows that the amplitude of the cavity mode (frequency $\omega_0 = 2\pi\nu_0$) has the time dependence

$$E(t) = E_0 e^{-\gamma_c t/2 - i\omega_0 t}, \quad t \geq 0, \quad \gamma_c = 1/\tau_c . \tag{4.88}$$

Taking the Fourier transform of the amplitude, we find that the spectral amplitude is

$$\widetilde{E}(\omega) = \int_0^\infty dt\, E(t) e^{i\omega t} = \frac{E_0}{i(\omega_0 - \omega) + \gamma_c/2} = \frac{E_0}{2\pi} \frac{1}{i(\nu_0 - \nu) + \gamma_c/4\pi} . \tag{4.89}$$

This leads to the cavity mode power spectrum

$$\left|\widetilde{E}(\omega)\right|^2 = \left|E_0\right|^2 \tau_c \left[\frac{1}{\pi} \frac{\gamma_c/4\pi}{(\nu_0 - \nu)^2 + (\gamma_c/4\pi)^2}\right] \equiv \left|E_0\right|^2 \tau_c S(\nu) , \tag{4.90a}$$

where
$$S(\nu) = \frac{1}{\pi} \frac{\gamma_c/4\pi}{(\nu_0 - \nu)^2 + (\gamma_c/4\pi)^2} . \tag{4.90b}$$

The function $S(\nu)$ is the normalized power spectral density, such that $S(\nu)d\nu$ represents the fraction of the total wave power contained in a narrow frequency interval of width $d\nu$ centered at ν. The normalized power spectral density $S(\nu)$ is plotted in Fig. 4.10 as a function of frequency ν. It has the form of a Lorentzian distribution (or lineshape). It peaks at $\nu = \nu_0$ and falls off symmetrically as ν deviates from ν_0, reaching half its maximum value when the frequency deviation from the center frequency is $|\nu - \nu_0| = \gamma_c/4\pi$. The two half maximum points are located symmetrically about the center frequency ν_0. The frequency interval between these points defines the width $\Delta\nu_c$ of the power spectrum. This width, called the full width at half maximum (FWHM), is given by

$$\Delta\nu_c = \frac{\gamma_c}{2\pi} = \frac{1}{2\pi\tau_c} = \frac{\mathcal{L}_c}{2\pi\tau_R} . \tag{4.91}$$

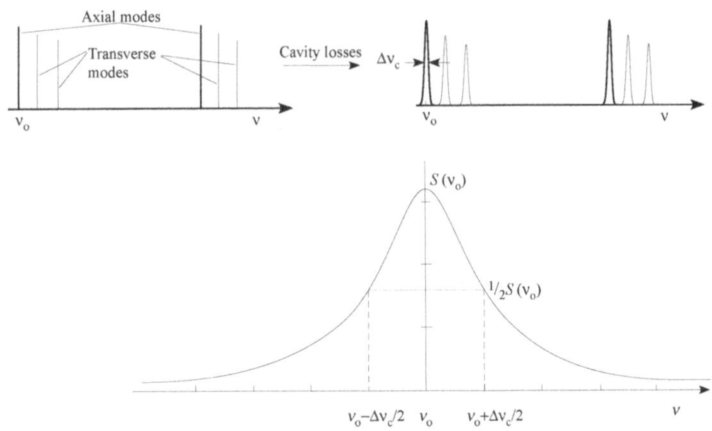

FIGURE 4.10
The decay of field amplitude caused by cavity losses imparts a width to each ideal (lossless) cavity mode frequency.

Thus, in the presence of loss, a cavity-mode frequency is no longer a sharp line (delta function) but acquires a width $\Delta\nu_c$. If the losses do not vary significantly over the range of cavity-mode frequencies of interest, $\Delta\nu_c$ may be referred to simply as the cavity linewidth. All cavity-mode frequencies of interest are then broadened by approximately the same amount, and the power in each mode may be considered to be distributed over a narrow band of frequencies centered on the corresponding ideal cavity-mode frequency ν_0. As the cavity losses decrease ($\mathcal{L}_c \to 0$), the cavity field amplitude retains its initial value for increasingly long times ($\tau_c \to \infty$), and the spectrum approaches a delta function centered at ν_0.

4.3.5 Cavity Quality Factors

Two related quantities are commonly used to characterize the effect of loss on the quality of the cavity-mode structure.

Cavity Finesse

The cavity finesse $\mathcal{F}$ is defined as the ratio of the axial-mode spacing $\Delta\nu_{ax} = 1/\tau_R$ to the cavity linewidth $\Delta\nu_c = 1/2\pi\tau_c$:

$$\mathcal{F} \equiv \frac{\Delta\nu_{ax}}{\Delta\nu_c} = 2\pi\frac{\tau_c}{\tau_R} = \frac{2\pi}{\mathcal{L}_c}. \tag{4.92}$$

The cavity finesse $\mathcal{F}$ characterizes the sharpness of a cavity resonance relative to the frequency spacing between adjacent axial modes. For a 90-cm-long standing-wave cavity with 4% loss per roundtrip, we find

$$\Delta\nu_c = \frac{3.00 \times 10^8\,\text{m/s}}{2(0.90\,\text{m})} \times \frac{0.04}{2\pi} = 1.06\,\text{MHz} \approx 1.1\,\text{MHz}. \tag{4.93}$$

The axial-mode separation is

$$\Delta\nu_{ax} = \frac{c}{2nL} = \frac{3.00 \times 10^8}{2 \times 0.90}\,\text{Hz} = 167\,\text{MHz} \tag{4.94}$$

In this case, $\Delta\nu_c$ is small compared with the axial-mode spacing and corresponds to a high cavity finesse of $\mathcal{F} \approx 157$. Thus, the free spectral range $\Delta\nu$ax is approximately 160 cavity linewidths wide.

Cavity Q-factor

The cavity Q-factor is defined in terms of the ratio of the cavity lifetime τ_c to the optical period τ_o:

$$Q \equiv 2\pi \times \frac{\tau_c}{\tau_o} = \omega_o\tau_c = \frac{\nu_o}{\Delta\nu_c} = 2\pi\frac{2nL}{\lambda\mathcal{L}_c}. \tag{4.95}$$

It is another useful quantity for characterizing the spectral response of the cavity. The cavity lifetime, although short by ordinary standards, is still long compared with an optical cycle. For λo = 600 nm, an optical cycle is $\tau_o = \lambda/c = (600 \times 10^{-9})/(3.00 \times 10^8) = 2.00 \times 10^{-15}$ s. The ratio τ_c/τ_o is therefore the number of optical cycles that occur during one cavity lifetime. For our example, this ratio is

$$\frac{\tau_c}{\tau_o} = \frac{150 \times 10^{-9}}{2.00 \times 10^{-15}} = 7.5 \times 10^7, \tag{4.96}$$

which is a very large number. Thus, if a certain amount of energy is deposited in the cavity, many millions of optical cycles occur before the stored energy decays appreciably.

A cavity that stores energy for a long time, corresponding to many optical cycles, is said to be a high-Q cavity. From the dependence of Q on cavity losses, we see that the lower the loss, the higher the quality factor of the cavity. Expressing τ_c in terms of Q,

$$\tau_c = Q\frac{\tau_o}{2\pi} = \frac{Q}{\omega_o}, \tag{4.97}$$

and noting that the field intensity $I \propto (W/V)c/n$, where W is the energy stored in the cavity mode and V is the effective cavity-mode volume, we can write an equation for the fractional rate of loss of energy stored in the cavity:

$$\frac{1}{W}\frac{dW}{dt} = \frac{1}{I}\frac{dI}{dt} = -\frac{1}{\tau_c} = -\frac{\omega_o}{Q}. \tag{4.98}$$

With the help of this equation, we find that the energy lost in one optical cycle $\tau_o = 2\pi/\omega_o$ is $\Delta W_0 = \tau_o\left|\frac{dW}{dt}\right| = \frac{2\pi W}{Q}$, so that Q can also be written as

$$Q = 2\pi\frac{W}{\Delta W_0}. \tag{4.99}$$

Thus, Q is 2π times the ratio of the energy stored in the resonator to the energy lost in one optical cycle. Equivalently, $Q/2\pi$ is the number of optical cycles required to lose an amount of energy equal to the stored energy if the energy were lost at the constant rate ΔW_0 per optical cycle.

Note that Q can be quite high even for relatively lossy cavities. For example, for a 50 cm long two-mirror standing-wave cavity with 30% transmission loss at each mirror at $\lambda = 600$ nm, the loss factor is $\mathcal{L}_c = -2\ln(0.70) = 0.71$, which gives a Q-factor of

$$Q = 2\pi\frac{2nL}{\mathcal{L}\lambda} = 2\pi\frac{2 \times 0.50}{(0.71)(600 \times 10^{-9})} = 1.5 \times 10^7. \tag{4.100}$$

Such large values of Q arise because, although the loss per roundtrip may be substantial, the fractional energy loss during a single optical cycle is still very small.

Thus, whereas the finesse measures the sharpness of a resonance relative to the spacing between neighboring axial resonances, the Q-factor measures its sharpness relative to the optical frequency itself.

4.4 Resonance Properties of Laser Resonators

Optical resonators can also be used as spectrum analyzers, filters, or buildup cavities. For these applications, the resonator mirrors must be partially transmitting to permit coupling of external fields into and out of the cavity. Consider a two-mirror resonator filled with a dielectric of refractive index n and absorption coefficient α. Such a resonator, when used as a spectrum analyzer, is referred to as a Fabry-Perot cavity or interferometer. A monochromatic field is incident on mirror M_1. We assume that the incident field is spatially

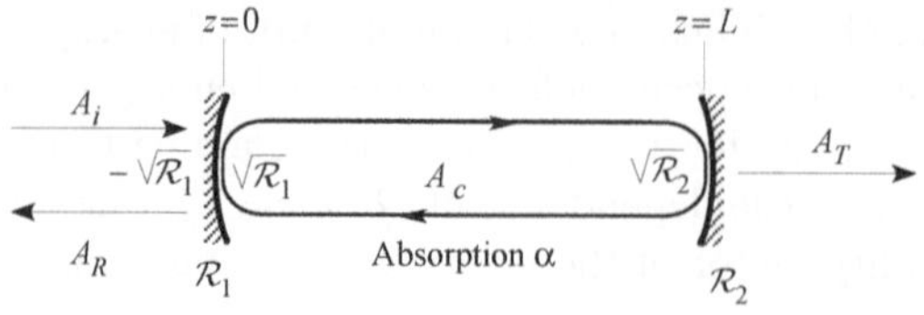

FIGURE 4.11
The right- and left-traveling waves are part of the same circulating field Ac inside the cavity.

mode-matched to one of the cavity modes. Let A_i, A_R, and A_T denote the amplitudes of the incident, reflected, and transmitted fields, respectively, and let A_c denote the amplitude of the circulating field inside the resonator. For mode-matched fields, the transverse envelope of the fields drops out of the boundary conditions. The remaining space-time dependence of the field is then $e^{-\alpha z/2+i(kz-\omega t)}$. In the steady state, the field amplitudes for a standing-wave cavity satisfy the following boundary conditions at $z=0$ and $z=L$:

$$A_c = \sqrt{\mathcal{T}_1}\, A_i + \sqrt{\mathcal{R}_1\mathcal{R}_2}\, e^{-\alpha L+i2\phi} A_c\,, \tag{4.101a}$$

$$A_R = -\sqrt{\mathcal{R}_1}\, A_i + \sqrt{\mathcal{T}_1} e^{-\alpha L+i2\phi} A_c\,, \tag{4.101b}$$

$$A_T = \sqrt{\mathcal{T}_2}\, A_c e^{-\alpha L/2+i\phi}\,. \tag{4.101c}$$

The phase ϕ is related to the wave frequency $\omega = 2\pi\nu$ and the refractive index n of the medium filling the resonator by

$$\phi \equiv kL = \frac{n\omega L}{c}\,. \tag{4.102}$$

We can solve Eqs. (4.101) for A_c, A_T, and A_R to find

$$A_T = \frac{\sqrt{\mathcal{T}_1\mathcal{T}_2}\, e^{-\alpha L/2} e^{i\phi}}{1-\sqrt{\mathcal{R}_1\mathcal{R}_2}\, e^{-\alpha L} e^{i2\phi}} A_i\,, \tag{4.103a}$$

$$A_R = -\frac{\sqrt{\mathcal{R}_1} - \sqrt{\mathcal{R}_2}\, e^{-\alpha L+i2\phi}}{1-\sqrt{\mathcal{R}_1\mathcal{R}_2} e^{-\alpha L} e^{i2\phi}} A_i\,, \tag{4.103b}$$

$$A_c = \frac{\sqrt{\mathcal{T}_1}}{1-\sqrt{\mathcal{R}_1\mathcal{R}_2}\, e^{-\alpha L} e^{i2\phi}} A_i\,. \tag{4.103c}$$

Using these equations, we can define the transmission and reflection coefficients of the cavity as the ratios of the transmitted and reflected intensities to the incident intensity $I_i = \frac{1}{2}\epsilon_o nc|A_i|^2$. These coefficients, together with the enhancement factor for the intensity of the circulating wave inside the cavity, are then given by

$$T_c \equiv \frac{I_T}{I_i} = \frac{T_{\rm max}}{1+F\sin^2\phi}\,, \tag{4.104a}$$

$$R_c \equiv \frac{I_R}{I_i} = \frac{R_{\rm min}+F\sin^2\phi}{1+F\sin^2\phi}\,, \tag{4.104b}$$

$$M_c \equiv \frac{I_c}{I_i} = \frac{M_{\rm max}}{1+F\sin^2\phi}\,, \tag{4.104c}$$

where we have taken the refractive index to be the same both inside and outside the cavity and introduced the following definitions:

$$F = \frac{4\sqrt{\mathcal{R}_1\mathcal{R}_2}e^{-\alpha L}}{(1-\sqrt{\mathcal{R}_1\mathcal{R}_2}e^{-\alpha L})^2}, \tag{4.105a}$$

$$T_{\max} = \frac{\mathcal{T}_1\mathcal{T}_2\, e^{-\alpha L}}{(1-\sqrt{\mathcal{R}_1\mathcal{R}_2}e^{-\alpha L})^2}, \tag{4.105b}$$

$$R_{\min} = \frac{(\sqrt{\mathcal{R}_1}-\sqrt{\mathcal{R}_2}e^{-\alpha L})^2}{(1-\sqrt{\mathcal{R}_1\mathcal{R}_2}e^{-\alpha L})^2}, \tag{4.105c}$$

$$M_{\max} = \frac{\mathcal{T}_1}{(1-\sqrt{\mathcal{R}_1\mathcal{R}_2}e^{-\alpha L})^2}. \tag{4.105d}$$

Here T_c and R_c represent, respectively, the overall power transmittance and reflectance of the cavity, and M_c is the enhancement factor by which the intracavity intensity is built up relative to the incident intensity. We neglect absorption and scattering in the mirror coatings, so that $\mathcal{R}_i + \mathcal{T}_i = 1$. The significance of the various quantities introduced in Eq. (4.105) will become clear in the following paragraphs. Equations (4.104) show that T_c, R_c, and M_c are periodic functions of ϕ. They pass through a series of maxima and minima as ϕ varies. In particular, the cavity transmission coefficient is not simply the product of the transmission coefficients of the two end mirrors.

The cavity transmission and reflection coefficients as functions of ϕ are shown in Figs. 4.12 and 4.13. The cavity transmission attains its maximum value $T_{\max}$, given by Eq. (4.105b), whenever ϕ is a multiple of π:

$$\phi \equiv \frac{n\omega L}{c} = p\pi, \tag{4.106}$$

where p is an integer. Successive transmission maxima are separated by $\Delta\phi = \pi$. This condition can be achieved in several different ways. For fixed frequency ω, we can change the cavity length L. To pass from one transmission maximum to the next, the cavity length must then change by $\Delta L = \frac{\pi c}{n\omega} = \frac{\lambda}{2n}$. On the other hand, for fixed cavity length L, the frequency ω can be tuned. In that case, the frequency must change by $\Delta\nu = \frac{c}{2nL}$ to pass from one maximum to the next. It is also possible to tune the refractive index n of the medium, for instance, by filling the cavity with a gas and varying the gas pressure.

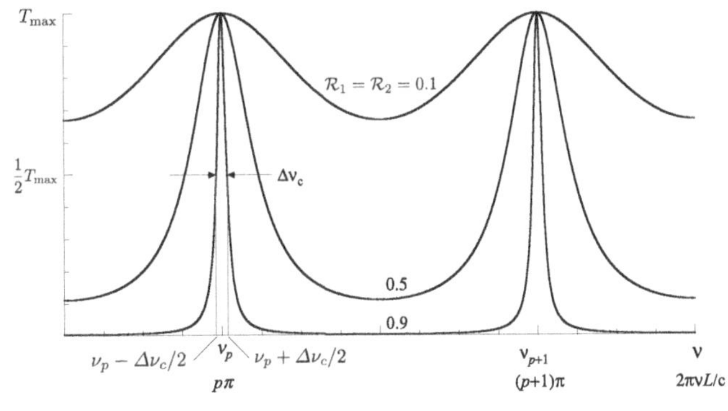

FIGURE 4.12
Transmission T_c [Eq. (4.104a)] from a Fabry-Perot cavity as a function of ϕ.

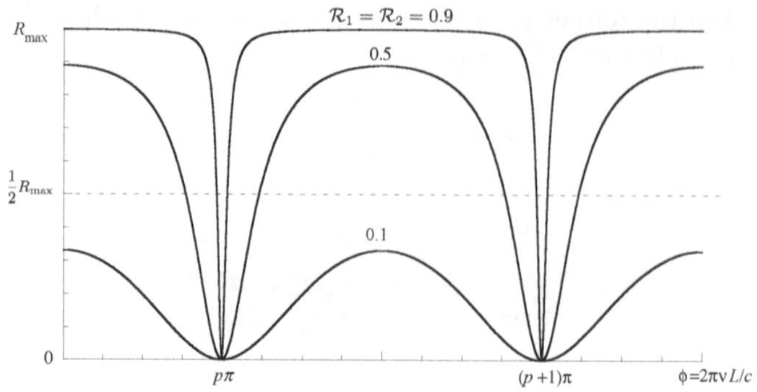

FIGURE 4.13
Reflection R_c [Eq. (4.104b)] from a Fabry-Perot cavity as a function of single-pass phase ϕ.

The condition for maximum transmission (4.106) implies that, in this longitudinal Fabry-Perot description, transmission is a maximum when the incident-field frequency coincides with a frequency

$$\nu_p = \frac{pc}{2nL} . \tag{4.107}$$

These frequencies have the same axial spacing as the cavity-mode frequencies given by Eq. (4.29). More generally, for a spatially mode-matched Gaussian beam, the absolute resonance frequencies also contain the transverse-mode-dependent Gouy-phase contribution appearing in Eq. (4.27). Thus, cavity transmission is a maximum whenever the incident-field frequency is in resonance with the corresponding cavity-mode frequency. For this reason, cavity transmission profiles are also referred to as cavity resonance profiles.

It is also remarkable to note from Fig. 4.12 that even when each mirror by itself has a small transmittance, the cavity as a whole may have a much larger transmittance. This is a dramatic consequence of wave interference. It can be understood by considering the transmitted field as a superposition of partial transmissions generated as the wave injected into the cavity bounces back and forth between the mirrors [Fig. 4.14]. When the resonance condition is satisfied, all the partially transmitted fields add in phase. The integer p is referred to as the order of a transmission maximum and serves to label a particular transmission peak. Assuming that the incident field is spatially mode-matched to a particular transverse mode of the cavity, the frequency interval between successive maxima is

$$\Delta\nu_{\mathrm{ax}} \equiv \nu_{p+1} - \nu_p = \frac{c}{2nL} , \tag{4.108}$$

which is equal to the axial-mode separation of the cavity. In the context of resonators as spectrum analyzers, this interval is known as the free spectral range[1]

$$\Delta\nu_{\mathrm{FSR}} = \frac{c}{2nL} = \Delta\nu_{\mathrm{ax}} . \tag{4.109}$$

The free spectral range is the maximum frequency interval over which a spectrum can be analyzed without ambiguity arising from transmission by adjacent resonance orders. If the input signal spectrum is wider than the free spectral range, $\frac{c}{2nL}$, more than one

[1]For a confocal resonator, the combined axial-transverse spectrum contains resonances separated by $c/4nL$ [see Sec. 4.2]; for a fixed transverse mode, the axial free spectral range remains $\Delta\nu_{\mathrm{FSR}} = \frac{c}{2nL}$.

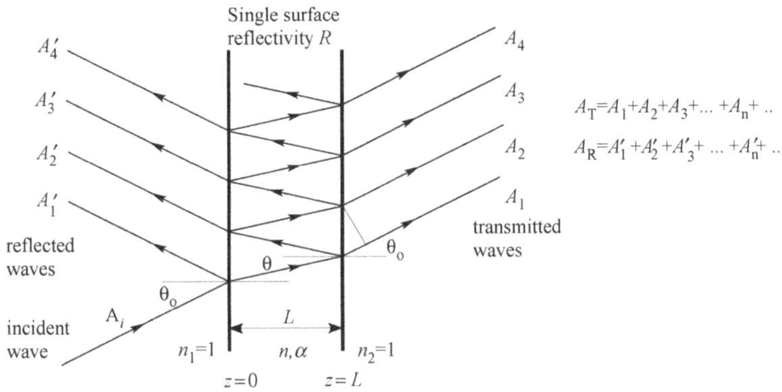

FIGURE 4.14
The reflected and transmitted beams can be written as a superposition of multiple beams generated by partial transmission and reflection of the beam entering the Fabry-Perot cavity. The angle of incidence has been exaggerated to show multiple reflections. For nonzero angle of incidence θ_0 and, therefore, nonzero internal angle θ, we can write the phase as $\phi = 2n\pi\nu L\cos\theta/c$.

frequency [signal spectrum I in Fig. 4.15] may be transmitted simultaneously, although the transmission orders of the two waves in Eq. (4.106) will be different. For example, frequencies ν_1 and ν_2 satisfying the resonance conditions $\nu_1 = pc/2nL$ and $\nu_2 = (p+1)c/2nL$ are transmitted simultaneously. If the cavity length is scanned, its resonances may similarly transmit two frequencies ν'_1 and ν'_2 at the same cavity setting. On the other hand, if the input signal spectrum is narrower than the free spectral range $\frac{c}{2nL}$ [signal spectrum II in Fig. 4.15], only one frequency within that spectral interval can satisfy the resonance condition for a given cavity setting. In other words, the relationship between the frequency ν and Fabry-Perot transmission $T(\nu)$ is unambiguous only if the input signal contains spectral components confined to a frequency interval smaller than $\frac{c}{2nL}$. Thus, the identification of spectral components of incident light by cavity-transmission resonances is free from ambiguity only if the input signal occupies a frequency interval smaller than the spectral range $\frac{c}{2nL}$. This interval is referred to as the free spectral range.

Halfway between two consecutive maxima of order p and $p+1$, the transmission has a minimum located at

$$\phi \equiv \frac{n\omega L}{c} = \left(p + \frac{1}{2}\right)\pi\,. \tag{4.110}$$

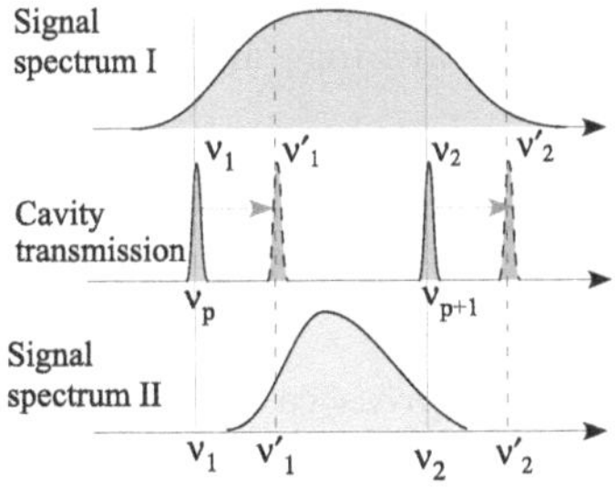

FIGURE 4.15
Cavity transmission $T(\nu)$, being a periodic function of frequency, can uniquely identify frequencies only if the extent of the signal spectrum or the separation between two discrete frequencies is less than the axial mode spacing, which in this context is referred to as the free spectral range.

This value of ϕ leads to the minimum transmission

$$T_{\min} = \frac{T_{\max}}{1+F}. \tag{4.111}$$

Note that the minima of the cavity transmission are not strictly zero but can be very small for $F \gg 1$.

From Fig. 4.12, we note that the transmission peaks become sharper as F increases. The coefficient F is determined by the mirror reflectances $\mathcal{R}_1$ and $\mathcal{R}_2$ and by the internal losses, represented here by $2\alpha L$. Other imperfections in an actual resonator may introduce additional losses and thereby reduce the sharpness of the resonances. When the transmission losses $\mathcal{T}_1 = 1-\mathcal{R}_1$, $\mathcal{T}_2 = 1-\mathcal{R}_2$, and the internal absorption loss $S_a = 1-e^{-2\alpha L}$ are small, we obtain a simpler expression for F:

$$F \equiv \frac{4\sqrt{\mathcal{R}_1\mathcal{R}_2}e^{-\alpha L}}{(1-\sqrt{\mathcal{R}_1\mathcal{R}_2}e^{-\alpha L})^2} \approx \frac{4}{(\mathcal{T}_1+\mathcal{T}_2+2\alpha L)^2/4} = \frac{16}{\mathcal{L}_c^2}. \tag{4.112}$$

The last approximation gives a value within a few percent of the exact value as long as the overall loss factor $\mathcal{L}_c \leq 0.1$.

A measure of the width of a transmission maximum is its full width at half maximum (FWHM). Let us consider a transmission maximum at $\nu = \nu_p$ and denote its FWHM by $\Delta\nu_c$. We wish to express $\Delta\nu_c$ in terms of the cavity parameters. The frequency $\nu_p + \Delta\nu_c/2$, at which the transmission has fallen to one-half of its maximum value at ν_p, is determined by $T(\nu_p + \frac{1}{2}\Delta\nu_c) = \frac{1}{2}T(\nu_p)$. With the help of Eq. (4.104a), this leads to

$$\frac{T_{\max}}{1+F\sin^2(p\pi + \pi\Delta\nu_c nL/c)} = \frac{1}{2}T_{\max},$$

or

$$F\sin^2(\Delta\nu_c \pi nL/c) = 1, \tag{4.113a}$$

where we have used the result $\sin^2(p\pi+\theta) = \sin^2\theta$. Thus, $\Delta\nu_c$ is determined exactly by

$$\Delta\nu_c = \frac{c}{\pi nL}\sin^{-1}\frac{1}{\sqrt{F}}, \tag{4.113b}$$

where the positive root has been retained because $\Delta\nu_c$, by definition, is positive.

For a large coefficient of finesse F, the argument of the sine function is small, and we can use the small-angle approximation $\sin\theta \approx \theta$ to obtain

$$\Delta\nu_c = \frac{2}{\pi\sqrt{F}}\cdot\frac{c}{2nL} \equiv \frac{\Delta\nu_{ax}}{\mathcal{F}}, \tag{4.114}$$

where $\Delta\nu_{ax} = c/2nL$ is the free spectral range and $\mathcal{F}$, a dimensionless quantity called the finesse of the resonator, is given by

$$\mathcal{F} = \frac{\pi\sqrt{F}}{2} = \frac{\pi(\mathcal{R}_1\mathcal{R}_2e^{-2\alpha L})^{1/4}}{1-\sqrt{\mathcal{R}_1\mathcal{R}_2e^{-2\alpha L}}}. \tag{4.115}$$

Here we have used Eq. (4.105a) to express cavity finesse directly in terms of the cavity parameters. It is clear that the width $\Delta\nu_c$ of a transmission peak is a small fraction of the separation between successive transmission peaks (the free spectral range) when the resonator finesse $\mathcal{F}$ is large.

The cavity finesse can also be interpreted in terms of spectral resolution when a Fabry-Perot interferometer (FPI) is used as a spectrum analyzer. Recall that the FWHM of a

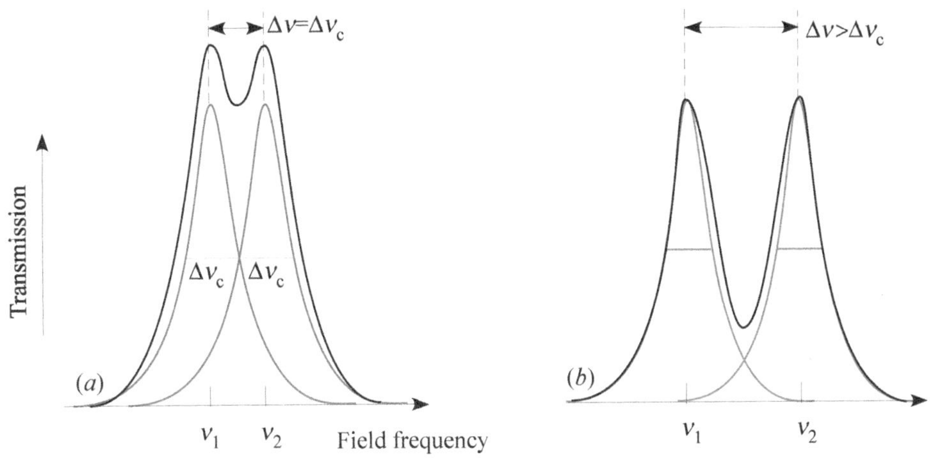

FIGURE 4.16
Cavity finesse and spectral resolution when the cavity is used as a spectrum analyzer.

transmission peak is the cavity linewidth $\Delta\nu_c$, and transmission occurs whenever a frequency component of the input field coincides with a resonance frequency of the FPI. If the separation between two transmission peaks corresponding to two frequency components in the input field is smaller than their linewidth, the output appears as a single broad peak [Fig. 4.16(a)], and the two frequencies cannot be readily distinguished. If the transmission peaks are separated by more than approximately one linewidth, but by less than the free spectral range, distinct peaks can be observed [Fig. 4.16(b)]. Thus, as a simple resolution criterion, two transmission peaks should be separated by approximately $\Delta\nu_c$ or more. The number of linewidth-sized resolution intervals contained within one free spectral range of the FPI is therefore

$$\frac{\Delta\nu_{_{FSR}}}{\Delta\nu_c} = \frac{\Delta\nu_{ax}}{\Delta\nu_c} .$$

This quantity is precisely the finesse of the FPI. Hence, finesse gives approximately the number of distinct frequency intervals that can be resolved unambiguously within one free spectral range and is therefore a useful measure of the spectral resolving capability of a resonator.

Another way of interpreting finesse follows from the relation between cavity linewidth and lifetime, $\Delta\nu_c = 1/2\pi\tau_c$, and between the free spectral range and the cavity roundtrip time, $\Delta\nu_{ax} = 1/\tau R$. We then find

$$\mathcal{F} = \frac{\Delta\nu_{_{ax}}}{\Delta\nu_c} = \frac{2\pi\tau_c}{\tau_{_R}} , \tag{4.116}$$

which shows that the finesse is 2π times the number of cavity roundtrips occurring during one cavity lifetime. For small losses [$\mathcal{T}_i$, αL, etc. $\ll 1$], the expression for finesse (4.115) simplifies to

$$\mathcal{F} = \frac{2\pi}{\mathcal{T}_1 + \mathcal{T}_2 + 2\alpha L} = \frac{2\pi}{\mathcal{L}_c} . \tag{4.117}$$

Using this expression for the finesse, the cavity resonance width can be written as

$$\Delta\nu_c \equiv \frac{\Delta\nu_{ax}}{\mathcal{F}} = \frac{\mathcal{L}_c}{2\pi\tau_{_R}} . \tag{4.118}$$

Thus, the cavity linewidth $\Delta\nu_c$ is proportional to the logarithmic power loss factor per roundtrip. For small losses, $\mathcal{L}_c$ is approximately equal to the fractional power loss per roundtrip.

In terms of finesse $\mathcal{F}$, the cavity transmission, reflection and buildup factors become

$$T_c = \frac{T_{\max}}{1 + (2\mathcal{F}/\pi)^2 \sin^2\phi}, \tag{4.119a}$$

$$R_c = \frac{R_{\min} + (2\mathcal{F}/\pi)^2 \sin^2\phi}{1 + (2\mathcal{F}/\pi)^2 \sin^2\phi}, \tag{4.119b}$$

$$M_c = \frac{M_{\max}}{1 + (2\mathcal{F}/\pi)^2 \sin^2\phi}. \tag{4.119c}$$

In the neighborhood of a transmission peak at $\nu_p = p\Delta\nu_{ax}$, we can write the frequency ν in terms of its deviation from the peak-transmission frequency ν_p as $\nu = \nu_p + (\nu - \nu_p)$, so that

$$\phi = [\nu_p + (\nu - \nu_p)]2\pi nL/c = p\pi + \pi(\nu - \nu_p)/\Delta\nu_{ax}, \tag{4.120a}$$

and $$\sin\phi = \sin[p\pi + \pi(\nu - \nu_p)/\Delta\nu_{ax}] = \sin[\pi(\nu - \nu_p)/\Delta\nu_{ax}]. \tag{4.120b}$$

For a high-finesse cavity, the field frequency need not deviate far from the resonance frequency before the transmission becomes negligibly small. Thus, for frequencies close to the transmission maximum, where the transmission remains appreciable, we can use the small-argument approximation for the sine term. The cavity transmission function then takes the form

$$\begin{aligned} T_c &= \frac{T_{\max}}{1 + (2\mathcal{F}/\pi)^2 \sin^2[\pi(\nu - \nu_p)/\Delta\nu_{ax}]} \approx \frac{T_{\max}}{1 + (2\mathcal{F}/\pi)^2[\pi(\nu - \nu_p)/\Delta\nu_{ax}]^2} \\ &= \frac{T_{\max}}{1 + [2(\nu - \nu_p)/\Delta\nu_c]^2} = \left(\frac{T_{\max}\pi\Delta\nu_c}{2}\right)\left[\frac{1}{\pi}\frac{\Delta\nu_c/2}{(\nu - \nu_p)^2 + (\Delta\nu_c/2)^2}\right] \\ &\equiv \left(\frac{T_{\max}\pi\Delta\nu_c}{2}\right) S(\nu). \end{aligned} \tag{4.121}$$

The function $S(\nu)$, which we encountered in the discussion of cavity losses, appears again here as the Lorentzian lineshape function. This form is particularly useful when a high-finesse resonator is used as a spectral filter.

We have derived three different expressions for $\Delta\nu_c$:

$$\Delta\nu_c = \begin{cases} \frac{c}{2nL}\frac{1}{\pi}\cdot\frac{1-\sqrt{\mathcal{R}_1\mathcal{R}_2}e^{-\alpha L}}{(\mathcal{R}_1\mathcal{R}_2)^{1/4}e^{-\alpha L/2}} = \frac{\Delta\nu_{ax}}{\mathcal{F}} \\ \frac{c}{2nL}\frac{1}{2\pi}(-\ln\mathcal{R}_1\mathcal{R}_2 e^{-2\alpha L}) = \frac{\Delta\nu_{ax}\mathcal{L}_c}{2\pi} = \frac{1}{2\pi\tau_c} \equiv \frac{\nu_p}{Q} \\ \frac{c}{2nL}\frac{1}{2\pi}(2\alpha L + \mathcal{T}_1 + \mathcal{T}_2) = \frac{\Delta\nu_{ax}}{\mathcal{F}} = \frac{1}{2\pi\tau_c} \end{cases} \tag{4.122}$$

The first expression $\Delta\nu_c = \Delta\nu_{ax}/\mathcal{F}$ comes from the Airy transmission function after applying the high-finesse approximation. The truly exact Airy FWHM is still Eq. (4.113b). The second expression $\Delta\nu_c = \mathcal{L}_c/2\pi\tau_R = 1/2\pi\tau_c$ comes from the coarse-grained exponential lifetime treatment, and the third expression replaces the logarithmic loss by its small-loss approximation $\mathcal{L}_c \approx 2\alpha L + \mathcal{T}_1 + \mathcal{T}_2$. In practice, the first and second expressions for $\Delta\nu_c$ give nearly the same result for sufficiently high-reflectance cavities, while both reduce to the third expression in the small-loss limit $|1 - \mathcal{R}_1|$, $|1 - \mathcal{R}_2|$, and $1 - e^{-2\alpha L} \leq 0.1$. The small differences among these expressions arise from the different approximations used in deriving them and from their corresponding ranges of validity. For high-quality, high-finesse cavities, these differences become negligible.

Case I: $\mathcal{R}_1 \approx \mathcal{R}_2 = 0.98$, $2\alpha L \approx 0$

$$\Delta\nu_c = \begin{cases} \frac{c}{2nL} \times \frac{1}{\pi} \times 0.02020305 = \frac{c}{2nL} \times 6.4308 \times 10^{-3} \\ \\ \frac{c}{2nL} \times \frac{1}{\pi} \times 0.020202071 = \frac{c}{2nL} \times 6.4307 \times 10^{-3} \\ \\ \frac{c}{2nL} \times \frac{1}{\pi} \times 0.02 = \frac{c}{2nL} \times 6.3662 \times 10^{-3} \end{cases}$$

Case II: $\mathcal{R}_1 \approx \mathcal{R}_2 = 0.5$, $2\alpha L \approx 0$

$$\Delta\nu_c = \begin{cases} \frac{c}{2nL} \times \frac{1}{\pi} \times 0.7071068 = \frac{c}{2nL} \times 0.225 \\ \\ \frac{c}{2nL} \times \frac{1}{\pi} \times 0.69314718 = \frac{c}{2nL} \times 0.221 \\ \\ \frac{c}{2nL} \times \frac{1}{\pi} \times 0.5 = \frac{c}{2nL} \times 0.159 \end{cases}$$

These examples illustrate the relative ranges of validity of the three expressions. For $\mathcal{R}_1 = \mathcal{R}_2 = 0.98$, all three give nearly the same result. Even for $\mathcal{R}_1 = \mathcal{R}_2 = 0.5$, the first two expressions differ by only about 2%, whereas the small-loss approximation represented by the third expression differs substantially. In practice, uncertainties in cavity losses may themselves be comparable to or larger than the small differences between the first two expressions, so either may provide an adequate estimate in many applications. The small-loss expression, however, should be used only when its underlying assumptions are satisfied.

For $L = 1.00$ m in air ($n \approx 1$), $\mathcal{R}_1 = \mathcal{R}_2 \approx 0.98$, and $\alpha L = 0$, we find the cavity linewidth

$$\Delta\nu_c = \frac{c}{2nL}\left(\frac{\mathcal{T}_1 + \mathcal{T}_2}{2\pi}\right) = \frac{3.00 \times 10^8 (0.04)}{2 \times 1.00 \times 2\pi} = 9.55 \times 10^5\,\text{Hz} \approx 1.0\,\text{MHz}\,.$$

The finesse is

$$\mathcal{F} \equiv \frac{\Delta\nu_{ax}}{\Delta\nu_c} = \frac{c/2nL}{(c/2nL)(\mathcal{L}_c/2\pi)} = \frac{2\pi}{\mathcal{L}_c} = 157\,.$$

4.4.1 Cavity Buildup

The intracavity field buildup is also a periodic function of ϕ, and at a transmission resonance the buildup factor attains its maximum value

$$M_{\max} = \frac{\mathcal{T}_1}{(1 - \sqrt{\mathcal{R}_1 \mathcal{R}_2 e^{-\alpha L}})^2}\,. \tag{4.123}$$

For a high-finesse cavity ($\mathcal{R}_1\,, \mathcal{R}_2 \approx 1$ and $2\alpha L \ll 1$)

$$M_{\max} \approx \frac{4\mathcal{T}_1}{(\mathcal{T}_1 + \mathcal{T}_2 + 2\alpha L)^2}\,. \tag{4.124}$$

From Eq. (4.105c), zero reflection at resonance requires $R_{\min} = 0$, or $\sqrt{\mathcal{R}_1} = \sqrt{\mathcal{R}_2}\, e^{-\alpha L}$, hence

$$\mathcal{R}_1 = \mathcal{R}_2 e^{-2\alpha L}\,. \tag{4.125}$$

At this condition, the field promptly reflected from the input mirror is canceled by destructive interference with the field leaking out of the cavity through the same mirror. Zero reflected power does not mean that the input mirror ceases to reflect the incident field.

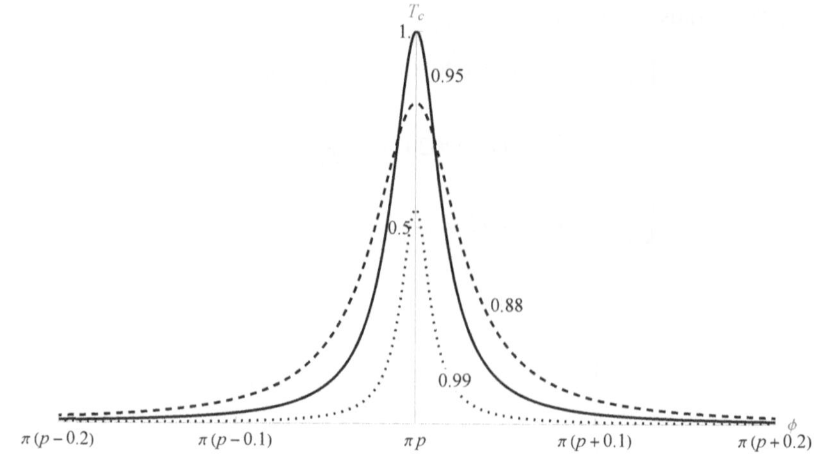

FIGURE 4.17
Cavity transmission for fixed input mirror reflectance $\mathcal{R}_1 = 0.95$ and several different output mirror reflectances $\mathcal{R}_2$.

Rather, at critical coupling the promptly reflected field and the field leaking back out of the cavity through the input mirror have equal amplitudes and opposite phases, so they interfere destructively.

For small losses, $\mathcal{R}_1 \approx 1 - \mathcal{T}_1$, $\mathcal{R}_2 \approx 1 - \mathcal{T}_2$, and $e^{-2\alpha L} \approx 1 - 2\alpha L$, the condition for zero reflection becomes

$$\mathcal{T}_1 \approx \mathcal{T}_2 + 2\alpha L\,, \tag{4.126}$$

corresponding to a maximum buildup factor of

$$M_{\max} = \frac{1}{\mathcal{T}_1}\,. \tag{4.127}$$

Thus, even a modest intracavity intensity buildup factor of 10 can produce approximately a 100-fold increase in a nonlinear optical signal whose strength scales as the square of the fundamental intensity, as in second-harmonic generation in the undepleted-pump regime.

Summary

In this chapter, we used the paraxial Gaussian beams of Chapter 3 to determine the preferred modes of open resonators. This was done by requiring that, after undergoing the transformations produced by the optical elements over one cavity roundtrip, a Gaussian beam recover its original Gaussian shape. We first illustrated this requirement for a two-mirror standing-wave resonator and then generalized it to an arbitrary stable resonator.

By further requiring that the phase of the Gaussian beam remain unchanged modulo 2π after one roundtrip, we found that this condition can be satisfied only for certain values of the wave number and, therefore, frequency, thereby determining the Gaussian-beam modes of stable resonators. This analysis yielded the location of the beam waist, the beam spot radius, and the spectrum of allowed frequencies in terms of cavity parameters such as mirror radii of curvature and mirror separations. These are the long-lived, preferred field modes of open cavities that provide the optical feedback crucial to laser operation and that we had set out to determine.

We also discussed how losses due to absorption, diffraction, scattering, and finite mirror transmission are specified and how they affect the cavity-mode spectrum. In doing so, we introduced the concepts of cavity lifetime, linewidth, finesse, and quality factor. Finally, just as electromagnetic energy can leak out of a cavity through mirror transmission, it can also be coupled into the cavity. We concluded the chapter by discussing applications of low-loss optical cavities for spectral analysis and for producing high intracavity intensities, for example, in nonlinear optical applications.

Problems

1. **Roundtrip matrix and beam parameters:**
Given the roundtrip ABCD matrix for a stable optical cavity, starting from some reference plane at z_1 (measured relative to the beam waist), show that the radius of curvature and spot radius of the trapped Gaussian beam at the reference plane are given by

$$R(z_1) = -\frac{2B}{A-D},$$

$$\frac{\pi n w^2(z_1)}{\lambda} = \frac{|B|}{\sqrt{1-\left(\frac{A+D}{2}\right)^2}}.$$

Why must the modulus $|B|$ appear in the expression for the spot radius? What condition must A and D satisfy if the reference plane z_1 coincides with a beam waist? What is the waist radius in that case?

Apply these results to a cavity consisting of a plane mirror facing a spherical concave mirror of radius of curvature R, with mirror separation $L = R/2$. Compute the roundtrip matrix using a reference plane just in front of the concave mirror and initially propagating toward the plane mirror. Use the matrix to determine the Gaussian-beam parameters at this reference plane.

2. The laser cavity shown here produces a fundamental Gaussian beam with its waist at $z = 0$, located at the plane mirror. The output beam propagates a distance d and is then incident on a thin lens of focal length f. Assume that $w_0 = 0.50$ mm, $\lambda = 633$ nm, $d = 1.0$ m, and $f = 0.25$ m.

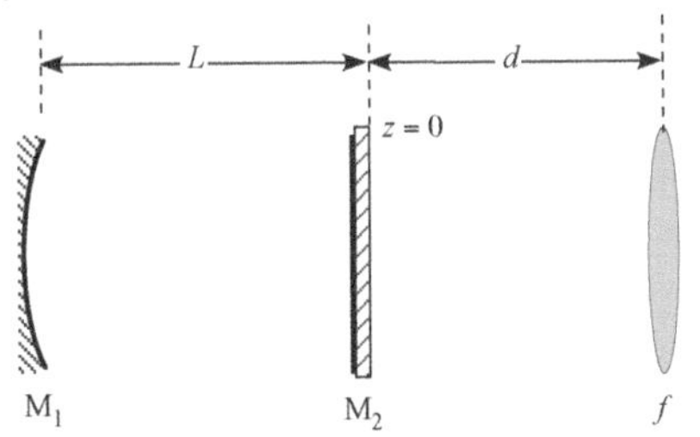

(a) Find the spot radius and phasefront radius of curvature of the beam incident on the lens.

(b) Find the location of the new waist after the lens and determine its waist radius. Also find the phasefront radius of curvature immediately after the lens.

3. The laser cavity shown here consists of a concave mirror and a convex mirror with radii of curvature $R_1 = R$ and $R_2 = -4R$, respectively, where $R > 0$.

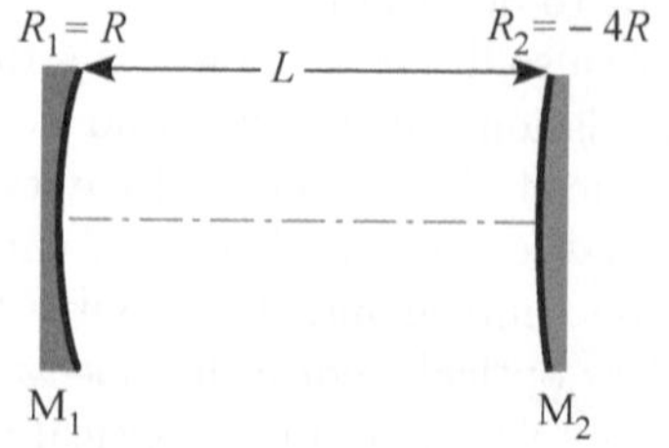

   (a) What is the maximum mirror separation L for which the cavity is stable?

   (b) If $L = R/2$, find the position of the Gaussian-mode waist. Determine w_0 and z_R in terms of R and the laser wavelength λ.

   (c) A thin lens with focal length $f = R$ is placed at the waist location found in part (b). Where does the new waist occur? Express its location and minimum spot radius in terms of the original waist radius w_0, the wavelength λ, and R as needed.

   (d) Write an expression for the electric field of the fundamental cavity mode as a function of space and time. Clearly specify where $z = 0$ is chosen. Leave your answer in terms of k, z_R, $w(z)$, and $R(z)$.

4. **Spot radius and mode spacing in symmetric resonators:** A symmetric resonator consists of two spherical concave mirrors, each with radius of curvature R, facing each other and separated by a distance L.

   (a) Find the stability condition in terms of R and L. For fixed $R = 50.0$ cm, $\lambda = 632.8$ nm, and $n = 1$, plot the waist radius w_0 and the mirror spot radius w_1 as functions of L/R over the stable range. Indicate the planar, confocal, and concentric limiting configurations on the plot. Which configuration gives the largest waist radius?

   (b) Now keep $L = 50.0$ cm fixed, so that the axial mode spacing $\Delta\nu_{ax}$ remains fixed. Compute the axial and transverse mode spacings for (i) $L/R = 0.1$ (near-planar), (ii) $L/R = 1$ (confocal), and (iii) $L/R = 1.9$ (near-concentric). Compare and comment on the results.

5. A Fabry-Perot cavity of length L with two plane mirrors is marginally stable. Placing a positive (focusing) thin lens of focal length f at the center of the cavity can make the cavity stable for an appropriate range of focal lengths.

   (a) Determine the range of focal lengths for which the cavity is stable. Identify any special marginal or degenerate cases that occur within or at the boundaries of this range. Sketch the cavity mode as the system approaches the stability boundaries.

   (b) Find the location of the waist (or waists) and the corresponding minimum spot radius. The situation considered in this problem provides a useful model for some lasers in which the gain medium has a cylindrical geometry. Absorption of pump power can make the center of the gain medium hotter than its periphery. Through the temperature dependence of the refractive index, this transverse temperature variation produces an effective focusing element known as a **thermal lens**.

6. **Near-hemispherical cavity:** Consider a Fabry-Perot cavity of length L in which mirror 1 is plane and mirror 2 is concave with radius of curvature R_2. The cavity is called near-hemispherical when $R_2 \approx L$. Consider a stable near-hemispherical cavity with $R_2 = L + \Delta L$, where $\Delta L > 0$ and $\Delta L \ll L$. As $\Delta L \to 0^+$, the cavity approaches the hemispherical stability limit.

(a) Sketch the cavity qualitatively, showing the expected variation of the Gaussian-beam spot radius and phasefront curvature along the cavity. Find the leading dependence of the spot radius on mirror 2 as a function of ΔL as $\Delta L \to 0^+$.

(b) For a cavity of length $L = 30$ cm operating at a wavelength of $1\,\mu$m, how small must ΔL be for the spot radius on mirror 2 to be 5 mm? What is the spot radius on mirror 1 for this design?

7. **Resonator for gravitational-wave detection:**

(a) For a symmetric two-mirror cavity, the spot radii of the fundamental Gaussian mode are equal at the two mirrors. For a symmetric cavity of length L, determine the mirror radius of curvature that minimizes the spot radius on the mirrors. What is the corresponding minimum mirror spot radius for light of wavelength λ?

(b) The LIGO gravitational-wave detector must measure extremely small changes in length over a 4-km baseline. It uses optical interferometry, with a 4-km-long Fabry-Perot interferometer (FPI) in each arm. Assuming a symmetric FPI, evaluate the results of part (a) for a wavelength of $1\,\mu$m. What is the smallest mirror spot radius that can be obtained, and what mirror radius of curvature is required?

(c) Assuming that the Gaussian beam is centered on each mirror, how large must the mirror radius be so that less than 10^{-6} of the total optical power extends beyond the mirror edge and is lost? Meeting the required tolerances for mirrors of this size and curvature presents a significant fabrication challenge.

8. **Effect of output mirror on exit beam waist:**
We have derived expressions for the location of the beam waist relative to the mirrors in a two-mirror resonator. As the beam exits the cavity through the output mirror, however, refraction by the mirror substrate changes the apparent waist location and waist radius of the emerging beam.

Consider, for concreteness, a symmetric confocal cavity of mirror separation L. Find the location of the waist of the beam emerging from the laser relative to the output mirror and determine its waist radius. Model the output mirror substrate as a thin plano-concave lens of refractive index n, with the curved surface having the same radius of curvature as the cavity mirror and the outer surface being plane. Assume that the medium outside the cavity is air. For $n = 1.5$, determine the shift of the output-beam waist relative to the intracavity waist in terms of L.

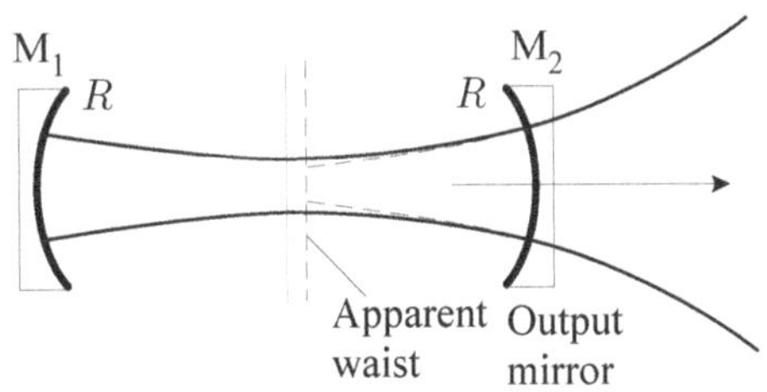

9. Suppose we wish to design a two-mirror cavity having the Gaussian-beam parameters shown in the figure. The waist is located $d = 20$ cm to the left of M_1 and has Rayleigh range $z_R = 100$ cm. Mirror M_2 is located $L = 60$ cm to the right of M_1.

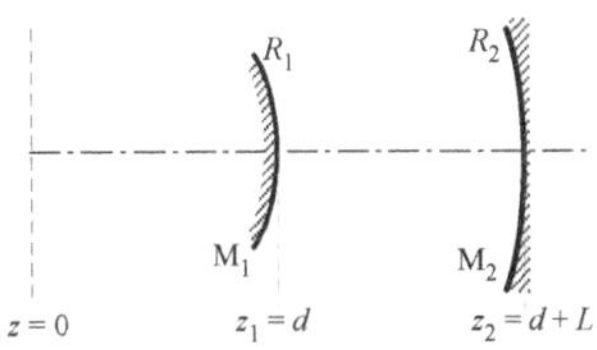

Determine the radii of curvature of mirrors M_1 and M_2, including their signs according to the convention used in this chapter. Also determine the axial and transverse mode-frequency separations.

10. A short section of the graded-index fiber considered in Sec. 2.8.1 [and Problem 9 of Chapter 2] is used as an optical cavity supporting Gaussian modes. Take the z-axis along the fiber axis. Determine the spot radii of the fundamental mode at the two end faces of the fiber. How does the beam spot radius vary with z inside the fiber? Write down the spatial dependence of the fundamental Gaussian mode supported by this cavity. [Hint: Compute the roundtrip matrix and use the ABCD law for Gaussian beams.]

    Note: The situation considered in this problem also provides a useful model for some optically pumped lasers in which the gain medium has a cylindrical geometry. Absorption of pump power can make the center of the gain medium hotter than its periphery. Through the temperature dependence of the refractive index, this transverse temperature variation can produce a **thermal lens** with a radial refractive-index profile of the type considered here.

11. In the stable optical cavity shown here, the waist plane, $z = 0$, occurs at a distance 25 cm to the left of M_1, with the beam parameter $z_R = 125$ cm. The mirrors are separated by 75 cm.

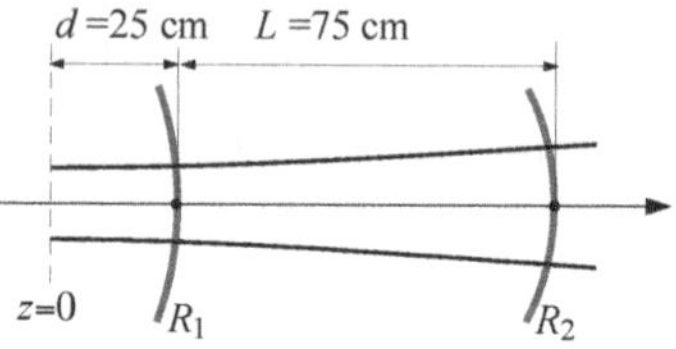

    (a) Find an expression for the resonant frequency of the $HG_{\ell mp}$ mode.

    (b) Find the separation between successive axial and transverse modes.

    (c) Find the radii of curvature R_1 and R_2 for the two mirrors.

12. (a) Evaluate the mode-volume integral, Eq. (4.54), for the fundamental Gaussian mode ($\ell = 0 = m$) of a traveling-wave cavity. What modification is required for a standing-wave cavity?

    (b) For a symmetric cavity ($R_1 = R = R_2$), determine the mirror separation L that maximizes the mode volume.

13. Consider the laser shown in the accompanying figure, which we wish to operate in the fundamental Gaussian mode at a wavelength of $0.4\,\mu$m. In this problem, rather than using the intensity-weighted mode volume defined in Eq. (4.54), we define a geometrical interaction volume as the volume of the gain medium ($n = 2.4$) lying within a radius $\rho \leq 2w(z)$ of the beam axis, where $w(z)$ is the Gaussian spot radius. Since the irradiance falls rapidly with distance from the beam axis, this provides a useful estimate of the volume of gain medium that interacts strongly with the laser mode.

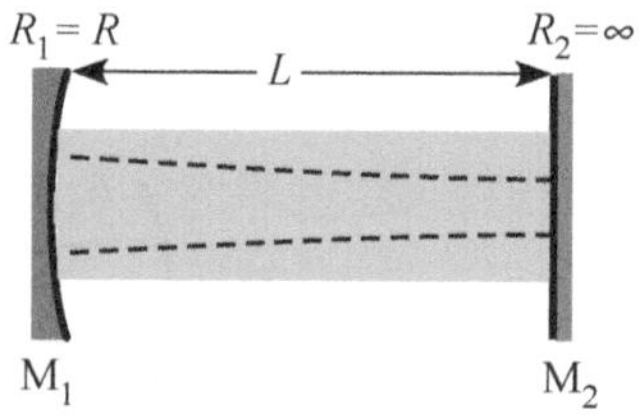

    Plot this geometrical interaction volume as a function of the mirror radius of curvature R for $L = 40$ cm, and discuss how its dependence on R might affect the efficiency of the laser.

14. The long arms of the bow-tie and z-shaped cavities shown in Fig. (a) are commonly used in lasers to accommodate modulators, frequency tuners, and other intracavity optical

elements. Both folded cavities can be represented by equivalent two-mirror cavities. Show this by following the steps outlined below.

For simplicity, assume symmetric fold angles and neglect astigmatism caused by non-normal incidence on the cavity mirrors and gain medium. [Astigmatism can be included by replacing (f_1, f_2, t) by $(f_{1t} = f_1 \cos\theta_1, f_{2t} = f_2 \cos\theta_2, t_t = t/n^3)$ for the tangential plane, and by $(f_{1s} = f_1/\cos\theta_1, f_{2s} = f_2/\cos\theta_2, t_s = t/n)$ for the sagittal plane.]

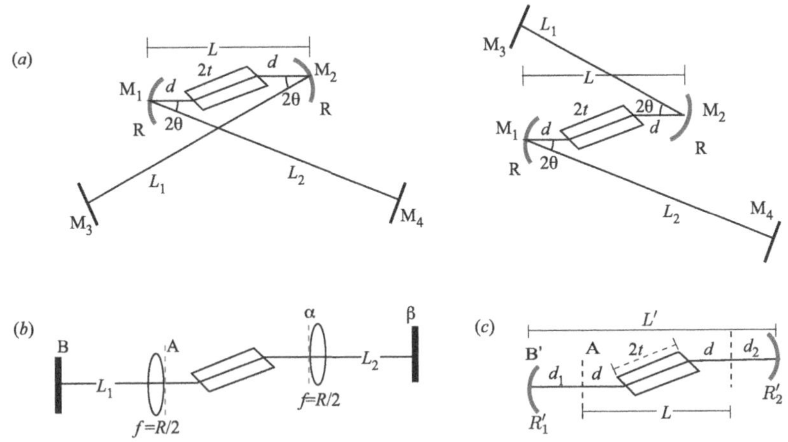

(a) Show that the folded cavities in (a) can be unfolded into the equivalent linear cavity shown in (b). Compute the ray-transfer matrix for the roundtrip path $A \to B \to A$ in (b), and for the corresponding symmetric path $A \to B' \to A$.

(b) By comparing the two matrices obtained in part (a), determine d_1 and R_1' in terms of R and L_1. Repeat the calculation for the other arm to determine d_2 and R_2' in terms of R and L_2.

The cavity in (c) can then be analyzed as an equivalent standing-wave two-mirror cavity with mirror radii of curvature R_1' and R_2' separated by $L' = d_1 + L + d_2$, using the results of Secs. 4.1 and 4.2 for two-mirror standing-wave cavities.

15. **Varying spot radius:** A 632.8 nm He-Ne laser cavity has a micrometer adjustment that varies the separation L between a plane mirror and an output mirror of radius of curvature $R = 50$ cm. If an output-beam spot radius between 1 mm and 2 mm is desired at the curved mirror, over what range must the mirror separation L be varied? Over what corresponding range does the waist radius at the plane mirror vary?

16. A laser resonator is formed by two mirrors with radii of curvature $R_1 = 20$ cm and $R_2 = 32$ cm, separated by a distance $L = 16$ cm. Take the operating wavelength to be $1.06\,\mu$m.

    (a) Find the location of the beam waist, the minimum spot radius, and the spot radii at the two mirrors.

    (b) Find the axial and transverse mode-frequency separations.

17. **Half-symmetric cavity parameters:** A Nd:YAG laser resonator ($\lambda = 1.06\,\mu$m) consists of a plane mirror and a concave mirror of radius of curvature 50.0 cm, separated by a distance of 25.0 cm.

    (a) Determine the location of the beam waist, the minimum spot radius, and the spot radii at the two mirrors.

(b) What are the axial and transverse mode-frequency spacings?

(c) If both mirrors have power reflectances of 98%, determine the cavity linewidth, neglecting all other losses. If this cavity is used as a spectrum analyzer, what is the minimum frequency separation it can resolve?

18. One technique for measuring small absorption is to place the absorber inside a high-Q resonator and inject a short laser pulse into the cavity. The injected pulse circulates around the cavity, and each time it reaches the output mirror, a transmitted pulse emerges. From the decay of the energies of successive output pulses, the cavity losses can be determined.

Consider a resonator consisting of two identical mirrors separated by a distance L, illuminated by a 1-ps pulse from an external source at a wavelength of 600 nm. The output consists of a regular sequence of 1-ps pulses separated by 10 ns. The pulse energy decays exponentially with a time constant of 100 ns. Determine the cavity length, the logarithmic cavity loss factor per roundtrip, and the corresponding fractional energy loss per roundtrip. If both mirrors have power reflectances of 99%, determine the absorption coefficient of a medium that uniformly fills the cavity.

19. A low-pressure gas cell ($L_a = 5.0$ cm, $\alpha = 0.002$ cm^{-1}, refractive index $n \approx 1$, the same as that of the rest of the cavity) is placed inside a two-mirror cavity of length $L = 10$ cm. Mirrors M_1 and M_2 both have power reflectances of 95% and radii of curvature $R = 10$ cm. In addition, there is a 2% scattering loss per cavity roundtrip.

For a fixed transverse cavity mode, determine (i) the frequency separation between successive transmission maxima, (ii) the cavity linewidth (FWHM), (iii) the cavity lifetime, and (iv) the number of roundtrips completed by the light in two cavity lifetimes.

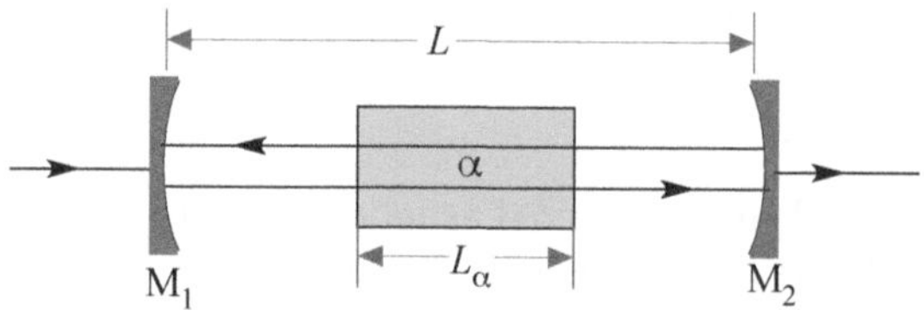

20. Three mirrors located at the vertices of a triangle form a ring cavity of perimeter L. Mirrors M_1 and M_2 are plane, while M_3 is concave with radius of curvature $R > 0$. Assume that the angle of incidence on the curved mirror is sufficiently small that astigmatism can be neglected and the mirror focal length may be taken as $R/2$ in both the tangential and sagittal planes.

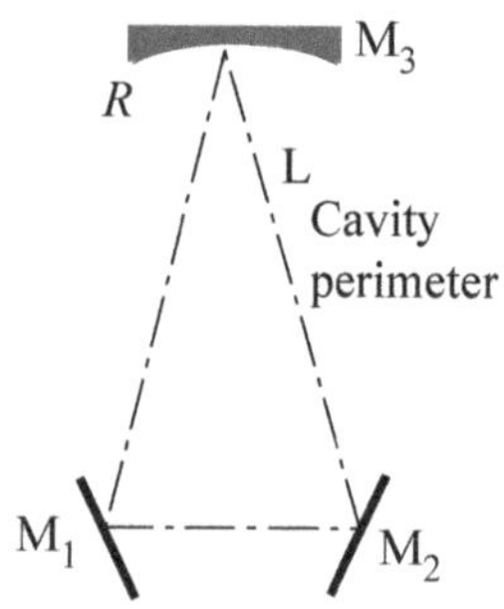

(a) Starting just in front of the curved mirror, construct a periodic lens waveguide equivalent to the cavity, identify a repeat unit, and write the roundtrip ray matrix as a product of the ABCD matrices of the optical elements encountered in one cavity roundtrip. Determine the range of values of L, in terms of R, for which the cavity is stable.

(b) Determine the resonance frequency $\nu_{\ell mp}$ of the HG$_{\ell mp}$ modes of the cavity. Express your result in terms of R, L, the mode indices, and c.

21. Consider again the ring cavity shown in the preceding problem. Suppose now that the angle of incidence θ at the curved mirror is sufficiently large that the effective mirror focal lengths differ in the two principal planes: $f_x = f_o \cos\theta$ in the x-z plane and $f_y = f_o / \cos\theta$

in the y-z plane, where $f_o = R/2$. Consequently, the conditions for bounded ray motion in the two planes are different, and the Gaussian mode is generally elliptical rather than circular. By symmetry, however, the beam waists in the two principal planes lie in the same longitudinal plane.

(a) Starting at the waist plane, construct the repeat cell for the ray coordinates in each principal plane and compute the corresponding roundtrip matrices. Determine the condition for bounded ray path in both planes, in terms of L, R, and θ.

(b) For $R = 25$ cm, $L = 50$ cm, and $\theta = 15^o$, determine whether the cavity is stable in both transverse planes. If it is not, identify which plane becomes unstable and explain why a stationary Gaussian mode with finite waist radii cannot exist for these parameters.

22. A cavity is excited by a 500 nm mode-matched laser beam and has resonances whose separation is 125 MHz and whose width is 2.5 MHz. Determine the following properties of the cavity: (a) cavity length, (b) cavity finesse, (c) cavity Q, and (d) photon lifetime.

23. A Fabry-Perot consisting of two identical mirrors is used for analyzing the emission spectrum of a laser. It is found to have a peak transmission of 99% and a finesse of 30. Find the reflectivity of the mirrors. If the resolution of the interferometer is 50 MHz, what is the separation between the mirrors?

24. A Fabry-Perot consisting of two identical mirrors separated by 2.5 cm is used for analyzing the emission spectrum of a 1.5 m long standing-wave dye laser. Find the minimum mirror reflectivity to just resolve two consecutive axial modes of the dye laser. If the same mirrors are used with a separation of 5.0 cm, what is the resolution of the Fabry-Perot? What is the cavity lifetime for this case?

25. A Fabry-Perot interferometer consisting of two identical air-spaced mirrors is illuminated by monochromatic light of tunable frequency. A measurement of the transmitted intensity versus the frequency of the input wave reveals a free spectral range of 3.0 GHz, resolution of 60 MHz, and peak transmission of 50%. Calculate the spacing, finesse, mirror reflectivity, and the internal loss of the interferometer.

5

Electron Oscillator Model of Light-Matter Interaction

In Chapters 2-4, we developed a comprehensive picture of the preferred long-lived electromagnetic wave modes of open resonators. Their spatial structure and frequency spectrum are determined by the boundary conditions imposed by the resonator walls, as well as by resonator losses. If a group of atoms capable of exchanging energy with these modes is introduced into the cavity, a mutual energy exchange occurs. Any energy emitted into these modes by excited atoms will be stored for millions of optical cycles—a duration sufficient to stimulate further emission from excited atoms into the same modes. Concurrently, unexcited atoms can absorb energy from these modes. To explore this energy exchange between atoms and fields further, we now describe how atoms respond to the fields of an electromagnetic wave.

Ideally, the atomic response to the electromagnetic field is described by the Schrödinger equation of quantum mechanics. However, it is much easier to derive and understand most aspects of atomic response using a classical model of light-matter interaction. The modifications required to align the resulting classical equations with a full quantum mechanical description can be motivated by relatively straightforward physical arguments. These include the incorporation of spontaneous emission, which is important for some aspects of laser operation. Therefore, we present here a classical model for the response of an atomic medium to a time-dependent electromagnetic field. According to this model, the fields of an electromagnetic wave drive the electronic charges inside the atoms, causing them to oscillate coherently with the driving fields. The atoms respond very much like classical driven charged oscillators. The fields reradiated by the atom oscillators combine coherently with the incident signal, resulting in absorption or amplification of the incident signal as well as a phase shift. For readers interested in a rigorous approach, a fully quantum mechanical treatment of the atomic response to an EM wave is presented in Chapter 6.

5.1 Harmonically Bound Charged Oscillator

In the classical model of light-matter interaction, the atomic medium is treated as a collection of bound electronic charges. Their behavior is described by Newton's second law of motion with a force given by the Lorentz force law. Under the influence of the fields of an applied electromagnetic wave, the bound electronic charges are perturbed from their equilibrium positions. For field strengths that are small compared to those exerted by the nucleus, these bound charges behave as simple harmonic oscillators driven by the applied field. Each oscillator is characterized by its own resonant frequency and energy damping constant. The overall response of the medium is the sum of these individual oscillator responses.

DOI: 10.1201/9781003584179-5

Consider the response of a single harmonically bound electron to the applied electromagnetic wave. Let $\boldsymbol{r}_o$ be the mean position and $\delta\boldsymbol{r}$ represent the complex[1] displacement of the electron from its mean position. Using the Lorentz force experienced by the electron in Newton's second law, and assuming small displacements from equilibrium, the equation of motion for the electron can be written as

$$\frac{d^2\delta\boldsymbol{r}}{dt^2}+\gamma_o\frac{d\delta\boldsymbol{r}}{dt}+\omega_o^2\delta\boldsymbol{r}=-\frac{e}{m}\left[\boldsymbol{E}(\boldsymbol{r}_o,t)+\frac{d\delta\boldsymbol{r}}{dt}\times\boldsymbol{B}(\boldsymbol{r}_o,t)\right], \tag{5.1}$$

where $-e$ is the charge of the electron, m is its rest mass, γ_o is the energy decay rate (accounting for both radiative and non-radiative processes), ω_o is the natural frequency (the eigenfrequency, corresponding to an atomic or molecular transition). $\boldsymbol{E}$ and $\boldsymbol{B}$ represent the complex electric and magnetic fields of the incident wave. Here we assume the displacement $\delta\boldsymbol{r}$ from equilibrium to be small compared with the wavelength of light (the dipole approximation) such that the variation of the field over atomic dimensions is negligible. This allows us to use fields evaluated at the equilibrium position $\boldsymbol{r}_o$ of the electron in the equation of motion.

Recalling that the magnitudes of the electric and magnetic fields of an electromagnetic wave are related by $|\boldsymbol{B}|=|\boldsymbol{E}|/c$, a comparison of the magnetic and electric force terms in Eq. (5.1) shows that their ratio is $\frac{1}{c}|\frac{d\delta\boldsymbol{r}}{dt}|\approx v/c$, where v is the speed of the atomic electron. For the non-relativistic motion of an atomic electron considered here, this ratio is 10^{-2} or smaller. Moreover, because the induced velocity and magnetic field are each first order in the applied field, the magnetic-force term is second order in field amplitude and does not contribute to the linear response considered here. The resulting equation for the electron displacement from equilibrium, and by extension the induced dipole moment $\boldsymbol{p}=-e\delta\boldsymbol{r}$ is given by

$$\frac{d^2\boldsymbol{p}}{dt^2}+\gamma_o\frac{d\boldsymbol{p}}{dt}+\omega_o^2\boldsymbol{p}=\frac{e^2}{m}\boldsymbol{E}(\boldsymbol{r}_o,t). \tag{5.2}$$

This is the equation of a charged damped harmonic oscillator driven by the electric field of an electromagnetic wave.

To analyze the frequency response of the oscillator, we consider a monochromatic driving field $\boldsymbol{E}(\boldsymbol{r},t)=\boldsymbol{E}(\boldsymbol{r})e^{-i\omega t}$. Recalling that the forced oscillations occur at the driving frequency, the steady-state induced dipole moment can be expressed as

$$\boldsymbol{p}(t)=\boldsymbol{p}_o e^{-i\omega t}, \tag{5.3}$$

where the amplitude $\boldsymbol{p}_o$ is to be determined. Substituting this into Eq. (5.2) and solving for $\boldsymbol{p}_o$ yields the steady-state induced dipole moment

$$\boldsymbol{p}(t)=\frac{(e^2/m)\boldsymbol{E}(\boldsymbol{r}_o,t)}{\omega_o^2-\omega^2-i\gamma_o\omega}. \tag{5.4}$$

Summing similar contributions from the remaining induced dipoles within the medium yields the macroscopic polarization (the dipole moment per unit volume), which ultimately characterizes the overall response of the bulk medium to an electromagnetic wave.

5.2 Induced Polarization and Susceptibility

Different electrons in an atom or molecule have different natural frequencies and energy damping constants. If a fraction f_j of the total number of electrons ($\sum_j f_j=1$) has natural

[1]The physical displacement is given by the real part of $\delta\boldsymbol{r}$.

frequency ω_j and a decay rate γ_j, then an electron belonging to this group, located at position $\boldsymbol{r}_j$, contributes a dipole moment

$$\boldsymbol{p}_j = \frac{(e^2/m)\boldsymbol{E}(\boldsymbol{r}_j, t)}{\omega_j^2 - \omega^2 - i\gamma_j\omega}. \tag{5.5}$$

By adding the contributions from all groups of electrons and carrying out a spatial average by introducing the electron number density $\mathcal{N}$ (the number of electrons per unit volume), the induced macroscopic polarization at a position $\boldsymbol{r}$ in the medium can be written as[2]

$$\boldsymbol{P}(\boldsymbol{r}, t) = \frac{\mathcal{N}e^2}{m}\sum_j \frac{f_j}{\omega_j^2 - \omega^2 - i\gamma_j\omega}\boldsymbol{E}(\boldsymbol{r})e^{-i\omega t} \equiv \epsilon_o\chi(\omega)\boldsymbol{E}(\boldsymbol{r})e^{-i\omega t}. \tag{5.6}$$

Here $\boldsymbol{E}(\boldsymbol{r})$ represents the spatially averaged electric field, and the dielectric susceptibility $\chi(\omega)$, which characterizes the frequency dependence of the medium response to the field, is given by

$$\chi(\omega) = \frac{\mathcal{N}e^2}{m\epsilon_o}\sum_j \frac{f_j}{\omega_j^2 - \omega^2 - i\gamma_j\omega}. \tag{5.7}$$

Using the definition of dielectric permittivity $\epsilon(\omega) = \epsilon_o[1 + \chi(\omega)]$ and the plasma frequency ω_p, defined by

$$\omega_p^2 = \frac{\mathcal{N}e^2}{m\epsilon_o}, \tag{5.8}$$

the permittivity of the medium can be written as

$$\epsilon(\omega) = \epsilon_o\left[1 + \frac{\mathcal{N}e^2}{m\epsilon_o}\sum_j \frac{f_j}{\omega_j^2 - \omega^2 - i\gamma_j\omega}\right]. \tag{5.9}$$

From Eq. (5.9), it is evident that the dielectric permittivity $\epsilon(\omega)$ is a complex quantity with real and imaginary parts given by

$$\epsilon'(\omega) = \epsilon_o\left[1 + \omega_p^2\sum_j \frac{f_j(\omega_j^2 - \omega^2)}{(\omega_j^2 - \omega^2)^2 + \gamma_j^2\omega^2}\right], \tag{5.10}$$

$$\epsilon''(\omega) = \epsilon_o\omega_p^2\sum_j \frac{f_j\gamma_j\omega}{(\omega_j^2 - \omega^2)^2 + \gamma_j^2\omega^2}. \tag{5.11}$$

The real part of the dielectric permittivity $\epsilon'(\omega)$ is an even function of frequency, whereas the imaginary part, $\epsilon''(\omega)$, is an odd function of frequency. This explicit frequency dependence of $\epsilon(\omega)$ illustrates how the macroscopic atomic response depends on both the driving field frequency and eigenfrequencies of the electron oscillators.

---

[2] In its simplest form, spatial averaging associates an averaging volume ΔV with every point in the medium. This volume is chosen to be large enough to contain many oscillators yet small compared to the cube of the optical wavelength. We then add the contributions of all oscillators within the averaging volume and divide by the volume. This averaged contribution, which is proportional to the density of oscillators, is then assigned to the specific point $\boldsymbol{r}$ around which the averaging volume is centered. Thus, $\frac{1}{\Delta V}\sum_j(-e)\delta\boldsymbol{r}_j \to \boldsymbol{P}(\boldsymbol{r}, t)$, where the sum over j includes only the oscillators within ΔV. To avoid a proliferation of symbols, the spatially averaged electric field is denoted by the same symbol as the applied field.

5.2.1 Refractive Index and Attenuation Index

In considering the effect of atomic response on the propagating field, it is convenient to separate the real and imaginary parts of the complex refractive index $\tilde{n}$, defined by

$$\sqrt{\epsilon(\omega)/\epsilon_o} \equiv \tilde{n}(\omega) = n(\omega) + i\kappa(\omega)\,, \tag{5.12}$$

where $n(\omega)$ and $\kappa(\omega)$ are the real and imaginary parts of $\tilde{n}$. The real part $n(\omega)$ is called the refractive index, while the imaginary part $\kappa(\omega)$ is termed the attenuation index of the medium. Both parameters are frequency-dependent quantities. Equating the real and imaginary parts on the two sides of Eq. (5.12) allows n and κ to be expressed in terms of the real and imaginary parts of $\epsilon(\omega)$ as

$$n(\omega) = \sqrt{\frac{1}{2\epsilon_o}\left[\epsilon'(\omega) + \sqrt{\epsilon'^2(\omega) + \epsilon''^2(\omega)}\right]}\,, \tag{5.13}$$

$$\kappa(\omega) = \sqrt{\frac{1}{2\epsilon_o}\left[-\epsilon'(\omega) + \sqrt{\epsilon'^2(\omega) + \epsilon''^2(\omega)}\right]}\,. \tag{5.14}$$

These general expressions for n and κ are too unwieldy for direct analytical application. However, they simplify under two limiting regimes of practical interest in the discussion of lasers. We consider these cases separately below.

5.2.1.1 Non-resonant Behavior

For a field frequency ω far from any of the resonance frequencies ω_j, such that the inequality $|\omega_j - \omega| \gg \gamma_j$ holds, the denominator of each term in Eqs. (5.9) can be approximated as $(\omega_j^2 - \omega^2)^2\left[1 + \gamma_j^2\omega^2/(\omega_j^2 - \omega^2)^2\right] \approx (\omega_j^2 - \omega^2)^2$. Under this condition, the dielectric permittivity simplifies to

$$\epsilon(\omega) \approx \epsilon_o\left[1 + \omega_p^2\sum_j \frac{f_j}{\omega_j^2 - \omega^2} + i\omega_p^2\sum_j \frac{f_j\gamma_j\omega}{(\omega_j^2 - \omega^2)^2}\right] \equiv \epsilon'(\omega) + i\epsilon''(\omega)\,. \tag{5.15}$$

Comparing the relative magnitudes of the real and imaginary parts of $\epsilon(\omega)$ in this limit reveals that the imaginary part of permittivity is negligibly small compared to the real part ($\epsilon''(\omega) \ll \epsilon'(\omega)$). This is because each term in the imaginary sum is smaller relative to its corresponding term in the real sum by a factor $\gamma_j\omega/|\omega_j^2 - \omega^2| \approx \gamma_j/2|\omega_j - \omega| \ll 1$. This means that as the wave propagates, its attenuation remains small, and the medium is considered transparent to light in this frequency range. It is clear that the transparency or opacity of a medium is a frequency-dependent property.

In this transparency frequency range, because the condition $\epsilon'' \ll \epsilon'$ holds, Eqs. (5.13) and (5.14) yield significantly simpler expressions for the refractive index and the attenuation index

$$n^2(\omega) \approx \frac{\epsilon'(\omega)}{\epsilon_o} = 1 + \omega_p^2\sum_j \frac{f_j}{\omega_j^2 - \omega^2} \tag{5.16}$$

$$\kappa(\omega) \approx \frac{\epsilon''(\omega)}{2\epsilon_o n(\omega)} = \frac{\omega_p^2}{2n(\omega)}\sum_j \frac{f_j\gamma_j\omega}{(\omega_j^2 - \omega^2)^2} \tag{5.17}$$

Many common materials are transparent at optical frequencies because their primary resonance frequencies ω_j lie in the ultraviolet or infrared regions. For these materials, the refractive index n is a monotonically increasing function of frequency throughout the optical range. This characteristic behavior of the refractive index is known as normal dispersion. The expression for the refractive index can also be formulated in terms of the free-space wavelength $\lambda = 2\pi c/\omega$ as

$$n^2 = 1 + \frac{\omega_p^2}{4\pi^2 c^2} \sum_j \frac{f_j \lambda_j^2 \lambda^2}{\lambda^2 - \lambda_j^2} \equiv \sum_j \frac{b_j \lambda^2}{\lambda^2 - \lambda_j^2}, \quad b_j = \frac{\omega_p^2 f_j \lambda_j^2}{4\pi^2 c^2}. \tag{5.18}$$

This very useful formula for the refractive index of a material is known as the Sellmeier dispersion formula. It describes the wavelength dependence of n with high fidelity within a material's transparency range. However, calculating the coefficients b_j analytically from first principles is often impractical due to large physical uncertainties in the underlying parameters. Instead, these coefficients are typically treated as empirical fitting parameters to model experimentally measured refractive index data. For example, the refractive index of fused silica at 20° C (with λ specified in microns), accurate to $\pm 3\times 10^{-5}$ in the wavelength range $0.2\,\mu$m—2.1μm, is given by

$$n^2 = 1 + \frac{0.6961663\lambda^2}{\lambda^2 - (0.0684043)^2} + \frac{0.4079426\lambda^2}{\lambda^2 - (0.1162414)^2} + \frac{0.8974794\lambda^2}{\lambda^2 - (9.896161)^2}. \tag{5.19}$$

As an additional example, the three-term Sellmeier equations for sapphire, where the refractive index depends on the polarization of light, yield the following expressions for the ordinary and extraordinary waves

$$n_o^2 = 1 + \frac{1.43134930\lambda^2}{\lambda^2 - (0.0726631)^2} + \frac{0.65054713\lambda^2}{\lambda^2 - (0.119324)^2} + \frac{5.3414021\lambda^2}{\lambda^2 - (18.0283)^2}, \tag{5.20a}$$

$$n_e^2 = 1 + \frac{1.5039759\lambda^2}{\lambda^2 - (0.0740298)^2} + \frac{0.55069141\lambda^2}{\lambda^2 - (0.121653)^2} + \frac{6.5927379\lambda^2}{\lambda^2 - (20.0722)^2}. \tag{5.20b}$$

These relations are valid over the wavelength range of 0.2 μm to 5 μm.

5.2.1.2 Resonant Behavior of Refractive Index

When the driving field frequency ω lies within a few γ_o of a specific resonance frequency ω_o, we can use the approximation $\omega_o^2 - \omega^2 \approx 2\omega(\omega_o - \omega)$ in the dominant resonant term. For example, for atomic electrons with $\omega_j \approx 2\pi(10^{14} - 10^{15})$ Hz, and $\gamma_j \approx 2\pi(10^8 - 10^{10})$ Hz [see Sec. 5.5], even when ω differs from ω_j by $10\gamma_j$, the relation $\omega + \omega_j \approx 2\omega$ remains accurate to within one part in 10^4. By separating the non-resonant (background) and resonant contributions, the real and imaginary components of the permittivity can be expressed as

$$\begin{aligned} \epsilon'(\omega) &= \epsilon_o \left[1 + \omega_p^2 {\sum_j}' \frac{f_j}{\omega_j^2 - \omega^2} + \left(\frac{\omega_p^2 f_o}{2\omega} \right) \frac{(\omega_o - \omega)}{(\omega_o - \omega)^2 + (\gamma_o/2)^2} \right] \\ &\equiv \epsilon_b' + \left(\frac{\epsilon_o \omega_p^2 f_o}{2\omega} \right) \frac{(\omega_o - \omega)}{(\omega_o - \omega)^2 + (\gamma_o/2)^2} \end{aligned} \tag{5.21}$$

$$\begin{aligned} \epsilon''(\omega) &= \epsilon_o \omega_p^2 {\sum_j}' \frac{f_j \gamma_j \omega}{(\omega_j^2 - \omega^2)^2} + \left(\frac{\epsilon_o \omega_p^2 f_o}{2\omega} \right) \frac{\gamma_o/2}{(\omega_o - \omega)^2 + (\gamma_o/2)^2} \\ &\equiv \epsilon_b'' + \left(\frac{\epsilon_o \omega_p^2 f_o}{2\omega} \right) \frac{\gamma_o/2}{(\omega_o - \omega)^2 + (\gamma_o/2)^2} \end{aligned} \tag{5.22}$$

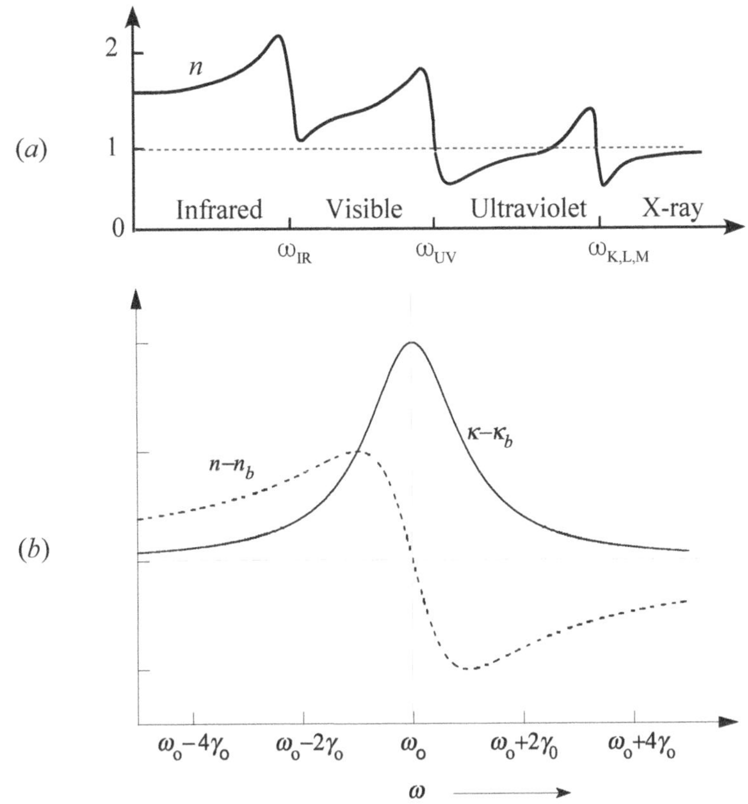

FIGURE 5.1
(a) Qualitative behavior of the refractive index n from infrared to ultraviolet frequencies; (b) behavior of refractive index n and attenuation index κ as a function of frequency near a resonance for a homogeneous collection of oscillators.

As noted earlier, for the non-resonant background terms, the inequality $\epsilon''_b \ll \epsilon'_b$ holds. By introducing a background refractive index $n_b \approx \sqrt{\epsilon'_b/\epsilon_o}$ and an associated background attenuation index $\kappa_b = \epsilon''_b/2\epsilon_o n_b$, we can write $\sqrt{\epsilon/\epsilon_o} = n + i\kappa$. This yields the simplified expressions

$$n(\omega) \approx n_b(\omega) + \frac{\omega_p^2 f_o}{4n_b\omega} \frac{(\omega_o - \omega)}{(\omega_o - \omega)^2 + (\gamma_o/2)^2} \tag{5.23}$$

$$\kappa(\omega) \approx \kappa_b(\omega) + \frac{\omega_p^2 f_o}{4n_b\omega} \frac{\gamma_o/2}{(\omega_o - \omega)^2 + (\gamma_o/2)^2} \tag{5.24}$$

The resonant terms vary rapidly with frequency ω in the immediate vicinity of the resonance frequency ω_o, whereas the background terms n_b and κ_b vary slowly. Dissipation, characterized by κ, is largest on resonance. A plot of n and κ in the neighborhood of the resonance frequency is shown in Fig. 5.1. Note that in the region of resonance, n can become a decreasing function of frequency. This behavior of n is termed anomalous dispersion, in contrast to its *normal* behavior between resonances where it monotonically increases with frequency. Away from a resonance, n reverts to its normal behavior. While we have considered a simple model of atomic response for an isolated resonance, it is clear that if multiple resonances overlap, a significantly more complicated frequency behavior can emerge.

5.3 Resonant Susceptibility and Atomic Response

We now return to Eq. (3.9), which governs the evolution of the electromagnetic field driven by the atomic response. Following the discussion of Sec. 2.2, we use the results of this chapter by identifying the non-resonant (background) polarization term with the host medium polarization $\boldsymbol{P}_m = \epsilon_0 \chi_m \boldsymbol{E}$. This allows us to equate $\epsilon = \epsilon_o(1+\chi_m) = \epsilon_b$, define the background index as $n_b = \sqrt{\epsilon_b/\epsilon_o}$ (representing the host medium refractive index n), and associate the resonant term with the polarization of the active atoms, $\boldsymbol{P}_{at} = \epsilon_0 \chi_{at} \boldsymbol{E}$. Consequently, from Eqs. (5.21) and (5.22), the resonant susceptibility $\chi_{\mathrm{at}}(\omega)$ can be expressed as

$$\chi_{at}(\omega) = \frac{\mathcal{N}_o e^2}{m\epsilon_o}\left[\frac{i/2\omega}{i(\omega_o-\omega)+(\gamma_o/2)}\right] = \frac{\mathcal{N}_o e^2}{2m\epsilon_o\omega}\left[\frac{1}{(\omega_o-\omega)-i(\gamma_o/2)}\right], \tag{5.25}$$

where $\mathcal{N}_o = \mathcal{N} f_o$ is the number density of the active atoms. This expression characterizes the steady-state polarization induced by a near-resonant incident field ($|\omega_o-\omega|, \gamma_o \ll \omega, \omega_o$). Approximating $\omega \approx \omega_o$ in the numerator, the quantity inside square brackets in the first equality can be recognized as the Fourier transform of the oscillator's impulse response function. The frequency dependence of susceptibility thus conforms to the well-known result that the steady-state frequency response of a linear oscillator is the Fourier transform of its impulse response function.

In deriving Eq. (5.25) for the collective response of active atoms, we assumed that the induced dipole moment of each active atom aligns with the direction of the applied field. This is generally not the case for real atoms. An atom possesses its own reference axes. Its induced dipole moment can point in any direction relative to the applied field. However, the magnitude of the induced dipole moment remains proportional to the cosine of the angle between the applied field and the induced dipole moment. In an ensemble of randomly oriented active atoms, for example, in a gas or impurity atoms embedded in a solid, the induced atomic dipoles point in random directions. Consequently, the collective atomic response in the direction of the applied field involves, on average, only a third of the total atomic population, $\mathcal{N}_o/3$. The same reduction factor applies if the individual atomic response involves a complete set of degenerate atomic transitions. Summing over these transitions amounts to an averaging over spatial orientations of the atomic axes, reducing the overall response in the direction of applied field to one third of the expression given in (5.25). To account for these common physical scenarios, we define the resonant atomic susceptibility as

$$\chi_{at}(\omega) = \frac{\mathcal{N}_o e^2}{6m\epsilon_o\omega}\left[\frac{1}{(\omega_o-\omega)-i(\gamma_o/2)}\right]. \tag{5.26}$$

This expression can be written in a more useful form by eliminating its explicit dependence on oscillator-specific constants, namely the mass m, charge e, resonance frequency ω_o, and the permittivity of free space ϵ_o. To this end, we recall that the radiative energy decay rate of the oscillator

$$A_{rad} = \frac{2}{3}\frac{e^2}{4\pi\epsilon}\frac{\omega_o^2}{m(c/n)^3} = \frac{e^2 n\omega_o^2}{6\pi m\epsilon_o c^3}, \tag{5.27}$$

depends on an identical set of parameters. Eliminating these parameters in favor of A_{rad}, the expression for resonant susceptibility becomes

$$\chi_{at}(\omega) = \frac{c\lambda_o^2 A_{rad}}{\omega 4\pi n}\left[\frac{1}{(\omega_o-\omega)-i(\gamma_o/2)}\right]\mathcal{N}_o, \tag{5.28}$$

where $\lambda_o = 2\pi c/\omega_o$ is the free-space wavelength corresponding to the natural frequency of the oscillator. The real and imaginary parts of atomic susceptibility, $\chi_{at} = \chi'_{at} + i\chi''_{at}$, can now be written as functions of the field frequency $\nu = \omega/2\pi$

$$\chi''_{at}(\nu) = \frac{c\lambda_o^2 A_{rad}}{\omega 8\pi n}\left[\frac{1}{\pi}\frac{\gamma_o/4\pi}{(\nu_o-\nu)^2+(\gamma_o/4\pi)^2}\right]\mathcal{N}_o \equiv \frac{c\lambda_o^2 A_{rad}}{\omega 8\pi n} S(\nu)\mathcal{N}_o\,, \tag{5.29a}$$

$$\chi'_{at}(\nu) = \frac{c\lambda_o^2 A_{rad}}{\omega 8\pi n}\left[\frac{1}{\pi}\frac{(\nu_o-\nu)}{(\nu_o-\nu)^2+(\gamma_o/4\pi)^2}\right]\mathcal{N}_o \equiv \frac{2(\nu_o-\nu)}{\Delta\nu_o}\chi''_{at}(\nu)\,. \tag{5.29b}$$

The function $S(\nu)$ is the lineshape function, which characterizes how the strength of atomic response to an applied field changes as the field frequency detunes from the oscillator resonance. For the homogeneous ensemble of atom oscillators considered here, it has the form of a Lorentzian profile

$$S(\nu) = \frac{1}{\pi}\frac{\Delta\nu_o/2}{(\nu_o-\nu)^2+(\Delta\nu_o/2)^2}\,, \tag{5.30}$$

with a peak value $S(\nu_o)$ and a full width at half maximum (FWHM) $\Delta\nu_o$ defined by

$$S(\nu_o) = \frac{2}{\pi\Delta\nu_o}\,, \qquad \Delta\nu_o = \frac{\gamma_o}{2\pi}\,. \tag{5.31}$$

The FWHM condition is given by $S(\nu_o \pm \Delta\nu_o/2) = \frac{1}{2}S(\nu_o)$, as shown graphically in Fig. 5.2. While the lineshape function does not always exhibit a Lorentzian profile—as discussed further in Sec. 5.5—the lineshape function $S(\nu)$ provides a normalized measure of the relative strength of the resonant response as a function of field frequency.

Equations (5.28) and (5.29) for the resonant susceptibility are advantageous because they are free of internal, oscillator-specific microscopic parameters. By expressing the collective macroscopic response purely in terms of experimentally measurable quantities, they are readily generalized to a broader class of systems, both classical and quantum mechanical, provided the appropriate radiative decay rates and natural frequencies are supplied. Furthermore, translating these formulations into a quantum mechanical framework is significantly more straightforward, bypassing the conceptual difficulties that arise because not all quantum mechanical variables possess classical analogs.

The expression for resonant susceptibility (5.28) can be derived directly by focusing on the group of electron oscillators of frequency ω_o embedded in a host medium of refractive index n. The macroscopic polarization induced by this subpopulation of dipoles is obtained by carrying out a spatial averaging of their individual contributions following the methodology detailed in Sec. 5.2. Using the single-dipole equation of motion (5.2) and averaging over the underlying spatial distribution, we find that the macroscopic polarization $\boldsymbol{P}_{at}(\boldsymbol{r},t)$

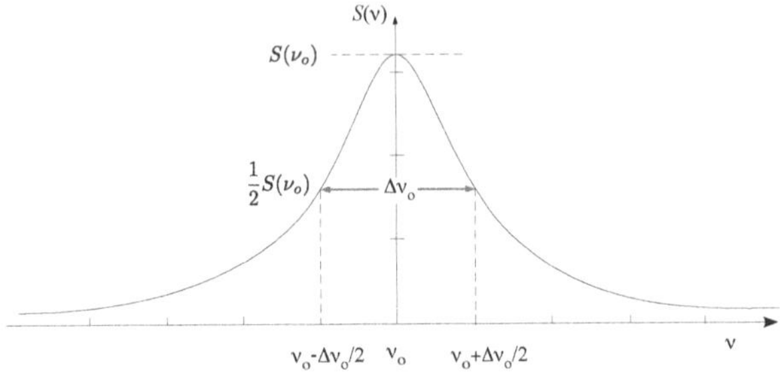

FIGURE 5.2
The Lorentzian lineshape function has a single peak centered at the atomic frequency ν_o, with full width at half maximum $\Delta\nu_o$.

satisfies the equation

$$\frac{d^2\boldsymbol{P}_{at}}{dt^2} + \gamma_o \frac{d\boldsymbol{P}_{at}}{dt} + \omega_o^2 \boldsymbol{P}_{at} = \frac{\epsilon_o c \lambda_o^2 A_{rad}}{2\pi n} \mathcal{N}_o \boldsymbol{E}(\boldsymbol{r},t)\,, \tag{5.32}$$

where the electronic charge and mass have been eliminated in favor of $A_{\rm rad}$ on the right side, mirroring the procedure employed when transitioning from Eq. (5.25) to Eq. (5.28). Solving this equation for the steady-state polarization $\boldsymbol{P}_{at}$ induced by a monochromatic field of frequency ω while using the near-resonance approximation, $\omega_o^2 - \omega^2 \approx 2\omega(\omega_o - \omega)$, we recover the expression (5.28) for the resonant atomic susceptibility.

This latter algebraic sequence can be bypassed entirely by operating directly on Eq. (5.32) within the resonance approximation framework. Recall that the resonance approximation implies that the driving field and natural oscillator frequencies are tightly coupled. Consequently, although their detuning can vary from zero to several linewidths (e.g., $10\gamma_o$ or more), both γ_o and $|\omega - \omega_o|$ remain orders of magnitude smaller than either frequency scale:

$$\omega \sim \omega_o \quad \text{and} \quad \gamma_o, |\omega_o - \omega| \ll \omega_o, \omega\,. \tag{5.33}$$

These conditions are very well satisfied for the optical atomic transitions typically encountered in laser physics.

To proceed further, we assume that the field and induced polarization are linearly polarized. This does not limit the general applicability of our results, as fields possessing other states of polarization can always be expressed in terms of a linear basis. We further generalize our results by considering an electromagnetic field with a more general time dependence of the form $E(t) = A(t)e^{-i\omega t}$ (where spatial variations are suppressed for clarity). Here, $A(t)$ represents a slowly varying amplitude corresponding to a narrow-band field, as introduced in Sec. 2.1. It is physically reasonable to assume that induced atomic polarization inherits this dynamic structure, taking the form $P_{at}(t) = P(t)e^{-i\omega t}$. Substituting these field configurations into Eq. (5.32) and canceling the overall $e^{-i\omega t}$ factor from both sides reveals that the polarization amplitude $\boldsymbol{P}(t)$ obeys the differential equation

$$\frac{d^2P(t)}{dt^2} + (-2i\omega + \gamma_o)\frac{dP(t)}{dt} + (\omega_o^2 - \omega^2 - i\omega\gamma_o)P(t) = \frac{\epsilon_o c \lambda_o^2 A_{rad}}{2\pi n}\mathcal{N}_o A(t)\,. \tag{5.34}$$

Invoking the resonance approximation,

$$\omega_o^2 - \omega^2 - i\omega\gamma_o \approx 2\omega(\omega_o - \omega - i\gamma_o/2) \text{ and } -2i\omega + \gamma_o = -2i\omega(1 + i\gamma_o/2\omega) \approx -2i\omega\,, \tag{5.35}$$

and executing the slowly varying envelope approximation,

$$\left|d^2P(t)/dt^2\right| \ll |2\omega dP(t)/dt|\,, \tag{5.36}$$

we can neglect the second-derivative contribution relative to the first derivative. Under these combined assumptions, the near-resonant polarization dynamics simplify to

$$\frac{dP(t)}{dt} + [i(\omega_o - \omega) + \gamma_o/2]P(t) = i\epsilon_o \frac{c\lambda_o^2 A_{rad}}{4\pi n\omega}\mathcal{N}_o A(t)\,. \tag{5.37}$$

For a strictly monochromatic field, $dP/dt = 0$, this equation immediately reduces to the linear susceptibility result (5.28). However, Eq. (5.37) remains far more versatile, accurately describing time-dependent, narrow-band field scenarios. As will be demonstrated in later chapters, this specific equation is ideally suited for formulating the quantum mechanical evolution of atomic polarization variables.

5.3.1 Effect of χ_{at} on Field

The effect of $\chi_{at}(\nu)$ on the field amplitude is described by Eq. (3.9)

$$\frac{\partial E_o(\boldsymbol{r})}{\partial z} = \frac{i}{2k}\nabla_\perp^2 E_o(\boldsymbol{r}) + i\frac{k\chi'_{at}}{2n^2}E_o(\boldsymbol{r}) - \frac{k}{2}\left(\frac{\chi''_{at}}{n^2} + \frac{\sigma}{\omega\epsilon_0 n^2}\right)E_o(\boldsymbol{r})\,. \tag{3.9*}$$

Having already accounted for the effect of the $\nabla_\perp^2$ transverse derivative term in Chapter 3, we write the overall field amplitude as $E(\boldsymbol{r},t) = E_o(\boldsymbol{r})e^{i(kz-\omega t)}$, where $E_o(\boldsymbol{r}) = Au(\boldsymbol{r})$. Here $u(\boldsymbol{r})$ represents a Hermite-Gauss (HG) or Laguerre-Gauss (LG) mode function satisfying Eq. (3.10), and the field amplitude A is allowed to vary explicitly with z. Substituting $E_o(\boldsymbol{r}) = A(z)u(\boldsymbol{r})$ into Eq. (3.9) and invoking Eq. (3.10), we find that the field amplitude $A(z)$ evolves according to

$$\frac{\partial A(z)}{\partial z} = -\frac{\sigma}{2\epsilon_0 nc}A(z) - \frac{\omega\chi''_{at}}{2nc}A(z) + i\frac{k\chi'_{at}}{2n^2}A(z)\,. \tag{5.38}$$

Integrating this differential relation yields the following expressions for the spatial evolution of the wave amplitude and the associated optical intensity

$$A(z) = A(0)e^{-\frac{1}{2}(\alpha+\alpha_{at})z}e^{ik(\chi'_{at}/2n^2)z}\,, \tag{5.39a}$$

$$I(z) = \frac{1}{2}\epsilon_o nc|A(0)|^2 e^{-(\alpha+\alpha_{at})z} \equiv I(0)e^{-(\alpha+\alpha_{at})z}\,, \tag{5.39b}$$

where the attenuation coefficients α and α_{at} are given by

$$\alpha = \frac{\sigma}{\epsilon_o nc}\,, \tag{4.59*}$$

$$\alpha_{at}(\nu) = \frac{\omega\chi''_{at}(\nu)}{nc} = \left(\frac{\lambda_o^2 A_{rad}}{8\pi n^2}S(\nu)\right)\mathcal{N}_o\,. \tag{5.40}$$

Here α represents the absorption coefficient arising from distributed background losses in the host medium (modeled via the fictitious conductivity σ), while α_{at} is the absorption coefficient due to the resonant oscillators. According to Eq. (5.29a), χ''_{at} is always a positive quantity, implying that the steady-state interaction of a propagating wave with a collection of classical oscillators invariably results in net attenuation.

The quantity inside the parentheses in Eq. (5.40) possesses the dimensions of a cross-section. We therefore define the absorption cross-section $\sigma(\nu)$ (not to be confused with conductivity σ) by

$$\sigma(\nu) = \frac{\lambda_o^2 A_{rad}}{8\pi n^2}S(\nu)\,. \tag{5.41}$$

Then the intensity decay as the wave passes through the ensemble of oscillators can be written in the form

$$\left[\frac{\partial I}{\partial z}\right]_{at} = -\sigma(\nu)\mathcal{N}_o I\,. \tag{5.42}$$

In this form, the attenuation of wave intensity can be physically interpreted as the wave power absorbed by a uniform spatial density $\mathcal{N}_o$ of effective absorbing disks per unit volume. Each disk represents an absorption cross-section $\sigma(\nu)$ and captures a total optical power equal to σI. Accordingly, $\sigma(\nu)$ is designated as the absorption cross-section of the oscillator. This cross-section inherits its frequency dependence from the underlying lineshape function, reflecting the characteristic resonance response of the atom.

The term containing the real part χ'_{at} in Eq. (5.38) contributes to the phase of the propagating wave. Because χ'_{at} can assume either positive or negative values depending on detuning [cf. Eq. (5.29b)], this contribution can either advance or retard the wave phase. Combining the phase factor $e^{ik(\chi'_{at}/2n^2)z}$ in Eq. (5.39a) with the primary spatial propagation e^{ikz} reveals that the effective propagation constant changes from k to

$$k + \frac{k\chi'_{at}}{2n^2} = \frac{\omega n}{c}\left(1 + \frac{\chi'_{at}}{2n^2}\right). \tag{5.43}$$

This perturbation directly alters the passive cavity mode spectrum. For example, the resonant frequencies of the $\mathrm{HG}_{\ell mp}$ modes in a two-mirror Fabry-Perot cavity are determined by enforcing the roundtrip phase boundary condition [cf. Eq. (4.32)]

$$\frac{\omega n}{c}\left(1 + \frac{\chi'_{at}}{2n^2}\right)L - 2(\ell + m + 1)\cos^{-1}\sqrt{g_1 g_2} = 2\pi p. \tag{5.44}$$

Consequently, the frequency $\nu = \omega/2\pi$ of an $\mathrm{HG}_{\ell mp}$ mode is now determined implicitly by the expression

$$\nu\left(1 + \frac{\chi'_{at}(\nu)}{2n^2}\right) = \frac{c}{2nL}\left[p + \frac{\ell + m + 1}{\pi}\cos^{-1}\sqrt{g_1 g_2}\right]. \tag{5.45}$$

This implicit relationship gives rise to rich dispersive phenomena because $\chi'_{at}(\nu)$ varies rapidly near the atomic resonance frequency. While the host medium refractive index n is also technically frequency-dependent, this background variation is assumed to be negligible over the narrow frequency interval of interest. It is evident that the modifications to the cavity mode spectrum can become substantial within the spectral window where χ_{at} undergoes rapid variation.

We will revisit this transcendental equation when analyzing laser operation in Chapter 7. For now, we conclude that a collection of classical atomic oscillators can only absorb power from an incident light wave, causing an exponential decay of its intensity during propagation. This poses a fundamental challenge, as laser realization requires field amplification and gain. Addressing this limitation requires a transition to a quantum mechanical framework of discrete atomic energy levels, which motivates a necessary modification of our classical expression for the susceptibility.

5.4 Quantum Mechanical Susceptibility

The relation between the absorption coefficient and the atomic susceptibility remains valid in a quantum mechanical framework, provided we interpret the resonance frequency ν_o as the transition frequency between two quantized atomic levels of energy E_2 and E_1, where $E_2 - E_1 = h\nu_o > 0$. Furthermore, we replace the classical oscillator number density $\mathcal{N}_o$ (the number of oscillators with eigenfrequency ω_o) with the population inversion density $\mathcal{N}_1 - \mathcal{N}_2 = \Delta\mathcal{N}_{12}$, where $\mathcal{N}_1$ and $\mathcal{N}_2$ represent the number densities of atoms occupying levels E_1 and E_2, respectively. This formulation is physically intuitive because quantum mechanically, atoms absorb or emit energy in discrete amounts during transitions between energy levels. An atom occupying the lower level E_1 can absorb a quantum of energy $h\nu_o$ from the electromagnetic wave and make a transition to E_2. Conversely, an atom initially in the upper level E_2 can undergo a transition to level E_1 by emitting a quantum of energy $h\nu_o$ into the electromagnetic wave. Thus, the net energy exchange between the field and the atoms is fundamentally governed by the population difference density $\Delta\mathcal{N}_{12}$. Motivated by the quantum-mechanical picture of absorption and stimulated emission, we anticipate that the classical oscillator density $\mathcal{N}_o$ should be replaced by the population difference $\Delta\mathcal{N}_{12} = \mathcal{N}_1 - \mathcal{N}_2$. Chapter 6 derives this result rigorously. A quantum mechanically accurate formula for the atomic susceptibility is therefore obtained by substituting $\Delta\mathcal{N}_{12}$ for $\mathcal{N}_o$ in Eq. (5.28):

$$\chi_{at}(\nu) = \frac{c\lambda_o^2 A_{21}}{\omega 4\pi n}\left[\frac{1}{\omega_o - \omega - i\gamma_o/2}\right]\Delta\mathcal{N}_{12}\,. \tag{5.46}$$

The absorption coefficient for this specific atomic transition takes the form

$$\alpha_{at} = \frac{\omega}{nc}\chi''_{at} = \frac{\lambda_o^2 A_{21}}{8\pi n^2}S(\nu)\Delta\mathcal{N}_{12}\,, \tag{5.47}$$

where the lineshape function $S(\nu)$ is still defined by Eq. (5.30) with a full width at half maximum of $\Delta\nu_o = \gamma_o/2\pi$. In the quantum description, however, γ_o represents the combined total decay rate of the upper and lower states $(\gamma_2 + \gamma_1)$, and A_{21} is the radiative decay rate from level E_2 to level E_1 via spontaneous photon emission. This parameter corresponds exactly to the Einstein-A coefficient, $A_{21} \equiv \gamma_{21\text{rad}}$ for the $E_2 \to E_1$ transition.

Equation (5.47) implies that the absorption coefficient is positive when $\Delta\mathcal{N}_{12} > 0$, but it can become negative if $\Delta\mathcal{N}_{12} < 0$. Indeed, when the upper-level population is negligible compared to the lower-level population ($\mathcal{N}_2 \ll \mathcal{N}_1$), as would be the case for atoms in thermal equilibrium, the population difference simplifies to $\Delta\mathcal{N}_{12} = \mathcal{N}_1 - \mathcal{N}_2 \approx \mathcal{N}_1$. Under these conditions, Eq. (5.47) reduces to the classical expression (5.40) for the absorption coefficient. Conversely, if an external excitation mechanism causes the upper-level population to exceed the lower-level population, the population difference becomes negative ($\Delta\mathcal{N}_{12} < 0$). Consequently, the absorption coefficient becomes negative, resulting in exponential growth of the wave amplitude. This condition, $\Delta\mathcal{N}_{12} < 0$, is known as a population inversion; it is a necessary condition for laser action. We shall see that this condition can be achieved in a wide variety of laser media through several distinct pump schemes. Ultimately, the quantized nature of atomic energy levels naturally introduces the possibility of optical amplification, provided a population inversion can be created in a medium.

Applying the same substitution rule that we used for susceptibility, Eq. (5.37) for the slowly varying polarization amplitude becomes

$$\frac{dP}{dt} + [i(\omega_o - \omega) + \gamma_o/2]P = i\epsilon_o\frac{c\lambda_o^2 A_{21}}{4\pi n\omega}\Delta\mathcal{N}_{12}(t)A(t)\,. \tag{5.48}$$

This represents the correct quantum-mechanical equation of motion for the induced polarization for a homogeneous group of atoms driven by an external field and a time-dependent population difference $\Delta\mathcal{N}_{12}$. In this quantum mechanical framework of light-matter energy exchange, the level populations themselves become dynamical variables driven by stimulated transitions, external pumping, and relaxation processes. This feedback mechanism makes the quantum mechanical model of atomic response intrinsically nonlinear. Consequently, the linear susceptibility $\chi_{at}(\omega)$ is often an inadequate description of the atomic response. To properly capture these dynamics, one must solve the polarization equation (5.48) together with an evolution equation for the population difference density, which can be derived by analyzing the quantum-mechanical power exchange between the field and the atomic medium.

The linear susceptibility result (5.46) for atomic response remains valid in the so-called rate-equation limit. This regime, also called the adiabatic limit, is defined by the condition that the rates of change of both the population difference and the field envelope are small compared to the linewidth γ_o, meaning they vary slowly on a timescale of order γ_o^{-1}. In this limit, we can neglect the first derivative term dP/dt compared to the remaining relaxation and detuning terms on the left-hand side of Eq. (5.48). Solving directly for the polarization amplitude yields

$$P(t) \approx \epsilon_o \frac{c\lambda_o^2 A_{21}}{4\pi n\omega} \left(\frac{\Delta\mathcal{N}_{12}(t)}{(\omega_o - \omega) - i\gamma_o/2} \right) A(t) \,. \tag{5.49}$$

This equation implies that the atomic polarization responds rapidly enough to instantaneously track the temporal variations of both the population difference and the driving field envelope. Consequently, the equations of motion for the field intensity and the underlying level populations can be cast entirely in terms of the rates of energy absorption, stimulated emission, and level relaxation rates. Most practical laser devices operate within this rate-equation regime, and we will utilize it extensively throughout this text. For monochromatic fields where both A and $\Delta\mathcal{N}_{12}$ are constants, this expression for polarization reduces to the linear resonant susceptibility result (5.46).

5.4.1 Power Transfer to Atoms and Level Population

The transfer of electromagnetic wave power to the atomic medium can be computed directly from the rate of work performed per unit volume by the field on the underlying dipoles. This rate of work done, derived from the Poynting theorem, Eq. (2.58), is given by

$$\frac{dw}{dt} = \frac{1}{2}\mathcal{R}e \left[E \cdot \frac{dP_{at}^*}{dt} \right] = \frac{i\omega}{4} \left[A \cdot P^* - A^* \cdot P \right] , \tag{5.50}$$

where A and P are the slowly varying field and polarization envelopes introduced by writing $E(t) = A(t)e^{-i\omega t}$ and $P_{\rm at}(t) = P(t)e^{-i\omega t}$, as in the derivation of Eq (5.48). Within the rate-equation limit, the macroscopic polarization amplitude simplifies to the form given in Eq. (5.49). Substituting this adiabatic polarization expression into Eq. (5.50), and recalling the definitions of the cross-section (5.41) and wave intensity (2.60), we find that the average power transferred to the atomic system per unit volume is

$$\begin{aligned} \frac{dw}{dt} &= \left[\frac{\lambda_o^2 A_{21}}{8\pi n^2} S(\nu) \right] \left(\frac{1}{2}\epsilon_o nc|A|^2 \right) \Delta\mathcal{N}_{12} \\ &= \sigma(\nu) I \Delta\mathcal{N}_{12} \,. \end{aligned} \tag{5.51}$$

This constitutes the quantum-mechanically correct expression for the field-to-atom power transfer on an $E_2 \to E_1$ transition in the rate-equation limit. The wave power absorbed by atoms occupying the lower level, $\sigma I \mathcal{N}_1$, appears as a positive contribution to the net power transfer. Conversely, power emitted into the field by atoms occupying the upper state, $\sigma I \mathcal{N}_2$, enters as a negative contribution. According to this expression, the net power transfer to the atomic medium is positive if $\Delta\mathcal{N}_{12} > 0$ (resulting in wave attenuation) and negative if $\Delta\mathcal{N}_{12} < 0$ (resulting in wave amplification).

5.4.2 Stimulated Transition Cross-section

As in the classical theory of light-atom interaction, the concept of cross-section remains highly useful in quantum mechanics, though its physical interpretation shifts. In the quantum picture, absorption entails an atom undergoing an upward transition from the lower level E_1 to the upper level E_2 by capturing an incident photon, whereas emission involves a downward transition from $E_2 \to E_1$ accompanied by the creation of a photon. The power absorbed by the population in level E_1 is conventionally written as $\sigma_{12}\mathcal{N}_1 I$, where σ_{12} is the cross-section for the $E_1 \to E_2$ transition. Similarly, the power emitted by the atomic population in the upper level E_2 is formulated as $\sigma_{21}\mathcal{N}_2 I$, where σ_{21} denotes the transition cross-section for the $E_2 \to E_1$ transition. Utilizing this notation, Eq. (5.51) can be partitioned as

$$\frac{dw}{dt} = \sigma_{12} I \mathcal{N}_1 - \sigma_{21} I \mathcal{N}_2\,, \tag{5.52a}$$

with $$\sigma_{21}(\nu) = \sigma_{12}(\nu) = \left[\frac{\lambda_o^2 A_{21}}{8\pi n^2} S(\nu)\right] \equiv \sigma(\nu)\,. \tag{5.52b}$$

While the transition cross-sections retain the dimensions of area, interpreting $\sigma_{12} I$ as the wave power intercepted and absorbed by an atomic disk, as in the classical picture, is conceptually meaningful for absorption, but a similar geometric view of $\sigma_{21} I$ is less intuitive for stimulated emission. Alternatively, we can introduce photon flux density Φ [photons $\cdot$ $\text{m}^{-2} \cdot \text{s}^{-1}$] and express the narrowband wave intensity as $I = \Phi h\nu$. Under this formulation, the product $\sigma_{12}\Phi$ takes the units of a transition rate [s^{-1}]. In quantum mechanics, this rate represents the probability per unit time for an individual atom to undergo the $E_1 \to E_2$ excitation process. Consequently, the product $\sigma_{12}\Phi\mathcal{N}_1$ yields the total rate at which atoms within the volume are excited from E_1 to E_2. Multiplying this rate by the energy of a single photon yields the rate of energy absorption, $(\sigma_{12}\Phi\mathcal{N}_1)h\nu \equiv \sigma_{12} I \mathcal{N}_1$. Similarly, $\sigma_{21}\Phi$ represents the downward transition probability per atom for the $E_2 \to E_1$ process, such that $(\sigma_{21}\Phi\mathcal{N}_2)h\nu \equiv \sigma_{21} I \mathcal{N}_2$ defines the power emitted by the upper-level atomic population.

In summary, the quantum description of a resonant electromagnetic field interacting with an atomic transition between levels, E_1 and E_2 (where $E_1 < E_2$), relies on an absorption cross-section $\sigma_{12} \equiv \sigma_{ab}$ and an emission cross-section $\sigma_{21} \equiv \sigma_{em}$. The product $\sigma_{ij}\Phi$ defines the quantum mechanical transition probability per unit time (transition rate) for an $E_i \to E_j$ transition by an individual atom.[3] Because this transition rate is proportional to the incident photon flux, it is referred to as the stimulated transition rate, and σ_{ij} is termed the stimulated transition cross-section.

The power balance described by Eq. (5.52) implies that $\sigma_{12} = \sigma_{21}$, meaning the transition probabilities for stimulated emission and absorption between the same two levels E_1 and E_2 are strictly equal.[4] This equality of transition cross-sections and their explicit dependence on the spontaneous emission rate A_{21} represent the exact quantum mechanical relations for two non-degenerate energy states. Expressing the transition cross-sections in terms of the spontaneous emission rate A_{21} (the Einstein A-coefficient) is very useful because it replaces internal microscopic variables with an empirically accessible parameter that can be measured in photoluminescence experiments.[5] Ultimately, two critical insights emerge from this quantum mechanical formulation of power transfer: first, the net power transfer to the

[3]See, for example, Chapter 10 in *Concepts in Quantum Mechanics*, V. S. Mathur and S. Singh (CRC Press, Boca Raton, FL) 2009.

[4]This statement must be modified if the upper or lower energy levels exhibit degeneracy. The reason for this modification is not fundamental but practical. It is often impossible or undesirable to resolve individual transitions between the upper and lower groups of sub-levels. Therefore, it is convenient to formulate total transition rates between levels with energy E_2 and E_1, which are directly accessible to measurements. If the lower energy level consists of a manifold of g_1 degenerate sub-levels (i.e., there are g_1 levels that have the same energy E_1) and the upper level consists of a manifold of g_2 degenerate sub-levels, the relation between the manifold-averaged absorption and emission cross-section becomes $\sigma_{12} = \frac{g_2}{g_1}\sigma_{21}$. See Sec. 7.5.

[5]First-principles calculations of A_{21} carry significant uncertainties due to a lack of detailed knowledge of the local environment of the atom. The rigorous quantum mechanical expression for radiative decay rate of an isolated atom is given by

$$A_{21} = \frac{8\pi^2\nu^3|\mu_{21}|^2}{3\epsilon_0 \hbar c^3}\,,$$

where $\mu_{21} = -e\langle 2|\boldsymbol{r}|1\rangle$ represents the electric dipole transition matrix element connecting states 2 and 1. See *Physics of atoms and molecules*, B.H. Bransdon and C.J. Joachain (Prentice Hall, 1983), Chapter 4.

atomic medium can be positive (absorption) or negative (stimulated emission) based entirely on the relative size of the level populations; second, the direct proportionality between the stimulated transition cross-section and the spontaneous emission rate implies that atomic transitions which fluoresce strongly exhibit correspondingly robust response via absorption or stimulated emission when subjected to an external field.

5.4.3 Stimulated Transition Rates for Narrowband Fields

The transfer of electromagnetic wave power to the atomic medium alters the net energy stored within the atomic system by inducing transitions between the energy levels E_1 and E_2 via absorption and stimulated emission of light. Therefore, the rate of change of the energy stored in the atomic system is directly proportional to the rate at which the atomic level populations change. We recall that the absorption of a single quantum of energy $\hbar\omega = h\nu$ by an individual atom increases the upper-level population density $\mathcal{N}_2$ by one and decreases the lower-level population density $\mathcal{N}_1$ by one. Assuming that the total number density of active atoms across both participating levels remains fixed ($\mathcal{N}_1 + \mathcal{N}_2 = \mathcal{N}$), a single absorption event increases the net population difference density, defined here as $\Delta\mathcal{N}_{21} = \mathcal{N}_2 - \mathcal{N}_1$ by two. Accordingly, the rate of increase of energy stored in the atomic system per unit volume can be written as

$$\frac{dw_a}{dt} = \frac{\hbar\omega}{2}\frac{d\Delta\mathcal{N}_{21}}{dt} = \hbar\omega\frac{d\mathcal{N}_2}{dt} = -\hbar\omega\frac{d\mathcal{N}_1}{dt}\,. \tag{5.53}$$

Equating this atomic power gain to the power transfer dw/dt from the driving field to the atoms [Eq. (5.50)], we obtain the equations governing level population changes brought about by stimulated transitions:

$$\frac{1}{2}\frac{d\Delta\mathcal{N}_{21}}{dt} = \frac{i}{4\hbar}\left[A\cdot P^* - A^*\cdot P\right], \tag{5.54a}$$

$$= -\left[\frac{\lambda_o^2 A_{21}}{8\pi n^2}S(\nu)\right]\Phi\left(\mathcal{N}_2 - \mathcal{N}_1\right) = -\sigma(\nu)\Phi\Delta\mathcal{N}_{21}\,. \tag{5.54b}$$

The last result follows directly in the rate-equation limit by invoking Eq. (5.49) to express the polarization amplitude in terms of the driving field envelope and the instantaneous population inversion. While these equations accurately capture level population changes due to stimulated transitions, spontaneous emission and the relaxation of levels E_2 and E_1 to other levels, and of other levels to them, also contribute to population dynamics. We revisit these population rate equations in Chapter 7 and incorporate these and other population-changing mechanisms to construct a comprehensive picture of level population evolution.

5.4.4 Wave Power Amplification and Attenuation

The discussion of the preceding sections demonstrates that, within a quantum mechanical framework, net power can flow from the radiation field to the atoms or, conversely, from the atoms to the radiation field. This two-way exchange of energy between the wave and the atomic medium allows for either the attenuation or the amplification of wave power as it propagates through a medium. To visualize this interaction, consider a narrowband optical wave with center frequency ν propagating along the z-axis through a medium containing a collection of active atoms with a characteristic transition frequency $\omega_o = 2\pi\nu_o$. The wave frequency is near-resonant with the atomic transition frequency ($\nu \approx \nu_o$). Let us consider the energy balance in a small volume in the form of a right cylinder of length δz and

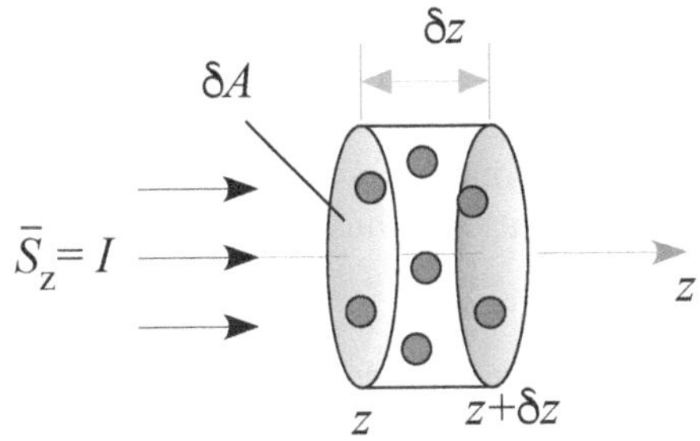

FIGURE 5.3
Power balance in a small volume $\delta V = \delta A\, \delta z$ for an electromagnetic wave propagating along the z-axis and interacting with a uniform distribution of oscillators. S_z represents the z-component of the Poynting vector, whose magnitude corresponds to the cycle-averaged optical intensity I.

end-face area δA, as shown in Fig. 5.3. The volume of the cylinder is $\delta V = \delta A \delta z$. The cycle-averaged electromagnetic energy stored in this cylinder can be written as $\bar{u}\delta V$, where $\bar{u}$ represents the average local energy density. This stored energy increases due to the net inflow of electromagnetic energy through the boundary facets and decreases due to the work done by the field on the oscillators within the volume δV. Applying the continuity equation, the rate of change of the electromagnetic energy contained within δV is given by

$$\delta V \frac{\partial \bar{u}}{\partial t} = \begin{Bmatrix} \text{net power flow into} \\ \delta V \end{Bmatrix} - \begin{Bmatrix} \text{average rate of work done by} \\ \text{the field on the oscillators in} \\ \delta V \end{Bmatrix} . \tag{5.55}$$

The first contribution on the right-hand side can be determined by evaluating the time-averaged Poynting vector, $\bar{\boldsymbol{S}}(z) = I(z)\hat{z}$ [W·m^{-2}], associated with the wave. The total wave power entering the volume δV through the facet at z is $\bar{\boldsymbol{S}}(z) \cdot \hat{z}\delta A = I(z)\delta A$, while the wave power exiting through the facet at position $z + \delta z$ is $\bar{\boldsymbol{S}}(z + \delta z) \cdot \hat{z}\delta A = I(z + \delta z)\delta A$. This evaluation yields the net power flowing into the volume δV:

$$I(z)\delta A - I(z + \delta z)\delta A \approx -\frac{\partial I}{\partial z}\delta z \delta A \equiv -\frac{\partial I}{\partial z}\delta V . \tag{5.56}$$

Substituting this result for the net wave power inflow along with Eq. (5.51) for the net wave power transfer to the atoms within the volume δV into Eq. (5.55), and canceling the common factor δV, we find that the energy density for the narrowband field increases at the rate

$$\frac{\partial \bar{u}}{\partial t} = -\frac{\partial I}{\partial z} - (\sigma_{12}\mathcal{N}_1 - \sigma_{21}\mathcal{N}_2)I . \tag{5.57}$$

Using the relation $I = \bar{u}v$ to express the wave intensity in terms of the average energy density $\bar{u}$ and the wave speed $v = c/n$ in the medium, the space-time evolution of the optical intensity is described by

$$\frac{1}{v}\frac{\partial I}{\partial t} + \frac{\partial I}{\partial z} = -(\sigma_{12}\mathcal{N}_1 - \sigma_{21}\mathcal{N}_2)I . \tag{5.58}$$

This equation represents an explicit differential form of Poynting's theorem. In steady state, or for a strictly monochromatic field, the wave intensity is independent of time t, so that $\partial I/\partial t = 0$, and Eq. (5.58) reduces to

$$\frac{\partial I}{\partial z} = (\sigma_{21}\mathcal{N}_2 - \sigma_{12}\mathcal{N}_1)\, I . \tag{5.59}$$

This equation describes the evolution of the wave intensity due to the processes of stimulated emission and absorption of radiation as the wave traverses the medium. The coefficient within the parentheses on the right-hand side has units of inverse length [m^{-1}]. Depending on the sign of the population inversion, $\Delta\mathcal{N}_{21} = \mathcal{N}_2 - \mathcal{N}_1$, the field intensity either grows or decays exponentially during propagation. Thus, the quantized nature of atomic energy levels naturally introduces the possibility of wave amplification, provided a state of population inversion can be achieved. If $\Delta\mathcal{N}_{21} < 0$, we define a positive absorption coefficient $\alpha(\nu)$, and if $\Delta\mathcal{N}_{21} > 0$, we define a positive gain coefficient $g(\nu)$ via

$$\alpha(\nu) = \sigma_{12}\Delta\mathcal{N}_{12} = \frac{\lambda_o^2 A_{21}}{8\pi n^2} S(\nu)(\mathcal{N}_1 - \mathcal{N}_2) \quad [\text{absorption coefficient, } \mathcal{N}_1 > \mathcal{N}_2]\,, \tag{5.60}$$

$$g(\nu) = \sigma_{21}\Delta\mathcal{N}_{21} = \frac{\lambda_o^2 A_{21}}{8\pi n^2} S(\nu)(\mathcal{N}_2 - \mathcal{N}_1) \quad [\text{gain coefficient, } \mathcal{N}_2 > \mathcal{N}_1]\,. \tag{5.61}$$

5.4.5 Transition Rates for Broadband Fields

The stimulated transition rates in population rate equations (5.54b) differ fundamentally from those presented in Eq. (1.5) of Chapter 1, which were originally employed by Einstein. The transition rates in Eq. (5.54b) depend directly on the lineshape function evaluated at the field frequency ν, whereas Einstein's formulation defines the transition rates using the same function evaluated at the atomic transition frequency ν_o. The difference can be traced to the spectral profile of the field. Recall that the lineshape function attains its maximum value when the field frequency is in resonance with the atomic transition frequency ($\nu = \nu_o$) and decreases as the field frequency detunes from the atomic transition frequency. Consequently, if the incident field possesses a distribution of frequencies, the overall absorption rate can exhibit a frequency dependence that differs from either extreme limit. To see this, we return to Eq. (5.52a) for the power transfer per unit volume and recall that the intensity I_ν [$W{\cdot}m^{-2}$] refers to a (narrowband) field with central (carrier) frequency ν. If the interacting field is instead characterized by a spectral intensity $I(\nu)$ [[$W \cdot m^{-2} \cdot Hz^{-1}$], such that $I(\nu)d\nu$ represents the power density within the frequency band $[\nu, \nu + d\nu]$, the total power transfer per unit volume is obtained by integrating the narrowband response over the entire frequency spectrum:

$$\frac{dw}{dt} = \int_0^\infty d\nu' \frac{\lambda_o^2 A_{21}}{8\pi n^2} S(\nu') I(\nu')(\mathcal{N}_1 - \mathcal{N}_2)\,. \tag{5.62}$$

Using the relation $I(\nu) = u(\nu)c/n$ to express the spectral power density $I(\nu)$ in terms of the spectral energy density $u(\nu)$ [$J \cdot m^{-3} \cdot Hz^{-1}$], the power transfer per unit volume can be written as

$$\frac{dw}{dt} = \int_0^\infty d\nu' \frac{c\lambda_o^2 A_{21}}{8\pi n^3} S(\nu') u(\nu')(\mathcal{N}_1 - \mathcal{N}_2)\,. \tag{5.63}$$

The integral can be evaluated if both the atomic lineshape function $S(\nu)$ and the field spectral energy density $u(\nu)$ are known. However, it simplifies dramatically in two extreme limits of practical importance in laser physics.

For a field with a very broad spectrum relative to the atomic lineshape function, such as blackbody radiation [see Fig. 5.4(a)], the variation of the spectral density $u(\nu')$ is negligible over the narrow frequency band where the atomic response $S(\nu')$ remains significantly different from zero. Under this condition, the field spectral density can be approximated by its value at the line center, $u(\nu_o)$, and factored out of the integral. The resulting integral, $\int_0^\infty S(\nu')d\nu' = 1$, then yields the following expression for the power absorption per unit

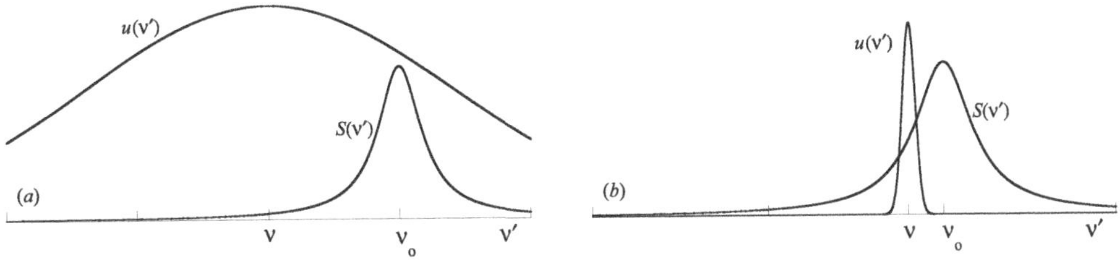

FIGURE 5.4
Field spectral density relative to the oscillator absorption profile for (a) broad-band and (b) narrow-band fields.

volume from a broadband field

$$\left.\frac{dw}{dt}\right|_{\text{broadband}} \approx \frac{c\lambda_o^2 A_{21}}{8\pi n^3} u(\nu_o) \left[\int_0^\infty d\nu' S(\nu')\right] (\mathcal{N}_1 - \mathcal{N}_2)$$
$$= \frac{c\lambda_o^2 A_{21}}{8\pi n^3} u(\nu_o)(\mathcal{N}_1 - \mathcal{N}_2)\,. \tag{5.64}$$

The absorption rate in this broad-band limit depends on the lineshape function evaluated at its resonance frequency $\nu = \nu_o$. To the extent that the background host medium refractive index n is frequency dependent, its value in the expressions above is likewise evaluated at ν_o. Einstein's derivation of the blackbody spectrum considered precisely this broadband limit. The resulting coefficient of the population difference density $(\mathcal{N}_1 - \mathcal{N}_2)$ in this equation corresponds exactly to the absorption and stimulated emission rates parametrized by the Einstein-B coefficient [cf. Eq. (1.9)].

In the opposite limit, the field energy is concentrated in a narrow spectral band centered around a carrier frequency ν (the quasi-monochromatic regime) while the lineshape function $S(\nu')$ [Fig. 5.4(b)] is broad compared to the bandwidth of the field. This situation is characteristic of light generated by a single-mode laser. In this case, we can neglect the variation of the lineshape function over the narrow spectral profile of the field and replace it with its value $S(\nu)$ evaluated at the carrier frequency of the spectral energy density $u(\nu')$. This yields the following expression for the power transfer from a narrowband field

$$\left.\frac{dw}{dt}\right|_{\text{narrowband}} \approx \frac{\lambda_o^2 A_{21}}{8\pi n^2} S(\nu) \left[\int_0^\infty d\nu' \frac{cu(\nu')}{n}\right] (\mathcal{N}_1 - \mathcal{N}_2)\,,$$
$$= \frac{\lambda_o^2 A_{21}}{8\pi n^2} S(\nu) I_\nu (\mathcal{N}_1 - \mathcal{N}_2)\,. \tag{5.65}$$

Here I_ν represents the integrated total intensity $[\mathrm{W\cdot m^{-2}}]$ for the narrow band of frequencies contained in $u(\nu')$ centered on ν. As expected, the power transfer to the atoms from a strictly monochromatic field where $u(\nu') = u_\nu \delta(\nu' - \nu)$ agrees with the narrow-band result in Eq. (5.52a).

The main conclusion from the preceding discussion is that for narrowband fields of interest in this book, the power absorption is directly proportional to the lineshape function $S(\nu)$ evaluated at the field carrier frequency ν. This function characterizes the relative strength of the atomic response to a field of frequency ν and its width determines the extent of field frequencies that will interact strongly with the atoms. It plays a central role in the energy exchange processes that govern the operation of a laser. Up to this point, our derivation of the lineshape function has incorporated only the effect of radiative and non-radiative energy loss processes (lifetime decay channels), which yields a symmetric Lorentzian spectral profile with an identical linewidth γ_o for every atom. In practical laser media, however, the atoms are subjected to numerous other processes that affect not just their energy lifetimes, but also the phase of their induced polarization and their localized resonance frequencies. For example, atoms collide frequently with each other in the gas phase or with phonons in a solid. These perturbations disrupt the phase coherence of the polarization and enhance the amplitude and phase coherence decay rates. Alternatively, atoms experience different local environments due to impurities or lattice defects in a solid that shift their individual resonance frequencies. When these additional mechanisms are taken into account, the lineshape function describing the collective response can deviate from a Lorentzian profile both in shape and in its spectral width. In the next section we review some important mechanisms and analyze how they affect the shape and width of the lineshape function.

5.5 Lineshape Function

Mechanisms affecting the lineshape function are classified as either homogeneous or inhomogeneous. Homogeneous mechanisms affect all atoms in the same way, whereas inhomogeneous mechanisms affect individual atoms differently. These two types of broadening mechanisms produce different lineshapes. Real atoms are, of course, subject to a combination of both types of mechanisms, leading almost always to intractable lineshapes. In such cases, we make an approximation by identifying the dominant mechanism and proceeding accordingly.

The prediction of the shape and width of the lineshape function from first principles is, in general, an unpleasant mathematical task that is beyond the scope of this book. In many cases, incomplete knowledge of the underlying processes may prevent such an undertaking altogether. Our approach is to provide analytic expressions for $S(\nu)$, where possible, without becoming entangled in mathematical details. In other cases, we aim to develop a qualitative understanding of how various processes affect the lineshape so that measurements of the lineshape function can be used to extract quantities such as its peak value and width.

In considering the effects of various broadening mechanisms on the lineshape function, the dipole oscillator will again serve as a simple analytical model because, as we have seen in the preceding sections, it is capable of reproducing the correct frequency response of an allowed atomic transition, provided that parameters such as ω_o, γ_o, and $A_{\rm rad}$ are interpreted in terms of the corresponding quantities for the atomic transition.

5.5.1 Homogeneous Line Broadening

5.5.1.1 Radiative and Lifetime Broadening

Radiative and nonradiative energy losses are homogeneous line-broadening mechanisms. They affect all oscillators the same way and result in damping of the amplitude of dipole oscillations described by the differential equation

$$\frac{d^2p}{dt^2} + \gamma_o \frac{dp}{dt} + \omega_o^2 p = 0\,, \tag{5.66}$$

where ω_o is the natural frequency and γ_o is the rate at which an excited oscillator loses energy through radiation and nonradiative processes such as inelastic collisions with other oscillators or quenching collisions, in which the excitation energy of an oscillator is converted into its kinetic energy. An isolated classical oscillator still loses energy by radiation at a rate given by Eq. (5.27).

In the quantum mechanical description, ν_o is the transition frequency between two quantized energy levels, say E_2 and E_1, with $h\nu_o = E_2 - E_1 > 0$, and γ_o is the sum of the energy or population relaxation rates (transition probabilities per unit time) associated with the upper and lower levels of the transition. An atom occupying level E_j can relax radiatively to a lower-level E_i by spontaneously emitting a photon of energy $h\nu_{ji} = E_j - E_i$ at a rate A_{ji} or non-radiatively at rate $\gamma_{ji\mathrm{nr}}$. The total transition rate (also called the decay rate or relaxation rate) of level E_j is then

$$\gamma_j = \sum_{i<j} (A_{ji} + \gamma_{ji\mathrm{nr}}) \equiv A_j + \gamma_{j\mathrm{nr}} \,. \tag{5.67a}$$

This means that the radiative decay rate A_2 of level E_2 includes A_{21}, the rate of spontaneous emission on the $E_2 \to E_1$ transition. If E_2 is the first excited state and state E_1 is the ground state of the atom, then $A_2 = A_{21}$ and $A_1 = 0$. (We will use the symbol A to denote spontaneous emission rates and γ to denote other relaxation rates, including the total decay rate.) The radiative decay rate A_j defines the radiative lifetime $\tau_{j\mathrm{rad}}$, and the total decay rate γ_j defines the lifetime τ_j of level j via

$$\frac{1}{\tau_{j\mathrm{rad}}} \equiv A_j = \sum_{i<j} A_{ji} \,, \tag{5.67b}$$

$$\frac{1}{\tau_j} = A_j + \gamma_{j\mathrm{nr}} \,. \tag{5.67c}$$

The energy-decay rate for the atomic dipole oscillator is therefore

$$\gamma_o = (A_2 + \gamma_{2\mathrm{nr}}) + (A_1 + \gamma_{1\mathrm{nr}}) = \gamma_2 + \gamma_1 \,, \tag{5.68}$$

where A_2 and A_1 are the spontaneous emission rates of the upper and lower levels, respectively, and $\gamma_{2\mathrm{nr}}$ and $\gamma_{1\mathrm{nr}}$ are the corresponding population-loss rates due to known and unknown nonradiative relaxation mechanisms.

Equation (5.66) accounts for energy loss by the dipole oscillator in a cycle-averaged sense over times much longer than an optical cycle, $2\pi/\omega_o$, but shorter than the decay time $1/\gamma_o$. Such a separation of time scales is possible because the radiative decay rate γ_o is much smaller than a typical optical transition frequency. The classical expression for radiative decay rate, Eq. (5.27) (with $n = 1$), for an optical wavelength of $\lambda = 500$ nm, corresponding to $\omega_o = 2\pi c/\lambda \approx 4 \times 10^{15}\,\mathrm{s}^{-1}$, yields

$$\gamma_o = A_{rad} = \frac{2}{3}\frac{e^2\omega_o^2}{4\pi\epsilon_o mc^3} = \frac{2}{3}\left(\frac{e^2}{4\pi\epsilon_o \hbar c}\right)\frac{\hbar c}{mc^2}\frac{4\pi^2 c}{\lambda} \approx 9 \times 10^7\,\mathrm{s}^{-1} \,. \tag{5.69}$$

This value is of the same order of magnitude as typical spontaneous-emission rates for optical transitions in atoms and is several orders of magnitude smaller than the corresponding optical frequency. From the Fourier transform of the impulse response of an oscillator, we find that its frequency response has a Lorentzian shape, as already shown in Eq. (5.30) of Sec. 5.3. Radiative decay affects all oscillators in the same way and is therefore an example of a homogeneous broadening mechanism. Consequently, the collective response of an ensemble of atoms undergoing an atomic transition at frequency ν_o also has a Lorentzian lineshape, with a full width at half maximum (FWHM), $\Delta\nu_{_H}$, given by

$$\varDelta\nu_o = \frac{1}{2\pi}\,(A_2 + \gamma_{2\mathrm{nr}} + A_1 + \gamma_{1\mathrm{nr}}) \,. \tag{5.70}$$

This broadening is referred to as lifetime broadening. If all nonradiative line-broadening mechanisms could be eliminated ($\gamma_{1nr} = 0 = \gamma_{2nr}$), the residual broadening due to spontaneous emission, $\Delta\nu_o = (A_2 + A_1)/2\pi$, would still remain. This is the smallest possible linewidth that can be realized and is referred to as natural broadening. For most laser transitions, natural broadening is small compared with other broadening mechanisms.

5.5.1.2 Collision Broadening

The most important homogeneous line-broadening mechanism is collisions of an atom with other atoms (in a gas) or with other entities, such as phonons (in a solid), in its environment. Inelastic collisions can increase the energy-decay rate of an oscillator by providing nonradiative decay channels for energy loss, for example, through the emission of one or more phonons in solids or by transferring energy to translational motion in a gas. Energy loss in inelastic collisions increases the decay rate from γ_o to $\gamma_o + \gamma_{\text{inel}}$ in the equation of motion of the atomic dipole oscillator. The net effect is to shorten the energy lifetime of the oscillator. This is another example of a homogeneous broadening mechanism. The resulting lineshape is Lorentzian but with an increased width. Energy-changing inelastic collisions are relatively rare; much more frequent are elastic collisions, which play a far more important role in the broadening of atomic spectral lines.

Elastic collisions do not change the energy (or the level populations) of the colliding atoms, but they do interrupt the phase of the dipole-moment oscillations. This means that after an elastic collision, the oscillator retains the same amplitude, but its oscillations resume with a random phase. To see how this affects the atomic response and, therefore, the lineshape, we begin with the solution of the polarization equation, (5.66) (with $\gamma_o \ll \omega_o$),

$$p(t) = p_o e^{i\phi_o} e^{-\gamma_o t/2} e^{-i\omega_o t}\,, \tag{5.71a}$$

where γ_o includes only the effects of energy-loss mechanisms. The constants p_0 and the initial phase ϕ_0 are determined by the initial conditions. This solution (more precisely, its real part) describes coherent, exponentially decaying oscillations in the absence of collisions, as shown in Fig. 5.5(a). If the oscillations are interrupted by a dephasing collision after a time T, they resume with the same amplitude they had immediately before the collision but with a completely random phase, as shown in Fig. 5.5(b). Thus, coherent dipole oscillations persist for only a time T and can be represented by the finite-duration signal

$$p(t) = p_o e^{i\phi_o} e^{-\gamma_o t/2} \epsilon^{-i\omega_o t}\,, \quad 0 \le t \le T\,. \tag{5.71b}$$

For random collisions occurring at a constant rate γ_{coll} (collisions per unit time), the time T between successive collisions is exponentially distributed,

$$w(T) = \gamma_{\text{coll}} e^{-\gamma_{\text{coll}} T}\,. \tag{5.72}$$

Taking the Fourier transform of Eq. (5.71b) and averaging it over the distribution of waiting times T, we find that the spectral response is Lorentzian [Eqs. (5.30) and (5.31)], with γ_o replaced by $\gamma_o + 2\gamma_{\text{coll}}$. Thus, averaging over dephasing collisions increases the decay rate of the amplitude of the coherent oscillations from γ_o to $\gamma_o + 2\gamma_{\text{coll}}$ [see Problem 5].

Collisional dephasing is a homogeneous line-broadening mechanism because it affects all oscillators in the same way. Its effect can be taken into account by replacing γ_o with $\gamma_o + 2\gamma_{\text{coll}}$ in Eq. (5.66). The resulting equation must then be interpreted as describing the average atomic dipole. With dephasing collisions, the equation for the collective polarization, Eq. (5.48), is modified to

$$\frac{dP(t)}{dt} + [i(\omega_o - \omega) + (\gamma_o + 2\gamma_{\text{coll}})/2]P = i\epsilon_o \frac{c\lambda_o^2 A_{21}}{4\pi n\omega} \Delta\mathcal{N}_{12}(t)E(t)\,, \tag{5.73}$$

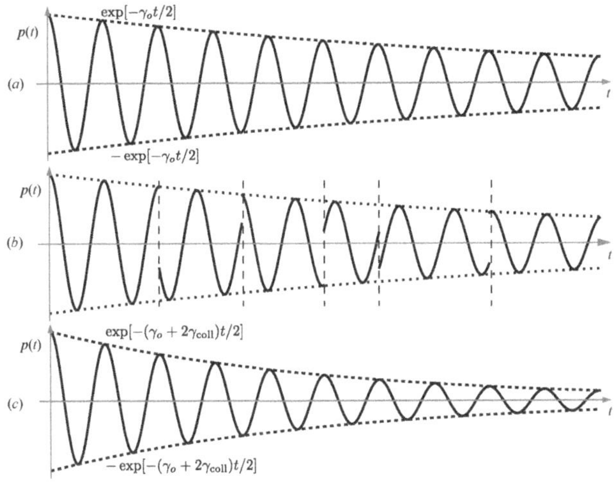

FIGURE 5.5
Broadening due to elastic collisions: (a) Exponential decay of oscillations due to energy loss; (b) Phase interruptions of the oscillations due to elastic (dephasing) collisions (amplitude unchanged); (c) Enhanced decay of oscillations averaged over dephasing collisions.

The collective polarization has the same resonance frequency as the individual atomic dipole oscillator, $p(t)$, but its oscillations decay more rapidly because of the dephasing of the individual dipole oscillators. The lineshape function for the collective response is again Lorentzian,

$$S(\nu) = \frac{1}{\pi} \frac{\Delta\nu_H/2}{(\nu - \nu_o)^2 + (\Delta\nu_H/2)^2}, \tag{5.74}$$

where the linewidth $\Delta\nu_H$, defined as the full width at half maximum (FWHM), together with the peak value $S(\nu_o)$ and the normalization of the Lorentzian, are given by

$$\Delta\nu_H = \frac{\gamma_o + 2\gamma_{\text{coll}}}{2\pi} \equiv \frac{\gamma_H}{2\pi}. \tag{5.75a}$$

$$S(\nu_o) = \frac{2}{\pi \Delta\nu_H} \approx \frac{0.637}{\Delta\nu_H}, \tag{5.75b}$$

$$\int_0^\infty d\nu S(\nu) = \frac{\Delta\nu_H}{2\pi} \int_{-\nu_o}^{\infty} \frac{d\Delta\nu}{(\Delta\nu)^2 + (\Delta\nu_H/2)^2} \approx \frac{\Delta\nu_H}{2\pi} \int_{-\infty}^{\infty} \frac{d\Delta\nu}{(\Delta\nu)^2 + (\Delta\nu_H/2)^2} = 1. \tag{5.75c}$$

In the normalization integral, the lower limit of integration, $-\nu_o$, has been extended to $-\infty$ with negligible error because $\Delta\nu_H \ll \nu_o$ for optical transitions, as the estimates below illustrate.

Collision Broadening in a Gas: Pressure Broadening

Collision broadening in a gas of atomic oscillators is referred to as pressure broadening. A simple model of collisions in a gas helps us understand how collision broadening depends on the parameters characterizing the gas. Consider a gas of identical atomic oscillators at temperature T, pressure P, and atomic density $\mathcal{N}$ (m^{-3}). From the ideal gas law, these quantities are related by

$$\mathcal{N} = \frac{P}{k_B T}, \tag{5.76}$$

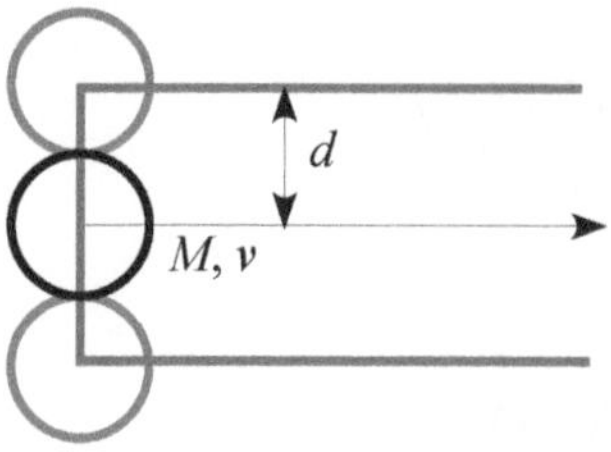

FIGURE 5.6
Atomic centers cannot come closer than atomic diameter d.

where k_B is the Boltzmann constant. Assuming each atom to be a hard sphere of diameter d, the number of collisions it suffers during a time interval δt can be estimated as follows. Two atoms cannot approach one another more closely than a center-to-center separation of d. During the time interval δt, an atom sweeps out a volume $\pi d^2 v \delta t$, where v is the average relative speed of the atoms. This volume contains, on average, $\pi d^2 v \delta \mathcal{N}$ atoms. Therefore, each atom undergoes, on average, $\pi d^2 v \delta t \mathcal{N}$ collisions during the interval δt. It follows that the collision frequency is

$$\gamma_{\text{coll}} = \frac{\#\text{ of collisions in time } \delta t}{\delta t} = \pi d^2 v \mathcal{N} \tag{5.77}$$

According to the kinetic theory of gases, the average relative speed for a gas of identical particles of mass M is

$$v = \sqrt{\frac{16 k_B T}{\pi M}}\,. \tag{5.78}$$

Using the ideal gas law, Eq. (5.76), for the density $\mathcal{N}$, we obtain

$$\gamma_{\text{coll}} = \pi d^2 \sqrt{\frac{16 k_B T}{\pi M}}\frac{P}{k_B T} = \pi d^2 P \sqrt{\frac{16}{\pi M k_B T}}\,. \tag{5.79}$$

The corresponding contribution of collisions to the linewidth is therefore

$$\Delta\nu_{\text{coll}} = \frac{\gamma_{\text{coll}}}{\pi} = d^2 P \sqrt{\frac{16}{\pi M k_B T}}\,. \tag{5.80}$$

Let us estimate the magnitude of collision broadening using this result. For atoms, the diameter d may be estimated from the approximate relation

$$r = r_o n^2\,, \tag{5.81}$$

where n is the principal quantum number of the outermost occupied shell and $r_o \approx 0.05$ nm. Expressing d in nanometers, P in Torr, and M in atomic mass units (amu), we obtain a form more convenient for order-of-magnitude estimates,

$$\begin{aligned}\Delta\nu_c &= \left[d(\text{nm})\frac{10^{-9}\,\text{m}}{\text{nm}}\right]^2 \left[P(\text{Torr})\frac{1.013\times 10^5\,\frac{\text{Pa}}{\text{atm}}}{760\,\frac{\text{Torr}}{\text{atm}}}\right]\sqrt{\frac{16}{\pi M(\text{amu})(1.66\times 10^{-27}\,\frac{\text{kg}}{\text{amu}})k_B T}} \\ &= 2.00\times 10^9 \frac{[d(\text{nm})]^2 P(\text{Torr})}{\sqrt{M(\text{amu})T(K)}}\,\text{Hz}\end{aligned} \tag{5.82}$$

For pure Ne, with $M(\text{amu}) \approx 20$ and $d = 2r_o n^2 = 2(0.05)4 = 0.40$ nm, at room temperature, $T = 300\text{K}$, we obtain

$$\Delta\nu_{\text{coll}} = 4.11\times 10^6 P(\text{Torr})\,\text{Hz}\,. \tag{5.83}$$

Thus, collision broadening in a gas is directly proportional to its pressure; for this reason it is commonly referred to as pressure broadening. Although this simple hard-sphere model is not expected to yield quantitatively accurate values for the collision broadening, it does explain its qualitative dependence on pressure and temperature. The actual values of $\Delta\nu_c$ can be larger, sometimes by an order of magnitude or more, than those predicted by Eq. (5.83).

Exact calculations that account for different particle masses and realistic interparticle interaction potentials are considerably more involved. Nevertheless, the simple model correctly predicts several general features of collision broadening. In particular, it predicts the linear dependence of the linewidth on gas pressure, a result that is well supported by experiment. Consequently, collision-broadened linewidths are often specified in units of MHz/Torr (at a fixed temperature). For example, the pressure dependence of the homogeneous linewidth of the 632.8-nm Ne transition in an electrical discharge containing a 7:1 He-Ne mixture is given by[6]

$$\Delta\nu_H = [2.6 \times 10^7 + 1.44 \times 10^8 P(\text{Torr})]\,\text{Hz}\,. \tag{5.84}$$

The first term in this equation represents the contribution of natural (radiative) broadening.

We also note that the collision frequency ($10^7 - 10^9$ Hz), though large, is still much smaller than a typical optical frequency ($10^{14} - 10^{15}$ Hz). Consequently, the homogeneous linewidth $\Delta\nu_H$ is also several orders of magnitude smaller than the optical frequency. As a result, the absorption profile of an atomic gas appears as a narrow spectral line on the optical frequency scale because $\Delta\nu_H \ll \nu_o$.

Collision broadening in solids: Phonon broadening

For active atoms embedded in a solid (typically in ionic form), the electrons of the active ions are affected by vibrations of the host lattice. Lattice vibrations are quantized and are described in terms of quanta of lattice vibration known as phonons. Elastic collisions with phonons can interrupt the phase of the radiating dipoles, leading to additional line broadening that is proportional to the collision rate. We see that phonon broadening is similar to collision broadening in a gas. It is a homogeneous broadening mechanism because all active ions in the lattice are affected by phonons in the same way.

5.5.2 Inhomogeneous Broadening

In addition to homogeneous broadening mechanisms, active atoms are also subject to inhomogeneous broadening mechanisms that shift their resonance frequencies by different amounts, depending on some property that distinguishes one group of atoms from another. Examples include the Doppler effect for atoms of the same species and mass but with different thermal velocities, the isotope effect for atoms of the same species but with different masses, and the Stark effect for atoms of the same species experiencing different local crystalline electric fields in a solid. These local field variations produce shifts of the energy levels and, consequently, of the transition frequencies of the ions. (Indeed, the term inhomogeneous broadening originated from this situation.) For random local field variations, the resulting distribution of transition frequencies, $S(\nu)$, is Gaussian, with a linewidth $\Delta\nu$ (FWHM) that depends on the extent of the transition-frequency variations and hence on the degree of field inhomogeneity within the crystal or glass. In practice, several of these mechanisms often act simultaneously, together with homogeneous broadening mechanisms, resulting in a complicated lineshape.

[6] P. W. Smith, "Linewidth and Saturation Parameters for the 6328A Transition in a Helium-Neon Laser," J. Appl. Phys. **37**, 2089–2093 (1966).

For active atoms in a gas, Doppler broadening due to thermal motion is the most important inhomogeneous broadening mechanism. Consider a gas of active atoms with transition frequency ν_o (in their rest frame) interacting with an electromagnetic wave of frequency ν propagating in the $+z$ direction. For a group of atoms moving with velocity component v_z along the z axis, the apparent (Doppler-shifted) frequency of the wave is

$$\nu' = \nu(1 - v_z/c)\,. \tag{5.85}$$

To be in resonance, the Doppler-shifted frequency ν' must equal ν_o; that is, $\nu(1-v_z/c) = \nu_o$. Hence, the wave frequency that is resonant with atoms moving at velocity v_z is

$$\nu = \frac{\nu_o}{(1 - v_z/c)} \approx \nu_o(1 + v_z/c)\,, \tag{5.86}$$

where the approximation assumes $|v_z| \ll c$. Thus, this group of atoms is resonant with radiation of frequency $\nu_o(1 + v_z/c)$, and we may regard

$$\nu_a = \nu_o(1 + v_z/c) \tag{5.87}$$

as the resonance frequency of atoms moving with velocity component v_z.

Different velocity groups in the gas therefore have different resonance frequencies. We recall that, for a gas in thermal equilibrium at temperature T, the fraction of atoms with velocity components between v_z and $v_z + dv_z$ along the direction of propagation is given by the Maxwell–Boltzmann distribution,

$$f(v_z) = \sqrt{\frac{M}{2\pi k_B T}}\, e^{-\frac{Mv_z^2}{2k_B T}}\,, \tag{5.88}$$

where M is the atomic mass and k_B is the Boltzmann constant. Using Eq. (5.87) as a change of variable, the distribution $G(\nu_a)$ of resonance frequencies is obtained from $G(\nu_a)d\nu_a = f(v_z)dv_z$, which yields

$$G(\nu_a) = f(v_z)\left|\frac{dv_z}{d\nu_a}\right| = \frac{c}{\nu_o}\sqrt{\frac{M}{2\pi k_B T}}\,\exp\left[-\left(\frac{\nu_a - \nu_o}{\nu_o}\right)^2 \frac{Mc^2}{2k_B T}\right]. \tag{5.89}$$

Thus, the Doppler-shifted resonance frequencies have a Gaussian distribution centered at $\nu_a = \nu_o$, with a peak value

$$G(\nu_o) = \frac{1}{\nu_o}\sqrt{\frac{Mc^2}{2\pi k_B T}}\,. \tag{5.90}$$

The full width at half maximum (FWHM), $\Delta\nu_D$, determined by the condition $G(\nu_o + \frac{1}{2}\Delta\nu_D) = \frac{1}{2}G(\nu_o)$, is

$$\Delta\nu_D = \nu_o\sqrt{\frac{k_B T 8\ln 2}{Mc^2}} = \frac{1}{\lambda_o}\sqrt{\frac{k_B T 8\ln 2}{M}}\,. \tag{5.91}$$

This quantity is referred to as the Doppler width. In terms of the FWHM $\Delta\nu_D$, we can rewrite the resonance frequency distribution function in a compact form as

$$G(\nu_a) = \frac{2}{\Delta\nu_D}\sqrt{\frac{\ln 2}{\pi}}\,\exp\left[-\left(\frac{\nu_a - \nu_o}{\Delta\nu_D}\right)^2 4\ln 2\right]. \tag{5.92}$$

The function $G(\nu_a)$ is normalized to unity,

$$\int_0^\infty d\nu_a G(\nu_a) = \frac{2}{\Delta\nu_D}\sqrt{\frac{\ln 2}{\pi}}\int_0^\infty d\nu_a \exp\left[-\left(\frac{\nu_a - \nu_o}{\Delta\nu_D}\right)^2 4\ln 2\right],$$

$$= \frac{1}{\sqrt{\pi}}\int_{-2\nu_o\sqrt{\ln 2}/\Delta\nu_D}^{\infty} dx \exp\left[-x^2\right] \approx \frac{1}{\sqrt{\pi}},\int_{-\infty}^{\infty} dx \exp\left[-x^2\right] = 1, \quad (5.93)$$

where, since $2\nu_o\sqrt{\ln 2}/\Delta\nu_D \gg 1$, we have replaced the lower limit by $-\infty$ with negligible error because the integrand is vanishingly small well before this limit is reached.

The expression for Doppler width $\Delta\nu_D$ can be cast in a form that is more convenient for estimating its numerical values

$$\Delta\nu_D = \frac{1}{\lambda_o}\sqrt{\frac{k_B T 8\ln 2}{M}} = \frac{1}{\lambda_o(\text{nm}) \times 10^{-9}\,\text{m/nm}}\sqrt{\frac{1.3807 \times 10^{-23}\,\text{J/K}T(\text{K}) \times 8\ln 2}{M(\text{amu}) \times 1.6605 \times 10^{-27}\,\text{kg/amu}}}$$

$$= 2.15 \times 10^{11}\left[\frac{1}{\lambda_o(\text{nm})}\sqrt{\frac{T(\text{K})}{M(\text{amu})}}\right]\ \text{Hz} \quad (5.94)$$

Similarly, for the relative width we obtain

$$\frac{\Delta\nu_D}{\nu_o} = \frac{1}{c}\sqrt{\frac{k_B T 8\ln 2}{M}\ln 2} = 7.16 \times 10^{-7}\sqrt{\frac{T(\text{K})}{M(\text{amu})}}. \quad (5.95)$$

Relative width is useful when the same atom can exhibit absorption at several transition frequencies. For a He-Ne laser, $M_{\text{Ne}} = 20$ amu, $T = 400$ K, and $\lambda_o = 632.8$ nm

$$\Delta\nu_D = 2.15 \times 10^{11} \times \frac{1}{632.8}\sqrt{\frac{400}{20}}\ \text{Hz} = 1518\,\text{MHz} \approx 1.52\,\text{GHz}$$

For the 10.6 μm line of CO_2, $M_{CO_2} = 44$ amu, $T = 400$ K

$$\Delta\nu_D = 2.15 \times 10^{11} \times \frac{1}{10.6 \times 10^3}\sqrt{\frac{400}{44}}\ \text{Hz} = 61\,\text{MHz}$$

Doppler broadening is an example of inhomogeneous broadening. Atoms in a collection have different resonance frequencies because of their different velocities. Other examples of inhomogeneous broadening include broadening of active atoms by impurities or defects in a crystal that cause different local crystal fields. These in turn have the effect of shifting the resonance frequencies of active atoms slightly differently. The distribution of these shifts is often approximately Gaussian, much like the Doppler distribution, and gives rise to an inhomogeneously broadened absorption line. This type of broadening is present in the Cr^{3+} line associated with the Ruby laser, for example.

5.6 The Voigt Profile

In general, both homogeneous and inhomogeneous mechanisms contribute to line broadening. Consequently, the lineshape function is, in general, neither purely Gaussian nor purely Lorentzian. Let us consider a case of considerable practical importance in gas lasers. Active atoms or ions with transition frequency ν_o interact with an electromagnetic wave of frequency ν (all frequencies are measured in the laboratory frame). Atoms moving with velocity component v_z (along the direction of propagation of the electromagnetic wave) have the resonance frequency

$$\nu_a = \nu_o(1 + v_z/c) \tag{5.96}$$

This velocity group, representing a fraction $G(\nu_a)\,d\nu_a$ of the total number of atoms, is subject to both collision broadening and natural broadening. Since both are homogeneous broadening mechanisms, its response has a Lorentzian lineshape

$$S(\nu, \nu_a) = \frac{1}{\pi} \frac{\Delta\nu_H/2}{(\nu_a - \nu)^2 + (\Delta\nu_H/2)^2} \tag{5.97}$$

where

$$\Delta\nu_H = \Delta\nu_o + \Delta\nu_{\text{coll}}\,. \tag{5.98}$$

The overall lineshape of the ensemble is obtained by summing the contributions from all resonance-frequency groups:

$$\begin{aligned} S(\nu) &= \int_o^\infty d\nu_a G(\nu_a) S(\nu, \nu_a) = \int_0^\infty d\nu_a \sqrt{\frac{4\ln 2}{\pi \Delta\nu_D^2}}\, e^{-\left(\frac{\nu_a - \nu_o}{\Delta\nu_D}\right)^2 4\ln 2} \frac{\Delta\nu_H/2\pi}{(\nu_a - \nu)^2 + (\Delta\nu_H/2)^2} \\ &= \frac{\Delta\nu_H \sqrt{\ln 2}}{\Delta\nu_D \pi^{3/2}} \int_0^\infty d\nu_a \frac{\exp\left[-(\nu_a - \nu_o)^2 \left(\frac{4\ln 2}{\Delta\nu_D^2}\right)\right]}{(\nu_a - \nu_o + \nu_o - \nu)^2 + (\Delta\nu_H/2)^2}\,. \end{aligned} \tag{5.99}$$

This line profile, which incorporates both homogeneous and inhomogeneous broadening mechanisms, is known as the Voigt profile. It is encountered frequently in laser spectroscopy and atomic physics. Values of the Voigt profile have traditionally been tabulated in handbooks, although with modern desktop computers it is straightforward to evaluate the function numerically.

Here we consider several limiting cases to gain insight into the behavior of the Voigt profile. We first rewrite the integral by introducing the variables

$$x = \frac{(\nu_a - \nu_o)2\sqrt{\ln 2}}{\Delta\nu_D}\,, \qquad w = \frac{(\nu - \nu_o)2\sqrt{\ln 2}}{\Delta\nu_D}\,, \qquad b = \frac{\Delta\nu_H \sqrt{\ln 2}}{\Delta\nu_D}\,, \tag{5.100a}$$

The variable x represents the Doppler shift of the resonance frequency from ν_o, w represents the detuning of the field frequency from ν_o, and $b = \sqrt{\ln 2}\,\Delta\nu_H/\Delta\nu_D$ is a dimensionless measure of the ratio of the homogeneous linewidth to the Doppler linewidth. In terms of these variables, the lineshape function becomes

$$\begin{aligned} S(\nu) &= \left(\frac{2\sqrt{\ln 2}}{\sqrt{\pi}\,\Delta\nu_D}\right) \frac{b}{\pi} \int_{-\left(\frac{2\nu_o\sqrt{\ln 2}}{\Delta\nu_D}\right)}^{\infty} \frac{dx\, e^{-x^2}}{(x-w)^2 + b^2} \\ &\approx G(\nu_0) \left[\frac{b}{\pi} \int_{-\infty}^{\infty} \frac{dx\, e^{-x^2}}{(x-w)^2 + b^2}\right] \equiv G(\nu_o) V(b, w)\,. \end{aligned} \tag{5.101}$$

In the last step, the lower limit of integration has been extended to $-\infty$ with negligible error. As we have seen, the optical frequency ν_o is much larger than the Doppler width, so that $\nu_o/\Delta\nu_D \gg 1$, and the integrand is already vanishingly small well before the original lower limit is reached. The expression in square brackets is the Voigt function, $V(b, w)$, which is implemented as a built-in function in software packages such as Mathematica.

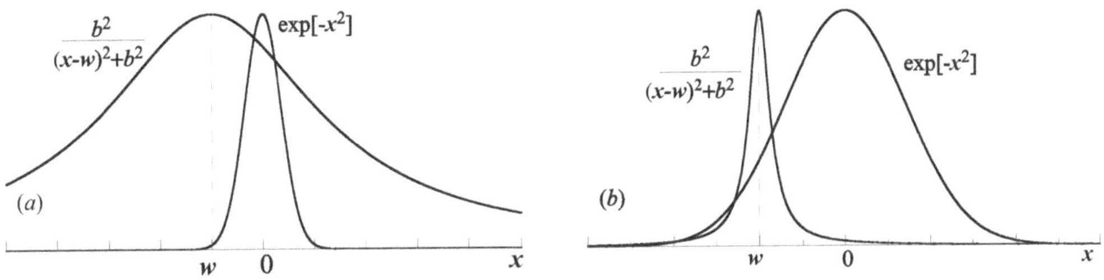

FIGURE 5.7
The integrand of Eq. (5.101) is the product of a Gaussian centered at $x = 0\,(\nu_a - \nu_o = 0)$ and a Lorentzian centered at $x = w\,[\nu_a - \nu_o = (\nu - \nu_o)]$. The figure illustrates these two factors in the limits of (a) strong collisional broadening, $b = 5$ ($\Delta\nu_H/\Delta\nu_D \approx 6$), and (b) strong Doppler broadening, $b = 0.14$ ($\Delta\nu_H/\Delta\nu_D \approx 1/6$).

The integrand is the product of a Gaussian and a Lorentzian. The Gaussian factor, e^{-x^2}, representing the distribution of Doppler-shifted resonance frequencies, is centered at $x = 0$ (corresponding to $\nu_a = \nu_o$), whereas the Lorentzian factor, representing the homogeneous response, is centered at $x = w$ (corresponding to $\nu_a - \nu_o = \nu - \nu_o$, or equivalently $\nu_a = \nu$). We now consider several limiting cases for which approximate analytic expressions for the lineshape function can be obtained.

5.6.1 Strong Collision Broadening

At sufficiently high gas pressures, collisional broadening increases the homogeneous linewidth so that the ratio $b \sim \Delta\nu_H/\Delta\nu_D \gg 1$. In this limit, the Lorentzian varies only slowly over the range of integration for which the Gaussian is appreciable [see Fig. 5.7(a)]. We may therefore replace the Lorentzian by its value at the peak of the Gaussian ($x = 0$) and take it outside the integral. The remaining integral is simply the normalization integral of the Gaussian, yielding

$$S(\nu) \approx \sqrt{\frac{4\ln 2}{\pi}}\frac{b}{\pi\Delta\nu_D}\frac{1}{w^2+b^2}\int_{-\infty}^{\infty} dx e^{-x^2} = \frac{1}{\pi}\frac{\Delta\nu_H/2}{(\nu-\nu_o)^2+(\Delta\nu_H/2)^2}. \tag{5.102}$$

Thus, in the limit of strong collisional broadening, the lineshape is well approximated by a Lorentzian with FWHM $\Delta\nu_H$.

5.6.2 Strong Inhomogeneous Broadening

At low gas pressure, inhomogeneous broadening due to the Doppler effect dominates, so that $\Delta\nu_H \ll \Delta\nu D$ and the ratio $b \ll 1$ [Fig. 5.7(b)]. In this case, the Lorentzian is a very narrow function, and, provided that w is not too large, we can evaluate the integral by replacing the Gaussian with its value at the peak of the Lorentzian ($x = w$) and factoring it out of the integral. We then obtain

$$\begin{aligned} S(\nu) &\approx G(\nu_o)e^{-w^2}\left[\frac{b}{\pi}\int_{-\infty}^{\infty}\frac{dx}{(x-w)^2+b^2}\right], \\ &= \frac{2\sqrt{\pi\ln 2}}{\pi\Delta\nu_D}\exp\left[-(\nu-\nu_o)^2\frac{4\ln 2}{\Delta\nu_D^2}\right]. \end{aligned} \tag{5.103}$$

In this limit, the lineshape function reduces to the pure Doppler-broadened Gaussian profile. We can develop corrections to the Doppler profile for finite but small homogeneous

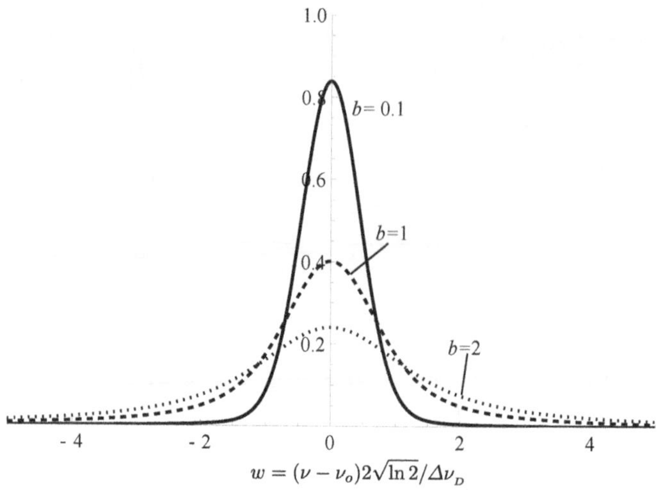

FIGURE 5.8
Voigt line profile for increasing homogeneous broadening relative to Doppler broadening, with $b = 0.1, 1, 2$.

broadening by using the identity

$$\int_0^\infty dy e^{-by} \cos[(x-w)y] = \frac{b}{b^2 + (x-w)^2}, \tag{5.104}$$

in the Voigt function integral. This gives

$$V(b,w) = \frac{b}{\pi}\int_{-\infty}^{\infty} \frac{dx e^{-x^2}}{b^2 + (x-w)^2} = \frac{1}{\pi}\int_{-\infty}^{\infty} dx e^{-x^2} \int_0^\infty dy e^{-by} \cos[(x-w)y]\,. \tag{5.105}$$

Expanding $\cos[(x-w)y] = \cos(xy)\cos(wy) + \sin(xy)\sin(wy)$, and interchanging the order of integration, we find that the contribution from the sine term vanishes because it is an odd function of x. This leads to

$$V(b,w) = \int_0^\infty dy e^{-by} \cos(wy) \left[\frac{1}{\pi}\int_{-\infty}^{\infty} e^{-x^2} \cos(xy) dx\right]. \tag{5.106}$$

The Gaussian Fourier integral evaluates to $\sqrt{\pi} e^{-y^2/4}$, so the quantity in square brackets is $e^{-y^2/4}/\sqrt{\pi}$, yielding

$$V(b,w) = \frac{1}{\sqrt{\pi}}\int_0^\infty dy\, e^{-by-y^2/4} \cos(wy)\,. \tag{5.107}$$

For small b, we may approximate $e^{-by} \approx 1 - by$ under the integral. Then, after changing the variable of integration to $y = 2t$, we obtain

$$\begin{aligned} V(b,w) &\approx \frac{2}{\sqrt{\pi}}\int_0^\infty e^{-t^2}\cos(2wt)dt - \frac{4b}{\sqrt{\pi}}\int_0^\infty te^{-t^2}\cos(2wt)dt\,, \\ &= \frac{2}{\pi}\frac{\sqrt{\pi}}{2}e^{-w^2} - \frac{2b}{\sqrt{\pi}}\left[1 - 2w\int_0^\infty dt\, e^{-t^2}\sin(2wt)\right] \\ &= e^{-w^2} - \frac{2b}{\sqrt{\pi}}\left[1 - 2wD(w)\right], \end{aligned} \tag{5.108a}$$

where

$$D(w) = \int_0^\infty dt\, e^{-t^2}\sin(2wt) = e^{-w^2}\int_0^w e^{t^2} dt \tag{5.108b}$$

is the Dawson function. The first term in Eq. (5.108a), when substituted into Eq. (5.101), reproduces the Doppler profile given by Eq. (5.103), while the second term provides the leading correction due to finite homogeneous broadening.

5.6.3 Far-wing Limit

Finally, when the field frequency ν is far from ν_0 such that the detuning $|\nu - \nu_0|$ is large compared with both $\Delta\nu_H$ and $\Delta\nu D$, we are in the so-called far-wing limit, for which $|w| \gg 1$. In this limit, the qualitative difference between the Lorentzian and Gaussian profiles becomes important. The Gaussian falls to zero much more rapidly than the Lorentzian. Consequently, the Gaussian is essentially zero near the peak ($x = w$) of the Lorentzian. On the other hand, the Lorentzian tail still makes a significant contribution near the peak of the Gaussian. We can therefore evaluate the integral approximately by replacing the Lorentzian with its far-wing value at the peak ($x = 0$) of the Gaussian:

$$\begin{aligned} S(\nu) &\approx \frac{b}{\pi}\frac{1}{w^2}\sqrt{\frac{4\ln 2}{\pi\Delta\nu_D^2}}\int_{-\infty}^{\infty} dx\, e^{-x^2} = \frac{b}{\pi}\frac{1}{w^2}\sqrt{\frac{4\ln 2}{\pi\Delta\nu_D^2}}\sqrt{\pi} \\ &= \frac{\Delta\nu_H}{2\pi}\frac{1}{(\nu-\nu_o)^2} \end{aligned} \tag{5.109}$$

This is the behavior expected for a homogeneously broadened line at large detuning. Thus, in the far-wing limit, the line behaves essentially as a homogeneously broadened line even when the dominant broadening mechanism is Doppler broadening. Within this isolated-resonance model, this provides a justification for treating the far-off-resonance response as essentially homogeneous.

We can summarize the results of this section as follows. Doppler broadening is most significant at low pressure, high temperature, and for short-wavelength transitions. Collision broadening is most significant at high pressure, low temperature, and for long-wavelength transitions. Under many conditions, both effects must be considered simultaneously, requiring the use of the Voigt profile. Doppler and pressure broadening are the dominant broadening mechanisms in gaseous media. We have presented them here to illustrate, within a simple analytical framework, how different broadening mechanisms determine the form of $S(\nu)$.

Our discussion of lineshapes applies primarily to gaseous media with isolated resonances, and even in such cases only the central portion of some spectral lines is adequately described

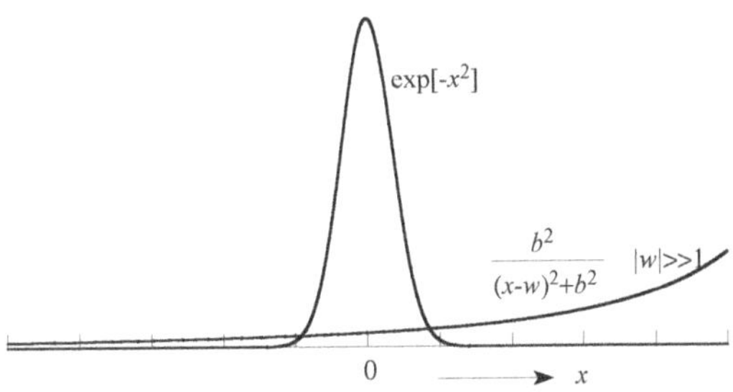

FIGURE 5.9
In the far-wing limit $|w| \gg 1$, only the tail of the Lorentzian contributes significantly near the peak of the Gaussian.

by the theory presented here. If atomic resonances overlap, if different isotopes are present, or if collisions with background gases become important, considerably more complicated lineshapes can arise. Such effects are common in molecular and solid-state lasers.

Doppler and pressure broadening do not apply in solids. Instead, the local electric fields (crystal fields) produced by neighboring atoms modify the energy levels of an active ion by shifting them and lifting the degeneracies that exist for the free ion. Dynamic fluctuations of the crystal field, caused by thermal vibrations of the host lattice, produce additional changes in the energy-level structure of the active ions. As a result, many experimentally observed lineshape functions are asymmetric. Because the underlying broadening mechanisms are often only partially understood and are mathematically intractable, it is common practice to use experimentally measured lineshape functions rather than attempt to derive them theoretically.

Regardless of its mathematical form, the lineshape function plays a fundamental role in light-matter interactions. The quantity $S(\nu)d\nu$ represents the relative probability that

1. an atom in the excited state E_2 spontaneously emits a photon with frequency between ν and $\nu + d\nu$ as it makes a transition to the lower state E_1;

2. an atom in the lower state E_1 absorbs a photon with frequency between ν and $\nu + d\nu$, making a transition to the excited state E_2; or

3. radiation in the frequency interval ν to $\nu + d\nu$ stimulates an atom in the excited state E_2 to emit a photon within the same frequency interval.

These three processes correspond, respectively, to spontaneous emission, absorption, and stimulated emission. Within the present isolated-transition model, the same normalized lineshape function characterizes the frequency dependence of spontaneous emission, stimulated emission, and absorption. A second important point is that an ensemble of atoms is subject to a variety of broadening mechanisms, each producing a characteristic lineshape that determines the collective optical response of the medium.

5.6.4 Lineshape Function and Cross-section

Modification of the lineshape function by broadening mechanisms is reflected in the frequency dependence of the absorption and emission cross-sections. In particular, the peak value of the cross-section for the $2 \to 1$ transition is given by $\sigma(\nu_o) = \frac{\lambda_o^2 A_{21}}{8\pi n^2} S(\nu_o)$ and away from line center it decreases according to the lineshape function $S(\nu)$. Thus, in the limit of strong homogeneous broadening, the absorption cross-section has a Lorentzian profile, whereas in the limit of strong inhomogeneous broadening it has a Gaussian profile:

$$\sigma_H(\nu) = \frac{\lambda_o^2 A_{21}}{4\pi^2 n^2 \Delta\nu_H} \left[\frac{1}{1 + [2(\nu - \nu_o)/\Delta\nu_H]^2} \right], \quad \text{homogeneously broadened line}, \tag{5.110}$$

$$\sigma_I(\nu) = \sqrt{\pi \ln 2} \frac{\lambda_o^2 A_{21}}{4\pi^2 n^2 \Delta\nu_D} \exp\left[-(4\ln 2) \left(\frac{\nu - \nu_o}{\Delta\nu_D} \right)^2 \right], \quad \text{inhomogeneously broadened line}. \tag{5.111}$$

Here we have used the near-resonance approximation, $\lambda \approx \lambda_o$, in the prefactor multiplying the Gaussian for an inhomogeneously broadened transition. Note that by defining an effective atomic linewidth, $\Delta\nu_{\text{at}} = \Delta\nu_H$ for homogeneous broadening and $\Delta\nu_{\text{at}} = \Delta\nu_D/\sqrt{\pi \ln 2} \approx 0.678\Delta\nu_D$, for inhomogeneous broadening, the peak cross-section can be written in the unified form

$$\sigma(\nu_o) = \frac{\lambda_o^2 A_{21}}{4\pi^2 n^2 \Delta\nu_{\text{at}}}. \tag{5.112}$$

The effective linewidth $\Delta\nu_{\text{at}}$ is defined specifically to unify the peak-cross-section expression; for an inhomogeneous Gaussian it is not the actual FWHM, which remains $\Delta\nu_{_D}$. Finally, recall that we have assumed $\sigma_{21} = \sigma_{12} = \sigma$, corresponding to nondegenerate upper and lower levels. For an arbitrary $i \to j$ transition, including the effects of degeneracy, the more general relation is

$$\sigma_{ji}(\nu_o) = \frac{g_i}{g_j}\sigma_{ij}(\nu_o) = \frac{\lambda_{ij}^2 A_{ji}}{4\pi^2 n^2 \Delta\nu_{\text{at}}}\,. \tag{5.113}$$

Cross-section Estimates

The cross-section is a key parameter in laser design because it determines the achievable gain and, consequently, influences both the laser threshold and the maximum output power. Larger cross-sections generally provide higher gain for a given population inversion and can reduce the lasing threshold; they also strongly influence saturation and power-extraction characteristics. Within this lifetime-limited two-level model, the maximum line-center cross-section is obtained when the transition is broadened only by its radiative lifetime. In this case, $\Delta\nu_{_H} = A_{21}/2\pi$, which gives

$$\sigma_{\max} = \frac{\lambda_o^2}{2\pi n^2}\,. \tag{5.114}$$

Thus, the maximum cross-section is of the order of one wavelength squared. For an optical transition with $\lambda = 500$ nm and $n = 1$, this corresponds to $(500\,\text{nm})^2/2\pi \approx (200\,\text{nm})^2$. By comparison, the physical size of an atom, as characterized by the radius of its outermost Bohr orbit, is only a few tenths of a nanometer. Consequently, the effective optical "size" of an atom can be thousands of times larger than its physical size because of the resonant enhancement of its interaction with light.

For real atoms, which are subject to additional broadening mechanisms, the line-center cross-section is reduced by the ratio of the radiative linewidth to the total linewidth. For example, under typical operating conditions, the 632.8-nm transition in a He-Ne laser is dominated by Doppler broadening with $\Delta\nu_{_D} = 1.5 \times 10^9$ Hz. The radiative decay rate for this transition is $A_{21} \approx 3.4 \times 10^6\ \text{s}^{-1}$ giving a peak emission cross-section

$$\begin{aligned}\sigma(\nu_o) &= \sqrt{\pi \ln 2}\frac{\lambda^2}{4\pi^2 n^2}\frac{A_{21}}{\Delta\nu_{_D}}\,,\\ &= \sqrt{\pi\ln 2}\frac{(632.8\times 10^{-9})^2(3.4\times 10^6)}{4\pi^2(1.5\times 10^9)} = 3.4\times 10^{-13}\,\text{cm}^2\,.\end{aligned}$$

Strong Doppler-broadened visible transitions in gas lasers typically have emission cross-sections in the range $10^{-11}\,\text{cm}^2$–$10^{-13}\,\text{cm}^2$.

Solid-state laser materials, in general, have much smaller emission cross-sections, typically in the range 10^{-18}–$10^{-20}\,\text{cm}^2$. The Nd:YAG laser is perhaps the best-known solid-state laser. It uses triply ionized neodymium ions as the active centers embedded in a yttrium aluminum garnet (YAG) host crystal with refractive index $n = 1.82$. The stronger of the two closely spaced laser transitions near 1064 nm has $A_{21} = 4.3 \times 10^3\ \text{s}^{-1}$ and is predominantly homogeneously broadened with linewidth $\Delta\nu_{_H} = 1.2 \times 10^{11}$ Hz. The corresponding emission cross-section is

$$\sigma(\nu_o) = \frac{\lambda^2}{4\pi^2 n^2}\frac{A_{21}}{\Delta\nu_{_H}} = \frac{(1064\times 10^{-9})^2(4.3\times 10^3)}{4\pi^2(1.82)^2(1.2\times 10^{11})} = 3.1\times 10^{-18}\,\text{cm}^2\,. \tag{5.115}$$

TABLE 5.1
Emission cross-section for selected laser transitions.

Laser	λ (nm)	σ_{em} (cm^2)
He-Ne	633	3.0×10^{-13}
Argon	488	2.5×10^{-12}
He-Cd	442	9.0×10^{-14}
Copper (CVL)	510	8.6×10^{-14}
CO_2	10,600	3.0×10^{-18}
Excimer	248	2.6×10^{-16}
Dye (Rh6G)	577	2.5×10^{-16}
Semiconductor	800	1.0×10^{-18}
Nd:YAG	1064	3.3×10^{-19}
Nd:Glass	1062	3.0×10^{-20}
Ti:Al_2O_3	800	3.4×10^{-19}
Cr:LiSrAlF	850	4.8×10^{-20}

Equation (5.115) is a simplified estimate ignoring such details as branching ratio and temperature dependence whereas Table 5.1 below lists typical cross-sections for some important laser transitions. Actual values may vary depending on isotopic composition, residual impurities, and growth conditions.

Summary

In this chapter, we presented the electron oscillator model of light–matter interaction by modeling the atomic response to the electric field of an electromagnetic wave as that of a classical driven charged oscillator. This model was used to calculate the steady-state atomic polarization and dielectric susceptibility, allowing us to revisit the separation of the atomic response into nonresonant and resonant components introduced in Chapter 2. It also provided a basis for introducing the refractive index, absorption coefficient, and absorption cross-section of a medium in terms of its susceptibility.

We also derived an equation of motion for the atomic polarization of an ensemble of atoms. By replacing the oscillator number density in this equation with the population density difference between the two atomic levels, identifying the oscillator frequency with the atomic transition frequency, and replacing the oscillator decay rate with the spontaneous emission rate of the transition, we obtained the quantum-mechanically correct equation for the atomic polarization. We then used Poynting's theorem for the energy exchange between an electromagnetic wave and the atomic medium to derive the quantum-mechanically correct rate equations for the level populations. Together with the equation for the atomic polarization, these rate equations provide a complete description of the atomic response to light.

Finally, we discussed modifications of the atomic response arising from finite excited-state lifetimes, collisions, atomic motion, local variations in the resonance frequency, and other broadening mechanisms in terms of the lineshape function and the associated emission and absorption cross-sections.

Problems

1. **Classical electron oscillator:** Classically, the response of a dielectric material to a weak external electric field (weak compared with the internal electric fields experienced by atomic electrons) can be modeled by treating the dielectric as a collection of electron oscillators. In this model, an atom consists of a heavy point nucleus of charge Ze surrounded by a rigid, concentric, uniformly charged spherical electron cloud of total charge $-Ze$, mass Zm_e, and radius a, where Z is the atomic number. When an external electric field is applied, the electron cloud is displaced slightly relative to the nucleus. Since the nucleus is much heavier than the electron cloud, its displacement may be neglected.

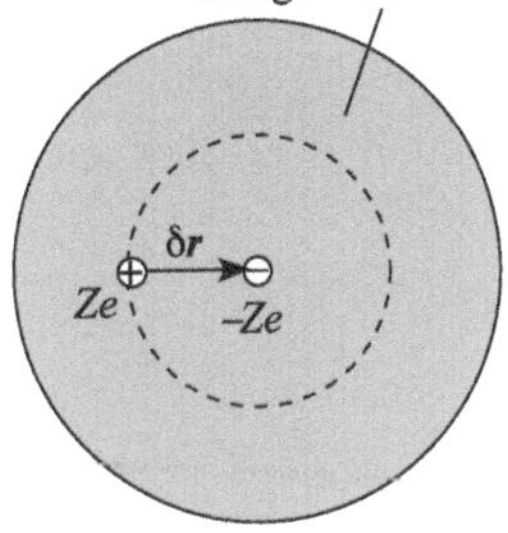

   (a) Find the restoring force on the displaced electron cloud and show that the atomic polarizability is $4\pi\epsilon_o a^3$, where the induced dipole moment and polarizability are related by $\boldsymbol{p}_{\text{induced}} = \alpha \boldsymbol{E}_{\text{ext}}$.

   (b) Find the natural angular frequency at which the electron cloud oscillates about the nucleus.

   (c) In the Bohr model of the hydrogen atom, the radius of the first electron orbit is $a_o = 5.3\times10^{-11}$ m $= 0.053$ nm. Using twice this value as an estimate of the electron-cloud radius, $a = 2a_o = 0.106$ nm, estimate the resonance frequency numerically. What wavelength does this frequency correspond to, and in what region of the electromagnetic spectrum does it lie?

2. Show that in the low-frequency limit, $\omega \ll \omega_o$, where ω_o is the nearest resonance frequency of the material, the dielectric susceptibility can be approximated by

$$\chi \approx \frac{e^2 \mathcal{N} f_o}{m\epsilon_o\omega_o^2}\,.$$

   Using the expression for ω_o obtained in Problem 1 and taking $a = 2a_o$, express χ in terms of the number density $\mathcal{N}$ and fraction f_o. Then estimate its numerical value for a representative atomic number density $\mathcal{N}$ and $f_o \approx 1$, and compare the result with the susceptibility corresponding to a typical transparent optical material with refractive index $n \sim 2$.

3. Estimate the speed of a bound electron in a hydrogen atom by assuming that the electron moves in a circular orbit of Bohr radius. Compare the magnitude of the magnetic part of the Lorentz force exerted by an electromagnetic wave on an electron moving at this speed with the magnitude of the electric force.

4. A linear electron oscillator in the absence of an applied field obeys the equation

$$\frac{d^2x}{dt^2} + \omega_o^2 x = 0\,.$$

   In the absence of radiation loss, its motion may be written as $x(t) = a_o\cos(\omega_o t + \phi)$, where a_o is the oscillation amplitude. When radiation loss is included, assume that the fractional energy loss per optical cycle is small, so that a_o varies slowly compared with the oscillation period. Using Larmor's formula for the instantaneous radiated power,

$$P(t) = \frac{2}{3}\frac{e^2}{4\pi\epsilon_o c^3}\ddot{x}^{\,2},$$

show that the total oscillator energy, averaged over a few optical cycles, is

$$\overline{W} = \overline{W}_{\rm KE} + \overline{W}_{\rm PE} = \frac{1}{4}m\omega_o^2 a_o^2 + \frac{1}{4}m\omega_o^2 a_o^2 = \frac{1}{2}m\omega_o^2 a_o^2,$$

and that it decreases according to

$$\frac{d\overline{W}}{dt} = -\gamma_o \overline{W}.$$

Find γ_o.

5. An oscillator of mass m, natural angular frequency ω_o, and damping rate γ_o is subjected to an impulsive force

$$F(t) = \Delta p\,\delta(t) = mv_o\delta(t),$$

where $\delta(t)$ is the Dirac delta function and $\Delta p = mv_o$ is the impulse delivered to the oscillator:

$$\text{Impulse} = \int_{-\infty}^{\infty} F(t)\,dt = \Delta p.$$

Assume that the oscillator is initially at rest at its equilibrium position and obeys

$$\ddot{x} + \gamma_o\dot{x} + \omega_o^2 x = \frac{F(t)}{m}.$$

(a) Show that, for $t \geq 0$, the oscillator response is

$$x(t) = \frac{v_o}{\omega_o'}e^{-\gamma_o t/2}\sin(\omega_o' t),$$

where

$$\omega_o' = \sqrt{\omega_o^2 - \frac{\gamma_o^2}{4}}.$$

(b) What form does the displacement take in the weak-damping limit $\gamma_o \ll \omega_o$?

(c) In the presence of collisions, suppose this coherent response persists until a collision occurs at a random time T. Assume that the waiting time between successive collisions is exponentially distributed,

$$P(T) = \gamma_{\rm coll}e^{-\gamma_{\rm coll}T},$$

where $\gamma_{\rm coll}$ is the mean collision rate. Show that the frequency response has a Lorentzian lineshape with homogeneous FWHM

$$\Delta\nu_{_H} = \frac{\gamma_o + 2\gamma_{\rm coll}}{2\pi} \equiv \frac{\gamma_{_H}}{2\pi}.$$

6. The phase of a monochromatic wave $E(t) = E_o e^{-i\omega_o t + i\phi_o}$ is interrupted at random times by some phase-randomizing mechanism. Suppose the waiting time τ between successive phase interruptions has the exponential distribution

$$P(\tau) = \frac{1}{\tau_c}e^{-\tau/\tau_c},$$

so that the mean interruption rate is $\gamma = 1/\tau_c$. Show that the resulting spectral distribution is Lorentzian with FWHM

$$\Delta\nu = \frac{\gamma}{2\pi} = \frac{1}{2\pi\tau_c}.$$

7. In fiber optics, an important quantity characterizing group-velocity dispersion is the wavelength dependence of the group velocity $v_g = [dk/d\omega]^{-1}$, where $k = n(\omega)\omega/c$. If the dispersion is large, the group velocity varies significantly with wavelength, whereas at a zero-dispersion wavelength it is locally insensitive to wavelength. Define the dispersion parameter by

$$D \equiv \frac{d}{d\lambda}\left(\frac{1}{v_g}\right).$$

   (a) Show that D is proportional to the second derivative of the refractive index with respect to wavelength.

   (b) Optical fibers are commonly made of silica glass. A useful approximation, simpler than the Sellmeier formula but reasonably accurate over a limited wavelength range, is

$$n(\lambda) = A - B\lambda^2 + \frac{B}{\lambda^2},$$

   where $A = 1.451$, $B = 0.003$, and the numerical value of the free-space wavelength λ is to be expressed in micrometers. Find the wavelength at which $D = 0$. This zero-dispersion wavelength lies near one of the important telecommunications wavelength regions.

8. The energy stored per unit volume in a collection of two-level atoms is $w_a = \mathcal{N}_1 E_1 + \mathcal{N}_2 E_2$, where $\mathcal{N} = \mathcal{N}_1 + \mathcal{N}_2$ is the total atomic number density and $E_2 - E_1 = h\nu_o$.

   (a) Assuming that the total number density $\mathcal{N}$ remains constant, express w_a in terms of $\mathcal{N}_2$, and hence obtain an expression for the rate of energy transfer per unit volume between the electromagnetic field and the atoms in terms of $d\mathcal{N}_2/dt$.

   (b) Simplify the result for the case $\mathcal{N}_1 \ll \mathcal{N}_2$.

9. **Sellmeier formula:** The Sellmeier formula is an empirical relation between refractive index and wavelength that accurately describes many transparent materials over broad wavelength ranges. It is commonly written in the form

$$n^2 = 1 + \frac{B_1\lambda^2}{\lambda^2 - C_1} + \frac{B_2\lambda^2}{\lambda^2 - C_2} + \frac{B_3\lambda^2}{\lambda^2 - C_3},$$

where B_1, B_2, B_3, C_1, C_2, and C_3 are material-dependent Sellmeier coefficients. For BK7 borosilicate glass, commonly used in optical components, the coefficients are

$$\begin{aligned} B_1 &= 1.03961212 & C_1 &= 6.00069867 \times 10^{-3}\ \mu\text{m}^2 \\ B_2 &= 2.31792344 \times 10^{-1} & C_2 &= 2.00179144 \times 10^{-2}\ \mu\text{m}^2 \\ B_3 &= 1.01046945 & C_3 &= 1.03560653 \times 10^{2}\ \mu\text{m}^2. \end{aligned}$$

   (a) Does light have a larger phase velocity in BK7 at 400 nm or at 700 nm? By what percentage do the two phase velocities differ?

   (b) What is the dispersion $dn/d\lambda$ of BK7 at the wavelength of a frequency-doubled Nd:YAG laser, $\lambda = 532$ nm? Give the result with appropriate units. Does the sign correspond to normal or anomalous dispersion?

(c) Using the Sellmeier formula, compute the phase velocity and group velocity of light from a He-Ne laser ($\lambda = 632.8$ nm) propagating in BK7 glass. Which velocity is larger?

10. Calculate the Doppler linewidth of a CO_2 laser transition at $\lambda = 10.6\,\mu$m and $T = 400$ K. Take the molecular mass of CO_2 to be 44 atomic mass units. The collision broadening of this transition is approximately 5 MHz/Torr. At what gas pressure do the Doppler and collision-broadening contributions to the linewidth become equal?

11. Consider an atomic transition centered at 500 nm with a linewidth of 0.10 nm in a laser cavity of volume 2.0 cm^3 and refractive index $n = 1$.

    (a) Assuming $\Delta\lambda \ll \lambda$, convert the wavelength linewidth $\Delta\lambda = 0.10$ nm to the corresponding frequency linewidth $\Delta\nu$. Using the electromagnetic mode density, estimate the total number of electromagnetic modes within the cavity volume that lie in this frequency interval.

    (b) Assume that the cavity is a circular cylinder of cross-sectional area 0.10 cm^2 and hence length 20.0 cm. How many longitudinal HG_{00p} cavity modes lie within the frequency interval found in part (a)? Include the two independent polarizations associated with each spatial mode.

    (c) Assuming, for this estimate, that spontaneous emission is distributed equally among all electromagnetic modes within the transition bandwidth and cavity volume, estimate the probability that an emitted photon enters one of the HG_{00p} modes counted in part (b).

12. **Doppler width and transition wavelength:** The He-Ne laser has several neon transitions that can support laser oscillation. Calculate the Doppler linewidth (FWHM) at $T = 400$ K for the following transitions:

    (i) $\lambda = 632.8$ nm ($3s_2$–$2p_4$ transition),

    (ii) $\lambda = 1.1523\,\mu$m ($2s_2$–$2p_4$ transition), and

    (iii) $\lambda = 3.39\,\mu$m ($3s_2$–$3p_4$ transition).

    Take the atomic mass of neon to be 20 atomic mass units. Comment on the dependence of the Doppler linewidth, expressed in frequency units, on the transition wavelength.

13. As a simple model of a lineshape in the presence of both inhomogeneous and homogeneous broadening, assume that the distribution of atomic resonance frequencies is a rectangular function of width $\Delta\nu_R$,

$$p(\nu_a) = \begin{cases} \dfrac{1}{\Delta\nu_R}, & |\nu_a - \nu_o| < \Delta\nu_R/2, \\ 0, & |\nu_a - \nu_o| > \Delta\nu_R/2. \end{cases}$$

    Assume that each atomic frequency group has a homogeneously broadened Lorentzian lineshape of FWHM $\Delta\nu_H$. Write the total lineshape as the convolution of the rectangular distribution with the homogeneous Lorentzian, and evaluate the convolution analytically. Plot the resulting lineshape as a function of frequency ν for

$$\frac{\Delta\nu_H}{\Delta\nu_R} = 0.01, \quad 0.10, \quad \text{and} \quad 0.50\,.$$

    Comment on how the shape changes as the relative importance of homogeneous broadening increases.

14. Estimate the line-center stimulated-emission cross-section for each of the following laser transitions:

    (i) a He-Ne laser operating at 632.8 nm with $\Delta\nu_H = 60$ MHz, $\Delta\nu_D = 1.5$ GHz, and $A_{21} = 1.7 \times 10^6\,\mathrm{s}^{-1}$;

    (ii) a He-Cd laser operating at 441.6 nm with $\Delta\nu_H = 45$ MHz, $\Delta\nu_D = 1.1$ GHz, and $A_{21} = 1.1 \times 10^6\,\mathrm{s}^{-1}$; and

    (iii) an Nd:YAG laser operating at 1.064 μm with $\Delta\nu_H = 120$ GHz, $A_{21} = 4.3 \times 10^3\,\mathrm{s}^{-1}$, and $n = 1.82$.

    For the gas lasers, assume that Doppler broadening dominates the observed lineshape. For Nd:YAG, assume that the transition is predominantly homogeneously broadened.

15. The first continuous-wave laser, which was also the first gas laser, operated on a neon transition near $\lambda = 1.15\,\mu$m [A. Javan, W. R. Bennett, Jr., and D. R. Herriott, "Population inversion and continuous optical maser oscillation in a gas discharge containing a He-Ne mixture," Phys. Rev. Lett. **6**, 106 (1961)]. The transition is predominantly Doppler broadened, and the upper-state lifetime, which we take to be approximately the radiative lifetime associated with the laser transition, is about 100 ns. Take the gas temperature to be $T = 500$ K and the atomic mass of neon to be 20 atomic mass units.

    (a) Calculate the Doppler linewidth (FWHM) and the line-center stimulated-emission cross-section for this transition.

    (b) Using the approximations developed in the text, estimate the saturation irradiance of the transition.

    (c) How many longitudinal modes of a 40-cm-long cavity lie within the FWHM of the Doppler-broadened gain profile?

16. **Absorption coefficient of ruby:** The gain medium used in the first laser, demonstrated in 1960, was a ruby rod pumped by a flash lamp. Ruby is Al_2O_3 doped with a small concentration of Cr^{3+} ions. A typical Cr^{3+} concentration is $2 \times 10^{19}\,\mathrm{cm}^{-3}$. Absorption in the red, near 694.3 nm, is due to resonant transitions of the Cr^{3+} ions. Take the radiative lifetime to be 3 ms, the homogeneous linewidth to be $\Delta\nu_H = 11\,\mathrm{cm}^{-1}$, and the refractive index of ruby to be 1.7. Assume that, before pumping, essentially all of the relevant Cr^{3+} ions occupy the lower laser level.

    (a) Calculate the line-center absorption coefficient of the unpumped ruby crystal.

    (b) What fraction of the irradiance of a 694.3-nm beam is transmitted through a 1-cm-thick ruby crystal of this type?

    Remark: The unit cm^{-1} is commonly used for spectroscopic wavenumber. In this notation, the spectroscopic wavenumber is $\tilde{\nu} = 1/\lambda$, where the vacuum wavelength λ is expressed in centimeters. A linewidth expressed in cm^{-1} can be converted to frequency units (Hz) by multiplying by the speed of light in cm/s.

17. **Homogeneous and inhomogeneous absorption:** The peak absorption coefficient of the $\lambda = 10.6\,\mu$m transition of CO_2 at room temperature is observed to increase approximately linearly with gas pressure at low pressures and to become nearly independent of pressure at high pressures [E. T. Gerry and D. A. Leonard, "Measurement of 10.6 μm CO_2 laser transition probability and optical broadening cross-sections," Appl. Phys. Lett. **8**, 227–229 (1966)]. Explain this behavior.

Hint: At low pressure, the line is predominantly Doppler-broadened and Gaussian, whereas at high pressure it is predominantly collision-broadened and Lorentzian. Natural broadening is negligible in both regimes.

18. **Absorption in Na vapor:** Estimate the line-center absorption coefficient for the 589 nm transition in sodium vapor containing 2.70×10^{12} atoms/cm^3 at 200° C. The radiative lifetime of the 589 nm transition is 16 ns [J. E. Bjorkholm and A. Ashkin, "cw self-Focusing and self-Trapping of light in sodium vapor," Phys. Rev. Lett. **32**, 129 (1974)]. Use the result of Problem 2 of Chapter 1 for the number density of an ideal gas at pressure P (in Torr) and absolute temperature T (in K). Take the atomic mass of sodium to be 23 atomic mass units.

    Hint: First estimate the gas pressure corresponding to the stated number density. At this very low pressure, assume that collision broadening is negligible compared with Doppler broadening. Determine whether natural broadening is also negligible.

19. **Doppler vs lifetime broadening:** A two-level atom of atomic mass 20 amu has lower- and upper-level radiative lifetimes $\tau_1 = 30$ ns and $\tau_2 = 1$ ns, respectively, and an energy separation $E_2 - E_1 = 0.5$ eV.

    (a) At what gas temperature does the natural linewidth of the transition equal its Doppler linewidth? [**Hint:** Use the approximation $k_B T \approx 1/40$ eV at $T = 300$ K.]

    (b) If $\tau_1 = 30$ ns remains fixed, what must the upper-level lifetime τ_2 be for the natural linewidth to equal the Doppler linewidth at $T = 300$ K?

20. The radiative lifetime of the upper level of the CO_2 laser transition at $\lambda = 10.6\,\mu$m is 5.38 s, and the collision-broadening contribution to the homogeneous linewidth is approximately 5 MHz/Torr. Take the molecular mass of CO_2 to be 44 atomic mass units.

    (a) Calculate the Doppler linewidth (FWHM) of this transition at $T = 400$ K.

    (b) Calculate the pressure at which the homogeneous and inhomogeneous contributions to the linewidth are equal. Show that natural broadening is negligible under these conditions.

    (c) In continuous-wave (cw) operation, CO_2 lasers commonly operate over gas pressures ranging from approximately 2 to 100 Torr. Describe how the lineshape changes as the pressure increases through this range. How does this change affect the frequency dependence and peak value of the absorption coefficient (or gain) at this wavelength?

6

Quantum Mechanics of Atomic Response

In the preceding chapter we established that the classical electron oscillator model of an atom correctly captures many fundamental aspects of light-matter interactions, including the frequency dependence and resonance enhancement of atomic response. However, a major limitation of this classical approach is its inability to predict the amplification of light propagating through an atomic medium. To address this limitation, this chapter presents a rigorous quantum-mechanical treatment of atomic response. This formulation naturally accounts for both optical absorption and coherent optical amplification under appropriate conditions.

The primary distinction between the classical and quantum-mechanical frameworks lies in the quantized nature of atomic and molecular energy structures. Within the quantum mechanical paradigm, a material medium consisting of atoms and molecules can exchange energy with an interacting electromagnetic field exclusively in discrete amounts corresponding to the difference in the energy of two atomic states. For complete consistency, the field energy itself must also be quantized; however, a vast class of optical phenomena can be accurately modeled by treating the atoms quantum mechanically while maintaining a classical representation of the electromagnetic field. This includes most aspects of laser light that can be understood without quantizing the field. This framework is designated as the semi-classical approximation. The quantized nature of light becomes necessary primarily for a complete understanding of the fundamental coherence properties, spontaneous initiation, and quantum noise statistics of laser light. We return to the details of cavity field quantization, photon statistics, and quantum coherence of laser light in Chapter 9.

Atoms are described spectroscopically by a discrete set of stationary energy levels. Atoms absorb and emit energy in discrete amounts corresponding to the energy difference between two atomic levels. A transition between an initial state of energy E_i and a final state of energy E_j requires the absorption or emission of a quantum of energy defined by

$$E_{ji} = E_j - E_i \equiv \hbar\omega_{ji} \,. \tag{6.1}$$

where $\hbar$ is the reduced Planck's constant and ω_{ji} represents the transition frequency. These atomic transitions are bound by strict selection rules dictated by conservation laws for angular momentum, parity, and magnetic quantum numbers. Furthermore, the real energy levels are not infinitely sharp because they possess finite lifetimes. Continuous random interaction with the surrounding environment, including coupling to the zero-point vacuum fluctuations of the electromagnetic field (which drives spontaneous emission) as well as collisions with neighboring atoms in a gas phase or phononic interactions within a host crystal lattice, can impart finite lifetimes and dephasing to atomic states. According to the energy-time uncertainty principle, this finite lifetime imparts a finite spectral width to each discrete energy level. This lifetime broadening is similar to the spectral broadening observed in passive optical resonators, where the cavity modes acquire a finite linewidth because of the finite lifetime of the stored energy resulting from cavity losses.

DOI: 10.1201/9781003584179-6

6.1 Two-level Atom in Electromagnetic Field

For analyzing the resonant interaction of an atomic system with a narrow-band optical field whose frequency is tuned close to a particular transition frequency, it is sufficient to focus on the two resonant states and neglect all other non-resonant levels. This is the so-called two-level approximation to a real atom. The effect of the omitted levels on this simplified two-level atom appears via the phenomenological level decay rates and external pumping rates that track transitions into and out of the two-level subspace. Concrete physical examples demonstrating how a complex, multi-level atomic system can be reduced to an effective two-level atom will be considered in detail in Chapter 7.

Consider an idealized two-level atom consisting of a single bound electron of mass m and charge $-e$ bound to an infinitely heavy nucleus. The isolated free atom is modeled by the unperturbed Hamiltonian $\hat{H}_A$, which admits two stationary eigenstates, $|\phi_1\rangle$ and $|\phi_2\rangle$ [see Fig. 6.1(a)] satisfying the time-independent Schrödinger equation:

$$\hat{H}_A\,|\phi_1\rangle = E_1\,|\phi_1\rangle\,, \tag{6.2a}$$

$$\hat{H}_A\,|\phi_2\rangle = E_2\,|\phi_2\rangle\,. \tag{6.2b}$$

The electron can execute an upward transition from the lower state $|\phi_1\rangle$ to the excited upper state $|\phi_2\rangle$ by absorbing a quantum of energy $\hbar\omega_{21} = E_2 - E_1$, or undergo a downward transition from level $|\phi_2\rangle$ to $|\phi_1\rangle$ by emitting a quantum of the same energy. While spontaneous downward transitions are driven by vacuum fluctuations, the processes of stimulated absorption and stimulated emission require the presence of an external driving field. Under the long-wavelength electric dipole approximation, the semi-classical Hamiltonian $\hat{H}$ of the atom-field system can be partitioned as

$$\hat{H} = \hat{H}_A + \hat{H}_{AF}\,, \tag{6.3a}$$

$$\hat{H}_A = \frac{\hat{p}^2}{2m} + \hat{V}\,, \tag{6.3b}$$

$$\hat{H}_{AF} = -\hat{\boldsymbol{\mu}}\cdot\boldsymbol{\mathcal{E}}(0,t) = e\hat{\boldsymbol{r}}\cdot\boldsymbol{\mathcal{E}}(0,t)\,, \tag{6.3c}$$

where $\hat{\boldsymbol{p}}$ is the electronic momentum operator and $\hat{V}(\hat{\mathbf{r}})$ is the binding potential. The interaction Hamiltonian $\hat{H}_{AF}$ models the potential energy of the active electronic dipole operator $\hat{\boldsymbol{\mu}} \equiv -e\hat{\boldsymbol{r}}$ in an external classical electric field $\boldsymbol{\mathcal{E}}(\boldsymbol{R},t)$. Here, $\hat{\boldsymbol{r}}$ represents the internal position operator of the bound electron relative to the atomic nucleus, which serves as the origin

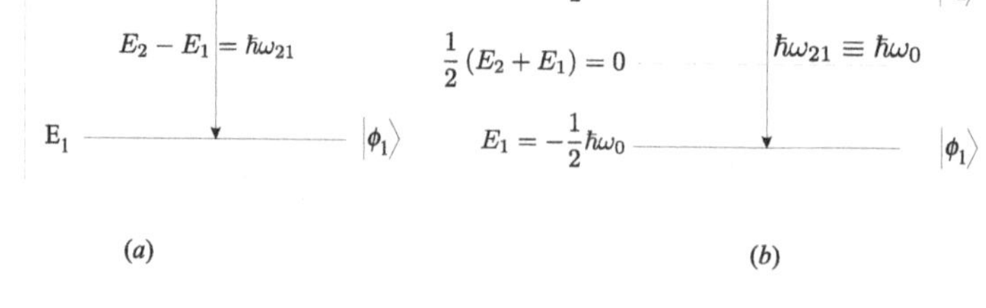

FIGURE 6.1
Energy levels for a resonant two-level atom: (a) basis states relative to an absolute energy reference and (b) symmetric energy representation of basis states relative to the mid-point of the transition.

of the internal coordinate system. This internal coordinate is distinct from the macroscopic position vector $\boldsymbol{R}$ representing the atomic center of mass within the laboratory frame. Under the electric dipole approximation, the spatial variation of the optical field across the sub-nanometer atomic volume is negligible ($\exp(i\boldsymbol{k}\cdot\boldsymbol{r}) \approx 1$), allowing the field to be evaluated at the center-of-mass coordinate $\boldsymbol{R}$. We model this driving field as a monochromatic traveling wave:

$$\boldsymbol{\mathcal{E}}(\boldsymbol{R},t) = \frac{1}{2}\left[\boldsymbol{E}_o\, e^{-i(\omega t - \boldsymbol{k}\cdot\boldsymbol{R})} + c.c.\right] \equiv \mathcal{R}e\, \boldsymbol{\mathcal{E}}_o\, e^{-i\omega t}\,, \tag{6.4}$$

where $\boldsymbol{\mathcal{E}}_o = \boldsymbol{E}_0 e^{i\boldsymbol{k}\cdot\boldsymbol{R}}$ is the complex field amplitude and c.c. denotes the complex conjugate.

In the presence of dipole interaction, the state of the atom evolves into a time-dependent superposition of the unperturbed eigenstates $|\phi_1\rangle$ and $|\phi_2\rangle$

$$|\Psi(t)\rangle = a_1(t)\,|\phi_1\rangle + a_2(t)\,|\phi_2\rangle\,, \tag{6.5}$$

where $a_1(t)$ and $a_2(t)$ are the complex probability amplitudes to be determined. The state vector $|\Psi(t)\rangle$ obeys the time-dependent Schrödinger equation

$$i\hbar\frac{\partial}{\partial t}\,|\Psi(t)\rangle = \hat{H}\,|\Psi(t)\rangle\,. \tag{6.6}$$

The equations of motion for the probability amplitudes are obtained by projecting the Schrödinger equation onto the orthogonal basis states. Projecting from the left with the bra vector $\langle\phi_1|$ yields

$$i\hbar\dot{a}_1 = \langle\phi_1|\left(\hat{H}_A + \hat{H}_{AF}\right)|\Psi(t)\rangle = E_1 a_1(t) - \boldsymbol{\mu}_{12}\cdot\boldsymbol{\mathcal{E}}(\boldsymbol{R},t)a_2(t)\,, \tag{6.7a}$$

Similarly, projecting from the left with the bra vector $\langle\phi_2|$ yields

$$i\hbar\dot{a}_2 = \langle\phi_2|\left(\hat{H}_A + \hat{H}_{AF}\right)|\Psi(t)\rangle = E_2 a_2 - \boldsymbol{\mu}_{21}\cdot\boldsymbol{\mathcal{E}}(\boldsymbol{R},t)a_1(t)\,. \tag{6.7b}$$

Here we have defined the transition electric dipole matrix elements as $\boldsymbol{\mu}_{12} \equiv -e\,\langle\phi_1|\,\hat{\boldsymbol{r}}\,|\phi_2\rangle = \boldsymbol{\mu}_{21}^*$ and used the fact that, because the unperturbed atomic states $|\phi_1\rangle$ and $|\phi_2\rangle$ have definite parity under inversion, the diagonal expectation values of the coordinate operator vanish identically ($\boldsymbol{\mu}_{11} = \boldsymbol{\mu}_{22} = 0$).

The differential equations for a_1 and a_2 can be cast into a symmetric form by shifting the zero-energy reference coordinate to the exact midpoint separating the two levels, enforcing the constraint $E_1 + E_2 = 0$ [see Fig. 6.1(b)]. Defining the line-center atomic transition frequency as $\omega_o \equiv \omega_{21} = (E_2 - E_1)/\hbar$, the shifted energy eigenvalues reduce to $E_1 = -\frac{1}{2}\hbar\omega_0$, and $E_2 = \frac{1}{2}\hbar\omega_0$. Substituting these into the equations for a_1 and a_2 yields the coupled equations of motion for the atomic probability amplitudes:

$$\dot{a}_1(t) = \frac{i}{2}\omega_o a_1(t) + \frac{i}{\hbar}\left[\boldsymbol{\mu}_{12}\cdot\boldsymbol{\mathcal{E}}(\boldsymbol{R},t)\right]a_2(t)\,, \tag{6.8a}$$

$$\dot{a}_2(t) = -\frac{i}{2}\omega_o a_2(t) + \frac{i}{\hbar}\left[\boldsymbol{\mu}_{21}\cdot\boldsymbol{\mathcal{E}}(\boldsymbol{R},t)\right]a_1(t)\,. \tag{6.8b}$$

In the absence of an external driving field, the uncoupled probability amplitudes exhibit a natural harmonic time dependence of the form $\sim a_{1,2}e^{\pm i(\omega_o/2)t}$. The presence of the time-dependent dipole interaction term modifies this. To track the slowly varying modifications to the state vector, we explicitly decompose the classical monochromatic field interaction terms from Eq. (6.8)

$$\frac{1}{\hbar}\left[\boldsymbol{\mu}_{12}\cdot\boldsymbol{\mathcal{E}}(\boldsymbol{R},t)\right] = \frac{\boldsymbol{\mu}_{12}\cdot\boldsymbol{\mathcal{E}}_o}{2\hbar}e^{-i\omega t} + \frac{\boldsymbol{\mu}_{12}\cdot\boldsymbol{\mathcal{E}}_o^*}{2\hbar}e^{i\omega t}\,, \tag{6.9a}$$

$$\frac{1}{\hbar}\left[\boldsymbol{\mu}_{21}\cdot\boldsymbol{\mathcal{E}}(\boldsymbol{R},t)\right] = \frac{\boldsymbol{\mu}_{21}\cdot\boldsymbol{\mathcal{E}}_o}{2\hbar}e^{-i\omega t} + \frac{\boldsymbol{\mu}_{21}\cdot\boldsymbol{\mathcal{E}}_o^*}{2\hbar}e^{i\omega t}\,. \tag{6.9b}$$

Utilizing these in the equations of motion (6.8) for a_1 and a_2 reveals that by factoring out the rapid time-dependence $e^{\pm i\omega t/2}$, slowly varying probability amplitudes can be introduced by writing

$$a_1(t) = c_1(t)e^{i(\omega/2)t}, \quad a_2(t) = c_2(t)e^{-i(\omega/2)t}. \tag{6.10}$$

Substituting this transformation back into the amplitude equations, Eq. (6.8), yields the coupled equations of motion for the slowly varying probability amplitudes c_1 and c_2

$$\dot{c}_1 = \frac{i}{2}\Delta\, c_1 + ic_2 \left[\frac{\boldsymbol{\mu}_{12}\cdot\boldsymbol{\mathcal{E}}_o}{2\hbar}e^{-2i\omega t} + \frac{\boldsymbol{\mu}_{12}\cdot\boldsymbol{\mathcal{E}}_o^*}{2\hbar}\right], \tag{6.11a}$$

$$\dot{c}_2 = -\frac{i}{2}\Delta\, c_2 + ic_1 \left[\frac{\boldsymbol{\mu}_{21}\cdot\boldsymbol{\mathcal{E}}_o}{2\hbar} + \frac{\boldsymbol{\mu}_{21}\cdot\boldsymbol{\mathcal{E}}_o^*}{2\hbar}e^{2i\omega t}\right], \tag{6.11b}$$

where

$$\Delta = \omega_o - \omega\,. \tag{6.11c}$$

The parameter Δ defines the frequency detuning of the classical driving field from the atomic transition frequency ω_o.

Equations (6.11) involve three distinct, widely separated physical timescales. The first timescale, $\tau_{\rm opt} \sim (2\omega)^{-1}$, corresponds to the rapid optical sub-cycle oscillations. The second timescale, $\tau_{\rm det} \sim \Delta\omega^{-1}$, is determined by the detuning. For near-resonant interactions, $|\Delta| \ll \omega$, so the detuning time scale is much longer than an optical period and may be comparable to the coherence-relaxation time when the detuning is of order the homogeneous linewidth. The third timescale, $\tau_{\rm int} \sim [|\boldsymbol{\mu}_{21}\cdot\boldsymbol{E}_o|/\hbar]^{-1}$ [$|\boldsymbol{\mu}_{21}\cdot\boldsymbol{E}_o|/\hbar$ has the dimensions of a frequency], stems from the coherent atom-field interaction. As long as the field of the light wave remains small compared to the internal electrostatic field produced by the nucleus, this interaction timescale is also long compared to the optical period.

We can filter out the rapid optical sub-cycle dynamics by integrating these equations for a coarse-grained time interval δt. This interval is chosen to be large compared to an optical period but short compared to the interaction and detuning timescales ($\omega^{-1} \ll \delta t \ll \tau_{\rm int}, \tau_{\rm det}$). Integrating Eq. (6.11) over this increment yields

$$c_1(t+\delta t) - c_1(t) = \frac{i}{2}\Delta\,\delta t c_1(t) + ic_2(t)\left[\frac{\boldsymbol{\mu}_{12}\cdot\boldsymbol{\mathcal{E}}_o}{2\hbar}\frac{e^{-2i\omega t}(e^{-2i\omega\delta t}-1)}{-2i\omega} + \frac{\boldsymbol{\mu}_{12}\cdot\boldsymbol{\mathcal{E}}_o^*}{2\hbar}\delta t\right]. \tag{6.12}$$

Dividing this equation by δt yields a coarse-grained time derivative

$$\begin{aligned} \dot{c}_1 \equiv \frac{c_1(t+\delta t)-c_1(t)}{\delta t} &= \frac{i}{2}\Delta c_1(t) + ic_2(t)\left[\frac{\boldsymbol{\mu}_{12}\cdot\boldsymbol{\mathcal{E}}_o}{2\hbar}\frac{e^{-2i\omega t}(e^{-2i\omega\delta t}-1)}{-2i\omega\delta t} + \frac{\boldsymbol{\mu}_{12}\cdot\boldsymbol{\mathcal{E}}_o^*}{2\hbar}\right] \\ &\approx i\frac{\Delta}{2}c_1(t) + i\frac{\boldsymbol{\mu}_{12}\cdot\boldsymbol{E}_o^*}{2\hbar}c_2(t)\,, \end{aligned} \tag{6.13}$$

where the high-frequency oscillating term is safely neglected because its integrated amplitude is suppressed by a large denominator ($|(e^{-2i\omega\delta t}-1)/(2\omega\delta t)| \ll 1$).

This averaging procedure defines the rotating-wave approximation (RWA). It effectively smooths out the rapidly varying phase terms that average to zero over a few optical cycles. Equation (6.13) describes the time evolution of c_1 over times long compared to an optical cycle. Applying an identical RWA simplification to the upper-state equation, the long-term coupled dynamics are governed by the set of differential equations

$$\dot{c}_1(t) = \frac{i}{2}\Delta\, c_1(t) + i\frac{\boldsymbol{\mu}_{12}\cdot\boldsymbol{\mathcal{E}}_o^*}{2\hbar}c_2(t)\,, \tag{6.14a}$$

$$\dot{c}_2(t) = -\frac{i}{2}\Delta\, c_2(t) + i\frac{\boldsymbol{\mu}_{21}\cdot\boldsymbol{\mathcal{E}}_o}{2\hbar}c_1(t)\,. \tag{6.14b}$$

These coupled equations can be decoupled by differentiating Eq. (6.14a) with respect to time and substituting Eq. (6.14b) to eliminate $\dot{c}_2(t)$:

$$\begin{aligned}\ddot{c}_1(t) &= \frac{i}{2}\Delta\,\dot{c}_1(t) + i\frac{\boldsymbol{\mu}_{12}\cdot\boldsymbol{\mathcal{E}}_o^*}{2\hbar}\dot{c}_2(t)\,,\\ &= \frac{i}{2}\Delta\left[\frac{i}{2}\Delta c_1(t) + i\frac{\boldsymbol{\mu}_{12}\cdot\boldsymbol{\mathcal{E}}_o^*}{2\hbar}c_2(t)\right] + i\frac{\boldsymbol{\mu}_{12}\cdot\boldsymbol{\mathcal{E}}_o^*}{2\hbar}\left[-\frac{i}{2}\Delta c_2(t) + i\frac{\boldsymbol{\mu}_{21}\cdot\boldsymbol{\mathcal{E}}_o}{2\hbar}c_1(t)\right]\,,\end{aligned} \tag{6.15}$$

which simplifies to a standard second-order linear differential equation,

$$\ddot{c}_1(t) + \left(\frac{\Delta^2}{4} + \frac{|\Omega_0|^2}{4}\right)c_1(t) = 0 \implies \ddot{c}_1(t) + \frac{\Omega^2}{4}c_1(t) = 0\,. \tag{6.16}$$

Here, the on-resonance Rabi frequency and generalized Rabi frequency are defined by

$$\Omega_o \equiv \frac{\boldsymbol{\mu}_{21}\cdot\boldsymbol{E}_0}{\hbar}\,, \quad \text{and} \quad \Omega \equiv \sqrt{\Delta^2 + |\Omega_o|^2}\,, \tag{6.17}$$

where $|\Omega_0|$ and Ω are referred to as the Rabi frequency and the generalized Rabi frequency, respectively. Eliminating c_1 proves that c_2 also satisfies the same second-order equation. The general solutions to Eq. (6.16) thus can be written as a superposition of harmonic functions

$$c_1(t) = A_1\cos\frac{\Omega}{2}t + B_1\sin\frac{\Omega}{2}t\,, \tag{6.18a}$$

$$c_2(t) = A_2\cos\frac{\Omega}{2}t + B_2\sin\frac{\Omega}{2}t\,. \tag{6.18b}$$

For an atom starting initially in the lower state at $t = 0$, the initial amplitude parameters are

$$\begin{aligned} c_1(0) &= 1, & c_2(0) &= 0\\ \dot{c}_1(0) &= i\frac{\Delta}{2}, & \dot{c}_2(0) &= i\frac{\Omega_o}{2}\,. \end{aligned}$$

With these initial conditions, the harmonic solutions (6.18) yield the following probabilities for the two-level atom to be found in the lower state $|\phi_1\rangle$ and the excited state $|\phi_2\rangle$,

$$\begin{aligned}P_1(t) \equiv |a_1(t)|^2 = |c_1(t)|^2 &= \cos^2\frac{\Omega}{2}t + \frac{\Delta^2}{\Omega^2}\sin^2\frac{\Omega}{2}t\\ &= \frac{1}{2}\left(1 + \frac{\Delta^2}{\Omega^2}\right) + \frac{1}{2}\frac{|\Omega_o|^2}{\Omega^2}\cos\Omega t\,,\end{aligned} \tag{6.19a}$$

$$P_2(t) \equiv |a_2(t)|^2 = \frac{|\Omega_o|^2}{\Omega^2}\sin^2\frac{\Omega}{2}t = \frac{1}{2}\frac{|\Omega_o|^2}{\Omega^2}\left(1 - \cos\Omega t\right)\,. \tag{6.19b}$$

It is easily verified that the overall probability for the atom to be in either of the two states remains unity for all times

$$\begin{aligned}P_1(t) + P_2(t) &= \frac{1}{2}\left(1 + \frac{\Delta^2}{\Omega^2}\right) + \frac{1}{2}\left(\frac{\Omega_o}{\Omega}\right)^2\cos\Omega t + \frac{1}{2}\left(\frac{\Omega_o}{\Omega}\right)^2 - \frac{1}{2}\left(\frac{\Omega_o}{\Omega}\right)^2\cos\Omega t\,,\\ &= \frac{1}{2}\left(1 + \frac{\Delta^2 + \Omega_o^2}{\Omega^2}\right) = 1\,.\end{aligned} \tag{6.20}$$

This is a statement of the conservation of probability, which holds because we have neglected the decay of states $\langle 1|$ or $\langle 2|$ to any other states.

The difference between the upper- and lower-state population probabilities is defined as the population inversion $w(t)$:

$$
\begin{aligned}
w(t) \equiv P_2(t) - P_1(t) &= \frac{1}{2}\left(\frac{\Omega_o}{\Omega}\right)^2 - \frac{1}{2}\left(\frac{\Omega_o}{\Omega}\right)^2 \cos \Omega t - \frac{1}{2}\left(1 + \frac{\Omega_o^2}{\Omega^2}\right) - \frac{1}{2}\left(\frac{\Omega_o}{\Omega}\right)^2 \cos \Omega t \\
&= -1 + \frac{\Omega_o^2}{\Omega^2} - \frac{\Omega_o^2}{\Omega^2} \cos \Omega t = -1 + \frac{\Omega_o^2}{\Omega^2}(1 - \cos \Omega t) \,. \qquad (6.21)
\end{aligned}
$$

Equation (6.21) shows that the population inversion oscillates dynamically between -1 (the completely unexcited ground state) and a maximum of $-1 + 2|\Omega_o|^2/\Omega^2$. For a perfectly resonant driving field ($\Delta = 0$), the generalized frequency equals the on-resonance value ($\Omega = |\Omega_o|$), allowing the population inversion to flip completely between -1 and $+1$.

6.2 Induced Electric Dipole Moment

The induced electric dipole moment is given by the expectation value of the dipole moment operator over the time-dependent state vector

$$
\begin{aligned}
\langle \Psi(t) | \hat{\boldsymbol{\mu}} | \Psi(t) \rangle &= [a_1^*(t) \langle \phi_1 | + a_2^*(t) \langle \phi_2 |] \, \hat{\boldsymbol{\mu}} \, [a_1(t) | \phi_1 \rangle + a_2(t) | \phi_2 \rangle] \\
&= -\left(|a_1(t)|^2 \boldsymbol{\mu}_{11} + a_1^*(t) a_2(t) \boldsymbol{\mu}_{12} + a_2^*(t) a_1(t) \boldsymbol{\mu}_{21} + |a_2(t)|^2 \boldsymbol{\mu}_{22} \right) . \qquad (6.22)
\end{aligned}
$$

Because the unperturbed atomic states have a definite parity, the diagonal dipole moments vanish identically ($\boldsymbol{\mu}_{11} = \boldsymbol{\mu}_{22} = 0$). Consequently, Eq. (6.22) reduces to

$$
\begin{aligned}
\langle \hat{\boldsymbol{\mu}} \rangle &= \boldsymbol{\mu}_{12} a_1^*(t) a_2(t) + \boldsymbol{\mu}_{21} a_2^*(t) a_1(t) = \boldsymbol{\mu}_{12} c_1^*(t) e^{-i\frac{\omega}{2}t} c_2(t) e^{-i\frac{\omega}{2}t} + \boldsymbol{\mu}_{21} c_2^*(t) e^{i\frac{\omega}{2}t} c_1(t) e^{i\frac{\omega}{2}t} \\
&= \boldsymbol{\mu}_{12} c_1^*(t) c_2(t) e^{-i\omega t} + \boldsymbol{\mu}_{21} c_2^*(t) c_1(t) e^{i\omega t} \,. \qquad (6.23)
\end{aligned}
$$

Thus the calculation of the induced dipole moment reduces to a calculation of the bilinear combinations of the underlying state amplitudes. It is mathematically more convenient to formulate equations for these bilinear quantities directly. To achieve this, we introduce the density matrix, whose elements are defined by

$$\rho_{11} = c_1 c_1^* \,, \qquad (6.24a)$$

$$\rho_{12} = c_1 c_2^* \,, \qquad (6.24b)$$

$$\rho_{21} = c_2 c_1^* = \rho_{12}^* \,, \qquad (6.24c)$$

$$\rho_{22} = c_2 c_2^* \,. \qquad (6.24d)$$

Using these definitions, the time-dependent expectation value of the induced dipole moment can be rewritten in terms of the complex off-diagonal elements

$$\langle \hat{\boldsymbol{\mu}} \rangle = \boldsymbol{\mu}_{12} \rho_{21}(t) e^{-i\omega t} + \boldsymbol{\mu}_{21} \rho_{12}(t) e^{i\omega t} \equiv \frac{1}{2}\boldsymbol{p}(t) + c.c. \Rightarrow \boldsymbol{p}(t) = 2\boldsymbol{\mu}_{12} e^{-i\omega t} \rho_{21}(t) \,, \qquad (6.25)$$

where $\boldsymbol{p}(t)$ is the slowly varying complex representation of the induced dipole moment as defined in Chapter 2.

The equations of motion for the density matrix elements $\rho_{ij}(t)$ are derived directly from Eqs. (6.14) for the amplitudes $c_1(t)$ and $c_2(t)$. Differentiating the off-diagonal element

$\rho_{12}(t) = c_1(t)c_2^*(t)$ with respect to time yields

$$\dot{\rho}_{12} = \dot{c}_1 c_2^* + c_1 \dot{c}_2^* = \left(\frac{i}{2}\Delta c_1 + \frac{i}{2}\Omega_o^* c_2\right) c_2^* + c_1 \left(\frac{i}{2}\Delta c_2^* - \frac{i}{2}\Omega_o^* c_1^*\right),$$

$$= i\Delta\rho_{12} + \frac{i}{2}\Omega_o^* \left(\rho_{22} - \rho_{11}\right), \tag{6.26}$$

and similarly,
$$\dot{\rho}_{21} = -i\Delta\rho_{21} - \frac{i}{2}\Omega_o \left(\rho_{22} - \rho_{11}\right), \tag{6.27}$$

$$\dot{\rho}_{11} = \dot{c}_1 c_1^* + c_1 \dot{c}_1^* = \left(\frac{i}{2}\Delta c_1 + \frac{i}{2}\Omega_o^* c_2\right) c_1^* + c_1 \left(-\frac{i}{2}\Delta c_1^* - \frac{i}{2}\Omega_o c_2^*\right),$$

$$= -\frac{i}{2}\left(\Omega_o \rho_{12} - \Omega_o^* \rho_{21}\right), \tag{6.28}$$

and
$$\dot{\rho}_{22} = \frac{i}{2}\left(\Omega_o \rho_{12} - \Omega_o^* \rho_{21}\right). \tag{6.29}$$

Thus the equations for ρ_{ij} are

$$\dot{\rho}_{12} = i\Delta\rho_{12} + \frac{i}{2}\Omega_o^* \left(\rho_{22} - \rho_{11}\right), \tag{6.30a}$$

$$\dot{\rho}_{21} = -i\Delta\rho_{21} - i\frac{\Omega_o}{2}\left(\rho_{22} - \rho_{11}\right), \tag{6.30b}$$

$$\dot{\rho}_{11} = -\frac{i}{2}\left(\Omega_o \rho_{12} - \Omega_o^* \rho_{21}\right), \tag{6.30c}$$

$$\dot{\rho}_{22} = \frac{i}{2}\left(\Omega_o \rho_{12} - \Omega_o^* \rho_{21}\right), \tag{6.30d}$$

where $\Delta = \omega_o - \omega$ is the detuning and Ω_0 is the on-resonance Rabi frequency. These equations describe the coherent, non-dissipative evolution of a closed two-level atom driven by a coherent classical field. They represent an incomplete physical description of the evolution of a real two-level atom open to its environment because they do not include the effects of relaxation processes into and out of the two-level manifold. A comprehensive model capable of predicting real laser behaviors requires these effects to be incorporated into the equations of motion. To include the effects of relaxation processes, we distinguish between elastic (energy-conserving) and inelastic (energy-changing) relaxation processes.

6.2.1 Incorporating Elastic Phase-Interrupting Collisions

Elastic collisions have a direct classical analog: they refer to interactions that interrupt only the phase of the induced dipole moment without inducing transitions between energy states. Such events modify only the off-diagonal coherence elements ρ_{12} and ρ_{21}. The dipole dephasing rate is γ_{coll}. The effect of these dephasing collisions is incorporated by adding an "elastic collision" term to the equation of motion for the complex coherence ρ_{12}:

$$\left[\dot{\rho}_{12}\right]_{\text{elastic}} = -\gamma_{\text{coll}}\, \rho_{12}\,. \tag{6.31}$$

The modified equation for ρ_{12} then becomes

$$\dot{\rho}_{12} = -\left(\gamma_{\text{coll}} - i\Delta\right)\rho_{12} + \frac{i}{2}\Omega_o^* \left(\rho_{22} - \rho_{11}\right). \tag{6.32}$$

It is critical to recognize that Eq. (6.32) no longer applies to a single atom; instead, it describes the evolution of a statistical ensemble-averaged atom, where the relaxation term represents a formal collision average. Similarly, the equation for ρ_{21} reads

$$\dot{\rho}_{21} = -\left(\gamma_{\text{coll}} + i\Delta\right)\rho_{21} - \frac{i}{2}\Omega_o \left(\rho_{22} - \rho_{11}\right). \tag{6.33}$$

6.2.2 Incorporating Inelastic Collisions and Spontaneous Decay

In contrast to elastic dephasing, inelastic collisions actively exchange energy with the environment, knocking populations out of levels $|\phi_1\rangle$ and $|\phi_2\rangle$ into auxiliary, nonresonant atomic levels at fixed rates designated by γ_1 and γ_2 , respectively. Furthermore, the fundamental process of spontaneous emission also transfers population from the excited state $|\phi_2\rangle$ to the lower state $|\phi_1\rangle$. Denoting the spontaneous emission rate by A_{21}, the equations of motion for the diagonal elements ρ_{11} and ρ_{22} become

$$\dot{\rho}_{11} = -\gamma_1\rho_{11} + A_{21}\rho_{22} - \frac{i}{2}\left(\Omega_o\rho_{12} - \Omega_o^*\rho_{21}\right), \tag{6.34a}$$

$$\dot{\rho}_{22} = -\gamma_2\rho_{22} - A_{21}\rho_{22} + \frac{i}{2}\left(\Omega_o\rho_{12} - \Omega_o^*\rho_{21}\right). \tag{6.34b}$$

Once again these equations must be understood as applying to an average atom.

Inelastic relaxation channels also implicitly affect the off-diagonal elements ρ_{12} and ρ_{21}. To see this, note that in the absence of relaxations, the absolute magnitude of the coherence element satisfies the algebraic identity

$$|\rho_{12}| = |c_1c_2^*| = |c_1||c_2^*| = \sqrt{c_1c_1^*}\,\sqrt{c_2c_2^*} = \sqrt{\rho_{11}\rho_{22}}\,. \tag{6.35}$$

This relation motivates the one-half contribution of each population-decay rate to the coherence-decay rate. Because inelastic collisions drive the exponential decay of the diagonal elements ρ_{11} and ρ_{22} according to

$$\left[\rho_{11}\right]_{\text{coll}} = \rho_{11}(0)e^{-\gamma_1 t}, \tag{6.36a}$$

$$\left[\rho_{22}\right]_{\text{coll}} = \rho_{22}(0)e^{-\gamma_2 t}, \tag{6.36b}$$

we conclude that inelastic collisions force off-diagonal elements to decay as

$$|\rho_{12}| = |\rho_{12}(0)|\,e^{-\frac{1}{2}(\gamma_1+\gamma_2)t}. \tag{6.37}$$

The spontaneous-emission term $A_{21}\rho_{22}$ contributes in the same way as an upper-state population-decay channel, adding $A_{21}\rho_{22}/2$ to the coherence-decay rate. Combining the contributions from elastic dephasing, inelastic population decay, and spontaneous emission, the ensemble-averaged equations of motion for the off-diagonal density matrix elements are

$$\dot{\rho}_{12} = -\left(\gamma_o/2 - i\Delta\right)\rho_{12} + i\frac{\Omega^*}{2}\left(\rho_{22} - \rho_{11}\right), \tag{6.38a}$$

$$\dot{\rho}_{21} = -\left(\gamma_o/2 + i\Delta\right)\rho_{21} - i\frac{\Omega}{2}\left(\rho_{22} - \rho_{11}\right), \tag{6.38b}$$

where γ_o is the homogeneous linewidth parameter, defined by:

$$\gamma_o \equiv 2\gamma_{\text{coll}} + \left(\gamma_1 + \gamma_2 + A_{21}\right), \tag{6.39}$$

so that the transverse coherence decays at the rate $\gamma_o/2$. In standard laser media and gases under practical operating conditions, the elastic phase-disrupting collision frequency vastly exceeds the inelastic transitions:

$$\gamma_{\text{coll}} \gg \frac{1}{2}\left(\gamma_1 + \gamma_2 + A_{21}\right). \tag{6.40}$$

Physically, this parameter inequality means that an active atom undergoes many elastic dephasing collisions before encountering an interaction energetic enough to cause population changes. Consequently, the transverse dephasing lifetime is typically orders of magnitude shorter than the longitudinal population lifetime.

6.3 Absorption and Emission of Radiation

Incorporating the phenomenological terms for both elastic dephasing and inelastic population decay, the ensemble-averaged equations of motion for the density matrix elements are given by:

$$\dot{\rho}_{12} = -\left(\gamma_o/2 - i\Delta\right)\rho_{12} + \frac{i}{2}\Omega_o^*\left(\rho_{22} - \rho_{11}\right), \tag{6.41a}$$

$$\dot{\rho}_{21} = -\left(\gamma_o/2 + i\Delta\right)\rho_{21} - \frac{i}{2}\Omega_o\left(\rho_{22} - \rho_{11}\right), \tag{6.41b}$$

$$\dot{\rho}_{11} = -\gamma_1\rho_{11} + A_{21}\rho_{22} - \frac{i}{2}\left(\Omega_o\rho_{12} - \Omega_o^*\rho_{21}\right), \tag{6.41c}$$

$$\dot{\rho}_{22} = -\gamma_2\rho_{22} - A_{21}\rho_{22} + \frac{i}{2}\left(\Omega_o\rho_{12} - \Omega_o^*\rho_{21}\right). \tag{6.41d}$$

where γ_o is the homogeneous linewidth parameter defined in Eq. (6.39), so that the off-diagonal coherence decays at the rate $\gamma_o/2$.

In many systems of practical interest, the off-diagonal coherence elements decay on a timescale that is orders of magnitude faster than the timescale for the diagonal population elements to change appreciably ($\gamma_o \gg \gamma_1, \gamma_2, A_{21}$). When the populations and driving field vary slowly compared with the coherence-relaxation rate $\gamma_o/2$, the off-diagonal elements follow their instantaneous values quasi-statically, and we may set $\dot{\rho}_{12} \approx 0$ and $\dot{\rho}_{21} \approx 0$. Setting the coherence derivatives to zero gives

$$\rho_{12}(t) \approx \frac{i}{2}\Omega_0^*\frac{\rho_{22}(t) - \rho_{11}(t)}{\gamma_o/2 - i\Delta}, \tag{6.42a}$$

$$\rho_{21}(t) \approx -\frac{i}{2}\Omega_0\frac{\rho_{22}(t) - \rho_{11}(t)}{\gamma_o/2 + i\Delta}. \tag{6.42b}$$

These equations provide the coherence elements needed to calculate the induced dipole moment, reducing the remaining dynamical problem to a calculation of the diagonal population elements ρ_{11} and ρ_{22}. Substituting these adiabatic expressions for ρ_{12} and ρ_{21} into the equations for ρ_{11} and ρ_{22} yields

$$\dot{\rho}_{11} = -\gamma_1\rho_{11} + A_{21}\rho_{22} + \frac{\left|\Omega_o\right|^2\gamma_o}{4\left[(\gamma_o/2)^2 + \Delta^2\right]}\left(\rho_{22} - \rho_{11}\right), \tag{6.43a}$$

$$\dot{\rho}_{22} = -\gamma_2\rho_{22} - A_{21}\rho_{22} - \frac{\left|\Omega_o\right|^2\gamma_o}{4\left[(\gamma_o/2)^2 + \Delta^2\right]}\left(\rho_{22} - \rho_{11}\right), \tag{6.43b}$$

where $|\Omega_o|$ is the on-resonance Rabi frequency.

Equations (6.43) constitute the population rate equations for a two-level atom. They express the rate of change of each level population in terms of rates at which the population in the two levels build up or decay. The terms proportional to γ_1 and γ_2 describe population loss from the two-level subsystem through nonresonant relaxation channels, while the terms proportional to A_{21} describe spontaneous emission from level 2 to level 1. The last terms, proportional to $\left|\Omega_o\right|^2$ – and thus to the intensity of light – represent the transitions induced via stimulated light-matter interactions. For an upward transition from the lower state to the excited state, which physically corresponds to the resonant absorption of light, the stimulated absorption rate is given by

$$W_{12} = \frac{\left|\Omega_o\right|^2\gamma_o}{4\left[(\gamma_o/2)^2 + \Delta^2\right]}, \tag{6.44a}$$

while for a downward transition from the excited state to the lower state, which corresponds to the stimulated emission of radiation, the stimulated emission rate is

$$W_{21} = \frac{|\Omega_o|^2 \gamma_o}{4\left[(\gamma_o/2)^2 + \Delta^2\right]}. \tag{6.44b}$$

A comparison of Eq. (6.44a) and Eq. (6.44b) yields the fundamental result that the field-induced stimulated transition rates for absorption and emission between an identical pair of quantum states are equal: $W_{12} = W_{21}$. The frequency dependence of the stimulated transition rate is Lorentzian, with angular-frequency FWHM γ_o in agreement with the lineshape obtained from the classical oscillator model in Chapter 5.

6.4 Classical Electron Oscillator Model: Lorentz Model

We can extend the semi-classical density matrix framework to derive the macroscopic polarization induced within a collection of two-level atoms under the influence of an external electromagnetic field. First, we calculate the complex dipole moment induced in a single atom located at an arbitrary position $\boldsymbol{R}$. Substituting the adiabatic off-diagonal element $\rho_{21}(t)$ from Eq. (6.42b) into the dipole definition, Eq. (6.25), yields

$$\begin{aligned} \boldsymbol{p}(\boldsymbol{R},t) &= 2e\boldsymbol{r}_{12}\rho_{21}(t)e^{-i\omega t+i\boldsymbol{k}\cdot\boldsymbol{R}} = 2e\boldsymbol{r}_{12}\left[-\frac{i}{2}\frac{\Omega_o\left(\rho_{22}-\rho_{11}\right)}{\gamma_o/2+i\Delta}\right]e^{-i\omega t+i\boldsymbol{k}\cdot\boldsymbol{R}} \\ &= e\boldsymbol{r}_{12}\frac{\Omega_o}{\Delta - i\gamma_o/2}\left(\rho_{11}-\rho_{22}\right)e^{-i\omega t+i\boldsymbol{k}\cdot\boldsymbol{R}}, \end{aligned} \tag{6.45a}$$

which, on substituting $\Omega_o \equiv \dfrac{e\boldsymbol{r}_{21}\cdot\boldsymbol{E}_0}{\hbar}$, simplifies to:

$$\boldsymbol{p}(\boldsymbol{R},t) = \frac{e^2}{\hbar}\frac{\boldsymbol{r}_{12}\left(\boldsymbol{r}_{21}\cdot\boldsymbol{E}(\boldsymbol{R},t)\right)}{\Delta - i\gamma_o/2}\left(\rho_{11}-\rho_{22}\right), \tag{6.45b}$$

where $\boldsymbol{E}(\boldsymbol{R},t) = \boldsymbol{E}_0 e^{-i\omega t+i\boldsymbol{k}\cdot\boldsymbol{R}}$ represents the electric field at the location of the atom. Here, $\boldsymbol{R}$ denotes the position of the nucleus in the laboratory frame to distinguish it explicitly from the electronic internal relative position $\boldsymbol{r}$. A critical physical property demonstrated by Eq. (6.45) is that the induced atomic dipole moment exhibits an intrinsic direction of its own. Because $\boldsymbol{p} \propto \boldsymbol{r}_{12}$, the induced dipole moment is not generally collinear with the applied electric field. For an isotropic ensemble, the transition-dipole matrix elements are randomly oriented. The macroscopic polarization is therefore obtained by averaging over both the positions and orientations of the atoms. The macroscopic polarization of a collection of atoms with a number density $\mathcal{N}$ is thus given by

$$\boldsymbol{P}_{\text{at}}(\boldsymbol{R},t) = \frac{\mathcal{N}e^2\left(\rho_{11}-\rho_{22}\right)}{\hbar\left(\Delta - i\gamma_o/2\right)}\left\langle \boldsymbol{r}_{12}\left(\boldsymbol{r}_{21}\cdot\boldsymbol{E}(\boldsymbol{R},t)\right)\right\rangle_{\text{orientations}}, \tag{6.46}$$

where $\boldsymbol{E}(\boldsymbol{R},t)$ is the macroscopic local electric field at position $\boldsymbol{R}$. To evaluate this orientational average, we express the macroscopic electric field as a linearly polarized monochromatic wave

$$\boldsymbol{E} \equiv \boldsymbol{\varepsilon}_1 E = \boldsymbol{\varepsilon}_1 E_o e^{-i\omega t+i\boldsymbol{k}\cdot\boldsymbol{R}}, \tag{6.47}$$

where $\boldsymbol{\varepsilon}_1$ is a unit polarization vector, and we introduce a right-handed triad of orthogonal unit vectors $\boldsymbol{\varepsilon}_1, \boldsymbol{\varepsilon}_2, \boldsymbol{\kappa} = \boldsymbol{k}/k$ with

$$\boldsymbol{\varepsilon}_1 \times \boldsymbol{\varepsilon}_2 = \boldsymbol{\kappa}. \tag{6.48}$$

Projecting the transition matrix vector $\boldsymbol{r}_{12}$ onto this basis yields:

$$\boldsymbol{r}_{12} = (\boldsymbol{r}_{12} \cdot \boldsymbol{\varepsilon}_1)\, \boldsymbol{\varepsilon}_1 + (\boldsymbol{r}_{12} \cdot \boldsymbol{\varepsilon}_2)\, \boldsymbol{\varepsilon}_2 + (\boldsymbol{r}_{12} \cdot \boldsymbol{\kappa})\, \boldsymbol{\kappa}\,, \tag{6.49a}$$

$$\Rightarrow \quad \boldsymbol{r}_{12}\, (\boldsymbol{r}_{21} \cdot \boldsymbol{E}) = [(\boldsymbol{r}_{12} \cdot \boldsymbol{\varepsilon}_1)\, \boldsymbol{\varepsilon}_1 + (\boldsymbol{r}_{12} \cdot \boldsymbol{\varepsilon}_2)\, \boldsymbol{\varepsilon}_2 + (\boldsymbol{r}_{12} \cdot \boldsymbol{\kappa})\, \boldsymbol{\kappa}]\, (\boldsymbol{r}_{21} \cdot \boldsymbol{\varepsilon}_1)\, E(\boldsymbol{R}, t)\,. \tag{6.49b}$$

On carrying out the orientational average over a random spatial distribution, only the component along the applied field survives, giving us

$$\langle \boldsymbol{r}_{12}\, (\boldsymbol{r}_{21} \cdot \boldsymbol{E}) \rangle_{\text{orientations}} = \langle |\boldsymbol{r}_{12} \cdot \boldsymbol{\varepsilon}_1|^2 \rangle_{\text{orientations}}\, \boldsymbol{\varepsilon}_1 E(\boldsymbol{R}, t) = \frac{\langle |\boldsymbol{r}_{12}|^2 \rangle}{3}\, \boldsymbol{E}(\boldsymbol{R}, \boldsymbol{t})\,. \tag{6.50}$$

Here we have invoked the equality of orientational averages

$$\langle |\boldsymbol{r}_{12} \cdot \boldsymbol{\varepsilon}_1|^2 \rangle = \langle |\boldsymbol{r}_{12} \cdot \boldsymbol{\varepsilon}_2|^2 \rangle = \langle |\boldsymbol{r}_{12} \cdot \boldsymbol{\kappa}|^2 \rangle = \frac{\langle |\boldsymbol{r}_{12}|^2 \rangle}{3}\,. \tag{6.51}$$

Substituting Eq. (6.50) into Eq. (6.46) yields the induced atomic polarization

$$\boldsymbol{P}_{\text{at}}(\boldsymbol{r}, t) = \frac{\mathcal{N} e^2 \langle |\boldsymbol{r}_{12}|^2 \rangle (\rho_{11} - \rho_{22})}{3\hbar(\Delta - i\gamma_o/2)}\, \boldsymbol{E}(\boldsymbol{r}, t)\,, \tag{6.52}$$

which is linearly proportional to the macroscopic applied field at position $\boldsymbol{r}$. Note that we have reverted to using $\boldsymbol{r}$ to denote atomic position in the lab frame because after performing the orientational averaging, only the directions relative to the fixed laboratory frame remain relevant. Utilizing this linear relationship, we define the quantum mechanical atomic susceptibility by writing $\boldsymbol{P}_{\text{at}}(\boldsymbol{r}, t) \equiv \epsilon_o \chi_{\text{at}} \boldsymbol{E}(\boldsymbol{r}, t)$. The resulting expression for quantum mechanical susceptibility can be written as

$$\chi_{\text{at}} = \frac{\mathcal{N} e^2}{2m\omega_o \epsilon_o} \left(\frac{2m\omega_o}{3\hbar} \langle |\boldsymbol{r}_{12}| \rangle|^2 \right) \left(\frac{1}{\Delta - i\gamma_o/2} \right) (\rho_{11} - \rho_{22})\,. \tag{6.53}$$

By defining the dimensionless quantum parameter known as the "atomic oscillator strength"

$$f_0 = \frac{2m\omega_o}{3\hbar}\, |\langle |\boldsymbol{r}_{12}| \rangle|^2\,, \tag{6.54}$$

the final quantum mechanical expression for the atomic susceptibility simplifies to the form

$$\chi_{\text{at}} = \frac{\mathcal{N} e^2 f_o}{2m\omega_o \epsilon_o} \left(\frac{1}{\Delta - i\gamma_o/2} \right) (\rho_{11} - \rho_{22})\,. \tag{6.55}$$

This quantum mechanical expression for the material susceptibility differs fundamentally from the classical Lorentz oscillator result (5.25), due to the presence of the population difference term $(\rho_{11} - \rho_{22})$. Under conditions of weak excitation of the medium by an optical field, where the atomic population stays trapped mostly in the ground state ($\rho_{11} \approx 1$, $\rho_{22} \approx 0$), the expression for atomic susceptibility reduces to

$$\chi_{\text{at}} = \frac{\mathcal{N} e^2 f_o}{2m\omega_o \epsilon_o} \left(\frac{1}{\Delta - i\gamma_o/2} \right)\,. \tag{6.56}$$

While Eq. (6.56) mimics the mathematical form of the classical Lorentz susceptibility, the parameter f_o has an entirely different physical meaning that has no classical analog. Whereas the classical model introduces f_o as a fraction representing a subpopulation of active electronic oscillators with natural frequency ν_o, the quantum mechanical definition, Eq. (6.54), connects f_o to the magnitude of the transition dipole matrix element. Under strong optical excitation, the quantum result permits population inversion and negative absorption, behavior that cannot arise from the passive classical oscillator model.

The imaginary part of susceptibility characterizes the cycle-averaged rate of mechanical work done by the field – or, equivalently, the power exchanged between the field and the medium. From the Poynting theorem, Eq. (2.57), this average power flow is given by

$$\frac{dw}{dt} = \left\langle \boldsymbol{\mathcal{E}}(\boldsymbol{r},t) \cdot \frac{\partial \boldsymbol{\mathcal{P}}(\boldsymbol{r},t)}{\partial t} \right\rangle = \frac{1}{2}\mathcal{R}e\left[\boldsymbol{E}(\boldsymbol{r},t) \cdot \frac{\partial \boldsymbol{P}^*(\boldsymbol{r},t)}{\partial t}\right] = \frac{1}{2}\omega\epsilon_0\chi''_{\text{at}}\left|\boldsymbol{E}_o\right|^2 , \tag{6.57}$$

where χ''_{at} is the imaginary part of the susceptibility, given from Eq. (6.55) by

$$\chi''_{\text{at}} = \frac{\mathcal{N}e^2 f_o}{2m\omega_o\epsilon_o} \frac{\gamma_o/2}{\Delta^2 + (\gamma_o/2)^2}(\rho_{11} - \rho_{22}) . \tag{6.58}$$

For the passive classical oscillator model of Chapter 5, $\chi''_{\text{at}} > 0$ under all conditions [Eq. (6.56)], implying a dissipation of optical energy into the medium via absorption. Quantum mechanically, however, Eq. (6.58) reveals that for a driven group of atoms such that $\rho_{22} > \rho_{11}$, the imaginary susceptibility becomes negative ($\chi''_{\text{at}} < 0$). Under this non-equilibrium condition, instead of absorbing energy from the field, the collection of atomic dipoles releases energy into the radiation field. This fundamental quantum mechanical result foretells the possibility of coherent optical amplification via stimulated emission and serves as the physical basis for laser action.

We now explore how a non-zero, steady-state population inversion ($\rho_{22} > \rho_{11}$) can be achieved. Introducing the level population densities $\mathcal{N}_1 \equiv \mathcal{N}\rho_{11}$ and $\mathcal{N}_2 \equiv \mathcal{N}\rho_{22}$, where $\mathcal{N}$ denotes the total number density of active atoms, their rate equations follow directly from the density matrix equations (6.43),

$$\dot{\mathcal{N}_1} = -\gamma_1\mathcal{N}_1 + A_{21}\mathcal{N}_2 + W_{21}\left(\mathcal{N}_2 - \mathcal{N}_1\right) , \tag{6.59a}$$

$$\dot{\mathcal{N}_2} = -\gamma_2\mathcal{N}_2 - A_{21}\mathcal{N}_2 - W_{12}\left(\mathcal{N}_2 - \mathcal{N}_1\right) , \tag{6.59b}$$

where the stimulated transition rates, Eqs. (6.44), satisfy $W_{12} = W_{21}$. The steady-state solution of these equations is the trivial state $\mathcal{N}_1 = 0 = \mathcal{N}_2$, which simply reflects the physical reality that due to the level relaxation rates γ_1 and γ_2, any populations initially trapped in active levels eventually leak out to other levels. To compensate for this population loss, there must be external pump mechanisms that inject populations into these states at rates Λ_1 and Λ_2. Incorporating these pump terms, the population rate equations become

$$\dot{\mathcal{N}_1} = \Lambda_1 - \gamma_1\mathcal{N}_1 + A_{21}\mathcal{N}_2 + W_{21}\left(\mathcal{N}_2 - \mathcal{N}_1\right), \tag{6.60a}$$

$$\dot{\mathcal{N}_2} = \Lambda_2 - \gamma_2\mathcal{N}_2 - A_{21}\mathcal{N}_2 - W_{21}\left(\mathcal{N}_2 - \mathcal{N}_1\right). \tag{6.60b}$$

As will be shown in Chapter 7, these equations can sustain a nonzero population inversion ($\mathcal{N}_2 > \mathcal{N}_1$) in the steady state.

6.4.1 Influence of Level Degeneracy

The population rate equations (6.60) are valid for a pair of nondegenerate levels. For degenerate upper or lower levels, with degeneracies g_1 and g_2, the population rate equations must be modified. Under these conditions, the material response includes contributions from all allowed sub-transitions. To track this fine-structure behavior, we introduce a set of secondary quantum indices, m_1 and m_2, to characterize the sublevels within groups 1 and 2, respectively,

$$\begin{array}{llll} ------ & m_2 & |\phi_2\rangle & g_2 \text{ discrete sublevels} \\ ---- & m_1 & |\phi_1\rangle & g_1 \text{ discrete sublevels} \end{array}$$

The rate equations for the individual sub-populations are written as

$$\begin{aligned} \dot{\mathcal{N}}_1(m_1) &= \Lambda_1(m_1) - \gamma_1(m_1)\mathcal{N}_1(m_1) + \sum_{m_2} A(m_2, m_1)\mathcal{N}_2(m_2) \\ &\quad + \sum_{m_2} W(m_2, m_1)\mathcal{N}_2(m_2) - \sum_{m_2} W(m_1, m_2)\mathcal{N}_1(m_1) \,, \end{aligned} \tag{6.61}$$

where $A(m_2, m_1)$ specifies the spontaneous emission rate, and $W(m_2, m_1)$ denotes the stimulated transition rate between specific sublevels. We introduce the total population density for group 1 by summing over its sublevels

$$\mathcal{N}_1 = \sum_{m_1} \mathcal{N}_1\left(m_1\right) . \tag{6.62a}$$

Under conditions where the relaxation rates among the sublevels within a degenerate group are fast compared to the optical transition rates, the sub-populations within a group equalize rapidly such that

$$\mathcal{N}_1(m_1) = \frac{\mathcal{N}_1}{g_1}. \tag{6.62b}$$

Summing the individual sub-population rate equations over the lower-level index m_1 yields

$$\begin{aligned} \dot{\mathcal{N}}_1 &= \Lambda_1 - \gamma_1\mathcal{N}_1 + \frac{1}{g_2}\sum_{m_1,m_2} A(m_2, m_1)\mathcal{N}_2 + \frac{1}{g_2}\sum_{m_1,m_2} W(m_2, m_1)\mathcal{N}_2 \\ &\quad - \frac{1}{g_1}\sum_{m_1,m_2} W(m_1, m_2)\mathcal{N}_1 \,, \end{aligned} \tag{6.63}$$

where $\Lambda_1 \equiv \sum_{m_1} \Lambda_1(m_1)$ is the total external pump rate and $\gamma_1 \equiv \sum_{m_1} \gamma_1(m_1)/g_1$ represents the average relaxation rate for group 1 population. To simplify Eq. (6.63), we define the statistically averaged spontaneous and stimulated emission rates via

$$A_{21} = \frac{1}{g_2}\sum_{m_1,m_2} A\left(m_2, m_1\right) , \tag{6.64a}$$

$$W_{21} = \frac{1}{g_2}\sum_{m_1,m_2} W\left(m_2, m_1\right) , \tag{6.64b}$$

$$W_{12} = \frac{1}{g_1}\sum_{m_1,m_2} W\left(m_1, m_2\right) . \tag{6.64c}$$

Because the microscopic quantum transition probabilities satisfy $W(m_2, m_1) = W(m_1, m_2)$, the group-averaged stimulated emission and absorption rates satisfy the relation

$$g_2 W_{21} = g_1 W_{12} \,. \tag{6.65}$$

Substituting these averaged coefficients into Eq. (6.63) yields the semi-classical population rate equations for a degenerate two-level atomic system:

$$\dot{\mathcal{N}}_1 = \Lambda_1 - \gamma_1\mathcal{N}_1 + A_{21}\mathcal{N}_2 + W_{21}\left(\mathcal{N}_2 - \frac{g_2}{g_1}\mathcal{N}_1\right) , \tag{6.66}$$

$$\dot{\mathcal{N}}_2 = \Lambda_2 - \gamma_2\mathcal{N}_2 - A_{21}\mathcal{N}_2 - W_{21}\left(\mathcal{N}_2 - \frac{g_2}{g_1}\mathcal{N}_1\right) . \tag{6.67}$$

The average spontaneous emission rate A_{21} and the average stimulated emission rate W_{21} are not independent; they are governed by the quantum mechanical structure of the transition. Using the quantum mechanical expression for spontaneous emission rate from the footnote 5 on page 183 and rewriting the stimulated emission rate with the help of Eq. (6.44b), the sums in Eqs. (6.64a) and (6.64b) can be expressed as

$$A_{21} = \frac{1}{g_2}\frac{e^2\omega_o^3 n}{3\pi c^3\hbar\epsilon_o}\sum_{m_1,m_2}|\boldsymbol{r}_{21}(m_2,m_1)|^2 \, , \tag{6.68}$$

$$W_{21} = \frac{1}{g_2}\frac{e^2}{2\hbar^2}\sum_{m_1,m_2}|\boldsymbol{r}_{21}(m_2,m_1)\cdot \boldsymbol{E}_o|^2\,\frac{\gamma_o/2}{\Delta^2+(\gamma_o/2)^2}. \tag{6.69}$$

To simplify the expression for W_{21}, we first invoke the orientational averaging [see Eq. 6.51] to resolve the dot product via

$$\sum_{m_1,m_2}|\mathbf{r}_{21}(m_2,m_1)\cdot\mathbf{E}_o|^2 = \frac{1}{3}\left(\sum_{m_1,m_2}|\mathbf{r}_{21}(m_2,m_1)|^2\right)|\boldsymbol{E}_o|^2 . \tag{6.70}$$

Then, recall the definition of optical intensity I in terms of the cycle-averaged energy density $\bar{u} = \frac{1}{2}\epsilon_o n^2\,|\boldsymbol{E}_o|^2$ inside a dielectric medium,

$$I \equiv \frac{1}{2}\epsilon_o nc\,|\boldsymbol{E}_o|^2 = \bar{u}\left(\frac{c}{n}\right). \tag{6.71}$$

and, finally, transform the angular frequency dependence of W_{21} into the cyclical frequency domain ($\nu \equiv \omega/2\pi$) by introducing the Lorentzian lineshape function, $S(\nu)$, via

$$\frac{\gamma_o/2}{\Delta^2+(\gamma_o/2)^2} = \frac{1}{2\pi}\frac{\gamma_o/4\pi}{(\Delta/2\pi)^2+(\gamma_o/4\pi)^2} = \frac{1}{2}S\left(\nu\right). \tag{6.72}$$

Substituting Eq. (6.70) and Eq. (6.72) into Eq. (6.69) yields:

$$W_{21} = \frac{1}{g_2}\frac{e^2}{6\hbar^2}\left(\sum_{m_1,m_2}|\mathbf{r}_{21}(m_2,m_1)|^2\right)|\mathbf{E}_0|^2\left[\frac{1}{2}S(\nu)\right]. \tag{6.73}$$

Isolating the explicit structure of the Einstein A-coefficient [Eq. (6.68)] in this expression then establishes the following relation between stimulated and spontaneous emission rates:

$$\begin{aligned} W_{21} &= \frac{\pi c^2}{2\hbar\omega_0^3 n^2}\left(\frac{1}{g_2}\frac{e^2\omega_0^3 n}{3\pi c^3\hbar\epsilon_0}\sum_{m_1,m_2}|\mathbf{r}_{21}(m_2,m_1)|^2\right)\left(\frac{1}{2}\epsilon_0 nc\,|\mathbf{E}_0|^2\right)S(\nu) \\ &= \frac{\pi c^2}{2\hbar\omega_0^3 n^2}A_{21}IS(\nu) = \frac{\pi c^2}{2(2\pi\nu_0)^3\hbar n^2}A_{21}IS(\nu) \\ &= \frac{\lambda_o^2}{8\pi n^2}A_{21}S(\nu)\left(\frac{I}{h\nu}\right), \end{aligned} \tag{6.74}$$

where we have used the near-resonance approximation $\nu_0 \approx \nu$ for the photon flux density $I/h\nu$. Using the degeneracy reciprocity factor from Eq. (6.65), the stimulated absorption rate is

$$W_{12} = \left(\frac{g_2}{g_1}\right)W_{21} = \left(\frac{g_2}{g_1}\right)\frac{\lambda_0^2}{8\pi n^2}A_{21}S(\nu)\left(\frac{I}{h\nu}\right). \tag{6.75}$$

These relations match the expressions derived for emission and absorption rates in Sec. 5.4 by modifying classical electron oscillator theory through physical arguments. The relation between the stimulated and spontaneous emission rates for a narrowband field derived here differs from the relation derived by Einstein for a broadband thermal field. The two results, in fact, represent opposite limits of a single, more general relationship, as was shown in Chapter 5.

6.4.2 Transition Cross-sections

Recalling that $I/h\nu$ is the photon flux density (the number of photons per unit area per second), the absorption rate W_{12} can be interpreted as the number of photons incident on an area $(g_2/g_1)(\lambda_o^2/8\pi n^2)A_{21}S(\nu)$ in one second. This effective area is referred to as the absorption cross-section $\sigma_{12} \equiv \sigma_{ab}$:

$$\sigma_{12} = \frac{g_2}{g_1}\frac{\lambda_o^2 A_{21}}{8\pi n^2}S(\nu)\ , \quad W_{12} = \sigma_{12}(\nu)\frac{I}{h\nu}\,. \tag{6.76}$$

In a similar manner, the stimulated emission rate W_{21} can be used to define the emission cross-section $\sigma_{21} \equiv \sigma_{em}$

$$\sigma_{21} = \frac{\lambda_o^2 A_{21}}{8\pi n^2}S(\nu) = \frac{g_1}{g_2}\sigma_{12}\,, \quad W_{21} = \sigma_{21}(\nu)\frac{I}{h\nu}\,. \tag{6.77}$$

We conclude this chapter by observing that the quantum-mechanical treatment of light-matter interaction confirms the conclusions of the preceding chapter, which were based on a modification of the expression for the dielectric susceptibility of a collection of classical electron oscillators. The modification was motivated by physical rather than mathematical aspects of quantum theory. By interpreting the oscillator frequency as the transition frequency between two specific quantum energy levels of an atom, and replacing the number density of classical oscillators with the difference in the population densities of the lower and upper levels, we were able to proceed relatively quickly and easily to a realistic theory of laser operation. This chapter shows that, in the weak-excitation limit and with the physically motivated identifications introduced in Chapter 5, the electron-oscillator model reproduces the principal form of the quantum-mechanical atomic response. An aspect that is still missing is the quantized nature of light energy. This feature is important in the initiation of laser radiation and the coherence properties of laser light, and it will be considered in Chapter 9.

Summary

In this chapter, we presented a quantum-mechanical treatment of the atomic response to an electromagnetic field. Absorption and emission of energy by atoms involve transitions from one energy level to another driven by an interaction Hamiltonian. These transitions must be allowed by the appropriate conservation laws. The state of the atom is described by the probability amplitudes for the atom to occupy these levels, and its time evolution, driven by the interaction Hamiltonian, is governed by the Schrödinger equation. The Schrödinger equation is used to derive the equations of motion for the atomic density matrix. The density matrix provides an alternative description of the atomic state and its time evolution. It is better suited to include the effects of processes not contained explicitly in the interaction Hamiltonian, such as collisions, level decay, and level shifts, which also affect the state of the atom and therefore its response to an EM wave.

Using the density matrix equation of motion and focusing on two atomic levels, the induced dipole moment for a single two-level atom, as well as the polarization and dielectric susceptibility for a collection of two-level atoms (including the effects of level decays and collisions), were calculated. Equations of motion for level populations were also calculated. These expressions naturally predict stimulated emission as well as absorption. Expressed in terms of various rates, including the spontaneous-emission rate, these results confirm the validity of the corresponding equations of motion introduced in Chapter 5 using the classical dipole-oscillator model of atomic response. The homogeneous linewidth parameter γ_o was shown to include both population-relaxation and pure dephasing contributions, while the optical coherence decays at the rate $\gamma_o/2$. We also showed that, in the weak-excitation limit, the quantum-mechanical susceptibility reduces to the Lorentz-oscillator form derived in Chapter 5.

It is worth mentioning that even in the quantum-mechanical approach, many processes, such as atomic collisions, are incorporated through phenomenological rates determined experimentally, exactly as was done in the classical approach. As pointed out in Chapter 1 and illustrated by Einstein's approach to the blackbody spectrum, this pragmatism avoids the mathematical complexity involved in first-principles calculations and keeps us focused on the physics of the problem.

7

Population Inversion, Saturation, and Laser Systems

In the preceding two chapters, we established that an electromagnetic wave interacting with an atomic medium at a frequency near an atomic transition induces a coherent polarization. This induced response is proportional to the instantaneous population difference between the two levels associated with that transition. We also found that the applied field alters the populations of the upper and lower levels through stimulated transitions. Chapter 5 already identified the rates of population change due to stimulated emission and absorption. However, other processes, such as energy-changing collisions and pumping, also modify these populations.

In this chapter, we present a heuristic derivation of atomic population rate equations that incorporate stimulated transitions alongside relaxation and pumping terms. This treatment provides an alternative framework that reinforces the concepts encountered in Chapter 5. The resulting atomic rate equations form the basis for analyzing population inversion, absorption, and amplification in laser systems.

Crucially, the inclusion of stimulated transition rates in the population equations naturally leads to saturation of the population inversion. This saturation is essential for understanding the steady-state power output, mode competition, and nonlinear behavior of laser oscillators and amplifiers. We will use these equations to discuss how light intensity saturates the population inversion and how this saturation depends on the gain medium, lineshape, and the nature of the line broadening. Furthermore, we will examine how saturation alters the lineshape itself, modifying gain and absorption to give rise to phenomena such as power broadening and spectral hole burning. We conclude this discussion by evaluating typical approximations used to describe laser level populations, their limits of validity, and how saturation behavior differs between homogeneous and inhomogeneous broadening.

The second part of this chapter brings together the equations of motion for atomic polarization, level population densities, and the wave equation for the cavity field. Together, they constitute a system of three dynamically coupled nonlinear differential equations known as the Maxwell–Bloch equations. These equations form the semiclassical foundation for describing atom-field dynamics and will be used in upcoming chapters to explore the rich and varied behavior of lasers. Finally, the third part of this chapter reviews selected gain media for various commonly used lasers.

7.1 Quantum Theory of Absorption and Emission

According to quantum mechanics, the energy of bound atomic states is restricted to discrete values, as illustrated in Fig. 7.1. These distinct values are referred to as the energy levels of the atom.

DOI: 10.1201/9781003584179-7

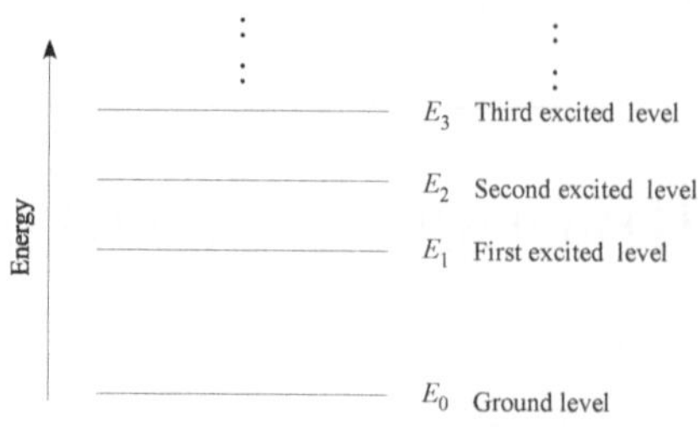

FIGURE 7.1
Energy levels of an atom.

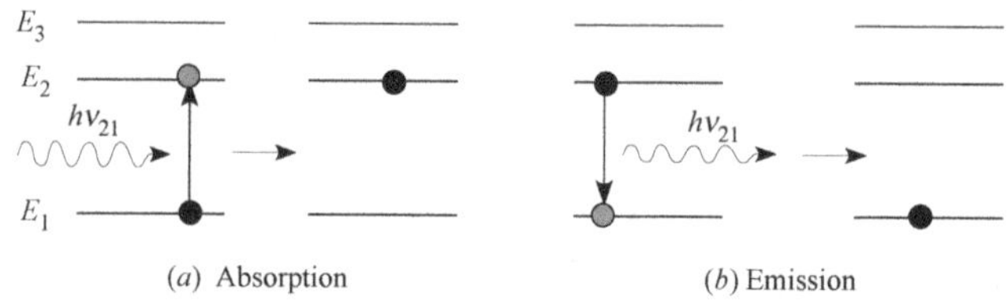

FIGURE 7.2
Absorption and emission of light by an atom.

Due to this quantization of energy, an atom exchanges energy with an electromagnetic field in discrete quanta (photons) and can absorb or emit energy only in certain discrete amounts. An atom, initially in a lower energy level E_i, can undergo a transition[1] to a higher energy level E_j by absorbing a photon of energy $h\nu_{ji} = E_j - E_i$ $(E_j > E_i)$. Conversely, the atom can undergo the reverse transition $E_j \to E_i$ by emitting a photon of the same energy. Thus the exchange of energy between the field and atom occurs either by absorption or emission of photons. Figure 7.2 illustrates these fundamental absorption and emission processes between two arbitrary levels, E_1 and E_2.

Throughout this chapter, we consider near-resonant light-matter interactions, consistent with the framework established in Chapter 5. This implies that the field frequency $\omega = 2\pi\nu$ and the atomic transition frequency $\omega_o = 2\pi\nu_o$ are closely matched $(\omega \approx \omega_o)$. However, the detuning $|\omega - \omega_o|$ can still be comparable to, or larger than, the population or polarization line widths. Crucially, these relaxation rates remain several orders of magnitude smaller than either ω or ω_o. In typical optical transitions, this condition is easily satisfied since $\nu, \nu_o \sim 10^{14} - 10^{16}$ Hz, whereas characteristic atomic and polarization line widths are typically only $\Delta\nu_a \sim 10^6 - 10^{12}$ Hz.

Population Change by Absorption

Semiclassical perturbation theory establishes that when an atomic system is exposed to radiation, the probability of absorbing a photon within a brief time interval δt is proportional to the incident photon flux density[2]. Consequently, the transition probability from level E_1

[1] Assuming the transition is dipole-allowed by quantum mechanical selection rules.

[2] The photon flux density $\Phi \equiv$ number of photons per unit area per unit time $(\mathrm{m^{-2}s^{-1}})$. For monochromatic light at frequency ν, the flux density relates to the optical intensity via $\Phi = I/h\nu$. We omit the frequency subscript hereafter for notational simplicity.

to E_2 is expressed as:

$$\text{differential transition probability} = \text{constant} \times \Phi_{12}\delta t\,, \tag{7.1}$$

where Φ_{12} represents the flux density of photons with energy $h\nu \approx E_2 - E_1 = h\nu_o$. This flux density relates directly to the field intensity by:

$$\Phi_{12} = \frac{I_\nu}{h\nu} = \frac{1}{2}\frac{\epsilon_0 nc|\mathbf{E}|^2}{h\nu}\,. \tag{7.2}$$

The proportionality expressed in Eq. (7.1) follows from Fermi's Golden Rule.

To connect this microscopic probability to a macroscopic population change, consider a sample with an initial population density of $\mathcal{N}_1$ atoms per unit volume in level E_1. The fraction of atoms excited to level E_2 via absorption during the interval δt is given by:

$$\left[\frac{|\delta\mathcal{N}_1|}{\mathcal{N}_1}\right]_{1\to 2} = \text{constant} \times \Phi_{12}\delta t\,, \tag{7.3}$$

where $\delta\mathcal{N}_1$ is the net change in population density. Since absorption depletes the lower state, the density of excited atoms is $|\delta\mathcal{N}_1| = -\delta\mathcal{N}_1$. Taking the continuous limit ($\delta t \to 0$), we obtain the time derivative of the population density:

$$\left[-\frac{d\mathcal{N}_1}{dt}\right]_{1\to 2} = \text{constant} \times \Phi_{12}\mathcal{N}_1\,. \tag{7.4}$$

Dimensional analysis reveals that the proportionality constant must carry the dimensions of area [m^2]. This parameter represents the transition cross-section, denoted by the Greek letter σ. While σ can be calculated using quantum mechanics if the atomic wave functions are known, in laser physics, we generally rely on values measured experimentally and compiled in handbooks and reference tables. Denoting the cross-section for the $E_1 \to E_2$ transition as σ_{12}, the depletion rate of level E_1 becomes:

$$\left[\frac{d\mathcal{N}_1}{dt}\right]_{1\to 2} = -\sigma_{12}\Phi_{12}\mathcal{N}_1\,. \tag{7.5}$$

The negative sign explicitly represents the depletion of $\mathcal{N}_1$ due to absorption. The product $\sigma_{12}\Phi_{12}$ represents the stimulated transition rate (the probability per unit time that a single atom undergoes an E_1 to E_2 transition). Because this process is driven by the external field, it is classified as a stimulated transition rate, and σ_{12} is called the stimulated absorption cross-section. The complete term, $\sigma_{12}\Phi_{12}\mathcal{N}_1$, defines the total rate of upward transitions per unit volume.

Absorption of photons by atoms in level E_1 also changes level population $\mathcal{N}_2$. Because total atomic density is conserved, the rate of population growth in the excited level E_2 due to $E_1 \to E_2$ transition must precisely balance the depletion rate of the lower level E_1:

$$\left[\frac{d\mathcal{N}_2}{dt}\right]_{1\to 2} = +\sigma_{12}\Phi_{12}\mathcal{N}_1\,. \tag{7.6}$$

This mathematical framework applies uniformly to all upward radiative transitions within the atom. For example, the population depletion rates due to E_1 to E_3 transitions can be written as

$$\left[\frac{d\mathcal{N}_1}{dt}\right]_{1\to 3} = -\sigma_{13}\Phi_{13}\mathcal{N}_1\,, \tag{7.7}$$

$$\left[\frac{d\mathcal{N}_3}{dt}\right]_{1\to 3} = +\sigma_{13}\Phi_{13}\mathcal{N}_1\,. \tag{7.8}$$

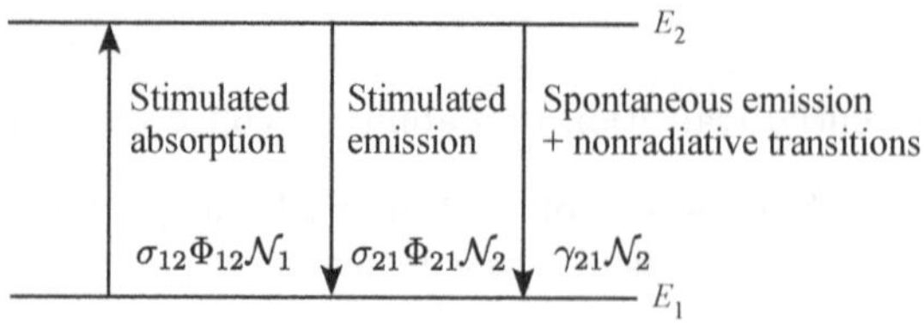

FIGURE 7.3
Various radiative and nonradiative rates mediating population transfer between levels 1 and 2 of a two-level system. Transitions involving auxiliary levels are omitted for clarity.

Population Change by Emission

In addition to absorption, radiative emission processes modify the populations of atomic energy levels. An atom in an excited state will eventually relax to a lower energy state by emitting a photon. This radiative transition from an upper level E_2 to a lower level E_1 can be driven by the presence of an external field, or it can occur spontaneously in the absence of a field [Fig. 7.3]. These processes are classified as stimulated (induced) emission and spontaneous emission, respectively. Their contribution to population rate equations can be written as follows.

For an upper-state population density $\mathcal{N}_2$, the rate of population depletion due to the $E_2 \to E_1$ transition is given by:

$$\left[\frac{d\mathcal{N}_2}{dt}\right]_{2\to 1} = -\sigma_{21}\Phi_{21}\mathcal{N}_2 - A_{21}\mathcal{N}_2\,. \tag{7.9}$$

The first term on the right-hand side, which depends linearly on the photon flux density Φ_{21}, represents the stimulated emission rate. Analogous to the absorption rate in Eq. (7.5) under an interchange of indices, this term accounts for field-induced downward transitions. Photons generated via stimulated emission are emitted into the same optical mode as the stimulating field and therefore have its frequency, direction, polarization, and phase coherence. Microscopic reversibility in quantum mechanics dictates that the stimulated emission cross-section σ_{21} is strictly equal to the absorption cross-section σ_{12} for non-degenerate states ($\sigma_{21} = \sigma_{12}$). Furthermore, since the absorbed and emitted photons share the same resonant frequency, their flux densities are identical ($\Phi_{21} = \Phi_{12}$).

The second term, $A_{21}\mathcal{N}_2$, is independent of the external radiation field. The coefficient A_{21} represents the spontaneous emission rate (the transition probability per unit time) for an isolated atom to decay from E_2 to E_1 ($E_1 < E_2$) via photon emission. This coefficient corresponds precisely to the Einstein A coefficient for the $E_2 \to E_1$ transition. In contrast to stimulated emission, spontaneously emitted photons lack coherence with respect to any external field.

By setting $\Phi_{21} = 0$ in Eq. (7.9), the differential equation yields the unperturbed exponential decay $\mathcal{N}_2(t) = \mathcal{N}_2(0)e^{-A_{21}t}$. The radiative lifetime $\tau_{21\,\mathrm{rad}}$ is defined as the time required for $\mathcal{N}_2$ to decay to $1/e \approx 0.37$ of its initial value $\mathcal{N}_2(0)$, establishing the fundamental reciprocal relationship:

$$\tau_{21\,\mathrm{rad}} = \frac{1}{A_{21}}\,. \tag{7.10}$$

By conservation of atomic density, every atom exiting level E_2 via photon emission must populate level E_1. Consequently, the corresponding rate of population growth in level E_1 is expressed as:

$$\left[\frac{d\mathcal{N}_1}{dt}\right]_{2\to 1} = \sigma_{21}\Phi_{21}\mathcal{N}_2 + A_{21}\mathcal{N}_2\,. \tag{7.11}$$

Rate equations of this form apply universally to any optical transition involving single-photon exchange.

Population Change by Nonradiative Transitions

Nonradiative processes, such as inelastic atomic collisions or energy quenching, also drive population transfer between atomic levels. While these interactions can induce both upward and downward transitions, the upward excitation rate relative to the corresponding downward rate is suppressed by a Boltzmann factor $\exp(-\hbar\omega_o/k_B T)$. For optical transitions where $\hbar\omega_o \gg k_B T$, this thermal excitation probability is negligible under typical conditions. We therefore ignore upward nonradiative relaxation under this optical approximation and consider only downward nonradiative decay of levels.

Nonradiative contributions to the population dynamics can be modeled phenomenologically by introducing a nonradiative transition probability per unit time, analogous to the spontaneous emission rate. The depletion rate of an upper level E_j due to a downward nonradiative transition to a lower level E_i is given by:

$$\left[\frac{d\mathcal{N}_j}{dt}\right]_{j\to i\,\mathrm{nr}} = -\gamma_{ji\,\mathrm{nr}}\mathcal{N}_j\,, \tag{7.12}$$

where $\gamma_{ji\,\mathrm{nr}}$ denotes the nonradiative relaxation rate for the $E_j \to E_i$ transition, representing the probability per unit time that an atom relaxes nonradiatively. The total relaxation rate γ_{ji} mediating the specific $E_j \to E_i$ channel and its corresponding lifetime τ_{ji} combine both radiative and nonradiative components:

$$\gamma_{ji} = A_{ji} + \gamma_{ji\,\mathrm{nr}}\,, \tag{7.13a}$$

$$\frac{1}{\tau_{ji}} = \gamma_{ji} = \frac{1}{\tau_{ji\,\mathrm{rad}}} + \frac{1}{\tau_{ji\,\mathrm{nr}}}\,. \tag{7.13b}$$

Summing over all allowed lower energy levels ($E_i < E_j$), the total decay rate γ_j and the aggregate lifetime τ_j of the excited level E_j are expressed as:

$$\gamma_j = \sum_{i<j}(A_{ji} + \gamma_{ji\,\mathrm{nr}}) \equiv A_j + \gamma_{j\,\mathrm{nr}}\,, \tag{7.13c}$$

$$\frac{1}{\tau_j} = \gamma_j = \sum_{i<j}\left(\frac{1}{\tau_{ji\,\mathrm{rad}}} + \frac{1}{\tau_{ji\,\mathrm{nr}}}\right)\,. \tag{7.13d}$$

These expressions provide a framework for tracking the energy dissipation pathways of an excited atomic system. While restricting nonradiative relaxation to the downward direction is a sound approximation for optical transitions with $\hbar\omega_o \gg k_B T$, this approximation can break down for sufficiently long-wavelength infrared, far-infrared, and submillimeter transitions, where $\hbar\omega_o \sim k_B T$ and upward thermally activated transitions must also be included.

7.2 Two-level Rate Equations

Using the stimulated transition rates and relaxation probabilities established in the preceding sections, we can construct the population rate equations for an arbitrary atomic transition. For a simple two-level atomic system defined by an upper-level E_2 and a lower-level E_1, the active population transfer pathways are illustrated in Fig. 7.3.

The population density of level E_1 decreases due to absorption-driven $E_1 \to E_2$ transitions [Eq. (7.5)] and increases via radiative and nonradiative $E_2 \to E_1$ decay [Eq. (7.11)]. Conversely, the population density of level E_2 undergoes depletion via $E_2 \to E_1$ transitions [Eq. (7.9)] and grows via $E_1 \to E_2$ absorption [Eq. (7.6)]. To simplify the notation without loss of generality, we define the cross-section $\sigma_{12} = \sigma_{21} \equiv \sigma$ and photon flux density $\Phi_{12} = \Phi_{21} = \Phi$. For an isolated two-level system, the upper-level total relaxation rate simplifies to $\gamma_{21} = A_{21} + \gamma_{21\,\mathrm{nr}} \equiv \gamma_2 \equiv 1/\tau_{21}$. Accounting exclusively for these single-photon radiative transitions and local environmental relaxation, the coupled time derivatives of the population densities are expressed as:

$$\frac{d\mathcal{N}_1}{dt} = \left[\frac{d\mathcal{N}_1}{dt}\right]_{1\to 2} + \left[\frac{d\mathcal{N}_1}{dt}\right]_{2\to 1} = \sigma(\mathcal{N}_2 - \mathcal{N}_1)\Phi + \gamma_{21}\mathcal{N}_2\,, \tag{7.14a}$$

$$\frac{d\mathcal{N}_2}{dt} = \left[\frac{d\mathcal{N}_2}{dt}\right]_{2\to 1} + \left[\frac{d\mathcal{N}_2}{dt}\right]_{1\to 2} = -\sigma(\mathcal{N}_2 - \mathcal{N}_1)\Phi - \gamma_{21}\mathcal{N}_2\,. \tag{7.14b}$$

Equations (7.14) constitute the core dynamical framework governing the evolution of level populations in a two-state system driven by an electromagnetic field. In a multi-level atom, these equations require additional coupling terms to account for radiative or collisional transitions involving auxiliary states, as explored in subsequent sections. Here, we use these idealized equations to introduce a number of useful concepts such as the saturation intensity and the nonlinear saturation behavior of the population difference.

7.2.1 Steady-state Inversion in a Two-level Medium

We now examine the feasibility of achieving a steady-state population inversion in this two-level framework. By assuming that the total atomic number density in the two levels remains fixed ($\mathcal{N}_1 + \mathcal{N}_2 = \mathcal{N}$) and decoupling the system from external states, we can reduce the coupled rate equations into a single differential equation for the population inversion.

Taking the sum and difference of Eqs. (7.14) yields the dynamical equations for the total atomic density $\mathcal{N}$ and the population inversion $\Delta\mathcal{N}_{21} \equiv \mathcal{N}_2 - \mathcal{N}_1$, respectively:

$$\frac{d(\mathcal{N}_1 + \mathcal{N}_2)}{dt} = 0\,, \tag{7.15a}$$

$$\frac{d\Delta\mathcal{N}_{21}}{dt} = -2\sigma\Phi\Delta\mathcal{N}_{21} - \gamma_{21}(\mathcal{N} + \Delta\mathcal{N}_{21})\,. \tag{7.15b}$$

Equation (7.15a) explicitly confirms that the total population density is a conserved quantity.

Under steady-state conditions ($d\Delta\mathcal{N}_{21}/dt = 0$), Eq. (7.15b) yields the steady-state population inversion:

$$\Delta\mathcal{N}_{\mathrm{ss}} = -\frac{\gamma_{21}}{\gamma_{21} + 2\sigma\Phi}\mathcal{N} = -\frac{\mathcal{N}}{1 + \frac{2\sigma I}{h\nu\gamma_{21}}}\,, \tag{7.16}$$

where we have expressed the photon flux density in terms of the light intensity, $\Phi = I/h\nu$. Physically, this steady-state population difference reflects a dynamic equilibrium between two competing mechanisms: stimulated transitions, which act symmetrically to equalize the state populations, and relaxation processes, which, within the optical approximation adopted here, tend to drive the population difference toward its equilibrium value $-N$, with essentially all atoms in the lower level.

Equation (7.15b) demonstrates that $\Delta\mathcal{N}_{\mathrm{ss}}$ is nonlinearly dependent on the incident optical intensity I. As the field intensity increases, the population difference is systematically driven toward zero, a phenomenon known as population saturation. The rate at which the population difference saturates as a function of intensity is governed by the coefficient of I in the denominator. This allows us to define a parameter known as the saturation intensity, I_{sat}:

$$I_{sat} = \frac{h\nu\gamma_{21}}{\sigma} = \frac{h\nu}{\sigma\tau_{21}} . \tag{7.17}$$

The quantity I_{sat} defines a characteristic intensity corresponding to one photon of energy incident per absorption cross-section per upper-state lifetime. For a specific transition and environment, I_{sat} establishes the natural scale to define weak and strong signal intensities. Utilizing this definition, the steady-state inversion can be compactly expressed as:

$$\Delta\mathcal{N}_{ss} = -\frac{\mathcal{N}}{1 + 2(I/I_{sat})} . \tag{7.18}$$

When the operating intensity is small relative to the saturation intensity ($I \ll I_{sat}$), the radiation field causes negligible perturbation to the population inversion. The factor of two in the denominator of Eq. (7.18) is specific to a two-state system and changes when analyzing complex multi-level systems. It stems from the fact that an individual field-induced transition decreases $\mathcal{N}_1$ by unity while simultaneously increasing $\mathcal{N}_2$ by unity, thereby altering the net population difference $\Delta\mathcal{N}_{21}$ by exactly two units.

The normalized two-level atomic inversion is plotted in Fig. 7.4 as a function of the normalized intensity ratio I/I_{sat}. In the asymptotic limits where $I \ll I_{sat}$ or $I \gg I_{sat}$, the inversion is relatively insensitive to intensity variations, whereas the intermediate regime ($I \sim I_{sat}$) exhibits a rapid, highly nonlinear suppression of the inversion. Crucially, this analytical result demonstrates that achieving a positive steady-state population inversion ($\Delta\mathcal{N}_{ss} > 0$) is impossible in a strictly two-level system. In the strong-signal limit ($I \gg I_{sat}$), the population difference asymptotically approaches $\Delta\mathcal{N}_{ss} \to 0$, representing complete saturation where the states are equally populated ($\mathcal{N}_1 = \mathcal{N}_2$).

7.2.2 Transient Response

Having established that a steady-state population inversion ($\Delta\mathcal{N}_{21} > 0$) cannot be maintained in an unpumped two-level system, we now investigate its transient dynamics when an arbitrary initial population difference $\Delta\mathcal{N}_{21}(0)$ is introduced. By defining a scaled time

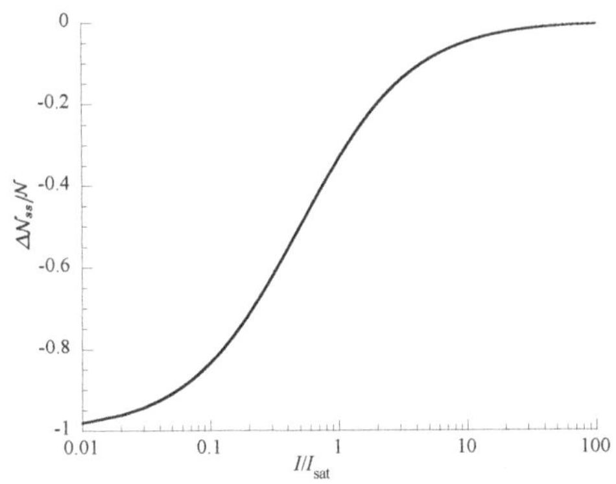

FIGURE 7.4
Normalized steady-state atomic inversion $\Delta\mathcal{N}_{ss}/\mathcal{N}$ for a two-level system plotted as a function of the normalized signal intensity I/I_{sat}.

variable $\tau = \gamma_{21}t$ and expressing the stimulated transition rates in terms of the optical intensity, the equation of motion (7.15b) and its analytical solution can be written as:

$$\frac{1}{\mathcal{N}}\frac{d\Delta\mathcal{N}_{21}}{d\tau} = -\left(1+\frac{2I}{I_{\text{sat}}}\right)\frac{\Delta\mathcal{N}_{21}}{\mathcal{N}} - 1\,, \tag{7.19a}$$

$$\begin{aligned}\Delta\mathcal{N}_{21}(t) &= -\frac{\mathcal{N}}{1+2I/I_{\text{sat}}} + \left[\Delta\mathcal{N}_{21}(0) - \left(\frac{-\mathcal{N}}{1+2I/I_{\text{sat}}}\right)\right] e^{-(1+2I/I_{\text{sat}})\tau} \\ &= \Delta\mathcal{N}_{\text{ss}} + \left[\Delta\mathcal{N}_{21}(0) - \Delta\mathcal{N}_{\text{ss}}\right] e^{-(1+2I/I_{\text{sat}})\gamma_{21}t}\,. \end{aligned} \tag{7.19b}$$

This solution implies that any deviation from the saturated steady-state inversion decays exponentially to zero. Figure 7.5 illustrates this transient response for a system prepared with an initial inversion of $\Delta\mathcal{N}_{21}(0)/\mathcal{N} = 0.5$. In the absence of an applied radiation field ($I = 0$), the population difference relaxes toward its thermal equilibrium value $-\mathcal{N}$ with a characteristic time constant $T_1 \equiv \gamma_{21}^{-1}$. This parameter is conventionally defined as the population recovery time. In this idealized framework, T_1 corresponds directly to the upper-level radiative and nonradiative energy lifetime τ_{21}. When a constant, non-zero signal intensity is applied, the inversion evolves monotonically toward its saturated steady-state value $\Delta\mathcal{N}_{\text{ss}}$ at an accelerated rate given by $\gamma_{21}(1+2I/I_{\text{sat}})$. This relaxation behavior can become oscillatory if the dynamic energy exchange between the atoms and the field is explicitly taken into account (i.e., when intensity is treated as a time-dependent variable), a phenomenon we will explore in the context of laser relaxation oscillations.

Saturation of Absorption and Gain

The saturation of the population difference directly manifests in the macroscopic behavior of the absorption and gain coefficients on a given transition. As laser oscillation builds up from spontaneous noise, the field amplitude grows until the intracavity intensity becomes large enough to saturate the gain medium. Steady-state laser oscillation is reached when the saturated gain precisely balances the total cavity losses. Consequently, gain saturation serves as the primary regulatory mechanism determining the steady-state output power of a laser oscillator. Similarly, the nonlinear absorption behavior of saturable absorbers is used for applications such as Q-switching and mode-locking. Because the real part of the complex susceptibility–and thus the phase shift contributed by the atomic medium–is also

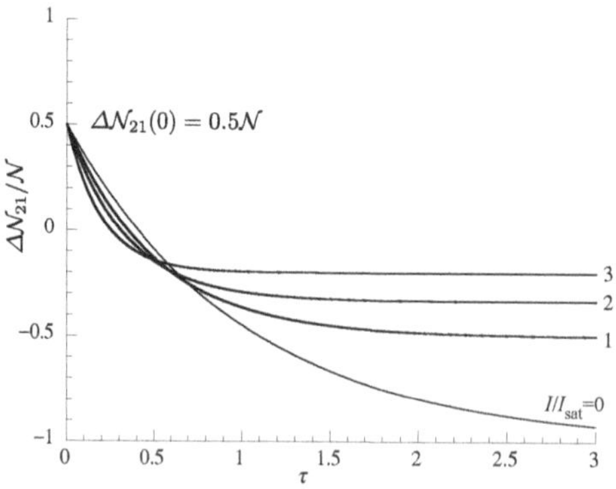

FIGURE 7.5
Transient response of the two-level population inversion as a function of scaled time $\tau = \gamma_{21}t$ for an initially positive inversion.

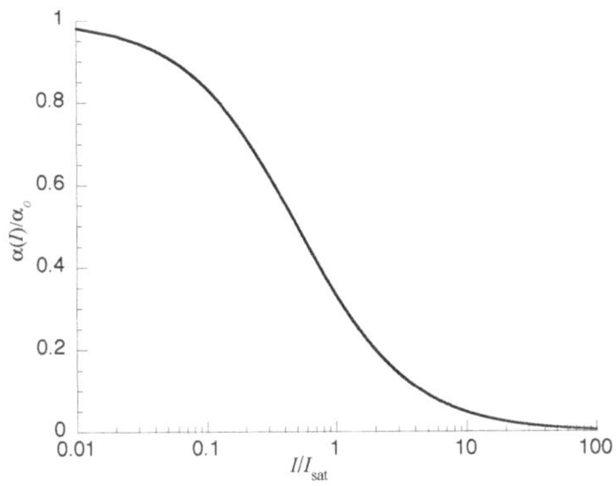

FIGURE 7.6
Absorption coefficient of a homogeneously broadened two-level medium as a function of normalized incident intensity.

linearly proportional to $\Delta\mathcal{N}_{21}$, an atomic transition can exhibit saturable gain or absorption together with a saturable phase shift under high optical intensities. The exact dependence of gain and absorption on signal intensity depends on the nature of the line broadening mechanism. Here, we analyze the basic case of a homogeneously broadened transition to highlight the underlying physical phenomena.

The steady-state absorption coefficient $\alpha(I) \equiv -\sigma\Delta\mathcal{N}_{21}$ is given by:

$$\alpha(I) = \frac{\sigma\mathcal{N}}{1 + 2(I/I_{\text{sat}})} = \frac{\alpha_0}{1 + 2(I/I_{\text{sat}})}, \tag{7.20}$$

where $\alpha_0 \equiv \sigma\mathcal{N}$ represents the unsaturated absorption coefficient in the zero-intensity limit ($I \to 0$). As the field intensity increases, the absorption coefficient decreases monotonically, asymptotically approaching zero for intense fields ($I \gg I_{\text{sat}}$). In this high-intensity regime, the medium becomes effectively transparent, a phenomenon known as optical bleaching.

Estimates of Saturation Intensity

Saturation intensity [W/m^2 or, in practice, W/cm^2] is an important parameter in determining the saturation behavior of gain as well as saturable absorption. In an amplifier, once the signal intensity becomes comparable to the saturation intensity, the gain begins to decrease appreciably with increasing signal intensity. Similarly, in a laser oscillator, the power builds until the saturated gain equals the total cavity loss. The saturation intensity thus sets the scale of power density that can be extracted from a laser.

- **Gas Lasers:** For a typical visible gas-laser transition (e.g., the HeNe laser at $\lambda =$ 633 nm), the photon energy is $h\nu = \frac{1240}{633}(1.60 \times 10^{-19}) \approx 3 \times 10^{-19}$ J. Given an atomic cross-section of $\sigma \approx 3 \times 10^{-13}$ cm^2 and an effective upper-state lifetime of $\tau_{\text{eff}} \approx 1.5 \times 10^{-7}$ s, the saturation intensity is quite low: $I_{\text{sat}} \approx 6\ \text{W/cm}^2$. For a typical beam radius of $w_0 = 1$ mm, this corresponds to a characteristic circulating saturation power of $P_{\text{sat}} = \frac{\pi w_0^2}{2} I_{\text{sat}} \approx 95$ mW. This low threshold explains why continuous-wave (cw) output powers from small gas lasers generally scale from milliwatts to a few watts.

- **Solid-State Lasers:** In contrast, a solid-state system such as a Nd:YAG laser operating at $\lambda = 1064$ nm features a significantly smaller cross-section, $\sigma \approx 10^{-19}$ cm^2, but a much longer upper-state storage lifetime, $\tau_{\text{eff}} \approx 10^{-3}$ s. These parameters yield a substantially higher saturation threshold: $I_{sat} \approx 2\ \mathrm{kW/cm^2}$. Consequently, a Nd:YAG crystal with a modest cross-sectional rod area of $(3\ \mathrm{mm})^2$ can routinely support cw output powers ranging from tens to several hundred watts.

- **Liquid Dye Lasers:** A liquid organic dye laser typically exhibits a very large cross-section, $\sigma \approx 10^{-16}\ \mathrm{cm^2}$, paired with an extremely short lifetime, $\tau_{eff} \approx 10^{-9}$ s. This combination results in an exceptionally high saturation intensity: $I_{sat} \approx 2.5\ \mathrm{MW/cm^2}$.

As these estimates show, the saturation intensity is determined by the photon energy $h\nu$, the transition cross-section σ, and the effective population recovery time τ_{eff}. While increasing the pump strength of a laser medium increases the small-signal gain (α_0), it leaves the saturation intensity I_{sat} unchanged.

7.3 Three- and Four-Level Population Inversion Schemes

As demonstrated in Section 7.2, maintaining a steady-state population inversion ($\Delta\mathcal{N}_{21} > 0$) is impossible within an isolated, two-level atomic system. Real atomic and molecular systems, however, possess a complex structure of multiple energy levels. By simultaneously driving transitions—either optically or via other excitation mechanisms—between different pairs of levels, one can dramatically alter the steady-state level populations. Radiative and nonradiative relaxation processes then populate specific intermediate states, enabling the realization of a population inversion on a specific transition. Remarkably, regardless of the number of energy levels involved, virtually all practical schemes for achieving population inversion can be reduced to an idealized three- or four-level system. In this section, we analyze these multi-level systems.

Figure 7.7 shows a simplified energy level diagram for a representative solid-state laser: triply ionized neodymium (Nd^{3+}) incorporated as the active "atom" within an yttrium aluminum garnet ($Y_3Al_5O_{12}$, or YAG) host crystal. Strong absorption of light from an optical pump excites ions from their ground state (the lower pump level) into a band of excited

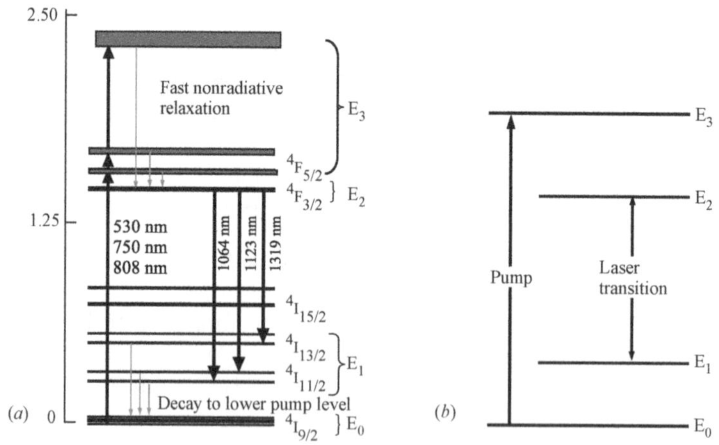

FIGURE 7.7
(a) Energy level structure, optical pumping channels, and primary laser transitions of the trivalent neodymium ion in a Nd^{3+}:YAG crystal. (b) Simplified four-level representation of the physical system customized for operation on a single laser transition. Auxiliary radiative and nonradiative relaxation pathways are omitted for clarity.

energy levels (the upper pump level). Ions in these highly excited levels undergo rapid, nonradiative relaxation into a lower-lying, metastable state that serves as the upper laser level. The relatively long lifetime of this state allows a substantial buildup of a population of excited ions. This upper laser level population relaxes primarily via the stimulated and spontaneous radiative transitions that provide the optical gain of interest. In real media, the upper and lower laser levels are typically clusters of closely spaced Stark-split sublevels, giving rise to multiple distinct laser lines. From the lower laser level, ions relax rapidly via phonon-assisted or radiative decay channels into lower-lying states, eventually returning to the initial ground state.

The intricate network of atomic states involved in laser gain media can be condensed into an idealized three-level or four-level laser system. In both cases, the upper pump level E_3 represents all high-lying states into which the pump mechanism excites population from the lower pump level. Level E_2 denotes the upper laser level—typically a long-lived metastable state separated from the lower laser level E_1 by an energy gap large compared to the thermal energy $k_B T$. If the lower laser level E_1 doubles as the ground state of the system, the medium is classified as a three-level laser. Conversely, if the lower laser level E_1 lies well above the ground level E_0, the medium operates as a four-level laser. In this case, auxiliary low-lying atomic states play no active role in the stimulated inversion dynamics and are neglected; they serve merely as rapid relaxation pathways back to the ground level E_0.

The ruby laser (Cr^{3+} : Al_2O_3) is the classic example of a three-level laser system, alongside many fiber-optic Erbium-doped amplifiers (EDFAs). In contrast, helium-neon (HeNe), argon-ion (Ar^+), carbon dioxide (CO_2), and organic liquid dye lasers are representative examples of four-level systems. The specific mechanism employed to supply the excitation energy required to excite atoms from the lower pump level (level E_1 in Fig. 7.8 and level E_0 in Fig. 7.9) to the upper pump level E_3 varies drastically across different lasers. For example, ruby and liquid dye lasers utilize optical pumping via flashlamps or secondary lasers; HeNe and argon-ion systems rely on electron-impact excitation and resonant collisional energy transfer within an electrical discharge; semiconductor lasers leverage electrical current injection across a p-n junction; and chemical lasers harvest the energy released during exothermic chemical reactions. Regardless of the underlying physical implementation, the pumping process can be modeled as an induced transition rate W_p (representing an excitation probability per unit time per atom) coupling the lower and upper pump levels. For an optical pump field, we explicitly assume its center frequency is sufficiently detuned from the laser transition ($E_2 \to E_1$) by several atomic linewidths to avoid any interference. The coupled rate equations governing these multi-level systems can be constructed by accounting for the field-induced, spontaneous radiative, and nonradiative transitions between all participating states, generalizing the procedure developed for the two-level atom.

7.3.1 Three-level Inversion Scheme

The population transition rates for a three-level pump scheme are shown in Fig. 7.8. These rates lead to the following equations for the population densities $\mathcal{N}_i$ (defined as the number of atoms per unit volume in level E_i):

$$\frac{d\mathcal{N}_3}{dt} = -W_p(\mathcal{N}_3 - \mathcal{N}_1) - \gamma_3 \mathcal{N}_3 \,, \tag{7.21a}$$

$$\frac{d\mathcal{N}_2}{dt} = \gamma_{32}\mathcal{N}_3 - \gamma_2 \mathcal{N}_2 - W(\mathcal{N}_2 - \mathcal{N}_1) \,, \tag{7.21b}$$

$$\frac{d\mathcal{N}_1}{dt} = W_p(\mathcal{N}_3 - \mathcal{N}_1) + \gamma_{31}\mathcal{N}_3 + \gamma_2 \mathcal{N}_2 + W(\mathcal{N}_2 - \mathcal{N}_1) \,, \tag{7.21c}$$

$$\mathcal{N} = \mathcal{N}_1 + \mathcal{N}_2 + \mathcal{N}_3 \,. \tag{7.21d}$$

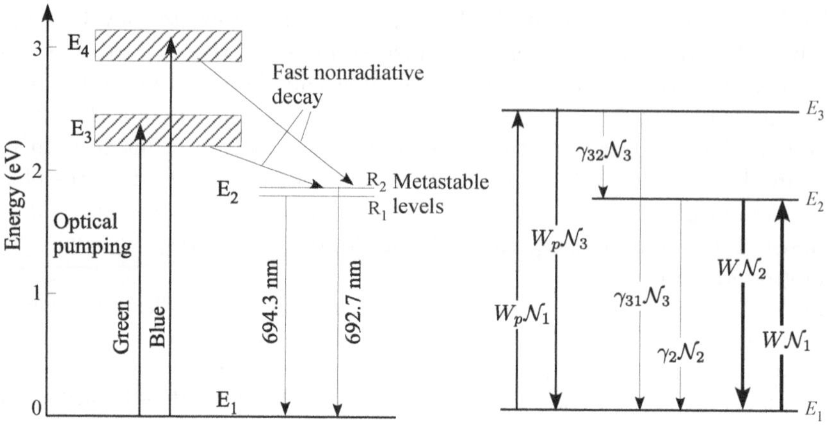

FIGURE 7.8
Three-level pump scheme operating with an external optical pump.

Here, $W_p = \sigma_{13}\Phi_p$ denotes the transition rate induced by an optical pump with photon flux density Φ_p between the ground level E_1 and the upper pump level E_3. The parameter $W = \sigma\Phi$ represents the stimulated transition rate between the upper and lower laser levels (E_2 and E_1, respectively) driven by the signal photon flux density Φ. As established in Eq. (7.13), each total relaxation rate γ_{ji} incorporates both radiative and nonradiative contributions, and γ_j represents the total decay rate out of level E_j.[3]

In formulating Eqs. (7.21), level E_1 is defined as the ground level, which precludes its downward decay. Equation (7.21d) implies that population transfer to omitted auxiliary levels is negligible, ensuring that the sum of the population densities equals the total density of active atoms $\mathcal{N}$. Furthermore, we assume that the energy states are non-degenerate and that there is no radiation field coupling the $2 \leftrightarrow 3$ transition. Although W_p is defined here for an optical pump, other excitation mechanisms, such as electron-impact excitation, current injection, or chemical pumping, can be incorporated through appropriate pump terms in the population equations.

Under steady-state conditions, the population densities become time-independent ($d\mathcal{N}_i/dt = 0$). In the small-signal regime where the saturating laser field on the $2 \leftrightarrow 1$ transition is absent ($W = 0$), the steady-state population equations simplify to:

$$0 = -W_p(\mathcal{N}_3 - \mathcal{N}_1) - \gamma_3\mathcal{N}_3\,, \tag{7.22a}$$
$$0 = \gamma_{32}\mathcal{N}_3 - \gamma_2\mathcal{N}_2\,, \tag{7.22b}$$
$$0 = W_p(\mathcal{N}_3 - \mathcal{N}_1) + \gamma_2\mathcal{N}_2 + \gamma_{31}\mathcal{N}_3\,. \tag{7.22c}$$

Solving these equations allows us to express the steady-state excited populations in terms of the ground-state population $\mathcal{N}_1$:

$$\mathcal{N}_3 = \frac{W_p/\gamma_3}{1 + W_p/\gamma_3}\mathcal{N}_1\,, \tag{7.23a}$$
$$\mathcal{N}_2 = \frac{\gamma_{32}}{\gamma_2}\mathcal{N}_3 = \frac{\gamma_{32}}{\gamma_2}\frac{W_p/\gamma_3}{1 + W_p/\gamma_3}\mathcal{N}_1\,. \tag{7.23b}$$

[3]Radiative decay is accompanied by the emission of a photon, whereas nonradiative decay dissipates energy through environmental interactions such as phonon emission or collisions.

Substituting Eq. (7.23b) into the definition for the population difference yields the unsaturated steady-state population inversion on the $2 \leftrightarrow 1$ transition, $\Delta\mathcal{N}^{(0)} \equiv [\mathcal{N}_2 - \mathcal{N}_1]_{\text{unsat}}$:

$$\Delta\mathcal{N}^{(0)} = \left[\frac{(\gamma_{32}/\gamma_2)(W_p/\gamma_3)}{1 + W_p/\gamma_3} - 1\right]\mathcal{N}_1 = \left[\frac{(1-\beta)\eta W_p/A_{21} - 1}{1 + W_p/\gamma_3}\right]\mathcal{N}_1 \,. \tag{7.24}$$

In the second step, the inversion is parameterized by dimensionless variables η and β, defined as:

$$\eta = \frac{\gamma_{32}}{\gamma_3}\frac{A_{21}}{\gamma_2}\,, \qquad \beta = \frac{\gamma_2}{\gamma_{32}}\,. \tag{7.25}$$

These variables serve as convenient combinations of level relaxation rates for comparing different laser transitions, capturing both the relative ease of establishing an inversion and the fundamental pump-to-laser quantum conversion efficiency. From the numerator of Eq. (7.24), it is evident that a positive steady-state population inversion ($\Delta\mathcal{N}^{(0)} > 0$) requires the pump transition rate W_p to exceed a threshold value:

$$W_{p,\min} = \frac{A_{21}}{(1-\beta)\eta} = \frac{\gamma_2\gamma_3/\gamma_{32}}{1 - \gamma_2/\gamma_{32}}\,. \tag{7.26}$$

Minimizing this threshold pump rate is key to optimizing the system for laser action. Inspection of Eq. (7.26) reveals that the physical conditions favoring a low threshold are:

$$\eta \approx 1 \quad \text{and} \quad \beta \ll 1\,. \tag{7.27}$$

Physical meaning of η and β

The parameter η is the product of two ratios. The first factor, γ_{32}/γ_3, represents the fraction of level 3 population that decays to the upper laser level 2. The second factor, A_{21}/γ_2, represents the fraction of level 2 population that transitions to level 1 via spontaneous radiative emission. In the limit $W_p \ll \gamma_3$, where saturation of the pump transition is negligible, the term $W_p\mathcal{N}_1$ represents the rate of pump photon absorption. Assuming a high-efficiency transfer where each absorbed pump photon excites one atom to the upper pump level, the rate of signal photon emission approaches $\eta W_p \mathcal{N}_1$. The ratio of the signal quantum emission rate to the pump quantum absorption rate corresponds exactly to η. Thus, η represents the fluorescent quantum efficiency with which pump photons are converted into signal photons. In an ideal system, $\eta \approx 1$, which implies that $\gamma_{32} \approx \gamma_3$ and $A_{21} \approx \gamma_2$. Physically, this means the upper pump level decays almost exclusively to the upper laser level, and the upper laser level subsequently de-excites to the lower laser level primarily via radiative transitions.

To establish the physical interpretation of β, consider a configuration where level 3 relaxes predominantly to level 2, ensuring that nearly all pumped atoms reach the upper laser level. In this regime, the total decay rate $\gamma_3 \approx \gamma_{32}$, so that the parameter $\beta \approx \gamma_2/\gamma_3 = \tau_3/\tau_2$. From the analytical expression for the unsaturated inversion $\Delta\mathcal{N}^{(0)}$ [Eq. (7.24)], a steady-state population inversion requires $\beta < 1$, and ideally $\beta \ll 1$. The condition $\beta \ll 1$ physically means that the upper laser level must decay much more slowly than the upper pump level through the $3 \to 2$ channel. This permits dynamic population accumulation in the upper laser level E_2. In terms of state lifetimes, $\beta \ll 1$ corresponds to $\tau_2 \gg \tau_3$.

In summary, the criteria for an optimized three-level laser medium [Eq. (7.27)] require:

- The upper pump level E_3 must decay primarily to the upper laser level E_2 ($\gamma_3 \approx \gamma_{32}$), ensuring that the overwhelming majority of pumped atoms populate the laser transition.
- Decays out of the upper laser level E_2 should occur predominantly through the radiative laser transition ($\gamma_2 \approx A_{21}$), maximizing the radiative quantum efficiency of the transition.

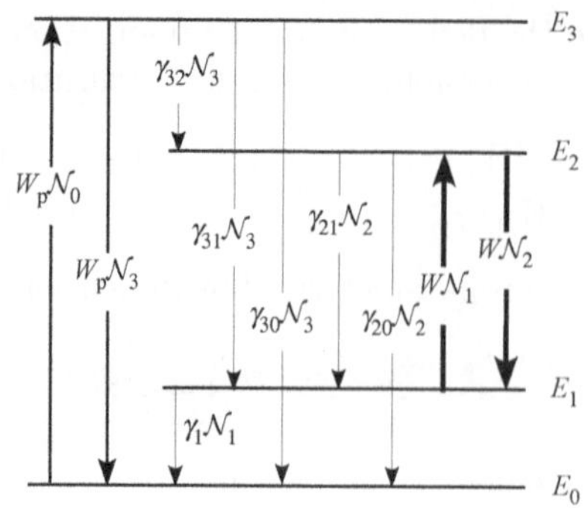

FIGURE 7.9
Four-level pump scheme using an optical pump.

- The upper laser level must be significantly longer-lived than the upper pump level ($\tau_2 \gg \tau_3$), allowing atoms to accumulate in level E_2 and create a net population inversion relative to the ground state.

7.3.2 Four-level Inversion Scheme

Figure 7.9 shows various transition rates for a four-level pump scheme. These rates lead to the following equations of motion for the population densities

$$\frac{d\mathcal{N}_3}{dt} = -W_p(\mathcal{N}_3 - \mathcal{N}_0) - \gamma_3\mathcal{N}_3\,, \tag{7.28a}$$
$$\frac{d\mathcal{N}_2}{dt} = \gamma_{32}\mathcal{N}_3 - \gamma_2\mathcal{N}_2 - W(\mathcal{N}_2 - \mathcal{N}_1)\,, \tag{7.28b}$$
$$\frac{d\mathcal{N}_1}{dt} = \gamma_{31}\mathcal{N}_3 + \gamma_{21}\mathcal{N}_2 - \gamma_1\mathcal{N}_1 + W(\mathcal{N}_2 - \mathcal{N}_1)\,, \tag{7.28c}$$
$$\frac{d\mathcal{N}_0}{dt} = \gamma_{30}\mathcal{N}_3 + \gamma_{20}\mathcal{N}_2 + \gamma_1\mathcal{N}_1 + W_p(\mathcal{N}_3 - \mathcal{N}_0)\,, \tag{7.28d}$$
$$\mathcal{N} = \mathcal{N}_0 + \mathcal{N}_1 + \mathcal{N}_2 + \mathcal{N}_3\,, \tag{7.28e}$$

where the last equation assumes that atomic decays to levels other than those shown in Fig. 7.9 are negligible so that the sum of population densities in the four levels equals the total density of active atoms in the medium. Level 0 is the ground state of the atom. We have also assumed that all levels are non-degenerate and no radiation fields are present on 2↔3 and 1↔0 transitions. In the absence of any saturating signal on 1 ↔ 2 transition ($W = 0$), the steady-state population densities are determined by the following equations

$$0 = -W_p(\mathcal{N}_3 - \mathcal{N}_0) - \gamma_3\mathcal{N}_3 \tag{7.29a}$$
$$0 = \gamma_{32}\mathcal{N}_3 - \gamma_2\mathcal{N}_2 \tag{7.29b}$$
$$0 = \gamma_{31}\mathcal{N}_3 + \gamma_{21}\mathcal{N}_2 - \gamma_1\mathcal{N}_1 \tag{7.29c}$$
$$0 = \gamma_{30}\mathcal{N}_3 + \gamma_{20}\mathcal{N}_2 + \gamma_1\mathcal{N}_1 + W_p(\mathcal{N}_3 - \mathcal{N}_0) \tag{7.29d}$$

By solving these equations, level populations can be expressed in terms of lower pump level population $\mathcal{N}_0$ as

$$\mathcal{N}_3 = \frac{W_p/\gamma_3}{W_p/\gamma_3 + 1}\mathcal{N}_0\,, \tag{7.30a}$$

$$\mathcal{N}_2 = \frac{\gamma_{32}}{\gamma_2}\mathcal{N}_3 = \frac{\gamma_{32}}{\gamma_2}\frac{W_p/\gamma_3}{W_p/\gamma_3 + 1}\mathcal{N}_0\,, \tag{7.30b}$$

$$\mathcal{N}_1 = \left[\frac{\gamma_{21}}{\gamma_1}\mathcal{N}_2 + \frac{\gamma_{31}}{\gamma_1}\mathcal{N}_3\right] = \left[\frac{\gamma_{21}}{\gamma_1}\frac{\gamma_{32}}{\gamma_2} + \frac{\gamma_{31}}{\gamma_1}\right]\frac{W_p/\gamma_3}{1 + W_p/\gamma_3}\mathcal{N}_0\,. \tag{7.30c}$$

Using these equations, we find that the steady-state unsaturated inversion on laser transition is given by

$$\Delta\mathcal{N}^{(o)} \equiv [\mathcal{N}_2 - \mathcal{N}_1]_{\text{unsat}} = \frac{(1-\beta)\eta W_p/A_{21}}{W_p/\gamma_3 + 1}\mathcal{N}_0\,, \tag{7.31}$$

where the parameters η and β are defined by

$$\eta = \left(\frac{\gamma_{32}}{\gamma_3}\right)\left(\frac{A_{21}}{\gamma_2}\right), \quad \beta = \frac{\gamma_2}{\gamma_1}\left(\frac{\gamma_{21}}{\gamma_2} + \frac{\gamma_{31}}{\gamma_{32}}\right). \tag{7.32}$$

Equation (7.31) for unsaturated inversion shows that a four-level system has positive inversion on the transition $2 \to 1$ for any $W_p > 0$ as long as $\beta < 1$. This means that there is no finite pump threshold for producing a positive population inversion in the ideal four-level model. It does not mean that there is no laser threshold. A real oscillator must still generate enough gain to overcome cavity loss.

The parameters η and β have the same significance for a four-level system as the parameters defined by Eq. (7.25) for a three-level system. Thus η is the quantum fluorescence efficiency. It is a product of two factors. The first factor, γ_{32}/γ_3, is the fraction of atoms pumped into level 3 that relaxes to the upper laser level, and the second factor, A_{21}/γ_2, is the fraction of atoms in level 2 that decays radiatively into level 1. Parameter β is the ratio of various relaxation rates. In an ideal system, the upper pump level 3 would decay primarily to the upper laser level 2 ($\gamma_{31} \ll \gamma_{32}$) and the upper laser level primarily to the lower laser level 1 so that the ratio $\beta \approx \gamma_2/\gamma_1 = \tau_1/\tau_2$ is essentially the ratio of the lifetimes of the lower and upper laser levels.

In terms of parameters η and β, the conditions favoring inversion are again given by

$$\eta \approx 1 \quad \text{and} \quad \beta \ll 1\,. \tag{7.27*}$$

The physical content of these conditions favoring steady-state inversion is similar to the case of the three-level pump scheme and can be stated as: (i) the upper laser level 2 relaxes predominantly to the lower laser level 1 via radiative decay ($A_{21}/\gamma_2 \approx 1$); and (ii) the upper laser level 2 has long lifetime compared with the lower laser level 1 ($\tau_1/\tau_2 = \gamma_2/\gamma_1 \ll 1$) or equivalently, the atoms relax out of the lower level at a much faster rate than the rate at which atoms relax into level 1 from levels above. In many practical four-level systems the upper pump level also relaxes rapidly, so that $W_p/\gamma_3 \ll 1$ and very little population accumulates in level 3.

7.3.3 Three- vs Four-level Inversions

We have seen that population inversion on a transition can be achieved using either three- or four-level pump schemes. Which scheme is relevant depends on the gain medium and the laser transition involved. Figure 7.10 shows the inversions for three-level and four-level

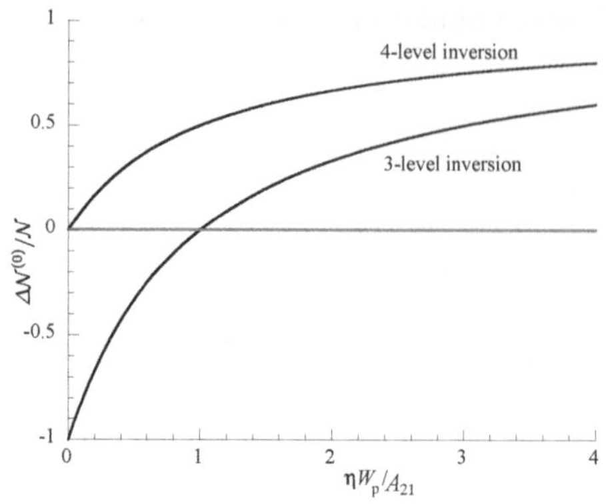

FIGURE 7.10
Three- and four-level inversions as functions of normalized pump $\eta W_p/A_{21}$ for $\beta = 0$.

pump schemes as functions of (normalized) pump rate. A comparison of the three- and four-level inversions [Eqs. (7.24) and (7.31)] reveals that it is easier to achieve inversion in a four-level pump scheme where any nonzero pumping can produce inversion provided $\beta < 1$; in the ideal four-level limit, this condition reduces approximately to requiring the upper laser level to be longer-lived than the lower laser level. In contrast, in the three-level scheme the pump must exceed the minimum given by Eq. (7.26) to achieve inversion. This difference between the two schemes can be traced to the fact that in the three-level scheme, the lower level is shared by both the pump and laser transitions. Since all the atoms are normally in the lower pump level 1, at least half of them must be pumped out to higher levels to achieve any population inversion between levels 2 and 1.

7.4 Effective Two-level Systems

In the previous section, we evaluated the conditions for achieving a steady-state population inversion in the limit when no light signal resonant with the laser transition is present. The population difference achieved under these conditions defines the unsaturated inversion, $\Delta\mathcal{N}^{(0)}$. In a laser amplifier or oscillator, however, a resonant light signal is inherently present. This signal drives stimulated transitions between the upper and lower laser levels at a rate $W = \sigma\Phi = \sigma I/h\nu$. Our current objective is to evaluate how this dynamic field coupling affects the population inversion and develop a simplified rate equation model that describes how the population inversion—and consequently the macroscopic optical gain—saturates with increasing signal intensity. In doing so, we re-examine our previous multi-level model by incorporating a resonant field. We will show that under several generally valid physical approximations, these complex multi-state systems can, under appropriate conditions, be reduced to effective two-level models. We analyze two distinct physical regimes that yield qualitatively different saturation behavior, which together provide useful analytical models to interpret the steady-state saturation and transient behavior of gain, absorption, and phase shifts across most real laser systems.

Case I: $\gamma_3 \gg W_p \gg \gamma_1$

This inequality dictates that the external pump excites atoms from the ground level E_0 to the pump manifold E_3 at a rate far exceeding the relaxation rate of the lower laser level (γ_1). Concurrently, any population entering the upper pump level E_3 undergoes near-instantaneous nonradiative decay to the lower levels. The condition $\gamma_3 \gg W_p \gg \gamma_1$ therefore introduces a severe bottleneck at the lower laser level E_1. Because level E_1 does not empty fast enough to keep pace with the intense pumping cycle, the states E_3 and E_0 reach a quasi-equilibrium state. Solving the corresponding rate equations [Eqs. (7.28a) and (7.28d)] for these quasi-steady values yields:

$$\mathcal{N}_3 = \frac{W_p}{W_p + \gamma_3}\mathcal{N}_0 \,, \tag{7.33a}$$

$$\mathcal{N}_0 = \frac{W_p + \gamma_3}{W_p}\left[\frac{\gamma_1}{\gamma_{32} + \gamma_{31}}\mathcal{N}_1 + \frac{\gamma_{20}}{\gamma_{32} + \gamma_{31}}\mathcal{N}_2\right] . \tag{7.33b}$$

Substituting these quasi-equilibrium conditions back into Eqs. (7.28b) and (7.28c), we find that the laser state populations obey a closed two-level system of rate equations:

$$\frac{d\mathcal{N}_2}{dt} = \Gamma_{12}\mathcal{N}_1 - \Gamma_{21}\mathcal{N}_2 - W(\mathcal{N}_2 - \mathcal{N}_1) \,, \tag{7.34a}$$

$$\frac{d\mathcal{N}_1}{dt} = -\Gamma_{12}\mathcal{N}_1 + \Gamma_{21}\mathcal{N}_2 + W(\mathcal{N}_2 - \mathcal{N}_1) \,, \tag{7.34b}$$

where the effective upward and downward relaxation rates are defined as:

$$\Gamma_{12} = \frac{\gamma_1\gamma_{32}}{\gamma_{32} + \gamma_{31}} \quad \text{and} \quad \Gamma_{21} = \gamma_{21}\left(1 + \frac{\gamma_{31}}{\gamma_{32} + \gamma_{31}}\frac{\gamma_{20}}{\gamma_{21}}\right) . \tag{7.34c}$$

These effective rate equations differ fundamentally from the isolated two-state system analyzed in Section 7.2. After adiabatic elimination of levels E_0 and E_3, the pump and relaxation cycle produces an effective upward transfer from level 1 to level 2 at rate Γ_{12} and from level 2 to level 1 at a rate $\Gamma_{21} \neq \Gamma_{12}$. This asymmetry arises because atoms relaxing to the ground level E_0 are promptly re-excited to the upper pump level E_3 by the intense pump field, followed by rapid decay into the upper laser level. Solving these reduced two-level equations under steady-state conditions ($d\mathcal{N}_i/dt = 0$) while recalling the stimulated transition rate $W = \sigma I/h\nu$, the steady-state population inversion $\Delta\mathcal{N} \equiv \mathcal{N}_2 - \mathcal{N}_1$ can be expressed as:

$$\Delta\mathcal{N} = \frac{\Delta\mathcal{N}^{(0)}}{1 + 2(I/I_{\text{sat}})} \,, \tag{7.35}$$

where the unsaturated inversion $\Delta\mathcal{N}^{(0)}$ and the saturation intensity I_{sat} are defined by:

$$\Delta\mathcal{N}^{(0)} = (\Gamma_{12} - \Gamma_{21})\frac{\mathcal{N}}{\gamma_{\text{eff}}} \,, \tag{7.36a}$$

$$I_{\text{sat}} = \frac{h\nu}{\sigma}\gamma_{\text{eff}} \,, \tag{7.36b}$$

$$\gamma 21\text{eff} \equiv \Gamma_{12} + \Gamma_{21} \,. \tag{7.36c}$$

The appearance of the factor of 2 in the denominator of Eq. (7.35) is a direct consequence of a closed, effective two-level system where ($\mathcal{N} = \mathcal{N}_1 + \mathcal{N}_2$) denotes the conserved population density of the effective two-level subsystem. Indeed, by taking the difference of Eqs. (7.34a) and (7.34b), we can reduce the entire multi-level framework into a single dynamic equation for the population inversion:

$$\frac{d\Delta\mathcal{N}}{dt} = -\gamma_{21\text{eff}}(\Delta\mathcal{N} - \Delta\mathcal{N}^{(0)}) - 2\gamma_{21\text{eff}}\left(\frac{I}{I_{\text{sat}}}\right)\Delta\mathcal{N}\,. \tag{7.37}$$

Equation (7.37) shares an identical mathematical form with the inversion rate equation for an isolated two-level atom, with the steady-state unsaturated inversion $\Delta\mathcal{N}^{(0)}$ given by Eq. (7.36a) and the population recovery rate governed by γ_{eff}, which corresponds to a recovery time constant of $T_1 = \gamma_{21\text{eff}}^{-1}$. To achieve a positive steady-state inversion, the effective pump rate must exceed the downward decay rate ($\Gamma_{12} > \Gamma_{21}$). In a well-optimized medium where the upper pump level decays almost entirely to the upper laser level and the upper laser level relaxes primarily to the lower laser level, these effective parameters reduce to $\Gamma_{12} \approx \gamma_1$, $\Gamma_{21} \approx \gamma_2$, and $\gamma_{21\text{eff}} \approx \gamma_1 + \gamma_2$. Under these ideal conditions, the unsaturated inversion simplifies to $\Delta\mathcal{N}^{(0)} \approx [(\gamma_1 - \gamma_2)/(\gamma_1 + \gamma_2)]\mathcal{N}$. This confirms that to achieve a positive inversion the lower laser level must empty faster than the upper laser level decays, which in this ideal limit corresponds to $\gamma_1 > \gamma_2$.

Case II: $\gamma_3 \gg W_p$

A qualitatively different two-level approximation arises when the combined population of the upper and lower laser levels can no longer be treated as a conserved quantity. Consider a regime where the upper pump level relaxation rate is fast relative to the pump rate ($W_p/\gamma_3 \ll 1$). In this limit, the upper pump-level population $\mathcal{N}_3$ remains negligible compared to the ground state population $\mathcal{N}_0$,

$$\mathcal{N}_3 = \left(\frac{W_p}{W_p + \gamma_3}\right)\mathcal{N}_0 \approx \left(\frac{W_p}{\gamma_3}\right)\mathcal{N}_0 \ll \mathcal{N}_0\,. \tag{7.38}$$

We further consider the low-excitation regime in which the populations of the laser levels [Eqs. (7.30a)–(7.30c)] also remain small compared with the ground-state population, $\mathcal{N}_1, \mathcal{N}_2 \ll \mathcal{N}_0$. Under these additional conditions, $\mathcal{N}_0 \simeq \mathcal{N}$ acts as an effectively undepleted reservoir. The rate equations for $\mathcal{N}_1$ and $\mathcal{N}_2$–explicitly incorporating induced transitions at the rate $W = \sigma I/h\nu$–can be written as:

$$\frac{d\mathcal{N}_2}{dt} = \frac{\gamma_{32}W_p}{\gamma_3}\mathcal{N}_0 - \gamma_2\mathcal{N}_2 - W(\mathcal{N}_2 - \mathcal{N}_1) \equiv R_2 - \gamma_2\mathcal{N}_2 - W(\mathcal{N}_2 - \mathcal{N}_1)\,, \tag{7.39a}$$

$$\frac{d\mathcal{N}_1}{dt} = \frac{\gamma_{31}W_p}{\gamma_3}\mathcal{N}_0 + \gamma_{21}\mathcal{N}_2 - \gamma_1\mathcal{N}_1 + W(\mathcal{N}_2 - \mathcal{N}_1) \equiv R_1 + \gamma_{21}\mathcal{N}_2 - \gamma_1\mathcal{N}_1 + W(\mathcal{N}_2 - \mathcal{N}_1)\,, \tag{7.39b}$$

where the parameters

$$R_2 = \frac{\gamma_{32}}{\gamma_3}W_p\mathcal{N}_0 \quad \text{and} \quad R_1 = \frac{\gamma_{31}}{\gamma_3}W_p\mathcal{N}_0 = \frac{\gamma_{31}}{\gamma_{32}}R_2 \tag{7.39c}$$

define the volumetric rates ($\text{m}^{-3}\text{s}^{-1}$) at which the pump populates the upper and lower laser levels, respectively. Because the ground state acts as an undepleted reservoir, these pump rates remain unaffected by the laser level populations. Solving Eqs. (7.39) under steady-state conditions ($d\mathcal{N}_i/dt = 0$), the saturated population inversion density on the laser transition is given by:

$$\Delta\mathcal{N} = \frac{R_2(\gamma_1 - \gamma_{21}) - R_1\gamma_2}{\gamma_1\gamma_2 + W(\gamma_1 + \gamma_{20})} = \frac{\eta(1-\beta)(W_p\mathcal{N}_0/A_{21})}{1 + W/\gamma_{2\text{eff}}}\,, \tag{7.40a}$$

where the parameters η and β match our previous definitions:

$$\eta = \frac{\gamma_{32}}{\gamma_3}\frac{A_{21}}{\gamma_2}, \quad \beta = \frac{\gamma_2}{\gamma_1}\left(\frac{\gamma_{21}}{\gamma_2} + \frac{\gamma_{31}}{\gamma_{32}}\right), \tag{7.40b}$$

and the effective upper-level relaxation rate is defined as:

$$\gamma_{21\mathrm{eff}} = \frac{\gamma_2}{1 + \gamma_{20}/\gamma_1}. \tag{7.40c}$$

The parameters η and β retain the physical significance established in Eq. (7.25). Expressing the stimulated transition rate in terms of the optical field intensity ($W = \sigma I/h\nu$), the population inversion can be cast into standard saturation form:

$$\Delta\mathcal{N} = \frac{\Delta\mathcal{N}^{(0)}}{1 + I/I_{\mathrm{sat}}}, \tag{7.41}$$

where the small-signal unsaturated inversion $\Delta\mathcal{N}^{(0)}$ and the frequency-dependent saturation intensity $I_{\mathrm{sat}}(\nu)$ are given by:

$$\Delta\mathcal{N}^{(0)} = \eta(1-\beta)\frac{W_p\mathcal{N}_0}{A_{21}}, \tag{7.42a}$$

$$I_{\mathrm{sat}}(\nu) = \frac{h\nu}{\sigma(\nu)}\gamma_{21\mathrm{eff}}. \tag{7.42b}$$

As in Case I, the unsaturated inversion increases linearly with the pump rate, and a positive steady-state inversion requires satisfying the condition $\beta < 1$. In a medium where the upper pump level decays exclusively to the upper laser level ($\gamma_{31} \to 0$), this parameter simplifies to $\beta \approx \gamma_{21}/\gamma_1$, reducing the unsaturated inversion to $\Delta\mathcal{N}^{(0)} \approx (1 - \gamma_{21}/\gamma_1)(R_2/\gamma_2)$. Consequently, generating a net inversion requires $\gamma_1 > \gamma_{21}$, implying that the lower laser level must empty out at a rate faster than it is replenished by upper-state relaxation. Crucially, Eq. (7.41) exhibits a more gradual saturation with intensity compared to Case I, missing the factor of two in the denominator's saturation term. This variation is a direct consequence of the shift from a closed system to an open system. When the joint population of the laser levels is conserved (a closed system), an individual stimulated transition simultaneously depletes one level while incrementing the other, changing the net population difference $\Delta\mathcal{N}$ by two units. In this open-system configuration, although each stimulated transition instantaneously changes $\mathcal{N}_2 - |\mathcal{N}_1$ by two units, rapid depletion of the lower laser level prevents population from accumulating in $\mathcal{N}_1$. In the limit $\mathcal{N}_1 \simeq 0$, the slowly varying inversion is therefore $\Delta\mathcal{N} =\simeq \mathcal{N}_2$, and the effective stimulated depletion term becomes $W\Delta\mathcal{N}$ rather than $2W\Delta\mathcal{N}$.

Consider the limiting regime where the lower-level decay rate γ_1 is sufficiently rapid to prevent any population buildup, $\mathcal{N}_1 \approx 0$ at all times. Under this condition, an individual stimulated emission event alters the population difference by exactly one unit ($\Delta\mathcal{N} \approx \mathcal{N}_2$). This changes the effective stimulated rate coefficient from $2W$ to W. In this limit, the coupled equations (7.39a) and (7.39c) yield a simplified, uncoupled differential equation for the population inversion:

$$\frac{d\Delta\mathcal{N}}{dt} = -\gamma_{21\mathrm{eff}}(\Delta\mathcal{N} - \Delta\mathcal{N}^{(0)}) - \gamma_{21\mathrm{eff}}\left(\frac{I}{I_{\mathrm{sat}}}\right)\Delta\mathcal{N}, \tag{7.43}$$

with the population recovery rate, saturation intensity, and unsaturated inversion given by:

$$\gamma_{21\mathrm{eff}} \approx \gamma_2 = \frac{1}{\tau_2}, \quad I_{\mathrm{sat}} \approx \frac{h\nu}{\sigma}\gamma_2, \quad \Delta\mathcal{N}^{(0)} \approx \mathcal{N}_2^{(0)} = \frac{R_2}{\gamma_2}. \tag{7.44}$$

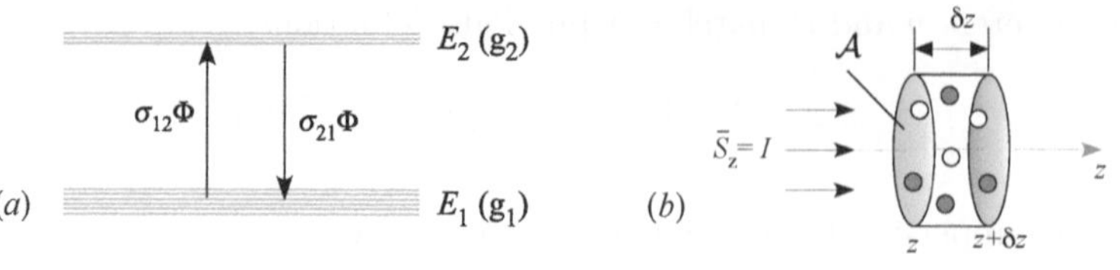

FIGURE 7.11
(a) Allowed transition pathways between degenerate atomic levels driven by a resonant signal field with photon flux density Φ. (b) Localized power exchange between the active atoms and an electromagnetic wave within a thin volumetric element $\mathcal{A}\delta z$.

Note that the approximation $\gamma_{21\text{eff}} \approx \gamma_2$ requires $\gamma_{20} \ll \gamma_1$. Equations (7.37) and (7.43) provide a simple and effective framework for understanding the nonlinear gain saturation and transient behavior across a broad class of laser media. The contributions of auxiliary states and external pump are absorbed into the pump rates R_i. This reduction yields a single, unified differential equation for the population inversion, where $\Delta\mathcal{N}^{(0)}$ represents the steady-state unsaturated inversion, $\gamma_{21\text{eff}}^{-1}$ defines the characteristic inversion recovery time constant, and the stimulated transition rate takes the limiting forms proportional to W or $2W$, depending on the population dynamics of the lower laser level.

Up to this point, our derivations have assumed non-degenerate laser states. If the upper and lower energy levels are degenerate, the atomic response at the driving field frequency represents a sum over all participating sublevels. While incorporating degeneracy modifies the quantitative expression for the unsaturated inversion, it leaves the structure of the dynamic rate equations and the behavior of gain saturation unchanged. These modifications are presented in the next section for completeness.

7.5 Laser Transitions Between Degenerate Levels

The rate equation formalism developed in the preceding sections can be generalized to atomic transitions where the upper and lower energy levels are degenerate by appropriately modifying the definition of the transition cross-section. This modification allows us to employ the same population equations to study how an optical field undergoes absorption or amplification as it propagates through a degenerate medium. Consider a laser transition coupling two degenerate energy levels, E_1 and E_2. As illustrated in Fig. 7.11(a), the lower level E_1 consists of g_1 degenerate sublevels, while the upper level E_2 possesses a degeneracy of g_2 sublevels. Let $\mathcal{N}_1$ and $\mathcal{N}_2$ represent the total volumetric population densities of the lower and upper levels, respectively. Radiative transitions can occur between allowed pairs of upper and lower sublevels, although the corresponding transition strengths need not be equal. We assume that rapid intra-level relaxation mechanisms equalize the populations within each degenerate group much faster than the inter-level transition rates. Under this condition, the sublevels within each group remain equally populated, yielding the sublevel population densities:

$$\mathcal{N}_{1i} = \frac{\mathcal{N}_1}{g_1}, \qquad \mathcal{N}_{2j} = \frac{\mathcal{N}_2}{g_2}, \tag{7.45}$$

where i and j denote specific sublevels belonging to the lower and upper groups, respectively.

The total population density of the upper state changes due to stimulated transitions from any of the upper sublevels to any of the lower sublevels as well as due to relaxation from the upper sublevels into the lower-level manifold. The rate equation for the total upper-level population density $\mathcal{N}_2$ is obtained by summing over these individual pathways:

$$\frac{d\mathcal{N}_2}{dt} = -\sum_{i=1}^{g_1}\sum_{j=1}^{g_2}\left[\left(\sigma_{2j1i}\mathcal{N}_{2j} - \sigma_{1i2j}\mathcal{N}_{1i}\right)\Phi + \gamma_{2j1i}\mathcal{N}_{2j}\right], \tag{7.46}$$

where $\sigma_{2j1i}\Phi$ represents the stimulated emission rate from sublevel j to sublevel i, and $\sigma_{1i2j}\Phi$ represents the corresponding stimulated absorption rate from sublevel i to sublevel j. The parameter $\gamma_{2j1i} \equiv A_{2j1i} + \gamma_{2j1i,\mathrm{nr}}$ defines the total relaxation rate out of sublevel j into sublevel i, combining the spontaneous radiative rate A_{2j1i} and the nonradiative decay rate $\gamma_{2j1i,\mathrm{nr}}$. The sums extend over all allowed sublevel pairs; forbidden transitions may equivalently be included with zero transition rate. Although the individual cross-sections can vary across different sublevel pairs, the microscopic principle of detailed balance dictates that the cross-sections for upward and downward induced transitions between any specific pair of states i and j are equal ($\sigma_{2j1i} = \sigma_{1i2j}$). Substituting the sublevel populations in terms of the total level populations from Eq. (7.45) into Eq. (7.46) and factoring out the population terms independent of the summation indices yields:

$$\frac{d\mathcal{N}_2}{dt} = -\frac{\mathcal{N}_2}{g_2}\left(\sum_{i,j}\sigma_{2j1i}\right)\Phi + \frac{\mathcal{N}_1}{g_1}\left(\sum_{i,j}\sigma_{1i2j}\right)\Phi - \frac{\mathcal{N}_2}{g_2}\sum_{i,j}\left(A_{2j1i} + \gamma_{2j1i,\mathrm{nr}}\right). \tag{7.47}$$

7.5.1 Effective Cross-sections and Spontaneous Emission Rate

We can now define the effective single-atom cross-section σ_{21} for induced transitions from the upper level to the lower level, alongside the effective cross-section σ_{12} for the absorption transition between the same two levels, as:

$$\sigma_{21} = \frac{1}{g_2}\sum_{i,j}\sigma_{2j1i}, \tag{7.48a}$$

$$\sigma_{12} = \frac{1}{g_1}\sum_{i,j}\sigma_{1i2j}. \tag{7.48b}$$

Similarly, we introduce an effective total decay rate γ_{21} for population relaxation from the upper level to the lower level:

$$\gamma_{21} = \frac{1}{g_2}\sum_{i,j}\left(A_{2j1i} + \gamma_{2j1i,\mathrm{nr}}\right) \equiv A_{21} + \gamma_{21,\mathrm{nr}}, \tag{7.48c}$$

where A_{21} represents the effective spontaneous emission rate (the Einstein A-coefficient) for the $2 \to 1$ transition, and $\gamma_{21,\mathrm{nr}}$ denotes the effective nonradiative transition rate between the same two energy levels. Evaluating Eqs. (7.48a) and (7.48b) together with the equality $\sigma_{2j1i} = \sigma_{1i2j}$ reveals that the effective cross-sections are related by the degeneracy ratio:

$$g_2\sigma_{21} = g_1\sigma_{12}. \tag{7.49}$$

Thus, although the induced cross-sections are equal for each corresponding pair of sublevels, the effective cross-sections averaged over the two degenerate manifolds need not be equal when $g_1 \neq g_2$. For non-degenerate energy levels ($g_1 = g_2 = 1$), this relation simplifies to $\sigma_{21} = \sigma_{12}$, matching our earlier result.

Recalling that the cross-sections for induced transitions between the upper and lower sublevels, $\sigma_{2j1i} = (\lambda^2 A_{2j1i}/8\pi n^2)S(\nu)$, the sum yields a direct relationship between the effective cross-section and the Einstein A_{21} coefficient for the $E_2 \to E_1$ transition:

$$\sigma_{21} \equiv \frac{1}{g_2}\sum_{i,j}\sigma_{2j1i} = \frac{\lambda^2}{8\pi n^2}\left(\frac{1}{g_2}\sum_{i,j}A_{2j1i}\right)S(\nu) = \frac{\lambda^2 A_{21}}{8\pi n^2}S(\nu) = \frac{g_1}{g_2}\sigma_{12}\,. \tag{7.50}$$

Utilizing these effective parameters, the equation of motion for the upper-level population density simplifies to the concise expression:

$$\frac{d\mathcal{N}_2}{dt} = -\left(\sigma_{21}\Phi\mathcal{N}_2 - \sigma_{12}\Phi\mathcal{N}_1\right) - \gamma_{21}\mathcal{N}_2\,. \tag{7.51}$$

This equation has the same mathematical form as the upper-level population rate equation for a nondegenerate transition. A similar equation for level E_1 population can be written down. These effective cross-sections and rates allow us to treat each degenerate sublevel group as a single level. Note that this equation incorporates only the direct transitions between states E_2 and E_1; decays of level E_2 to outside levels can be incorporated by replacing γ_{21} with the total decay rate γ_2.

7.5.2 Gain and Absorption Coefficients

The gain and absorption coefficients can similarly be expressed using the effective cross-sections. Consider an incident optical beam of intensity I traversing a thin slab of thickness δz and cross-sectional area $\mathcal{A}$ containing total population densities $\mathcal{N}_1$ and $\mathcal{N}_2$ [Fig. 7.11(b)]. The total population within this slice is $\mathcal{N}_1\mathcal{A}\delta z$ for the lower level and $\mathcal{N}_2\mathcal{A}\delta z$ for the upper level. The total power extracted from the wave via lower-level absorption is $\sigma_{12}I\mathcal{N}_1\mathcal{A}\delta z$, while the power added back into the field via stimulated emission from the upper level is $\sigma_{21}I\mathcal{N}_2\mathcal{A}\delta z$. The net power absorbed by the medium within this localized segment is:

$$\delta P_{\text{absorption}} = \left(\sigma_{12}\mathcal{N}_1 - \sigma_{21}\mathcal{N}_2\right)I\mathcal{A}\delta z\,. \tag{7.52}$$

By conservation of energy, the change in beam power in traversing the slab must balance this net absorption, $(\delta I)\mathcal{A} = -\delta P_{\text{absorption}}$:

$$(\delta I)\mathcal{A} = -\left(\sigma_{12}\mathcal{N}_1 - \sigma_{21}\mathcal{N}_2\right)I\mathcal{A}\delta z\,. \tag{7.53}$$

Canceling the common area factor $\mathcal{A}$, dividing by δz, and taking the continuous limit ($\delta z \to 0$) yields the spatial differential equation for the beam intensity:

$$\frac{dI}{dz} = -\left(\sigma_{12}\mathcal{N}_1 - \sigma_{21}\mathcal{N}_2\right)I\,. \tag{7.54}$$

We can now define gain and absorption coefficients from Eq. (7.54) based on the population difference. If the level populations satisfy $\mathcal{N}_1/g_1 > \mathcal{N}_2/g_2$, we use the cross-section ratio from Eq. (7.49) to express the intensity variation as

$$\frac{dI}{dz} = -\sigma_{12}\left(\mathcal{N}_1 - \frac{g_1}{g_2}\mathcal{N}_2\right)I = -\sigma_{12}\Delta\mathcal{N}_{12}I\,. \tag{7.55}$$

Because the degenerate population difference $\Delta\mathcal{N}_{12} \equiv (\mathcal{N}_1 - \frac{g_1}{g_2}\mathcal{N}_2)$ is positive, the wave intensity decays exponentially as it propagates. This defines the absorption coefficient α_{12}:

$$\alpha_{12} = \sigma_{12}\Delta\mathcal{N}_{12} \quad \text{when} \quad \Delta\mathcal{N}_{12} = \left(\mathcal{N}_1 - \frac{g_1}{g_2}\mathcal{N}_2\right) > 0\,. \tag{7.56}$$

Conversely, if the medium is inverted such that $\mathcal{N}_2/g_2 > \mathcal{N}_1/g_1$, we cast the intensity equation in terms of the downward cross-section:

$$\frac{dI}{dz} = \sigma_{21}\left(\mathcal{N}_2 - \frac{g_2}{g_1}\mathcal{N}_1\right) I = \sigma_{21}\Delta\mathcal{N}_{21} I \,. \tag{7.57}$$

Here, the value $\Delta\mathcal{N}_{21} \equiv (\mathcal{N}_2 - \frac{g_2}{g_1}\mathcal{N}_1)$ is positive, causing the field to grow exponentially with distance. This defines the optical gain coefficient g:

$$g = \sigma_{21}\left(\mathcal{N}_2 - \frac{g_2}{g_1}\mathcal{N}_1\right) \quad \text{when} \quad \Delta\mathcal{N}_{21} = \left(\mathcal{N}_2 - \frac{g_2}{g_1}\mathcal{N}_1\right) > 0 \,. \tag{7.58}$$

Note that for degenerate levels the condition for inversion is not simply $\mathcal{N}_2 > \mathcal{N}_1$ but rather

$$\frac{\mathcal{N}_2}{g_2} > \frac{\mathcal{N}_1}{g_1} \,. \tag{7.59}$$

Thus, for degenerate levels, optical gain requires the population per upper-state sublevel to exceed the population per lower-state sublevel. When the lower level is overwhelmingly populated in the absence of pumping, $\mathcal{N}_1 \gg \mathcal{N}_2$, the attenuation simplifies to the standard expression:

$$\alpha \approx \sigma_{12}\mathcal{N}_1 \,. \tag{7.60}$$

For a near-ideal four-level laser, rapid clearing of the lower level typically ensures that $\mathcal{N}_2 \gg \mathcal{N}_1$, reducing the gain coefficient to:

$$g \approx \sigma_{21}\mathcal{N}_2 \,. \tag{7.61}$$

These limiting expressions are identical in form to those obtained for nondegenerate systems, demonstrating that, under the equal-sublevel-population assumption, degeneracy can be incorporated through effective cross-sections and appropriately defined population differences.

The degeneracy factors scale such that the term $(\sigma_{12}\mathcal{N}_1 - \sigma_{21}\mathcal{N}_2)$ converts cleanly into the product $\sigma_{ij}\Delta\mathcal{N}_{ij}$, where the population difference accommodates the level populations via $\Delta\mathcal{N}_{ij} = \mathcal{N}_i - (g_i/g_j)\mathcal{N}_j$. This redefinition is particularly useful because many measurements and laser models characterize transitions in terms of effective quantities associated with entire degenerate manifolds rather than individual sublevel transitions. This uniform sublevel population model can break down under specialized conditions, such as polarization-selective optical pumping, magnetic-field-induced Zeeman splitting, or atomic orientation and alignment. Such configurations require a full density-matrix or uncoupled sublevel treatment. However, because these special cases do not alter the basic logic of gain saturation, we will proceed through the remainder of this chapter using our standard nondegenerate rate equations, knowing that degeneracy adjustments can be applied to the final results via the effective cross-sections and modified population difference.

7.6 Gain Saturation

In the preceding sections, we analyzed the atomic response to external fields in terms of polarization and the population inversion established on specific transitions. We now examine how the optical gain associated with an inverted atomic transition amplifies an incident optical signal, focusing on the mechanics of gain saturation, its spectral profile, and its operational bandwidth. In subsequent chapters, we will place this inverted gain medium inside an optical resonator to discuss the threshold conditions for laser oscillation, identify which cavity modes achieve net gain, and predict the steady-state output power.

Our initial treatment of gain saturation assumes that all active atoms exhibit an identical frequency response; that is, the transition is homogeneously broadened. In practical media, however, an atomic ensemble is subject to both homogeneous and inhomogeneous broadening mechanisms. The resulting gain saturation dynamics under these two regimes exhibit qualitatively different physical characteristics. To appreciate these differences, consider an atomic transition in a collection of active atoms. The transition undergoes homogeneous broadening driven by finite state lifetimes and phase-disrupting collisions, alongside an inhomogeneous broadening mechanism that causes the specific resonance frequency to shift from atom to atom due to local environmental variations or Doppler shifts. Crucially, each individual atom interacts with the radiation field according to its homogeneous line profile centered at its unique resonance frequency ν_a. Let $G(\nu_a)d\nu_a$ define the fraction of atoms possessing a resonance frequency within the interval $[\nu_a, \nu_a + d\nu_a]$. This subgroup of atoms interacts with a monochromatic signal field at frequency ν via its homogeneous lineshape, saturating according to the laws derived previously. The gain profile of the entire ensemble is the integral sum of these individual homogeneous line profiles $S(\nu_a, \nu)$, weighted by the probability distribution $G(\nu_a)d\nu_a$. The resulting integral for the net optical gain encountered by the signal field is expressed as:

$$g(I, \nu_o, \nu) = \int_0^\infty \sigma(\nu_a, \nu)\Delta\mathcal{N}(\nu_a, I, \nu)G(\nu_a)d\nu_a\,, \tag{7.62}$$

where $\sigma(\nu_a, \nu)$ is the transition cross-section and $\Delta\mathcal{N}(\nu_a, I, \nu)$ is the saturated population inversion density of the frequency subgroup ν_a in the presence of a laser signal with frequency ν and intensity I. The functional arguments are written explicitly here to emphasize the parameter space governing the interaction; in subsequent derivations, only the variables of immediate relevance are explicitly retained. The integral in Eq. (7.62) is similar to that encountered in the discussion of line shape functions and generally requires numerical integration. However, as in the discussion of line shape functions, we will approximate and simplify the integral to gain important physical insights.

7.6.1 Gain Saturation on a Homogeneously Broadened Line

For a purely homogeneously broadened transition, the resonance frequencies of all atoms within the ensemble are identical, collapsing the frequency distribution into a Dirac delta function: $G(\nu_a) = \delta(\nu_a - \nu_o)$. Under this condition, the integral in Eq. (7.62) reduces to $g(\nu_o, I, \nu) = \sigma(\nu_o, \nu)\Delta\mathcal{N}(\nu_o, I, \nu)$. Recalling the expression for the steady-state population inversion on an open four-level transition [Eq. (7.41)], the saturated gain coefficient can be written as:

$$g(\nu) = \frac{g_o(\nu)}{1 + I/I_{\text{sat}}(\nu)}\,, \tag{7.63}$$

where $g_o(\nu) \equiv \sigma(\nu)\Delta\mathcal{N}^{(0)}$ represents the unsaturated gain, and $I_{\text{sat}}(\nu) = h\nu\gamma_{21\text{eff}}/\sigma(\nu)$ denotes the frequency-dependent saturation intensity. Equation (7.63) demonstrates that gain saturation on a homogeneously broadened line follows the exact nonlinear intensity dependence established for the population inversion discussed in Secs. 7.3 and 7.4. To analyze how this saturation behavior varies across the line profile, we note that the frequency dependence of both the unsaturated gain and the saturation intensity is governed by the transition cross-section $\sigma(\nu)$. For a homogeneous Lorentzian line shape, the cross-section can be written as:

$$\sigma(\nu) = \frac{\sigma(\nu_o)}{1 + \left[\frac{2(\nu-\nu_o)}{\Delta\nu_H}\right]^2}, \tag{7.64}$$

where $\Delta\nu_H$ is the homogeneous full width at half maximum (FWHM) linewidth, and the peak line-center cross-section $\sigma(\nu_o)$ is given by:

$$\sigma(\nu_o) = \frac{\lambda_o^2 A_{21}}{4\pi^2 n^2 \Delta\nu_H}. \tag{7.65}$$

Using this profile, the unsaturated gain coefficient and the saturation intensity are expressed as:

$$g_o(\nu) = \frac{g_o(\nu_o)}{1 + \left[\frac{2(\nu-\nu_o)}{\Delta\nu_H}\right]^2}, \tag{7.66a}$$

$$I_{\text{sat}}(\nu) = I_s \left[1 + \left(\frac{2(\nu-\nu_o)}{\Delta\nu_H}\right)^2\right], \tag{7.66b}$$

where the peak line-center parameters are defined by:

$$g_o(\nu_o) = \sigma(\nu_o)\Delta\mathcal{N}^{(0)} = \left(\frac{\lambda_o^2 A_{21}}{4\pi^2 n^2 \Delta\nu_H}\right)\Delta\mathcal{N}^{(0)}, \tag{7.66c}$$

$$I_s \approx \frac{h\nu_o\gamma_{21\text{eff}}}{\sigma(\nu_o)} = \frac{\hbar c\gamma_{21\text{eff}} 8\pi^3 n^2 \Delta\nu_H}{\lambda_o^3 A_{21}}. \tag{7.66d}$$

In defining the line-center saturation intensity I_s, the photon energy is approximated via the line-center frequency ($h\nu \approx h\nu_o$), consistent with the near-resonance approximation. Equation (7.66b) demonstrates that the saturation intensity reaches its minimum value at the line center and increases quadratically off-resonance. Physically, this implies that for a fixed signal intensity I, it is easier to saturate the medium at the line center than in the detuned wings of the transition. Substituting the profiles from Eqs. (7.66) into the gain saturation equation (7.63) yields the explicit spectral dependence of the saturated gain:

$$g(\nu) = g_o(\nu)\left[\frac{(\nu-\nu_o)^2 + (\Delta\nu_H/2)^2}{(\nu-\nu_o)^2 + (\Delta\nu_H/2)^2(1+I/I_s)}\right], \tag{7.67}$$

$$= \frac{g_o(\nu_o)}{1+I/I_s}\left[\frac{1}{1+\left(\frac{2(\nu-\nu_o)}{\Delta\nu_b}\right)^2}\right], \tag{7.68}$$

where $\Delta\nu_b$ defines the intensity-dependent, power-broadened linewidth:

$$\Delta\nu_b = \Delta\nu_H\sqrt{1+I/I_s}. \tag{7.69}$$

This field-induced broadening of the gain bandwidth is known as power broadening. The bracketed term in Eq. (7.67) reveals that the degree of gain suppression, $g(\nu)/g_o(\nu)$, depends strongly on the signal frequency. The gain reduction is most severe at the line center, where it is suppressed by the factor $1/(1+I/I_s)$, and becomes less pronounced as the signal frequency is detuned into the wings. Because all atoms in a homogeneously broadened ensemble share the same resonance frequency, a monochromatic field interacts with the entire atomic ensemble. As the field frequency is scanned across the transition, however, the degree

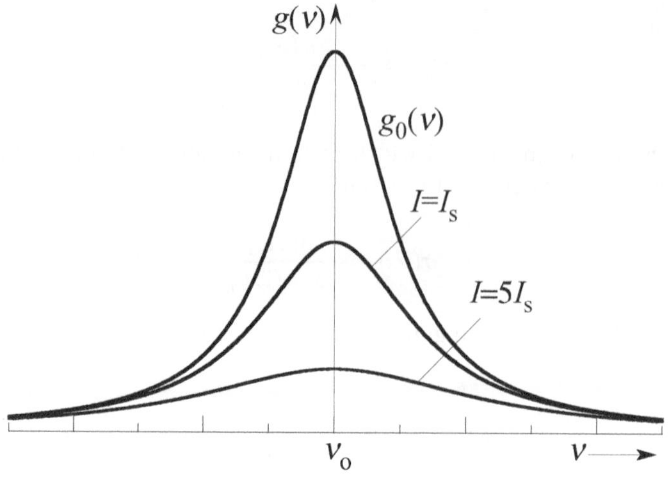

FIGURE 7.12
The saturated gain curve for a homogeneously broadened line is lower and broader than the unsaturated gain curve.

of saturation varies with detuning because the transition cross-section is frequency dependent. Consequently, the saturated gain profile is reduced most strongly near line center and becomes power broadened as described by $\Delta\nu_b$. This global gain compression is illustrated in Fig. 7.12, which plots the saturated gain profile $g(\nu)$ for several values of incident intensity I. Equation (7.68) confirms that the saturated gain retains a Lorentzian profile, whose full width at half maximum increases nonlinearly with optical power according to Eq. (7.69).

7.6.2 Gain Saturation on an Inhomogeneously Broadened Line

As an example of inhomogeneous broadening of an atomic transition, we consider the Doppler shift of the atomic resonance frequency due to the random thermal motion of atoms in a gas. The resulting distribution of the shifted resonance frequencies ν_a has a Gaussian profile centered at the rest-frame resonance frequency ν_o:

$$G(\nu_a) = \frac{1}{\Delta\nu_D}\sqrt{\frac{4\ln 2}{\pi}}\exp\left[-\left(\frac{\nu_a-\nu_o}{\Delta\nu_D}\right)^2 4\ln 2\right]. \tag{7.70}$$

Substituting Eq. (7.70) into the gain integral [Eq. (7.62)], alongside the homogeneous cross-section $\sigma(\nu_a,\nu)$ and the open four-level saturated inversion density [Eq. (7.41)], the net saturated gain coefficient yields:

$$\begin{aligned} g(\nu) &= \int_0^\infty d\nu_a G(\nu_a)\frac{\lambda_a^2 A_{21}\Delta\mathcal{N}^{(0)}}{8\pi n^2}\frac{\Delta\nu_H}{2\pi}\frac{1}{(\nu-\nu_a)^2+(\Delta\nu_H/2)^2(1+I/I_s)}, \\ &\approx \frac{\lambda_o^2 A_{21}\Delta\mathcal{N}^{(0)}\sqrt{\ln 2}}{4\pi^{3/2}n^2\Delta\nu_D}\frac{b}{\pi}\int_{-\infty}^{\infty}\frac{dx\, e^{-x^2}}{(w-x)^2+b^2(1+I/I_s)}, \end{aligned} \tag{7.71a}$$

where the dimensionless variables and parameters are defined by:

$$x = \frac{(\nu_a-\nu_o)2\sqrt{\ln 2}}{\Delta\nu_D}, \quad w = \frac{(\nu-\nu_o)2\sqrt{\ln 2}}{\Delta\nu_D}, \quad b = \frac{\Delta\nu_H\sqrt{\ln 2}}{\Delta\nu_D}. \tag{7.71b}$$

In evaluating Eq. (7.71a), the wavelength λ_a is approximated by its line-center value λ_o, consistent with the near-resonance approximation. This simplification is well justified for optical transitions because all spectral linewidths remain several orders of magnitude smaller than the absolute frequencies ν_o and ν. The integrand in Eq. (7.71a) is structured as the product of a Gaussian centered at $x = 0$ ($\nu_a = \nu_o$) and a field-dependent Lorentzian peaked at $x = w$ ($\nu_a = \nu$). This convolution matches the Voigt profile encountered in Chapter 5, with the crucial distinction that the external radiation field nonlinearly broadens the core width of the Lorentzian component.

We can extract the underlying physics of this saturation behavior without explicit numerical integration by considering two limiting cases. If the homogeneous mechanism dominates, the Lorentzian width is broad compared to the Gaussian distribution. In this regime, the Lorentzian can be factored out of the integral by evaluating it at the peak ($x = 0$) of the Gaussian. Increasing the radiation intensity broadens the homogeneous response and can therefore drive the system toward the homogeneous-broadening limit when the power-broadened homogeneous linewidth becomes comparable to or exceeds the inhomogeneous width. Under this condition, the integrated expression reduces exactly to the homogeneous gain formula [Eq. (7.67)], as expected. Conversely, if Doppler broadening dominates, the Gaussian profile is much broader than the power-broadened Lorentzian ($\Delta\nu_D \gg \Delta\nu_H\sqrt{1+I/I_s}$). Provided the signal lies within the central portion of the Doppler profile, where the Gaussian varies slowly over the power-broadened homogeneous linewidth, the Gaussian can be evaluated at the peak ($x = w$) of the narrow Lorentzian and factored out of the integration. Evaluating the remaining Lorentzian integral yields the analytical expression for gain in the strong inhomogeneous broadening limit:

$$g(\nu) = \frac{g_o(\nu)}{\sqrt{1+I/I_s}}, \tag{7.72}$$

where the unsaturated gain profile $g_o(\nu)$ has the Doppler shape:

$$g_o(\nu) = g_o(\nu_o)\exp\left[-\left(\frac{\nu-\nu_o}{\Delta\nu_D}\right)^2 4\ln 2\right], \tag{7.73a}$$

$$\text{with} \quad g_o(\nu_o) = \sqrt{\frac{\ln 2}{\pi}}\left(\frac{\lambda_o^2 A_{21}}{4\pi^2 n^2 \Delta\nu_D}\right)\Delta\mathcal{N}^{(0)}. \tag{7.73b}$$

Comparing Eq. (7.72) to the homogeneously broadened case [Eq. (7.67)], we find that the gain coefficient of an inhomogeneously broadened transition is suppressed by the square-root factor $1/\sqrt{1+I/I_s}$. This saturation factor $g(\nu)/g_o(\nu)$ is independent of the signal frequency ν, and indicates that inhomogeneous lines undergo gain suppression at a significantly slower rate than their homogeneous counterparts as a function of signal intensity. In this extreme Doppler limit, the power broadening of individual velocity groups is masked within the overall gain coefficient expression. However, this localized broadening is physically present for the specific subgroups of atoms whose thermal velocities shift their resonances close to the driving field frequency. The consequences of this localized interaction become apparent when considering multi-mode laser operation inside a resonant cavity. Once threshold conditions are met, photons spontaneously emitted into the cavity modes can circulate for many roundtrips, stimulating further emission and allowing the intracavity intensity to build. This intensity growth saturates the available gain. A critical question then arises: how does gain saturation driven by a single oscillating mode affect the available gain seen by a neighboring cavity mode? In other words, how does field intensity at one frequency perturb the gain profile at an independent frequency? It is vital to recognize that Eq. (7.72) cannot answer this question, because the variable ν in these single-field equations is the frequency of the saturating signal itself. This cross-saturation dynamic is investigated via the pump-probe formalism described in the following section.

7.7 Saturation Spectroscopy

Saturation spectroscopy relies on a strong laser field (the saturating beam) at a fixed frequency ν_L interacting with an atomic collection to modify its population inversion. This field-induced modification alters the optical gain (or loss) profile of the atomic collection, which is subsequently probed by a second, much weaker beam (the probe beam) as its frequency ν is continuously tuned. The spectral modification depends on the nature of the line broadening mechanism and the frequency detuning of the probe field relative to the saturating laser frequency $(\nu - \nu_L)$. A knowledge of this interaction provides critical insights into multi-mode laser dynamics, where the conditions for simultaneous multi-mode oscillation are governed by how gain saturation caused by one mode affects the gain available to other cavity modes.

Consider a subgroup of active atoms with resonance frequencies within the interval $[\nu_a, \nu_a + d\nu_a]$. This subgroup constitutes a fraction $G(\nu_a)d\nu_a$ of the total atomic number density. The gain coefficient $g(\nu)$ for the weak probe field at frequency ν in the presence of the intense saturating field at frequency ν_L can be expressed as:

$$g(\nu) = \frac{\lambda_o^2 A_{21}}{8\pi n^2} \int_0^\infty d\nu_a G(\nu_a) S(\nu_a, \nu) \Delta\mathcal{N}(\nu_a, \nu_L) , \tag{7.74}$$

where $\Delta\mathcal{N}(\nu_a, \nu_L)$ is the steady-state population inversion determined primarily by the strong saturating field. This is because the probe beam is assumed to be sufficiently weak $(I \ll I_L)$ that it causes a negligible population saturation and therefore can be omitted in the derivation of the inversion density. Under this condition, the saturated population inversion is:

$$\Delta\mathcal{N}(\nu_a, \nu_L) = \frac{\Delta\mathcal{N}^{(0)}}{1 + I_L / I_{\text{sat}}(\nu_a, \nu_L)} = \Delta\mathcal{N}^{(0)} \left[\frac{(\nu_a - \nu_L)^2 + (\Delta\nu_H/2)^2}{(\nu_a - \nu_L)^2 + (\Delta\nu_H/2)^2 (1 + I_L/I_s)} \right] , \tag{7.75}$$

where I_L represents the intensity of the saturating beam and the peak line-center saturation intensity I_s is defined by Eq. (7.66d). With this inversion density and the homogeneous line-shape function $S(\nu_a, \nu)$, Eq. (7.74) for the gain experienced by the probe beam can be expressed as:

$$g(\nu) = \left(\frac{\lambda_o^2 A_{21} \Delta\mathcal{N}^{(0)}}{8\pi n^2} \right) \left(\frac{2\sqrt{\ln 2}\, b}{\pi^{3/2} \Delta\nu_D} \right) \int_{-\infty}^{\infty} \frac{dx\, e^{-x^2}}{(x - w)^2 + b^2} \left[\frac{(x - w_L)^2 + b^2}{(x - w_L)^2 + b^2 (1 + I_L/I_s)} \right] , \tag{7.76}$$

where the dimensionless scaled parameters are defined by:

$$x = \frac{(\nu_a - \nu_o)}{\Delta\nu_D} 2\sqrt{\ln 2} , \quad w = \frac{(\nu - \nu_o)}{\Delta\nu_D} 2\sqrt{\ln 2} , \tag{7.77a}$$

$$w_L = \frac{(\nu_L - \nu_o)}{\Delta\nu_D} 2\sqrt{\ln 2} , \quad b = \frac{\Delta\nu_H}{\Delta\nu_D} \sqrt{\ln 2} . \tag{7.77b}$$

In writing Eq. (7.76), the lower integration limit has been extended to $-\infty$ with negligible error because the Gaussian integrand approaches zero rapidly well before the variable x reaches its lower physical limit $-\nu_o/\Delta\nu_D$.

7.7.1 Spectral Hole in an Inhomogeneous Gain Profile

In the extreme Doppler limit where the homogeneous linewidth is very narrow ($b \ll 1$) [$\Delta\nu_H \ll \Delta\nu_D$, Eq. (7.77)], the broad Gaussian distribution is essentially constant compared to the narrow Lorentzians. Because the first Lorentzian (the probe interaction) is sharply peaked at $x = w$, we can evaluate the Gaussian at $x = w$ and pull it out of the integral. The remaining integral involving the product of the two Lorentzian factors can be integrated exactly to yield

$$g(\nu) \approx g_o(\nu)\left[1 - \left(1 - \frac{1}{\Gamma}\right)\frac{\left[\frac{1}{2}(1+\Gamma)\Delta\nu_H\right]^2}{(\nu - \nu_L)^2 + \left[\frac{1}{2}(1+\Gamma)\Delta\nu_H\right]^2}\right], \tag{7.78}$$

where the constant Γ and the small-signal inhomogeneous gain coefficient are given by

$$\Gamma = \sqrt{1 + \frac{I_L}{I_s}}, \tag{7.79a}$$

$$g_o(\nu) = \sqrt{\frac{\ln 2}{\pi}}\left(\frac{\lambda_o^2 A_{21}}{4\pi^2 n^2 \Delta\nu_D}\right)\Delta\mathcal{N}^{(0)} e^{-4\ln 2\left(\frac{\nu-\nu_o}{\Delta\nu_D}\right)^2}. \tag{7.79b}$$

This immediately yields

$$\frac{g(\nu_L)}{g_o(\nu_L)} = \frac{1}{\Gamma} = \frac{1}{\sqrt{1 + I_L/I_s}}, \tag{7.80}$$

$$\Delta\nu_{\text{hole}} \approx (1+\Gamma)\Delta\nu_H = \left(1 + \sqrt{1 + I_L/I_s}\right)\Delta\nu_H. \tag{7.81}$$

If the probe beam frequency ν is far from the frequency ν_L of the saturating beam ($|\nu - \nu_L| \gg \Delta\nu_H$), the factor inside the square brackets becomes equal to unity. This means that the small-signal gain remains unchanged far away from the frequency of the strong laser. However, when the probe frequency ν is tuned close to ν_L, this factor drops below unity, reaching its smallest value of $1/(1 + I_L/I_s)$ when the probe frequency matches the saturating frequency ($\nu = \nu_L$). The gain curve therefore exhibits a sharp dip centered at $\nu = \nu_L$, as shown in Fig. 7.13. This phenomenon is called spectral hole burning. If the same strong field is used both to saturate the medium and to sample the gain as its frequency is scanned, the gain experienced by that field follows the scaled self-saturated profile of Eq. (7.72); the localized modification of the underlying gain distribution is not separately resolved. Revealing the spectral hole requires an independent weak probe.

Physics of the Spectral Hole

Despite the mathematical complexity surrounding the saturation of an inhomogeneously broadened transition, the underlying physics is easy to grasp if we recall that an inhomogeneous medium is simply a collection of atomic groups distinguished by their unique resonance frequencies. For example, a Doppler-broadened gas is a collection of atoms of different velocity groups. Each velocity group experiences a different Doppler-shifted resonance frequency, but within each group, the response is entirely homogeneous and described by a Lorentzian line shape centered at its shifted resonance frequency. The overall sample response is the sum of these individual homogeneous group responses, where each group's contribution is weighted by its probability in the frequency distribution (Fig. 7.14(a)). An intense laser field at frequency ν_L will interact strongly only with those atoms whose transition frequencies lie within a homogeneous linewidth $\Delta\nu_H$ of the laser frequency. Consequently, only this narrow spectral group of atoms undergoes significant population saturation. Atoms

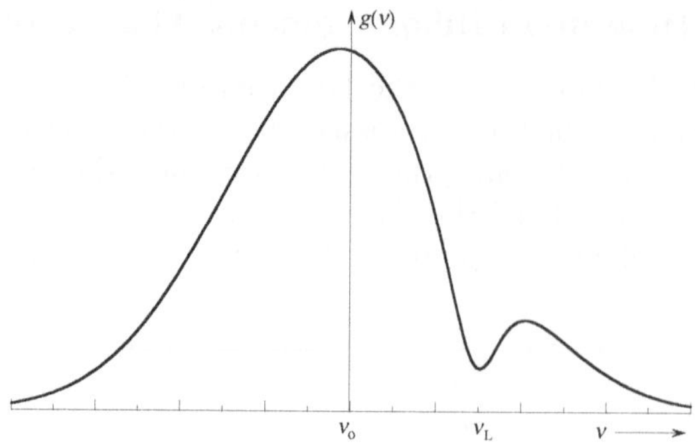

FIGURE 7.13
Gain coefficient for an inhomogeneously broadened transition as a function of probe frequency ν in the presence of a strong saturating beam of fixed frequency ν_L.

whose resonance frequencies satisfy ($|\nu_a - \nu_L| \gg \Delta\nu_H$) remain largely unaffected. As a result, the gain on an inhomogeneously broadened transition is reduced significantly only for frequencies near ν_L, while remaining undisturbed elsewhere. This localized gain saturation creates a distinct dip, or "hole," in the overall gain profile centered at the saturating signal frequency ν_L (Fig. 7.14(b)).

The gain (or loss) encountered by the saturating signal right at the center of this spectral feature saturates as $1/\sqrt{1 + I_L/I_s}$. The depth of the hole relative to the small-signal gain is given by:

$$\frac{g_o(\nu_L) - g(\nu_L)}{g_o(\nu_L)} = 1 - \frac{1}{\sqrt{1 + I_L/I_s}} \approx \begin{cases} \frac{I_L}{2I_s} & \text{for } I_L \ll I_s\,, \\ 1 - \sqrt{\frac{I_s}{I_L}} & \text{for } I_L \gg I_s\,. \end{cases} \tag{7.82}$$

Thus, the hole depth increases linearly as $I_L/2I_s$ for weak fields, but asymptotically approaches unity under intense fields when the local gain approaches complete depletion. The total spectral width of this feature can be approximated as:

$$\Delta\nu_{\text{hole}} \approx \left[1 + \sqrt{1 + I_L/I_s}\right] \Delta\nu_H\,, \tag{7.83}$$

which yields an unperturbed width of $\approx 2\Delta\nu_H$ under weak saturation, and $\approx \sqrt{I_L/I_s}\,\Delta\nu_H$ for intense saturating fields. The factor of two in the weak-saturation limit results from the overlap of the homogeneous response widths of the saturating and probe interactions.

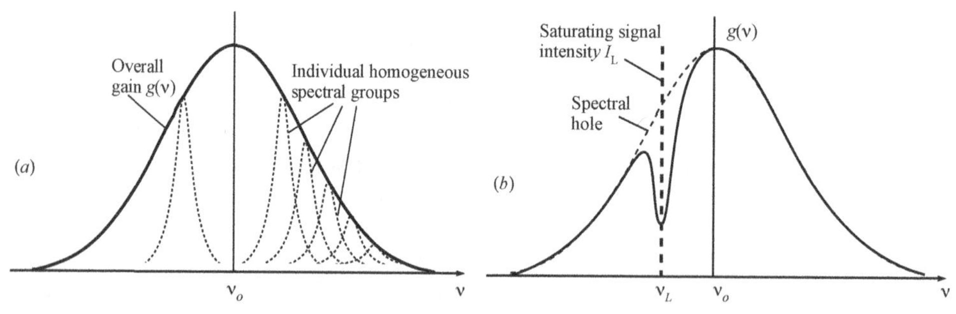

FIGURE 7.14
(a) The overall gain profile is a superposition of the responses of individual homogeneous spectral groups. (b) A strong signal depletes the gain of the specific spectral group centered at the field frequency ν_L, burning a localized "hole" into the broader gain profile.

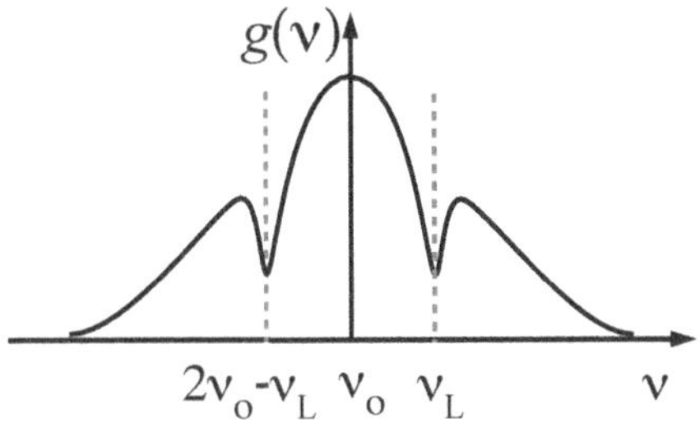

FIGURE 7.15
A weak probe beam counter-propagating to an intense saturating beam (frequency ν_L) inside a Doppler-broadened medium encounters a mirror-image spectral hole burned at the symmetric frequency $\nu = 2\nu_o - \nu_L$.

In summary, the core features of spectral hole burning in a strongly inhomogeneous line are: (i) the hole possesses a symmetric Lorentzian profile; (ii) its width is roughly twice the homogeneous linewidth at low intensities, increasing slowly as the square root of intensity under strong saturation; and (iii) the gain right at the center of the hole saturates as $1/\sqrt{1+I/I_s}$, which is slower than the $1/(1+I/I_s)$ scaling characteristic of homogeneous transitions.

An interesting situation arises when the saturating beam and the probe beam are counter-propagating. An atom that is Doppler-shifted by an amount $+\delta\nu$ relative to the forward-propagating saturating beam will experience a mirrored shift of $-\delta\nu$ relative to the counterpropagating probe beam. Because of this direction-dependent shift, the probe beam does not see the hole at the saturating frequency ν_L. To first order in the Doppler shift, the corresponding probe frequency is at a mirror-image frequency reflected across the line center: $\nu = 2\nu_o - \nu_L$ (Fig. 7.15). This cross-saturation behavior has important implications for laser oscillators operating with standing wave cavities, where fields naturally propagate in both directions simultaneously, a topic we will analyze in the next chapter.

7.7.2 Homogeneously Broadened Gain Profile

In the limit of strong homogeneous broadening where the homogeneous linewidth far exceeds the Doppler width ($\Delta\nu_H \gg \Delta\nu_D$, or $b \gg 1$), the first Lorentzian inside the probe gain integral [Eq. (7.76)] is much broader than the Gaussian distribution. The second Lorentzian is broader still due to the added effect of power broadening. Under these conditions, we can approximate the two slowly varying Lorentzians by evaluating them at the peak of the narrow Gaussian ($x = 0$) and factor them out of the integration. Evaluating the remaining Gaussian integral yields a factor of $\sqrt{\pi}$, which simplifies the expression for probe gain to:

$$\begin{aligned} g(\nu) &\approx \frac{\lambda_o^2 A_{21}\Delta\mathcal{N}^{(0)}}{8\pi n^2}\frac{2\sqrt{\ln 2}\,b}{\pi^{3/2}\Delta\nu_D}\frac{1}{w^2+b^2}\left[\frac{w_L^2+b^2}{w_L^2+b^2(1+I_L/I_s)}\right]\sqrt{\pi} \\ &= \frac{\lambda_o^2 A_{21}\Delta\mathcal{N}^{(0)}}{8\pi n^2}\left(\frac{1}{\pi}\frac{(\Delta\nu_H/2)}{(\nu-\nu_o)^2+(\Delta\nu_H/2)^2}\right)\left[\frac{(\nu_L-\nu_o)^2+(\Delta\nu_H/2)^2}{(\nu_L-\nu_o)^2+(\Delta\nu_H/2)^2(1+I_L/I_s)}\right] \\ &= g_o(\nu)\left[\frac{(\nu_L-\nu_o)^2+(\Delta\nu_H/2)^2}{(\nu_L-\nu_o)^2+(\Delta\nu_H/2)^2(1+I_L/I_s)}\right], \end{aligned} \tag{7.84}$$

where $g_o(\nu)$ represents the small-signal homogeneous gain profile, Eq. (7.66). The influence of the strong saturating field on the gain profile (as experienced by the weak probe beam) is contained within the bracketed factor of Eq. (7.84). Crucially, this factor is independent of the probe frequency ν. This independence means that the gain on a homogeneously broadened transition saturates uniformly across the entire line profile when a strong signal is applied anywhere within the atomic linewidth. As the intensity of the saturating signal increases, the magnitude of the gain decreases uniformly at the driving frequency ν_L and at all other probe frequencies ν, without any distortion to the original Lorentzian line shape or linewidth. The factor by which the gain is reduced depends on the frequency of the saturating field relative to the line center $(\nu_L - \nu_o)$. Within the spectral range over which the present approximation applies, the saturation remains uniform with respect to probe frequency; however, the required saturating intensity increases rapidly as ν_L is detuned from line center. This global, uniform reduction of gain across all probe frequencies on a homogeneous line stands in sharp contrast to our findings for a strongly Doppler-broadened line [Eq. (7.78)], where the saturation ratio $g(\nu)/g_o(\nu)$ depends on the frequency offset between the probe and saturating fields. Importantly, this analysis implicitly assumes that both the probe and saturating beam frequencies, ν and ν_L, are located near the line center relative to the total width $\Delta\nu_H$. For a Gaussian inhomogeneous distribution combined with a Lorentzian homogeneous profile, the far wings eventually become dominated by the Lorentzian contribution and therefore exhibit predominantly homogeneous behavior.

7.8 Maxwell-Bloch Equations of the Laser

Up to this point, we have analyzed optical radiation fields and atomic media as isolated systems. We are now ready to construct the coupled dynamic equations that govern an interacting light-atom system, which forms the core of laser theory. As established in Chapters 2–4, light is described by Maxwell's equations. These can be distilled into a three-dimensional wave equation for the electric field, where the macroscopic atomic polarization serves as the driving source term. In Chapter 5 and the preceding sections of this chapter, we found that the atomic response to an optical field is macroscopically characterized by its polarization and its level population densities. Although the quantized nature of atomic energy states is fundamental to calculating these responses accurately, we demonstrated that a simple electron oscillator model augmented by quantized energy levels can yield equations of motion consistent with the quantum-mechanical response required here. In a real light-matter system, the field and the atoms evolve together in both space and time; the atoms respond dynamically to the driving field, and the field simultaneously adapts to the atomic polarization. While this coupled system may initially appear complex, we can make significant progress by following our core methodology: approximate, simplify, and solve, while always validating our results for physical plausibility.

As in previous chapters, we work with complex analytic representations of the fields [see Sec. 2.1]. The physical electric field and macroscopic atomic polarization are obtained by taking the real parts of the corresponding complex fields, which we write as

$$E(\mathbf{r},t) = A(z,t)u(x,y)e^{ikz-i\omega t}\,, \tag{7.85}$$

$$P_{\text{at}}(\mathbf{r},t) = P(z,t)u(x,y)e^{ikz-i\omega t}\,, \tag{7.86}$$

where we have assumed that both the electric field and the atomic polarization are linearly polarized, allowing us to treat them as scalar quantities. Equations (7.85) and (7.86) are formulated within the paraxial approximation, taking the z-axis as the predominant direction of propagation. The variable $A(z,t)$ represents the slowly varying field amplitude, which changes in space and time much more slowly than the rapid carrier phase factor $e^{ikz-i\omega t}$.

The function $u(x, y)$ captures the transverse spatial profile of the beam due to diffraction and can represent, for example, a specific Hermite-Gauss (HG) or Laguerre-Gauss (LG) mode. Similar definitions apply to the macroscopic atomic polarization and its slowly varying polarization envelope, $P(z, t)$. We assume that the polarization is projected onto the same transverse mode as the field and that the level populations possess the corresponding transverse dependence induced by the local field intensity. We also assume that the system dynamics are dominated by a single optical frequency ω that is close to both a specific longitudinal cavity mode frequency ω_p and the resonant atomic transition frequency ω_o. Alternatively, the radiation field can be expanded into a complete set of orthogonal cavity modes (as discussed in Chapter 3), and the atomic response can be projected onto this basis. This modal projection yields a net polarization of the form shown in Eq. (7.86), where $P(z, t)$ acts as an expansion coefficient corresponding to the specific mode function $u(x, y)$. We will explore an important example of this multi-mode expansion in Chapter 10 when we analyze laser mode competition dynamics.

7.8.1 Equation of Motion for the Polarization

As established in Chapter 5, the microscopic atomic polarization is a dynamical variable that obeys Eqs. (5.48) and (5.73). We can generalize Eq. (5.73) to include all relevant homogeneous broadening and dephasing mechanisms that damp the atomic response. The collective macroscopic polarization on a laser transition connecting atomic energy levels E_1 and E_2, with a resonant transition frequency $\nu_o = (E_2 - E_1)/h$, obeys the following equation of motion:

$$\frac{dP(z,t)}{dt} + \left[i(\omega_o - \omega) + \frac{\gamma_H}{2}\right] P(z,t) = -i\epsilon_o \frac{c\lambda_o^2 A_{21}}{4\pi n\omega} \Delta\mathcal{N}(z,t) A(z,t)\,, \tag{7.87}$$

where $\Delta\mathcal{N}(z,t) \equiv \mathcal{N}_2(z,t) - \mathcal{N}_1(z,t)$ defines the population inversion density. The homogeneous linewidth parameter γ_H is related to the FWHM linewidth by $\gamma_H = 2\pi\Delta\nu_H$; correspondingly, the polarization envelope decays at the rate $\gamma_H/2$.

7.8.2 Equation of Motion for the Population Inversion

We established the rate equations for the population inversion density in the rate equation limit in Section 7.4. In general, when the macroscopic atomic polarization is treated as a dynamic variable, the governing equation of motion for the inversion density is derived by equating the volumetric optical power transfer [Eq. (5.50)] to the net power gained or lost by the atomic system on the laser transition. For a closed two-level system where the combined population of the upper and lower states remains conserved ($\mathcal{N}_1 + \mathcal{N}_2 = \mathcal{N}$), the field-induced population change is given by Eq. (5.54a). Incorporating external pumping and state relaxation terms from Eq. (7.37), the complete equation of motion for the population inversion is expressed as:

$$\frac{d\Delta\mathcal{N}}{dt} = -\gamma_{21\text{eff}}(\Delta\mathcal{N} - \Delta\mathcal{N}^{(0)}) + \frac{2i}{4\hbar}(AP^* - A^*P)\,, \tag{7.88}$$

where the effective population recovery rate $\gamma_{21\text{eff}}$ and the unsaturated population inversion $\Delta\mathcal{N}^{(0)}$ are given by Eq. (7.36). The factor of 2 in the stimulated interaction term is characteristic of a closed two-level system, corresponding directly to Case I analyzed in Section 7.4.

Most practical laser media are better modeled by the nearly ideal four-level system discussed under Case II in Section 7.4. In this regime, the lower laser level experiences negligible direct pumping ($R_1 \ll R_2$) and empties via rapid nonradiative decay ($\gamma_1 \gg$

γ_2), ensuring that it remains effectively unpopulated ($\mathcal{N}_1 \approx 0$). Under this open-system condition, the net population inversion is essentially the upper-level population density ($\Delta\mathcal{N} \equiv \mathcal{N}_2 - \mathcal{N}_1 \approx \mathcal{N}_2$), and the volumetric power gained by the atomic system is given by:

$$\frac{dw_a}{dt} = \hbar\omega\frac{d\mathcal{N}_2}{dt} \approx \hbar\omega\frac{d\Delta\mathcal{N}}{dt}\,. \tag{7.89}$$

Setting this equal to the volumetric wave power transfer dw/dt [Eq. (5.50)] and appending the pump and relaxation terms from Eq. (7.43), we find that the population inversion density obeys:

$$\frac{d\Delta\mathcal{N}}{dt} = -\gamma_{21\text{eff}}(\Delta\mathcal{N} - \Delta\mathcal{N}^{(0)}) + \frac{i}{4\hbar}(AP^* - A^*P)\,, \tag{7.90}$$

where the effective population recovery rate $\gamma_{2\text{eff}}$ and the unsaturated inversion $\Delta\mathcal{N}^{(0)}$ are given by Eq. (7.44). The dynamic inversion profiles of real lasers typically span the continuum between these two extreme modeling limits. Equation (7.90) provides an excellent approximation for a broad class of four-level and effectively four-level lasers and will be our primary tool for modeling transient inversion dynamics. Examples where the closed effective-two-level form (7.88), or a closely related quasi-three-level treatment, becomes relevant include ruby and ytterbium-doped crystals and glasses.

7.8.3 Equation of Motion for the Field

The circulating intracavity electric field E is governed by Maxwell's equations and satisfies the three-dimensional wave equation:

$$\nabla^2 E - \frac{n^2}{c^2}\frac{\partial^2 E}{\partial t^2} - \sigma\mu\frac{\partial E}{\partial t} = \mu\frac{\partial^2 P_{\text{at}}}{\partial t^2}\,, \tag{2.30*}$$

where distributed internal cavity losses are introduced via a fictitious medium conductivity σ (not to be confused with the transition cross-section). Substituting the slowly varying complex envelopes for the electric field and the atomic polarization from Eqs. (7.85) and (7.86) into Eq. (2.30), the field envelope amplitude $A(z,t)$ reduces to the following paraxial evolution equation:

$$\frac{n}{c}\frac{\partial A}{\partial t} + \frac{\partial A}{\partial z} + \frac{\alpha}{2}A = i\frac{\omega}{2\epsilon_o nc}P\,, \tag{7.91}$$

where $\alpha \equiv \sigma/\epsilon_o nc$ is the linear absorption coefficient representing the distributed cavity losses. Multiplying Eq. (7.91) by its complex conjugate and simplifying, we arrive at the corresponding transport equation for the field intensity envelope, $I(z,t) \equiv \frac{1}{2}\epsilon_o nc|A(z,t)|^2$:

$$\frac{n}{c}\frac{\partial I}{\partial t} + \frac{\partial I}{\partial z} + \alpha I = i\frac{\omega}{4}(PA^* - P^*A)\,. \tag{7.92}$$

It is important to remember that the complete spatial intensity distribution of the beam, accounting for diffraction, is given by $I(\mathbf{r},t) = \frac{1}{2}\epsilon_o nc|E|^2 = \frac{1}{2}\epsilon_o nc|A(z,t)|^2|u(x,y)|^2$. Carefully maintaining these definitions is essential for the consistency of the reduced-dimensional models.

Equations (7.87), (7.90), and (7.91) constitute the Maxwell-Bloch equations; Eq. (7.92) is the corresponding field-intensity equation. Together, they form a self-consistent semiclassical framework that models the interaction between electromagnetic fields and quantized atomic transitions. The laser is characterized as a dynamical system governed by three coupled nonlinear differential equations describing the dynamical behavior of the field envelope, the population inversion, and the macroscopic atomic polarization:

- Equation (7.87) shows that the field drives a macroscopic polarization in the medium whenever a nonzero population difference exists between the two resonant energy states. The induced polarization is coherent with the driving field, with an amplitude and phase determined by the population difference and detuning.
- Equation (7.92) demonstrates that this induced polarization generates a secondary field component that either amplifies the original wave via stimulated emission (if $\Delta\mathcal{N} > 0$) or damps it via resonant absorption (if $\Delta\mathcal{N} < 0$).
- Equation (7.90) confirms that the phase-coherent energy exchange between the field and the macroscopic polarization modifies the level populations, governing the energy stored within the atomic medium.

Because this coupled system of equations is inherently nonlinear, it can exhibit highly complex dynamical phenomena under specific operating parameters, including periodic self-pulsing, quasi-periodic oscillations, and deterministic chaos. Under appropriate single-mode and spatially uniform approximations, the Maxwell–Bloch equations can be transformed into the Lorenz equations—the classic hydrodynamic convection model that led to the discovery of deterministic chaos. While the parameter regimes required for intrinsic chaotic instabilities are difficult to satisfy in most laser systems, chaotic dynamics can be induced by subjecting the laser system to external perturbations, such as pump modulation, optical feedback, or injected signals.

7.9 Laser Dynamic Classes

Each dynamic variable in the Maxwell-Bloch framework possesses a characteristic relaxation time constant governing its return to steady-state following a perturbation. For the intracavity field, the population inversion, and the macroscopic atomic polarization, these time scales correspond to the photon lifetime τ_c, the population recovery time T_1, and the polarization decoherence time T_2 defined by

$$\tau_c = \gamma_c^{-1}, \quad T_1 = \gamma_{21\text{eff}}^{-1}, \quad T_2 \equiv \gamma_\perp^{-1} = \frac{2}{\gamma_H} = \frac{1}{\pi\Delta\nu_H}. \tag{7.93a}$$

The quantity $\gamma_\perp$ represents the transverse relaxation rate. The transient and dynamical behavior of a laser depends explicitly on the relative magnitude of these three time scales. Based on the relationships between these relaxation rates, lasers are classified into three distinct categories.

Class-A Lasers

A laser system is categorized as a Class-A laser if the cavity lifetime is significantly longer than all other relaxation times. In this case, the characteristic relaxation rates satisfy the condition:

$$\gamma_{21\text{eff}}\,, \gamma_\perp \gg \gamma_c \quad \text{or} \quad T_1\,, T_2 \ll \tau_c\,. \tag{7.93b}$$

Physically, this means that both the polarization P and the population inversion $\Delta\mathcal{N}$ respond sufficiently rapidly to adiabatically follow the relatively slow evolution of the cavity field. This separation of time scales allows us to express P and $\Delta\mathcal{N}$ algebraically in terms of the field amplitude. Consequently, the system of three coupled differential equations reduces to a single nonlinear differential equation for the field variable alone. Representative Class-A systems include helium-neon (He-Ne), argon-ion, and organic liquid dye lasers.

Class-B Lasers

In a Class-B laser, the atomic polarization P responds quickly enough to instantaneously follow the slower variations of both the population inversion and the cavity field envelope. The polarization rapidly approaches its local quasi-steady value, meaning its time derivative can be set to zero ($dP/dt \approx 0$). The characteristic relaxation rates in this class satisfy the condition:

$$\gamma_\perp \gg \gamma_c\,, \gamma_{21\text{eff}} \quad \text{or} \quad T_2 \ll \tau_c\,, T_1\,. \tag{7.93c}$$

Many important practical lasers, including Nd:YAG and other solid-state lasers and, under typical operating conditions, several semiconductor and gas laser systems, exhibit Class-B dynamics. Because the dephasing rate γ_H is dominant, we can express the polarization envelope P purely as an algebraic function of the population inversion $\Delta\mathcal{N}$ and the field amplitude A, eliminating it from the remaining differential equations. This mathematical reduction, known as adiabatic elimination, is derived in the following section. It reduces the Maxwell-Bloch equations to the standard, highly useful coupled laser rate equations.

Class-C Lasers

A laser is classified as a Class-C laser if all three time scales are comparable:

$$\gamma_\perp \sim \gamma_{21\text{eff}} \sim \gamma_c \quad \text{or} \quad T_2 \sim T_1 \sim \tau_c\,. \tag{7.93d}$$

Under this condition each dynamic variable retains a unique, essential role in driving the system's time evolution. A classic example of a Class-C system is the ammonia maser operating in the microwave region. Because all three dynamical variables remain coupled, Class-C lasers are particularly susceptible to complex dynamical instabilities, including deterministic chaos.

Within the same underlying Maxwell–Bloch model, the three dynamic classes obey the same steady-state conditions, but their transient behaviors can differ substantially. The subsequent analysis focuses on Class-A and Class-B lasers, which encompass nearly all lasers of practical interest, to discuss their steady-state operation and transient behavior.

7.10 Rate Equations of Laser Physics

According to the classification criteria for Class-A and Class-B lasers, the macroscopic atomic polarization exhibits the fastest temporal response, relaxing quickly to a local equilibrium dictated by the instantaneous field amplitude and population inversion density. Let us recall that the cavity photon lifetime $\tau_c = \gamma_c^{-1}$ is determined entirely by passive cavity losses, whereas the population recovery time is governed by the lifetimes of the atomic levels. These atomic level lifetimes naturally affect the polarization response as well. In an isolated atom free from external perturbations, the population inversion and the polarization response times would be of the same order. In many real atomic ensembles, additional dephasing processes shorten the polarization coherence time without producing a corresponding change in the level populations. Consequently, the polarization relaxes rapidly to reach a local steady-state with the field and inversion density. To see this quantitatively, we begin with the formal solution to the polarization equation of motion [Eq. (7.87)]:

$$P(t) = -i\epsilon_o \frac{c\lambda_o^2 A_{21}}{4\pi n\omega} \int_{-\infty}^{t} \Delta\mathcal{N}(t')A(t')e^{-\left[i(\omega_o-\omega)+\gamma_H/2\right](t-t')}dt' ,$$
$$= -i\epsilon_o \frac{c\lambda_o^2 A_{21}}{4\pi n\omega} \int_{0}^{\infty} \Delta\mathcal{N}(t-\tau)A(t-\tau)e^{-\left[i(\omega_o-\omega)+\gamma_H/2\right]\tau}d\tau , \tag{7.94}$$

where we have transformed the variable of integration to a time delay $\tau = t - t' > 0$ in the second step. The final expression makes the causal nature of atomic response explicit by expressing the polarization at time t exclusively in terms of the past values of the population inversion and field amplitude. Furthermore, the presence of the exponentially decaying factor $e^{-\gamma_H\tau/2}$ in the integrand dictates that only past values extending over a time of order T_2 contribute significantly to the integration. If A and $\Delta\mathcal{N}$ vary slowly on the time scale T_2, as is the case for Class-A and Class-B lasers, we can replace $\Delta\mathcal{N}(t-\tau)$ and $A(t-\tau)$ with their instantaneous values at time t and pull them out of the integration. Evaluating the remaining exponential integral yields:

$$P(t) \approx -i\epsilon_o \frac{c\lambda_o^2 A_{21}}{4\pi n\omega} \Delta\mathcal{N}(t)A(t) \int_{0}^{\infty} e^{-\left[i(\omega_o-\omega)+\gamma_H/2\right]\tau}d\tau$$
$$= \epsilon_o \frac{c\lambda_o^2 A_{21}}{4\pi n\omega} \frac{\Delta\mathcal{N}(t)}{(\omega-\omega_o)+i\gamma_H/2} A(t) . \tag{7.95}$$

Equation (7.95) implies that on time scales longer than the dephasing time γ_H^{-1}, the polarization envelope adiabatically follows the time variations of $A(t)$ and $\Delta\mathcal{N}(t)$. Any transient contributions related to initial conditions decay exponentially and are omitted from Eq. (7.94). This mathematical reduction is defined as the adiabatic approximation. In practice, it is implemented by neglecting the time derivative dP/dt relative to the damping term $(\gamma_H/2)P$ in Eq. (7.87), which converts the differential equation for polarization into a simple algebraic equation. Setting $dP/dt = 0$ directly inside the Maxwell-Bloch equations also yields an identical algebraic expression. Substituting the adiabatic expression (7.95) into the four-level inversion equation [Eq. (7.90)], and using the definition of optical intensity $I = \frac{1}{2}\epsilon_o nc|A|^2$ together with the transition cross-section $\sigma(\nu)$, the equation of motion for the population inversion reduces to:

$$\frac{d\Delta\mathcal{N}}{dt} = -\gamma_{21\text{eff}}(\Delta\mathcal{N} - \Delta\mathcal{N}^{(0)}) - \gamma_{2\text{eff}}\left(\frac{I}{I_{\text{sat}}}\right)\Delta\mathcal{N} , \tag{7.96}$$

where $I_{\text{sat}}(\nu) = h\nu\gamma_{21\text{eff}}/\sigma(\nu)$ represents the saturation intensity. As expected, Eq. (7.96) is identical to Eq. (7.43), which we derived directly using level population rate equations. Similarly, substituting the adiabatic polarization envelope (7.95) into the field amplitude equation [Eq. (7.91)] yields the updated evolution equation:

$$\frac{n}{c}\frac{\partial A}{\partial t} + \frac{\partial A}{\partial z} = \frac{1}{2}(\sigma\Delta\mathcal{N} - \alpha)A + i\frac{(\omega-\omega_o)}{\gamma_H}\sigma\Delta\mathcal{N}A . \tag{7.97}$$

Note that the gain coefficient $g(\nu) \equiv \sigma(\nu)\Delta\mathcal{N}$ is itself time dependent because it follows the dynamical inversion density. Recalling the definition of light intensity in terms of field amplitude, $I(z,t) = \frac{1}{2}\epsilon_o nc|A(z,t)|^2$, or alternatively substituting the adiabatic polarization directly into the intensity equation [Eq. (7.92)], yields the equation for space-time evolution of the field intensity:

$$\frac{n}{c}\frac{\partial I}{\partial t} + \frac{\partial I}{\partial z} = (\sigma\Delta\mathcal{N} - \alpha)I . \tag{7.98}$$

Equations (7.96) and (7.97) [or (7.98)] show that within the adiabatic approximation, the laser is described by two coupled dynamical variables—the population inversion density and the field intensity (or amplitude). With this reduction the atomic medium can be characterized dynamically in terms of its level populations without retaining the polarization as an independent variable. This approximation is known as the rate equation approximation. The introductory sections of this chapter used this framework to model how external fields saturate atomic inversion.

Equations (7.97) and (7.98) describe a purely homogeneously broadened medium or a system where homogeneous broadening dominates its macroscopic response. If inhomogeneous broadening mechanisms are significant, the expression for gain must be averaged over the distribution of some appropriate parameter that distinguishes one homogeneous group from another as was done in Eq. (7.62) for atoms with Doppler broadening. Taking Doppler broadening as an example, and recalling that the cross-section, saturation intensity and inversion density all depend on atomic resonance frequency ν_a, the expression for gain, $\sigma(\nu)\Delta\mathcal{N}$, on the right-hand side of Eq. (7.98) is replaced by the ensemble integral

$$g(\nu) \equiv \sigma(\nu)\Delta\mathcal{N} \rightarrow g(\nu) = \int_0^\infty \sigma(\nu_a, \nu)\Delta\mathcal{N}(\nu_a, \nu, I)G(\nu_a)d\nu_a\,, \tag{7.99}$$

where $G(\nu_a)$ represents the distribution of atomic resonance frequencies, $\sigma(\nu_a, \nu)$ is the transition cross-section of the frequency group ν_a, and $\Delta\mathcal{N}(\nu_a, \nu, I)$ is its corresponding saturated population inversion. For dynamic Class-B operation, the inversion $\Delta N(\nu_a, t)$ of each homogeneous subgroup must in general be evolved from its own population rate equation. In steady state, or when the subgroup inversions can themselves be adiabatically eliminated, the algebraic saturation expressions derived earlier may be inserted into Eq. (7.99).

In the limit of strong inhomogeneous broadening and steady-state operation of the laser, this integral evaluates to the square-root saturation law of Eq. (7.72). A similar ensemble integral can be written for the imaginary dispersion term on the right-hand side of the field amplitude equation [Eq. (7.97)]. However, no simple closed-form analytical approximations exist for it, even in the strong Doppler limit. This term does not alter the intensity equation; instead, its influence manifests as an intensity-dependent phase shift that perturbs the final operating frequency of the laser, as we will explore in the next chapter.

For Class-A lasers, both the macroscopic atomic polarization and the population inversion relax at rates much faster than the cavity field decay rate. Consequently, both atomic variables can be adiabatically eliminated simultaneously, leaving the intracavity field as the sole dynamical variable. This operational regime is characteristic of many common low-pressure gas lasers and liquid organic dye lasers. Following the same adiabatic elimination procedure used for the polarization, we set $d\Delta\mathcal{N}/dt \approx 0$ in Eq. (7.96), which allows the instantaneous population inversion to be expressed algebraically as:

$$\Delta\mathcal{N}(t) \approx \frac{\Delta\mathcal{N}^{(0)}}{1 + I(t)/I_{\rm sat}}\,. \tag{7.100}$$

While Eq. (7.100) formally mirrors the steady-state formula derived in Section 7.4 [Eq. (7.41)], here it has a broader dynamical interpretation. It establishes that on time scales long compared to the atomic relaxation times ($T_1, T_2 \ll \tau_c$), the population inversion adiabatically follows the time-dependent optical intensity. Substituting this adiabatic population inversion into the field intensity equation [Eq. (7.98)] yields the final equation governing the time evolution of the field intensity in a Class-A laser:

$$\frac{n}{c}\frac{\partial I}{\partial t} + \frac{\partial I}{\partial z} = (g - \alpha)I\,. \tag{7.101}$$

For Class-A lasers, the saturated gain coefficient $g \equiv g(I)$ can be computed using the ensemble integral from Eq. (7.99) for both homogeneous and inhomogeneous transitions, yielding the analytical form of Eq. (7.63) or Eq. (7.72) at any given instant. For Class-B and Class-C lasers, the algebraic steady-state saturation formulas cannot generally be used instantaneously; the gain must instead be obtained from the dynamically evolving population variables.

7.10.1 Roundtrip-averaged Time-evolution of Cavity Field

In the discussion of time-dependent phenomena in lasers, it is highly practical to treat localized losses–such as output coupler mirror transmission and scattering–alongside the distributed internal absorption coefficient α. For high-finesse cavities where the localized gain and loss parameters per pass are relatively small, these localized losses can be successfully integrated into the equation of motion following the averaging procedure introduced in Section 4.3. By a coordinate transformation to the moving frame, defined by the variables $\tau = t - z/v$ and $\zeta = z$, the field transport equation [Eq. (7.97)] transforms into

$$\frac{\partial A(\zeta,\tau)}{\partial \zeta} = \frac{1}{2}(g-\alpha)A(\zeta,\tau) + i\frac{(\nu-\nu_o)}{\Delta\nu_H}gA(\zeta,\tau)\,. \tag{7.102}$$

Assuming that g and α are approximately uniform over the propagation segment, integration of Eq. (7.102) over a propagation distance ζ yields the spatial evolution of the field amplitude envelope:

$$A(\tau,\zeta) = A(\tau,0)e^{(g-\alpha)\zeta/2}e^{i(\nu-\nu_o)g\zeta/\Delta\nu_H} \tag{7.103}$$

Let us now apply this spatial integration to a standard linear Fabry-Pérot cavity of length L filled with an active medium of background refractive index n. The total physical distance traversed during a single roundtrip is $2L$, corresponding to a roundtrip transit time of $\tau_R = 2nL/c$. Over one complete roundtrip, the field encounters distributed gain over the active medium length L_g, distributed background loss over the physical length L_α, and localized attenuation at the mirrors and scattering centers. Consequently, the field envelope at a fixed spatial reference point z inside the cavity at time $\tau + \tau_R$ can be related to the field envelope at that same location one roundtrip earlier by:

$$\begin{aligned} A(\tau+\tau_R, z) &= \sqrt{\mathcal{R}_1\mathcal{R}_2 e^{-\mathcal{L}_{\rm loc}}}\, e^{(gL_g-\alpha L_\alpha)}e^{i(\omega-\omega_o)g2L_g/\gamma_H}A(\tau,z) \\ &= e^{(gL_g-\alpha L_\alpha-\mathcal{L}_{\rm loc}/2)+\ln\sqrt{\mathcal{R}_1\mathcal{R}_2}}e^{i(\omega-\omega_o)g2L_g/\gamma_H}A(\tau,z)\,, \end{aligned} \tag{7.104}$$

where $\mathcal{R}_1$ and $\mathcal{R}_2$ represent the power reflectances of the cavity mirrors, and $e^{-\mathcal{L}_{\rm loc}}$ accounts for localized roundtrip insertion losses due to diffraction and scattering. The imaginary exponential phase factor $e^{i(\omega-\omega_o)g2L_g/\gamma_H}$ in Eq. (7.104) combines with the primary spatial carrier phase e^{ik2L} from Eq. (7.85) to satisfy the overall cavity roundtrip resonance condition, which requires that the net roundtrip phase shift must equal an integer multiple of 2π. This self-consistent phase condition defines the operating frequency of the laser via the frequency pulling equation. For a high-Q cavity where the single-pass gain and loss coefficients are small, the remaining real amplitude multiplier on the right-hand side can be accurately approximated via a first-order Taylor expansion ($e^x \approx 1 + x$):

$$e^{(gL_g-\alpha L_\alpha-\mathcal{L}_{\rm loc}/2)+\ln\sqrt{R_1R_2}}A(\tau,z) \approx \left[1+\frac{1}{2}\left(2gL_g-\mathcal{L}_c\right)\right]A(\tau,z)\,, \tag{7.105a}$$

where

$$\mathcal{L}_c \equiv 2\alpha L_\alpha - \ln(\mathcal{R}_1\mathcal{R}_2) + \mathcal{L}_{\rm loc}\,, \tag{7.105b}$$

defines the total roundtrip cavity loss parameter. Because the gain and loss parameters are small, the absolute change in the field envelope amplitude A over the span of a single roundtrip is also small. We can therefore approximate the left-hand side of Eq. (7.104) as:

$$A(\tau + \tau_R, z) \approx A(\tau, z) + \tau_R \frac{\partial A(\tau, z)}{\partial \tau} \,. \tag{7.106}$$

Equating the first-order approximations in Eqs. (7.105a) and (7.106) allows us to isolate the temporal evolution of the field envelope. For a fixed point z inside the cavity, the partial derivative with respect to the moving frame time τ reduces to the standard laboratory time derivative. The resulting equation of motion for the slowly varying field amplitude is given by:

$$\frac{dA}{dt} + \frac{1}{2}\gamma_c A \approx \frac{1}{2}\left(\frac{2gL_g}{\tau_R}\right) A = \frac{1}{2}\left(\frac{c \cdot 2gL_g}{2nL}\right) A \,, \tag{7.107}$$

where $\gamma_c \equiv \mathcal{L}_c/\tau_R = c\mathcal{L}_c/2nL$ is the lumped cavity photon decay rate.

Equation (7.107) governs the long-term temporal evolution of the electromagnetic field at a fixed location inside a high-Q cavity over time scales long compared to a single roundtrip transit time τ_R. This formulation simplifies the complex spatial propagation dynamics by expressing the field evolution in terms of gain and loss parameters averaged over a complete cavity roundtrip.

7.11 Specific Laser Systems

This section summarizes the properties of some common laser media. The discussion in this book focuses primarily on atomic and molecular lasers. Semiconductor and free-electron lasers differ considerably from atomic and molecular lasers and, because of their importance in certain applications, are discussed in more detail than the other laser systems considered here.

7.11.1 Gas Lasers

7.11.1.1 Helium-Neon (He-Ne) Laser

The He-Ne gas mixture, typically with a He:Ne ratio of about 7:1, consists mostly of He atoms. Energetic electrons produced in an electrical discharge collide with and excite primarily the He atoms to their metastable energy levels 2^3S_1 and 2^1S_0, which cannot readily return to the ground state by spontaneous emission. The metastable He atoms transfer their excitation energy to Ne atoms through collisions, returning the He atoms to the ground state. This energy transfer is particularly efficient because excited energy levels (4s and 5s) of Ne atoms have nearly the same energies as the metastable levels of He atoms. In this way, the 4s and 5s levels of neon become populated. These serve as the upper laser levels for a number of laser transitions.

The excited Ne atoms in the upper laser levels return to lower energy levels through stimulated emission at several wavelengths, including 1.15 μm (corresponding to a 4s-to-3p transition), 3.39 μm (corresponding to a 5s-to-4p transition), and 632.8 nm (corresponding to a 5s-to-3p transition). Among these, the 632.8-nm transition is the one most commonly used, while operation at other wavelengths can be obtained with suitable cavity mirror

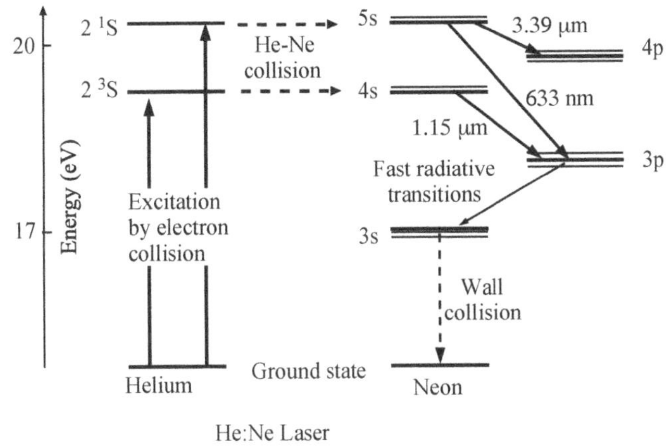

FIGURE 7.16
Simplified energy-level diagram for He-Ne laser.

coatings. The 3p levels are efficiently emptied by rapid radiative decay to the 3s levels. The 3s levels are metastable, and collisions with the walls of the discharge tube assist in returning the Ne atoms to the ground state. The small-signal gain is typically of order 1.5×10^{-3} cm^{-1}. The total gas pressure is typically 1–3 Torr, with He:Ne ratios ranging from about 5:1 to 10:1.

7.11.1.2 Argon-ion Laser

The most common ionized-gas lasers are based on noble gases such as argon and krypton. Many metals (mercury, selenium, gold, cadmium, etc.), when vaporized, can also serve as active media for ion lasers. Ion lasers produce a large number of relatively high-power laser wavelengths ranging from the ultraviolet through the visible and into the near-infrared portions of the spectrum. Ion lasers differ from neutral-atom gas lasers primarily in that the atoms must first be ionized by the electrical discharge. The ions thus created are subsequently excited to higher energy levels by further electron collisions. Lasing then occurs between excited states of the ions when a population inversion is established.

The argon-ion laser is a representative ion laser. Ar atoms are first ionized by electron collisions in a high-current electrical discharge. Further collisions with energetic electrons excite the argon ions, Ar^+, to a group of 4p energy levels about 35 eV above the ground state of the neutral Ar atom, as shown in Fig. 7.17. This pumping process creates a population inversion between the 4p levels and the 4s levels, which lie about 33.5 eV above the neutral-Ar ground state. Laser transitions from the 4p levels to the 4s levels produce a series of wavelengths ranging from about 351 nm to 529 nm. Ar^+ ions in the lower laser levels (4s) decay radiatively to lower ionic levels and eventually reach the Ar^+ ground state. Recombination with an electron then returns the ion to the neutral atomic state, completing the cycle.

The strongest argon-ion laser transitions are the green and blue lines at wavelengths of 514.5 nm and 488 nm, respectively. Multiline operation in the visible, including lines at 514.5, 501.7, 488, and 476.5 nm, as well as ultraviolet operation in the approximate range 334–364 nm, is possible. The krypton-ion laser, with multiline operation extending over approximately 520–676 nm, is another important ion-gas laser.

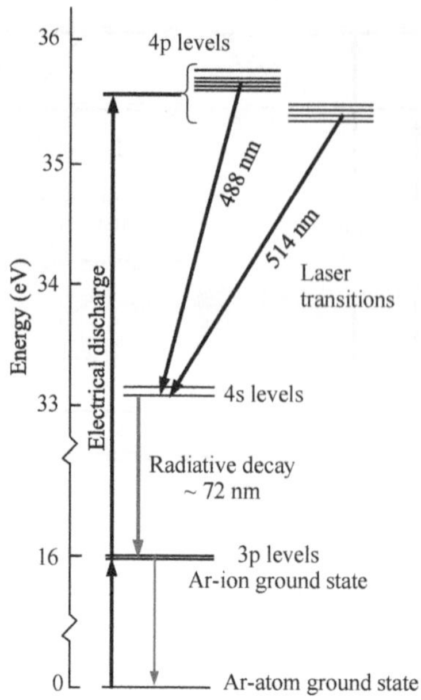

FIGURE 7.17
Schematic energy levels and laser transitions in argon-ion laser.

7.11.1.3 Carbon Dioxide Laser

One of the most important molecular lasers uses carbon dioxide gas as its active medium. The CO_2 molecule is a linear molecule and has three different vibrational modes, each having a sequence of vibrational energy levels. In addition, each vibrational level is associated with a large number of closely spaced rotational sublevels. The simplified energy-level diagram in Fig. 7.18 shows the vibrational levels relevant to the operation of the CO_2 laser. These are all vibrational levels within the ground electronic state of the CO_2 molecule.

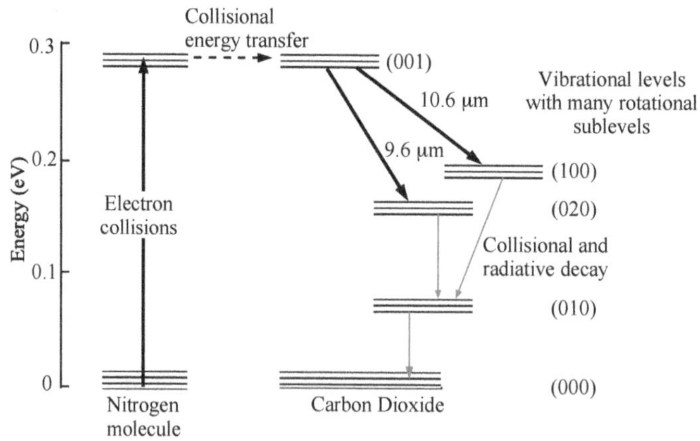

FIGURE 7.18
Energy levels relevant to the operation of a CO_2-N_2 laser. The numbers refer to the vibrational states (each with numerous rotational sublevels) of the CO_2 molecule.

The upper laser level is populated through collisions with vibrationally excited nitrogen molecules, which in turn are excited by collisions with electrons in an electrical discharge in a CO_2-N_2 gas mixture. Energy transfer from the first excited vibrational state of N_2 to the nearly resonant (001) vibrational level of the CO_2 molecule is highly efficient. This process produces population inversions between the (001) level and the lower (100) and (020) vibrational levels, giving rise to laser emission near 10.6 μm and 9.6 μm, respectively. Because these two groups of transitions share a common upper vibrational level and the gain is generally higher for the 10.6-μm transition, operation near 10.6 μm usually suppresses operation near 9.6 μm. If operation near 9.6 μm is desired, a wavelength-selective element, such as a grating, can be used in the resonator. The lower laser levels relax through collisions and radiative processes, eventually returning the molecules to the ground (000) vibrational level.

The table below summarizes useful data for some common atomic and ionic gas lasers.

7.11.1.4 Chemical Lasers

Chemical lasers emit radiation primarily in the infrared region, typically at wavelengths of about 1.3–11 μm. They use the energy released in chemical reactions to produce a population inversion. Laser action in many chemical lasers occurs between vibrational-rotational levels of molecules. One of the most prominent examples is the hydrogen-fluoride (HF) laser. In this laser, vibrationally excited HF molecules are produced through a sequence of exothermic reactions.

The reaction proceeds by first producing atomic fluorine, either by collisional dissociation of F_2 in a high-temperature chamber or by flash photolysis. The nascent F atoms react with H_2 according to

$$F + H_2 \rightarrow HF^* + H\,,$$

producing vibrationally excited HF molecules, predominantly in vibrational states $v = 1$–3, together with atomic hydrogen. The nascent H atoms then react with F_2 according to

$$H + F_2 \rightarrow HF^* + F\,,$$

producing additional vibrationally excited HF molecules and regenerating atomic fluorine. The chain reaction therefore continues, resulting in the production of a large population of vibrationally excited HF molecules. By placing the active medium in a suitable optical resonator, laser action can be obtained on a number of vibrational-rotational transitions of HF.

7.11.2 Solid-State Lasers

7.11.2.1 Titanium-Sapphire (Ti:sapphire) Laser

Solid-state lasers include optically pumped lasers in which the gain medium is a solid. Conventionally, semiconductor lasers are considered a separate group even though their

TABLE 7.1
Spectroscopic data for some common atomic and ionic gas lasers.

Laser type	He-Ne	Copper Vapor	Argon-ion	He-Cd
Laser wavelength [nm]	632.8	510.5	514.5	441.6
σ_{em} [10^{-14} cm^2]	30	9	25	9
Upper-level lifetime [ns]	150	500	6	700
Lower-level lifetime [ns]	10	$\sim 10^4$	~ 1	1
Doppler linewidth [GHz]	1.5	2.5	3.5	1
Partial pressure of gases [Torr]	7 He:1Ne	40 He:0.1-1 Cu	0.1 Ar	10 He:0.1 Cd

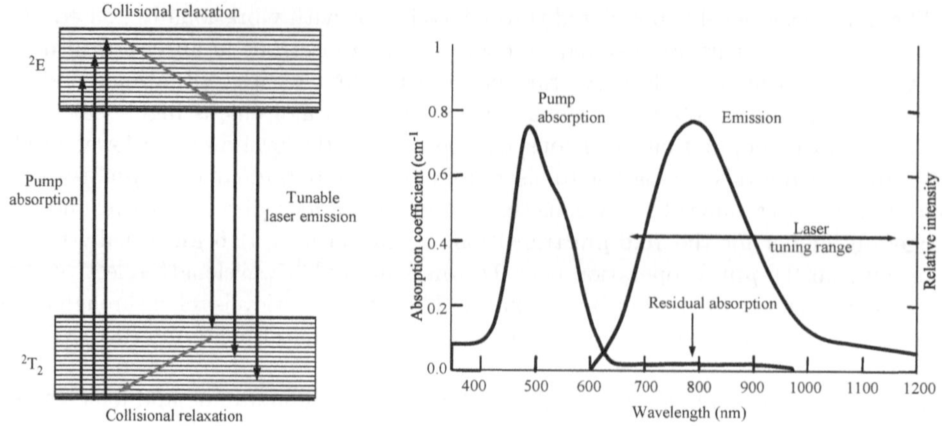

FIGURE 7.19
Schematic energy levels involved in laser operation in a titanium-sapphire laser.

gain medium is also a solid. Solid-state lasers can be divided broadly into two groups: (i) lasers that emit at one or a few discrete wavelengths and (ii) tunable lasers that can emit continuously over a broad spectral range.

The phrase "tunable solid-state laser" is nearly synonymous with the Ti:sapphire laser. Ti:sapphire is an example of a vibronic laser. The term "vibronic" reflects the strong coupling of the electronic states of the Ti^{3+} ion to vibrations of the surrounding crystal lattice. This coupling broadens the absorption and emission bands and makes laser operation possible over a wide range of wavelengths.

A simplified energy-level diagram of the Ti^{3+} ion is shown in Fig. 7.19. The electronic configuration of the free Ti^{3+} ion consists of an argon-like closed shell plus a single 3d electron. The crystal field of the sapphire lattice removes the fivefold degeneracy of the free-ion 3d level, splitting it into a lower $^2T_2$ state and an upper ^2E state. Coupling of these electronic states to lattice vibrations produces broad vibronic bands. Consequently, Ti:sapphire has a broad optical absorption band extending approximately from 380 to 620 nm and an exceptionally broad laser tuning range, approximately from 660 to 1100 nm.

The stimulated-emission cross-section is approximately 3.9×10^{-19} cm^2, the spontaneous-emission bandwidth (FWHM) is of order 1×10^{14} Hz, and the upper-state lifetime is approximately 3.2 μs. Typical values of other parameters are a small-signal gain coefficient of about 0.2 cm^{-1}, an inversion density of order 5×10^{17} cm^{-3}, a Ti^{3+} doping density of order 4.6×10^{19} cm^{-3}, and a refractive index of approximately $n = 1.76$.

7.11.2.2 Nd:YAG Lasers

Neodymium lasers and other rare-earth lasers are examples of non-tunable solid-state lasers that emit radiation at one or a few discrete wavelengths. YAG denotes yttrium aluminum garnet ($Y_3Al_5O_{12}$), which is a widely used host for many laser-active ions. The term YAG laser often refers to a laser based on neodymium-doped YAG (Nd:YAG or, more precisely, Nd^{3+}:YAG). However, YAG crystals can also be doped with other rare-earth ions, such as ytterbium, erbium, thulium, and holmium, to form other laser gain media. Yttrium ions in the YAG lattice can be partially replaced by laser-active rare-earth ions without strongly perturbing the lattice because of their similar ionic sizes.

For its principal 1064-nm transition, Nd^{3+}:YAG behaves as a four-level gain medium and can therefore provide substantial laser gain at moderate pump intensities. Its gain band-

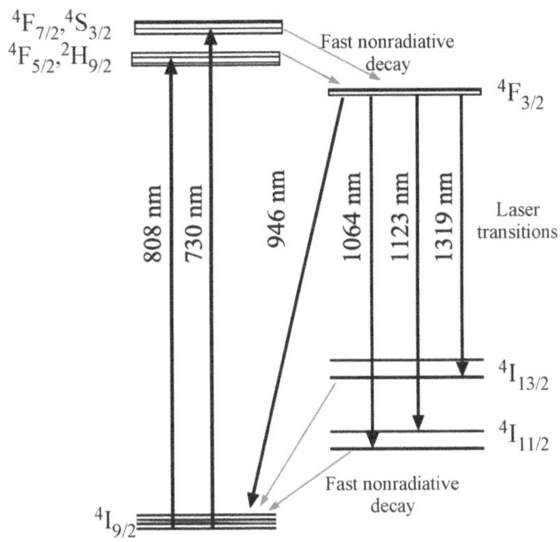

FIGURE 7.20
Pump and laser transitions in the neodymium laser.

width is relatively narrow compared with that of vibronic gain media such as Ti:sapphire. Nd:YAG lasers can be optically pumped by laser diodes or flash lamps. Efficient diode pumping is commonly achieved using strong absorption bands near 808 nm, while the relatively broad absorption bands of Nd^{3+} also permit flash-lamp pumping.

The most common Nd:YAG laser wavelength is 1064 nm. Radiation at 532, 355, and 266 nm can be generated from the 1064-nm output by second-, third-, and fourth-harmonic generation, respectively. Other laser transitions occur near 946, 1123, 1319, 1338, 1415, and 1444 nm. The 946-nm transition behaves as a quasi-three-level laser transition and therefore requires substantially higher pump intensities than the four-level 1064-nm transition.

Neodymium ions can also be incorporated into other crystalline hosts and into glasses. Important crystalline hosts include YLF ($YLiF_4$, yttrium lithium fluoride) and YVO_4 (yttrium orthovanadate). Neodymium-doped glasses are used as the active media in Nd:glass lasers.

7.11.3 Erbium-doped Fiber Lasers and Amplifiers

Erbium-doped glasses are widely used as gain media for bulk lasers, fiber lasers, and fiber amplifiers. The most common laser transition, also widely used in erbium-doped fiber amplifiers, is from the $^4I_{13/2}$ manifold to the ground-state manifold $^4I_{15/2}$ [Fig. 7.21]. Depending on the glass composition, laser emission can occur over a broad wavelength range, approximately from 1.53 μm to 1.6 μm. Because the lower laser level belongs to the ground-state manifold, this is a quasi-three-level transition. Consequently, a significant fraction of the erbium ions must be excited to overcome reabsorption at the laser wavelength. The wavelength range around 1.55 μm is particularly important in optical fiber communications because silica fibers exhibit low propagation loss in this region.

A common pumping scheme uses the $^4I_{15/2} \rightarrow {}^4I_{11/2}$ absorption band near 980 nm. In-band pumping directly into the $^4I_{13/2}$ manifold, commonly at wavelengths near 1.48 μm, is also widely used. Following excitation to the $^4I_{11/2}$ manifold, relatively rapid multiphonon relaxation transfers the ions to the $^4I_{13/2}$ manifold. The lifetime of the $^4I_{13/2}$ upper laser level is typically of order 8–10 ms, whereas the higher-lying manifolds generally have much shorter lifetimes because of rapid multiphonon relaxation.

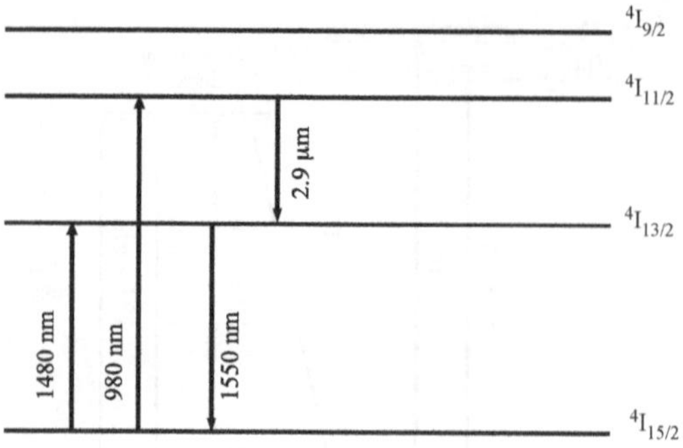

FIGURE 7.21
Energy-level structure of the trivalent erbium ion and some common optical transitions.

Pump absorption by Er^{3+} ions near 980 nm can be relatively weak because of the modest absorption cross-section and practical limits on the erbium concentration. One way to increase pump absorption is to co-dope the gain medium with ytterbium (Yb^{3+}) ions, producing an erbium-ytterbium-doped gain medium. Yb^{3+} ions have strong absorption in the approximate 900–1000-nm region and can efficiently transfer the absorbed energy to Er^{3+} ions, exciting them from the ground-state manifold into the ${}^4I_{11/2}$ manifold. The Er^{3+} ions then rapidly relax into the ${}^4I_{13/2}$ upper laser level. Erbium-ytterbium-codoped fibers are widely used in fiber lasers and amplifiers, particularly when strong pump absorption over a relatively short fiber length is desirable.

The behavior of Er^{3+} ions can be quite different in host glasses with lower phonon energies, such as fluoride glasses. The reduced multiphonon relaxation rates allow higher-lying energy levels to retain their excitation for longer times, making a variety of additional pumping and lasing schemes possible. For example, green-emitting upconversion fiber lasers and lasers emitting near 2.9 μm can be realized. Radiation near 2.9 μm is strongly absorbed by water, making lasers operating in this wavelength region useful for a variety of medical applications.

Erbium-doped crystals also serve as useful laser materials; highly doped Er:YAG can lase on the ${}^4\mathrm{I}_{11/2} \rightarrow {}^4\mathrm{I}_{13/2}$ transition, producing emission near 2.9 μm. At low erbium concentrations, this transition tends to be self-terminating because the lower laser level has a substantially longer lifetime than the upper laser level, resulting in a population bottleneck. At high erbium concentrations, however, ion-ion energy-transfer processes can substantially modify the population dynamics and alleviate this bottleneck.

Er:YAG lasers can also operate on the ${}^4I_{13/2} \rightarrow {}^4I_{15/2}$ transition, producing emission near 1.645 μm. Such lasers can, for example, be in-band pumped by an erbium-doped fiber laser operating near 1.53 μm. Erbium is also used as the active ion in other crystalline hosts, including Er:YLF, Er:YALO, tungstates, and vanadates.

7.11.4 Quasi-three-level Systems

In our discussion, we have distinguished between four-level and three-level laser gain media, which exhibit different characteristics, for example, in terms of the threshold pump intensity required for laser operation. There are also systems that exhibit intermediate behavior. In such systems the lower laser level lies only slightly above the lowest populated state [Fig. 7.22]. Despite the presence of four levels, the system does not exhibit typical four-level behavior because the lower laser level has a significant thermal population, as determined

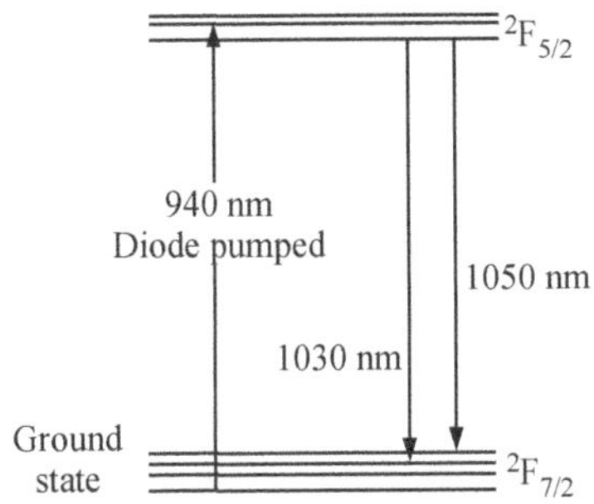

FIGURE 7.22
Energy levels of Yb^{3+} ions in Yb^{3+}:YAG, and the usual pump and laser transitions.

by the Boltzmann distribution. Consequently, the unpumped gain medium exhibits reabsorption at the laser wavelength, and transparency (zero net gain) is reached only after a finite fraction of the active ions has been excited. At still higher pump intensities, net gain is obtained, as required for laser operation or signal amplification. Such a system is referred to as a quasi-three-level gain medium.

One example of such behavior is the Yb^{3+}:YAG laser crystal, in which the laser-active Yb^{3+} ions have only two electronic manifolds, $^2F_{7/2}$ and $^2F_{5/2}$ [Fig. 7.22]. The crystal field splits each of these manifolds into a number of Stark sublevels. The commonly used laser transitions terminate on Stark sublevels of the ground-state manifold and have wavelengths only moderately longer than the pump wavelength. This small difference between the pump and laser photon energies, commonly referred to as a small quantum defect, allows high power-conversion efficiency, relatively low heat generation, and reduced thermal-lensing effects. On the other hand, the thermal population of the lower laser level causes reabsorption at the laser wavelength, so that a significant excitation density is required to achieve net gain.

Ytterbium ions can also be incorporated into glasses to form Yb-doped fiber gain media. In glasses, inhomogeneous broadening causes substantial overlap of transitions between the Stark sublevels, resulting in broad absorption and emission bands rather than well-resolved individual transitions. A beneficial consequence is a relatively broad gain bandwidth. The gain and reabsorption depend strongly on wavelength. At wavelengths near 1060 nm and longer, reabsorption is relatively weak, and the gain medium approaches four-level behavior. At shorter wavelengths, reabsorption becomes increasingly important and quasi-three-level behavior becomes more pronounced. Near 975 nm, where absorption and emission strongly overlap, the system approaches three-level behavior. Similar quasi-three-level behavior occurs in erbium-doped fiber lasers and amplifiers operating in the 1.5-μm wavelength region.

7.11.5 Liquid Dye Lasers

In liquid dye lasers, the gain medium consists of an organic dye dissolved in a liquid solvent. With laser emission covering a broad wavelength range, approximately 300–1100 nm depending on the dye, and operation in either pulsed or cw modes, dye lasers were once workhorses of scientific laboratories. Organic dyes are characterized by broad absorption and emission bands. The broad gain bandwidth can encompass many longitudinal modes of an optical resonator and, with a suitable wavelength-selective element, allows the laser wavelength to be tuned continuously over a substantial range.

The levels participating in absorption and laser emission are vibrational-rotational sublevels belonging to different electronic states of the dye molecule. A typical energy-level

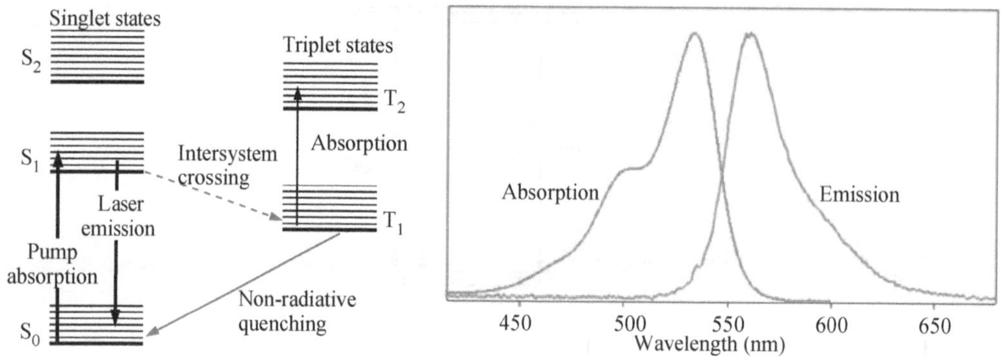

FIGURE 7.23
Energy levels of an organic dye molecule and the absorption and emission spectra of Rhodamine 6G.

diagram is shown in Fig. 7.23. Here S_0 is the singlet ground electronic state, S_1 is the first excited singlet state, and T_1 and T_2 are excited triplet states. Each electronic state contains a large number of closely spaced vibrational and rotational sublevels. Strong interactions with the solvent broaden these closely spaced transitions, producing broad absorption and emission bands.

The dye molecules are optically pumped with visible or ultraviolet radiation, exciting them from S_0 to various vibrational levels of the S_1 state. They then relax rapidly, typically in 10^{-12}–10^{-11} s, toward the lowest vibrational levels of S_1. From there, the molecules can undergo radiative transitions to various vibrational levels of S_0. Because some of the excitation energy has already been transferred to vibrational motion and subsequently to the surrounding medium before fluorescence occurs, the fluorescence spectrum is generally shifted toward longer wavelengths relative to the absorption spectrum.

From the S_1 state, molecules can also undergo intersystem crossing to the triplet state T_1. Molecules accumulated in T_1 can absorb radiation through the optically allowed $T_1 \rightarrow T_2$ transition. This triplet-state absorption may overlap the laser emission band and can therefore introduce significant optical loss. Accumulation of molecules in the relatively long-lived triplet state also removes molecules from the singlet-state laser cycle and can impede laser operation. The severity of these effects depends on the particular dye, solvent, concentration, and operating conditions. Dye lasers can be operated in either pulsed or cw modes.

7.12 Semiconductor Lasers

Semiconductor lasers are one of the most important classes of lasers. With their high electrical-to-optical power-conversion efficiency, compact size, and the ability to engineer their optical and electronic properties, semiconductor lasers have found widespread use in consumer products, telecommunications, and many other applications. They are also widely used as pumps for solid-state lasers. For this reason, and because most of our discussion in this book uses atomic and molecular lasers as examples, we devote relatively more space to semiconductor lasers here.

Semiconductor lasers (also called diode lasers) are quite different from the solid-state lasers we have encountered so far, and conventionally they are considered a separate class, even though the gain medium is a solid. In conventional solid-state lasers, such as Nd:YAG, the laser transitions occur between discrete electronic energy levels of impurity ions (in this case, triply ionized neodymium) embedded in a host (YAG) lattice. In a semiconductor laser, by contrast, the gain medium is a many-electron system, and the relevant electronic states are characteristic of the entire crystal lattice.

In a crystal lattice, the discrete energy levels of the constituent atoms broaden into energy bands containing a very large number of closely spaced states. Two bands of particular interest for semiconductor lasers are the valence band and the conduction band. At zero temperature in an intrinsic semiconductor, the valence band is the highest occupied energy band and the conduction band, lying higher in energy, is empty. The two bands are separated by an energy gap, called the bandgap and denoted by E_g, within which there are no allowed electronic states. Electrons excited into the conduction band can act as mobile charge carriers. A vacancy ("hole") in an otherwise filled valence band also behaves as a mobile positive charge carrier.

Absorption and emission of light in a semiconductor involve electronic transitions between states in the valence and conduction bands. An electron in the valence band may be excited into the conduction band by absorption of a photon having sufficient energy, leaving behind a hole in the valence band. Conversely, an electron in the conduction band may recombine with a hole in the valence band, releasing energy. For efficient radiative recombination, the semiconductor should have a direct bandgap, for which transitions between the conduction- and valence-band extrema can satisfy both energy and crystal-momentum conservation without the participation of a phonon. In an indirect-bandgap semiconductor, radiative recombination generally also requires the participation of a phonon and is therefore much less probable. Such materials are consequently inefficient light emitters and are generally unsuitable as gain media for conventional semiconductor lasers. This level of description is sufficient for gaining a basic understanding of semiconductor-laser operation.

For laser action, we need sufficiently large populations of electrons in the conduction band and holes in the valence band within the same active region. These populations can be produced by carrier injection across a forward-biased p-n junction. In an intrinsic semiconductor at sufficiently low temperature, the valence band is essentially filled and the conduction band essentially empty. Adding suitable donor impurities to an intrinsic semiconductor produces an n-type semiconductor, in which electrons are the majority charge carriers and holes are the minority carriers. Similarly, adding suitable acceptor impurities produces a p-type semiconductor, in which holes are the majority carriers and electrons are the minority carriers.

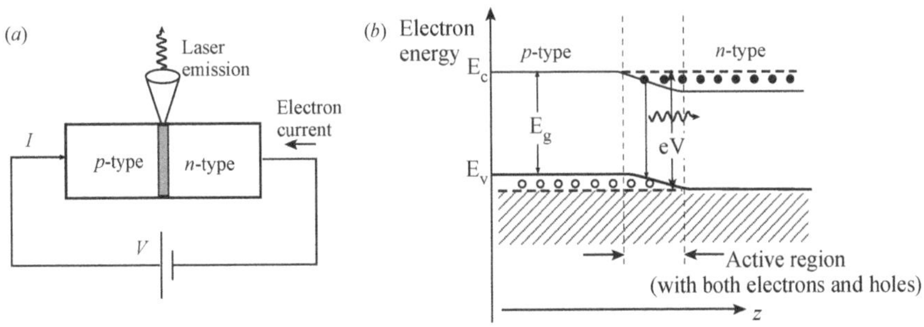

FIGURE 7.24
(a) Basic structure of a p-n junction (diode) laser with forward bias V. The p and n layers are about 1 μm thick and a few hundred microns in each transverse direction. (b) Energy-level structure (band structure) of a p-n junction laser with forward voltage V applied to the junction.

The simplest semiconductor-laser structure is a p-n junction formed between p-type and n-type semiconductor regions [Fig. 7.24(a)]. When the junction is forward biased, electrons are injected from the n-type region into the p-type region and holes are injected from the p-type region into the n-type region. In a thin active region near the junction [Fig. 7.24(b)], the injected electrons and holes coexist and can recombine radiatively. At sufficiently high injection current, carrier populations can be established for which stimulated emission produces sufficient optical gain to overcome the losses. In a semiconductor, the corresponding inversion condition is expressed in terms of the occupation probabilities of conduction- and valence-band states, or equivalently the separation of the electron and hole quasi-Fermi levels. Although semiconductor inversion and therefore gain is described using carrier occupation probabilities rather than populations of two discrete atomic levels, the same basic requirement remains: stimulated emission must exceed absorption and cavity loss.

Because of the large refractive index of semiconductor materials, substantial Fresnel reflection occurs at a semiconductor-air interface. For example, for GaAs with $n \simeq 3.6$, the normal-incidence Fresnel reflectance is approximately 32%. Thus, two cleaved parallel end faces of the semiconductor can themselves provide sufficient optical feedback to form a laser resonator. Loss mechanisms in semiconductor lasers include optical absorption, scattering, incomplete confinement of the optical field to the active region, and transmission through the end facets.

The active region is thin compared with the lateral dimensions of the device, so that the optical field is strongly confined in one transverse direction. Light propagating parallel to the junction can therefore interact with the injected carriers over a long distance, and the laser emission emerges from an edge of the structure (edge emission) [Fig. 7.24(a)]. Because the emitting aperture is much narrower in the direction perpendicular to the junction than in the direction parallel to it, the output beam is strongly divergent and astigmatic. Typical divergence angles are of the order of 10 degrees in the plane of the junction and 40 degrees perpendicular to it. Suitable optics are therefore generally employed to collimate and, when necessary, circularize the beam.

An important alternative to the edge-emitting laser is the vertical-cavity surface-emitting laser (VCSEL), in which the laser cavity is perpendicular to the plane of the semiconductor layers. The cavity mirrors are typically multilayer distributed Bragg reflectors fabricated as part of the semiconductor structure. Because the cavity and gain region are very short, highly reflecting mirrors are required. The short cavity gives a large longitudinal-mode spacing, which facilitates single-longitudinal-mode operation. The approximately symmetric transverse geometry also produces an output beam that is generally more circular than that of an edge-emitting diode laser.

The simple p-n homojunction laser considered above is useful for understanding the basic principles of semiconductor-laser operation but has largely been replaced by more advanced structures. It requires a relatively large threshold current because neither the injected carriers nor the optical field is strongly confined to a thin active region. In addition, a substantial fraction of the optical field can extend into surrounding regions, increasing optical losses.

Many of the deficiencies of the simple homojunction laser are overcome by using a double-heterostructure laser, in which a lower-bandgap semiconductor such as GaAs is sandwiched between higher-bandgap semiconductor layers such as AlGaAs [Fig. 7.25]. The lower-bandgap active material generally also has a higher refractive index than the surrounding layers. The resulting refractive-index profile forms an optical waveguide that confines the laser field primarily to the active region. This optical confinement reduces the overlap of the laser field with the surrounding p- and n-type cladding layers and thereby reduces optical losses. At the same time, the band offsets at the heterojunctions confine the injected

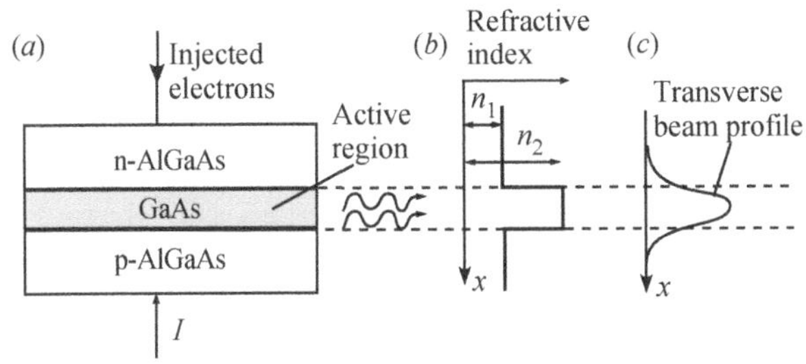

FIGURE 7.25
(a) An example of a double-heterostructure diode laser. (b) Refractive-index profile. (c) Transverse beam profile.

electrons and holes to the active layer. The combination of optical and carrier confinement greatly reduces the threshold current compared with that of a homojunction laser.

The diode lasers discussed so far operate through stimulated emission associated with electron-hole recombination between the conduction and valence bands. The approximate emission wavelength is therefore determined primarily by the bandgap energy of the active semiconductor. It can be varied to a limited extent by temperature and injection current, but much larger changes are obtained by changing the composition of semiconductor alloys, subject to constraints such as lattice matching and material quality.

A further means of engineering the electronic states and emission wavelength is provided by a quantum-well laser (QWL). By reducing the thickness of the active layer in a double heterostructure to typically a few nanometers, comparable to the de Broglie wavelength of the charge carriers, quantum-confinement effects become important. The resulting structure is referred to as a quantum-well (QW) heterostructure. Figure 7.26(a) shows an example of a quantum-well structure based on GaAs.

Confinement of the carriers in the direction perpendicular to the quantum-well layer quantizes their motion in that direction, producing discrete electron and hole subbands. The carriers remain free to move in the plane of the well. Optical transitions between the quantized conduction- and valence-band subbands determine the gain spectrum, so that the emission wavelength depends not only on the bulk bandgaps of the constituent materials but also on the thickness and composition of the quantum well. The electronic and optical properties of the laser can therefore be engineered by controlling the dimensions and composition of the semiconductor layers.

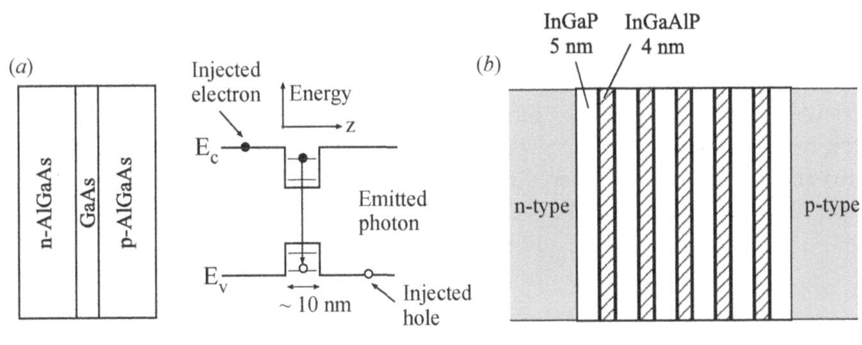

FIGURE 7.26
(a) An example of a quantum-well diode-laser structure. (b) A multiple-quantum-well structure.

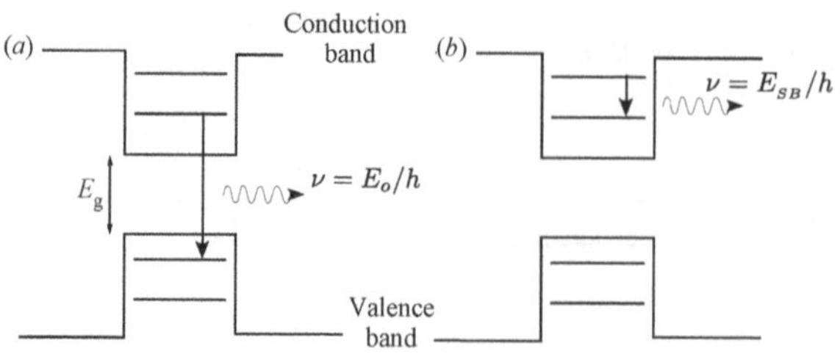

FIGURE 7.27
Comparison of (a) an interband transition in a conventional quantum-well laser and (b) an intersubband transition in a quantum-cascade laser.

Quantum-well lasers generally provide stronger carrier confinement and favorable gain characteristics compared with bulk active-layer lasers, allowing reduced threshold current and improved efficiency. A multiple-quantum-well laser contains several quantum wells separated by barrier layers [Fig. 7.26(b)]. Such structures provide additional freedom to optimize optical gain, carrier confinement, threshold current, and output power.

7.12.1 Quantum-Cascade Lasers (QCL)

A quantum-cascade laser (QCL) is a semiconductor laser that operates by a physical mechanism quite different from that of a conventional semiconductor diode laser.

In a QCL, quantum confinement is used to engineer a sequence of subbands within the conduction band. Stimulated emission occurs between these conduction-band subbands and is therefore known as an intersubband transition. In contrast to an ordinary diode laser, no electron-hole recombination is involved, and an electron remains in the conduction band after emitting a photon [Fig. 7.27]. The emission wavelength is therefore determined primarily by the engineered energy separation between the subbands rather than by the bandgap energy of the semiconductor. This design flexibility is one of the distinguishing features of QCLs.

A QCL consists of many repeated semiconductor heterostructure stages containing quantum wells and barriers. When a voltage is applied across the structure, the conduction-band profile is tilted so that electrons can be transported successively through these stages. In each active stage, an electron can undergo an intersubband transition and emit a photon before being injected into the next stage. Thus, a single electron can generate several photons as it cascades through the device, as illustrated schematically in Fig. 7.28. QCLs are therefore unipolar devices: electrons are the only charge carriers involved in the laser transition, and no *p*-*n* junction is required.

Because the intersubband energy separations can be engineered over a wide range, QCLs are particularly important sources in the mid-infrared and terahertz spectral regions. QCL operation has been demonstrated over wavelengths spanning approximately a few micrometers into the far-infrared/terahertz region, with much practical and commercial activity concentrated in the mid-infrared.

7.13 Free-electron Laser

The free-electron laser is an unusual type of laser, very different from all the lasers we have considered so far, including semiconductor lasers. There is a broad conceptual similarity

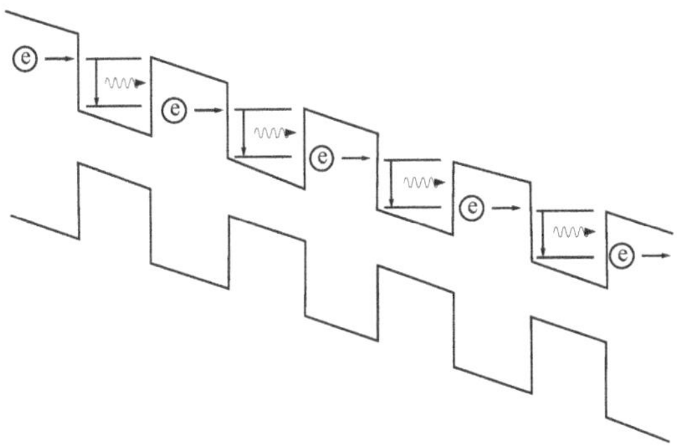

FIGURE 7.28
Schematic representation of the cascade process in a biased quantum-cascade structure, in which intersubband transitions occur in successive active stages.

to the Smith–Purcell effect, in which an energetic beam of electrons passing close to the surface of a periodic metallic grating generates electromagnetic radiation. However, the mechanism of a free-electron laser is quite different: relativistic electrons oscillate in the periodic magnetic field of an undulator, and the resulting radiation interacts collectively with the electron beam to produce coherent amplification. The term “free-electron” refers to electrons in a beam in vacuum, rather than to electrons bound within a gas, liquid, or solid as in the other lasers considered so far.

The free-electron laser is powered by a beam of relativistic ($v_z \approx c$) electrons passing through a periodic array of magnets with alternating polarity [Fig. 7.29]. The magnetic array, called an undulator, produces a periodic magnetic field transverse to the electron-beam direction, with period Λ_u. The Lorentz force causes the electrons to undergo transverse oscillations and, because they are accelerated charges, to emit electromagnetic radiation. The interaction between this radiation and the electron beam can cause the electrons to become bunched on the scale of the optical wavelength, producing coherent emission whose intensity grows rapidly along the undulator. Thus, the relativistic electron beam provides the energy source and gain mechanism of the FEL.

A simple estimate of the emitted wavelength can be obtained by considering the electromagnetic field seen in the rest frame of an electron. In this frame, the periodic magnetic structure is transformed into an oscillating electromagnetic field. Because of Lorentz contraction, the undulator period appears shortened to approximately Λ_u/γ, where $\gamma = 1/\sqrt{1 - v^2/c^2}$ is the Lorentz factor. The corresponding frequency in the electron rest

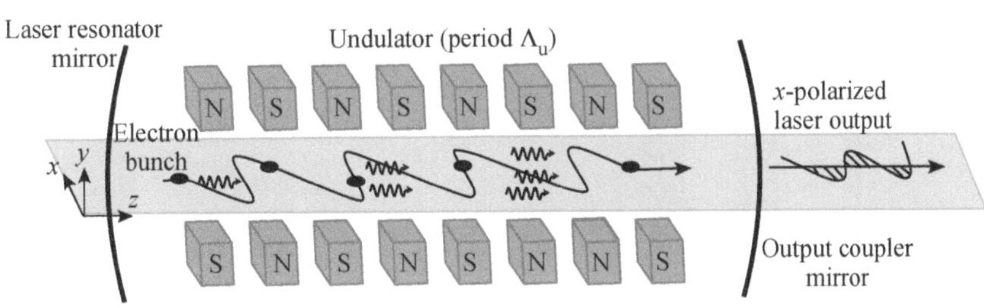

FIGURE 7.29
In a free-electron laser, high-speed electrons pass through a periodic array of magnets, causing the electrons to oscillate in a plane transverse to the direction of the beam. The accelerated electrons emit electromagnetic radiation.

frame is therefore of order $\nu' \approx \gamma c/\Lambda_u$. Radiation emitted in the forward direction is Doppler shifted by approximately another factor of 2γ when transformed back to the laboratory frame, giving

$$\nu \approx \frac{2\gamma^2 c}{\Lambda_u}\,. \tag{7.108}$$

Thus, in the limit of weak transverse electron motion, the emitted wavelength is approximately

$$\lambda \approx \frac{\Lambda_u}{2\gamma^2}\,. \tag{7.109}$$

Using the relativistic energy relation $E = \gamma mc^2$, where m is the electron rest mass, this can be written as

$$\lambda = \frac{\Lambda_u}{2}\left(\frac{mc^2}{E}\right)^2\,. \tag{7.110}$$

This result shows how an undulator with a period of centimeters can generate radiation with wavelengths many orders of magnitude shorter. For example, using $mc^2 = 0.511$ MeV for an electron [Table 1.1], a 3-GeV electron beam passing through an undulator of period $\Lambda_u = 20$ mm gives, in this approximation, a wavelength of approximately 0.3 nm.

A more complete calculation takes account of the transverse oscillatory motion of the electrons, which reduces their average longitudinal velocity. For radiation observed along the undulator axis, the resonance wavelength for a planar undulator is

$$\lambda = \frac{\Lambda_u}{2\gamma^2}\left(1+\frac{K^2}{2}\right) = \frac{\Lambda_u}{2}\left(\frac{mc^2}{E}\right)^2\left(1+\frac{K^2}{2}\right), \tag{7.111a}$$

where $$K = \frac{eB_o\Lambda_u}{2\pi mc} = 0.934\,B_o[\text{tesla}]\,\Lambda_u[\text{cm}]\,. \tag{7.111b}$$

The dimensionless parameter K, called the undulator parameter (or sometimes the wiggler parameter), characterizes the strength of the transverse electron motion and depends on the peak undulator magnetic field B_o and the undulator period Λ_u.

The wavelength of radiation from an FEL can range from the infrared to X-ray wavelengths, although no single FEL spans this entire range. The emission wavelength can be tuned by changing the electron-beam energy E and, in some systems, the undulator magnetic field and hence K. For example, for a fixed undulator period, increasing the electron energy strongly decreases the radiation wavelength because $\lambda \propto 1/\gamma^2$.

At visible, infrared, and some ultraviolet wavelengths, FELs can operate with optical resonators in which the radiation makes repeated passes through the electron beam. At X-ray wavelengths, where high-reflectivity normal-incidence cavity mirrors are generally unavailable, high-gain FELs commonly operate without a conventional optical resonator. In self-amplified spontaneous emission (SASE), spontaneous undulator radiation initiates the interaction, and the radiation is amplified during a single passage through a long undulator, producing intense, coherent ultrashort X-ray pulses.

Summary

This chapter began with a heuristic derivation of the atomic population rate equations, including population changes due to stimulated transitions, level relaxation, and pumping. This treatment provides an alternative perspective on many of the concepts introduced in Chapter 5 and offers additional physical insight into the corresponding level-population equations. These atomic rate equations provide the basis for analyzing population inversion, absorption, and amplification in laser systems.

The presence of stimulated transition rates in the population rate equations naturally leads to saturation of the population inversion, which is crucial for understanding the steady-state power output, mode competition, and saturation behavior of laser oscillators and amplifiers. These equations were used to examine how the population inversion saturates with increasing light intensity; how this saturation depends on the gain medium, including three- and four-level systems; how homogeneous and inhomogeneous broadening affect saturation; and how saturation modifies the spectral lineshape. The discussion also included power broadening in homogeneously broadened transitions and spectral hole burning in inhomogeneously broadened media. Common approximations used to describe population inversion in lasers and the limits of their validity were also discussed.

The second part of the chapter summarized the equations of motion for the atomic polarization and level population densities, together with the wave equation for the cavity field. Together, the equations for the field amplitude, atomic polarization, and population inversion constitute a system of three coupled nonlinear differential equations known as the Maxwell–Bloch equations. The relative relaxation times of these three dynamical variables provide the basis for classifying lasers into Class A, Class B, and Class C systems. These equations will be used in subsequent chapters to explore the rich dynamical behavior of lasers.

The third part of the chapter presented a brief survey of selected gain media and the operating principles of representative gas, solid-state, dye, semiconductor, and free-electron laser systems.

Problems

1. The maximum output power obtainable from a laser is related to its saturation irradiance. Estimate the saturation intensity for each of the following lasers:

   (a) a He-Ne laser with $\lambda = 632.8$ nm, peak stimulated-emission cross-section $\sigma_o = 5.8 \times 10^{-13}$ cm^2, and upper-state lifetime 30 ns;

   (b) an Ar-ion laser with $\lambda = 514.5$ nm, peak stimulated-emission cross-section $\sigma_o = 2.5 \times 10^{-13}$ cm^2, and upper-state lifetime 6 ns;

   (c) an Nd:YAG laser with $\lambda = 1.064\,\mu$m, peak stimulated-emission cross-section $\sigma_o = 2.8 \times 10^{-19}$ cm^2, and upper-state lifetime 230 μs; and

   (d) a Rhodamine 6G dye laser with $\lambda = 570$ nm, peak stimulated-emission cross-section $\sigma_o = 3.2 \times 10^{-16}$ cm^2, and upper-state lifetime 5.5 ns.

2. **Relaxing the optical approximation:** In our discussion of the two-level system, we neglected the upward transition rate $\gamma_{12}\mathcal{N}_1$. This is an excellent approximation for optical transitions when $h\nu_{21} \gg k_B T$, and we referred to it as the optical approximation.

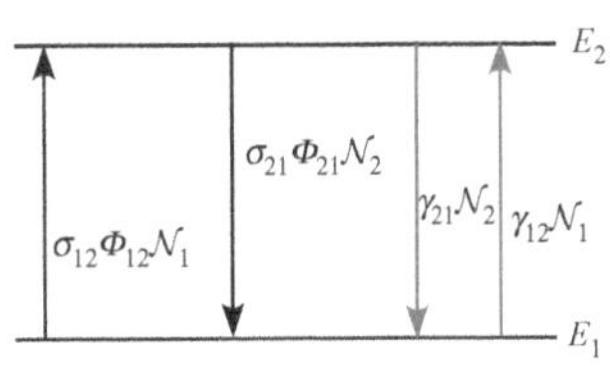

   (a) Include the upward transition term $\gamma_{12}\mathcal{N}_1$ in the population rate equations and find the steady-state population inversion $\Delta\mathcal{N} = \mathcal{N}_2 - \mathcal{N}_1$ for the modified two-level system.

(b) What must the ratio γ_{12}/γ_{21} be for the equilibrium population ratio $\mathcal{N}_2/\mathcal{N}_1$ to satisfy the Boltzmann distribution? Hence obtain an expression for the population inversion in thermal equilibrium. Assume that both levels are nondegenerate.

3. **A different kind of three-level system:** In the text, we considered a three-level system in which atoms were pumped to level 3 and population inversion was produced between levels 2 and 1. Consider instead the system shown in the figure, in which atoms are pumped directly from the ground state 0 to the upper laser level 2, and laser action occurs on the $2 \to 1$ transition. Assume that level 2 has a relatively long lifetime, while level 1 decays rapidly to the ground state. Let W_p denote the induced transition rate on the pump transition $0 \leftrightarrow 2$, and let γ_{21}, γ_{20}, and γ_{10} denote the indicated relaxation rates. Assume that upward relaxation from level 1 to level 2 is negligible.

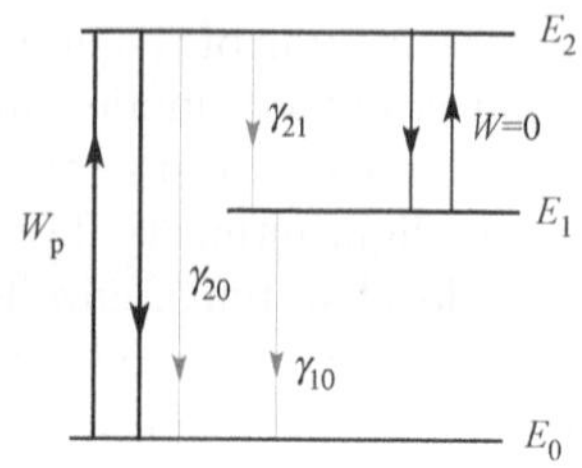

   (a) Find the steady-state population inversion $\Delta\mathcal{N}_{21} = \mathcal{N}_2 - \mathcal{N}_1$ as a function of the total atomic number density $\mathcal{N}$, the pump rate W_p, and the relaxation rates γ_{21}, γ_{20}, and γ_{10}.

   (b) Find the population difference $\Delta\mathcal{N}_{10} = \mathcal{N}_1 - \mathcal{N}_0$. Compare the pumping required to obtain inversion on the $2 \to 1$ transition with that required in the conventional three-level laser system discussed in the text.

4. **Transient gain:** Some laser transitions can exhibit population inversion during a transient interval even though no inversion is possible in the steady state. Such transitions may therefore be useful for pulsed operation but not for cw laser operation. Consider the following simplified three-level model. Atoms are pumped from level 0 into level 2 at a constant rate R_p per unit volume. Assume that level 0 remains essentially undepleted and that the only important decay processes are from level 2 to level 1 at rate γ_{21} and from level 1 to level 0 at rate γ_{10}. Assume also that stimulated transitions are negligible and that upward thermal transitions may be ignored.

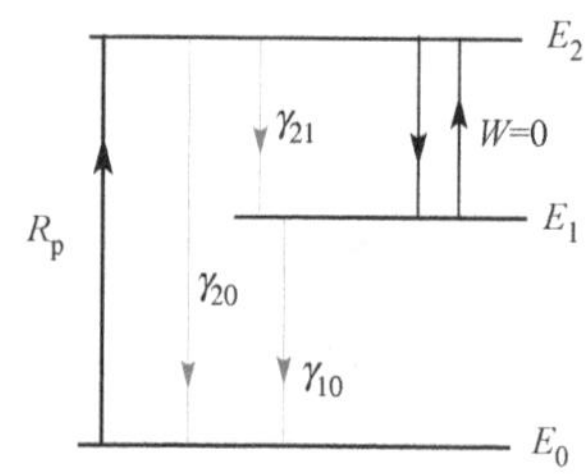

   (a) Write the rate equations for the population densities $\mathcal{N}_2$ and $\mathcal{N}_1$. Solve them in the steady state and find the steady-state population inversion $\Delta\mathcal{N}_{21} = \mathcal{N}_2 - \mathcal{N}_1$. Under what condition can $\Delta\mathcal{N}_{21} > 0$ be achieved in the steady state?

   (b) Solve the time-dependent rate equations for $\mathcal{N}_2(t)$ and $\mathcal{N}_1(t)$ assuming $\mathcal{N}_2(0) = \mathcal{N}_1(0) = 0$. Show that the inversion approaches the steady-state result obtained in part (a) as $t \to \infty$.

   (c) For $\gamma_{21} = 2\gamma_{10}$, no steady-state inversion is possible. Determine the time interval after the pump is turned on during which $\Delta\mathcal{N}_{21}(t) > 0$.

5. The $\lambda = 10.6\,\mu$m laser transition in CO_2 is homogeneously broadened with FWHM $\Delta\nu_H = 1.0$ GHz. The spontaneous-emission rate for the transition is $A_{21} = 0.35$ s^{-1}, and the lifetimes of the upper and lower laser levels are $10\,\mu$s and $0.1\,\mu$s, respectively. Assume $n \approx 1$.

   (a) Calculate the line-center stimulated-emission cross-section. Estimate the saturation irradiance, taking advantage of the fact that the lower laser level relaxes much more rapidly than the upper level.

(b) What population inversion density $\Delta\mathcal{N}_{21}$ is required to produce a line-center gain coefficient of 0.05 cm^{-1}?

6. A simple method for estimating the saturation irradiance I_{sat} of a laser transition is to measure the side fluorescence from a small volume of the gain medium. Assume that the laser transition is homogeneously broadened and that the lower laser-level population is negligible compared with the upper-level population. Let P_o denote the side-fluorescence power when laser oscillation is suppressed by blocking one of the cavity mirrors, and let P denote the side-fluorescence power when the laser is oscillating.

   Derive an expression relating P/P_o to the intracavity irradiance I and the saturation irradiance I_{sat}. If laser oscillation reduces the side fluorescence by 50% when the intracavity irradiance is 100 W/cm^2, determine I_{sat}.

7. The upper level of a laser transition is pumped from the ground state at a rate R_2 per unit volume. The total population-decay rates from level 2 to levels 0 and 1 are $\gamma_{20} = 5 \times 10^6$ s^{-1} and $\gamma_{21} = 10^7$ s^{-1}, respectively. The radiative contribution to the $2 \rightarrow 1$ decay rate is $A_{21} = 10^6$ s^{-1}. The lifetime of level 1 is 50 ns. The level degeneracies are $g_1 = 3$ and $g_2 = 1$.

   (a) Assuming a homogeneously broadened line of width $\Delta\nu_H = 2\,\text{cm}^{-1}$ and wavelength $\lambda = 400$ nm, calculate the line-center stimulated-emission cross-section.

   (b) What pump rate density R_2 is required to produce a small-signal gain coefficient of 0.003 cm^{-1} (approximately 0.3% gain per cm)? Take proper account of the level degeneracies in the population difference that determines the net gain.

8. An Ar-ion laser operating at 514.5 nm uses a cavity of length 100 cm. The upper laser level has a lifetime of 8 ns, which we assume to be dominated by spontaneous emission on the laser transition. The lower laser level relaxes approximately 25 times faster. The transition is predominantly Doppler broadened by the thermal motion of the atoms at a gas temperature of 3000 K. Take the atomic mass of argon to be 40 atomic mass units and $n \approx 1$.

   (a) Calculate the Doppler linewidth (FWHM) $\Delta\nu_D$ of the transition.

   (b) Estimate the number of electromagnetic modes within the Doppler linewidth and within a volume of 1 cm^3.

   (c) How many longitudinal HG$_{00p}$ cavity modes lie within $\pm\Delta\nu_D/2$ of line center? Include the two independent polarizations if appropriate to the mode-counting convention used in the text. Assuming that spontaneous emission is distributed among the available electromagnetic modes according to the simple mode-counting argument developed in the text, estimate the fraction emitted into these HG$_{00p}$ cavity modes.

   (d) Calculate the line-center stimulated-emission cross-section. If a population inversion density of $10^{10}\,\text{cm}^{-3}$ fills the cavity, what is the corresponding small-signal gain coefficient at line center?

9. The neon laser transition at $\lambda = 1.15\,\mu$m is predominantly Doppler broadened, with $\Delta\nu_D = 900$ MHz. The upper-state lifetime is 10^{-7} s. Assuming that this lifetime is determined entirely by spontaneous emission on the laser transition, calculate the line-center stimulated-emission cross-section. Take $n \approx 1$.

10. The R_1 laser transition of ruby has a homogeneously broadened Lorentzian lineshape with FWHM 330 GHz at room temperature. The measured line-center stimulated-emission

cross-section is $\sigma_o = 2.5 \times 10^{-20}\,\text{cm}^2$. Estimate the radiative lifetime of the transition. Take the refractive index of ruby to be $n = 1.76$.

11. **Laser amplifier:** The irradiance in a homogeneously broadened laser amplifier obeys

$$\frac{dI}{dz} = \frac{g_o I}{1 + I/I_s} - \alpha I\,,$$

where g_o is the unsaturated gain coefficient, α is the linear absorption coefficient, and I_s is the saturation irradiance.

Integrate this equation for an amplifier of length L. For $g_o = 0.05\ \text{cm}^{-1}$ and $\alpha = 0.02\ \text{cm}^{-1}$, plot the normalized output irradiance I_{out}/I_s, where $I_{\text{out}} \equiv I(L)$, as a function of $g_o L$ for the two input irradiances $I(0) = 0.1 I_s$ and $I(0) = 0.01 I_s$.

(a) Explain qualitatively how the irradiance changes with propagation distance z. Show that for $g_o > \alpha$ the irradiance approaches a finite asymptotic value as $z \to \infty$. For the parameters given above, estimate a useful amplifier length by defining it as the length at which the irradiance reaches 90% of this asymptotic value.

(b) Discuss qualitatively how the behavior would change for an inhomogeneously broadened gain medium. In particular, comment on how the different saturation behavior of the gain modifies the approach to the large-z irradiance. You may find it useful to plot $I(z)$ as a function of z.

8

Laser Threshold and Oscillations

In the preceding chapters, we analyzed the macroscopic atomic response to resonant optical signals, various pumping schemes for creating population inversion on an atomic transition, and the saturation of this population inversion as the signal power increases. We also discussed how a propagating light wave interacts with a collection of resonant atoms and developed the concepts of gain, absorption, loss, and optical feedback. By combining these fundamental elements, we arrived at a self-consistent system of coupled nonlinear differential equations governing the evolution of the population inversion, the macroscopic polarization, and the circulating cavity field–the three dynamical variables that describe the physical behavior of a laser oscillator.

In this chapter, we examine the steady-state solutions of these coupled equations to analyze the basic features of laser operation: the population inversion required to reach oscillation threshold; the steady-state laser output power and its dependence on pump power and output coupling; the differences between homogeneously and inhomogeneously broadened gain media; and frequency pulling, which shifts the laser oscillation frequencies from the corresponding passive-cavity mode frequencies. We also discuss the frequency dependence of the output power and the conditions for single-mode and multimode steady-state laser operation. In the next chapter, we will use these equations to explore the buildup of laser oscillations and analyze the dynamical transition that a laser oscillator undergoes as it crosses the oscillation threshold.

8.1 Laser Oscillation Threshold

An oscillator is an amplifier with positive feedback in which a nonzero steady-state field intensity can be continuously sustained without an input field, provided certain conditions on the material gain and cavity losses are satisfied.[1]

In a laser oscillator, amplification, or gain, is provided by stimulated emission from a population of inverted atoms or molecules in the gain medium, while optical feedback is provided by arranging a set of mirrors around the gain medium to form a closed optical path [see Fig. 8.1]. Light emitted spontaneously by the active atoms in directions close to the cavity axis can be trapped between the mirrors, making repeated passes through the gain medium. With each passage through the medium, the light is amplified by stimulated emission. At the same time, the circulating intensity is reduced by absorption and scattering within the medium (distributed losses) and by localized losses, principally transmission through the mirrors.

[1]The formal descriptive acronym for this device is Light Oscillation by Stimulated Emission of Radiation. The resulting acronym, however, would be linguistically inappropriate for a device which, by all technical benchmarks, is an outstanding winner! We therefore stick with the universally accepted term laser for a coherent light oscillator.

DOI: 10.1201/9781003584179-8

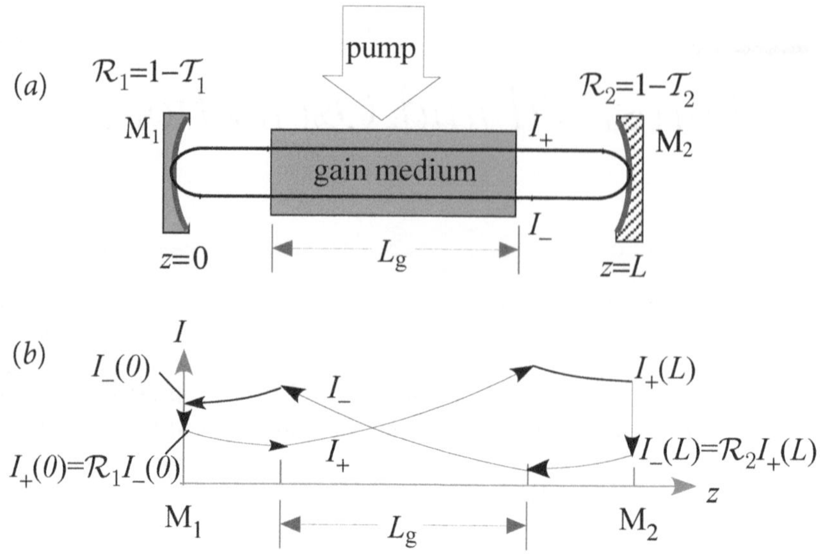

FIGURE 8.1
(a) A laser is an oscillator formed by a gain medium surrounded by mirrors to provide optical feedback. (b) Spatial variation of the counter-propagating wave intensities inside the cavity.

For a nonzero steady-state intensity to be sustained, the gain accumulated over one cavity roundtrip must exactly compensate the total roundtrip loss. At threshold, the small-signal gain is just sufficient to satisfy this condition. Below threshold, the roundtrip gain is insufficient to compensate the losses, and a self-sustained intracavity field cannot be maintained. The minimum gain required to satisfy the roundtrip gain-loss balance is called the threshold gain. An expression for the threshold gain can be derived as follows.

Consider a standing-wave laser cavity enclosing a saturable gain medium, as shown in Fig. 8.1(a), operating near its oscillation threshold in the steady state. The standing-wave inside the laser cavity can be described in terms of the intensities of a right-traveling and a left-traveling wave, each satisfying the paraxial transport equation (7.98). Near the threshold, the circulating intracavity field intensity is exceptionally small, allowing nonlinear gain saturation to be neglected ($I \ll I_{\text{sat}}$) and the material gain to be approximated by its small-signal unsaturated value g_o. Under these steady-state conditions, the spatial evolution of the counter-propagating wave intensities is governed by

$$\frac{dI_+}{dz} = (g_o - \alpha) I_+ \,, \tag{8.1a}$$

$$\frac{dI_-}{dz} = -\,(g_o - \alpha) I_- \,, \tag{8.1b}$$

where I_+ and I_- denote the intensities of the waves traveling in the $+z$ and $-z$ directions, respectively. Assuming that the gain medium fills the entire cavity, integrating these differential equations from $z = 0$ to $z = L$ for the right-traveling wave intensity, and from $z = L$ to 0 for the left-traveling wave intensity yields

$$I_+(z) = I_+(0) e^{(g_o - \alpha) z} \,, \tag{8.2a}$$

$$I_-(z) = I_-(L) e^{(g_o - \alpha)(L - z)} \,. \tag{8.2b}$$

The right-traveling wave is reflected by mirror 2 at $z = L$, transforming into the left-traveling wave. Similarly, the left-traveling wave is reflected by mirror 1 to regenerate the right-traveling wave. Consequently, the intensities of the two waves at the boundaries are related by the mirror power reflectances

$$I_+(0) = \mathcal{R}_1 I_-(0)\,, \tag{8.3a}$$
$$I_-(L) = \mathcal{R}_2 I_+(L)\,, \tag{8.3b}$$

where $\mathcal{R}_1$ and $\mathcal{R}_2$ denote the power reflectances of the end mirrors, as shown in Fig. 8.1(b). We use Eqs. (8.2) and (8.3) to trace the spatial evolution of the optical intensity through one complete roundtrip around the cavity. Figure 8.1(b) sketches the intensities of the counter-propagating waves as they traverse the cavity.

Suppose the right-traveling wave begins with an initial intensity $I_+(0)$ at $z = 0$. According to Eq. (8.2a), its intensity just before striking mirror M_2 is given by e $I_+(L) = I_+(0)e^{(g_o-\alpha)L}$. After reflection from mirror M_2, this determines the initial value of the left-traveling wave: $I_-(L) = \mathcal{R}_2 I_+(L) = \mathcal{R}_2 I_+(0)e^{(g_o-\alpha)L}$. According to Eq. (8.2b), the left-traveling wave arrives at mirror M_1 with an intensity $I_-(0) = I_-(L)e^{(g_o-\alpha)L} = \mathcal{R}_2 I_+(0)e^{(g_o-\alpha)2L}$. Finally, reflection from mirror M_1 produces the right-traveling wave for the next roundtrip, $I_{+1}(0) = \mathcal{R}_1 I_-(0) = \mathcal{R}_1\mathcal{R}_2 e^{2(g_o-\alpha)L}$. Therefore, the intensity $I_{+1}(0)$ of the right-traveling wave at $z = 0$ after completing one roundtrip is related to its initial value $I_+(0)$ by

$$I_{+1}(0) = \mathcal{R}_1\mathcal{R}_2 e^{2(g_o-\alpha)L} I_+(0)\,. \tag{8.4}$$

Such a relation can, indeed, be established for any starting point z within the cavity. If the single roundtrip multiplier satisfies $\mathcal{R}_1\mathcal{R}_2 e^{2(g_o-\alpha)L} < 1$, the wave intensity decreases with each successive roundtrip. Under this condition, no self-sustained coherent oscillation can build up, and the oscillator is said to be operating below threshold. For laser oscillations to build up from an initially small field, the small-signal gain must be sufficiently large that $\mathcal{R}_1\mathcal{R}_2 e^{2(g_o-\alpha)L} > 1$. At the oscillation threshold, the roundtrip multiplier is exactly unity, $\mathcal{R}_1\mathcal{R}2e^{2(g_{\text{th}}-\alpha)L} = 1$, or, equivalently,

$$2g_{\text{th}}L = 2\alpha L - \ln(\mathcal{R}_1\mathcal{R}_2)\,. \tag{8.5}$$

Recognizing that the quantity on the right-hand side of this equation is the cavity loss factor $\mathcal{L}_c$ defined in Eq. (4.70), the threshold condition can be written

$$2g_{\text{th}}L = \mathcal{L}_c \equiv 2\alpha L - \ln \mathcal{R}_1\mathcal{R}_2\,. \tag{8.6}$$

Thus, at threshold, the integrated unsaturated roundtrip gain exactly compensates the total roundtrip cavity loss.

If auxiliary localized loss channels are present that cause the wave to lose a fraction S of its power in each roundtrip, and if the gain and background absorption are confined to cavity segments of lengths L_g and L_a, respectively, the threshold condition becomes

$$2g_{\text{th}}L_g = \mathcal{L}_c = 2\alpha L_a - \ln[\mathcal{R}_1\mathcal{R}_2(1-S)]\,. \tag{8.7}$$

This threshold gain implies that the population inversion must likewise reach a threshold value before laser action can occur. Recalling the definition of the small-signal gain coefficient in terms of the population inversion density,

$$g_o(\nu) = \sigma(\nu)\Delta\mathcal{N}^{(o)} = \frac{\lambda_o^2 A_{21}}{8\pi n^2} S(\nu)\Delta\mathcal{N}^{(o)}\,, \tag{5.61*}$$

we find that the population inversion required to reach the oscillation threshold is

$$\Delta\mathcal{N}_{\rm th} \equiv \Delta\mathcal{N}^{(o)}_{\rm th} = \frac{\mathcal{L}_c}{2L_g}\frac{1}{\sigma(\nu)} = \frac{\mathcal{L}_c}{2L_g}\frac{8\pi n^2}{\lambda_o^2 A_{21} S(\nu)}\,, \tag{8.8}$$

and the necessary condition for laser action becomes

$$2g_o(\nu)L_g > \mathcal{L}_c \quad \Longleftrightarrow \quad \Delta\mathcal{N}^{(0)} > \frac{\mathcal{L}_c}{2L_g\sigma(\nu)}\,. \tag{8.9}$$

A wealth of fundamental physics is contained within Eqs. (8.6)–(8.9). These threshold criteria apply broadly to laser oscillators described by the present gain model, independent of the particular atomic line-broadening mechanism, provided the appropriate lineshape function $S(\nu)$ is used. Inspection of Eq. (8.8) shows that minimizing the threshold inversion requires minimizing the cavity loss factor, increasing the gain length, and choosing a transition with a large stimulated-emission cross section $\sigma(\nu)$. From the expression for $\sigma(\nu)$ above, a large cross section is favored by a large value of $\lambda_o^2 A_{21} S(\nu)$; thus, other factors being equal, narrow spectral lines, large radiative transition rates, and longer wavelengths tend to reduce the threshold inversion.

The condition for laser action (8.9) is inherently frequency dependent. If the small-signal unsaturated gain is sufficiently high above threshold, there may be an extended spectral range over which this inequality is satisfied. This raises several important questions: Which cavity mode frequencies will oscillate? What physical mechanisms limit their amplitudes? How do laser oscillations build up from noise and reach a stable steady state? We will explore these questions, together with other related phenomena, as they naturally arise throughout this chapter.

Example 8.1 Threshold Gain and Inversion in a He-Ne Laser: A Four-Level Laser

Consider a two-mirror cavity employing a high reflector ($\mathcal{R}1 = 99.8\%$) and an output coupler ($\mathcal{R}2 = 98.0\%$), with the gain medium filling the entire cavity length ($L_g = L = 50$ cm). The gain medium consists of a He-Ne gas mixture with a ratio of 7:1 and a neon partial pressure of $P_{\rm Ne} = 0.3$ Torr at an absolute temperature of $T = 400$ K. The operating wavelength of the laser is $\lambda_o = 632.8$ nm. Take the radiative decay rate to be $A21 = 6.56 \times 10^6\ s^{-1}$, and assume that the oscillating cavity mode coincides with the center of the gain line.

Under these low-density operating conditions, the gain profile is predominantly Doppler broadened, with an inhomogeneous linewidth

$$\Delta\nu_{D} = \frac{2.15\times 10^{11}}{632.8}\sqrt{\frac{400}{20}}\ \text{Hz} = 1.52\,\text{GHz}\,,$$

and $$\frac{1}{S(\nu_o)} = \Delta\nu_{D}\sqrt{\frac{\pi}{4\ln 2}} = (1.52\,\text{GHz})\times 1.064 \approx 1.62\times 10^9\,\text{Hz}\,.$$

Assuming a gas-phase refractive index of $n \approx 1$, negligible internal absorption losses $\mathcal{L}_i \approx 0$, and a total mirror transmission loss $\mathcal{L}_{T} \approx \mathcal{T}_1 + \mathcal{T}_2 = 2.2\%$, the roundtrip cavity loss factor is approximately $\mathcal{L}_c = 0.022$. Substituting these quantities into Eq. (8.8), the threshold inversion density is

$$\Delta\mathcal{N}_{\rm th} = \frac{\mathcal{L}_c}{2L_g\sigma} = \frac{0.022}{2\times 0.50}\cdot\frac{8\pi(1.62\times 10^9)}{(632.8\times 10^{-9})^2(6.56\times 10^6)}\,\text{m}^{-3} = 3.5\times 10^8\,\text{cm}^{-3}\,.$$

This seemingly large number density represents an exceptionally small fraction of the total neon atom density, which, from the Ideal Gas Law, is

$$\mathcal{N} = (9.65\times 10^{18})\,\frac{0.3}{400}\,\text{cm}^{-3} = 7.24\times 10^{15}\,\text{cm}^{-3}\,.$$

Consequently, the fractional inversion required to reach the threshold of laser oscillation is only

$$\frac{\Delta\mathcal{N}_{\text{th}}}{\mathcal{N}} \approx 4.8 \times 10^{-8}\,.$$

Thus, only about five out of every hundred million neon atoms need to contribute to the population inversion required to reach threshold.

This threshold inversion density corresponds to a threshold gain coefficient

$$g_{\text{th}} = \frac{0.022}{2 \times 50}\,\text{cm}^{-1} = 2.2 \times 10^{-4}\,\text{cm}^{-1}\,.$$

In laser engineering, gain is also sometimes expressed in logarithmic decibel units. The integrated roundtrip threshold gain expressed in decibels is

$$G_{\text{th}}(\text{dB}) = 4.34 \times (2g_{th}L_g) = 4.34 \times 2(2.2 \times 10^{-4}\,\text{cm}^{-1})50\,\text{cm} = 0.095\,\text{dB}\,.$$

The factor of 4.34 follows from the identity $10\log_{10} e = 4.343$.

For comparison, another neon laser transition at 3.39 μm shares the same upper laser level with the red 632.8 nm line but has a spontaneous radiative rate of 2.87×10^6 s^{-1} and a considerably smaller Doppler width, $\Delta\nu_D$ =(1.52 GHz)(0.6328/3.39)≈ 284 MHz.

Example 8.2 Threshold Gain and Inversion in a Ruby Laser: A Three-Level Laser

Ruby is Al_2O_3 (corundum) in which a fraction of the Al^{3+} ions are replaced by trivalent chromium (Cr^{3+}) ions. For practical laser applications, synthetic crystals are grown from a molten mixture of Al_2O_3 containing a small percentage of Cr_2O_3 (typically about 0.05% by weight, corresponding to approximately 1.6×10^{19} cm^{-3} Cr^{3+} ions). Undoped Al_2O_3 is optically colorless and is known as sapphire. The active lasing centers are the Cr^{3+} ions embedded as impurities in the sapphire host crystal ($n = 1.77$).

At room temperature (300 K), the $\lambda = 694$ nm laser transition is predominantly homogeneously broadened because of strong ion-phonon coupling, resulting in a broad Lorentzian gain profile with linewidth $\Delta\nu_H = 330$ GHz. The metastable upper laser level has a radiative lifetime of 3 ms, corresponding to a spontaneous emission rate $A21 = 333$ s^{-1}.

Consider a standing-wave Fabry-Perot resonator enclosing a 5-cm long ruby laser rod. Assume an internal single-pass distributed scattering and diffraction loss factor of $\mathcal{L}_i = 0.03$, mirror power reflectances of $\mathcal{R}_1 \approx 100\%$ and $\mathcal{R}2 = 96.0\%$, and line-center modal operation such that the normalized lineshape peak value is $S(\nu_o) = 2/\pi\Delta\nu_H$. The roundtrip cavity loss factor is then $\mathcal{L}_c = 2\mathcal{L}_i - \ln(\mathcal{R}_1\mathcal{R}_2) \approx 0.06 - \ln(0.96) \approx 0.1008 \approx 0.10$. Substituting these variables into Eq. (8.8), the population inversion density required to reach the oscillation threshold is:

$$\Delta\mathcal{N}_{\text{th}} = \frac{\mathcal{L}_c}{2L_g}\frac{8\pi n^2}{\lambda^2 A_{21}}\frac{\pi\Delta\nu_H}{2} = \frac{0.1008}{(2 \times 5)}\frac{8\pi(1.77)^2(\pi \times 3.3 \times 10^{11})}{(0.694 \times 10^{-4})^2 \times 333 \times 2}\,\text{cm}^{-3}\,,$$
$$\approx 2.6 \times 10^{17}\,\text{cm}^{-3}\,.$$

This threshold inversion represents a much larger fraction of the total Cr^{3+} ion concentration (1.6×10^{19} ions/cm^3) than in the He-Ne laser example. In fact, $\Delta\mathcal{N}_{\text{th}}/\mathcal{N}_{Cr} \approx 1.6 \times 10^{-2}$, so the required threshold inversion is of order one percent of the total active-ion density.

The underlying reason for this striking difference is that the He-Ne laser transition has a relatively large stimulated-emission cross section because of its much larger radiative decay rate and much narrower gain profile, whereas the ruby transition has a slow spontaneous radiative decay rate and a very broad phonon-broadened gain profile. The required threshold inversion density corresponds to an unsaturated threshold gain coefficient $g_{\text{th}} = \mathcal{L}_c/2L_g = [0.1008/(2 \times 5)]\,\text{cm}^{-1} \approx 1.0 \times 10^{-2}\,\text{cm}^{-1}$.

8.2 Laser Power Output

Once the material gain exceeds the threshold value, the intracavity laser intensity can grow with each consecutive cavity roundtrip. However, this exponential growth cannot continue indefinitely because, as the light intensity increases, it begins to saturate the material gain, systematically reducing the gain until the growth rate is exactly balanced by the loss rate. When this dynamic equilibrium is established, the system reaches a stable steady state capable of sustaining a nonzero intracavity field intensity in the absence of an external input. Under these conditions, the system operates as an oscillator and produces a steady coherent optical output.

In this operating regime, gain saturation must be included in the equations governing the spatial evolution of the intracavity light intensity. Under steady-state conditions, the equations for the left- and right-traveling wave intensities in a standing-wave laser are

$$\frac{dI_+}{dz} = \left(\frac{g_o(\nu)}{1 + (I_+ + I_-)/I_{\text{sat}}} - \alpha\right) I_+ \,, \tag{8.10a}$$

$$\frac{dI_-}{dz} = -\left(\frac{g_o(\nu)}{1 + (I_+ + I_-)/I_{\text{sat}}} - \alpha\right) I_- \,. \tag{8.10b}$$

In writing these coupled differential equations, we assume a homogeneously broadened gain medium and explicitly account for cross-saturation of the gain by the simultaneous presence of the two counter-propagating waves. At this stage, we neglect the spatial interference pattern of the counter-propagating waves and describe saturation in terms of the sum of their traveling-wave intensities. The consequences of the standing-wave intensity modulation will be considered in Sec. 8.2.2.

If the end-mirror power reflectances are high (a high-Q cavity) and the small-signal gain remains moderate, the sum of the counter-propagating intensities, $I_+(z) + I_-(z)$, varies only weakly along the resonator axis. This slow spatial variation motivates the standard spatial-average approximation

$$I_+(z) + I_-(z) \approx 2I_c \tag{8.11}$$

within the saturated gain coefficient, which can then be written as

$$g = \frac{g_o}{1 + (I_+ + I_-)/I_{\text{sat}}} = \frac{g_o}{1 + 2I_c/I_{\text{sat}}} \tag{8.12}$$

The parameter I_c is defined as the intracavity circulating intensity, representing the characteristic intensity of either traveling-wave component in the spatial-average approximation. The concept of an intracavity circulating wave will also prove useful in the discussion of laser dynamics.

Using this expression for the gain in Eqs. (8.10) and integrating the equation for I_+ from $z = 0$ to $z = L$ and that for I_- from $z = L$ to $z = 0$, we obtain

$$\ln \frac{I_+(L)}{I_+(0)} = (g - \alpha)L \,, \tag{8.13a}$$

$$\ln \frac{I_-(L)}{I_-(0)} = -(g - \alpha)L \,. \tag{8.13b}$$

These equations differ from Eqs. (8.2) only in that the unsaturated gain coefficient g_o has been replaced by the saturated gain coefficient g. Subtracting these and using the boundary conditions (8.3), we obtain

$$\ln \frac{I_+(L)I_-(0)}{I_-(L)I_+(0)} = 2(g-\alpha)L$$
$$\Rightarrow \qquad -\ln \mathcal{R}_1\mathcal{R}_2 = 2(g-\alpha)L\,. \tag{8.14}$$

Applying this to a homogeneously broadened line and assuming that internal loss and gain are confined to cavity segments of lengths L_a and L_g, respectively, we obtain

$$2\alpha L_a - \ln \mathcal{R}_1\mathcal{R}_2 = \frac{2g_o L_g}{1+2I_c/I_{\text{sat}}}\,. \tag{8.15}$$

The quantity on the left-hand side of this equation is the total cavity loss factor $\mathcal{L}_c$, and the quantity on the right is the saturated roundtrip gain.

Equation (8.15) shows that in the steady state, the saturated roundtrip laser gain is clamped at its threshold value $\mathcal{L}_c$. The laser adjusts its circulating intensity until the saturated roundtrip gain exactly compensates the losses encountered in one roundtrip. As discussed in Sec. 4.3.2, other losses, such as localized scattering at a mirror, can be included by adding terms such as $-\ln(1-S)$ to $\mathcal{L}_c$, where S is the fraction of incident power lost to scattering in one roundtrip, so that $1-S$ is the corresponding survival factor.

Solving Eq. (8.15) for the circulating intensity I_c, we obtain

$$I_c = \frac{I_{\text{sat}}}{2}\left(\frac{2g_o L_g}{\mathcal{L}_c}-1\right) = \frac{I_{\text{sat}}}{2}\left(\frac{2g_o L_g}{2g_{\text{th}}L_g}-1\right) = \frac{I_{\text{sat}}}{2}(r-1)\,, \tag{8.16}$$

where r is the pump ratio given by

$$r = \frac{2g_o L_g}{\mathcal{L}_c} = \frac{g_o}{g_{\text{th}}} = \frac{\Delta\mathcal{N}^{(o)}}{\Delta\mathcal{N}_{\text{th}}}\,. \tag{8.17}$$

The pump ratio r is a very useful parameter that can be expressed in several equivalent ways: as the ratio of the unsaturated roundtrip gain to the cavity loss factor $\mathcal{L}_c$, the unsaturated gain coefficient to the threshold gain coefficient, or the unsaturated population inversion to the threshold inversion. In operating regimes where the unsaturated population inversion is proportional to the chosen pump strength, the ratio r may also be expressed as the ratio of the pump strength to its threshold value. The pump strength may be specified as optical pump power, semiconductor injection current, gas-discharge current, or another quantity characterizing the excitation process. Once the threshold value of the pump strength has been determined, the pump ratio r specifies how far above threshold the laser is operating and provides a convenient parameter for calculating the laser output power without explicitly retaining all the details of the pumping and excitation processes.

The circulating intensity depends linearly on r, and since laser intensity cannot be a negative quantity, Eq. (8.16) requires $r > 1$ for a nonzero steady-state laser field. Once the circulating intensity is known, the output from the laser can be obtained from the fraction of the circulating intensity transmitted through the cavity mirrors. If essentially all useful output coupling occurs through a single output mirror of power transmission T, then, in the small-loss approximation, $\mathcal{L}_T \simeq T$ and

$$I_{\text{out}} = (\text{transmission loss}) \times (\text{circulating intensity}) = \mathcal{L}_{_T} I_c$$
$$= \mathcal{T}\frac{I_{\text{sat}}}{2}\left(\frac{2g_o L_g}{\mathcal{T}+\mathcal{L}_i}-1\right) = \mathcal{T}\frac{I_{\text{sat}}}{2}(r-1)\,. \tag{8.18}$$

Laser cavities are usually designed with one mirror having a relatively large transmission while the other mirrors have negligible transmission. The principal laser output then emerges through the more transmissive mirror, which is referred to as the output mirror or output coupler.

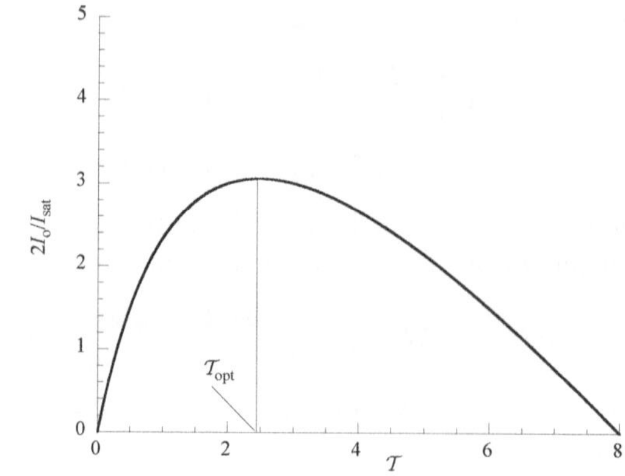

FIGURE 8.2
Laser output intensity as a function of output coupling for fixed gain $2g_oL_g = 10\%$ and internal losses $\mathcal{L}_i = 2\%$.

Optimum Power Output

The dependence of the output intensity on mirror transmission is not linear. We expect the output to increase initially as the output-coupler transmission increases. However, increasing the transmission also increases the overall cavity loss and, therefore, the minimum gain required to reach threshold. For a fixed unsaturated gain, this increased threshold gain reduces the pump ratio and eventually reduces the output. Hence, for a given gain $2g_oL_g$, there is an optimum output coupling that maximizes the laser output [see Fig. 8.2]. This optimum can be found from Eq. (8.18) by differentiating I_{out} with respect to $\mathcal{T}$.

$$\begin{aligned}\frac{dI_{out}}{d\mathcal{T}} = 0 &= \frac{I_{sat}}{2}\left(\frac{2g_oL_g}{\mathcal{L}_i + \mathcal{T}} - 1\right) - \mathcal{T}\frac{I_{sat}}{2}\frac{2g_oL_g}{(\mathcal{L}_i + \mathcal{T})^2}, \\ 0 &= \frac{I_{sat}}{2(\mathcal{L}_i + \mathcal{T})^2}\left[2g_oL_g(\mathcal{L}_i + \mathcal{T}) - (\mathcal{L}_i + \mathcal{T})^2 - \mathcal{T}2g_oL_g\right], \\ &= \frac{I_{sat}}{2(\mathcal{L}_i + \mathcal{T})^2}\left[2g_oL_g\mathcal{L}_i - (\mathcal{L}_i + \mathcal{T})^2\right].\end{aligned}$$

Solving for the optimum output coupling, we find

$$\mathcal{T}_{opt} = \sqrt{2g_oL_g\mathcal{L}_i} - \mathcal{L}_i. \tag{8.19}$$

Rewriting this result as

$$\mathcal{L}_i + \mathcal{T}_{opt} = \sqrt{2g_oL_g\mathcal{L}_i},$$

and substituting it in Eq. (8.18), the optimum output intensity from the laser is

$$[I_{out}]_{opt} = \frac{I_{sat}}{2}\left(\sqrt{2g_oL_g} - \sqrt{\mathcal{L}_i}\right)^2. \tag{8.20}$$

Example 8.3 *Find the output intensity for a high-Q single-mode unidirectional ring laser.*

Assuming homogeneous broadening and that the spatial variation of the gain is small, the steady-state gain condition is

$$\frac{g_o(\nu)L_g}{1 + I_c/I_{sat}} = \mathcal{L}_i + \mathcal{L}_{\mathcal{T}}.$$

It follows that

$$I_c = I_{\text{sat}}\left(\frac{g_o(\nu)L_g}{\mathcal{L}_i + \mathcal{L}_T} - 1\right) = I_{\text{sat}}(r-1)\,.$$

The useful output transmitted through the output coupler is therefore

$$I_{\text{out}} = \mathcal{T}_{\text{out}} I_{\text{sat}}\left(\frac{g_o(\nu)L_g}{\mathcal{L}_i + \mathcal{L}_T} - 1\right) = \mathcal{T}_{\text{out}} I_{\text{sat}}(r-1)\,,$$

where $\mathcal{T}_{\text{out}}$ is the transmission of the output coupler and $\mathcal{L}_T$ represents the total transmission loss from all cavity mirrors. In the high-Q, small-loss limit, $\mathcal{L}_T \approx \sum_j \mathcal{T}_j$.

Example 8.4 *For a strongly inhomogeneous line, the saturated gain is given by*

$$g(\nu) = \frac{g_o(\nu)}{\sqrt{1 + I/I_s}}\,,$$

where $g_o(\nu)$ is the unsaturated gain. Find the output intensity from the laser as a function of the laser pump ratio for an inhomogeneously broadened gain medium.

For a unidirectional traveling-wave cavity, the steady-state condition requires the saturated gain to equal the threshold gain:

$$\frac{g_o(\nu)}{\sqrt{1 + I_c/I_s}} = g_{\text{th}}\,.$$

Since $r = g_o(\nu)/g_{\text{th}}$, it follows that

$$\frac{r}{\sqrt{1 + \frac{I_c}{I_s}}} = 1\,,$$

and hence

$$I_c = I_s(r^2 - 1)\,.$$

The output intensity is therefore

$$I\text{out} = \mathcal{T}_{\text{out}} I_s(r^2 - 1)\,. \tag{8.21}$$

Example 8.5 *How does the pump ratio vary with frequency for (i) a homogeneously broadened gain medium and (ii) an inhomogeneously broadened gain medium?*

$$\text{Homogeneous} \quad r(\nu) = \frac{g_o(\nu_o)}{g_{\text{th}}}\frac{1}{1 + [2(\nu - \nu_o)/\Delta\nu_H]^2} = \frac{r_o}{1 + [2(\nu - \nu_o)/\Delta\nu_H]^2}\,,$$

$$\text{Inhomogeneous} \quad r(\nu) = \frac{g_o(\nu_o)}{g_{\text{th}}}\exp\left[-\left(\frac{\nu - \nu_o}{\Delta\nu_D}\right)^2 4\ln 2\right] = r_o \exp\left[-\left(\frac{\nu - \nu_o}{\Delta\nu_D}\right)^2 4\ln 2\right],$$

where $r_o = r(\nu_o) = g_o(\nu_o)/g_{\text{th}}$ is the pump ratio at the line center. These expressions for $r(\nu)$ apply directly to a traveling-wave laser. For a standing-wave laser, the same expressions apply to a stationary ensemble of active atoms, as in a solid-state gain medium. In a gas, however, the interaction of the two counter-propagating waves with different atomic velocity classes makes the gain of a standing-wave laser more complex. This leads to characteristic Doppler-dependent behavior, which will be discussed next.

8.2.1 Lamb Dip

The Lamb dip refers to a characteristic frequency dependence of the laser output power from a single-mode standing-wave gas laser. If the frequency of a single-mode standing-wave cavity is tuned across a Doppler-broadened gain profile, the laser output power exhibits a narrow dip when the oscillation frequency coincides with the center of the gain profile.

The field inside a standing-wave laser can be regarded as a superposition of two waves traveling in opposite directions, $\pm z$. Recall that a light wave of frequency ν interacts strongly only with a group of atoms whose velocity Doppler shifts the field into resonance with the atomic transition frequency ν_o in their rest frame. Thus, the $+z$-traveling wave interacts with atoms having velocity $v/c = (\nu - \nu_o)/\nu_o$, whereas the $-z$-traveling wave interacts with atoms having velocity $v/c = -(\nu - \nu_o)/\nu_o$. When the cavity frequency is sufficiently detuned from line center, the two counter-propagating waves burn two separate spectral holes in the Doppler-broadened gain profile, corresponding to the velocity classes $v/c = \pm(\nu - \nu_o)/\nu_o$. As the detuning is reduced toward zero, the two holes approach one another and eventually overlap. At line center, both traveling-wave components interact with the same group of atoms corresponding to $v = 0$.

Thus, away from line center, the two counter-propagating waves draw gain from two different velocity classes, whereas at line center they compete for gain from the same velocity class. As the oscillation frequency approaches line center, the laser output initially increases because of the Doppler gain profile, but once the two saturated holes begin to overlap, the available gain is reduced and the output drops, reaching a relative minimum at line center. The resulting dip in the output power of a single-mode standing-wave gas laser is known as the *Lamb dip*[2]. The center of the Lamb dip marks the center of the atomic transition and can therefore be used as a frequency reference in stabilized lasers. In addition, the width of the dip provides a means of measuring the homogeneous linewidth of an atomic transition even in the presence of strong inhomogeneous broadening.

For an oscillation frequency ν more than approximately a homogeneous linewidth away from the center of the Doppler line, the two counter-propagating waves burn separate holes in the velocity classes $v/c = \pm(\nu - \nu_o)/\nu_o$. If these holes are separated by several hole widths, there is essentially no cross-saturation between them. The gain experienced by each wave is then saturated primarily by its own intensity. If the output coupling is small, the two counter-propagating waves have approximately the same circulating intensity I_c, and the gain experienced by either wave at the center of its own hole saturates as $\frac{1}{\sqrt{1+I_c/I_s}}$.

If the oscillation frequency ν coincides with the line-center frequency ν_o, however, the two oppositely traveling waves burn a single hole at the center of the line in the $v = 0$ velocity class. This common hole is saturated by the total intensity $2I_c$ of the two counter-propagating waves. The gain coefficient experienced by either wave therefore saturates as $\frac{1}{\sqrt{1+2I_c/I_s}}$, with an additional factor of 2 in the saturation denominator.

The changeover from $\frac{1}{\sqrt{1+I_c/I_s}}$ to $\frac{1}{\sqrt{1+2I_c/I_s}}$ saturation will occur when the oscillation frequency ν is offset from line center by approximately one half of the homogeneous linewidth. This corresponds to normalized detuning $\delta \equiv 2(\nu - \nu_o)/\Delta\nu_H = \pm 1$. A simple phenomenological interpolation between these limiting behaviors is to modify the gain saturation by incorporating a detuning-dependent factor given by

$$f(\delta) = \frac{2 + \delta^2}{1 + \delta^2}\,. \tag{8.22}$$

[2]W.E. Lamb, Jr., "Theory of an optical maser," Phys. Rev. **134**, A1429–A1450 (1964); R.A. McFarlane, W. R. Bennett, Jr., and W. E. Lamb, Jr., "Single mode tuning dip in the power output of an He-Ne optical maser," Appl. Phys. Lett. **2**, 189–190 (1963).

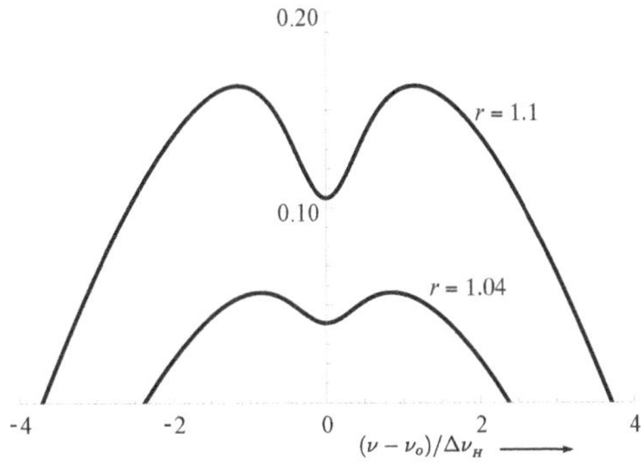

FIGURE 8.3
Lamb dip profiles (I_c/I_s as a function of normalized detuning) for a single-mode standing-wave laser at two different pump ratios above threshold (as indicated) for strong Doppler broadening $\Delta\nu_D = 20\Delta\nu_H$.

This factor has the desired property that it changes from a value of 1 for $|\delta| \gg 1$ to a value of 2 for $|\delta| \ll 1$. With this modification, the oscillation condition that the saturated roundtrip power gain just equal the roundtrip power losses can be written as

$$\frac{2g_{\text{th}}L_g r e^{-4\ln 2\left[(\nu-\nu_o)/\Delta\nu_D\right]^2}}{\sqrt{1+f(\delta)I_c/I_s}} = \frac{2g_{\text{th}}L_g r e^{-\ln 2\left(\Delta\nu_H/\Delta\nu_D\right)^2\delta^2}}{\sqrt{1+f(\delta)I_c/I_s}} = \mathcal{L}_c\,. \tag{8.23}$$

Recalling $2g_{\text{th}}L_g = \mathcal{L}_c$ and solving this equation yields the circulating intensity as a function of detuning

$$I_c(\delta) = \frac{r^2 e^{-2\ln 2\left(\Delta\nu_H/\Delta\nu_D\right)^2\delta^2} - 1}{f(\delta)} I_s\,. \tag{8.24}$$

A plot of I_c as a function of normalized detuning $(\nu_{-}\nu_o)/\Delta\nu_H$ for a strongly Doppler-broadened single-mode standing-wave laser is shown in Fig. 8.3. A dip in the circulating wave intensity as mode frequency is tuned to line center is clearly visible and becomes more pronounced as the pump ratio increases. To observe the Lamb dip clearly, it is important that the inhomogeneously broadened laser oscillates on only a single axial mode within the inhomogeneous gain profile. This can be achieved either by using a sufficiently short laser cavity so that only one axial mode lies within the relevant gain bandwidth, or by employing mode-control techniques to suppress all but one selected axial mode. Multimode oscillation, overlapping gain lines, or line-splitting mechanisms can smear out or eliminate the Lamb dip even for a strongly inhomogeneously broadened transition.

8.2.2 Effect of Spatial Hole Burning

In calculating laser power output, we have made several simplifying approximations. If these approximations do not hold, more sophisticated approaches to calculating the laser power output are required. For example, in a standing-wave laser, the field inside the cavity is a superposition of left- and right-traveling waves. For approximately equal counter-propagating amplitudes, and choosing the origin and relative phase so that the standing-wave has a node at $z = 0$, the field may be written as

$$E = E_+ e^{i(kz-\omega t)} + iE_- e^{-i(kz+\omega t)} \approx 2E_c \sin(kz)\, e^{-i\omega t}\,, \tag{8.25}$$

with the intensities of the right- and left-traveling waves given by

$$I_+ = \frac{1}{2}\epsilon_o nc|E_+|^2 \approx \frac{1}{2}\epsilon_o nc|E_c|^2 \approx I_- \,. \tag{8.26}$$

The overall intensity of the standing-wave can therefore be written as

$$I = \frac{1}{2}\epsilon_o nc|E|^2 \approx \frac{1}{2}\epsilon_o nc\left(4|E_c|^2 \sin^2 kz\right) = 4I_c \sin^2 kz \,. \tag{8.27}$$

Thus, the overall intensity is a function of z, which means that the saturated gain g is also a function of z, given by

$$g(\nu, z) = \frac{g_o(\nu)}{1 + 4(I_c/I_{\text{sat}})\sin^2 kz} \,. \tag{8.28}$$

Consequently, we can no longer represent the saturated gain by a spatially uniform gain coefficient throughout the gain medium. Replacing $\sin^2 kz$ by its spatial average $1/2$ recovers the uniform-field approximation leading to Eq. (8.15).

Even when the gain depends on z, however, the overall rate at which the cavity field gains energy must equal the rate at which it loses energy. A generalization of Eq. (8.15) therefore gives

$$\begin{aligned}\mathcal{L}_c I_c &= \int_0^{L_g} \frac{4g_o I_c \sin^2 kz}{1 + 4(I_c/I_{\text{sat}})\sin^2 kz}\, dz = 4g_o I_c \int_0^{L_g} \frac{\sin^2 kz}{1 + 4(I_c/I_{\text{sat}})\sin^2 kz}\, dz \\ &\approx g_o L_g I_{\text{sat}} \left[1 - \frac{1}{\sqrt{1 + 4I_c/I_{\text{sat}}}}\right], \qquad kL_g \gg 1 \,. \end{aligned} \tag{8.29}$$

Here the last result follows by averaging over many standing-wave periods within the gain medium. Using $r = 2g_o L_g/\mathcal{L}_c$ for the pump ratio, this equation can be manipulated into the form

$$\sqrt{1 + 4I_c/I_{\text{sat}}} = 2r - (1 + 4I_c/I_{\text{sat}}) \,, \tag{8.30}$$

Squaring both sides leads to a quadratic equation in the variable $1 + 4I_c/I_{\text{sat}}$. The physical solution that satisfies $I_c = 0$ at $r = 1$, the threshold condition, is

$$I_c = \frac{I_{\text{sat}}}{2}\left(r - \frac{1}{4} - \sqrt{\frac{r}{2} + \frac{1}{16}}\right) . \tag{8.31}$$

The output intensity is then $\mathcal{T}_{\text{out}} I_c$, giving

$$I_{\text{out}} = \frac{1}{2}\mathcal{T}_{\text{out}} I_{\text{sat}}\left(r - \frac{1}{4} - \frac{1}{4}\sqrt{8r + 1}\right) . \tag{8.32}$$

Comparing this result with Eq. (8.18), we see that accounting for the standing-wave intensity pattern reduces the output intensity relative to the uniform-field approximation [Fig. 8.4]. This is physically reasonable because the gain medium is utilized nonuniformly: atoms near the antinodes experience strong saturation, whereas atoms near the nodes interact only weakly with the standing-wave field. This periodic depletion of the population inversion at the standing-wave antinodes is referred to as spatial hole burning.

8.3 Large Output Coupling Analysis

If the transmission of the output mirror is large, the variation of the wave intensity during propagation through the gain medium can no longer be neglected. With substantial output

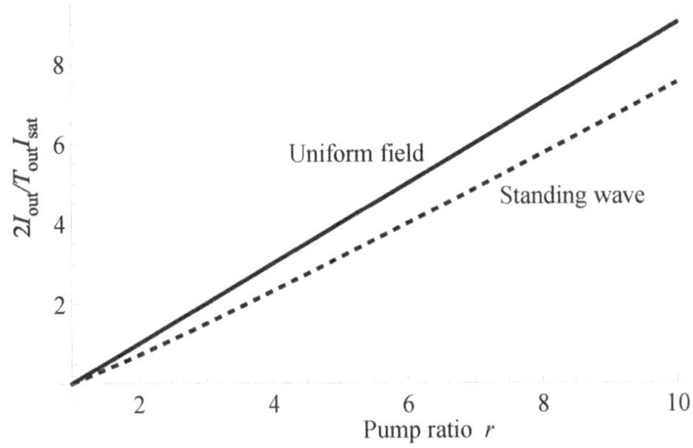

FIGURE 8.4
Effect of spatial hole burning on the output intensity.

coupling, the intracavity intensity changes appreciably at the output mirror, and the resulting difference between the forward- and backward-traveling intensities persists throughout the gain medium. In this case, provided that distributed internal losses are negligible, Rigrod obtained an analytic solution of the coupled intensity equations:[3]

$$\frac{dI_+}{dz} = \frac{g_o}{1 + (I_+ + I_-)/I_{\text{sat}}}, \tag{8.33}$$

$$\frac{dI_-}{dz} = -\frac{g_o}{1 + (I_+ + I_-)/I_{\text{sat}}}. \tag{8.34}$$

Note that this treatment neglects both the standing-wave interference pattern and distributed losses. In terms of the scaled variables $u = I_+/I_{\text{sat}}$ and $v = I_-/I_{\text{sat}}$, these equations become

$$\left.\begin{aligned} \frac{du}{dz} &= \frac{g_o u}{1 + (u + v)} \\ \frac{dv}{dz} &= -\frac{g_o v}{1 + (u + v)} \end{aligned}\right\} \quad \Rightarrow \quad \ln(uv) = \text{constant}. \tag{8.35}$$

Thus, although both I_+ and I_- depend on z, their product is independent of z. This invariant, together with the boundary conditions at the mirrors, makes it possible to integrate the coupled equations analytically.

Integrating the equations for u and v from one end of the gain medium to the other ($z = 0$ to $z = L_g$), and applying the mirror boundary conditions, we find

$$u(L_g) = \frac{\sqrt{\mathcal{R}_1}\left(g_o L_g + \ln\sqrt{\mathcal{R}_1\mathcal{R}_2}\right)}{(1 - \sqrt{\mathcal{R}_1\mathcal{R}_2})(\sqrt{\mathcal{R}_1} + \sqrt{\mathcal{R}_2})}, \tag{8.36}$$

$$v(0) = \frac{\sqrt{\mathcal{R}_2}\left(g_o L_g + \ln\sqrt{\mathcal{R}_1\mathcal{R}_2}\right)}{(1 - \sqrt{\mathcal{R}_1\mathcal{R}_2})(\sqrt{\mathcal{R}_1} + \sqrt{\mathcal{R}_2})}. \tag{8.37}$$

[3]W. W. Rigrod, "Saturation effects in high gain lasers," J. Appl. Phys. **36**, 2487-2490 (1965); W. Rigrod, "Homogeneously broadened CW lasers with uniform distributed loss," IEEE J. Quantum Electron. **14**, 377–381 (1978); T.S. Mansuripur, G.D. Naurois, A. Belyanin, and F. Capasso, "Lasers with distributed loss have a sublinear output power characteristic," Optica **2**, 48–55 (2014).

The total output power from the laser is $P_{\text{out,total}} = A[\mathcal{T}_2 I_+(L_g) + \mathcal{T}_1 I_-(0)]$ for common beam area A. The single-ended output intensity through mirror M_2 is therefore

$$I_{\text{out}} \equiv \mathcal{T}_2\, I_+(L_g) = I_{\text{sat}} \frac{\mathcal{T}_2\sqrt{\mathcal{R}_1}\,(g_o L_g + \ln\sqrt{\mathcal{R}_1\mathcal{R}_2})}{(1-\sqrt{\mathcal{R}_1\mathcal{R}_2})(\sqrt{\mathcal{R}_1}+\sqrt{\mathcal{R}_2})}\,. \tag{8.38}$$

For a perfectly reflecting mirror $\mathcal{R}_1 = 1$, the cavity has a single output port. Multiplying both sides of Eq. (8.38) by the beam area A, the output power is

$$P_{\text{out}} = I_{\text{sat}} A \frac{\mathcal{T}_2\,(g_o L_g + \ln\sqrt{\mathcal{R}_2})}{(1-\sqrt{\mathcal{R}_2})(1+\sqrt{\mathcal{R}_2})}\,. \tag{8.39}$$

Since $(1-\sqrt{\mathcal{R}_2})(1+\sqrt{\mathcal{R}_2}) = 1 - \mathcal{R}_2 = \mathcal{T}_2$ when mirror absorption is negligible, this expression may also be written as

$$P_{\text{out}} = I_{\text{sat}} A \left(g_o L_g + \ln\sqrt{\mathcal{R}_2}\right) = I_{\text{sat}} A \left(g_o L_g + \frac{1}{2}\ln \mathcal{R}_2\right)\,. \tag{8.40}$$

For a high-Q cavity with high-reflectivity mirrors, $\mathcal{R}_1 = 1 - \mathcal{T}_1$ and $\mathcal{R}_2 = 1 - \mathcal{T}_2$, where $\mathcal{T}_1, \mathcal{T}_2 \ll 1$. In this limit, Eq. (8.38) reduces to

$$I_{\text{out}} = \frac{1}{2} I_{\text{sat}} \mathcal{T}_2 \left(\frac{2 g_o L_g}{\mathcal{T}_2 + \mathcal{T}_1} - 1\right)\,. \tag{8.41}$$

This agrees with the standing-wave result derived earlier when the spatial standing-wave interference pattern is neglected. The agreement is expected because neither the large-output-coupling analysis presented here nor the earlier uniform-field approximation includes spatial hole burning. Inclusion of standing-wave effects and additional phenomena arising, for example, from atomic motion requires a more advanced treatment[4]. For many lasers with modest single-pass gain and a high-Q cavity, however, the uniform-field approximation provides reasonably accurate results.

8.4 An Exactly Solvable Model

For a unidirectional ring laser with a homogeneously broadened active medium, the cavity intensity propagation equation inside the gain medium is

$$\frac{dI_c}{dz} = \frac{g_o I}{1 + I_c/I_{\text{sat}}}\,, \tag{8.42}$$

where the gain medium occupies a length L_g. As the wave propagates through the gain medium, it is amplified, whereas elsewhere in the cavity its intensity is reduced by partial transmission at the mirrors and by scattering and absorption in the various intracavity elements. For a nonzero steady-state intensity to exist, the intensity at the output of the gain medium, multiplied by all the survival factors encountered during the remainder of the roundtrip, must reproduce the intensity at the input of the gain medium.

[4] See, for example, M. Sargent, III, M.O. Scully, and W.E. Lamb, Jr., *Laser Physics* (Addison-Wesley Publishing, Reading, MA, 1974).

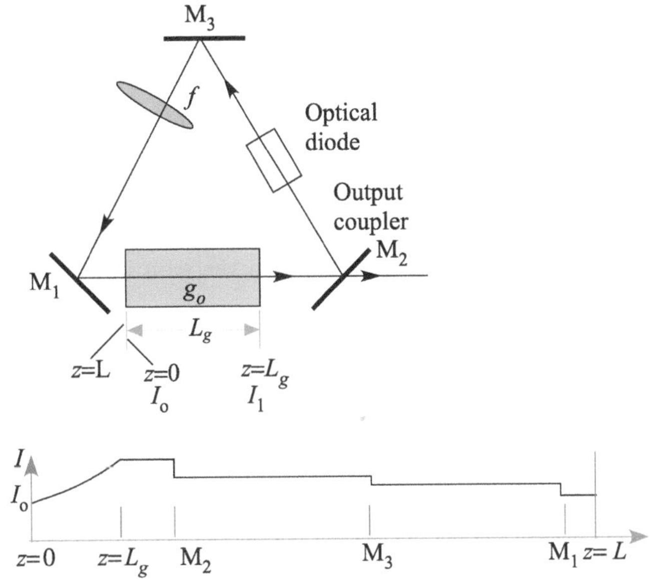

FIGURE 8.5
A ring laser supports traveling waves in both clockwise and counter-clockwise directions. An optical diode introduces greater loss in one direction, allowing the gain to exceed threshold only in the lower-loss direction (counter-clockwise in the figure).

Starting with an intensity I_o at the input of the gain medium, which extends over a length L_g, let I_1 denote the intensity at its output. Integrating the propagation equation through the gain medium gives

$$g_o L_g = \ln\left(\frac{I_1}{I_o}\right) + \frac{1}{I_{\rm sat}}(I_1 - I_o)\,. \tag{8.43}$$

Following the intensity I_1 around the cavity back to the input of the gain medium, we find that the returning intensity is $\mathcal{R}_1\mathcal{R}_2\mathcal{R}_3 e^{-\alpha L_a} I_1$. Scattering and other losses can be included by multiplying this expression by the corresponding survival factors. In steady state, the returning intensity must equal the initial intensity I_o, giving

$$\mathcal{R}_1\mathcal{R}_2\mathcal{R}_3 e^{-\alpha L_a} I_1 = I_o \Rightarrow \frac{I_o}{I_1} = \mathcal{R}_1\mathcal{R}_2\mathcal{R}_3 e^{-\alpha L_a} \equiv f_s\,, \tag{8.44}$$

where $f_s < 1$ is the overall fraction of the wave intensity that survives in traveling from the output of the gain medium around the remainder of the cavity and back to its input. Using this result in Eq. (8.43), we find

$$\ln\left(\frac{1}{f_s}\right) + \frac{I_1}{I_{\rm sat}}(1 - f_s) = g_o L_g \tag{8.45a}$$

or

$$I_1 = I_{\rm sat}\frac{g_o L_g + \ln f_s}{1 - f_s}\,. \tag{8.45b}$$

This result can be expressed in a more physically useful form by recalling that the survival factor f_s is related to the cavity loss factor by $-\ln(f_s) = \alpha L_a - \ln \mathcal{R}_1\mathcal{R}_2\mathcal{R}_3 \equiv \mathcal{L}_c$. Thus,

$$I_1 = I_{\text{sat}} \frac{\mathcal{L}_c}{1 - f_s} \left(\frac{g_o L_g}{\mathcal{L}_c} - 1 \right) . \tag{8.45c}$$

Equation (8.45c) vanishes at $g_o L_g = \mathcal{L}_c$, recovering the oscillation-threshold condition for the unidirectional ring cavity. The output intensity is obtained by propagating the intensity I_1 at the output of the gain medium to the output mirror. Taking M_2 to be the output mirror, and assuming that there is no loss between the gain medium and mirror M_2, the output intensity $I_{\text{out}} = \mathcal{T}_2 I_1$ is

$$I_{\text{out}} = \mathcal{T}_2 I_{\text{sat}} \frac{\mathcal{L}_c}{1 - f_s} \left(\frac{g_o L_g}{\mathcal{L}_c} - 1 \right) . \tag{8.46}$$

For a high-Q cavity, the overall survival factor f_s is close to unity, so that $1 - f_s$ is small. Assuming negligible mirror absorption, the survival factor in Eq. (8.46) can then be approximated as

$$\begin{aligned} f_s &\equiv \mathcal{R}_1\mathcal{R}_2\mathcal{R}_3 e^{-\alpha L_a} = (1 - \mathcal{T}_1)(1 - \mathcal{T}_2)(1 - \mathcal{T}_3) e^{-\alpha L_a} \\ &\approx 1 - (\alpha L_a + \mathcal{T}_1 + \mathcal{T}_2 + \mathcal{T}_3) = 1 - (\mathcal{L}_a + \mathcal{L}_T) , \end{aligned} \tag{8.47a}$$

and

$$1 - f_s \approx \alpha L_a + \mathcal{T}_1 + \mathcal{T}_2 + \mathcal{T}_3 = \mathcal{L}_a + \mathcal{L}_T , \tag{8.47b}$$

where in this low-loss approximation, $\mathcal{L}_a \equiv \alpha L_a$ and $\mathcal{L}_T \equiv \mathcal{T}_1 + \mathcal{T}_2 + \mathcal{T}_3$ represent the absorptive and mirror-transmission contributions to the roundtrip loss. In the same low-loss limit, $\mathcal{L}_c = -\ln f_s \approx 1 - f_s \approx \mathcal{L}_a + \mathcal{L}_T$. The output intensity can therefore be written as

$$I_{\text{out}} \approx \mathcal{T}_2 I_{\text{sat}} \left(\frac{g_o L_g}{\mathcal{L}_a + \mathcal{L}_T} - 1 \right) . \tag{8.48}$$

This result differs from the corresponding standing-wave result by the absence of the factor $\frac{1}{2}$. This difference arises because, in the unidirectional ring laser, the gain is saturated by a single traveling wave rather than by the two counter-propagating components used in the corresponding uniform-field standing-wave model. Also, in one roundtrip around a ring cavity, the gain medium is traversed once, and any distributed absorptive region is likewise traversed once. Thus, the integrated gain and absorption factors involve $g_o L_g$ and αL_a, respectively, whereas in a standing-wave linear cavity these regions are traversed twice per roundtrip, giving factors $2g_o L_g$ and $2\alpha L_a$.

We have derived several analytic expressions for the output power of a laser using different models and approximations. Some assumed that the intracavity intensity varies little during propagation, whereas others explicitly accounted for its spatial variation. The Rigrod analysis neglected distributed absorption but allowed arbitrary mirror losses, while the present ring-laser model treats the spatial evolution of a single traveling wave exactly within the assumptions of homogeneous saturation and a uniform gain coefficient. These models are all approximations to real lasers, and for physical effects common to the models, their predictions approach one another in the low-loss, high-Q limit. They illustrate how analytic expressions for laser output can take different forms depending on which physical aspects of a real laser are retained. They still omit effects that can influence laser output, including the transverse Gaussian beam profile, nonlinear loss mechanisms, spatially nonuniform pumping, and nonlinearities or saturation in the pumping process. Nevertheless, within their regimes of validity, these relatively simple models can provide surprisingly good descriptions of experimental observations.

8.5 Laser Operating Frequency

The dispersive response associated with gain or absorption modifies the operating frequencies of a laser cavity. This problem was briefly considered in Sec. 5.3.1 for a collection of classical oscillators. Equation (7.97) for the cavity field amplitude shows that the imaginary term on the right-hand side contributes an additional phase shift beyond the usual plane-wave propagation phase.

Consider a laser with a cavity of length L filled with a medium of refractive index n and containing a gain medium of length L_g. Suppose the laser operates at a frequency ν_p (to be determined) close to the pth longitudinal mode of the corresponding passive cavity, whose frequency is $\nu_p^{(0)} = pc/2nL$. From Eq. (7.97), the steady-state field-amplitude equation can be written as

$$\frac{\partial A}{\partial z} = \frac{1}{2}(g - \alpha)A + i\Delta k A\,, \tag{8.49}$$

where
$$g = \sigma\Delta\mathcal{N}, \quad \Delta k = \frac{(\nu - \nu_o)}{\Delta\nu_H}g(\nu)\,. \tag{8.50}$$

The expression for Δk in Eq. (8.49) holds for a homogeneously broadened transition. As pointed out in the discussion following Eqs. (7.97) and (7.98), such a simple expression for the correction Δk in terms of the gain is not generally possible for an inhomogeneously broadened transition. Nevertheless, even for a more complex lineshape, the phase correction for axial modes near the center of the atomic line varies approximately linearly with the frequency offset $\nu - \nu_o$. To gain insight into how the gain affects the operating frequency without introducing a more complicated analysis, we retain this near-resonance linear dependence of phase on $\nu - \nu_o$ and define an effective atomic width $\Delta\nu_{\text{at}}$ by

$$\Delta\nu_{\text{at}} = \begin{cases} \Delta\nu_H & \text{Homogeneous gain broadening} \\ \Delta\nu_D\sqrt{\frac{\pi}{4\ln 2}}\,, & \text{Inhomogeneous gain broadening} \end{cases} \tag{8.51}$$

Now recalling Eqs. (7.85) and (7.86), and noting that A is the slowly varying field amplitude after the rapid axial variation e^{ikz} and the transverse variation described by the mode function have been factored out, we find that the effective propagation constant in the gain medium is $k + \Delta k$. The operating frequency of the laser is then determined by the resonance condition that the total roundtrip phase be an integer multiple of 2π. Thus, $\Delta\phi \equiv k2L + (\Delta k)2L_g = 2\pi p$, where p is an integer. With $k = n2\pi\nu/c$, this condition leads to the following equation for the operating frequency ν_p of the laser, close to the pth axial mode of the passive cavity:

$$\Delta\phi \equiv \frac{n2\pi\nu_p 2L}{c}\left[1 + \frac{(\nu_p - \nu_o)}{\Delta\nu_{\text{at}}}\frac{cgL_g}{2\pi n\nu_p L}\right] = 2\pi p\,, \tag{8.52}$$

or
$$\nu_p\left[1 - \left(\frac{\nu_o - \nu_p}{\Delta\nu_{\text{at}}}\right)\frac{cg(\nu_p)L_g}{2\pi n\nu_p L}\right] = \frac{pc}{2nL} \equiv \nu_p^{(0)}\,. \tag{8.53}$$

For a given p, this is still, in general, an implicit equation for ν_p. However, for most lasers the integrated roundtrip gain $2gL_g$ is small, typically of the order of a few percent. The operating frequency is therefore expected to lie close to the passive-cavity frequency $\nu_p^{(o)}$. Replacing ν_p by $\nu_p^{(o)}$ in the small correction term, including the gain coefficient, we find to leading order

$$\nu_p \approx \frac{\nu_p^{(o)}}{1-\left(\frac{\nu_o-\nu_p^{(o)}}{\Delta\nu_{\rm at}}\right)\frac{cg(\nu_p^{(0)})L_g}{2\pi n\nu_p^{(o)}L}} \approx \nu_p^{(o)} + \left(\frac{\nu_o-\nu_p^{(o)}}{\Delta\nu_{\rm at}}\right)\frac{cg(\nu_p^{(o)})L_g}{2\pi nL} . \tag{8.54}$$

Subtracting this result from the line-center frequency ν_o, we can express the laser-frequency offset $\nu_o - \nu_p$ in terms of the passive-cavity frequency offset $\nu_o - \nu_p^{(o)}$ as

$$\nu_o - \nu_p = (\nu_o - \nu_p^{(o)})\left[1 - \frac{cg(\nu_p^{(o)})L_g}{\Delta\nu_{\rm at}2\pi nL}\right] . \tag{8.55}$$

This is a first-order perturbative result. The factor inside the square brackets on the right-hand side is less than unity for gain ($g > 0$) and greater than unity for absorption ($g < 0$). Thus, if the field of a passive-cavity mode $\nu_p^{(0)}$ experiences gain, its operating frequency ν_p (the dressed-cavity frequency) is shifted toward the atomic resonance frequency ν_o. This phenomenon is known as frequency pulling [Fig. 8.6a].

On the other hand, if the cavity mode field experiences absorption ($g < 0$), its dressed-cavity frequency is shifted away from the atomic resonance frequency relative to its passive-cavity frequency. This phenomenon is known as frequency pushing [Fig. 8.6b].

For a laser pumped above threshold, the unsaturated gain satisfies $g_o > g_{\rm th}$, while in the steady state the saturated gain is clamped at its threshold value, $g = g_{\rm th}$. Thus,

$$2gL_g = 2g_{\rm th}L_g = \mathcal{L}_c = \left[\frac{c\mathcal{L}_c}{2nL}\right]\frac{2nL}{c} = \frac{1}{\tau_c}\frac{2nL}{c} = 2\pi\Delta\nu_c \times \frac{2nL}{c} ,$$

where $\Delta\nu_c$ is the FWHM of the cavity transmission profile. Using this result in the frequency equation and assuming that the gain medium fills the entire cavity ($L_g = L$), we obtain

$$\nu_p\left[1 - \frac{(\nu_o-\nu_p)}{\nu_p}\frac{\Delta\nu_c}{\Delta\nu_{\rm at}}\right] = \nu_p^{(o)} . \tag{8.56}$$

This equation can be rewritten as

$$\nu_p = \frac{\nu_p^{(o)}/\Delta\nu_c + \nu_o/\Delta\nu_{\rm at}}{1/\Delta\nu_c + 1/\Delta\nu_{\rm at}} . \tag{8.57}$$

Thus, the laser operating frequency is a linewidth-weighted average of the passive-cavity frequency $\nu_p^{(o)}$ and the atomic resonance frequency ν_o. It lies between the two frequencies and is pulled more strongly toward the resonance having the narrower linewidth.

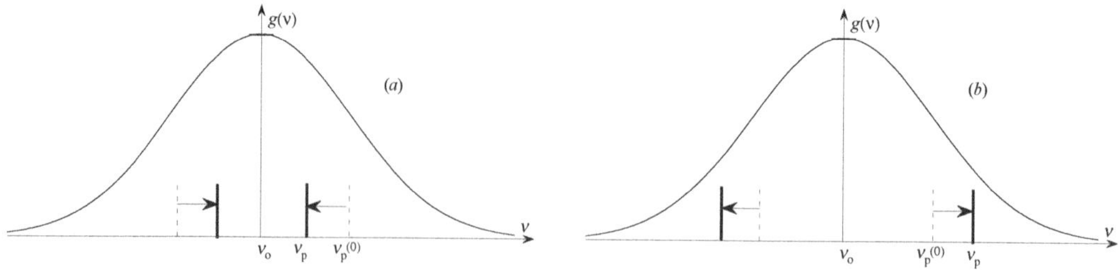

FIGURE 8.6
(a) Cavity frequency pulling in the presence of gain ($g > 0$) and (b) cavity frequency pushing in the presence of absorption ($g < 0$). Dashed vertical lines show bare cavity axial modes and solid lines show the operating frequencies of axial laser modes.

In terms of the Q-factors for the atomic transition and the cavity,

$$Q_{\text{at}} = \nu_o/\Delta\nu_{\text{at}}\,, \quad Q_c = \nu_p^{(o)}/\Delta\nu_c \approx \frac{\nu_o}{\Delta\nu_c}\,, \tag{8.58}$$

and using $\nu_p^{(o)} \approx \nu_o$ in the slowly varying frequency factors, Eq. (8.57) can be written approximately as

$$\nu_p = \frac{\nu_p^{(o)} Q_c + \nu_o Q_{\text{at}}}{Q_c + Q_{\text{at}}}\,. \tag{8.59}$$

According to this equation, the dressed-cavity frequency is approximately a Q-weighted average of the passive-cavity frequency $\nu_p^{(o)}$ and the atomic frequency ν_o. It lies between $\nu_p^{(o)}$ and ν_o, being closer to the frequency associated with the larger Q.

For many conventional lasers, $Q_{\text{at}} \ll Q_c$, so that the operating frequency is approximately

$$\nu_p \approx \nu_p^{(o)} + (\nu_o - \nu_p^{(o)})\frac{Q_{\text{at}}}{Q_c} = \nu_p^{(o)} + (\nu_o - \nu_p^{(o)})\frac{\Delta\nu_c}{\Delta\nu_{\text{at}}} \tag{8.60}$$

in agreement with Eq. (8.54).

Example 8.6 *Estimates of frequency pulling*

For a He-Ne laser with threshold gain $g_{\text{th}} = 0.001$ cm^{-1}, $\Delta\nu_D = 1500$ MHz, $\lambda = 633$ nm, $n = 1$, and the gain medium filling the entire cavity ($L_g = L$), recall that the cavity and effective atomic linewidths are given by

$$\Delta\nu_c = \frac{g_{\text{th}} c}{2\pi n}\,, \quad \Delta\nu_{\text{at}} = \Delta\nu_D\sqrt{\frac{\pi}{4\ln 2}}\,.$$

We then obtain

$$\frac{\Delta\nu_c}{\Delta\nu_{\text{at}}} = \frac{g_{\text{th}} c}{2\pi n \Delta\nu_D}\sqrt{\frac{4\ln 2}{\pi}} = \frac{0.001(3.00 \times 10^{10})}{2\pi \times 1.5 \times 10^9}\sqrt{\frac{4\ln 2}{\pi}} = 3.0 \times 10^{-3} = 0.3\%\,.$$

Thus, the frequency-pulling correction is only about 0.3% of the bare cavity detuning $|\nu_o - \nu_p^{(o)}|$.

For the $\lambda = 3.39\,\mu$m transition with $\Delta\nu_D = 280$ MHz and $g = 0.03$ cm^{-1}, we obtain

$$\frac{\Delta\nu_c}{\Delta\nu_{\text{at}}} = \frac{3 \times 10^{-2} \times (3.00 \times 10^{10})}{2\pi \times 2.8 \times 10^8} \times \sqrt{\frac{4\ln 2}{\pi}} = 0.48\,.$$

The first-order estimate corresponds to a correction of approximately 48% of the bare cavity detuning, indicating that frequency pulling is substantial and that higher-order corrections may no longer be negligible.

Finally, for a high-gain laser such as a He:Xe laser operating at $\lambda = 3.51\,\mu$m, with $\Delta\nu_D = 100$ MHz and $g = 0.5$ cm^{-1}, we obtain

$$\frac{\Delta\nu_c}{\Delta\nu_{\text{at}}} = \frac{0.5 \times (3.00 \times 10^{10})}{2\pi \times 10^8} \times \sqrt{\frac{4\ln 2}{\pi}} \approx 22\,.$$

For such a large ratio, the cavity-mode structure is strongly perturbed, and the perturbative analysis presented above is no longer applicable. We must then return to Eq. (8.53) and solve the full frequency equation to determine the mode structure. Both strongly absorbing and strongly amplifying media can substantially alter the cavity-mode structure and give rise to more complex resonance behavior. For example, the resonance equation can have several

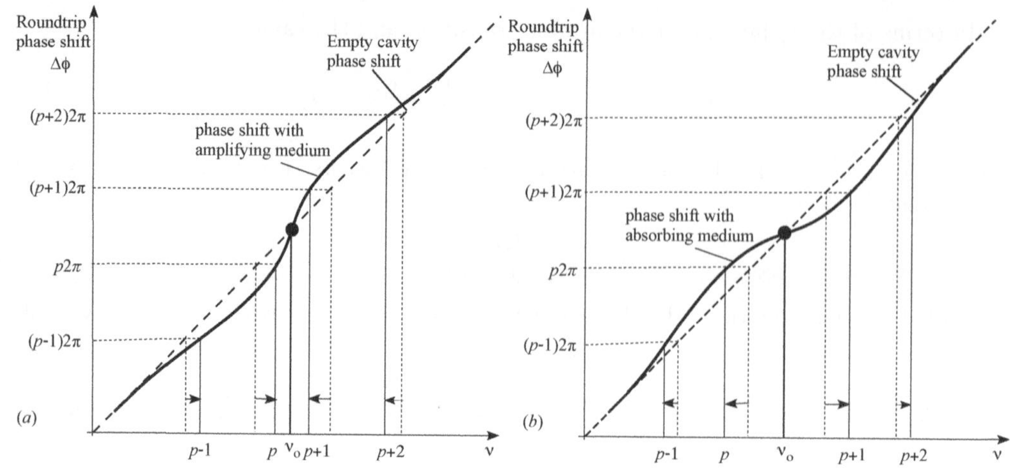

FIGURE 8.7
Graphical solutions illustrating frequency pulling and pushing when the dispersive medium fills the entire cavity.

frequency solutions associated with a single longitudinal mode index p. In such cases, the analysis is further complicated by spectral hole burning and mode-competition effects.[5]

For the more complex situations described in the preceding paragraph, a graphical analysis of frequency pulling and pushing is particularly instructive. Recall that the roundtrip phase shift $\Delta\phi$ in a laser cavity must satisfy the condition $\Delta\phi(\nu) \equiv (k + \Delta k)2L = 2\pi p$. Without the gain medium, the $\Delta\phi(\nu)$ versus frequency curve is the dashed straight line in Fig. 8.7, and its intersections with the $2\pi p$ resonance values define the bare cavity modes shown by the dotted vertical lines. The solid curve gives an exaggerated representation of how the additional contribution Δk to the wave number shifts the frequencies at which $\Delta\phi(\nu)$ intersects the $2\pi p$ resonance values. An amplifying transition tends to pull the cavity-mode frequencies toward the atomic transition frequency ν_o [Fig. 8.7]. In contrast, for an absorbing transition, Δk has the opposite sign and tends to push the cavity resonances away from line center [Fig. 8.7(b)]. Modes within the central part of the atomic line are generally affected more strongly by frequency pulling or pushing than modes farther from line center because the additional phase contribution is appreciable primarily in the neighborhood of the atomic transition frequency ν_o, as determined by the lineshape function and saturation effects.

8.6 Single-mode or Multimode?

In our discussion so far, we have assumed that only one cavity mode interacts strongly with the active atoms. Lasers can be designed or controlled to operate so that only a single cavity mode oscillates, and for many applications, such as holography, imaging, and interferometry, this is the desired mode of operation. Other applications, such as the generation of trains of short pulses, require the simultaneous oscillation of many modes, and multimode operation is then desirable.

[5]Lee Casperson and A. Yariv, "Longitudinal modes in a high gain laser," Appl. Phys. Lett. **17**, 259–261 (1970).

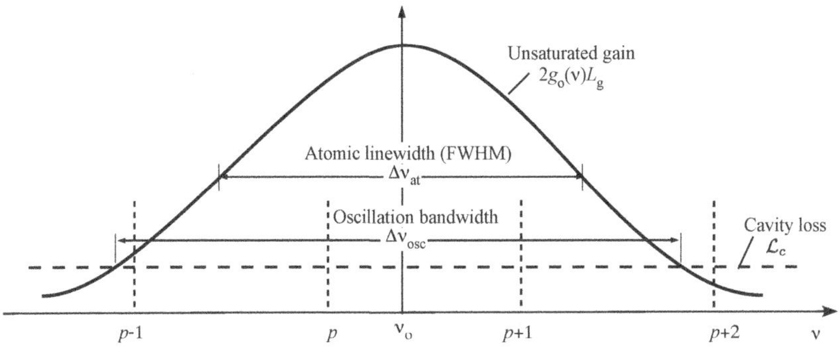

FIGURE 8.8
Unsaturated gain, atomic linewidth, oscillation bandwidth and axial cavity modes of a laser.

Recalling that the frequency dependence of the unsaturated gain is contained in the lineshape function $S(\nu)$, the unsaturated roundtrip gain $2g_o(\nu)L_g$ as a function of frequency may have the form shown schematically in Fig. 8.8. If the roundtrip cavity loss $\mathcal{L}_c$ is represented on the same graph by the dashed line, the gain exceeds the loss over a frequency interval of width $\Delta\nu_{\text{osc}}$ centered at ν_o within which laser oscillation *can* occur. As an example, consider a homogeneously broadened line centered at ν_o, with FWHM $\Delta\nu_{at} = \Delta\nu_H$, and suppose that the unsaturated line-center gain $2g_o(\nu_o)L_g$ is r times the cavity loss. The gain exceeds the loss over the frequency interval centered at frequency ν_o.

$$\Delta\nu_{\text{osc}} = (r-1)^{1/2}\Delta\nu_H\,, \tag{8.61}$$

Laser oscillation, however, can occur only at the discrete frequencies corresponding to cavity modes. We already understand the reason for this from our discussion of laser modes in Chapters 3 and 4. Once the pump creates a population inversion, spontaneous emission from the inverted atoms can couple into any one of the approximately $\left(\frac{8\pi n^3\nu^2\Delta\nu_{\text{at}}}{c^3}\right)V$ electromagnetic modes within the gain bandwidth and the volume of the gain medium. We estimated this number [see Eq. (2.78)] to be of order $10^8 - 10^{12}$. Radiation emitted into most of these directions and modes is not retained by the resonator, whereas emission into directions corresponding to cavity modes is redirected by the cavity mirrors and repeatedly passes through the gain medium, stimulating further emission into those modes and thereby amplifying their fields.

Because the initial frequency dependence of the available gain is governed by $g_o(\nu)$, cavity modes close to line center are generally amplified more strongly than those in the wings. As the mode intensities grow, however, gain saturation becomes important and, as discussed in Sec. 7.6, changes the gain profile. What, then, are the amplitudes or powers of the cavity modes for which the unsaturated gain exceeds threshold? We explore this question next.

8.6.1 Homogeneous Lasers: Single-Mode Oscillation

In a homogeneously broadened laser transition, the atomic lineshape is identical for all atoms in the gain medium. Suppose the gain on such a transition is increased until the axial mode closest to line center, say the pth mode, is the first to reach threshold, as shown in Fig. 8.9. This mode begins to oscillate while all other cavity modes remain below threshold. As its

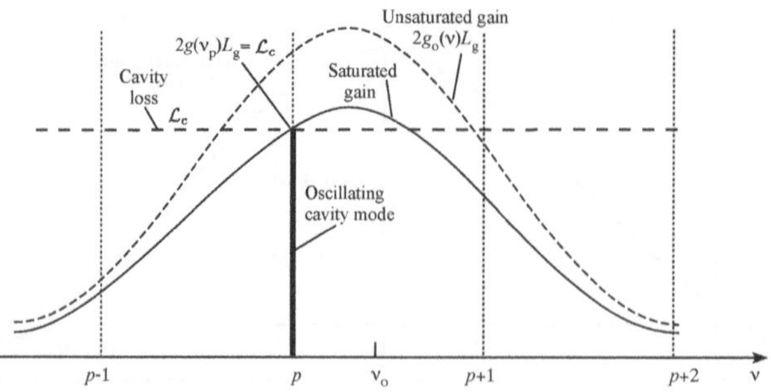

FIGURE 8.9
In a homogeneously broadened laser, the first mode to reach threshold saturates the entire gain curve, without changing its shape [Eq. (7.84)], to clamp it to cavity loss at its operating frequency (solid curve), keeping all other modes below threshold.

intensity grows, it saturates the gain at its own frequency down to the cavity-loss level and, in the process, reduces the gain across the entire homogeneous line, as discussed in Sec. 7.6. The saturated gain curve is shown by the solid line in Fig. 8.9. Further increases in pump power increase the intensity of the oscillating mode, but within this ideal homogeneous-saturation model they do not raise the saturated gain curve sufficiently to bring another axial mode to threshold. Here the oscillating mode remains at a fixed frequency while the gain available to other weak cavity modes is considered, so the relevant behavior is homogeneous cross-saturation rather than the self-saturation power broadening discussed earlier. Note that the saturated gain can exceed the cavity loss at line center if no cavity mode is located exactly at that frequency.

It is possible for several modes to experience gain exceeding their threshold values transiently during initial turn-on or pulsed operation of the laser. In steady-state operation of an ideal homogeneously broadened laser with complete cross-saturation, however, the mode with the largest net gain dominates, while the other modes fall to levels sustained primarily by spontaneous emission.

The axial mode closest to line center need not be the first or preferred mode to oscillate if mode-control techniques are used to increase its loss relative to that of another axial mode farther from the center of the atomic gain profile. In a Nd:YAG laser, for example, the atomic transition has a linewidth of ∼ 120 GHz and strongly homogeneous character, so that hundreds of axial cavity modes may fall within the atomic gain profile. Under transient conditions, such a laser can exhibit strongly multimode oscillation. Under continuous-wave (cw) conditions, however, it commonly oscillates in only one or a few simultaneous axial modes.

Thus, under ideal steady-state conditions, a laser with homogeneous gain tends to oscillate in a single preferred mode, namely the mode with the highest net gain. The gain in the laser medium is clamped at the level required to reduce the saturated gain at that mode frequency to the cavity-loss level. Increasing the pump then increases the intensity of the preferred mode rather than increasing the saturated gain sufficiently to bring additional modes to threshold. Several practical effects, including spatial hole burning, a nonuniform spatial distribution of gain, polarization effects, and unequal modal losses, can weaken this conclusion in real lasers. Nevertheless, homogeneously broadened lasers generally have a strong tendency toward single-frequency operation on the most favored axial and transverse mode.

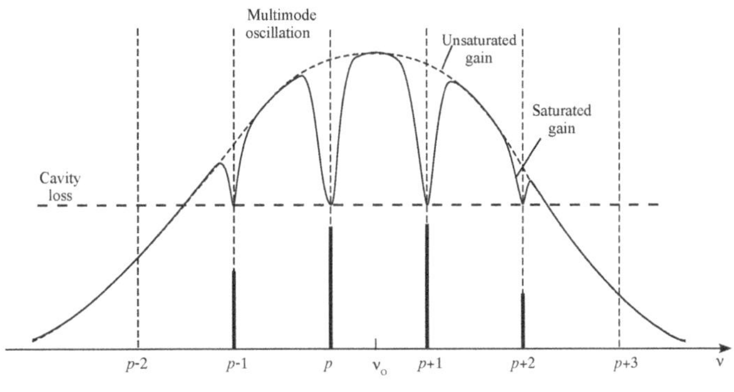

FIGURE 8.10
In a laser with an inhomogeneously broadened gain profile, each oscillating mode predominantly saturates a different subgroup of atoms, burning a hole in the gain profile near its frequency.

8.6.2 Inhomogeneous Lasers: Multimode Oscillation

In contrast to lasers based on homogeneously broadened lines, Doppler-broadened gas lasers and other lasers with strongly inhomogeneously broadened transitions readily allow simultaneous oscillation on multiple axial modes within the laser gain profile. When the gain on an inhomogeneously broadened transition exceeds the cavity loss, each axial mode for which this condition is satisfied can grow. However, each mode predominantly saturates only the subgroup of atoms whose resonance frequencies lie within a homogeneous linewidth of its oscillation frequency. As a result, different axial modes can saturate different subgroups of atoms, each "burning a hole" in the gain profile near its own frequency, as shown in Fig. 8.10.

Doppler-broadened gas lasers can therefore oscillate simultaneously on many axial modes. For example, in a He-Ne laser with a Doppler linewidth of $\Delta\nu_D \approx 1500$ MHz and an axial-mode spacing of 150–500 MHz, typically several axial modes lie within the gain linewidth, and the laser can oscillate simultaneously on several of them. On the other hand, a low-pressure CO_2 laser with a much smaller Doppler linewidth of approximately 60–100 MHz may have only one axial mode within the gain linewidth, depending on the cavity length. Far-infrared molecular gas lasers also tend to have narrow Doppler linewidths because they operate at relatively low gas pressures and because Doppler broadening in frequency units decreases with increasing operating wavelength. Consequently, such lasers may also have only one axial mode within the atomic linewidth.

8.6.3 Multimode Oscillations and Spatial Hole Burning

One of the most important mechanisms leading to multimode operation even in homogeneously broadened lasers is spatial inhomogeneity of gain saturation. Spatial hole burning is an important example. A standing-wave laser oscillating in the pth axial mode has a periodic intensity profile along the z-axis, with adjacent intensity maxima separated by approximately one-half of the wavelength in the medium, as shown in Fig. 8.11(a). Consequently, the population inversion is saturated in a similarly periodic fashion, producing spatial holes in the inversion, as illustrated in Fig. 8.11(b). The standing-wave pattern of the $(p+1)$th mode—which fits one additional half-wavelength into the cavity length—is displaced relative to that of the pth mode. Consequently, over substantial portions of the gain medium, the $(p+1)$th mode has appreciable intensity where the inversion has

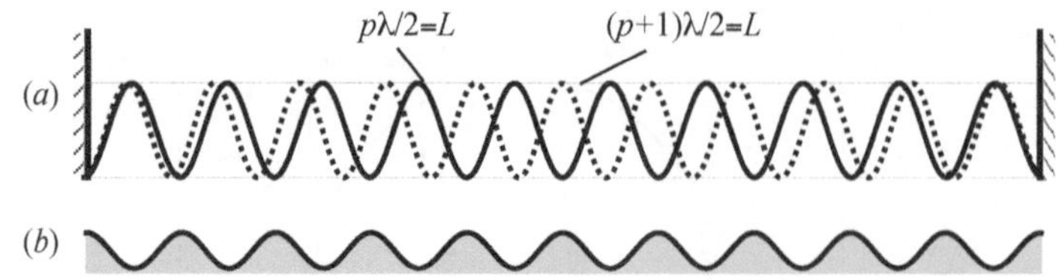

FIGURE 8.11
Spatial hole burning in a standing-wave laser: (a) Intensity profiles of pth and $(p+1)$th modes; (b) Spatial holes in the inverted population.

not been strongly saturated by the pth mode, as shown schematically in the figure. The two adjacent axial modes therefore sample the spatially varying inversion differently. Their cross-saturation is reduced, and both modes may be able to oscillate simultaneously even in a strongly homogeneously broadened laser medium.

Spatial-hole-burning effects can also be reduced in suitable resonator geometries by placing short gain sections where the intensity patterns of competing axial modes have greater spatial overlap. Unidirectional oscillation in a ring laser provides another important way of eliminating standing-wave spatial hole burning and thereby promoting single-frequency operation.

In addition to the longitudinal modes of a cavity, higher-order transverse modes can oscillate, depending on the spatial distribution of gain and loss. Their resonant frequencies $\nu_{\ell mp}$ differ from the frequencies ν_{00p} of the fundamental transverse modes and may therefore experience sufficient net gain to reach threshold. In addition, higher-order transverse modes sample different regions of the gain medium because of their different transverse intensity distributions, as can be seen from the profiles of the HG_{mn} and $\mathrm{LG}_{p\ell}$ modes in Figs. 3.13 and 3.14. For example, the ν_{00p} mode interacts most strongly with excited atoms near the axis of the gain medium, whereas a higher-order mode such as ν_{11p} samples regions farther from the axis. Thus, even a homogeneously broadened transition can support multimode oscillation through spatially nonuniform gain saturation. Multimode oscillation becomes particularly likely when the gain distribution itself is nonuniform in ρ or φ.

8.6.4 Single-Mode Selection

Many applications of laser light require single-transverse-mode operation, most commonly in the fundamental transverse mode ν_{00p}. For this purpose, mode-dependent losses can be introduced into the laser cavity, forcing unwanted modes to experience sufficiently high losses that they cannot reach threshold. From our discussion of diffraction losses in Sec. 4.3, it is clear that higher-order transverse modes generally suffer greater diffraction losses at the mirrors. However, this effect may not provide sufficient loss discrimination between the fundamental transverse modes ν_{00p} and the lower-order transverse modes $\nu_{\ell mp}$ with $\ell + m \leq 4$. Additional mode-dependent diffraction losses can be introduced by inserting an aperture inside the cavity at a location where the difference between the spot radii of the fundamental and higher-order transverse modes is large (a hard aperture). Because the higher-order modes generally have larger transverse extent than the fundamental mode, such an aperture introduces greater diffraction loss for the higher-order modes, thereby favoring operation in the fundamental transverse mode.

In the case of an inhomogeneously broadened medium, even after the transverse modes have been suppressed by an aperture, several longitudinal modes may have unsaturated gain exceeding the threshold value determined by the cavity loss $\mathcal{L}_c$. Moreover, as discussed

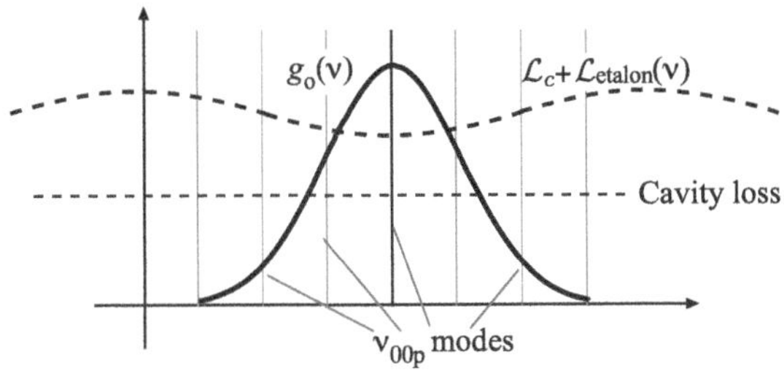

FIGURE 8.12
Gain profile and axial cavity modes in an inhomogeneously broadened laser in the presence of an etalon for axial mode selection.

earlier, different longitudinal modes can predominantly saturate different groups of atoms and can therefore oscillate simultaneously. Single-longitudinal-mode operation can be promoted by inserting a frequency-selective element into the cavity that increases the loss for all but one selected mode.

This can be achieved by inserting an etalon, i.e., a thin plane-parallel transparent plate, into the cavity. The two partially reflecting surfaces of the etalon form a low-Q Fabry-Perot resonator with a frequency-dependent transmission. At frequencies corresponding to an etalon transmission resonance, multiple reflections within the plate interfere constructively in transmission, and the insertion loss is minimized. At other frequencies, the etalon transmission is reduced, thereby increasing the effective cavity loss experienced by the corresponding laser modes [Fig. 8.12]. The etalon can be tuned so that a transmission maximum coincides with the desired laser mode by changing the angle of incidence. Its thickness is chosen so that the next coincidence between a cavity mode and an etalon transmission maximum occurs outside the frequency range over which the laser gain can exceed threshold.

8.7 Special Techniques for CW Operation

8.7.1 Wavelength Tuning

Certain laser systems, such as dye lasers and vibronic solid-state lasers (e.g., Ti:sapphire), have inherently broad gain bandwidths, allowing laser action over a broad, continuous range of wavelengths. Numerous other laser media exhibit gain on several discrete transitions (e.g., argon-ion, He-Ne, or CO_2 lasers). Consequently, for applications requiring laser light at a specific wavelength or tunability over a range of wavelengths, wavelength-selective elements are incorporated into the cavity to select the operating wavelength and, where possible, tune it over the available gain bandwidth.

8.7.1.1 Diffraction Grating Tuner

Diffraction gratings serve as important wavelength-selective elements in laser systems. A common method used in many dye and CO_2 lasers is to replace a conventional resonator mirror with a reflective diffraction grating. The grating is used in the so-called Littrow configuration [Fig. 8.13(a)]. In this configuration, the grating is oriented so that, for a given angular setting, a particular wavelength is diffracted exactly back along the incident direction and, therefore, back into the laser cavity. The selected wavelength is determined

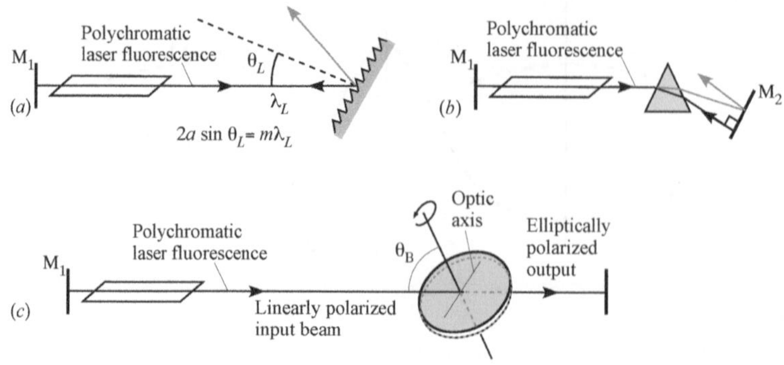

FIGURE 8.13
Laser wavelength tuning using the dispersive behavior of a diffraction grating (a), a dispersive prism (b), and the wavelength-dependent retardation of a birefringent plate (c).

by the Littrow condition[6]

$$\lambda_L = \frac{2a \sin \theta_L}{m}, \tag{8.62}$$

where θ_L is the angle of incidence (Littrow angle), a is the grating period (groove spacing), and m is the diffraction order, often $m = 1$. The grating is often blazed to maximize diffraction efficiency in the desired order. To tune the laser wavelength, the angle of incidence θ_L is varied by rotating the grating about an axis parallel to its grooves. Configurations other than the Littrow configuration are sometimes used in which an external mirror returns a diffracted beam to the grating. In this case, the grating no longer serves as an end reflector, and the selected beam interacts with the grating twice in the wavelength-selective section of the cavity. This double interaction generally provides greater wavelength discrimination than a single interaction with the grating, although it also introduces additional loss.

8.7.1.2 Dispersive Prism Tuner

Another wavelength-tuning method employs a dispersive prism placed directly within the laser cavity [Fig. 8.13(b)]. To minimize reflection losses, the prism is commonly arranged so that the intracavity beam is incident at or near Brewster's angle at its surfaces. As the broadband emission from the gain medium passes through the prism, angular dispersion spatially separates its constituent wavelengths. The angularly dispersed wavelengths propagate toward the cavity end mirror M_2. For a given angular setting of the prism, only a particular wavelength is directed so that reflection from M_2 returns it along the appropriate optical path through the prism and back into the laser cavity, thereby providing optical feedback at that wavelength. Wavelength tuning is achieved by rotating the prism about an axis parallel to its refracting edge, thereby changing the wavelength that satisfies this retroreflection condition.

[6] T. M. Hard, "Laser wavelength selection and output coupling by a grating," Appl. Opt. **9**, 1825-1830 (1970); M. G. Littman, "Single-mode operation of grazing incidence pulsed dye laser," Opt. Lett. **3**, 138-140 (1978).

8.7.1.3 Birefringent Tuner

A third wavelength-selective method uses a birefringent tuner inserted within the laser cavity [Fig. 8.13(c)]. The tuner consists of one or more plates of a suitable birefringent crystal (for example, quartz), with the optic axis lying in the plane of the plate. The plate is oriented with its surface normal inclined so that the intracavity beam is incident at Brewster's angle (θ_B), thereby minimizing reflection loss for the selected linear polarization. Upon entering the birefringent plate, the electric field is resolved into orthogonally polarized ordinary and extraordinary components, which propagate with different phase velocities and therefore acquire a relative phase difference. In a simplified description, the phase difference acquired in passing through the plate can be written as

$$\Delta\phi = \frac{2\pi|n_e - n_o|t}{\lambda}, \tag{8.63}$$

where n_o and n_e are the effective refractive indices experienced by the ordinary and extraordinary components, respectively, and t is the propagation length through the plate. A more exact treatment of the Brewster-angle geometry must take into account the internal propagation angle and the angular dependence of the extraordinary refractive index.

If $\Delta\phi = 2\pi m$, where m is an integer, the polarization state is restored after passage through the plate, and the beam experiences minimum polarization-dependent loss upon subsequent propagation through the Brewster-oriented elements of the cavity. For other values of $\Delta\phi$, the recombined beam is generally elliptically polarized and therefore contains a component orthogonal to the low-loss Brewster polarization, increasing the cavity loss. The birefringent plate therefore produces a periodic wavelength dependence of the intracavity loss.

For a sufficiently broad emission bandwidth, the condition for minimum loss can be satisfied at several wavelengths. Assuming that the birefringence $|n_e - n_o|$ does not vary appreciably between adjacent minima, the frequency separation between consecutive minimum-loss orders of the birefringent tuner–its free spectral range–is approximately

$$\Delta\nu_{\mathrm{FSR}} \approx \frac{c}{|n_e - n_o|t}. \tag{8.64}$$

Thus, the plate thickness determines the free spectral range: the thinner the plate, the larger the free spectral range and, consequently, the broader the wavelength interval over which a single minimum-loss order can be selected. Tuning over one such interval is achieved by rotating the plate, thereby changing the angle between the propagation direction and the optic axis and hence the effective extraordinary refractive index n_e and the birefringence $\Delta n = |n_e - n_o|$. Multiple plates of different thicknesses are often combined to provide progressively finer wavelength discrimination and, when required, to select a single longitudinal mode. Polarization-selective elements in the cavity, including Brewster-angle surfaces, provide the loss discrimination that converts the wavelength-dependent retardation of the birefringent plate into wavelength-dependent cavity loss.

The grating, prism, or birefringent tuner primarily selects a wavelength region within the broad gain profile; additional fine spectral discrimination may still be required to isolate a single longitudinal cavity mode.

8.8 Laser Frequency Stabilization

Methods for wavelength tuning and mode selection alone are not sufficient for a laser to produce a stable single-frequency output. In fact, the frequency of a single-mode laser can fluctuate because of mirror vibrations, thermal expansion of the laser cavity, fluctuations in the refractive index, and other disturbances that affect the optical path length of the resonator. For applications requiring light with a well-defined and highly stable frequency, a widely used method for laser frequency stabilization—or frequency locking—is the Pound-Drever-Hall (PDH) technique.[7] This method is based on the phase-sensitive reflection of laser light from a stable Fabry-Perot (FP) reference cavity [Fig. 8.14]. We present a simplified treatment here to explain the principle underlying this method.

In the PDH technique, a small fraction of the output beam from the laser is phase-modulated and sent to an FP reference cavity. If a Pockels cell is used to phase modulate the laser field $E_o e^{-i\omega t}$, the output from the phase modulator [see Sec. 12.7.2] can be written as[8]

$$E_i(t) = E_o e^{-i\omega t} e^{iM\cos(\omega_M t)} \approx E_o e^{-i\omega t} + i\frac{M}{2} E_o e^{-i(\omega-\omega_M)t} + i\frac{M}{2} E_o e^{-i(\omega+\omega_M)t}, \quad (8.65)$$

where we have assumed $M \ll 1$. The weak sinusoidal phase modulation thus produces, in addition to the carrier at frequency ν, two first-order sidebands at frequencies $\nu - \nu_M$ and $\nu + \nu_M$, where $\omega = 2\pi\nu$ and $\omega_M = 2\pi\nu_M$, as usual.

If this modulated wave is incident on the reference cavity, the reflected wave amplitude is

$$E_R(t) = E_o e^{-i\omega t} \left[\rho_c(\nu) + i\frac{M}{2}\rho_c(\nu - \nu_M) e^{i\omega_M t} + i\frac{M}{2}\rho_c(\nu + \nu_M) e^{-i\omega_M t} \right], \quad (8.66)$$

where $\rho_c(\nu)$ is the amplitude reflection coefficient of the FP reference cavity [see Sec. 4.4]. Taking, for simplicity, the power reflectances of the reference-cavity mirrors to be the same, $\mathcal{R}_1 = \mathcal{R}_2 = \mathcal{R}$, the refractive index $n = 1$, and the internal losses to be negligible ($\alpha = 0$), the amplitude reflection coefficient $\rho_c(\nu)$ is given by Eq. (4.103) as

$$\rho_c(\nu) \equiv \frac{E_R}{E_i} = \frac{\sqrt{\mathcal{R}}\left(e^{i4\pi\nu d/c} - 1\right)}{1 - \mathcal{R}e^{i4\pi\nu d/c}}. \quad (8.67)$$

Assume that the modulation frequency is much larger than the reference-cavity linewidth but sufficiently smaller than the cavity free spectral range that both sidebands remain well away from neighboring resonances. When the carrier lies close to the selected cavity resonance, the sidebands are then almost completely reflected and their reflection coefficients may, over the small locking range, be treated as approximately constant. Choosing the overall phase convention so that

$$\rho_c(\nu - \nu_M) \approx 1 \approx \rho_c(\nu + \nu_M), \quad (8.68)$$

the reflected field can be written approximately as

$$E_R(t) \approx E_o e^{-i\omega t} \left[\rho_c(\nu) + iM\cos\omega_M t\right]. \quad (8.69)$$

[7] R. W. P. Drever, J. L. Hall, F. V. Kowalski, J. Hough, G. M. Ford, A. J. Munley, and H. Ward, "Laser phase and frequency stabilization using an optical resonator," Appl. Phys. B **31**, 97–105 (1983).

[8] The exact expansion $\exp[iM\cos\phi] = J_0(M) + 2\sum_{n=1}^{\infty} i^n J_n(M)\cos(n\phi)$, for $M \ll 1$ [see Sec. 11.2], reduces to Eq. (8.65) in the limit $M \ll 1$.

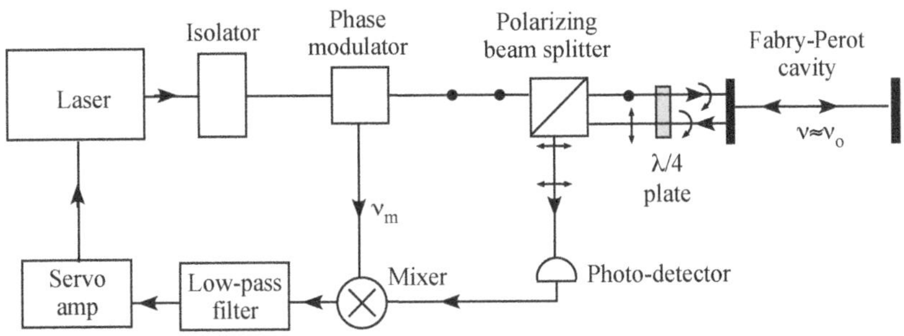

FIGURE 8.14
Simplified diagram (top view) of a Pound-Drever-Hall laser-frequency-stabilization scheme. • ≡ vertical polarization, ↕ ≡ horizontal polarization.

The photocurrent $i_R(t)$ from a detector illuminated by the reflected light is proportional to $|E_R(t)|^2$ and is therefore

$$i_R(t) = \eta |E_o|^2 \left[|\rho_c(\nu)|^2 + \frac{M^2}{2} + \frac{M^2}{2} \cos 2\omega_M t + 2M\rho_c''(\nu) \cos \omega_M t \right] , \tag{8.70}$$

where η is the detector sensitivity and $\rho_c''(\nu)$ is the imaginary part of the amplitude reflection coefficient. The overall sign of the last term depends on the phase convention adopted for the reflected sidebands and has no physical significance for the locking procedure because the electronic demodulation phase can be chosen accordingly.

If the laser carrier frequency ν is offset from the reference-cavity resonance frequency ν_{ref}, we write

$$\nu = \nu_{\text{ref}} + \Delta\nu \equiv p\frac{c}{2d} + \Delta\nu\,, \tag{8.71}$$

for some fixed integer p, with $|\Delta\nu| \ll c/2d$. Near resonance,

$$\begin{aligned} \rho_c''(\nu) &= \mathcal{I}m \left[\frac{\sqrt{\mathcal{R}}\left(e^{i4\pi\Delta\nu d/c} - 1\right)}{1 - \mathcal{R}e^{i4\pi\Delta\nu d/c}} \right] \\ &\approx \frac{4\pi d\sqrt{\mathcal{R}}}{c(1-\mathcal{R})}\Delta\nu = \frac{2\Delta\nu}{\Delta\nu_c}\,, \end{aligned} \tag{8.72}$$

where we have used the high-finesse expression for the cavity intensity linewidth $\Delta\nu_c$ from Eqs. (4.114) and (4.115). Thus, sufficiently close to resonance, the imaginary part of the cavity reflection coefficient is proportional to the frequency detuning and changes sign as the carrier passes through resonance.

The photocurrent component at frequency ω_M in Eq. (8.70), proportional to $\rho_c''(\nu)$, is the term of interest in the Pound-Drever-Hall stabilization technique. Its crucial property is that it is proportional to the laser frequency offset $\Delta\nu$ from the reference-cavity resonance and changes sign according to whether ν lies above or below ν_{ref}. After phase-sensitive demodulation, this component therefore provides a bipolar error signal that can be used in an electronic feedback loop to force the laser carrier frequency toward the reference-cavity resonance. The stability that can be achieved depends on such factors as the stability of the reference cavity, the signal-to-noise ratio, and the gain and bandwidth of the feedback system.

Figure 8.14 shows a simplified schematic diagram of a Pound-Drever-Hall laser frequency stabilization scheme. Vertically polarized light from the laser passes through an isolator and is phase-modulated at frequency ν_M. The resulting carrier and sidebands transmitted by the polarizing beam splitter (PBS) are incident on a quarter-wave plate (QWP; $\lambda/4$ plate), with its fast axis at 45° to the incident linear polarization. The circularly polarized light exiting the QWP reflects from the reference cavity and passes through the QWP again, emerging with linear polarization rotated by 90°. It is therefore reflected by the PBS onto a photodetector whose output current contains the component at ω_M described by Eq. (8.70).

The photocurrent component at frequency ω_M is retrieved by combining the photodetector signal with a phase-adjusted portion of the modulation signal in a mixer, whose output contains a term proportional to the product of the two signals. Passing the mixer output through a low-pass filter removes the rapidly oscillating components and produces an error-signal voltage which, near resonance, has the form

$$V_s = \text{const} \times M|E_o|^2 \frac{\Delta\nu}{\Delta\nu_c} . \tag{8.73}$$

The proportionality constant includes the detector response, electronic gain, demodulation phase, and numerical factors. This error signal is applied through a servo system to an appropriate laser-frequency actuator so as to bring the laser frequency into coincidence with the selected reference-cavity resonance ν_{ref}. For laser frequency stabilization, only the region in which the error signal varies approximately linearly with frequency offset is normally used.

Outside this linear region, the complete PDH error signal must retain the complex reflection coefficients of both sidebands. Apart from an overall phase and normalization determined by the demodulation convention, its frequency dependence is obtained from the appropriate quadrature of

$$\rho_c(\nu)\rho_c^*(\nu + \nu_M) - \rho_c^*(\nu)\rho_c(\nu - \nu_M) . \tag{8.74}$$

Instead of a reference cavity, an atomic or molecular absorption line can also provide a frequency reference for laser stabilization. In this case, laser light passing through an absorption cell experiences a frequency-dependent phase shift. The total propagation phase is $\phi = 2\pi n\ell/\lambda$, where ℓ is the length of the absorption cell and n is the refractive index of the medium. For a homogeneously broadened Lorentzian transition, the dispersive contribution to the refractive index is related to the absorption coefficient α_{at} by [Eqs. (5.29) and (5.40)]

$$n(\nu) = n_o + \frac{c}{2\pi\nu} \frac{\nu_o - \nu}{\Delta\nu_H} \alpha_{\text{at}}(\nu) , \tag{8.75}$$

where n_o is the refractive index of the background medium, ν is the frequency of the light wave, ν_o is the transition frequency of the absorption line, and $\Delta\nu_H$ is the homogeneous linewidth. The resonant contribution to $n(\nu)$ produces a frequency-dependent dispersive phase shift $\Delta\phi_{\text{disp}} = 2\pi[n(\nu) - n_o]\ell/\lambda$. Near line center, the dispersive phase shift is approximately linear in the frequency offset $\nu - \nu_o$ and changes sign at resonance.

For an inhomogeneously broadened line, the susceptibility must be obtained by summing, or integrating, the contributions at frequency ν from atomic groups having different resonance frequencies ν_a, weighted by the appropriate distribution function. The resulting expression is more complicated, but near the center of a symmetric isolated line the dispersive contribution to the refractive index—and therefore the corresponding phase shift—is again approximately linear in the laser-frequency offset $\nu - \nu_o$. This dispersive response can consequently be used to generate an error signal for frequency stabilization.

Summary

In this chapter, we used the laser rate equations developed in Sec. 7.10 to discuss the continuous-wave (cw) operation of a laser. We derived the threshold population inversion, or, equivalently, the threshold gain, required for laser oscillation and showed that, above threshold, gain saturation clamps the steady-state gain to the cavity loss. We then obtained expressions for the intracavity intensity and laser output power and examined their dependence on pump strength, cavity losses, and output coupling, including the existence of an optimum output coupling that maximizes the extracted power.

Because of the nonlinear nature of the laser equations, we considered several models and approximations to examine how the laser output depends on the spatial evolution of the intracavity field, on whether the cavity supports standing or traveling waves, and on whether the gain is homogeneously or inhomogeneously broadened. These considerations led naturally to phenomena such as spatial hole burning and the Lamb dip. We also examined how the dispersive response of the active medium modifies the cavity resonance frequencies, giving rise to frequency pulling in an amplifying medium and frequency pushing in an absorbing medium.

The frequency dependence of the gain and its saturation also determines which cavity modes can oscillate. We therefore examined the conditions for single-mode and multimode operation in homogeneously and inhomogeneously broadened lasers, including the important role of spatial hole burning in permitting multimode oscillation even in a homogeneously broadened medium. Finally, we reviewed techniques for transverse- and longitudinal-mode selection, wavelength tuning, and single-mode laser frequency stabilization.

In the next chapter, we will use a simplified form of the laser rate equations to explore the temporal buildup of laser oscillations and the dynamical transition that occurs as a laser passes through oscillation threshold.

Problems

1. Write down the threshold condition for laser oscillation in a cavity containing a gain medium, and derive an expression for the threshold population inversion density. Using this expression, discuss the physical conditions that favor laser action. Express your discussion in terms of cavity loss, gain-medium length, transition wavelength, atomic linewidth, and radiative and nonradiative decay rates. In particular, explain the relation between a narrow transition linewidth and a large radiative transition rate.

2. The energy-level diagram shown here is representative of certain three-level systems, such as molecular nitrogen, N_2. The pumped level E_2, which is also the upper laser level, decays predominantly to the lower laser level E_1 with a lifetime $\tau_2 = 20$ ns. The lower laser level E_1 decays to the ground state E_0 with a lifetime $\tau_1 = 1\,\mu$s. Assume that atoms are pumped into level E_2 at a constant rate R_p per unit volume and neglect stimulated transitions.

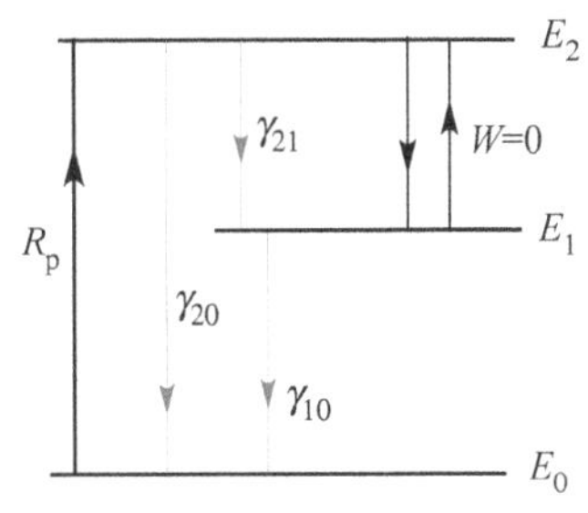

(a) If $R_p = 10^{23}$ cm^{-3}s^{-1}, determine the steady-state populations of levels E_2 and E_1. Is population inversion possible in the steady state? Is this system suitable for cw laser operation?

(b) Suppose the pump is turned on suddenly at $t = 0$ and thereafter remains constant at the value given in part (a). Assuming $\mathcal{N}_2(0) = \mathcal{N}_1(0) = 0$, determine the time dependence of the small-signal gain coefficient $g(t) = \sigma\,[\mathcal{N}_2(t) - \mathcal{N}_1(t)]$ for a stimulated-emission cross section $\sigma = 10^{-12}$ cm^2. Plot $g(t)$ as a function of time and identify the time interval, if any, during which the medium provides gain.

3. The first continuous-wave laser, which was also the first gas laser, operated on a neon transition at $\lambda = 1.15\,\mu$m [A. Javan, W. R. Bennett, Jr., and D. R. Herriott, "Population inversion and continuous optical maser oscillation in a gas discharge containing a He-Ne mixture," Phys. Rev. Lett. **6**, 106–110 (1961)]. The transition is predominantly Doppler broadened. The upper-state lifetime, which we assume to be approximately the radiative lifetime associated with the laser transition, is about 100 ns. Take the gas temperature to be $T = 500$ K, the atomic mass of neon to be 20 atomic mass units, and $n \approx 1$.

   (a) Calculate the Doppler linewidth (FWHM) and the line-center stimulated-emission cross section for this transition.

   (b) Using the saturation model developed in the text, estimate the saturation irradiance of the laser transition.

   (c) If the standing-wave laser cavity is 40 cm long, estimate how many longitudinal cavity modes can lie within the FWHM of the Doppler-broadened gain profile. Explain why the exact number depends on the position of the cavity-mode frequencies relative to line center.

4. **Determining internal laser losses:** One method for estimating the internal losses of a laser is to measure the threshold pump power using two output couplers with different reflectances. Explain the principle of this method and derive an expression for the internal roundtrip loss in terms of the two threshold pump powers and the two output-coupler reflectances. Here, the internal loss is to include all roundtrip losses other than transmission through the output coupler, such as absorption, scattering, and transmission losses at the remaining cavity mirrors and optical elements.

   For a diode-pumped solid-state laser, the threshold pump powers are measured to be $P_{\text{th}1} = 1.0$ W and $P_{\text{th}2} = 600$ mW when output couplers with power reflectances $\mathcal{R}_{o1} = 90\%$ and $\mathcal{R}_{o2} = 95\%$, respectively, are used. Assuming that the small-signal gain is proportional to pump power and that changing the output coupler does not otherwise alter the laser, determine the internal logarithmic roundtrip loss factor.

5. **Optimum output coupling:** A high-power CO_2 laser has a small-signal line-center gain coefficient $g_o(\nu_o) \approx 0.005$ cm^{-1}. The laser transition is homogeneously broadened with a Lorentzian lineshape of FWHM $\Delta\nu_{_H} \approx 2.0$ GHz. The gain medium fills nearly the entire 50-cm separation between the cavity mirrors. Internal losses due to scattering and absorption amount to approximately 2% per cavity roundtrip. One of the mirrors is perfectly reflecting.

   (a) Determine the output-coupler transmission that maximizes the laser output.

   (b) The saturation irradiance I_s for this laser is estimated to be 100 kW/cm^2. For the optimum output coupling found in part (a), estimate the intracavity irradiance and the output irradiance. If the laser beam has cross-sectional area A, express the corresponding output power in terms of A.

6. **Threshold inversion and saturation irradiance for an Nd:YAG laser:**

   (a) Estimate the threshold inversion density for a four-level Nd^{3+}:YAG laser with a 20-cm-long cavity, a total roundtrip loss of 4%, refractive index $n = 1.82$, wavelength $\lambda = 1.064\,\mu$m, radiative transition rate $A_{21} = 1.82 \times 10^3$ s^{-1}, peak stimulated-emission cross section $\sigma_o = 3.5 \times 10^{-19}$ cm^2, and linewidth (FWHM) $\Delta\nu_H = 120$ GHz at room temperature. Assume that the gain medium fills the entire cavity.

   Compare the threshold inversion density with the density of Nd^{3+} ions in the crystal. The density of YAG, $Y_3Al_5O_{12}$, is 4.56 g/cm^3, and approximately 1% of the Y^{3+} sites are occupied by Nd^{3+} ions. Take the atomic weights of Y, Al, and O to be 88.91, 26.98, and 16, respectively. [Data from *Solid State Laser Engineering*, W. Koechner (Springer-Verlag, Berlin, 1988), 2nd ed., p. 49.]

   (b) Assuming that the effective upper-laser-level population lifetime is 550 μs and that the lower laser level empties rapidly, estimate the saturation irradiance. If the 4% roundtrip cavity loss is assumed to arise entirely from the output coupler, and the intracavity beam has a uniform radius of 0.5 mm, estimate the unsaturated inversion (or equivalently the pump level) required to obtain an output power of 1.0 W.

7. **Intracavity photon number and output photon flux:** A standing-wave He-Ne laser operating at $\lambda = 632.8$ nm has a cavity length of 50 cm. The mirror power reflectances are $\mathcal{R}_1 = 1.00$ and $\mathcal{R}_2 = 0.98$, and all losses other than output coupling are negligible. The laser operates in a single mode and produces an output power of 10 mW through mirror M_2.

   Calculate the average number of photons stored in the cavity. Determine the circulating photon flux incident on the output mirror and the photon flux and optical power transmitted through it.

8. **Single-mode operation:** An argon-ion laser operating at $\lambda = 514.5$ nm has a predominantly Doppler-broadened gain profile with $\Delta\nu_D = 3.5$ GHz. The standing-wave cavity has length $L = 120$ cm, the gain tube has length $L_g = 100$ cm, and the total roundtrip cavity loss is 10%. The line-center stimulated-emission cross section is 2.5×10^{-13} cm^2, and the upper-state lifetime is approximately 5 ns.

   (a) Assuming that the lower laser level empties much faster than the upper laser level and that one longitudinal cavity mode coincides exactly with line center, calculate the threshold inversion density for this mode and the corresponding threshold pump rate density.

   (b) Assuming that the unsaturated gain profile scales linearly with pump rate and that different velocity groups saturate independently, determine how far above the line-center threshold the pump must be increased before either of the two adjacent longitudinal modes can reach threshold.

9. **Number of oscillating modes:** An argon-ion laser operating at 514 nm uses a standing-wave two-mirror resonator of length 0.50 m. The gas discharge fills the entire cavity and has an estimated temperature of 2000 K, producing a predominantly Doppler-broadened gain profile. The gas density is sufficiently low that the refractive index may be taken as unity. Take the atomic mass of argon to be 40 atomic mass units.

   If the small-signal line-center gain is 1.5 times the threshold gain, estimate the number of longitudinal modes for which the unsaturated gain exceeds threshold. Comment on how the exact integer number depends on the placement of the cavity-mode frequencies relative to line center.

10. Consider the laser transition between the levels shown in the figure to the right. The pump excites atoms from level E_0 to level E_2 at a rate R_2 per unit volume. The nonradiative decay rates from level E_2 to levels E_1 and E_0 are $\gamma_{21} = 10^7\ \text{s}^{-1}$ and $\gamma_{20} = 5 \times 10^6\ \text{s}^{-1}$, respectively. Level E_1 relaxes sufficiently rapidly that the approximation $\mathcal{N}_1 \approx 0$ is valid. The spontaneous-emission rate for the $E_2 \to E_1$ transition is $A_{21} = 6 \times 10^6\ \text{s}^{-1}$, and the transition has a homogeneous Lorentzian linewidth (FWHM) of 10 GHz. Assume steady-state operation.

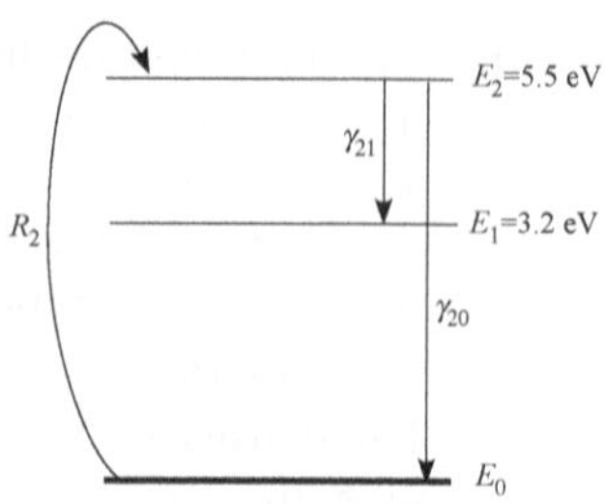

(a) Calculate the line-center stimulated-emission cross section and the saturation irradiance for the $E_2 \to E_1$ transition.

(b) What pump rate density R_2 is required to produce a small-signal line-center gain coefficient of 0.01 cm^{-1} (1.0% gain per cm)?

(c) What minimum pump power density is required to produce the gain coefficient found in part (b), assuming that each pump excitation requires an energy $E_2 - E_0$?

11. A thin quartz plate used as an intracavity etalon can assist in longitudinal-mode selection [see Sec. 8.6.4]. The etalon is operated near normal incidence and tuned by rotating it about an axis in the plane of the plate and perpendicular to the plane of incidence. Take the refractive index to be $n = 1.5$ and the plate thickness to be $d = 2$ mm.

(a) Determine the free spectral range of the etalon near normal incidence.

(b) Derive an expression for the frequency ν_q of the qth transmission maximum as a function of the external angle of incidence θ. Use Snell's law to express your result in terms of θ.

(c) Obtain the small-angle dependence of ν_q on θ. Is the frequency shift linear in the tilt angle near normal incidence?

12. A cylindrical Nd:YAG rod of length $L_g = 5.0$ cm and cross-sectional area 4 mm^2 is placed in a unidirectional ring cavity of roundtrip length 30 cm. The laser operates at $\lambda = 1064$ nm, with the cavity parameters indicated in the accompanying figure. Assume a four-level gain medium with an upper-laser-level lifetime of 230 μs and a negligibly short lower-level lifetime. The stimulated-emission cross section is $\sigma = 2.8 \times 10^{-19}\ \text{cm}^2$.

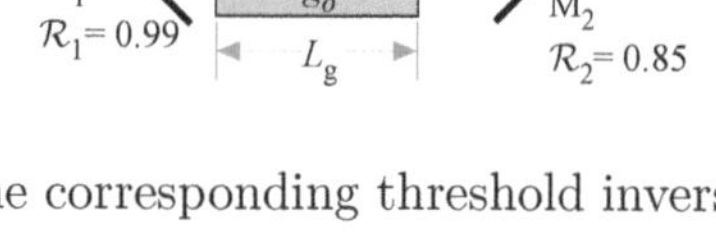

(a) Determine the threshold gain coefficient and the corresponding threshold inversion density.

(b) If the unsaturated gain coefficient is $g_o = 2g_{\text{th}}$, determine the output power through mirror M_2, assuming a uniform intracavity beam of radius 1 mm.

13. **Frequency dependence of laser power in a homogeneously broadened laser:** Consider a standing-wave laser with a homogeneously broadened gain profile operating at a fixed pump rate above threshold. To enforce single-mode operation, a narrowband frequency-selective element is inserted into the cavity and tuned across the gain profile so that only one cavity mode, at the selected frequency, can oscillate.

(a) Using the small-gain and small-output-coupling approximation, determine the output irradiance as a function of the selected oscillation frequency. Express your result in terms of the line-center saturation irradiance I_s, the line-center pump ratio r, the output-coupler transmission $\mathcal{T}$, and the homogeneous linewidth $\Delta\nu_H$.

(b) Determine the frequency range over which laser oscillation is possible. What is the total tuning range?

The situation modeled here is an idealization of that encountered in cw tunable lasers such as dye lasers.

14. **Frequency dependence of laser power in an inhomogeneously broadened laser:** In the text, we discussed the frequency dependence of the single-mode output power of a standing-wave laser with a Doppler-broadened gain medium.

Discuss the corresponding behavior for a unidirectional ring laser with a Doppler-broadened gain medium. Do you expect a Lamb dip at line center? Explain the physical basis for your answer.

Next, consider a laser whose gain medium has a strongly inhomogeneously broadened transition arising from a static distribution of atomic resonance frequencies, as in an inhomogeneously broadened glass. Discuss the frequency dependence of the single-mode output power for both standing-wave and unidirectional ring configurations. Compare this behavior with that of the Doppler-broadened gas laser, emphasizing the physical origin of any differences.

15. **Littrow condition:**

(a) Light with a wavelength equal to the grating period is incident on a reflection grating operated in the first-order Littrow configuration. Find the Littrow angle.

(b) A grating with 1200 lines/mm is used in the first-order Littrow configuration for 580-nm light from a dye laser. At what angle should the grating normal be oriented relative to the cavity axis?

9

Buildup of Oscillations and Photon Statistics of Laser Light

In this chapter we use the laser rate equations derived in Sec. 7.10 to develop a simpler set of rate equations that couple the cavity photon number to the population of laser-active atoms in the gain medium. These equations are then modified to include the contribution of spontaneous emission to the cavity photon number. This contribution, though small, plays a significant role near threshold and is critical to the buildup of laser oscillations, the photon statistics of laser light, and the ultimate limit to the laser linewidth. Using these modified equations, we explore the buildup of laser oscillations from noise in a laser cavity and the transformation of laser light as the system passes through the threshold of laser oscillation. The insights gained from these equations are then used to derive rate equations for the cavity photon-number probabilities and to discuss the photon statistics of laser light.

We begin with the population and intensity rate equations (7.96) and (7.98) for the laser and rewrite them in a simpler form by reinterpreting some of the rates that appear in them. We consider a standing-wave laser cavity in which the field is described in terms of the intensities I_+ and I_- of right- and left-traveling waves, respectively. We assume that the phase part of the amplitude equation (7.97) has been solved and that the mode frequencies have been adjusted to account for mode-pulling effects. The field inside the cavity can then be described in terms of its intensity alone. The field interacts with a four-level gain medium in which atoms are pumped primarily to the upper laser level, while the lower laser level decays sufficiently rapidly that it remains essentially empty. Equations (7.96) and (7.98), together with Eqs. (7.43) and (7.44), then lead to the following equations for this system:

$$\frac{d\Delta\mathcal{N}}{dt} = R_2 - \gamma_2\Delta\mathcal{N} - \sigma[(I_+ + I_-)/h\nu]\Delta\mathcal{N}\,, \tag{9.1}$$

$$\frac{\partial I_+}{\partial z} + \frac{n}{c}\frac{\partial I_+}{\partial t} = (\sigma\Delta\mathcal{N} - \alpha)I_+\,, \tag{9.2}$$

$$-\frac{\partial I_-}{\partial z} + \frac{n}{c}\frac{\partial I_-}{\partial t} = (\sigma\Delta\mathcal{N} - \alpha)I_-\,. \tag{9.3}$$

The population rate equation (9.1) is written for a nearly ideal four-level gain medium for which the inversion density is essentially the population density of the upper laser level, $\Delta\mathcal{N} \equiv \mathcal{N}_2 - \mathcal{N}_1 \approx \mathcal{N}_2$, and the population recovery time is effectively the upper-level lifetime, $\gamma_{21\text{eff}} \approx \gamma_2$ [see Eqs. (7.43) and (7.44)]. Here R_2 is the rate per unit volume at which the pump introduces atoms into the upper laser level, with units of atoms·m^{-3}s^{-1}. The stimulated transition rate per atom has been written as $\sigma(\nu)(I_+ + I_-)/h\nu$, since both the right- and left-traveling waves contribute to stimulated emission from the atoms. Equations (9.2) and (9.3), describing the evolution of the right- and left-traveling wave intensities I_+ and I_-, respectively, follow from Eq. (7.98). Equations (9.1)–(9.3) show that the atomic variable $\Delta\mathcal{N}$ couples to the field intensity through the gain coefficient $\sigma(\nu)\Delta\mathcal{N} = g(\nu)$, while the field variables I_+ and I_- couple to the atomic inversion $\Delta\mathcal{N}$ through the stimulated

DOI: 10.1201/9781003584179-9

transition rate. The coupling is nonlinear because these interaction terms involve products of the atomic and field variables.

9.1 Simplified Laser Rate Equations

First consider the stimulated transition rate $\sigma(\nu)(I_+ + I_-)/h\nu$. The intensities I_+ and I_- can be written in terms of the photon number q inside the cavity, the cavity volume V_c, and the speed $v = c/n$ of light inside the cavity as

$$\frac{I_\pm}{h\nu} = \frac{1}{2}\frac{q}{V_c}\frac{c}{n}. \tag{9.4}$$

Here q/V_c represents the cavity photon-number density, one half of which contributes to the photon flux associated with the right-traveling wave and the other half to the photon flux associated with the left-traveling wave. Using Eq. (9.4), the stimulated transition rate can be written as

$$\sigma(I_+ + I_-)/h\nu = \sigma(\nu)\left(\frac{q}{2V_c} + \frac{q}{2V_c}\right)\frac{c}{n} = \left(\frac{c\sigma(\nu)}{nV_c}\right)q \equiv Kq. \tag{9.5}$$

Recalling the expression for the stimulated transition cross-section $\sigma(\nu)$ [Eq. (5.52)], the constant K can be written as

$$K = \frac{c\sigma(\nu)}{nV_c} = \frac{c}{nV_c}\frac{\lambda^2 A_{21} S(\nu)}{8\pi n^2} = A_{21}\left(\frac{c^3}{8\pi\nu^2 n^3}\frac{S(\nu)}{V_c}\right) \equiv \frac{A_{21}}{p}, \tag{9.6}$$

where $S(\nu)$ is the homogeneous lineshape of the laser transition[1] and p is a dimensionless number given by

$$p = \left[\frac{8\pi\nu^2 n^3}{c^3}\right]\frac{V_c}{S(\nu)}. \tag{9.7}$$

The quantity inside square brackets is the density of electromagnetic modes (number of modes per unit volume per unit frequency interval) in a uniform medium of refractive index n, in the vicinity of the atomic transition frequency $\nu_o \sim \nu$ [Eqs. (2.77) and (2.78)]. Thus, within this mode-counting interpretation, p represents the effective number of electromagnetic modes available within the homogeneous transition bandwidth and cavity volume, and $K = A_{21}/p$ represents the spontaneous-emission rate into one such mode per excited atom.

For the normalized Lorentzian lineshape, the line-center value is $[1/S(\nu_o)] = \pi\Delta\nu_H/2$. Since V_c is the cavity volume, p is approximately the number of electromagnetic modes within the homogeneous linewidth of the atomic transition and within the volume V_c. The number p can vary from about 10^8 for gas lasers to 10^{12} for dye lasers. For example, for a He-Ne laser with $\lambda = 633$ nm, $\Delta\nu_H = 250$ MHz at normal operating pressures, a cavity length of 50 cm, a beam radius of 1.0 mm, and $n = 1$, the number p is given by

$$\begin{aligned} p &= \left[\frac{8\pi\nu^2 n^3}{c^3}\right]\frac{V_c}{S(\nu)} = \left[\frac{8\pi n^3}{\lambda^2 c}\right]\frac{V_c\pi\Delta\nu_H}{2} \\ &= \left[\frac{8\pi}{(0.633\times10^{-6})^2(3.00\times10^8)}\right]\frac{(0.50\times\pi(10^{-3})^2)\pi(250\times10^6)}{2} = 1.3\times10^8. \end{aligned}$$

[1]Note that the stimulated transition cross-section σ is always defined with respect to the homogeneous line.

Thus, spontaneous emission on the 633-nm transition is distributed among approximately 10^8 electromagnetic modes within the homogeneous linewidth and the volume considered here. The long-lived cavity modes discussed earlier in Chapters 2 and 4 constitute only a small subset of these modes.

Multiplying Eq. (9.1) for the inversion density by V_c and using Eq. (9.5) for the stimulated transition rate, we obtain the equation of motion for the atomic inversion, $\Delta\mathcal{N}V_c = \Delta N$, as

$$\frac{d\Delta N}{dt} = R_p - \gamma_2\Delta N - K\Delta N q\,, \tag{9.8}$$

where $R_p = R_2V_c$ is the pump rate (number of atoms per unit time) at which atoms are introduced into the upper laser level. Note that the atomic inversion ΔN represents the total population inversion of the atoms interacting with the field within the cavity volume V_c. For the nearly ideal four-level system considered here, this is approximately the total population of the upper laser level within that volume.

The equations for I_+ and I_- can be combined to obtain a single equation for the cavity photon number q. Adding Eqs. (9.2) and (9.3), we obtain

$$\frac{\partial(I_+ - I_-)}{\partial z} + \frac{n}{c}\frac{\partial(I_+ + I_-)}{\partial t} = (\sigma\Delta\mathcal{N} - \alpha)(I_+ + I_-)\,. \tag{9.9}$$

For a low-loss laser cavity, the right- and left-traveling wave intensities are approximately equal, $I_+ \approx I_-$. Within the spatially uniform, low-loss approximation, the spatial variation of their difference can therefore be neglected, while Eq. (9.4) gives $I_+ + I_- = qc/(nV_c)$. Using these results in Eq. (9.9), we find that the cavity photon number obeys the rate equation

$$\frac{dq}{dt} = \frac{c}{n}(\sigma\Delta\mathcal{N} - \alpha)q\,. \tag{9.10}$$

This equation describes the effects of distributed gain and loss on the cavity photon number. We can include the effects of mirror transmission and other localized losses by adding to this equation a photon-loss rate $q(-\ln\mathcal{R}_1\mathcal{R}_2)/\tau_R \equiv q\mathcal{L}_T/\tau_R$, where $\tau_R = 2nL/c$ is the cavity roundtrip time and $\mathcal{L}_T = -\ln(\mathcal{R}_1\mathcal{R}_2)$ is the roundtrip transmission loss, as in the discussion of cavity lifetime in Sec. 4.3.3. Furthermore, the gain term in the preceding equation can be rewritten as

$$\frac{c\sigma}{n}\Delta\mathcal{N} = \frac{c\sigma}{nV_c}(V_c\Delta\mathcal{N}) = K\Delta N\,, \tag{9.11}$$

where $\Delta N = \Delta\mathcal{N} \times V_c$ is the total population inversion within the cavity volume V_c (i.e., the population inversion of the atoms interacting with the cavity field), and we have used Eq. (9.6) for K. Incorporating these changes into Eq. (9.10), the equation of motion for the cavity photon number q takes the form

$$\begin{aligned}\frac{dq}{dt} &= \frac{c}{n}(\sigma\Delta\mathcal{N} - \alpha)q - \frac{\mathcal{L}_T}{\tau_R}q = K\Delta N q - \frac{c(2\alpha L + \mathcal{L}_T)}{2nL}q \\ &= K\Delta N q - \gamma_c q\,.\end{aligned} \tag{9.12}$$

Here $\gamma_c = \mathcal{L}_c/\tau_R$ is the cavity photon decay rate, where the total roundtrip cavity loss is $\mathcal{L}_c = 2\alpha L + \mathcal{L}_T$.

For the nearly ideal four-level system considered here, the population difference ΔN is essentially the population of the upper laser level. We therefore denote it simply by N to simplify the notation. With this change, the simplified rate equations can be written as

$$\dot{N} = R_p - \gamma_2 N - KNq\,, \tag{9.13}$$

$$\dot{q} = KNq - \gamma_c q\,. \tag{9.14}$$

These coupled nonlinear rate equations describe the laser in terms of two dynamical variables—the atomic population N, representing the upper laser level population for the nearly ideal four-level system, and the photon number q, representing the number of photons in the laser cavity. Despite the simplifying assumptions and approximations used in their derivation, these equations retain much of the essential physics of laser operation and prove remarkably useful for understanding many aspects of laser behavior.

These equations describe how the atomic inversion N and the cavity photon number q influence one another and are affected by external processes such as pumping and relaxation to other atomic levels. The upper laser level population increases at the rate R_p (atoms/s) as atoms are pumped into the upper level, decreases at the rate $\gamma_2 N$ because of relaxation from the upper level, and also decreases at the rate KNq because of stimulated transitions from the upper laser level to the lower laser level. The cavity photon number increases at the rate KNq because of stimulated emission into the cavity mode and decreases at the rate $\gamma_c q$ because of photon losses due to absorption, transmission, and other loss mechanisms.

In steady state, $\dot{N} = 0 = \dot{q}$, these equations lead to

$$0 = R_p - \gamma_2 N_{ss} - K N_{ss} q_{ss} \,, \tag{9.15a}$$

$$0 = K N_{ss} q_{ss} - \gamma_c q_{ss} \,, \tag{9.15b}$$

where the subscript ss denotes the values of the dynamical variables in steady state. Solving Eq. (9.15a) for the steady-state inversion N_{ss}, we can express it in terms of the steady-state photon number q_{ss} as

$$N_{ss} = \frac{R_p/\gamma_2}{1 + K q_{ss}/\gamma_2} = \frac{N^{(o)}}{1 + q_{ss}/q_s} \,, \tag{9.16}$$

where $N^{(o)}$ represents the unsaturated inversion and q_s represents the saturation photon number, given by

$$N^{(o)} \equiv \frac{R_p}{\gamma_2} \,, \qquad q_s \equiv \frac{\gamma_2}{K} = \frac{\gamma_2}{A_{21}/p} = \frac{\gamma_2}{A_{21}} p \,. \tag{9.17}$$

This result for the unsaturated inversion is the same as Eq. (7.44), and q_s has the expected relation to the saturation intensity:

$$\frac{q_s}{V_c} \frac{c}{n} h\nu = \frac{\gamma_2}{K V_c} \frac{c}{n} h\nu = \gamma_2 \frac{h\nu}{\sigma} = I_{\text{sat}} \,.$$

For the nearly ideal four-level laser considered here, if radiative decay on the laser transition is the dominant relaxation mechanism for the upper laser level, then $\gamma_2 \approx A_{21}$. Thus, in order of magnitude, the saturation photon number q_s is approximately equal to the number of free-space modes p that couple strongly to the active atoms.

Using the steady-state inversion in Eq. (9.15b), we obtain an equation involving only q_{ss},

$$\left[\frac{K N^{(o)}}{1 + q_{ss}/q_s} - \gamma_c \right] q_{ss} = 0 \,. \tag{9.18}$$

This equation gives two possible steady-state values for the cavity photon number:

$$(i) \quad q_{ss} = 0 \qquad \text{and} \qquad (ii) \quad q_{ss} = q_s \left(\frac{K N^{(o)}}{\gamma_c} - 1 \right) . \tag{9.19}$$

For the second solution to be physically meaningful, the cavity photon number q_{ss} must be nonnegative. Thus, the unsaturated inversion $N^{(o)}$ must satisfy $K N^{(o)} \geq \gamma_c$; that is, the unsaturated inversion $N^{(o)}$ must reach or exceed the threshold inversion

$$N_{th} = \frac{\gamma_c}{K} = \frac{\gamma_c n V_c}{c\sigma} \,, \tag{9.20}$$

which is set by the cavity loss $\mathcal{L}_c$ through the cavity decay rate $\gamma_c = \mathcal{L}_c/\tau_R = c\mathcal{L}_c/2nL$. This condition yields $N_{th}/V_c = \mathcal{L}_c/2L\sigma$, which is precisely the threshold condition given in Eq. (8.8).

Below threshold, $q_{ss} = 0$ is the only physically meaningful solution, and the corresponding inversion is $N_{ss} = N^{(o)} = R_p/\gamma_2$. Above threshold, the nonzero photon-number solution is the stable steady-state solution, whereas the solution $q_{ss} = 0$ is unstable with respect to any small positive photon-number perturbation. Thus, a small deviation from the stable steady-state decays with time, returning the system to its steady-state values, whereas a small positive photon-number perturbation away from the zero-photon solution above threshold grows and drives the system toward the nonzero steady-state. Notice, however, that within these deterministic rate equations a system starting with exactly $q = 0$ remains at $q = 0$, since Eq. (9.14) then gives $\dot{q} = 0$. The spontaneous-emission term introduced below removes this deficiency and provides the photons from which laser oscillations can build up.

Introducing the pump ratio

$$r = \frac{N^{(o)}}{N_{th}}, \tag{9.21}$$

which measures the pump strength relative to its threshold value, the steady-state photon number inside the cavity can be written in the compact form

$$q_{ss} = \begin{cases} 0, & r \leq 1 \quad \text{(below threshold)} \\ q_s(r-1), & r > 1 \quad \text{(above threshold)} \end{cases}. \tag{9.22}$$

Once the cavity photon number is known, the intracavity photon flux toward the output mirror is given by $\Phi_+ = \frac{1}{2}(q_{ss}/V_c)(c/n)$. If the output mirror transmission is $\mathcal{T}$, the output intensity from the laser operating above threshold is therefore

$$I_{out} = \frac{1}{2}\frac{q_{ss}}{V_c}\frac{c}{n}h\nu\mathcal{T} = \frac{1}{2}\mathcal{T}\frac{q_s h\nu}{V_c}\frac{c}{n}(r-1) = \frac{1}{2}\mathcal{T}I_{\text{sat}}(r-1). \tag{9.23}$$

Here $I_{\text{sat}} = q_s h\nu c/nV_c = \gamma_2 h\nu/\sigma$, where Eqs. (9.6) and (9.17) have been used. These expressions for the laser output intensity and saturation intensity agree with our earlier results.

The results of the preceding paragraphs show that our simplified laser rate equations are capable of reproducing the results derived in Chapter 8 from the original laser rate equations of Sec. 7.10. They also inherit a deficiency of those equations: they fail to reproduce an important dynamical feature—the turn-on of the laser. To see this, we consider these equations in the limit $\gamma_2 \gg r\gamma_c$. This condition is invoked only to make the problem analytically tractable and does not affect the qualitative conclusion reached below. This assumption means that the population can respond sufficiently rapidly to follow the variation of the field. Thus, N can adjust rapidly to remain approximately in instantaneous equilibrium with the cavity photon number, so that in this limit we can write

$$N(t) \approx \frac{N^{(o)}}{1 + q(t)/q_s}. \tag{9.24}$$

Substituting this expression into the equation of motion (9.14) for the cavity photon number, we find

$$\begin{aligned} \frac{dq}{dt} &= \frac{\gamma_c r q}{1 + q/q_s} - \gamma_c q \approx \gamma_c r q(1 - q/q_s) - \gamma_c q, \\ &= \gamma_c(r-1)q - \gamma_c r q^2/q_s, \end{aligned} \tag{9.25}$$

where we have expanded the saturation factor under the assumption that q has not grown to become comparable to q_s. This approximation is particularly appropriate near threshold, where $r \approx 1$ and the steady-state photon number is small compared with q_s. Separating the variables and integrating from the initial value $q(0) = q_o$ at $t = 0$ to $q(t)$ at time t, we obtain

$$q(t) = q_{ss} \frac{q_o e^{(r-1)\gamma_c t}}{q_{ss} + q_o[e^{(r-1)\gamma_c t} - 1]}, \tag{9.26}$$

where q_{ss} is the steady-state solution of the approximate Eq. (9.25) above threshold, given by

$$q_{ss} = q_s \frac{r-1}{r} \qquad r > 1. \tag{9.27}$$

Notice that this expression differs slightly from the exact steady-state result $q_{ss} = q_s(r-1)$ obtained earlier because Eq. (9.25) was derived by expanding the saturation factor to first order in q/q_s. Near threshold, where $r \approx 1$, the two expressions agree to leading order.

As a check, we note that for $t = 0$ the solution reduces to $q(0) = q_o$, while for $t \to \infty$ it approaches q_{ss}, as it should. The evolution of the cavity photon number as a function of time is shown in Fig. 9.1 for two different initial values, as indicated. The time required for the photon number to reach a significant fraction—say, 10%—of its steady-state value is shorter for the larger initial photon number. Once it reaches this level, however, the subsequent rise toward the steady-state value takes approximately the same time in both cases.

The solution (9.26) for the cavity photon number has two unsatisfactory features. First, if the initial photon number is $q_o = 0$, the cavity photon number remains zero for all times. Thus, if a laser is initially off, a common initial condition, it will never turn on according to these equations, even when it is pumped above threshold ($r > 1$). This conclusion can also be seen directly from the photon rate equation: its right-hand side is proportional to q, so that $q(t) = 0$ is an exact solution. Consequently, in the absence of a source term independent of q, the deterministic rate equation provides no mechanism for generating the first photons from which laser oscillations can build up.

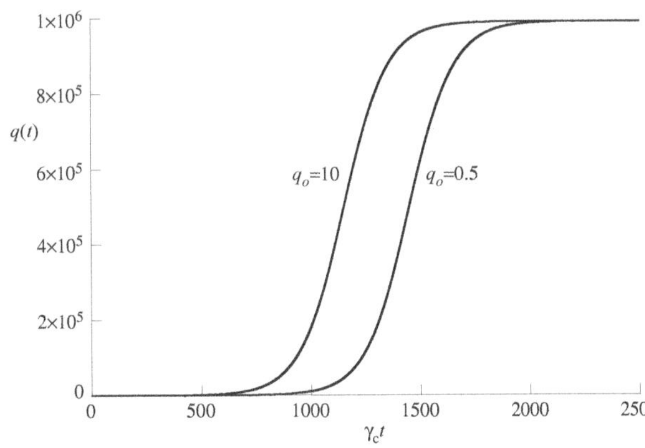

FIGURE 9.1
Buildup of laser radiation (cavity photon number) for initial photon numbers $q_o = 0.5$ and $q_o = 10$ for the pump ratio $r = 1.01$.

Second, for $r < 1$, the cavity photon number relaxes to zero regardless of its initial value q_o. Thus, according to these equations, there can be no steady-state photon flux escaping from the cavity below threshold. This is contrary to observation, since an excited laser medium emits spontaneous radiation, a small fraction of which is coupled into the cavity mode and escapes through the output coupler even below threshold. To be sure, this photon flux is generally very small compared with the coherent laser output well above threshold, but it is nonzero. Both deficiencies have the same origin: spontaneous emission into the laser cavity mode has not yet been included in the photon rate equation.

The origin of these problems can be traced to the fact that we have not included the contribution of spontaneous emission in the photon rate equation. The term KNq represents only the stimulated-emission rate. However, atoms in the upper laser level can also emit photons spontaneously, independent of q, at a rate KN into the cavity mode (recall that K is the spontaneous-emission rate per mode). This contribution is crucial for initiating the buildup of laser oscillations.

With the addition of spontaneous emission, the term KNq is replaced by $KN(q+1)$ in the photon rate equation (9.14). There is no corresponding change to the photon-loss term, since photon loss is proportional to the number of photons present in the cavity. Equation (9.13) for N remains unchanged because γ_2 already includes spontaneous emission from the upper laser level, including spontaneous emission into the cavity mode. The equations of motion for q and N can then be written as

$$\frac{dN}{dt} = R_p - \gamma_2 N - KNq = \gamma_2 N_{th}\left[r - \frac{N}{N_{th}} - \frac{N}{N_{th}} q/q_s\right], \tag{9.28a}$$

$$\frac{dq}{dt} = KN(q+1) - \gamma_c q = \gamma_c\left[\frac{N}{N_{th}}(q+1) - q\right], \tag{9.28b}$$

where, in the first equation, the unsaturated inversion $N^{(o)} \equiv R_p/\gamma_2$ has been expressed in terms of the pump ratio r and the threshold inversion N_{th} as $N^{(o)} = rN_{\text{th}}$.

9.2 Steady-State with Spontaneous Emission

In steady state, $\dot{N} = 0 = \dot{q}$, Eqs. (9.28a) and (9.28b) lead to the following equations for the inversion and cavity photon number:

$$r - \frac{N}{N_{th}} - \frac{N}{N_{th}} q/q_s = 0\,. \tag{9.29a}$$

$$\frac{N}{N_{th}}(q+1) - q = 0\,. \tag{9.29b}$$

Solving the first equation for N and substituting the result into the second equation, we find that the steady-state cavity photon number is determined by

$$r(q_{ss}+1) - q_{ss}[1 + q_{ss}/q_s] = 0\,. \tag{9.30}$$

Introducing $x = q_{ss}/q_s$, this becomes the quadratic equation

$$x^2 - (r-1)x - r/q_s = 0\,. \tag{9.31}$$

Solving for x, we obtain

$$x_\pm = \frac{1}{2}\left[(r-1) \pm \sqrt{(r-1)^2 + 4r/q_s}\,\right]. \tag{9.32}$$

Only the + sign gives a physically meaningful, nonnegative value for the steady-state photon number. Denoting this solution by q_c, we find

$$q_c \equiv q_s x_+ = \frac{q_s}{2}\left[(r-1) + \sqrt{(r-1)^2 + 4r/q_s}\,\right]. \tag{9.33}$$

Figure 9.2 shows the steady-state cavity photon number as a function of the pump ratio. We see that the cavity photon number below threshold ($r < 1$) is nonzero and increases rapidly as the laser approaches threshold. As the laser crosses the threshold of oscillation, the cavity photon number increases by several orders of magnitude over a very narrow range of pump ratios around $r = 1$. Well above threshold, the cavity photon number increases approximately linearly with pump ratio, as predicted by the laser rate equations both with and without spontaneous emission.

The presence of the term $4r/q_s$ in the radicand, whose origin can be traced directly to the inclusion of spontaneous emission in the photon rate equation, plays a crucial role in producing the observed behavior of the cavity photon number near threshold. Recalling that q_s for most lasers ranges from 10^8–10^{12} and r is typically less than 5, we see that $4r/q_s$ is ordinarily very small. In the radicand, however, it occurs alongside $(r-1)^2$, which vanishes at threshold. Thus, the term $4r/q_s$ cannot be neglected unless $(r-1)^2$ exceeds it by roughly two orders of magnitude. Near $r = 1$, this requires, to order of magnitude, $|r-1| > 10/\sqrt{q_s}$. Outside this narrow threshold region, the term under the radical sign can be approximated by

$$\sqrt{(r-1)^2 + \frac{4r}{q_s}} \approx |r-1|\left(1 + \frac{2r}{q_s|r-1|^2}\right) = |r-1| + \frac{2r}{q_s|r-1|}. \tag{9.34}$$

With the help of this result, Eq. (9.33) yields the following simple expressions for the cavity photon number as the laser pump ratio changes from below to above threshold:

$$q_c = \begin{cases} \dfrac{r}{1-r} & (1-r) > 10/\sqrt{q_s} : \text{Below threshold}, \\ \sqrt{q_s} & r = 1 : \text{At threshold}, \\ q_s(r-1) & (r-1) > 10/\sqrt{q_s} : \text{Above threshold}. \end{cases} \tag{9.35}$$

It is clear that, with the inclusion of spontaneous emission, the steady-state cavity photon number is nonzero and therefore produces a nonzero photon flux from the laser below threshold, in agreement with experimental observations. Spontaneous emission plays an especially important role in the threshold region, approximately where $|r-1| < 10/\sqrt{q_s}$. Outside this region, its direct contribution to the mean cavity photon number rapidly becomes negligible. Nevertheless, spontaneous emission continues to play a significant role above threshold in determining the ultimate linewidth of laser light, as will be seen later in this chapter. As a numerical estimate of the threshold region for the $\lambda = 633$ nm transition of a He-Ne laser, for which $q_s \approx 10^8$, the threshold region may be defined approximately by the pump-ratio interval $0.999 < r < 1.001$.[2]

Equation (9.35) also reveals an important change in the behavior of atomic emission at laser threshold. To see this, consider the expressions for the cavity photon number below and above threshold. Recalling the definitions of the threshold inversion (9.20) and pump ratio (9.21), the cavity photon number well below threshold can be written as

$$q_c = \frac{r}{1-r} \xrightarrow{r \ll 1} r = \frac{N^{(0)}}{N_{th}} = \frac{KN^{(o)}}{\gamma_c} = \frac{A_{21}N^{(o)}}{p\gamma_c}. \tag{9.36}$$

[2] S. Singh and L. Mandel, "Method for determining absolute photon numbers in a laser," J. Opt. Soc. Am. **72**, 304–305 (1982).

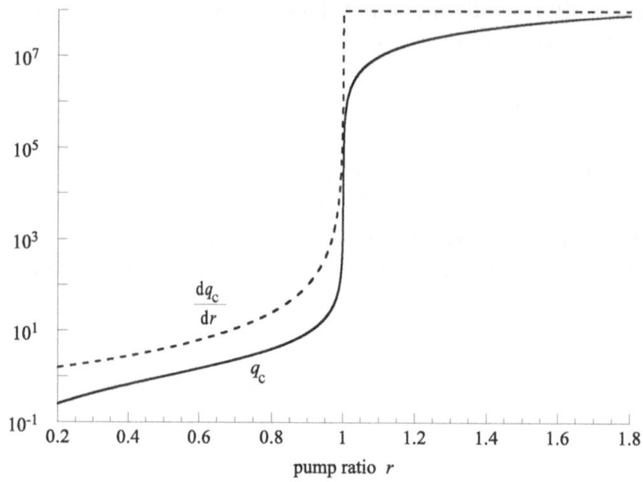

FIGURE 9.2
Steady-state cavity photon number as a function of pump ratio.

Here we have used the definition of K in Eq. (9.6) as the spontaneous-emission rate into one mode per excited atom. The quantity $A_{21}N^{(o)}/\gamma_c$ is the number of photons that would be produced by spontaneous emission from the unsaturated excited-state population during one cavity lifetime. Only a fraction $1/p$ of this spontaneous emission enters the selected cavity mode. Thus, well below threshold, the cavity mode behaves essentially as just one among the p electromagnetic modes available to the atomic transition.

Above threshold, recalling the definition of the saturation photon number (9.17), the cavity photon number can be written as

$$q_c = q_s(r-1) \xrightarrow{r\gg 1} q_s r = \frac{\gamma_2}{K}\frac{KN^{(o)}}{\gamma_c} = \frac{\gamma_2 N^{(o)}}{\gamma_c} \approx \frac{A_{21}N^{(o)}}{\gamma_c}. \tag{9.37}$$

In the last step we have assumed $\gamma_2 \approx A_{21}$, so that radiative decay on the laser transition dominates the upper-level relaxation. The physical interpretation of this result requires some care. Above threshold, the actual upper-level population is strongly affected by stimulated emission and is no longer equal to the unsaturated population $N^{(o)}$. The result therefore does not mean that essentially all spontaneously emitted photons enter the cavity mode. Rather, well above threshold, stimulated emission into the lasing mode becomes the dominant channel through which the pump-induced excitation is converted into radiation. Thus, as the laser passes through threshold, the selected cavity mode changes from being one among a very large number of modes receiving spontaneous emission to becoming the preferred mode into which an increasingly large fraction of the radiative energy flow is directed by stimulated emission.

The rate at which the steady-state cavity photon number changes with pump ratio follows from Eq. (9.33):

$$\frac{dq_c}{dr} = \frac{q_s}{2} + \frac{q_s(r-1)+2}{2[(r-1)^2+4r/q_s]^{1/2}}. \tag{9.38}$$

Well below threshold, where $q_c \approx r/(1-r)$, this rate becomes $dq_c/dr \approx 1/(1-r)^2$. At threshold, $\left.\frac{dq_c}{dr}\right|_{r=1} = \frac{q_s}{2} + \frac{\sqrt{q_s}}{2} \approx \frac{q_s}{2}$, where the last approximation holds for $q_s \gg 1$.

Well above threshold, $dq_c/dr \approx q_s$. Thus, as r increases through threshold, the response of the preferred cavity mode to additional pumping increases enormously. The lasing mode consequently captures an increasing fraction of the energy supplied to the active atoms. Another important change occurs in the photon statistics of the emitted light, as will be discussed later in this chapter.

9.3 Buildup of Laser Radiation

With spontaneous emission included in Eqs. (9.28a) and (9.28b), we now revisit the buildup of laser radiation from an initially empty cavity when the laser is turned on. To make the problem analytically tractable while retaining the essential physics, we again consider the limit in which the inversion dynamics is sufficiently fast to follow the temporal variation of q. In this approximation, which is useful near threshold for many gas and solid-state lasers, the inversion remains approximately in instantaneous equilibrium with the cavity photon number, so that

$$N(t) \approx \frac{rN_{th}}{1 + q/q_s}. \tag{9.39}$$

Substituting this result into the equation of motion for q, and using the fact that q/q_s is small near threshold, we obtain

$$\frac{dq}{dt} = \frac{KN_{th}r(q+1)}{1 + q/q_s} - \gamma_c q \approx KN_{th}r(q+1)(1 - q/q_s) - \gamma_c q. \tag{9.40}$$

Using $KN_{th}/\gamma_c = 1$ and introducing the scaled photon number x and scaled time τ,

$$x = \frac{q}{q_s} \quad \text{and} \quad \tau = \gamma_c t, \tag{9.41}$$

the equation for the scaled cavity photon number can, within the near-threshold approximation, be written as

$$\frac{dx}{d\tau} \approx -rx^2 + (r-1)x + \frac{1}{q_s}.$$

Here terms of relative order $1/q_s$ have been retained only where they provide the spontaneous-emission source responsible for initiating the buildup. Factoring the right-hand side gives

$$\frac{dx}{d\tau} = -r(x - x_+)(x - x_-), \tag{9.42}$$

where $x_\pm$ are the two steady-state solutions of this approximate equation,

$$x_\pm = \frac{1}{2r}\left[(r-1) \pm \sqrt{(r-1)^2 + 4r/q_s}\right]. \tag{9.43}$$

These steady-state solutions agree with Eq. (9.32) near threshold ($r \approx 1$), which is the regime considered here. The physical solution, which gives a positive cavity photon number, is x_+. In arriving at these expressions, terms of relative order r/q_s in comparison with unity have been neglected, while the spontaneous-emission source term of order $1/q_s$ has necessarily been retained. Integrating Eq. (9.42) by separation of variables, using the initial

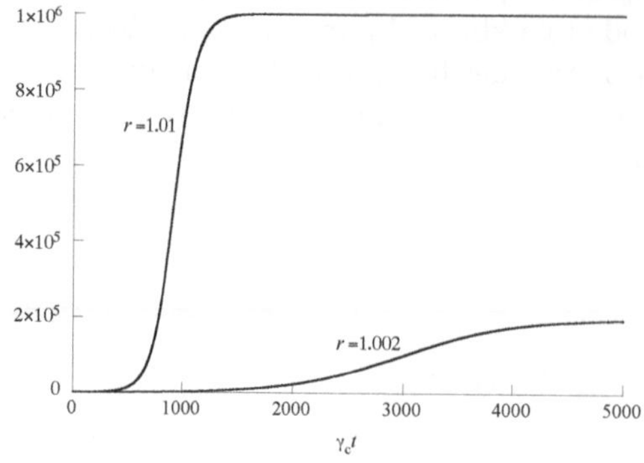

FIGURE 9.3
Growth of the cavity photon number from an initially empty cavity, $q(0) = 0$, for pump ratios $r = 1.002$ and $r = 1.01$.

condition $q(0) = 0$, and then reverting to the cavity photon number $q = xq_s$, we obtain

$$q(t) = \frac{q_c(1 - e^{-\Gamma_c t})}{1 + rq_c^2/q_s e^{-\Gamma_c t}}, \tag{9.44}$$

where $\quad \Gamma_c = \gamma_c \sqrt{(r-1)^2 + 4r/q_s}\,,$ $\qquad$ (9.45)

and $q_c = q_s x_+$ is the steady-state cavity photon number which, from Eq. (9.43), is given by

$$q_c \approx \frac{q_s}{2r}\left[r - 1 + \sqrt{(r-1)^2 + 4r/q_s}\right]. \tag{9.46}$$

Equation (9.44) describes the buildup of the cavity photon number after the laser is turned on. Starting from an initially empty cavity, spontaneous emission provides a nonzero source of photons, which are subsequently amplified by stimulated emission. The cavity photon number approaches the steady-state value q_c given by Eq. (9.46) as $t \to \infty$. Thus, once spontaneous emission is included, the cavity rate equations correctly describe the initiation and buildup of laser oscillations after the laser is turned on.

Figure 9.3 shows the growth of the cavity photon number as a function of the scaled time $\gamma_c t = t/\tau_c$ for two pump ratios close to threshold. Recall that, without the spontaneous-emission contribution, the transient solution given by Eq. (9.26) predicts that if the laser starts from an empty cavity $[q_o \equiv q(0) = 0]$, no buildup of laser radiation can occur. This deficiency of the deterministic laser equations is removed by including spontaneous emission, which supplies the initial photons that seed the subsequent stimulated-emission process.[3]

[3]These papers describe experiments on the buildup of laser light from noise.
D. Meltzer and L. Mandel, "Dynamics of a Q-Switched Laser near Threshold," Phys. Rev. A **3**, 1763–1772 (1971); F. T. Arecchi and V. DiGiorgio, "Statistical Properties of Laser Radiation During a Transient Buildup," Phys. Rev. A **3**, 1108–1124 (1971); M. R. Young and S. Singh, "Observation of the 'Brownian motion' of the electric field in a laser," Phys. Rev. A **35**, 1453–1456 (1987); M. R. Young and S. Singh, "Measurements of first-passage-time distributions in laser transients near threshold," J. Opt. Soc. Am. **5**, 1011–1019 (1988).

9.4 Photon Statistics of Laser Light

Another important change that takes place at laser threshold is the transformation of the photon statistics of laser light. In fact, almost all properties of laser light, including directionality, linewidth, intensity, and coherence, can, in principle, be attained by light from natural sources. Ultimately, it is the photon statistics that provide one of the most fundamental distinctions between laser light and light from thermal sources.

To consider this aspect of laser light, we rewrite the equation of motion for the inversion by explicitly separating the spontaneous-emission rate into the cavity mode from γ_2 and combining it with the stimulated-emission term by replacing q by $q+1$ in Eq. (9.28a). To compensate for this change, we replace the population decay rate γ_2 by $\gamma_2' = \gamma_2(1-\beta)$, where β is the fraction of the total upper-level relaxation rate associated with spontaneous emission into the cavity mode. Thus, γ_2' represents the population decay rate through all channels other than spontaneous emission into the cavity mode.

For most lasers, spontaneous emission into the cavity mode constitutes only a very small fraction of the overall upper-level decay rate ($\beta \ll 1$), so we can neglect the difference between γ_2' and γ_2 and continue to use γ_2 for the population decay rate. To estimate the magnitude of β, consider an ideal four-level laser in which the upper laser level decays solely to the lower laser level by spontaneous emission. It then follows from Eq. (9.6) that spontaneous emission into the selected cavity mode constitutes a fraction $1/p$ of the total spontaneous emission. For real lasers, additional decay channels can make this fraction still smaller. Thus, for the conventional macroscopic lasers considered here, β is typically of the order of 10^{-8}–10^{-12} or less.[4]

The equations of motion for q and N can then be written as

$$\frac{dN}{dt} = R_p - \gamma_2 N - KN(q+1) = \gamma_2 N_{th}\left[r - \frac{N}{N_{th}} - \frac{N}{N_{th}}(q+1)/q_s\right], \tag{9.47a}$$

$$\frac{dq}{dt} = KN(q+1) - \gamma_c q = \gamma_c\left[\frac{N}{N_{th}}(q+1) - q\right]. \tag{9.47b}$$

Here, in the first equation, the unsaturated inversion $N^{(o)} \equiv R_p/\gamma_2$ has been expressed in terms of the pump ratio $r \equiv N^{(o)}/N_{\text{th}}$ and the threshold inversion $N_{\text{th}} = \gamma_c/K$ as $N^{(o)} = rN_{\text{th}}$.

Assuming again that the atomic dynamics are fast compared with the field dynamics ($\gamma_2 \gg \gamma_c$), we can solve for the population N in terms of the cavity photon number q,

$$N(t) = \frac{rN_{th}}{1 + (q(t)+1)/q_s}. \tag{9.48}$$

Thus, within this adiabatic approximation, the time dependence of the inversion follows the time dependence of the cavity photon number q. Substituting this result into Eq. (9.47b), we obtain an effective rate equation for the cavity photon number alone,

$$\dot{q} = \frac{\gamma_c r(q+1)}{1 + (q+1)/q_s} - \gamma_c q\,. \tag{9.49}$$

9.4.1 Photon Number Probabilities

Equation (9.49) identifies two competing processes. When there are q photons in the cavity, emission by the atoms, including both spontaneous and stimulated emission, increases the

[4]This approximation can break down for wavelength-sized microcavity lasers, where spontaneous emission into the lasing mode can constitute a significant fraction of the total decay rate. In such lasers, the distinction between γ_2' and γ_2 must be retained. See, for example, P. R. Rice and H. J. Carmichael, "Photon statistics of a cavity-QED laser: A comment on the laser phase-transition analogy," Phys. Rev. A **50**, 4318-4329 (1994).

photon number by one at the rate

$$\frac{\gamma_c r(q+1)}{1+(q+1)/q_s},$$

whereas cavity loss decreases the photon number by one at the rate $\gamma_c q$. These emission and loss events occur at random times. Consequently, the number of photons in the cavity is not a deterministic quantity but a discrete stochastic variable that fluctuates in time.

It is therefore appropriate to introduce the probability $p(q)$ that the cavity contains q photons. By definition, $p(q)$ is a number between 0 and 1, whereas q is a nonnegative integer, $q = 0, 1, 2, \ldots$. The probabilities must satisfy

$$\sum_{q=0}^{\infty} p(q) = 1 \,. \tag{9.50}$$

This relation is known as the normalization condition.

As photons are emitted into the cavity or lost through absorption, transmission, and other loss mechanisms, the probabilities $p(q)$ evolve in time while always satisfying the normalization condition. Their time evolution is described by a photon-number rate equation, or master equation. The emission and loss terms appearing in Eq. (9.49) provide the corresponding transition rates for constructing this master equation.

9.4.2 Master Equation for Photon Number Probabilities

Consider the rate equation for the probability $p(q)$ that the cavity contains q photons at time t. The probability $p(q)$ decreases if a photon is emitted into the cavity, because the cavity will then contain $q+1$ photons. Conversely, $p(q)$ increases through emission if the cavity previously contained $q-1$ photons. Similarly, $p(q)$ increases if a cavity containing $q+1$ photons loses one photon, and decreases if the cavity contained q photons and lost one.

We can regard photon-emission and loss rates as transition rates between different "levels" representing different photon numbers in the cavity, with $p(q)$ representing the probability of occupying each level. Thus, the transition rate from level q to level $q+1$ due to emission by the atoms is $\gamma_c r(q+1)/[1+(q+1)/q_s]$. Multiplying this transition rate by the probability $p(q)$ of occupying the initial level gives the probability current from level q to level $q+1$. Similarly, the transition rate from level q to level $q-1$ due to cavity loss is $\gamma_c q$, so that $\gamma_c q p(q)$ is the probability current from level q to level $q-1$. Figure 9.4 illustrates the levels corresponding to different photon numbers in the cavity and the probability currents between them.

We can then write the rate equation for $p(q)$ as

$$\begin{array}{rcll}
\dfrac{d}{dt}p(q) & = & -\dfrac{\gamma_c r(q+1)}{1+(q+1)/q_s}p(q) & + \quad \gamma_c(q+1)p(q+1) \\
 & & \left\{\begin{array}{l}\text{rate of decrease due to}\\ \text{emission of a photon into}\\ \text{a cavity containing } q\\ \text{photons}\end{array}\right\} & \left\{\begin{array}{l}\text{rate of increase due to}\\ \text{loss of a photon from a}\\ \text{cavity containing } q+1\\ \text{photons}\end{array}\right\} \\
 & & +\dfrac{\gamma_c r q}{1+q/q_s}p(q-1) & - \quad \gamma_c q p(q) \\
 & & \left\{\begin{array}{l}\text{rate of increase due to}\\ \text{emission of a photon into}\\ \text{a cavity containing } q-1\\ \text{photons}\end{array}\right\} & \left\{\begin{array}{l}\text{rate of decrease due to}\\ \text{loss of a photon from a}\\ \text{cavity containing } q \text{ pho-}\\ \text{tons}\end{array}\right\} .
\end{array} \tag{9.51}$$

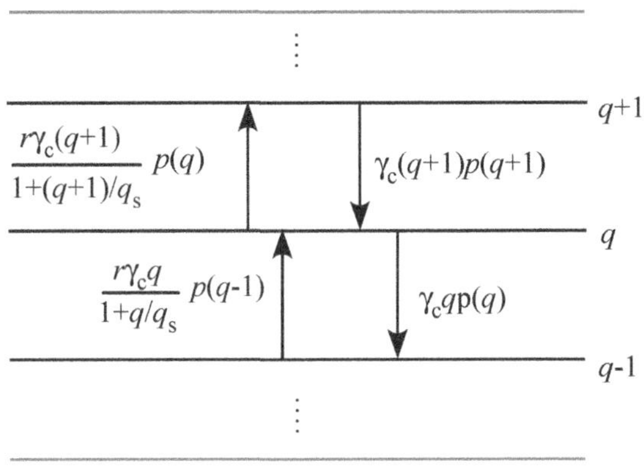

FIGURE 9.4
Probability currents between levels with $q+1$, q, and $q-1$ photons in the laser cavity.

An equation of this type is called a master equation. More specifically, Eq. (9.51) is a birth–death master equation associated with the simplified Scully–Lamb laser model.[5] Equations for $p(q+1)$, $p(q-1)$, etc., can be obtained from Eq. (9.51). For example, the equation for $p(q-1)$ is obtained by replacing $q \to q-1$ in Eq. (9.51), which yields

$$\frac{d}{dt}p(q-1) = -\frac{\gamma_c r q}{1+q/q_s}p(q-1)+\gamma_c q p(q)+\frac{\gamma_c r(q-1)}{1+(q-1)/q_s}p(q-2)-\gamma_c(q-1)p(q-1)\,. \quad (9.52)$$

In steady state, the probability current between neighboring photon-number states is independent of q. Because the photon number is bounded below by $q=0$ and the stationary probability distribution is normalizable, this constant current must vanish. Thus, the upward and downward probability currents between each pair of neighboring states balance. Considering the transitions between states containing q and $q-1$ cavity photons, we obtain

$$\gamma_c q p(q) = \frac{\gamma_c r q}{1+q/q_s}p(q-1)\,. \quad (9.53)$$

The steady-state probabilities therefore satisfy the recurrence relation

$$p(q) = \frac{r}{1+q/q_s}p(q-1)\,. \quad (9.54)$$

Iterating this relation, the steady-state photon-number probability $p(q)$ can be expressed in terms of $p(0)$ as

$$\begin{aligned} p(q) &= \frac{r}{1+q/q_s}p(q-1) = \frac{r^2}{(1+q/q_s)(1+(q-1)/q_s)}p(q-2) = \cdots \\ &= p(0)r^q\prod_{k=1}^{q}\frac{1}{1+k/q_s} = Cr^q\frac{q_s^q}{(q_s+q)!}\,, \end{aligned} \quad (9.55)$$

where the normalization constant $C = p(0)q_s!$ is determined by the requirement

$$\sum_{q=0}^{\infty} p(q) = 1\,. \quad (9.50^*)$$

[5]M. O. Scully and W. E. Lamb, "Quantum Theory of an Optical Maser. I. General Theory," Phys. Rev. **159**, 208–226 (1967).

Thus, the steady-state probability is the product of two factors, r^q and $q_s^q/(q_s+q)!$. Below threshold ($r<1$), the factor r^q decreases monotonically with increasing q, whereas above threshold ($r>1$), it increases monotonically with increasing q. The second factor, $q_s^q/(q_s+q)!$, always decreases with increasing q. Consequently, $p(q)$ exhibits distinctly different behavior below and above threshold. We now examine these two limiting regimes.

9.4.2.1 Far Below Threshold ($r \ll 1$)

In this regime, r^q decreases rapidly with increasing q. Moreover, for the small values of q that carry significant probability, $q \ll q_s$, so that

$$\prod_{k=1}^{q} \frac{1}{1+k/q_s} \approx 1 .$$

The photon-number probability can therefore be approximated as

$$p(q) \approx C r^q . \tag{9.56}$$

The constant C is determined by the normalization condition (9.50),

$$\sum_{q=0}^{\infty} p(q) = C \sum_{q=0}^{\infty} r^q = C \frac{1}{1-r} = 1$$

$$\Rightarrow \qquad C = 1 - r . \tag{9.57}$$

Using this expression for the normalization constant, we can write the photon-number distribution below threshold as

$$p(q) = (1-r) r^q . \tag{9.58}$$

This distribution depends only on the pump ratio r. Its mean photon number is

$$\langle q \rangle = \sum_{q=0}^{\infty} q p(q) = \frac{r}{1-r} . \tag{9.59}$$

We can therefore express the pump ratio r in terms of $\langle q \rangle$ as

$$r = \frac{\langle q \rangle}{\langle q \rangle + 1} . \tag{9.60}$$

Finally, the photon-number distribution below threshold can be written as

$$p(q) = \frac{\langle q \rangle^q}{[1+\langle q \rangle]^{q+1}} . \tag{9.61}$$

This is a monotonically decreasing function of q, with $q=0$ as the most probable value. Expressed in terms of the mean cavity photon number, this distribution has the Bose–Einstein, or thermal, form for a single electromagnetic mode. Thus, far below threshold, the laser cavity field exhibits thermal photon statistics.

9.4.2.2 Far Above Threshold ($r \gg 1$)

In this regime, the cavity photon number becomes large. For sufficiently large r, the values of q for which $p(q)$ is appreciable satisfy $q \gg q_s$. The product in Eq. (9.55) can then be approximated by

$$\prod_{k=1}^{q} \frac{1}{1+k/q_s} \approx \prod_{k=1}^{q} \frac{q_s}{k} = \frac{q_s^q}{q!} .$$

The cavity photon-number distribution therefore takes the asymptotic form

$$p(q) \approx \text{const} \times \frac{(rq_s)^q}{q!} \,. \tag{9.62}$$

The normalization constant is determined from

$$\sum_{q=0}^{\infty} p(q) = 1 = \text{const} \times \sum_{q=0}^{\infty} \frac{(rq_s)^q}{q!} = \text{const} \times e^{rq_s} \,, \tag{9.63}$$

which gives

$$p(q) = \frac{(rq_s)^q}{q!} e^{-rq_s} \,. \tag{9.64}$$

The mean photon number is therefore

$$\langle q \rangle = \sum_{q=0}^{\infty} q p(q) = e^{-rq_s} \sum_{q=0}^{\infty} q \frac{(rq_s)^q}{q!} = rq_s \,. \tag{9.65}$$

Thus, the photon-number distribution can finally be written as

$$p(q) = \frac{\langle q \rangle^q}{q!} e^{-\langle q \rangle} \,, \tag{9.66}$$

which is a Poisson distribution with mean photon number $\langle q \rangle$.

Recall that the more accurate expression for the mean photon number well above threshold is $q_s(r-1)$. In the present far-above-threshold limit, $r \gg 1$, this becomes $q_s(r-1) \approx q_s r$, in agreement with the mean obtained from the asymptotic Poisson distribution.

9.4.2.3 Near-Threshold Region: $|r - 1| < 10/\sqrt{q_s}$

In this case, we write the photon-number distribution as

$$p(q) = p(0) \frac{(rq_s)^q}{(q_s + q)!} \,. \tag{9.67}$$

Recall that in this regime the mean photon number $\langle q \rangle$ is of the order of $\sqrt{q_s}$. The transition from $\langle q \rangle \sim r/(1-r)$ below threshold to $\langle q \rangle = (r-1)q_s$ above threshold occurs over a region defined approximately by $|r-1| < 10/\sqrt{q_s}$. To examine this threshold region more closely, we introduce the pump parameter

$$a = (r - 1)\sqrt{2q_s} \,. \tag{9.68}$$

This pump parameter expands the narrow threshold region by a factor of order $\sqrt{q_s} \approx 10^4$, allowing us to examine the threshold transition in greater detail. In terms of a, the laser threshold corresponds to $a = 0$, the region above threshold to $a > 0$, and the region below threshold to $a < 0$. The range $|r - 1| < 10/\sqrt{q_s}$ corresponds to $|a| < 10\sqrt{2}$. The near-threshold region may then be represented approximately by the convenient range $-10 < a < 10$. Within this range, we can express the parameter r as

$$r = 1 + (r - 1) = 1 + \frac{a}{\sqrt{2q_s}} \approx e^{a/\sqrt{2q_s}} \,. \tag{9.69}$$

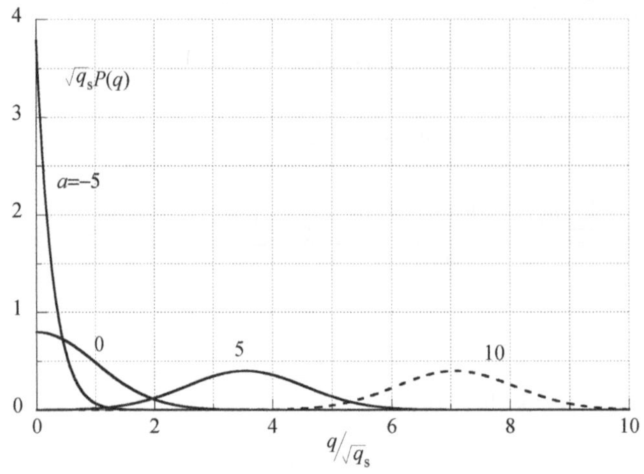

FIGURE 9.5
The cavity photon-number distribution changes as the laser passes through threshold. Below threshold, the distribution peaks at $q = 0$, whereas above threshold it peaks at a nonzero value of q.

Since q_s and the relevant values of q are both large, while q/q_s remains small near threshold, we can evaluate the denominator using Stirling's approximation[6]

$$\begin{aligned}\ln(q_s + q)! &= (q_s + q + 1/2)\ln(q_s + q) - (q_s + q) + \frac{1}{2}\ln(2\pi) \\ &= (q_s + q + 1/2)\left[\ln q_s + \ln(1 + q/q_s)\right] - (q_s + q) + \frac{1}{2}\ln(2\pi) \\ &\approx \text{const} + q\ln q_s + \frac{1}{2}\frac{q^2}{q_s} + \mathcal{O}\left(q^3/q_s^2\right) . \end{aligned} \tag{9.70}$$

Substituting this expansion into Eq. (9.67) and using the approximation $\ln r = \ln(1 + a/\sqrt{2q_s}) \approx a/\sqrt{2q_s}$ from Eq. (9.69), we find that the distribution $p(q)$ takes the form

$$p(q) = \text{const} \times e^{q(\ln r + \ln q_s) - \ln(q_s+q)!} = \text{const} \times \exp\left[\frac{a}{\sqrt{2}}\frac{q}{\sqrt{q_s}} - \frac{1}{2}\frac{q^2}{q_s}\right] . \tag{9.71}$$

The shape of this distribution is shown in Fig. 9.5 for pump parameters just below, at, and just above threshold. As the pump moves sufficiently below threshold, a becomes negative with sufficiently large magnitude that the linear term in the exponent dominates over the quadratic term for the values of q carrying appreciable probability. The distribution then approaches the exponential form

$$p(q) \approx \frac{|a|}{\sqrt{2q_s}} e^{-|a|q/\sqrt{2q_s}} . \tag{9.72}$$

[6]Stirling's approximation gives the factorial function as $m! \approx \sqrt{2\pi}m^{m+1/2}e^{-m}$ for $m \gg 1$. This formula gives accurate results even for relatively small values of m. For example, the fractional error between the exact and approximate values for $m = 10$ is only about 0.008.

This is the large-mean-photon-number limit of the thermal distribution obtained earlier below threshold.

As the pump moves sufficiently above threshold, while remaining within the range of validity of the near-threshold approximation, the distribution approaches the Gaussian form

$$p(q) = \text{const} \times \exp\left[-\frac{1}{2q_s}\left(q - a\sqrt{\frac{q_s}{2}}\right)^2\right], \tag{9.73}$$

apart from a factor independent of q. The distribution is centered at $q = \langle q \rangle = a\sqrt{q_s/2} = (r-1)q_s$, with variance $\langle(\Delta q)^2\rangle = q_s$. Its relative variance is therefore

$$\frac{\langle(\Delta q)^2\rangle}{\langle q \rangle^2} = \frac{2}{a^2}. \tag{9.74}$$

This result applies in the near-threshold regime, where $r \approx 1$. It should not be extrapolated to $r \gg 1$, where the photon-number distribution approaches the Poisson distribution derived earlier, with variance $\langle(\Delta q)^2\rangle = \langle q \rangle$.[7] Thus, as the laser passes through threshold, its photon-number distribution evolves continuously from a thermal-like distribution peaked at $q = 0$ to a narrower distribution peaked at a large nonzero photon number.

Within the near-threshold range $-10 < a < 10$, a more careful treatment is required. From the steady-state photon-number distribution $p(q)$, we find that the mean photon number is

$$\langle q \rangle = \sum_{q=0}^{\infty} q p(q) \to \int_0^{\infty} q p(q) dq = \sqrt{\frac{q_s}{2}}\left[a + \frac{2}{\sqrt{\pi}} \frac{e^{-\frac{1}{4}a^2}}{1 + \operatorname{erf}(a/2)}\right], \tag{9.75}$$

where $\operatorname{erf}(x)$ is the error function of argument x.[8] In certain limits, simpler expressions for the mean cavity photon number can be derived:

$$\langle q \rangle = \begin{cases} \sqrt{\frac{q_s}{2}} \frac{2}{|a|} = \frac{1}{1-r} & a < 0,\ |a| \gg 1, \\ \sqrt{\frac{q_s}{2}} \frac{2}{\sqrt{\pi}} \equiv q_{th} & a = 0 \\ \sqrt{\frac{q_s}{2}}\, a = q_s(r-1) & a \gg 1. \end{cases} \tag{9.76}$$

The variance of q is

$$\langle(\Delta q)^2\rangle \equiv \int_0^{\infty} dq (q - \langle q \rangle)^2 p(q) = \frac{q_s}{2}\left[2 - \frac{2a}{\sqrt{\pi}} \frac{e^{-\frac{1}{4}a^2}}{1 + \operatorname{erf}(a/2)} - \frac{4}{\pi} \frac{e^{-a^2/2}}{[1 + \operatorname{erf}(a/2)]^2}\right]. \tag{9.77}$$

---

[7] Near threshold: the expansion of the Scully–Lamb distribution gives a Gaussian with $\langle(\Delta q)^2\rangle = q_s$ and $\langle q \rangle = (r-1)q_s$. Far above threshold: the distribution becomes Poissonian, and for $\langle q \rangle \gg 1$ that Poisson distribution can itself be approximated by a Gaussian, but now $\langle(\Delta q)^2\rangle = \langle q \rangle \simeq r q_s$.

[8] The error function is defined by $\operatorname{erf}(x) = \frac{2}{\sqrt{\pi}} \int_0^x dy\, e^{-y^2}$. Some of its important properties are

$$\begin{aligned} \operatorname{erf}(x) &= \frac{2}{\sqrt{\pi}} \sum_{k=0}^{\infty} \frac{(-1)^k x^{2k+1}}{k!(2k+1)} = \frac{2}{\sqrt{\pi}}\left(x - \frac{x^3}{3} + \cdots\right) \\ \operatorname{erf}(-x) &= -\operatorname{erf}(x) \\ \operatorname{erf}(x) &\xrightarrow{x \gg 1} 1 - \frac{e^{-x^2}}{x\sqrt{\pi}}\left(1 - \frac{1}{2x^2} + \frac{1 \cdot 3}{2^2 x^4} - \cdots\right). \end{aligned}$$

Using the small- and large-argument expansions of the error function, the following limiting forms of the variance can be derived:

$$\langle(\Delta q)^2\rangle = \begin{cases} \dfrac{2q_s}{a^2} & a < 0,\ |a| \gg 1\,, \\ q_s\left(1 - \dfrac{2}{\pi}\right) & a = 0\,, \\ q_s & a \gg 1\,. \end{cases} \tag{9.78}$$

We note that sufficiently above threshold, while still within the range of validity of the near-threshold approximation, the mean photon number increases as $\langle q\rangle = q_s(r-1)$, while the variance approaches the constant value q_s. Thus, in a relative sense, the distribution narrows. Indeed, the relative variance is given by

$$\frac{\langle(\Delta q)^2\rangle}{\langle q\rangle^2} = \begin{cases} 1 & a < 0,\ |a| \gg 1\,, \\ \dfrac{\pi}{2} - 1 = 0.57 & a = 0\,, \\ \dfrac{2}{a^2} & a \gg 1\,. \end{cases} \tag{9.79}$$

From these expressions, we see that below threshold the laser has large photon-number fluctuations, of the same order as the mean, characteristic of thermal light. As the laser approaches threshold, the relative photon-number fluctuations decrease. Above threshold, they decrease rapidly, and the photon-number distribution becomes increasingly narrow relative to its mean. The result $\langle(\Delta q)^2\rangle \to q_s$ applies within the near-threshold approximation and should not be extrapolated to $r \gg 1$, where the photon-number distribution approaches the Poisson distribution derived earlier, for which $\langle(\Delta q)^2\rangle = \langle q\rangle$.

Figure 9.6 shows the scaled mean photon number $\langle q\rangle\sqrt{2/q_s}$, the normalized variance κ_2, and the normalized third central moment κ_3, which measures the asymmetry of the distribution about its mean, as functions of the pump parameter a. This behavior has been confirmed in a number of experiments.[9] For example, Fig. 9.7 shows the experimentally measured behavior of the normalized photon-number variance as a function of the mean photon number (normalized to the mean at threshold). Note that the rapid transformation of the photon-number distribution from an exponential form below threshold to a distribution peaked about a nonzero photon number above threshold takes place over the very narrow threshold region, represented approximately by $|a| < 10$. For most realistic laser systems, the saturation photon number is of the order of $q_s \sim 10^8$, so that this threshold region corresponds to changes in the pump ratio of a few parts in 10^4. Thus, the statistical properties of laser light change rapidly over an extremely narrow range of pump strength. Farther above threshold, the photon-number distribution approaches the Poisson distribution characteristic of coherent laser light, as discussed earlier.

9.5 Single-mode Laser Linewidth

We have seen in Chapter 4 that in an empty cavity, the electric field amplitude decays according to

$$E(\boldsymbol{r}, t) = E_0(\boldsymbol{r})\, e^{-t/2\tau_c - i\omega t + i\phi_0}\,. \tag{9.80}$$

[9] R. F. Chang, V. Korenman, C. O. Alley, and R. W. Detenbeck, "Correlations in light from a laser at threshold," Phys. Rev. **178**, 612–621 (1969); Y. Qu, S. Singh, and C. D. Cantrell, "Measurements of Higher Order Photon Bunching of Light Beams," Phys. Rev. Lett. **76**, 1236–1239 (1996).

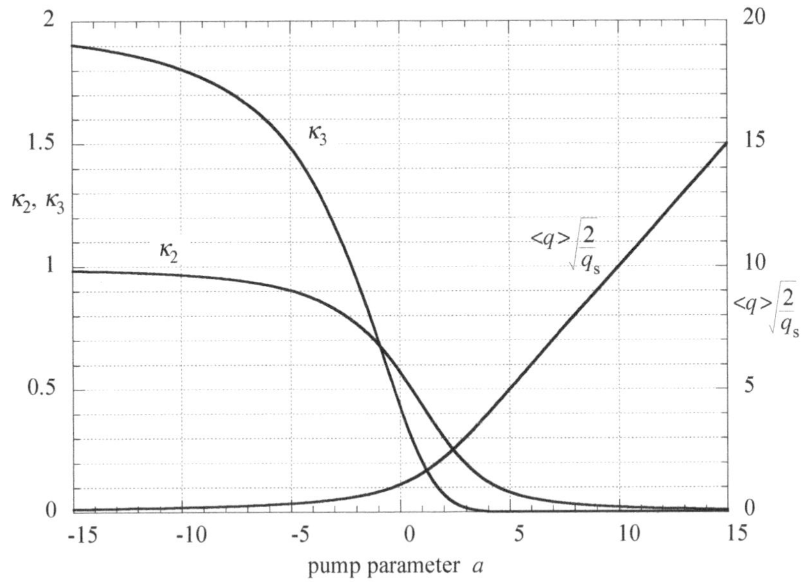

FIGURE 9.6
Behavior of scaled mean photon number $\langle q\rangle\sqrt{2/q_s}$, normalized variance $\kappa_2 \equiv \langle(\Delta q)^2\rangle/\langle q\rangle^2$, and normalized third central moment $\kappa_3 \equiv \langle(\Delta q)^3\rangle/\langle q\rangle^3$ of the photon-number distribution in the threshold region.

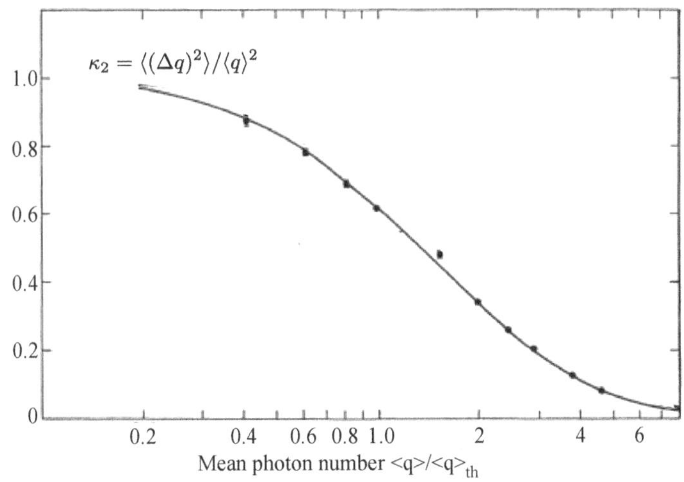

FIGURE 9.7
Normalized photon-number variance κ_2 as a function of mean photon number in a laser. The solid curve is derived from Eqs. (9.77) and (9.75). [Figure adapted from Phys. Rev. A **27**, 381–391 (1983)].

This leads to a Lorentzian spectral density with a FWHM given by

$$\Delta\nu_c = \frac{1}{2\pi\tau_c} = \frac{\gamma_c}{2\pi} = \frac{1}{2\pi}\frac{c\mathcal{L}_c}{2nL}, \tag{9.81}$$

where $\mathcal{L}_c$ is the loss per cavity roundtrip. When gain is present, the cavity field amplitude does not decay. Does this mean that the linewidth vanishes? This would indeed be the case for an idealized laser without spontaneous-emission noise. In a real laser, however, spontaneous emission of photons into the cavity mode introduces field contributions with random phases. These random contributions cause phase diffusion and give the laser field a small but finite spectral width that decreases as the cavity photon number increases.

A simple pictorial argument illustrates this behavior. As shown in the previous section, the magnitude of the electric field amplitude increases rapidly above threshold and stabilizes at a large, well-defined value, while the relative photon-number fluctuations $\langle(\Delta q)^2\rangle/\langle q\rangle^2$ decrease with increasing pump parameter. The phase, however, is not fixed by the gain and continues to perform a random walk due to spontaneous-emission events. This leads to the situation depicted in Fig. 9.8. We represent the electric field of the laser by a phasor of large amplitude proportional to $\sqrt{\langle q\rangle}$. A spontaneously emitted photon adds to this field a phasor with a random phase, represented in the figure as having unit magnitude. The phase change $\delta\phi$ due to a single photon emitted spontaneously into the cavity mode can then be written as

$$\delta\phi = \frac{\sin\theta}{\sqrt{\langle q\rangle}}. \tag{9.82}$$

Taking the statistical average, we find

$$\langle(\delta\phi)^2\rangle = \frac{\langle\sin^2\theta\rangle}{\langle q\rangle} = \frac{1}{2\langle q\rangle}. \tag{9.83}$$

In the last step, we used the fact that the phase θ of the spontaneously emitted field contribution is uniformly distributed between 0 and 2π, so that $\langle\sin^2\theta\rangle = \langle\cos^2\theta\rangle = \frac{1}{2}$. The

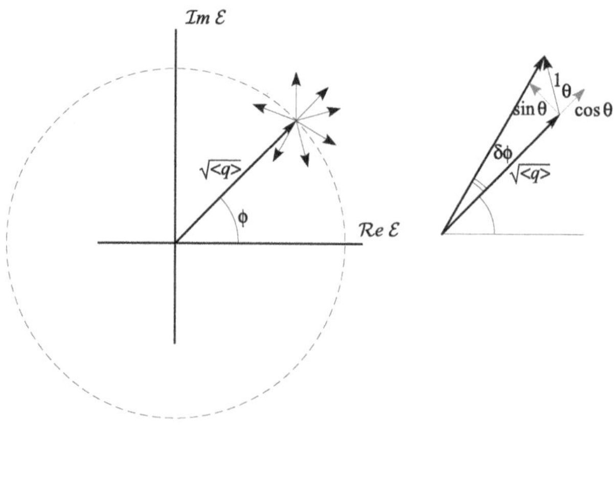

FIGURE 9.8
Each spontaneous-emission event is represented by the addition of a unit-magnitude phasor with random phase θ to a field phasor of magnitude $\sqrt{\langle q\rangle}$ and phase ϕ. This changes the field phase by $\delta\phi = \sin\theta/\sqrt{\langle q\rangle}$.

rate at which photons are spontaneously emitted into the cavity mode is

$$R_{sp} = \frac{N_2 A_{21}}{p}\,, \tag{9.84}$$

where N_2 is the number of atoms in the upper laser level, A_{21} is the spontaneous-emission rate for the laser transition, and p is the number of modes coupled to the atomic transition. Thus, during a time interval t, the mean-square phase displacement of the cavity field is

$$\langle(\Delta\phi)^2\rangle = \frac{1}{2\langle q\rangle}\frac{N_2 A_{21}}{p}\,t \equiv \Delta\omega_{_L}\,t\,, \tag{9.85}$$

where we have identified the coefficient of t as the phase-diffusion rate $\Delta\omega_{_L}$. As shown below, this phase-diffusion rate determines the laser linewidth:

$$\Delta\nu_{_L} = \frac{\Delta\omega_{_L}}{2\pi} = \frac{N_2 A_{21}}{4\pi\langle q\rangle p}\,. \tag{9.86}$$

We can see explicitly how the mean-square phase displacement $\langle(\Delta\phi)^2\rangle$ gives rise to a linewidth by considering the temporal coherence of the laser field. Write the complex electric field at time $t = 0$ as

$$E(0) = E_o(\boldsymbol{r})e^{i\phi_o}\,, \tag{9.87}$$

and at a later time t as

$$E(t) = E_o(\boldsymbol{r})e^{-i\omega t + i\phi(t)}\,. \tag{9.88}$$

Apart from the uniform rotation of the electric field phasor represented by the factor $e^{-i\omega t}$, there is a random contribution to its phase through the factor $e^{i\phi(t)}$. The temporal field correlation therefore contains the factor[10]

$$\langle E^*(0)E(t)\rangle = |E_o(\boldsymbol{r})|^2 e^{-i\omega t}\langle e^{i(\phi(t)-\phi_o)}\rangle \equiv |E_o(\boldsymbol{r})|^2 e^{-i\omega t}\langle e^{i\Delta\phi(t)}\rangle\,. \tag{9.89}$$

Here the average is taken over an ensemble of phase-diffusion trajectories and can be evaluated using a Taylor expansion:

$$\langle e^{i\Delta\phi(t)}\rangle = \sum_{n=0}^{\infty}\frac{i^n\langle(\Delta\phi(t))^n\rangle}{n!}\,. \tag{9.90}$$

Because the phase change $\Delta\phi(t)$ is the cumulative result of many independent spontaneous-emission events, it approaches Gaussian statistics according to the central limit theorem. Thus, the odd-order moments of $\Delta\phi(t)$ vanish, and the even-order moments can be expressed in terms of the second-order moment:

$$\langle(\Delta\phi(t))^{2m+1}\rangle = 0\,, \tag{9.91a}$$

$$\langle(\Delta\phi(t))^{2m}\rangle = (2m-1)!!\langle(\Delta\phi(t))^2\rangle^m\,. \tag{9.91b}$$

Using these results in Eq. (9.90), we find

$$\begin{aligned}\langle e^{i\Delta\phi(t)}\rangle &= \sum_{m=0}^{\infty}\frac{(-1)^m\langle(\Delta\phi(t))^2\rangle^m}{m!2^m} = e^{-\langle(\Delta\phi(t))^2\rangle/2}\,,\\ &= e^{-\frac{1}{2\langle q\rangle}\frac{N_2A_{21}}{p}\frac{t}{2}} \equiv e^{-t/2\tau_L}\,.\end{aligned} \tag{9.92}$$

[10]Spontaneous emission also produces fluctuations in the length of the field phasor. However, for the large photon numbers characteristic of a laser sufficiently above threshold, these fluctuations are relatively small, being of order $1/\sqrt{\langle q\rangle}$.

Thus, the temporal field correlation decays exponentially, leading to a Lorentzian lineshape with a full width at half maximum (FWHM) of

$$\Delta\nu_{L} = \frac{1}{2\pi\tau_{L}} = \frac{N_2 A_{21}}{4\pi p\langle q\rangle}. \tag{9.93}$$

Unlike the field in an empty cavity or a laser operating below threshold, the laser field amplitude above threshold does not itself decay in time. Rather, its phase ϕ performs a random walk under the influence of spontaneously emitted photons. It is the temporal coherence, represented by the field correlation $\langle E^*(0)E(t)\rangle$, that decays because of this phase diffusion. The laser field continues to oscillate with an essentially constant amplitude, but each spontaneous-emission event introduces a small random phase perturbation. The cumulative effect of many such perturbations produces a loss of phase coherence and, consequently, a finite laser linewidth.

The situation is analogous to an atom undergoing coherent Rabi oscillations while interacting with a driving field. A spontaneous-emission event disrupts the phase coherence of the atomic oscillation, and the cumulative effect of many such events leads to decay of the ensemble-averaged coherent oscillation. Similarly, spontaneous-emission events in a laser progressively randomize the field phase, with the resulting phase-diffusion rate determining the laser linewidth.

We can relate the average number of photons $\langle q\rangle$ in the laser cavity to the average optical power P lost from the cavity,

$$P = \gamma_c\langle q\rangle\hbar\omega\,. \tag{9.94}$$

Here P represents the total power associated with all cavity losses included in γ_c. This allows us to rewrite the linewidth as[11]

$$\Delta\nu_{L} = \frac{\gamma_c\hbar\omega}{4\pi P}\frac{N_2 A_{21}}{p} = \Delta\nu_c\frac{\hbar\omega}{2P}\frac{N_2 A_{21}}{p}. \tag{9.95}$$

The ratio A_{21}/p, which represents the spontaneous-emission rate into the cavity mode per excited atom, can be expressed in terms of the laser threshold inversion. Recalling that the threshold inversion is defined by

$$K\,\Delta N_{\text{th}} \equiv \frac{A_{21}}{p}\,\Delta N_{\text{th}} = \gamma_c\,, \tag{9.96}$$

we can write the laser linewidth, historically known as the Schawlow-Townes linewidth, as

$$\Delta\nu_{L} = \Delta\nu_c\frac{\hbar\omega}{2P}\frac{N_2\gamma_c}{\Delta N_{\text{th}}} = \Delta\nu_c\frac{\hbar\omega}{P}\frac{N_2}{\Delta N_{\text{th}}}\pi\frac{\gamma_c}{2\pi} = \pi(\Delta\nu_c)^2\frac{\hbar\omega}{P}\frac{N_2}{\Delta N_{\text{th}}}. \tag{9.97}$$

As P increases rapidly through threshold, the laser linewidth correspondingly decreases. For many four-level laser systems operating sufficiently above threshold, the last factor $N_2/\Delta N_{\text{th}}$ is approximately unity.

Estimates of the Schawlow-Townes Linewidth

Using typical parameters for a He-Ne laser, $\lambda = 633$ nm, $\mathcal{R}_1 \approx 1.0$, $\mathcal{R}_2 \approx 0.97$, $L = 30$ cm, and $P = 1$ mW, we find a Schawlow–Townes linewidth of approximately $\Delta\nu_{L} = 0.07$ Hz.

For a GaAs/GaAlAs laser with $\lambda = 850$ nm, corresponding to $\nu = c/\lambda = 3.53\times10^{14}$ Hz, cavity length $L = 3\times10^{-2}$ cm, $P = 1$ mW, $\mathcal{R} = 31\%$ (Fresnel reflectance), and $n = 3.5$, the cavity linewidth is of order 10^{10} Hz, giving a Schawlow-Townes linewidth of a few MHz.

[11] A. L. Schawlow and C. H. Townes, "Infrared and Optical Masers," Phys. Rev. **112**, 1940–1949 (1958).

These Schawlow–Townes linewidths are generally much smaller than experimentally observed linewidths. In practical lasers, mechanical vibrations, thermal fluctuations, pump fluctuations, and other environmental perturbations can produce frequency fluctuations substantially larger than the fundamental spontaneous-emission contribution. The Schawlow–Townes linewidth therefore represents the fundamental spontaneous-emission limit for an ideal single-mode laser.

9.6 Semiclassical Theory of Laser Light Fluctuations

The buildup of laser light during turn-on and its intensity and phase fluctuations can also be viewed from a classical-noise perspective. In this approach, the effect of spontaneous emission is taken into account by adding a random contribution (noise) to the equation of motion for the electric field. With the addition of noise, the equation of motion for the electric field becomes a Langevin equation. We can then use methods of statistical mechanics to discuss the statistical properties of laser light.

We present a simple treatment here. Consider the complex electric field of a single-mode laser, which can be written as

$$E(\boldsymbol{r},t) = A(t)u(\boldsymbol{r})e^{-i\omega t}\,, \tag{9.98}$$

where ω is the carrier (or average) field frequency, $u(\boldsymbol{r})$ describes the spatial variation of the field, and $A(t)$ represents the slowly varying complex field amplitude, normalized such that the cycle-averaged intensity is $I = |A(t)|^2$. Because the photon number $q \propto |A(t)|^2$, it follows from Eq. (9.25) for the cavity photon number that the amplitude $A(t)$ obeys the following equation of motion[12]

$$\frac{dA}{dt} = \frac{1}{2}\gamma_c \left[r - 1 - |A(t)|^2/I_{\text{sat}}\right] A(t)\,. \tag{9.99}$$

Here I_{sat} is the saturation intensity and γ_c is the cavity decay rate. The solution for the laser intensity with the initial condition $|A(0)|^2 = I(0) = I_o$ is

$$I(t) = \frac{I_{\text{sat}}(r-1)I_o e^{(r-1)\gamma_c t}}{I_{\text{sat}}(r-1) + I_o(e^{(r-1)\gamma_c t} - 1)} \xrightarrow{t\to\infty} \begin{cases} 0\,, & r < 1\,; \\ I_{\text{sat}}(r-1)\,, & r > 1\,. \end{cases} \tag{9.100}$$

Thus, when losses exceed gain ($r < 1$), the laser approaches a steady state with zero intensity. On the other hand, when gain exceeds losses ($r > 1$), it approaches a steady state with nonzero intensity. Notice that both asymptotic ($t \to \infty$) limits—zero for $r < 1$ and $I_{\text{sat}}(r-1)$ for $r > 1$—are reached independently of the particular nonzero initial value I_o. Thus, the stable steady state is $I_{ss} = 0$ for $r < 1$ and $I_{ss} = I_{\text{sat}}(r-1)$ for $r > 1$.

The expression for the light intensity derived above also shows that if the laser is initially off ($I_o = 0$), it will remain at zero intensity, no matter how far the pump is raised above threshold ($r > 1$). This limitation was already noted in Sec. 9.1. In a phenomenological approach, we can remedy this deficiency by adding a noise term to the equation of motion for the electric field. Physically, this noise term represents the contribution to the electric field from spontaneous emission. With the addition of such a term, the equation of motion

[12]This equation also follows from Eq. (7.106) by using the definition of the pump ratio $r = 2gL_g/\mathcal{L}_c$ and expanding the gain to first order in light intensity, appropriate for near-threshold operation of a laser.

for the field amplitude becomes

$$\dot{A}(t) = \frac{1}{2}\gamma_c \left[r - 1 - |A(t)|^2/I_{\text{sat}}\right] A(t) + \eta(t)\,, \tag{9.101}$$

where $\eta(t)$ is a complex random process. We expect spontaneous-emission contributions to the field to have zero mean and successive contributions to be statistically independent. Mathematically, these properties can be represented by modeling $\eta(t)$ as a delta-correlated (white), complex Gaussian random process:

$$\langle \eta(t) \rangle = 0\,, \qquad \langle \eta^*(t)\eta(t') \rangle = S\delta(t - t')\,, \tag{9.102}$$

where S specifies the noise strength. This phenomenological representation is convenient because it bypasses a detailed microscopic description of the noise source. It provides qualitative insight into the effect of noise on the laser equations without describing the precise microscopic mechanism. From the discussion of Sec. 9.5, we may regard it as representing the cumulative effect of spontaneously emitted photons, each changing the electric field of the laser by a small random amount. We may therefore expect S to depend on the inversion (or gain), since the spontaneous-emission rate depends on the number of atoms in the excited state. Photon loss from the cavity also produces fluctuations in the intracavity field. Other sources of noise, such as gain fluctuations, can instead produce random fluctuations of the gain coefficient and hence multiplicative, rather than additive, noise. Although all of these processes can contribute to laser fluctuations, we will not pursue these distinctions here and will represent the effect of spontaneous emission solely by an additive random contribution to the electric field.

The equation of motion for the field amplitude can be simplified by introducing the scaled (dimensionless) variables

$$E' = \left(\frac{2\gamma_c}{I_{\text{sat}}S}\right)^{1/4} A\,, \quad t' = \left(\frac{\gamma_c S}{8I_{\text{sat}}}\right)^{1/2} t\,, \quad a = (r-1)\sqrt{\frac{2\gamma_c I_{\text{sat}}}{S}}\,. \tag{9.103}$$

For notational simplicity, we now drop the primes and use E and t to denote the scaled field amplitude and scaled complex time. Then the equation for the scaled complex field amplitude is

$$\frac{dE(t)}{dt} = \left(a - |E(t)|^2\right) E(t) + \xi(t)\,, \tag{9.104}$$

where a is the pump parameter, which is positive above threshold, negative below threshold, and zero exactly at threshold. The parameter a plays the same role as the pump parameter introduced in Eq. (9.68). The scaled complex Gaussian white-noise term $\xi(t)$ has the properties

$$\langle \xi(t) \rangle = 0\,, \qquad \langle \xi^*(t)\xi(t') \rangle = 4\delta(t - t')\,. \tag{9.105}$$

Equation (9.104) for $E(t)$ has the form of a Langevin equation describing the overdamped motion of a Brownian particle in a two-dimensional potential

$$U(a, E) = -\frac{a}{2}|E|^2 + \frac{1}{4}|E|^4\,. \tag{9.106}$$

If the noise term $\xi(t)$ is ignored, the stationary field amplitudes are determined by

$$2\frac{\partial U}{\partial E^*} = 0 = E(|E|^2 - a)\,. \tag{9.107}$$

Below threshold ($a < 0$), the only stationary solution is $E = 0$, which is stable. Above threshold ($a > 0$), $E = 0$ becomes unstable and the stable stationary states satisfy

$$|E| = \sqrt{a}\,. \tag{9.108}$$

Since E is complex, these states form a circle $E = \sqrt{a}\, e^{i\phi}$ with arbitrary phase ϕ. Thus, above threshold the field acquires a preferred nonzero magnitude, while its phase remains undetermined.

If the most probable magnitude of the field amplitude is treated as an order parameter, its behavior is

$$|E|_m = \begin{cases} 0\,, & a < 0\,; \\ \sqrt{a}\,, & a > 0\,, \end{cases} \tag{9.109}$$

which is analogous to the behavior of the magnetization magnitude in the Curie–Weiss model of a ferromagnet, with $-a$ playing the role of a temperature-like control parameter. This analogy can be extended further by regarding $U(a, E)$ as an effective "free energy" for the laser. Because of this close mathematical correspondence, the laser threshold transition is frequently likened to a second-order phase transition.[13]

The Brownian-motion analogy also allows us to obtain the steady-state probability distribution for E using the classical Boltzmann form

$$\mathcal{P}_s(E) = \text{const} \times \exp[-U(a, E)] = \text{const} \times \exp\left[\frac{a}{2}|E|^2 - \frac{1}{4}|E|^4\right], \tag{9.110}$$

where we have used the fact that the scaled noise intensity (the analog of $k_B T$) in our system is unity. The distribution in Eq. (9.110) is the exact steady-state probability distribution for the complex field amplitude within this Langevin model.[14] By writing the field amplitude in terms of intensity and phase, $E = \sqrt{I}e^{i\phi}$, where $I = |E|^2$ now denotes the dimensionless scaled intensity, we see that the steady-state distribution $\mathcal{P}_s(\sqrt{I}e^{i\phi})$ is independent of phase. Integrating over the phase, the normalized distribution for the scaled intensity can be written as

$$P_s(I) = \frac{e^{-\frac{1}{4}(I-a)^2}}{\sqrt{\pi}[1 + \text{erf}(a/2)]}\,. \tag{9.111}$$

Comparison with the photon-number distribution in (9.73) shows that, upon identifying the scaled photon number $\sqrt{2}\,(q/\sqrt{q_s})$ with the scaled intensity I, the two distributions have exactly the same form.

Using the asymptotic form of the error function from Sec. 9.4, we find that far below threshold, $a < 0$ and $|a| \gg 1$, the intensity distribution reduces to the exponential form

$$P_s(I) = \frac{|a|}{2} e^{-\frac{1}{2}|a| I}\,, \tag{9.112}$$

which is characteristic of thermal light. Far above threshold ($a \gg 1$), the intensity distribution becomes

$$P_s(I) = \frac{1}{2\sqrt{\pi}} e^{-\frac{1}{4}(I-a)^2}\,, \tag{9.113}$$

which is a Gaussian centered at $I = a$. As a increases, the variance of the intensity remains constant at 2, while its mean increases approximately as a. Thus, the relative intensity variance decreases as $2/a^2$, and the laser approaches a state with increasingly small relative intensity fluctuations, while its phase remains undetermined and undergoes diffusion. In the threshold region, the laser undergoes a smooth transition from thermal-like fluctuations below threshold to the intensity statistics characteristic of a coherent state above threshold.

---

[13] R. Graham and H. Haken, "Laser light: first example of a second-order phase transition far away from thermal equilibrium," Z. Phys. **237**, 31–46 (1970); V. DiGiorgio and M. O. Scully, "Analogy between the Laser Threshold Region and a Second-Order Phase Transition," Phys. Rev. A **2**, 1170–1177 (1970).

[14] H. Risken in *Progress in Optics VIII*, ed. E. Wolf (North Holland, Amsterdam, 1970).

The mean and variance computed from $P_s(I)$ agree with the corresponding results for the scaled photon number obtained from Eqs. (9.76) and (9.77).

It is also instructive to consider the laser far above threshold ($a \gg 1$), where we can linearize the Langevin equation (9.104) about the steady-state field magnitude $\sqrt{a}$. We write

$$E = (\sqrt{a} + u)e^{i(\phi_o + \delta\phi)}, \tag{9.114}$$

where u and $\delta\phi$ represent small amplitude and phase fluctuations, respectively. Resolving the complex noise into locally radial and tangential components, $\xi(t) = \xi_R(t) + i\xi_T(t)$, the linearized equations become

$$\frac{du}{dt} = -2au + \xi_R(t), \tag{9.115}$$

$$\frac{d\delta\phi}{dt} = \frac{1}{\sqrt{a}}\xi_T(t), \tag{9.116}$$

where ξ_R and ξ_T are two independent real, zero-mean Gaussian white-noise processes satisfying

$$\langle \xi_R(t) \rangle = 0, \qquad \langle \xi_R(t)\xi_R(t') \rangle = 2\delta(t - t'), \tag{9.117}$$

$$\langle \xi_T(t) \rangle = 0, \qquad \langle \xi_T(t)\xi_T(t') \rangle = 2\delta(t - t'). \tag{9.118}$$

Since the intensity fluctuation is $\delta I \approx 2\sqrt{a}\,u$, it follows that

$$\frac{d\delta I}{dt} = -2a\,\delta I + 2\sqrt{a}\,\xi_R(t). \tag{9.119}$$

Thus, the intensity fluctuations constitute an Ornstein–Uhlenbeck process, whereas the phase fluctuations constitute a Wiener process. The phase equation can be integrated directly, giving

$$\langle [\delta\phi(t) - \delta\phi(0)]^2 \rangle = \frac{2t}{a}. \tag{9.120}$$

This linear growth of the phase variance with time, analogous to Eq. (9.85), represents phase diffusion and therefore produces a finite laser linewidth. Since a is proportional to the steady-state scaled intensity far above threshold, this result reproduces the essential feature of the Schawlow–Townes linewidth formula (9.97): the linewidth varies inversely with laser power.

Summary

In this chapter we introduced the cavity photon number and the population of active atoms inside the laser cavity as the key dynamical variables describing the behavior of a laser. We used the laser rate equations to derive a set of coupled nonlinear equations linking the cavity photon number to the excited-state population of atoms (inversion) in the gain medium. Because these equations cannot describe the growth of laser light from an initially off state, we modified them phenomenologically to include the contribution of spontaneous emission to the cavity photon number. The resulting modified rate equations were used to analyze the buildup of laser oscillations from an initially zero-intensity state and the rapid growth of the cavity photon number as the laser crosses the threshold of oscillation.

The insights gained from these modified rate equations allowed us to derive a master equation for the cavity photon-number probability distribution. The steady-state solution of this equation was used to describe the transformation of the photon statistics of laser light near threshold. The ultimate limit placed on the laser linewidth by spontaneous emission was also discussed.

Finally, we explored an alternative approach to laser photon statistics that models the dynamics of the complex electric field of a laser as a Brownian particle moving within a two-dimensional potential well under the influence of spontaneous-emission noise. The mathematical agreement between the master-equation and Langevin-equation approaches confirms that this semiclassical treatment successfully captures the essential physics of laser-light fluctuations.

Problems

1. **Laser turn-on:** With spontaneous emission included, the cavity photon number q in a single-mode laser operating near threshold builds up according to Eq. (9.42),

$$\frac{dq}{dt} = \gamma_c r \left(1 - \frac{q}{q_s}\right)(q+1) - \gamma_c q\,,$$

where the pump ratio is $r = KR_p/(\gamma_2\gamma_c) = N^{(0)}/N_{th}$, and the saturation photon number is $q_s = \gamma_2/K$. This equation assumes $q/q_s \ll 1$, as appropriate near threshold. In terms of the dimensionless variables $x = q/q_s$ and $\tau = \gamma_c t$, it becomes

$$\frac{dx}{d\tau} = r(1-x)\left(x + \frac{1}{q_s}\right) - x\,.$$

For the numerical plots in this problem, take $q_s = 10^8$.

(a) Find the steady-state cavity photon number predicted by this equation. Show that, in the threshold region $|r-1| < 10/\sqrt{q_s} \ll 1$, it agrees with the exact steady-state result derived earlier. Plot the steady-state photon number as a function of pump ratio over a range sufficiently narrow to display the threshold transition clearly, for example $0.998 < r < 1.002$.

(b) Show that, for the initial condition $x(0) = 0$, the time evolution of the cavity photon number after the pump is turned on is

$$q(t) = \frac{q_{ss}(1 - e^{-\gamma_c' t})}{1 + \dfrac{q_{ss}^2}{q_s} e^{-\gamma_c' t}}\,,$$

where

$$q_{ss} = \frac{q_s}{2r}\left[r - 1 - \frac{r}{q_s} + \sqrt{\left(r - 1 - \frac{r}{q_s}\right)^2 + \frac{4r^2}{q_s}}\right],$$

and

$$\gamma_c' = \gamma_c\sqrt{\left(r - 1 - \frac{r}{q_s}\right)^2 + \frac{4r^2}{q_s}}\,.$$

Verify that the solution has the correct limits as $t \to 0$ and $t \to \infty$.

Plot $x = q/q_s$ as a function of the scaled time $\gamma_c t$ for representative pump ratios below, near, and above threshold, for example $r = 0.98$, 1.005, and 1.02. Depending on r, many cavity lifetimes may be required before the photon number approaches its steady-state value. Define a reasonable buildup time, derive an analytic expression for it, and discuss its dependence on the pump ratio r.

2. **Photon number variance for thermal and Poisson distributions:**

   (a) Show that for a Poisson photon-number distribution the variance is

$$\langle(\Delta q)^2\rangle \equiv \langle(q - \bar{q})^2\rangle = \bar{q}\,.$$

   (b) Show that for a thermal (Bose-Einstein) photon-number distribution the variance is

$$\langle(\Delta q)^2\rangle \equiv \langle(q - \bar{q})^2\rangle = \bar{q}^2 + \bar{q}\,.$$

3. **Quantum-limited laser linewidth:** Using the Schawlow-Townes linewidth expression derived in the text, estimate the quantum-limited linewidth for each of the following lasers. Assume that the population-inversion factor $N_2/\Delta N_{\text{th}} \approx 1$, and neglect technical sources of frequency noise.

   (a) A He-Ne laser operating at $\lambda = 632.8$ nm with cavity length 20 cm, total logarithmic roundtrip loss corresponding to a 1% fractional roundtrip power loss, and output power 1 mW.

   (b) A semiconductor diode laser operating at $\lambda = 850$ nm with uncoated facets, refractive index $n = 3.5$, cavity length 120 μm, internal intensity-loss coefficient $\alpha = 10$ cm^{-1}, and output power 5 mW. Calculate the facet reflectance from the refractive index and include both mirror and internal losses in determining the cavity linewidth.

   (c) A unidirectional ring dye laser operating at $\lambda = 590$ nm with roundtrip path length 75 cm contains a dye jet of thickness 0.2 mm having an internal intensity-loss coefficient $\alpha = 0.05$ cm^{-1}. The output coupling is 3%, and the output power is 100 mW. Take $n \approx 1$.

4. **Intracavity photon number and saturation photon number:** A 1.0-m-long Nd:YAG laser operating at $\lambda = 1.064\,\mu$m has an internal fractional power loss of 5% per roundtrip and end mirrors with power reflectances $\mathcal{R}_1 = 99.9\%$ and $\mathcal{R}_2 = 95.0\%$. Mirror M_2 is the output coupler, and its losses other than transmission may be neglected. The cw output power through M_2 is 1.0 W. Take the refractive index throughout the cavity to be unity.

   (a) Estimate the average number of photons stored in the cavity while the laser is oscillating.

   (b) If the pump ratio is $r = 6$, estimate the saturation photon number q_s for the laser.

5. **Photon number and output flux of a semiconductor laser:** A semiconductor laser operating at $\lambda = 850$ nm has a cavity length of 300 μm and an effective transverse mode area of $1\,\mu\text{m} \times 5\,\mu\text{m}$. The two facets have power reflectances $\mathcal{R}_1 \approx \mathcal{R}_2 \approx 35\%$. Take the refractive index to be $n = 3.5$ and the effective upper-state population lifetime to be 3 ns. The stimulated-emission cross section is 3.6×10^{-16} cm^2. Within the simplified rate-equation model developed in the text, the laser operates at a pump ratio $r = 1.5$.

   (a) Estimate the average number of photons stored in the laser cavity.

   (b) Calculate the photon flux and optical power emitted through one facet. What are the total photon flux and optical power emitted through both facets?

10

Laser Transients and Q-Switched Operation

Lasers exhibit a wide range of transient behavior when their operating parameters are perturbed or modulated. These dynamics are described by coupled nonlinear differential equations and can become even richer when nonlinear optical elements, such as saturable absorbers, are introduced into the cavity. Such elements can destabilize steady-state continuous-wave (cw) operation and produce new forms of laser dynamics. A complete treatment is beyond the scope of this book; here we consider several phenomena that are particularly important in laser operation and coherent-light generation.

The time-dependent phenomena considered here fall naturally into two groups. The first involves essentially a single cavity mode and includes relaxation oscillations, Q-switching, and injection-locking. The second involves several modes simultaneously and includes mode competition and mode-locking. Multimode behavior may be described either through the amplitudes of the individual oscillating modes in the frequency domain or through a self-consistent circulating pulse in the time domain.

In this chapter we consider spiking, relaxation oscillations, mode competition, injection-locking, and Q-switching. These examples introduce several important concepts and mathematical techniques for describing transient laser behavior and each illustrates a **fundamental principle** of laser dynamics. Mode-locking and short-pulse generation, which involve a large number of cavity modes and require a different approach, are discussed in the next chapter.

10.1 Spiking and Relaxation Oscillations

Many lasers exhibit oscillatory transients when they are turned on or when an operating parameter, such as the gain, loss, or cavity frequency, is perturbed. The terminology used for these transients is not uniform. Here we use spiking for the relatively large, narrow pulses that commonly occur during laser turn-on, and relaxation oscillations for the small-amplitude, exponentially damped oscillations about the steady state that follow a weak perturbation of a CW laser. During turn-on, the initial spiking generally evolves into smooth relaxation oscillations as the laser approaches steady state.

Spiking and relaxation oscillations are characteristic of many solid-state and semiconductor lasers, in which the population-inversion recovery time is substantially longer than the cavity decay time ($\gamma_2^{-1} \gg \gamma_c^{-1}$). Most gas lasers do not satisfy this condition, and therefore pronounced spiking and relaxation oscillations are generally not observed in gas lasers.

DOI: 10.1201/9781003584179-10

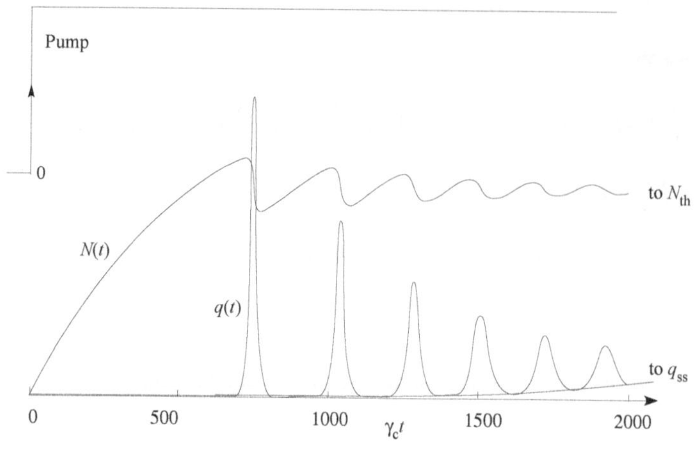

FIGURE 10.1
The growth of laser radiation following a sudden turn-on of the laser pump is, in general, oscillatory.

10.1.1 Relaxation Oscillations

When a laser is suddenly turned on or perturbed from steady state, it may approach its steady state either monotonically or through damped oscillations. Which behavior occurs depends mainly on the relative magnitudes of the cavity decay rate γ_c, the population decay rate γ_2, and the pump parameter r. The basic behavior is illustrated in Fig. 10.1.

During laser turn-on, pumping first builds the population inversion. Once the inversion exceeds threshold, the intracavity photon number begins to grow rapidly. The growing field then saturates the inversion and reduces the gain, slowing the buildup of radiation. Depending on the population and cavity decay rates and on the pump level, both the photon number and the inversion may overshoot their steady-state values and undergo a sequence of damped oscillations.

For a quantitative understanding, we return to the simplified coupled rate equations,

$$\frac{dq}{dt} = KNq - \gamma_c q = K(N - N_{th})q\,, \tag{10.1a}$$

$$\frac{dN}{dt} = R_p - \gamma_2 N - KNq = \gamma_2\left[N^{(0)} - N(1 + q/q_s)\right]\,. \tag{10.1b}$$

These equations describe the cavity photon number $q(t)$ and the instantaneous population inversion $N(t)$, which is approximated by the upper-laser-level population. The cavity photon decay rate is γ_c, with cavity lifetime $\tau_c = 1/\gamma_c$. The inversion is pumped at R_p and decays at the rate γ_2. We are primarily interested in operation well above threshold, where $q \gg 1$. We therefore neglect the additional "+1" spontaneous-emission term and retain only the stimulated-emission term KNq in the photon rate equation. As discussed in Chapter 9 , spontaneous emission is important mainly in the narrow threshold region $|r - 1| \leq 10/\sqrt{q_s}$. Outside this region, it is needed to initiate the buildup from the off-state, but it quickly becomes negligible once the intracavity photon number begins to grow.

Because Eqs. (10.1a) and (10.1b) contain the product $KN(t)q(t)$, they form a pair of coupled nonlinear differential equations. The inversion and photon number respond on different time scales and therefore need not follow one another instantaneously. This delayed response can produce spiking or relaxation oscillations as the laser approaches steady state.

For class-A lasers, where $\gamma_2 \gg \gamma_c$, we have already discussed an approximate analytic solution. No comparably simple closed-form solution exists during strong spiking, when both N and q change rapidly. Nevertheless, Eqs. (10.1) provide a unified description of both spiking and small-amplitude relaxation oscillations.

Equation (10.1a) shows that the cavity photon number grows only when the inversion exceeds the threshold value $N_{th} \equiv \gamma_c/K$. Equation (10.1b), in turn, shows that the growing photon number depletes the inversion from its unsaturated value $N^{(0)}$. Using N_{th}, unsaturated inversion $N^{(o)}$, saturation photon number q_s, and pump ratio r defined by

$$N_{\text{th}} = \frac{\gamma_c}{K}, \quad N^{(o)} = \frac{R_p}{\gamma_2}, \tag{10.2}$$

$$q_s = \frac{\gamma_2}{K}, \qquad r = \frac{N^{(o)}}{N_{\text{th}}} = \frac{KR_p}{\gamma_2\gamma_c}, \tag{10.3}$$

the steady-state solutions of Eqs. (10.1) above threshold can be written as

$$q_{ss} = \frac{R_p}{KN_{ss}} - \frac{\gamma_2}{K} = \frac{\gamma_2}{K}\left(\frac{N^{(o)}}{N_{\text{th}}} - 1\right) = q_s(r-1), \tag{10.4a}$$

$$N_{ss} = \frac{N^{(0)}}{1+q/q_s} = \frac{rN_{\text{th}}}{1+q_{ss}/q_s} = N_{\text{th}}. \tag{10.4b}$$

It is useful to express these equations in dimensionless form by defining

$$x = \frac{q}{q_s}, \quad y = \frac{N}{N_{\text{th}}}, \quad \tau = \gamma_c t, \quad s = \frac{\gamma_2}{\gamma_c}. \tag{10.5}$$

The rate equations then become

$$\frac{dx}{d\tau} = x(y-1), \tag{10.6a}$$

$$\frac{dy}{d\tau} = s(r - y - xy). \tag{10.6b}$$

By taking the ratio of these two equations, we eliminate the time variable τ, and obtain a single autonomous differential equation

$$\frac{dx}{dy} = \frac{x(y-1)}{s(r-y-xy)}. \tag{10.7}$$

This equation describes the trajectory of the laser in the two-dimensional phase space spanned by the scaled photon number x and the scaled inversion y. Two examples of this trajectory are shown in Fig. 10.2 for different values of the lifetime ratio $s = \gamma_2/\gamma_c = \tau_c/\tau_2$ at a fixed pump ratio $r = 1.25$. As a general rule, for $s \ll 1$, typical of many solid-state lasers, the system approaches steady state in a spiraling, oscillatory manner [Fig. 10.2(a)]. Conversely, for $s \gg 1$, typical of class-A dye or gas lasers, the system approaches steady state monotonically without overshooting [Fig. 10.2(b)].

The full time dependence of q and N must generally be obtained by solving Eqs. (10.1) numerically as illustrated in Fig. 10.1. For small perturbations, however, the equations can be linearized and solved analytically. We write

$$q = q_{ss} + \epsilon, \tag{10.8a}$$

$$N = N_{ss} + \eta, \tag{10.8b}$$

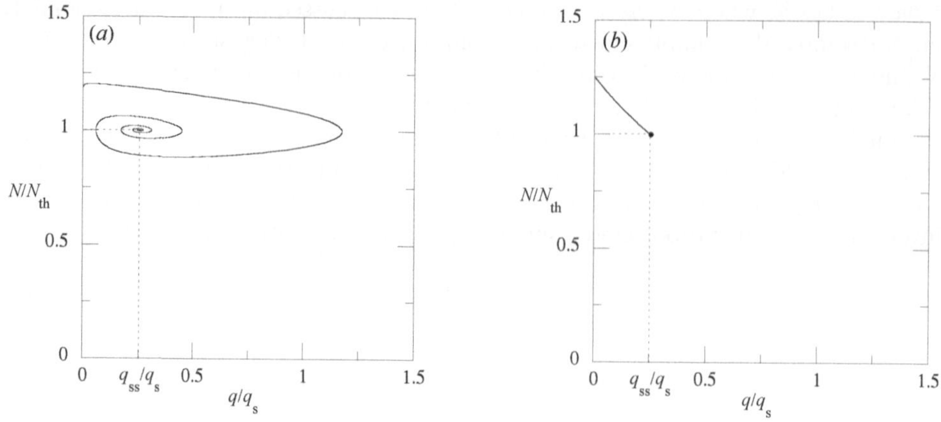

FIGURE 10.2
Phase space trajectory of a laser during its turn-on for pump ratio $r = 1.25$ and (a) a population lifetime longer than the cavity lifetime: $s = \gamma_2/\gamma_c = 0.02$ and (b) a population lifetime shorter than the cavity lifetime: $s = \gamma_2/\gamma_c = 10$.

where $|\epsilon| \ll q_{ss}$ and $|\eta| \ll N_{ss}$ represent small perturbations of photon number and population inversion from their steady-state values. Substituting these into Eqs. (10.1) and neglecting the product $\epsilon\eta$ yields the linearized equations of motion for ϵ and η

$$\begin{aligned}\frac{d\epsilon}{dt} &= K(N_{ss}+\eta)(q_{ss}+\epsilon) - \gamma_c(q_{ss}+\epsilon)\,,\\ &\approx \gamma_2(r-1)\eta\,, \qquad (10.9a)\\ \frac{d\eta}{dt} &= R_p - \gamma_2(N_{ss}+\eta) - K(N_{ss}+\eta)(q_{ss}+\epsilon)\,,\\ &\approx -\gamma_c\epsilon - \gamma_2 r\eta\,, \qquad (10.9b)\end{aligned}$$

where we used the steady-state conditions $KN_{ss} = \gamma_c$ and $Kq_{ss} = \gamma_2(r-1)$ to obtain the linearized equations. By differentiating Eq. (10.9a) and using Eq. (10.9b) to eliminate η, we obtain a second-order, linear differential equation for ϵ

$$\ddot{\epsilon} + r\gamma_2\dot{\epsilon} + (r-1)\gamma_c\gamma_2\epsilon = 0\,. \qquad (10.10)$$

This is mathematically identical to the equation of motion for a damped harmonic oscillator characterized by the natural oscillation frequency ω_o and the amplitude decay rate β

$$\omega_o \equiv \sqrt{(r-1)\gamma_c\gamma_2}\,, \qquad (10.11a)$$
$$2\beta \equiv r\gamma_2\,. \qquad (10.11b)$$

Rewriting Eq. (10.10) in terms of ω_o and β, we obtain

$$\ddot{\epsilon} + 2\beta\dot{\epsilon} + \omega_o^2\epsilon = 0\,. \qquad (10.12)$$

We seek a solution of the form

$$\epsilon(t) = Ae^{-i\omega t}\,, \qquad (10.13)$$

where A is a constant determined by the initial perturbation and ω may be complex. Substituting this solution into (10.12) yields the algebraic equation satisfied by ω

$$-\omega^2 - 2i\beta\omega + \omega_o^2 = 0\,. \tag{10.14}$$

This quadratic equation for ω yields two values,

$$\omega_\pm = -i\beta \pm \sqrt{\omega_o^2 - \beta^2} \equiv -i\beta \pm \omega_r\,. \tag{10.15}$$

When real, $\omega_r \equiv \sqrt{\omega_0^2 - \beta^2}$ represents the relaxation-oscillation frequency. The general solution for the photon number perturbation $\epsilon(t)$ can then be written as a linear combination of solutions corresponding to these two values of ω,

$$\epsilon(t) = A_+ e^{-i\omega_+ t} + A_- e^{-i\omega_- t} = e^{-\beta t}\left[A\cos\omega_r t + B\sin\omega_r t\right]\,. \tag{10.16}$$

Expressing the integration constants A and B explicitly in terms of the initial perturbation parameters, the solution takes the form

$$\epsilon(t) = e^{-\beta t}\left[\epsilon(0)\cos\omega_r t + \frac{\dot{\epsilon}(0)}{\omega_r}\sin\omega_r t\right] \equiv \mathrm{C}\,e^{-\beta t}\cos(\omega_r t + \phi_o)\,, \tag{10.17}$$

where C and ϕ_o represent the amplitude and phase offset of the initial disturbance. Equation (10.17) shows that small perturbations decay as damped sinusoidal oscillations with frequency ω_r and lifetime τ given by

$$\omega_r = \sqrt{\omega_o^2 - \beta^2} = \sqrt{(r-1)\gamma_c\gamma_2}\,\sqrt{1 - \frac{r^2}{4(r-1)}\frac{\gamma_2}{\gamma_c}}\,, \tag{10.18a}$$

$$\tau \equiv \frac{1}{\beta} = \frac{2}{r\gamma_2}\,. \tag{10.18b}$$

Figure 10.3 shows an example of relaxation oscillations.

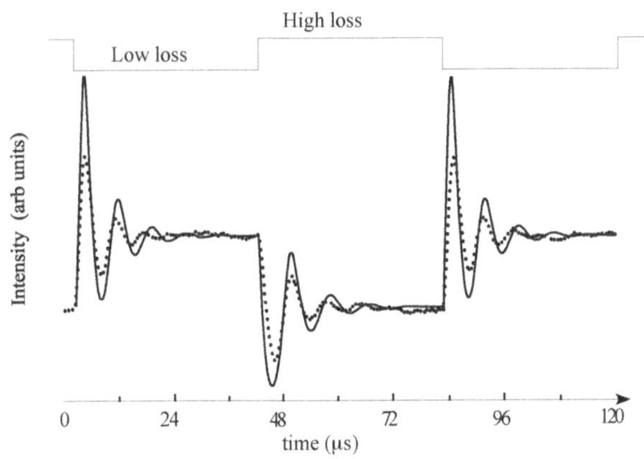

FIGURE 10.3
For small perturbations of gain or loss, the laser approaches steady state in a (damped) sinusoidal manner. The dotted curve shows the measured intensity, and the solid curve shows the calculated intensity for a Ti:sapphire laser whose loss is periodically switched between two levels differing by 0.3%. [From W. L. Erikson, M.S. Thesis, University of Arkansas (unpublished, 1992).]

Example 10.1 *Relaxation oscillations in solid-state lasers.*

1. For a ruby laser ($n = 1.77$) operating at $r = 2.0$ with $\mathcal{R}_1 = 1.0$, $\mathcal{R}_2 = 0.94$, $L = 5.0$ cm, $\gamma_2 = 500$ Hz, we have

$$\gamma_c = \frac{c\mathcal{L}_c}{2nL} = \frac{3.00 \times 10^8[-\ln(0.94)]}{2(1.77)(0.05)} \text{ Hz} = 1.0 \times 10^8 \text{ Hz}$$
$$T_r = \frac{2\pi}{\omega_r} = \frac{2\pi}{\sqrt{(r-1)\gamma_c\gamma_2 - (r\gamma_2/2)^2}} \approx \frac{2\pi}{\sqrt{1.0 \times 10^8 \times 500}} \text{ s} = 28\ \mu\text{s}$$
$$\tau = \frac{2}{r\gamma_2} = \frac{1}{\gamma_2} = 2 \text{ ms}$$

2. Consider a unidirectional Ti:sapphire ring laser with a roundtrip length of 117 cm and a loss of 3.6% per pass. Let $\gamma_2 = 288$ kHz (corresponding to $\tau_2 = 3.47\ \mu$s), and assume operation 40% above threshold ($r = 1.4$). Then

$$\gamma_c = \frac{c\mathcal{L}_c}{L} = \frac{(3.00 \times 10^8)[-\ln(0.964)]}{1.17} \text{ Hz} = 9.40 \times 10^6 \text{ Hz}$$
$$T_r = \frac{2\pi}{\sqrt{(1.4-1)(288 \times 10^3)(9.40 \times 10^6) - (1.4(288 \times 10^3)/2)^2}} \text{ s}$$
$$= 6.2\ \mu\text{s},$$
$$\tau = \frac{2}{r\gamma_2} = \frac{2}{4.03 \times 10^5} = 4.96\ \mu\text{s} \approx 5.0\ \mu\text{s}.$$

In many lasers, particularly gas lasers, the atomic decay rate γ_2 is much larger than the cavity decay rate γ_c (class-A lasers), and the laser does not operate too far above threshold. In such cases

$$\omega_\pm = -i\beta \pm i\sqrt{\beta^2 - \omega_o^2} = -i[\beta \mp \sqrt{\beta^2 - \omega_o^2}] \equiv -i\gamma_\pm, \tag{10.19}$$

so that the perturbations decay exponentially

$$\epsilon(t) = A_+ e^{-\gamma_+ t} + A_- e^{-\gamma_- t}, \tag{10.20}$$

where the constants $A_\pm$ are determined by the initial perturbation. Note that both decay rates $\gamma_\pm$ are positive, as they should be, and correspond to two decay time constants

$$\tau_\pm \equiv \frac{1}{\gamma_\pm} = \frac{1}{\beta \mp \sqrt{\beta^2 - \omega_o^2}}. \tag{10.21}$$

This case describes an overdamped oscillator. For $\beta \gg \omega_o$, the expressions for decay rates and decay times simplify to

$$\gamma_\pm \equiv \beta \mp \sqrt{\beta^2 - \omega_o^2} \approx \beta \mp (\beta - \omega_o^2/2\beta) \tag{10.22}$$

which leads to two decay times

$$\tau_+ \approx \frac{2\beta}{\omega_o^2} = \frac{r}{(r-1)\gamma_c}, \tag{10.23}$$
$$\tau_- \approx \frac{1}{2\beta} = \frac{1}{r\gamma_2}. \tag{10.24}$$

The first time constant τ_+ is essentially the modified cavity decay time, and the second time constant τ_- is the modified population decay time.

Example 10.2 *Possibility of relaxation oscillations in a gas laser:*

Consider a He-Ne laser operating at 632.8 nm and pumped 50% above threshold. It has an upper level lifetime of 50 ns. Assume a cavity of length 50.0 cm, output coupling of 1% and negligible other losses.

Laser operation 50% above threshold corresponds to a pump parameter $r = 1.50$. The upper-level decay rate is $\gamma_2 = 1/(50 \times 10^{-9})$ Hz. The cavity decay rate is (assuming $n = 1$)

$$\gamma_c = \frac{c\mathcal{L}_c}{2L} = \frac{3.00 \times 10^8 \times 0.01}{2 \times 0.50}\,\text{Hz} = 3.00 \times 10^6\,\text{Hz}\,.$$

This gives

$$\begin{aligned}\omega_r &= \sqrt{\frac{(1.50 - 1) \times 3.00 \times 10^6}{50 \times 10^{-9}} - \left(\frac{1.50}{2 \times 50 \times 10^{-9}}\right)^2}\ \text{rad/s} \\ &= \sqrt{-1.95} \times 10^7\,\text{rad/s} = i1.40 \times 10^7\,\text{rad/s} \equiv i\gamma_r\end{aligned}$$

Thus ω_r is pure imaginary and we do not have relaxation oscillations. Instead, the perturbations from the steady state decay exponentially with two decay rates,

$$\begin{aligned}\gamma_- &= \frac{r\gamma_2}{2} - \gamma_r = (1.50 \times 10^7 - 1.40 \times 10^7)\,\text{Hz} = 1.0 \times 10^6\,\text{Hz}\,, \\ \gamma_+ &= \frac{r\gamma_2}{2} + \gamma_r = (1.50 \times 10^7 + 1.40 \times 10^7)\,\text{Hz} = 2.9 \times 10^7\,\text{Hz}\,.\end{aligned}$$

These rates correspond to two decay time scales

$$\tau_- = \frac{1}{1.0 \times 10^6}\,\text{s} = 1.0\,\mu\text{s}\,, \quad \tau_+ = \frac{1}{2.90 \times 10^7}\,\text{s} = 34\,\text{ns}\,.$$

Relaxation oscillations are commonly observed during the turn-on of solid-state lasers as the intensity approaches its steady-state value. Pump fluctuations and other noise sources can also excite them during steady operation. When their decay time is relatively long, even mechanical or thermal perturbations may produce observable oscillations. A continuously pumped ruby laser, for example, may emit a sequence of irregular spikes as thermal and mechanical noise repeatedly excites the relaxation response.

For large perturbations, such as sudden laser turn-on, the response remains oscillatory, but is no longer sinusoidal [Fig. 10.4]. Even so, the small-signal results remain useful: the dominant oscillation frequency is still of order ω_r, and the decay time remains close to $1/\beta = 2/r\gamma_2$. Both the period and the duration decrease as the pump level increases. At sufficiently high pump rates, the oscillations become a sequence of narrow, high-amplitude spikes before the laser finally settles to a nearly steady output.

10.2 Two-mode Coupling and Mode Competition

Most lasers can potentially oscillate in several modes—for example, different axial or transverse modes, opposite propagation directions in a ring cavity, or different polarizations. These modes generally experience different gain, loss, and saturation. Because they draw gain from the same population inversion, oscillation in one mode can reduce the gain available to another. This mode competition can produce a variety of dynamical behavior.

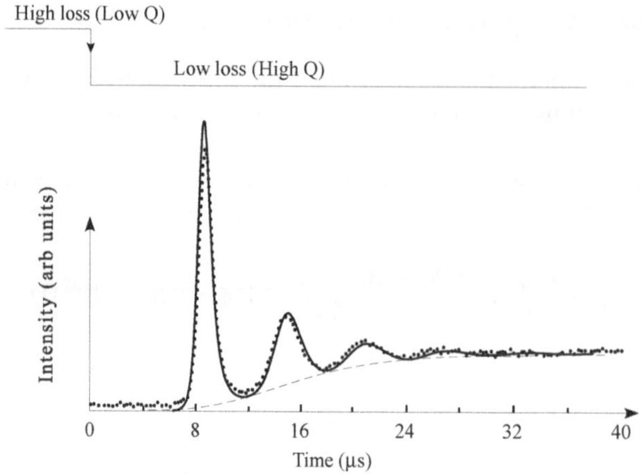

FIGURE 10.4
Time evolution of the laser intensity for a Ti:sapphire laser under steady pumping when the cavity Q is suddenly increased. The dotted curve is the measured intensity, whereas the solid curve is the theoretical prediction based on Eqs. (10.1). From W. L. Erikson, M.S. Thesis, University of Arkansas (1992, unpublished).

We illustrate the basic physics by considering two modes coupled through self-saturation and cross-saturation. Independent scattering mechanisms, such as mirror or defect backscattering, will be neglected, while the coupling produced by a population grating in the gain medium will be retained.

For mode intensities small compared with the saturation intensity, the gain may be expanded to first order in intensity. The two mode intensities then obey

$$\frac{dI_1}{dt} = I_1\left[p_1 - \beta_1 I_1 - \theta_{12} I_2\right] , \tag{10.25a}$$

$$\frac{dI_2}{dt} = I_2\left[p_2 - \beta_2 I_2 - \theta_{21} I_1\right] , \tag{10.25b}$$

where p_1 and p_2 represent the unsaturated net growth rates. The coefficients β_1 and β_2 represent self-saturation, while θ_{12} and θ_{21} describe cross-saturation of one mode by the other. Their values depend on the saturation intensity of the gain medium, the spatial and spectral overlap of the modes, the type of broadening, and the detuning from line center.

To see how Eqs. (10.25) arise, consider two counter-propagating modes of a ring laser. Let both modes have the same optical frequency ν and propagate through a homogeneously broadened gain medium with a transition frequency ν_o [Fig. 10.5]. For simplicity, let the gain medium fill the entire cavity and take both waves to be resonant with line center, $\nu = \nu_o$. The complex electric field of the two waves traveling in the $+z$ and $-z$ directions can be written as

$$E(z,t) = \left[A_1(z,t)e^{ikz} + A_2(z,t)e^{-ikz}\right] e^{-i\omega t} , \tag{10.26}$$

where the mode amplitudes $A_i(z,t)$ are slowly varying functions of z and t. In writing the field in this form, we assume that the transverse variation of the mode function over the extent of the gain medium is negligible. The total light intensity inside the cavity, defined

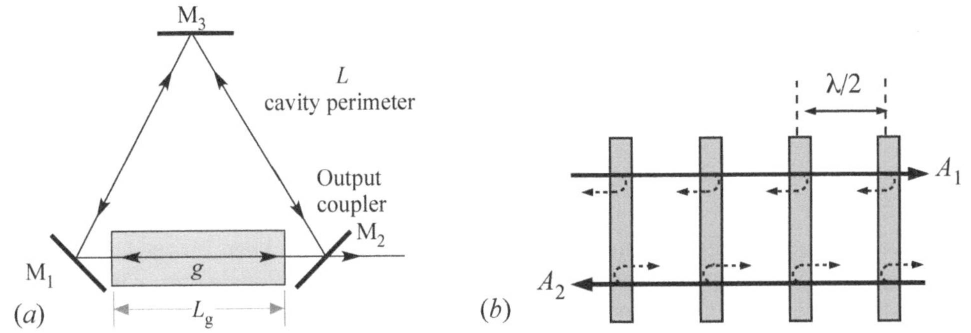

FIGURE 10.5
(a) Two oppositely traveling-wave modes of a ring cavity interacting with a gain medium. In the text we assume that the gain fills the entire cavity, making the gain length equal to the ring cavity perimeter. (b) Interference between two counter-propagating waves of the same frequency creates a standing-wave intensity grating, inducing spatial hole-burning within the population inversion. This grating can backscatter each wave into the direction of the other, as shown schematically here.

as $I(z,t) = \frac{1}{2}\epsilon_o nc|E|^2$, is then given by

$$I(z,t) = I_1(t) + I_2(t) + 2\sqrt{I_1 I_2}\cos 2kz = I_1(t) + I_2(t) + \sqrt{I_1 I_2}\left(e^{i2kz} + e^{-i2kz}\right) , \quad (10.27)$$

where I_1 and I_2 are the intensities of the two waves separately. With this total intensity, the inversion density can be expressed as

$$\Delta\mathcal{N}(z,t) = \frac{\Delta\mathcal{N}^{(0)}}{1 + I(z,t)/I_s} , \quad (10.28)$$

where $\Delta\mathcal{N}^{(0)}$ is the unsaturated inversion. Near threshold, the total intensity is small compared with the saturation intensity, $I(z) \ll I_s$. Expanding Eq. (10.28) to first order in I/I_s gives

$$\Delta\mathcal{N} \approx \Delta\mathcal{N}^{(0)}\left[1 - \frac{1}{I_s}\left(I_1 + I_2 + \sqrt{I_1 I_2}\,(e^{2ikz} + e^{-2ikz})\right)\right] . \quad (10.29)$$

Equation (10.29) shows that interference between the counter-propagating waves produces a standing-wave population grating, or spatial hole-burning, with a period equal to one-half the optical wavelength in the medium.

Substituting Eq. (10.26) into the wave equation (2.30) and following the envelope-propagation procedure of Sec. 7.8 gives

$$\begin{aligned}&\left[\frac{n}{c}\frac{\partial A_1}{\partial t} + \frac{\partial A_1}{\partial z}\right]e^{ikz} + \left[\frac{n}{c}\frac{\partial A_2}{\partial t} + \frac{\partial A_2}{\partial z}\right]e^{-ikz}\\ &\quad \approx \frac{1}{2}\left[g_o - \alpha - \frac{g_o}{I_s}\left(I_1 + I_2 + \sqrt{I_1 I_2}(e^{2ikz} + e^{-2ikz})\right)\right]\left(A_1 e^{ikz} + A_2 e^{-ikz}\right) , \quad (10.30)\end{aligned}$$

where $g_o = \sigma(\nu_o)\Delta\mathcal{N}^{(0)}$ is the line-center gain coefficient, and α represents distributed cavity loss. In obtaining this equation, we neglect second derivatives of the slowly varying field amplitudes and use the rate-equation approximation to express the polarization in terms of the local inversion. Equating the coefficients of e^{ikz} and e^{-ikz} then gives two coupled propagation equations,

$$\frac{\partial A_1}{\partial t} + \frac{c}{n}\frac{\partial A_1}{\partial z} = -\frac{c\alpha}{2n}A_1 + \frac{cg_o}{2n}\left[1 - \frac{1}{I_s}(I_1 + 2I_2)\right] A_1\,, \tag{10.31}$$

$$\frac{\partial A_2}{\partial t} - \frac{c}{n}\frac{\partial A_2}{\partial z} = -\frac{c\alpha}{2n}A_2 + \frac{cg_o}{2n}\left[1 - \frac{1}{I_s}(I_2 + 2I_1)\right] A_2\,. \tag{10.32}$$

There are also mode-pulling terms, which are assumed to have been taken into account by adjusting the mode frequency. Integrating these equations over one cavity roundtrip, as in the derivation from Eq. (7.97) to Eq. (7.107), gives

$$\frac{dA_1}{dt} = \frac{1}{2}\left[\frac{cg_o}{n} - \gamma_{c1} - \frac{cg_o}{nI_s}(I_1 + 2I_2)\right] A_1\,, \tag{10.33a}$$

$$\frac{dA_2}{dt} = \frac{1}{2}\left[\frac{cg_o}{n} - \gamma_{c2} - \frac{cg_o}{nI_s}(I_2 + 2I_1)\right] A_2\,. \tag{10.33b}$$

The photon decay rate γ_{ci} is determined by the roundtrip loss of mode i via $\gamma_{ci} = \mathcal{L}_{ci}/\tau_R = c\mathcal{L}_{ci}/nL$. We allow the two modes to have slightly different losses, for example because of their interaction with the gain medium, scattering centers, or output coupler. However, the difference is usually small unless a nonreciprocal loss element is deliberately introduced into the cavity. The nonlinear terms in Eqs. (10.33) are cubic in the field amplitudes, so these equations are often called third-order amplitude equations.

For laser operation close to threshold, the first two terms inside the square brackets in Eqs. (10.33) can be approximated as

$$\frac{cg_o}{n} - \gamma_{ci} = \frac{cg_o}{n}\left(1 - \frac{\mathcal{L}_{ci}}{g_o L}\right) = \frac{cg_o}{n}\left(1 - \frac{1}{r_i - 1 + 1}\right) \approx \frac{cg_o}{n}(r_i - 1)\,, \tag{10.34}$$

where r_i is the pump ratio for wave i. Using this result in Eq. (10.33) and recalling the definition of mode intensity, $I_i = \frac{1}{2}\epsilon_o nc|A_i|^2 = \frac{1}{2}\epsilon_o ncA_iA_i^*$, we obtain the coupled equations for the two mode intensities

$$\frac{dI_1}{dt} = \left[\frac{cg_o}{n}(r_1 - 1) - \frac{cg_o}{nI_s}(I_1 + 2I_2)\right] I_1\,, \tag{10.35a}$$

$$\frac{dI_2}{dt} = \left[\frac{cg_o}{n}(r_2 - 1) - \frac{cg_o}{nI_s}(I_2 + 2I_1)\right] I_2\,. \tag{10.35b}$$

These equations have the form of Eqs. (10.25). For this example, the self-saturation coefficients are equal, $\beta_1 = \beta_2 = cg_o/nI_s \equiv \beta$, while the cross-saturation coefficients are twice as large, $\theta_{12} = \theta_{21} = 2\,(cg_o/nI_s) = 2\beta$. Thus, for two counter-propagating modes, one mode suppresses the gain available to the other twice as strongly as it suppresses its own gain. The additional cross-saturation arises from scattering by the standing-wave population grating shown in Fig. 10.5(b). Similar standing-wave grating effects play a critical role in saturable-absorber mode-locking of lasers.

The population-grating picture also shows why the coupling is weaker in a strongly Doppler-broadened medium. Rapid atomic motion washes out the grating, suppressing the associated backscattering. Detailed calculations then give equal self- and cross-saturation coefficients, $\theta_{12} = \theta_{21} = \beta$ in Eqs. (10.25), so the saturation depends simply on the total intensity $I(z) = I_1 + I_2$.

If the common mode frequency is detuned from line center, the two counter-propagating waves interact with different velocity groups. Their overlap decreases with increasing detuning, further reducing the cross-saturation [see Fig. 7.15 for hole burning by two counter-propagating waves of identical frequency]. The dependence of the cross-saturation coefficient on detuning $\omega - \omega_o$ is given by $\theta_{12} = \theta_{21} = \beta/[1 + (\omega - \omega_o)^2/\gamma_2^2]$, where β is the self-saturation coefficient. For large detuning $\theta_{12} = \theta_{21} \to 0$ and the modes become decoupled.

There are other ways in which the spatial interference effects can be washed out, justifying the approximation $I(z) = I_1 + I_2$ in the atomic saturation. For example, if the modes have different frequencies, their interference fringes move through the medium at the beat frequency $|\omega_1 - \omega_2|$. If this beat frequency is large relative to the population recovery rate, $|\omega_1 - \omega_2| \gg \gamma_2$, the population cannot follow the moving pattern and the spatial saturation averages away.

More generally, time-dependent saturation or other nonlinear mixing at the beat frequency can generate sidebands on both waves. Such effects can lead to much richer mode coupling, but we will not consider them here.

Equating the coefficients of $e^{\pm ikz}$ from the two sides in Eq. (10.30) is equivalent to projecting the total polarization, or the gain in the rate-equation approximation, onto the two mode functions. In practice, calculating the exact gain coefficient for each mode can be more complex. So far we have neglected transverse variation across the gain medium. If the transverse mode profile $u(\boldsymbol{r}, z)$ is retained, the gain calculation contains overlap integrals involving both $|u|^2$ and higher powers such as $|u|^4$. These factors modify the numerical values of the self- and cross-saturation coefficients but not the basic physics of mode competition.

The saturation coefficients can vary widely from one laser to another, but the essential mode-competition behavior depends on only a few dimensionless parameters. We therefore introduce scaled time $t' = (cg_o/n)t$ and scaled intensities $I_i' = I_i/I_s$. Dropping the primes after the rescaling, Eqs. (10.35) then become

$$\frac{dI_1}{dt} = I_1 \left[a_1 - I_1 - \xi I_2\right] , \tag{10.36}$$

$$\frac{dI_2}{dt} = I_2 \left[a_2 - I_2 - \xi I_1\right] . \tag{10.37}$$

Here the pump parameters $a_i = r_i - 1$ characterize pump excess above individual mode threshold $r_i = 1$. The dimensionless coupling parameter ξ is the ratio of cross-saturation to self-saturation and ranges from 0 to 2 in the examples considered here. The limit $\xi = 0$ corresponds to independent modes, whereas $\xi = 2$ corresponds to strongly coupled modes, frequency-degenerate counter-propagating modes of a homogeneously broadened ring laser.

10.2.1 Steady States of a Two-mode Laser

We next find the possible steady states of the two-mode system as the coupling parameter ξ varies from 0 to 2. Physical solutions require $I_i \geq 0$. If both pump parameters a_1 and a_2 are negative, both modes are below threshold, and the only allowed state is $I_1 = I_2 = 0$. We therefore focus on the case a_1, $a_2 > 0$, for which both modes would independently lie above threshold.

Setting the time derivatives to zero for the steady state ($\dot{I}_1 = 0 = \dot{I}_2$) gives the algebraic system of equations

$$0 = I_1 \left[a_1 - I_1 - \xi I_2\right] , \tag{10.38a}$$

$$0 = I_2 \left[a_2 - I_2 - \xi I_1\right] . \tag{10.38b}$$

These equations imply that either a mode intensity is zero or its bracketed net gain expression vanishes. Combining these conditions yields four distinct steady-state solutions

$$S_0 \equiv (I_1 = 0, I_2 = 0); \quad S_1 \equiv (I_1 = a_1, I_2 = 0); \quad S_2 \equiv (I_1 = 0, I_2 = a_2); \tag{10.39a}$$

$$S_3 = \left(I_1 = \frac{a_1 - \xi a_2}{1 - \xi^2} \equiv \frac{a_1'}{1 - \xi^2}, I_2 = \frac{a_2 - \xi a_1}{1 - \xi^2} \equiv \frac{a_2'}{1 - \xi^2}\right) . \tag{10.39b}$$

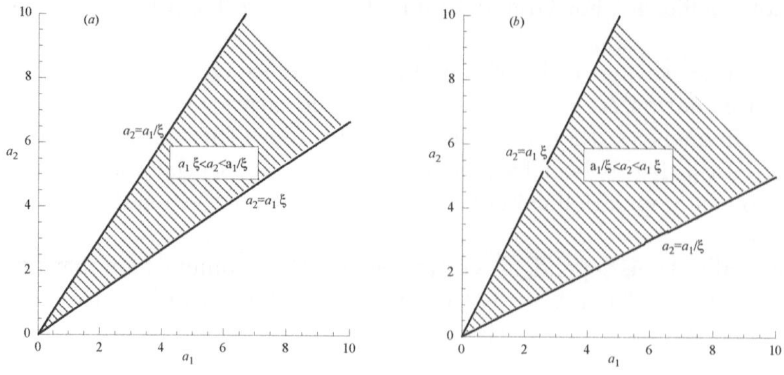

FIGURE 10.6
Regions in the $a_1 - a_2$ plane where a two-mode steady state, stable or unstable, can exist: (a) Weak coupling, $\xi = \frac{2}{3}$; (b) Strong coupling, $\xi = 2$.

For $a_1, a_2 > 0$, the single-mode states S_1, S_2, as well as the zero intensity state S_0 are mathematically allowed. The two-mode coexistence steady-state S_3 is physical only when both mode intensities are positive. This leads naturally to two cases.

For weak coupling, $0 < \xi < 1$, S_3 exists when $\xi a_1 < a_2 < a_1/\xi$. For strong coupling, $\xi > 1$, the solution S_3 is acceptable when $a_1/\xi < a_2 < \xi a_1$. These regions are shown as the hatched regions in Fig. 10.6. The case $\xi = 1$ represents neutral mode coupling and must be treated separately.

Geometrically, a physically acceptable two-mode state S_3 corresponds to the intersection of two straight lines, $a_1 - I_1 - \xi I_2 = 0$ and $a_2 - I_2 - \xi I_1 = 0$, within the positive quadrant of the $I_1 - I_2$ plane. For $\xi = 1$, the two straight lines defining S_3 are parallel. If $a_1 \neq a_2$, they do not intersect and no two-mode steady state exists. In the symmetric case, $a_1 = a_2 = a$, the two lines coincide, so any combination satisfying $I_1 + I_2 = a$ is a steady state. Because exact symmetry is difficult to maintain in a real laser, this degenerate case is usually of limited practical importance.

10.2.2 Stability of Steady States

A mathematically allowed steady state need not be stable. To determine which states can actually persist, we perturb each steady state slightly and ask whether the perturbation decays or grows. If these perturbation decay over time, the laser returns to the steady state under consideration, demonstrating its stability. This is the same linear-stability procedure used earlier for relaxation oscillations. For definiteness, we take $a_1 > a_2$ and $0 < \xi < 2$. Introduce small perturbations of mode intensities from their steady-state values by writing $I_1 = I_{1ss} + \epsilon_1$, $I_2 = I_{2ss} + \epsilon_2$, where $|\epsilon_1|$, $|\epsilon_2| \ll a_1, a_2$. Substituting these into the equations of motion and retaining only terms linear in the perturbations gives the linearized system of equations governing the temporal evolution of perturbations.

First, consider the zero-intensity solution $S_0 \equiv (I_{1ss} = 0,\, I_{2ss} = 0)$. In this case, the deviations from the steady state obey

$$\frac{d\epsilon_1}{dt} = a_1\epsilon_1\,, \tag{10.40a}$$

$$\frac{d\epsilon_2}{dt} = a_2\epsilon_2\,. \tag{10.40b}$$

Because both pump parameters a_1 and a_2 are positive above threshold, these equations show that ϵ_1 and ϵ_2 grow exponentially over time; any small perturbation will drive the laser away from S_0. Solution S_0 is thus unstable above threshold.

Next, consider the single-mode steady states S_1 and S_2. Working with $S_1 \equiv (I_{1ss} = a_1,\ I_{2ss} = 0)$, the linearized differential equations for the small deviations about this steady-state are

$$\frac{d\epsilon_1}{dt} = -a_1\epsilon_1 - \xi a_1 \epsilon_2\,, \tag{10.41a}$$

$$\frac{d\epsilon_2}{dt} = (a_2 - \xi a_1)\,\epsilon_2\,. \tag{10.41b}$$

From Eq. (10.41b), if $(a_2 - \xi a_1) < 0$, the small deviation ϵ_2 decays to zero and the laser returns to stable single-mode operation in state S_1. If instead, $(a_2 - \xi a_1) > 0$; that is, a_2 is sufficiently large to overcome the cross-saturation of its gain by mode 1, the perturbation grows and S_1 is unstable. An identical analysis applies to state $S_2 \equiv (I_{1ss} = 0,\ I_{2ss} = a_2)$.

For the two-mode coexistence state $S_3 \equiv \left(I_{1ss} = \frac{a_1 - \xi a_2}{1-\xi^2},\ I_{2ss} = \frac{a_2 - \xi a_1}{1-\xi^2}\right)$, we substitute $I_i = I_{iss} + \epsilon_i$, into the intensity equations of motion. Defining the effective pump parameters $a_1' = a_1 - \xi a_2$ and $a_2' = a_2 - \xi a_1$, the perturbations evolve according to the linearized coupled differential equations

$$\left.\begin{aligned}\frac{d\epsilon_1}{dt} &= -\frac{a_1'}{1-\xi^2}\left[\epsilon_1 + \xi\epsilon_2\right]\\ \frac{d\epsilon_2}{dt} &= -\frac{a_2'}{1-\xi^2}\left[\epsilon_2 + \xi\epsilon_1\right]\end{aligned}\right\} \Rightarrow \frac{d}{dt}\begin{bmatrix}\epsilon_1\\ \epsilon_2\end{bmatrix} = M\begin{bmatrix}\epsilon_1\\ \epsilon_2\end{bmatrix}, \tag{10.42}$$

where the stability matrix M is defined as

$$M = -\frac{1}{1-\xi^2}\begin{bmatrix}a_1' & \xi a_1'\\ \xi a_2' & a_2'\end{bmatrix}. \tag{10.43}$$

Solution S_3 is stable if the eigenvalues of M have negative real parts. The eigenvalues of M are given by

$$\lambda_\pm = -\frac{a_1' + a_2'}{2(1-\xi^2)} \pm \sqrt{\left(\frac{a_1' + a_2'}{2(1-\xi^2)}\right)^2 - \frac{a_1' a_2'}{1-\xi^2}}\,. \tag{10.44}$$

The eigenvalues in Eq. (10.44) lead to two qualitatively different regimes.

1. Weak coupling, $\xi < 1$. When $a_1',\ a_2' > 0$, both eigenvalues are negative and the two-mode coexistence state S_3 is stable.

2. Strong coupling, $\xi > 1$. When $a_1', a_2' < 0$, at least one eigenvalue has a positive real part and S_3 is unstable, while both single-mode steady states S_1 and S_2 are stable. The laser is then bistable; it can operate in state $S_1 \equiv (I_1 = a_1, I_2 = 0)$ or in state $S_2 \equiv (I_1 = 0, I_2 = a_2)$. Which mode oscillates depends on the initial conditions, and hysteresis may occur as an operating parameter is varied.

Linear stability analysis describes only the behavior close to a steady state. To see the global dynamics, Eqs. (10.36) and (10.37) can be integrated numerically for many initial conditions. The resulting trajectories in the $I_1 - I_2$ plane show which stable states attract

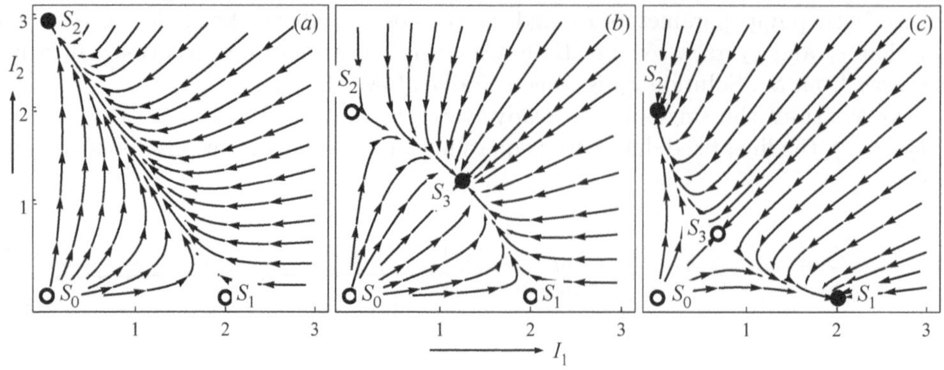

FIGURE 10.7
Two-mode laser trajectories in the $I_1 - I_2$ plane: (a) single-mode oscillation at S_2, (b) two-mode oscillation at S_3 (weak coupling), (c) bistable (S_1 or S_2) single-mode oscillation (strong coupling). Filled circles represent stable steady states, and open circles represent unstable steady states. Note that S_3 does not exist in case (a).

the system and which unstable states repel it. Figure 10.7 illustrates the three characteristic cases: stable single-mode operation, stable two-mode operation for weak coupling, and bistable single-mode operation for strong coupling. In each case, various trajectories converge toward the stable points (indicated by filled circles) and away from unstable points found in the preceding analysis.

These results have direct implications for real lasers. A homogeneously broadened ring laser with two counter-propagating modes corresponds approximately to the strongly coupled case $\xi = 2$. It therefore tends to oscillate in only one direction at a time. Backscattering, departures from homogeneous broadening, or other coupling mechanisms can modify this behavior. A homogeneous saturable absorber can even allow simultaneous oscillation in both directions, as exploited in colliding-pulse mode-locking discussed in the next chapter.

Mode competition is much weaker in a strongly Doppler-broadened medium. A strong mode saturates mainly the narrow velocity group with which it interacts resonantly. A second mode separated by more than about one homogeneous linewidth therefore interacts largely with a different group of atoms, reducing cross-saturation and allowing simultaneous multimode oscillation. As the frequency separation decreases, the overlap between the velocity groups increases and so does the coupling.

Mode competition can be still more complicated in Zeeman lasers. These lasers support different polarization states whose gains depend on Zeeman splitting, polarization-selection rules, Doppler broadening, and several coupled atomic subtransitions. Although the detailed saturation coefficients are more complicated, the resulting competition can still be described within the same intensity-coupling framework.

The discussion in this section has been restricted to intensity coupling within the rate-equation approximation. Direct coupling of the mode amplitudes—for example, through backscattering in a ring cavity—can also produce phase and frequency locking. If the rate-equation approximation itself is relaxed, still richer nonlinear dynamics, including deterministic chaos, can occur. We do not pursue these effects here.

10.3 Injection-Locking

Injection-locking provides another important example of laser dynamics. A weak external signal injected into a more powerful laser can force the laser to oscillate at the frequency of the injected signal. The same phenomenon occurs in many other physical systems, including electronic oscillators, phase-locked loops, pendulum clocks, and tuning forks.

Consider a laser operating at a free-running oscillation frequency ω_o and producing a steady-state output intensity I_o. A weak external signal at frequency ω_i, close to but not exactly equal to ω_o, is injected through one of the cavity mirrors. We write the injected field $E_i(t) = A_{in}(t)e^{-i\omega_i t}$ and the intracavity field as $E_c(t) = A_c(t)e^{-i\omega_i t}$. Both field amplitudes are defined at the same position and use the injected frequency ω_i as the common carrier frequency. Any difference between the actual laser frequency and ω_i manifests as a time-dependent phase of $A_c(t)$.

Within the rate-equation approximation, the polarization follows the cavity field adiabatically [Eq. (7.95)]. For a weak injected field, the cavity amplitude obeys

$$\frac{dA_c}{dt} + \frac{\gamma_c}{2}A_c = \frac{cg}{2n}A_c + i(\omega_i - \omega_o)A_c + \frac{\gamma_e}{2}A_{\text{in}}\,. \tag{10.45}$$

Equation (10.45) has the same form as Eq. (7.107). The second term on the right represents the detuning $\omega_i - \omega_o$ while the third describes injection through the cavity mirror. Small frequency-pulling effects from the gain medium are assumed to be included in the definition of the free-running frequency ω_o. For simplicity, we also take the gain medium to fill the entire cavity length, $L_g = L$. The injection rate γ_e is related to the output coupler transmittance $\mathcal{T}_{\text{oc}}$ and the cavity roundtrip time $\tau_{_R}$ by

$$\gamma_e = \frac{\mathcal{T}_{\text{oc}}}{\tau_{_R}}\,, \tag{10.46}$$

Writing the two complex amplitudes in terms of their real magnitudes and phases as

$$A_c = |A|e^{i\phi}, \quad A_{\text{in}} = |A_i|e^{i\phi_i}\,, \tag{10.47}$$

substituting them into Eq. (10.45), and separating the real and imaginary parts, we obtain the coupled equations for amplitude and phase evolution

$$\frac{d|A|}{dt} = \frac{1}{2}\left(\frac{cg}{n} - \gamma_c\right)|A| + \frac{1}{2}\gamma_e|A_i|\cos(\phi_i - \phi)\,, \tag{10.48}$$

$$\frac{d\phi}{dt} = (\omega_i - \omega_o) + \frac{\gamma_e}{2}\frac{|A_i|}{|A|}\sin(\phi_i - \phi)\,. \tag{10.49}$$

Equations (10.48) and (10.49) govern the amplitude and phase dynamics of the injection-locked laser.

For constant injected signal, A_i and ϕ_i are constant. Because $|A_i| \ll |A_o|$, the injected field produces only a small change in the steady-state laser amplitude. Thus, to a good approximation, the steady-state laser amplitude $|A| \approx |A_o|$, while the phase contains the essential injection-locking dynamics. Shifting the phase origin so that $\phi_i = 0$, Eq. (10.49) reduces to

$$\frac{d\phi}{dt} = \omega_i - \omega_o - \frac{\gamma_e}{2}\frac{|A_i|}{|A_o|}\sin\phi \equiv \omega_i - \omega_o - \omega_{_L}\sin\phi\,, \tag{10.50}$$

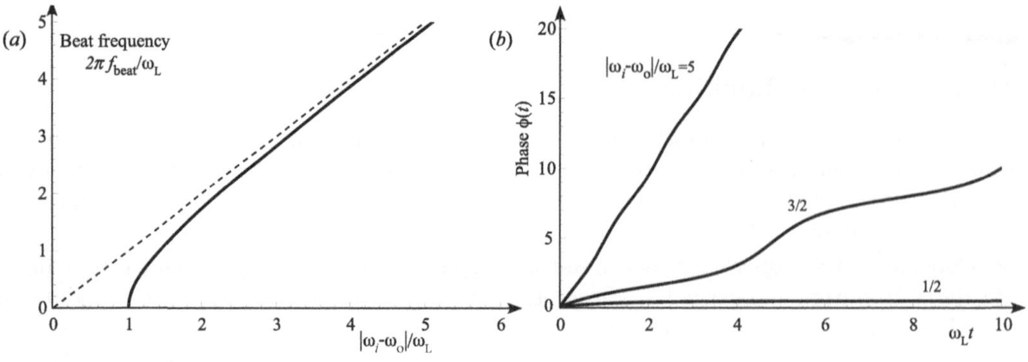

FIGURE 10.8
(a) Normalized beat frequency $2\pi f_{\text{beat}}/\omega_L$ versus normalized detuning $|\omega_i - \omega_o|/\omega_L$. The dashed line shows the beat frequency in the absence of injection locking. (b) Time evolution of the relative phase ϕ for several detunings.

where the semi-locking range parameter ω_L is given by

$$\omega_L = \frac{\gamma_e}{2}\frac{|A_i|}{|A_o|}. \tag{10.51}$$

Equation (10.50) is the classical Adler equation[1] for a weakly injected oscillator

10.3.1 Locked Solution

In the locked state, the relative phase is constant, so $(d\phi/dt = 0)$. Equation (10.50) then gives

$$\omega_i - \omega_o = \omega_L \sin\phi \Rightarrow \sin\phi = \frac{\omega_i - \omega_o}{\omega_L}. \tag{10.52}$$

This yields a real physical solution for ϕ when the injected signal frequency ω_i remains within the range

$$0 \leq |\omega_o - \omega_i| \leq \omega_L. \tag{10.53}$$

When $|\omega_o - \omega_i| \leq \omega_L$ the relative phase settles to a constant value [Fig. 10.8]. The laser then ceases to oscillate at its free-running frequency and instead follows the injected frequency. The full locking bandwidth is

$$\Delta\omega_{\text{lock}} = 2\omega_L = \gamma_e\frac{|A_i|}{|A_o|} = \gamma_e\sqrt{\frac{I_i}{I_o}}. \tag{10.54}$$

Thus, the locking bandwidth grows as the square root of the injected intensity and decreases as the square root of the free-running laser intensity. Outside this range, the phase cannot remain stationary and the laser enters an unlocked regime.

10.3.2 Unlocked or Free-running Solution

Outside the locking region, $|\omega_o - \omega_i| > \omega_L$, the relative phase ϕ drifts continuously, producing a beat note between the laser and the injected signal. The time-averaged beat frequency

[1] R. Adler, "A study of locking phenomena in oscillators," Proc. IRE **34**, 351–357 (1946).

can be computed by separating variables in Eq. (10.50), yielding

$$\frac{d\phi}{1 - \frac{\omega_L}{\omega_o - \omega_i} \sin\phi} = (\omega_o - \omega_i)dt\,. \tag{10.55}$$

Integrating this relation over one complete beat oscillation period T, which corresponds to ϕ changing from 0 to 2π, gives

$$\int_0^{2\pi} \frac{d\phi}{1 - \frac{\omega_L}{\omega_o - \omega_i} \sin\phi} = (\omega_o - \omega_i)T\,. \tag{10.56}$$

Using the standard integral identity (for $a < 1$)

$$\int_0^{2\pi} \frac{d\phi}{1 - a\sin\phi} = \frac{2\pi}{\sqrt{1 - a^2}}\,, \tag{10.57}$$

the beat frequency between the injected signal and the laser is found to be

$$f_{\text{beat}} = \frac{1}{2\pi}\sqrt{(\omega_o - \omega_i)^2 - \omega_L^2}\,. \tag{10.58}$$

Figure 10.8(a) shows the two regimes clearly. Inside the locking range, $|\omega_i - \omega_o| < \omega_L$, the laser follows the injected frequency and the beat note vanishes. Outside the locking range, the beat frequency is given by [Eq. (10.58)] and approaches the free-running detuning as $|\omega_i - \omega_o|$ becomes large.

Figure 10.8(b) shows the corresponding phase evolution. Inside the locking range the phase approaches a constant value. Just outside the range it drifts nonuniformly, producing an anharmonic beat note. Far outside the range the phase increases almost linearly with time, corresponding to the ordinary difference frequency between the two oscillators.

Injection locking is widely used to transfer the spectral purity of a low-power master laser to a high-power oscillator. If the master frequency lies within the locking range, the high-power laser follows its frequency and phase while retaining its much larger output power. This provides an effective method of power scaling with low phase noise.

The same principle appears in active mode-locking, where an intracavity amplitude or phase modulator generates sidebands that couple neighboring axial modes and establish a fixed phase relationship among them.

Injection locking can also be undesirable. Light reflected from external surfaces can reenter the laser cavity and act as an unintended injected signal. Because its frequency is naturally close to the laser frequency, even a very weak reflection can couple strongly to the oscillator. Mechanical motion, air currents, or other path-length fluctuations then convert directly into phase and frequency noise on the laser output.

An important example of injection-locking occurs in the ring-laser gyroscope (RLG).[2] In a stationary ring cavity, the clockwise (CW) and counterclockwise (CCW) modes are frequency-degenerate. Rotation of the cavity about an axis normal to the ring plane breaks this degeneracy through the Sagnac effect: one propagation direction experiences a slightly longer effective path and the other a slightly shorter one. The resulting frequency difference

$$f_{\text{ccw}} - f_{\text{cw}} = \text{C} \times \Omega_{\text{rot}}\,,$$

is proportional to the rotation rate and provides the gyroscope signal. Accurate operation therefore requires the two counter-propagating modes to remain independent. In practice,

[2]F. Aronowitz, "The Laser Gyro," in *Laser Applications*, edited by M. Ross (Academic, New York), pp. 113–120 (1971).

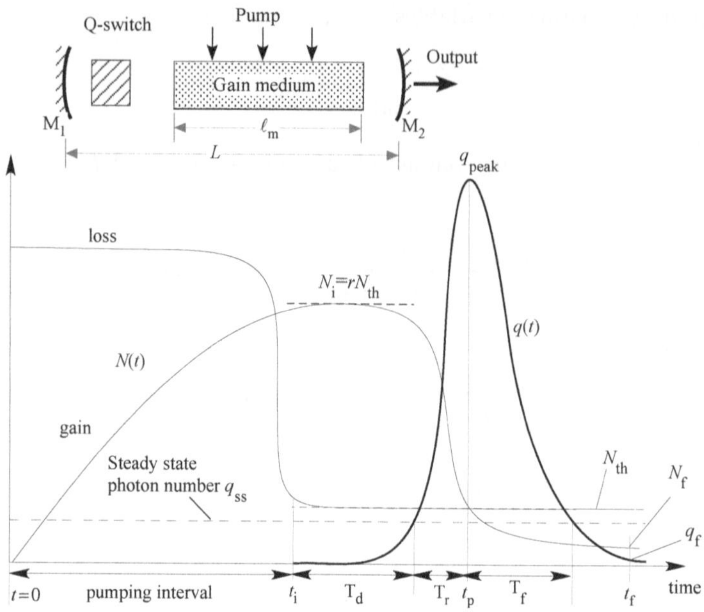

FIGURE 10.9
Development of a Q-switched laser pulse.

however, weak mirror backscattering couples them. At sufficiently small rotation rates the two frequencies are pulled together and can lock to a common value, producing a dead zone in which the beat frequency vanishes. Avoiding this lock-in requires techniques such as mechanical dithering or magneto-optic biasing. Because the two coupled waves now have comparable amplitudes, this regime is more complicated than the weak-injection case treated above.

10.4 Q-Switching

Q-switching is a powerful technique for generating short, intense laser pulses. During the pumping interval, the cavity is held in a low-Q state by suppressing optical feedback or introducing a large loss. Laser oscillation is therefore prevented while the pump builds a population inversion well above the normal steady-state value [Fig. 10.9]. Once sufficient energy has been stored in the gain medium, the cavity loss is rapidly reduced and the high-Q state is restored. The roundtrip gain then greatly exceeds the loss, and the intracavity photon number grows rapidly from spontaneous-emission noise. Stimulated emission quickly depletes the stored inversion, producing a short, intense burst known as a giant pulse. As the inversion falls below threshold, the pulse reaches its maximum and then decays as photons leave the cavity through the output coupler.

The result is a pulse much shorter than the pumping interval used to store the energy. Thus, energy accumulated over a relatively long pumping period can be released in a very short time, giving a peak power many orders of magnitude larger than the CW output power of the same laser at the same pump rate. Different stages in the evolution of a Q-switched laser pulse are illustrated in Fig. 10.9. Several features of Q-switching can be understood quantitatively from the coupled rate equations for the cavity photon number q and the population inversion N,[3]

$$\frac{dq}{dt} = KNq - \gamma_c q = \gamma_c\left(\frac{N}{N_{\text{th}}} - 1\right)q\,, \tag{10.59a}$$

$$\frac{dN}{dt} = R_p - \gamma_2 N - KNq = \gamma_2\left[N^{(o)} - \left(1 + \frac{q}{q_s}\right)N\right], \tag{10.59b}$$

where $\gamma_c = 1/\tau_c$ is the cavity decay rate, R_p is the pumping rate, γ_2 is the population decay rate, and K is the atom-field stimulated-transition coupling constant. The unsaturated inversion is $N^{(o)} = R_p/\gamma_2 = rN_{\text{th}}$, where r is the pump ratio and $N_{\text{th}} = \gamma_c/K$ is the threshold inversion of the high-Q cavity.

During the pumping interval, cavity feedback is suppressed and the photon number remains negligible, $q(t) \ll q_s$. For a constant pump turned on at $t = 0$, the population rate equation reduces to

$$\frac{dN}{dt} = R_p - \gamma_2 N\,, \tag{10.60}$$

which yields the analytical solution

$$N(t) = \frac{R_p}{\gamma_2}(1 - e^{-\gamma_2 t})\,. \tag{10.61}$$

The inversion approaches its maximum value R_p/γ_2. Efficient Q-switching requires this value to lie well above the threshold inversion of the high-Q cavity.

After two upper-state lifetimes, the inversion has reached about 86% of its limiting value. Pumping much longer therefore stores relatively little additional energy because excited atoms are already being lost through spontaneous decay. In solid-state lasers, where upper-state lifetimes are typically hundreds of microseconds to about a millisecond, useful pumping intervals are correspondingly long. In most visible gas and dye lasers, however, the population lifetime is only a few to a few tens of nanoseconds. These short storage times make conventional Q-switching much less useful.

10.4.1 Pulse Buildup or Delay Time

At the end of the pumping interval, the cavity loss is suddenly reduced and the laser is switched from low Q to high Q. Oscillation does not appear immediately, however; a finite time is required for the intracavity field to grow from spontaneous-emission noise. We estimate this delay by assuming that the loss changes nearly instantaneously at $t = t_i$. The inversion immediately after switching is $N_i \equiv N(t_i) = (R_p/\gamma_2)(1 - e^{-\gamma_2 t_i}) \equiv rN_{\text{th}}$, which is a factor r times the threshold inversion $N_{\text{th}} = \gamma_c/K$ for the high-Q cavity.

During the early buildup, the cavity photon number is much smaller than the saturation photon number, $q \ll q_s$, so saturation of the inversion is negligible. If the buildup time is also short compared with the population lifetime, pumping and spontaneous population

[3] Recall that inversion $N = N_2 - N_1 \approx N_2$ for four-level lasers, where the lower laser level decays sufficiently fast so that $N_1(t) \approx 0$ at all times, even during the Q-switched pulse. In some cases, for example, in a Q-switched ruby laser, the lower level cannot empty out sufficiently fast during the laser pulse. In such cases, transfer of an atom between the upper and lower levels changes inversion by 2, and the inversion equation is modified to read

$$\frac{dN}{dt} = R_p - \gamma_2 N - 2KNq\,,$$

while the cavity photon number equation remains unchanged.

decay can be neglected. The inversion therefore remains approximately fixed at $N_i = rN_{\text{th}}$. The cavity photon number evolution is thus governed by

$$\frac{dq}{dt} \approx \gamma_c \left(\frac{N_i}{N_{\text{th}}} - 1 \right) q = \gamma_c (r-1) q \tag{10.62}$$

with the solution

$$q(t) = q_i \exp[\gamma_c (r-1)(t - t_i)] \,, \tag{10.63}$$

where $q_i \equiv q(t_i)$ is the initial photon number in the cavity due to spontaneous emission. Typically q_i is of order unity. As an example, for a pump ratio of $r = 3$, the photon flux will be amplified by $e^{(r-1)\times 10} \approx 5 \times 10^8$ in ten roundtrips.

We define the delay time T_d as the time required for the photon number to grow from q_i to the CW steady-state value $q_{ss} = q_s(r-1)$. The delay time T_d is then determined by the condition

$$q(t = t_i + T_d) = q_{ss} \quad \Rightarrow \quad \frac{q_{ss}}{q_i} = \exp[\gamma_c (r-1) T_d] \,. \tag{10.64a}$$

Solving this yields the delay time T_d at which the laser pulse begins to appear

$$T_d = \frac{\ln(q_{ss}/q_i)}{\gamma_c (r-1)} = \frac{\tau_c}{(r-1)} \ln \left(\frac{q_s (r-1)}{q_i} \right) , \tag{10.64b}$$

where τ_c is the cavity lifetime of the high-Q cavity. Depending on q_i and r, the ratio $q_{ss}/q_i = q_s(r-1)/q_i$ may range from $10^8 - 10^{12}$. However, because T_d depends only logarithmically on q_{ss}/q_i, it is relatively insensitive to the precise value of the initial spontaneous photon number q_i unless r is very close to threshold.

Example 10.3 *Estimate the pulse delay time for a 60-cm-long, standing-wave Nd:YAG laser. The cavity contains a 10-cm-long rod with an effective cross-sectional area of* 0.3 *cm*^{2} *and a refractive index* $n = 1.83$. *The laser parameters are: pump ratio* $r = 3$, *mirror reflectances* $\mathcal{R}_1 = 1, \mathcal{R}_2 = 0.35$, *upper-state lifetime* $\tau_2 = 230\,\mu s$, *and emission cross-section* $\sigma = 3.5 \times 10^{-19}$ *cm*^{2}.

From the given data

$$\mathcal{L}_c = -\ln \mathcal{R}_1 \mathcal{R}_2 = 1.05$$

$$\tau_c = \frac{2(L + (n-1)\ell_g)}{c\mathcal{L}_c} \approx \frac{2(60 + 8.3)\,\text{cm}}{(3.00 \times 10^{10}\,\text{cm/s})1.05} = 4.34\,\text{ns}$$

$$q_s = \frac{\gamma_2}{K} = \frac{\gamma_2 V_c n}{c\sigma} = \frac{[(0.3 \times 68.3)\,\text{cm}^3]1.83}{(230 \times 10^{-6}\,\text{s})(3.00 \times 10^{10}\,\text{cm/s})(3.5 \times 10^{-19}\,\text{cm}^2)} = 1.55 \times 10^{13}$$

$$T_d = \tau_c \frac{\ln[q_s(r-1)/q_i]}{r-1} = 4.34 \times \frac{\ln[(1.55 \times 10^{13})(3-1)]}{(3-1)}\,\text{ns} = 67\,\text{ns} \,, \quad q_i \approx 1$$

For comparison, consider a 2-m-long Nd:YAG cavity. Its roundtrip time is approximately 14 ns, and its larger mode volume gives $q_s = 4.73 \times 10^{13}$. With 20% output coupling, $\mathcal{L}_c = 0.223$, giving a cavity lifetime of 63 ns. If this laser is 50% above threshold, $r = 1.5$, the delay time is

$$T_d = \tau_c \frac{\ln[q_s(r-1)/q_i]}{r-1} = (63\,\text{ns}) \frac{\ln[(4.73 \times 10^{13})(1.5-1)]}{(1.5-1)} = (63\,\text{ns})61.6 = 3.9\,\mu\text{s} \,.$$

10.4.2 Pulse Width and Pulse Energy

During the Q-switched pulse, the stored inversion is depleted on a timescale much shorter than the pumping or population-relaxation times. We can therefore neglect pumping and spontaneous population decay during the pulse, reducing the rate equations to

$$\frac{dq}{dt} = K(N - \gamma_c/K)q = K(N - N_{\text{th}})q \tag{10.65a}$$

$$\frac{dN}{dt} \approx -KNq \tag{10.65b}$$

subject to the initial conditions $N(t_i) \equiv N_i = rN_{\text{th}}$ and $q(t_i) \equiv q_i \approx 1$ at the switching time t_i. The initial photon number corresponds to a single spontaneous photon per mode far below threshold. These equations can be combined by dividing the photon rate equation by the population rate equation, yielding

$$\frac{dq}{dN} = \frac{N_{\text{th}} - N}{N}\,. \tag{10.66}$$

This equation relates the intracavity photon number directly to the changing population inversion during the pulse. Integrating Eq. (10.66) from the switching instant t_i to an arbitrary time $t > t_i$ during the pulse, we obtain

$$q(t) - q_i = N_i - N(t) + N_{\text{th}} \ln \left[\frac{N(t)}{N_i}\right]\,. \tag{10.67}$$

This equation leads to several useful results for the Q-switched laser pulse. Recall that the initial photon number $q_i \approx 1$ is negligible compared to the value of $q(t)$ at any instant during the main pulse.

Q-Switched Pulse Energy

To find the pulse energy, we evaluate Eq. (10.67) after the pulse has ended, when the photon number has returned to a negligible value $q(t_f) \equiv q_f \approx 1$ and the inversion has fallen to $N(t_f) \equiv N_f$. Since both q_i and q_f are negligible compared with the photon number during the pulse, Eq. (10.67) yields the final inversion implicitly in terms of N_i and N_{th}

$$N_i - N_f + N_{\text{th}} \ln\left(\frac{N_f}{N_i}\right) \approx 0\,. \tag{10.68}$$

The energy transferred from the inversion to the circulating pulse is the difference in energy stored in the initial and final inversions

$$U_c \equiv (N_i - N_f)\hbar\omega_o = \left(1 - \frac{N_f}{N_i}\right) N_i \hbar\omega_o\,. \tag{10.69}$$

The ratio N_f/N_i can be computed from Eq. (10.68), and used in Eq. (10.69) to calculate the pulse energy. It is useful to express this result in terms of an energy-extraction efficiency.

Energy Extraction Efficiency

The energy-extraction efficiency η is the fraction of the initially stored inversion energy transferred to the Q-switched pulse:

$$U_c \equiv \eta U_i = \eta N_i \hbar\omega_o = \eta r N_{\text{th}} \hbar\omega_o\,. \tag{10.70}$$

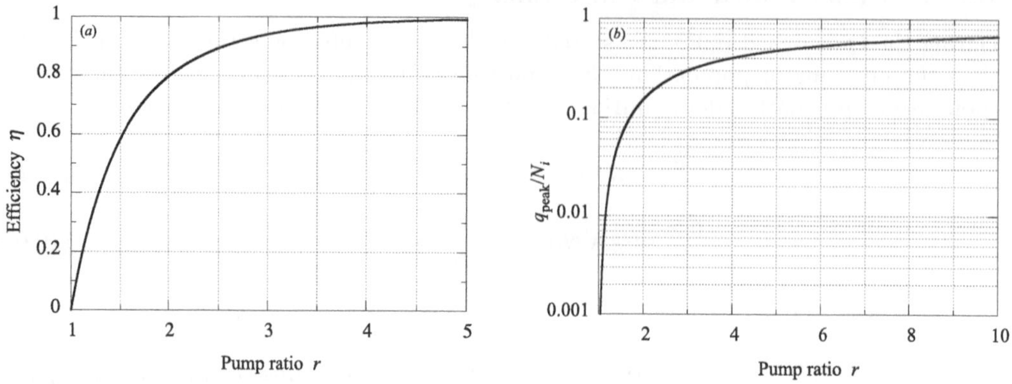

FIGURE 10.10
Variation of energy extraction efficiency η and peak photon number q_{peak} (as a fraction of initial inversion rN_{th}) with pump ratio r.

To determine η, we divide Eq. (10.68) by N_i and use Eqs. (10.69) and Eq. (10.70) to obtain

$$
\begin{aligned}
1-\frac{N_f}{N_i}+\frac{1}{r}\ln\left(\frac{N_f}{N_i}\right) &= 1-\frac{N_f}{N_i}+\frac{1}{r}\ln\left[1-\left(1-\frac{N_f}{N_i}\right)\right], \\
&= \eta+\frac{1}{r}\ln(1-\eta)=0 \\
\Rightarrow \qquad \eta &= 1-e^{-\eta r}\,. \qquad (10.71)
\end{aligned}
$$

This equation shows that the extraction efficiency depends only on the pump ratio r. At threshold, $r = 1$, no energy can be extracted into a Q-switched pulse and $\eta = 0$. As r increases, the efficiency rises rapidly toward unity [Fig. 10.10]; for example, $\eta = 0.797$ at $r = 2$ and 0.941 at $r = 3$. Thus, efficient Q-switching requires an inversion appreciably above the high-Q threshold.

The total Q-switched pulse energy is largely independent of cavity geometry and losses, provided that the losses are not so large as to reduce initial pump ratio r too close to unity. A Q-switched laser with r marginally above threshold will suffer from poor energy extraction efficiency, leaving a substantial fraction of the stored inversion unextracted.

Peak Pulse Power

The cavity photon number q increases while the population inversion stays above its threshold value [Eq. (10.66)] and reaches its maximum when the inversion falls through threshold $N = N_{\text{th}}$. Equivalently, the maximum occurs when $dq/dN = 0$,

$$\frac{dq}{dN}=0=-1+\frac{N_{\text{th}}}{N} \quad \Rightarrow \quad N=N_{\text{th}}\,. \qquad (10.72)$$

Thus, when the inversion $N(t)$ drops to its threshold value, photon number reaches its peak. Substituting $N(t) = N_{\text{th}}$ into Eq. (10.67) and neglecting the initial photon number $q_i \approx 1$ relative to N_i, we find that the peak photon number inside the cavity is

$$q_{\text{peak}}=N_i-N_{\text{th}}+N_{\text{th}}\ln\left(\frac{N_{\text{th}}}{N_i}\right)=N_i\left(1-\frac{1}{r}-\frac{\ln r}{r}\right). \qquad (10.73)$$

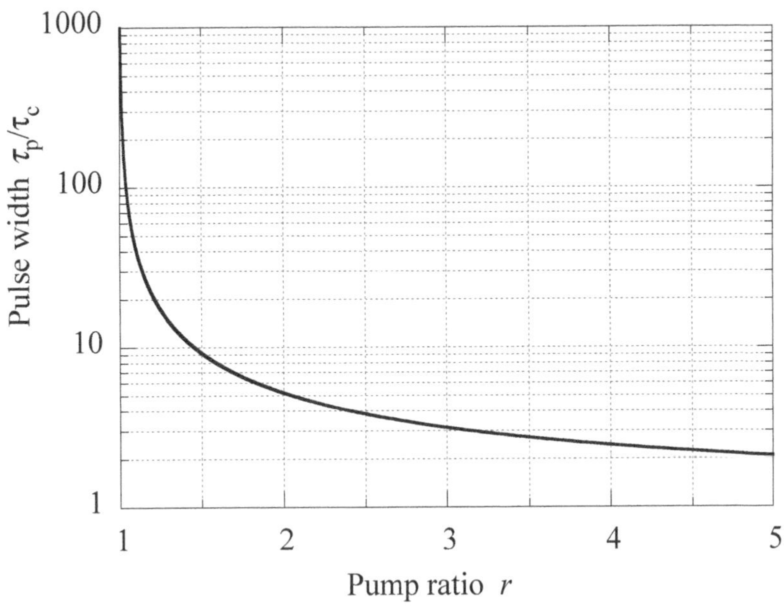

FIGURE 10.11
Q-switched pulse duration [Eq. (10.75)], in units of cavity lifetime, as a function of pump ratio r.

The peak power output from the Q-switched laser is therefore given by

$$P_{\text{peak}} = \gamma_c q_{\text{peak}} \hbar\omega_o = \gamma_c N_i \hbar\omega_o \left(1 - \frac{1}{r} - \frac{\ln r}{r}\right) . \tag{10.74}$$

Equation (10.73) shows that q_{peak} scales with the initial inversion N_i multiplied by the factor $(1 - 1/r - \ln r/r)$. This factor as a function of r is plotted in Fig. 10.10(b); it vanishes at $r = 1$ and approaches unity for $r \gg 1$. Thus, a large initial inversion produces a large intracavity photon population and a correspondingly high peak power. If internal losses are present, the useful output peak power is reduced by the output-coupling efficiency η_{coupl}.

Q-Switched Pulse Duration

The exact shape of the pulse can be determined by integrating Eqs. (10.65) numerically. A numerical solution gives the full pulse shape and hence any chosen width measure, such as the FWHM. A simple estimate is obtained by dividing the pulse energy by the peak power:

$$\tau_p \approx \frac{U_c}{P_{\text{peak}}} = \frac{\eta}{(1 - 1/r - \ln r/r)\gamma_c} = \frac{\eta\tau_c}{(1 - 1/r - \ln r/r)} , \tag{10.75}$$

where $\tau_c = 1/\gamma_c$ is the high-Q cavity lifetime. This top-hat estimate treats the pulse as having constant power P_{peak} for a time τ_p. It typically yields a pulse width roughly 20% smaller than the FWHM value. As r increases, the leading edge becomes faster, while the decay remains limited by the empty-cavity lifetime. The pulse therefore develops a sharp rise and a longer tail, with its width approaching τ_c at high pump ratio [see Fig. 10.11].

Internal cavity loss prevents all of the circulating pulse energy from reaching the useful output. The output pulse energy is therefore

$$U_{\text{pulse}} \equiv \eta_{coupl}\, U_c = \eta_{coupl} \left(1 - \frac{N_f}{N_i}\right) N_i \hbar\omega_o \,, \tag{10.76}$$

where $\eta_{coupl} = \mathcal{L}_T/\mathcal{L}_c$ is the fraction of total cavity loss associated with output coupling. In a well-designed Q-switched laser, output coupling is usually made larger than the internal loss, so η_{coupl} can approach unity.

Example 10.4 Q-switched pulse in Nd:YAG laser: *An 18-cm-long standing-wave cavity contains a 50-mm-long Nd:YAG rod ($n = 1.83$) with an effective cross-sectional area of 0.20 cm^2. The laser is operated in a Q-switched mode. The threshold inversion density is $\mathcal{N}_{th} = 4.35 \times 10^{17}$ cm^{-3} and the laser is pumped to $r = 2.9$. The cavity parameters are $\mathcal{R}_1 \approx 1$, $\mathcal{R}_2 = 0.70$, and the roundtrip internal loss of $2\alpha\ell_\alpha = 0.30$. Estimate the energy extraction efficiency, peak power, pulse duration, and total output pulse energy.*

Nd:YAG is a four-level laser system. The energy extraction efficiency η for Q-switched operation is determined by $\eta - 1 + e^{-\eta r} = 0$. For $r = 2.9$, successive iteration of $\eta = 1 - e^{-\eta r}$, starting with $\eta_0 = 1$, gives $0.945, 0.935, 0.933 \cdots$. Thus $\eta = 0.93$.
The threshold population in the gain volume is

$$N_{th} = \mathcal{N}_{th} \times \text{volume} = 4.35 \times 10^{17} \times (0.20 \times 5.0) = 4.35 \times 10^{17}\,\text{ions}\,,$$

and the energy of a single quantum at $\lambda = 1060$ nm is [recall $\hbar c = 197$ eV-nm]

$$\begin{aligned}\hbar\omega_o &= \frac{2\pi\hbar c}{\lambda} = \frac{2\pi(197)(1.60 \times 10^{-19})}{1060}\,\text{J}\,,\\ &= 1.868 \times 10^{-19}\,\text{J}.\end{aligned}$$

From the given cavity data, the effective cavity length is $L_{\text{eff}} = 18 + (1.83 - 1)5 \approx 22$ cm and the total roundtrip cavity loss is $\mathcal{L}_c = \mathcal{L}_i + \mathcal{L}_T = 2\alpha\ell_a - \ln(\mathcal{R}_1\mathcal{R}_2) = 0.30 - \ln(0.7) = 0.6567$. Using these, we obtain the cavity lifetime and decay rate as

$$\tau_c = \frac{1}{\gamma_c} = \frac{2L_{\text{eff}}}{c\mathcal{L}_c} \approx \frac{2 \times 22}{(3.0 \times 10^{10})0.6567}\,\text{s} = 2.23\,\text{ns}$$

Then the peak output power is

$$\begin{aligned}P_{\text{peak}} &= \left(\frac{\mathcal{L}_T}{\mathcal{L}_c}\right)\gamma_c(r - 1 - \ln r)N_{\text{th}}\hbar\omega_o\\ &= \left(\frac{0.3567}{0.6567}\right)\frac{0.835}{2.23 \times 10^{-9}\,\text{s}}(4.35 \times 10^{17})(1.868 \times 10^{-19}\,\text{J})\\ &= 1.65 \times 10^7\,\text{W} \approx 1.7 \times 10^7\,\text{W}\end{aligned}$$

The pulse duration τ_p under the top-hat approximation is given by

$$\tau_p = \frac{r\eta\,\tau_c}{r - 1 - \ln r} = \frac{2.9 \times 0.93(2.23\,\text{ns})}{0.835} = 7.2\,\text{ns}\,.$$

For comparison, the CW output power at the same pump ratio is $P_o = \gamma_c q_s(r-1)\hbar\omega_o$. Given the emission cross-section $\sigma = 3.5 \times 10^{-19}$ cm^2 for the laser transition from earlier examples and an upper-state population lifetime of $\tau_2 = 230\,\mu$s, the saturation photon number q_s is given by

$$q_s = \frac{\gamma_2}{K} = \frac{\gamma_2 n V_c}{c\sigma} = \frac{1.8(0.2 \times 5)}{(230 \times 10^{-6})(3.00 \times 10^{10})(3.5 \times 10^{-19})} = 7.45 \times 10^{11}\,.$$

The cw output from the laser will be

$$\begin{aligned} P_{\text{out}}^{cw} &= \left(\frac{\mathcal{L}_T}{\mathcal{L}_c}\right) \gamma_c q_s (r-1)\hbar\omega_o\,, \\ &= \left(\frac{0.3567}{0.6567}\right) \frac{(7.45 \times 10^{11})(2.9-1)(1.868 \times 10^{-19}\,\text{J})}{2.23 \times 10^{-9}\,\text{s}}\,, \\ &= 64\,\text{W}\,. \end{aligned}$$

The Q-switched peak power is therefore more than five orders of magnitude larger than the CW output, although it lasts for only about 7 ns.

Finally, the total (useful) pulse energy is given by

$$\begin{aligned} U_{\text{out}} &= \left(\frac{\mathcal{L}_T}{\mathcal{L}_c}\right) \eta r N_{\text{th}} \hbar\omega_o = \left(\frac{0.3567}{0.6567}\right) (0.93)(2.9)(4.35 \times 10^{17})(1.868 \times 10^{-19}\,\text{J}) \\ &= 0.118\,\text{J} \approx 120\,\text{mJ}\,. \end{aligned}$$

This agrees with the product of the top-hat peak power and pulse width $P_{\text{peak}} \times \tau_p$.

10.5 Methods for Q-switching

Several devices and techniques used for laser Q-switching are shown in Fig. 10.12. The first three are active methods because the elements responsible for switching the cavity Q require an external drive for their activation. The last method relies on the intrinsic nonlinear absorption of a saturable absorber triggered by the internal dynamics of the laser and is an example of passive Q-switching.

10.5.1 Rotating Mirror Q-switch

A simple mechanical method for Q-switching is to mount a cavity mirror or prism on a rapidly rotating shaft [Fig. 10.12]. In this configuration, the cavity feedback is restored only during the brief period when the rotating element aligns precisely with other cavity elements. This historically important method is rarely used today because the switching time is relatively long and mechanical vibration introduces noise and instability. For example, if the high-Q condition extends over an angular range of 1 mrad and the mirror rotates at 400 Hz, the high-Q interval is about 400 ns–too long for most applications."

10.5.2 Electro-optic Q-switch: Pockels Cell

Electro-optic Q-switches use the Pockels effect [Sec. 12.7] to switch the cavity loss rapidly. A Pockels cell contains a noncentrosymmetric crystal whose birefringence changes when a voltage is applied. Common materials include KDP and lithium niobate in the visible and near-infrared, and cadmium telluride in the mid-infrared.

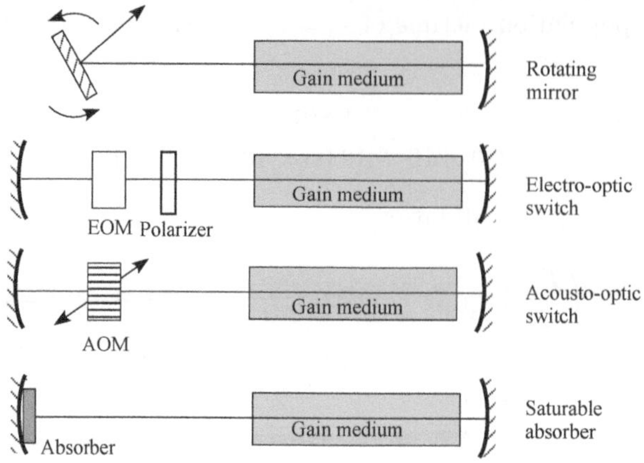

FIGURE 10.12
Methods for laser Q-switching.

The Pockels cell is oriented and biased in such a way that the x and y axes of the induced birefringence are in the plane perpendicular to the resonator axis. The cavity also contains a polarizer, such as a thin-film polarizing plate or Glan-Thompson prism. Its transmission axis is oriented at 45° to the electrically induced birefringence axes of the Pockels cell.

Light transmitted by the polarizer enters the Pockels cell with equal field components along the induced x- and y- axes. Inside the crystal, the x and y components of the optical field experience different refractive indices. Because these components experience different refractive indices, they acquire a relative phase difference

$$\Delta\varphi = k\Delta n L_p = 2\pi \frac{(n_y - n_x)L_p}{\lambda}, \tag{10.77}$$

where n_y and n_x are the indices experienced by the x and y components of the field, and L_p is the physical beam path length through the Pockels cell.

The DC voltage applied to the Pockels cell is chosen so that $\Delta\varphi = \pi/2$ on a single pass. This converts the linearly polarized wave into a circularly polarized wave [Fig. 10.13(a)]. The linearly polarized field therefore becomes circularly polarized. After reflection from the end mirror and a second pass through the cell, the total retardation is π, rotating the linear polarization by 90° relative to that of the incident wave. This orthogonally polarized beam is either reflected out of the cavity or absorbed. Removing the voltage removes the induced birefringence, so the returning beam retains its original polarization and remains in the cavity. The switch is then open and the cavity returns to its high-Q state.

The required voltage is called the quarter-wave voltage because it produces an optical path difference of $\lambda/4$ on a single pass. Typical values are 1–5 kV, depending on the crystal, cell length, and laser wavelength. The voltage must be switched on or off within a time short compared with the Q-switched pulse buildup, typically less than about 10 ns. Electro-optic Q-switches therefore provide fast, accurately timed, and highly repeatable switching, although they are relatively expensive and introduce additional intracavity optical surfaces subject to optical damage at the high intensities inside the Q-switched laser.

10.5.3 Acousto-optic Q-Switch

An acousto-optic Q-switch uses a traveling ultrasonic wave to create a refractive-index grating inside a transparent crystal. A piezoelectric transducer, driven typically at 20–80

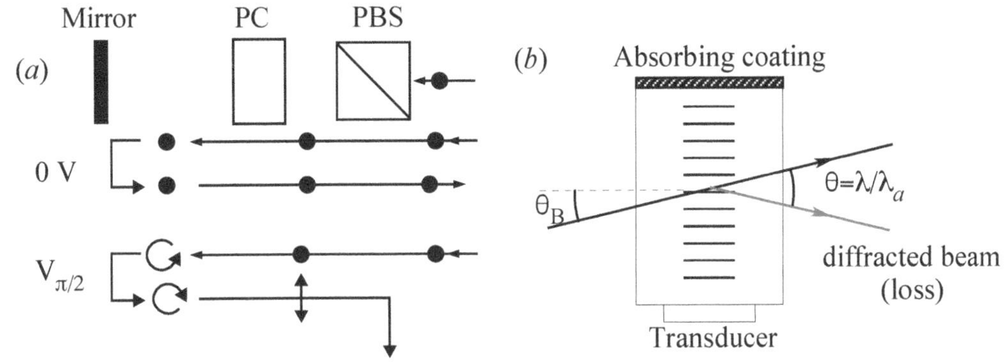

FIGURE 10.13
(a) Pockels cell (PC) Q-switch. Dots represent field polarization perpendicular to the plane of the paper and double-headed arrows represent polarization parallel to the plane. (b) Acousto-optic Q-switch.

MHz, launches the acoustic wave through the material. The resulting strain modulates the refractive index through the photoelastic effect and diffracts part of the laser beam out of the cavity mode [Fig. 10.13(b)]. An absorbing coating on the opposite face suppresses acoustic reflections and maintains a traveling-wave grating. Fused quartz is commonly used in the visible and near-infrared, while germanium and cadmium selenide are suitable for longer wavelengths.

For sufficiently thick modulators, the internal deflection angle of the diffracted wave is given by $\theta = \lambda/n\lambda_a$, where λ and λ_a are the wavelengths of light and acoustic wave, respectively. For an acoustic wave with a frequency of 50 MHz and speed of 3.75×10^5 cm/s, the acoustic wavelength is about $\lambda_a = 75\,\mu$m. Taking $n \approx 1.45$ for quartz at the YAG laser wavelength $\lambda = 1.06\,\mu$m, the internal diffraction angle $\theta \approx 0.56^o$, corresponding to about 0.8^o in air relative to the incident beam direction.

When the RF drive is applied, diffraction introduces an additional cavity loss. At sufficiently high drive power, this loss suppresses laser oscillation and places the cavity in its low-Q state. Turning off the acoustic drive removes the diffraction loss and restores high Q.

Acousto-optic Q-switches have low insertion loss, but their modulation depth is limited and their switching time is set by the acoustic transit time across the laser beam. They are therefore especially useful for repetitive Q-switching of low-gain, CW-pumped lasers

10.5.4 Passive Saturable Absorber Q-switch

Passive Q-switching requires no externally driven switching element. Instead, an intracavity saturable absorber provides a loss that decreases as the laser intensity rises. At low intensity the absorber keeps the cavity in a low-Q state. As the field builds, the absorber bleaches and the loss falls rapidly. If the absorber saturates at a much lower intensity than the gain medium, the cavity Q rises before the gain is significantly depleted, allowing a giant pulse to develop.

A saturable absorber may be modeled as a collection of two-level atoms with a large absorption cross section ($\sim 10^{-16}$ cm^2 for typical saturable dyes) at the laser wavelength. Optical excitation depletes the lower-state population and reduces the absorption; rapid relaxation subsequently restores the absorber. Thus, the switching occurs automatically as part of the intracavity field dynamics."

To describe this process quantitatively, we add a saturable absorber to the laser rate equations. LetN_a denote the lower-minus-upper population difference of the absorber, K_a its coupling constant, and γ_{2a} its population-relaxation rate [see Eq.(7.37)]. The coupled equations become

$$\frac{dq}{dt} = KNq - K_a N_a q - \gamma_c q\,, \tag{10.78a}$$

$$\frac{dN}{dt} = R_p - \gamma_2 N - KNq = \gamma_2\left[N^{(0)} - N - KNq/\gamma_2\right], \tag{10.78b}$$

$$\frac{dN_a}{dt} = \gamma_{2a}\left[N_a^{(0)} - N_a - 2K_a N_a q/\gamma_{2a}\right]. \tag{10.78c}$$

where $N^{(0)}$ and $N_a^{(0)}$ represent the unsaturated gain and unsaturated absorber populations, respectively. For many absorbers, the response time is short compared with the cavity lifetime, gain-recovery time, and pulse-buildup time. The absorber population can then be replaced by its steady-state value

$$N_a(t) \approx \frac{N_a^{(0)}}{1 + 2K_a q/\gamma_{2a}}. \tag{10.79}$$

The remaining laser equations therefore reduce to

$$\dot{q} = KNq - \gamma_c q - \frac{K_a N_a^{(0)} q}{1 + 2K_a q/\gamma_{2a}}, \tag{10.80a}$$

$$\dot{N} = \gamma_2\left(N^{(0)} - N - \frac{K}{\gamma_2}Nq\right). \tag{10.80b}$$

To simplify these equations, we introduce scaled dimensionless variables and parameters defined by

$$\begin{aligned} &\tau = \gamma_c t\,, \quad q_s = \frac{\gamma_2}{K}, \quad q_{sa} = \frac{\gamma_{2a}}{2K_a}, \quad s = \frac{q_s}{q_{sa}}, \quad N_{\text{th}} = \frac{\gamma_c}{K}, \\ &x = \frac{q}{q_s}, \quad y = \frac{N}{N_{\text{th}}}, \quad r = \frac{KN^{(0)}}{\gamma_c}, \quad r_a = \frac{K_a N_a^{(0)}}{\gamma_c}. \end{aligned} \tag{10.81}$$

The parameter $s = q_s/q_{sa}$ measures how much more easily the absorber saturates than the gain medium. In these variables, the equations become

$$\dot{x} = (y-1)x - \frac{r_a x}{1 + sx}, \tag{10.82a}$$

$$\dot{y} = \frac{\gamma_2}{\gamma_c}\left[r - (1+x)y\right]. \tag{10.82b}$$

Suppose the initial inversion is $y_i = r$ and exceeds the combined cavity and unsaturated-absorber loss, $r > 1 + r_a$. Under this condition, the cavity photon number begins to grow from spontaneous emission noise. During the initial growth phase, both gain saturation and absorber bleaching are negligible, so $y \approx r$ and the growth equation for the cavity photon number becomes

$$\dot{x} \approx (r - 1 - r_a)x \equiv \bar{r}x\,. \tag{10.83}$$

The cavity photon number thus grows exponentially as $x = x_i e^{\bar{r}\tau}$ provided $\bar{r} > 0$. This is the first threshold condition for passive Q-switching. As x grows, the absorber begins to bleach. Because $s = q_s/q_{sa} \gg 1$, the absorber saturates well before the gain medium. This falling loss accelerates the photon buildup and initiates the giant pulse. For most practical laser systems, the ratio $\gamma_2/\gamma_c \ll 1$, so the inversion changes little from its initial value r during early pulse development, allowing us to approximate $y \approx r$ within the gain equation (10.82b). The equation for the inversion then becomes

$$\dot{y} \approx -\frac{\gamma_2}{\gamma_c} xy \,. \tag{10.84}$$

Substituting the early exponential solution $x = x_i e^{\bar{r}\tau}$ into this equation and integrating from the initial condition $y_i = r$ up to an arbitrary early time τ, we find the initial depletion behavior of the atomic inversion y is given by

$$y = re^{-\frac{\gamma_2}{\gamma_c}(x-x_i)/\bar{r}} \approx r\left(1 - \frac{\gamma_2}{\gamma_c \bar{r}} x\right) , \tag{10.85}$$

where we utilized the fact that $x \gg x_i$ during buildup. This linear approximation is consistent with an expansion of $y(x)$ obtained by evaluating $dy/dx = (dy/d\tau)/(dx/d\tau)$ directly at early times (using Eqs. (10.83) and (10.84)). Substituting this linear inversion response into Eq. (10.82a), the instantaneous growth rate for the cavity photon number can be expressed as

$$\frac{\dot{x}}{x} = \left(r - \frac{\gamma_2 r}{\gamma_c \bar{r}} x - 1 - \frac{r_a}{1 + sx}\right) \approx (r - 1 - r_a) + \left(-\frac{\gamma_2 r}{\gamma_c \, \bar{r}} + r_a s\right) x \,. \tag{10.86}$$

The first term, $\bar{r} = r - 1 - r_a > 0$, is the ordinary small-signal growth rate. The second term describes the competition between gain saturation and absorber bleaching. If it is negative, the gain saturates faster than the absorber bleaches and the pulse growth slows. If it is positive, the absorber bleaches first, the net gain increases, and the field grows faster than exponentially into a giant pulse.

The two conditions for passive Q-switching are therefore

1. $r > 1 + r_a > 0$, and
2. $\frac{\gamma_2 r}{\gamma_c \bar{r}} < r_a s$.

The first condition simply requires the initial gain to exceed the combined cavity and unsaturated-absorber losses to allow initial exponential growth of the cavity photon number. The second condition requires the absorber to saturate faster than the gain medium. This is favored by $s = q_s/q_{sa} \gg 1$, meaning that the absorber saturation photon number is much smaller than that of the gain medium. Because the ratio γ_2/γ_c is usually very small in solid-state lasers, these conditions can be satisfied over a useful range of pump ratio r.

Figure 10.14(a) summarizes the pulse evolution. Initially, the gain exceeds the combined cavity and absorber loss, and the field begins to grow. As the absorber bleaches, the loss falls and the net gain rises, causing a rapid increase in intensity. Eventually the gain medium itself saturates; the pulse reaches its maximum as the inversion falls through threshold and then decays. Numerical solutions of Eqs. (10.82a) and (10.82b) shown in Fig. 10.14(a) reproduce this behavior and, for suitable parameters, also predict repetitive passive Q-switching [Fig. 10.14(b)].

In addition to design simplicity, passive Q-switches frequently generate output pulses concentrated within a single longitudinal mode. Because the absorber requires several hundred cavity roundtrips to fully saturate and thus increase the cavity Q, multiple longitudinal modes initially begin building up from background spontaneous-emission noise. Over these hundreds of roundtrips prior to bleaching, even microscopic differences in the loss of different modes are amplified exponentially. Consequently, only the single lowest-loss mode survives to dominate the resulting giant pulse. In contrast, active Q-switching features an exceptionally rapid transition to a high-Q state that typically unfolds within mere tens of roundtrips. This short timescale provides insufficient multi-pass discrimination to filter out competing modes. As a result, the output frequency spectrum of a passively Q-switched

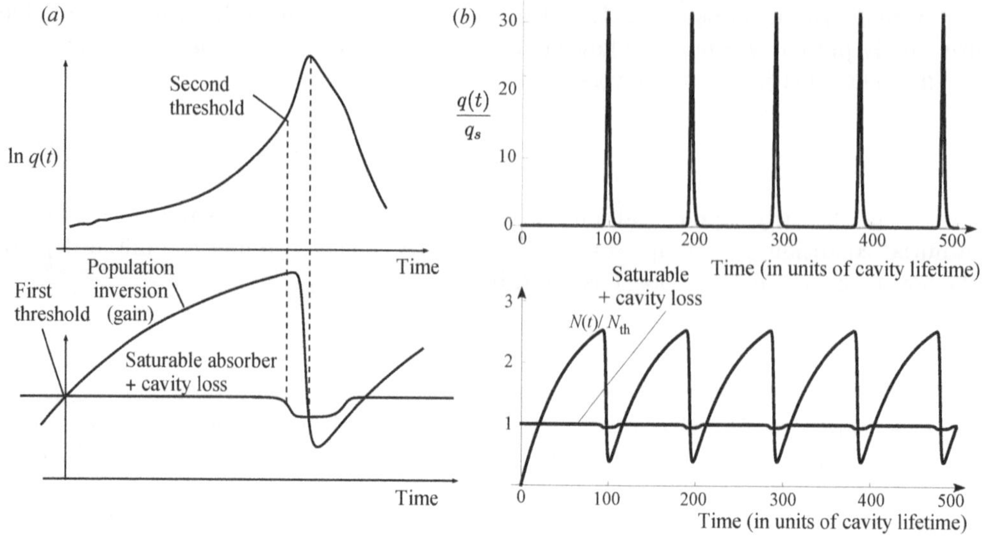

FIGURE 10.14
(a) Evolution of gain, loss, and cavity photon number during passive Q-switching; (b) Repetitive Q-switching with an intracavity absorber.

pulse is characteristically narrower than that produced via an active electro-optic Pockels cell or acousto-optic modulator.

Summary

In this chapter, we discussed several examples of transient and nonlinear dynamical behavior exhibited by lasers. These included spiking and relaxation oscillations, mode competition, injection locking, and Q-switching. These phenomena are important in understanding the operation and dynamics of lasers. They were used to illustrate fundamental principles and concepts of laser dynamics and to introduce techniques for analyzing them.

Transient phenomena such as spiking and relaxation oscillations were analyzed in Sec. 10.1 using simplified coupled nonlinear rate equations for the cavity photon number and the upper-laser-level population. Spiking and relaxation oscillations characterize a laser's response when operating parameters such as gain, loss, or pumping are perturbed.

Mode competition between simultaneously oscillating modes (Sec. 10.2) illustrates another aspect of laser dynamics, in which the modes become coupled through their competition for gain from the same atomic population. This behavior was analyzed using coupled-mode equations. Although the possibility of phase coupling is inherent in these equations, phase locking was considered separately through injection locking (Sec. 10.3). Injection locking illustrates the principle by which a weak signal injected into a more powerful laser can force the latter to lock its frequency and phase to those of the injected signal.

Finally, Q-switching was discussed in some detail. It was introduced as a technique for generating short, intense pulses of light from a laser. The process was analyzed using coupled rate equations for the cavity photon number and the atomic population inversion. Several experimental techniques for implementing Q-switching were also described. In particular, we discussed the use of an intracavity saturable absorber to enhance the nonlinearity, destabilize CW operation, and force the laser into a pulsed regime.

Problems

1. **Relaxation oscillations:** For each of the following laser systems, determine whether small perturbations about the steady state produce damped relaxation oscillations or purely exponential relaxation. If relaxation oscillations occur, calculate their angular frequency, oscillation frequency, period, and damping time. If they do not occur, determine the relevant exponential decay time constants.

   (a) An Nd:YAG laser operating at $\lambda = 1.064\,\mu$m has a pump ratio $r = 1.50$. The cavity is 50 cm long and has a total logarithmic roundtrip loss factor $\mathcal{L}_c = 0.12$. The upper-laser-level lifetime is 0.23 ms, and the lower-laser-level lifetime is much shorter.

   (b) A dye laser operates at a pump ratio $r = 2.5$. The cavity is 5 cm long, and the fractional roundtrip transmission loss is 5%. Treat the dye laser as a four-level system with an effective upper-laser-level lifetime $\gamma_2^{-1} = 1$ ns and a negligibly short lower-level lifetime.

   (c) A He-Ne laser operates at a pump ratio $r = 1.50$. The upper-laser-level lifetime is 50 ns. The cavity length is 50.0 cm, the output coupling is 1%, and all other losses are negligible.

2. A single-mode Nd:YAG NPRO (nonplanar ring oscillator) operates at a pump ratio $r = 5$. The roundtrip optical path has a physical length of 11.5 mm in material of refractive index $n = 1.82$. The output-coupler transmission is 0.4%, and the internal fractional roundtrip cavity loss is 0.5%. The upper-laser-level lifetime is 0.23 ms, and the lower-laser-level lifetime is much shorter. Determine whether relaxation oscillations occur and, if so, calculate their frequency and damping time.

3. A typical GaAs injection laser has a cavity length of 300 μm. Its cleaved end faces act as mirrors with power reflectance $\mathcal{R} = 30\%$, and the refractive index of the semiconductor is $n = 3.35$. The distributed intensity-loss coefficient is $\alpha = 60\ \text{cm}^{-1}$. Assuming operation at a pump ratio $r = 1.50$ and an effective upper-state population lifetime of 3 ns, determine whether relaxation oscillations occur and, if so, calculate their frequency, period, and damping time.

4. **Relaxation oscillations in a semiconductor laser:** The rate equations for a semiconductor laser are

$$\frac{dq}{dt} = \kappa(n - n_o)q - \gamma_c q\,,$$
$$\frac{dn}{dt} = J - \gamma n - \kappa(n - n_o)q\,,$$

   where q is the cavity photon number, n is the number of charge carriers in the active region, and J is the carrier injection rate produced by the steady pump current through the p-n junction. The term κn_o represents an additional optical loss rate associated with absorption by carriers in the lower states.

(a) Determine the steady-state values of the carrier population and cavity photon number, and obtain the threshold injection rate J_{th}.

(b) Linearize the rate equations about the above-threshold steady state and determine the condition for relaxation oscillations. Obtain expressions for the relaxation-oscillation angular frequency and damping time.

(c) For a GaAs injection laser, take $\gamma^{-1} = 3$ ns, $\gamma_c^{-1} = 1.1$ ps, and $n = 3.35$. Assume that the lower-state absorption represented by κn_o corresponds to an intensity-loss coefficient of 300 cm^{-1} in the optical waveguide. Calculate the relaxation-oscillation angular frequency ω_r at a pump ratio $J/J_{\text{th}} = 1.25$.

5. **Conditions for Q-switched operation:** What types of laser systems are good candidates for Q-switched operation? Discuss your answer in terms of the upper-level population decay rate γ_2, the high-Q cavity decay rate γ_c, and the pump ratio r.

(a) What requirements should the Q-switch satisfy for efficient generation of a short pulse?

(b) Are laser systems well suited to Q-switching generally more or less likely to exhibit relaxation oscillations? Explain the connection in terms of the relative time scales of the population and field dynamics.

(c) How does the efficiency with which the energy initially stored in the gain medium is converted into Q-switched pulse energy vary as the pump ratio r increases?

(d) How does the Q-switched pulse width vary as r increases? Describe its limiting behavior for $r \gg 1$.

6. **Q-switched Nd:YAG laser:** An 18-cm-long standing-wave laser cavity contains a 5.0-cm-long Nd:YAG rod operating at $\lambda = 1.064\,\mu$m. The rod has refractive index $n = 1.8$ and an effective cross-sectional area of 0.20 cm^2. The laser is operated in the Q-switched regime. The threshold inversion density is $\mathcal{N}_{th} = 4.35 \times 10^{17}$ cm^{-3}, and immediately before the Q-switch is opened the initial inversion corresponds to a pump ratio $r = 3$. The cavity mirrors have $\mathcal{R}_1 \approx 1$ and $\mathcal{R}_2 = 85\%$. The internal logarithmic roundtrip loss factor is $\mathcal{L}_i = 0.33$. For $r = 3$, take the energy-extraction efficiency to be $\eta = 0.941$.

(a) How much energy is initially stored in the population inversion?

(b) Estimate (i) the peak photon flux emitted through mirror M_2, (ii) the total number of photons emitted through M_2 during the pulse, and (iii) the pulse duration.

7. **Q-switching in a three-level gain medium:**

(a) Consider a three-level laser described during pumping and decay by

$$
\begin{aligned}
\dot{N} &= W_p(N_o - N) - 2KNq - \gamma_2(N_o + N)\,, \\
\dot{q} &= KNq - \gamma_c q\,,
\end{aligned}
$$

where q is the cavity photon number, N_o is the total number of active atoms, and N is the population inversion. During the short Q-switched pulse, neglect the pump and population-decay terms. Let N_i denote the inversion immediately before the Q-switch is opened, define the threshold inversion by $N_{\text{th}} = \gamma_c/K$, and define the pump ratio by $r = \frac{N_i}{N_{\text{th}}}$. Derive expressions for the energy-extraction efficiency, energy released into the cavity mode, peak intracavity photon number, peak intracavity power, and pulse duration. Express your results in terms of r and the energy-extraction efficiency η as far as possible.

(b) A ruby rod, treated as a three-level gain medium, has diameter 6.3 mm and length 7.5 cm and is placed in a 50-cm-long resonator with $\mathcal{R}_1 = 1.00$, $\mathcal{R}_2 = 0.50$, and internal logarithmic roundtrip loss factor $\mathcal{L}_i = 0.10$. The line-center stimulated-emission cross section is $\sigma = 2.5 \times 10^{-20}$ cm^2, the refractive index of ruby is 1.76, and the total active-ion density is $n_o = 1.60 \times 10^{19}$ ions/cm^3.

Assume that immediately before Q-switching the initial inversion is $N_i = 2N_{\text{th}}$, corresponding to $r = 2$. Determine the energy emitted through mirror M_2, the peak output power through M_2, and the pulse duration.

11

Mode-locking and Short Pulse Generation in Lasers

In the preceding chapter, we studied the generation of short, intense pulses of light using the technique of Q-switching. Q-switching can also be used to generate periodic pulse trains by continuously pumping the laser medium and repeatedly Q-switching the cavity at a repetition rate that may range from a few hundred to several thousand pulses per second. Q-switched laser pulses may involve a single mode or multiple modes of the laser cavity. As discussed in the previous chapter, their width is limited by the cavity lifetime. The relationship between pulse width and laser spectrum is governed by the time–bandwidth product: a short pulse duration requires a broad spectral bandwidth, and vice versa. This is a fundamental consequence of the Fourier-transform relation between the temporal and spectral representations of a physical signal. Consequently, the temporal and spectral widths of a pulse cannot both be made arbitrarily narrow.

Lasers can also produce repetitive pulse trains. Because of the properties of Fourier series, this periodic temporal behavior implies a comb-like spectral distribution, as sketched in Fig. 11.1. The spectral lines, or longitudinal modes, are separated by $\Delta\nu_{\text{ax}} = 1/\tau_R$, where τ_R is the cavity roundtrip time. The mode spacing, typically a few hundred MHz, is small compared with the carrier frequency ν_c. However, a large number of modes must participate in order to represent very short pulses. For ultrashort pulses with very small τ_p, the spectral width $\Delta\nu$ occupied by these modes can become significant compared with ν_c.

For example, consider a pulse with $\tau_p = 10$ fs at $\lambda = 750$ nm, corresponding to a carrier frequency $\nu_c \approx 4.0 \times 10^{14}$ Hz. The estimate $\Delta\nu \sim 1/(2\pi\tau_p)$ gives $\Delta\nu \approx 1.6 \times 10^{13}$ Hz. Thus, the bandwidth over which the pulse interacts with the gain medium is about 4% of the carrier frequency and corresponds to a wavelength interval of approximately 30 nm. Such a pulse has a spatial extent $c\tau_p \approx 3.0$ μm and contains only about four optical cycles.

Such pulses cannot be obtained by Q-switching. Under suitable circumstances, they can occur naturally; in general, however, their generation requires multimode laser operation with the longitudinal-mode phases locked together. This operating regime is called mode-locking. In a mode-locked laser, many longitudinal modes oscillate with fixed phase relationships, producing a periodic train of ultrashort optical pulses. We first introduce the general principles of laser mode-locking and then discuss several techniques for achieving it.

11.1 Mode-locked Laser Pulses

Consider a laser with a large number of oscillating axial modes, $N \gg 1$, all having the same polarization. If $\Delta\nu_{\text{osc}}$ is the oscillation bandwidth—defined as the frequency range over which gain exceeds loss (Fig. 11.1)—and $\Delta\nu_{\text{ax}}$ is the axial mode separation, the number of

DOI: 10.1201/9781003584179-11

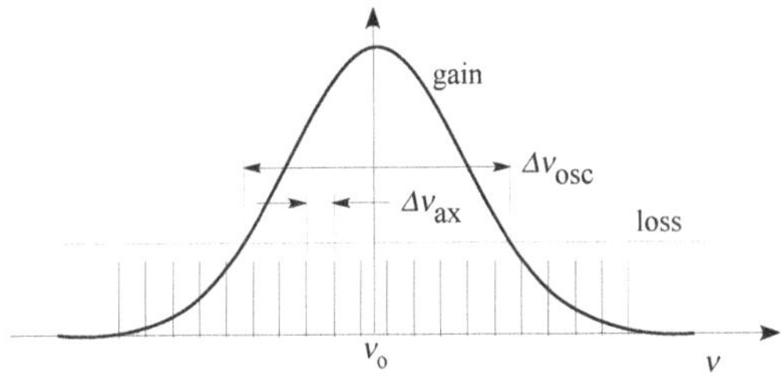

FIGURE 11.1
The number of oscillating modes N in a laser depends on the nature of the gain medium (homogeneous or inhomogeneous), the spatial distributions of gain and loss, the axial-mode separation, and the excess of gain over loss.

oscillating modes N is given by

$$N = \frac{\Delta\nu_{\rm osc}}{\Delta\nu_{\rm ax}}\,. \tag{11.1}$$

For simplicity, assume that all oscillating modes have nearly the same amplitude and that N is an odd integer so that $(N-1)/2$ is also an integer. Recall that axial mode frequencies are separated by $\Delta\nu_{\rm ax} = c/2L = 1/\tau_R$. If $\nu_c = Mc/2L$ (where $M \sim 10^7$ is a large integer) represents the central mode frequency (average frequency), then the frequencies ν_p and wave numbers k_p for the modes can be written as

$$\left.\begin{aligned} \nu_p &= \nu_c + p\Delta\nu_{\rm ax} = \nu_c + \frac{p}{\tau_R} \\ k_p &\equiv \frac{2\pi\nu_p}{c} = \frac{2\pi\nu_c}{c} + \frac{2\pi p}{c\tau_R} \end{aligned}\right\}, \qquad -\frac{(N-1)}{2} \le p \le \frac{(N-1)}{2}\,. \tag{11.2}$$

The electric field of the laser can then be written as a superposition of the standing-wave fields of these different modes[1]

$$A(z,t) = iA_o \sum_{p=-(N-1)/2}^{(N-1)/2} e^{-i2\pi\nu_p t + i\phi_p} \sin k_p z\,, \tag{11.3}$$

where ϕ_p is the initial phase of the pth mode. In general, ϕ_p varies randomly from mode to mode. However, for phase-locked modes, the fundamental mode-locking condition $\phi_{p+1} - \phi_p = \phi_o$ holds, where ϕ_o is a constant.[2]

The laser intensity has very different temporal behavior in these two cases. To illustrate this difference, Figure 11.2 shows the time-dependent intensity of a beam generated by a laser oscillating in $N = 21$ modes, each with the same amplitude, for both cases. Figure 11.2(a) shows the output intensity when the modes have random initial phases, while Figure 11.2(b) shows the output when the mode phases are locked. With random initial mode phases, the output intensity consists of a chaotic sequence of light pulses (bursts of light) of varying heights and durations. While most bursts are small, intense spikes of short duration ($\approx 1/\Delta\nu_{\rm osc}$) occasionally occur. The average intensity remains close to N ($A_o = 1$ for these

---

[1] In writing this field amplitude, we assume that all modes have nearly the same transverse structure, which can then be factored out from both sides of this equation [See Eq. (7.85)].

[2] This is only one possible phase-locking condition and is referred to as the fundamental mode-locking condition. $\phi_{p+1} - \phi_p = \phi_p - \phi_{p-1} + \pi$ corresponds to second-harmonic locking; other phase-locking relations are also possible.

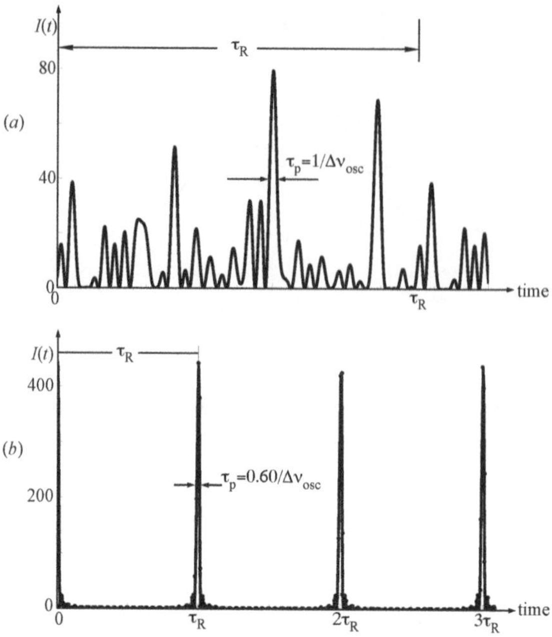

FIGURE 11.2
Time behavior of the squared amplitude of the total electric field for 21 oscillating modes, all with the same amplitude and with (a) random initial phases, and (b) locked phases.

figures). Notably, despite the random initial mode phases, the intensity waveform is periodic with a period of $1/\Delta\nu_{ax} = \tau_R$ because mode frequencies are harmonics of $\Delta\nu_{ax}$. This is a general property of Fourier series. If, more realistically, we allow the mode phases to evolve as random functions of time rather than remaining fixed, this waveform periodicity will be lost. Note that Figure 11.2(a) shows the waveform for a time window slightly larger than τ_R to illustrate its periodic nature and its short-term time evolution.

In the following analysis, we set $\phi_o = 0$ for convenience. By writing $\sin k_p z = (e^{ik_p z} - e^{-ik_p z})/2i$, we can express the standing-wave electric field of each mode as a superposition of left- and right-traveling waves

$$\begin{aligned}
A(z,t) &= \frac{A_o}{2} \sum_{p=-(N-1)/2}^{(N-1)/2} \left[e^{-i2\pi\nu_p(t-z/c)} - e^{-i2\pi\nu_p(t+z/c)} \right], \\
&= \frac{A_o}{2} \sum_{p=-(N-1)/2}^{(N-1)/2} \left[e^{-i2\pi(\nu_c+p/\tau_R)(t-z/c)} - e^{-i2\pi(\nu_c+p/\tau_R)(t+z/c)} \right], \\
&= \frac{A_o}{2} \left[e^{-i2\pi\nu_c(t-z/c)} \left(\frac{\sin N\pi(t-z/c)/\tau_R}{\sin \pi(t-z/c)/\tau_R} \right) \right. \\
&\qquad \left. - e^{-i2\pi\nu_c(t+z/c)} \left(\frac{\sin N\pi(t+z/c)/\tau_R}{\sin \pi(t+z/c)/\tau_R} \right) \right]. \qquad (11.4)
\end{aligned}$$

The first term inside the square brackets represents an optical wave of carrier frequency ν_c traveling to the right, with its amplitude modulated by the factor inside the parentheses (the modulation envelope). The frequency ν_c is the carrier frequency, and the modulation envelope is called the pulse envelope (or carrier envelope). Similarly, the second term represents an optical wave of frequency ν_c with a modulated amplitude traveling to the left.

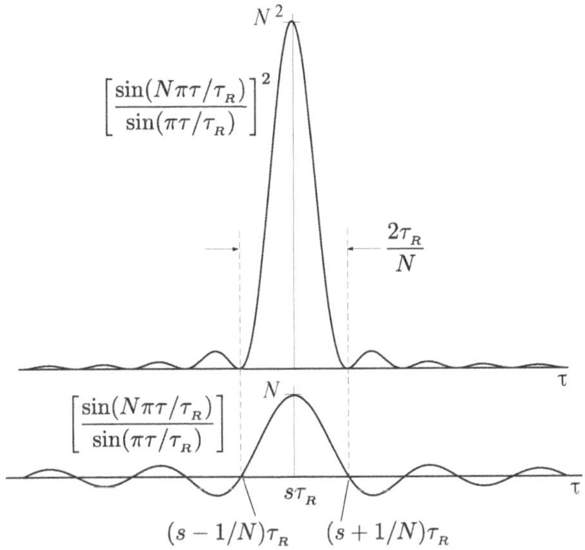

FIGURE 11.3
Pulse envelope and intensity profiles as functions of $\tau = t - z/c$ at a fixed point z inside the cavity. Here, τ_R represents the cavity roundtrip time.

The wave intensity inside the cavity, averaged over an optical cycle, $\frac{1}{2}\epsilon_o nc|A|^2$, is

$$I_c(z,t) = I_o\left[\left(\frac{\sin N\pi(t-z/c)/\tau_R}{\sin \pi(t-z/c)/\tau_R}\right)^2 + \left(\frac{\sin N\pi(t+z/c)/\tau_R}{\sin \pi(t+z/c)/\tau_R}\right)^2\right]. \tag{11.5}$$

The interference term drops out because, as seen from Eq. (11.6), right- and left-traveling waves are nonzero at two different times and exhibit negligible spatial overlap. Modulation envelope and intensity have sharp peaks at times (at a fixed position z inside the cavity) given by

$$\pi(t_s \mp z/c)/\tau_R = s\pi \quad \Rightarrow \quad (t_s \mp z/c) = s\tau_R \tag{11.6}$$

where s is an integer. Envelope peaks have a height of N, meaning the peak wave amplitude rises to NA_o and intensity scales to $N^2 I_o$. On either side of a peak at time t_s, the intensity falls to zero at times $t_s \pm \delta t$ such that

$$\pi(t_s \mp z/c \pm \delta t)/\tau_R = s\pi \pm \pi/N \quad \Rightarrow \quad \delta t = \tau_R/N\,. \tag{11.7}$$

There are also $N-2$ secondary intensity peaks between successive principal maxima. These secondary peaks are substantially weaker than the principal peak and are further suppressed when the mode amplitudes decrease away from line center. In the remainder of this chapter, we neglect these weaker secondary peaks.

11.1.1 Mode-locked Pulse Width

The pulses may be defined to last for a duration Δt, spanning the interval between the first zeros on either side of the peak

$$\Delta t_0 = 2\delta t = 2\tau_R/N = \frac{2}{\Delta\nu_{osc}}\,. \tag{11.8}$$

This interval, however, does not represent the full width at half maximum (FWHM). The FWHM width is instead obtained by solving

$$\left[\frac{\sin(N\pi\delta t_{1/2}/\tau_R)}{\sin(\pi\delta t_{1/2}/\tau_R)}\right]^2 = \frac{N^2}{2}. \tag{11.9}$$

For large values of N, this expression yields the approximate solution $N\pi\delta t_{1/2}/\tau_R \approx 0.443\pi$. Hence, a good approximation to the FWHM pulse width τ_p is

$$\tau_p = 2\delta t_{1/2} = 0.89\frac{\tau_R}{N} = \frac{0.89}{\Delta\nu_{\text{osc}}}. \tag{11.10}$$

As expected, this relation shows that mode-locked pulses become narrower as the oscillation bandwidth $\Delta\nu_{\text{osc}}$ (and, therefore, the number of oscillating modes N) increases.

This result is a direct consequence of the fundamental properties of Fourier transforms. In practice, the oscillation bandwidth is of the order of the gain linewidth $\Delta\nu_{at}$. This implies that mode-locked solid-state or semiconductor lasers can typically generate light pulses as short as a few picoseconds. For certain tunable dye and solid-state lasers with gain linewidths at least 100 times larger, ultrashort pulses with widths of order 10 fs are possible and have been experimentally observed. Gas lasers, on the other hand, with much narrower linewidths (up to a few GHz), generate relatively long pulses down to approximately 100 ps.

Note also that the peak power of a mode-locked pulse is proportional to $N^2A_o^2$, whereas for modes with random phases the average power is proportional to NA_o^2. Thus, the ratio of the peak pulse power in the mode-locked case to the average power in the non-mode-locked case is equal to the number of oscillating modes, N. This value can vary from 10^2 to 10^4 depending on the gain medium. Thus, mode-locking is useful not only for producing pulses of very short duration but also for achieving immense peak power.

11.1.2 Space-time Picture of Mode-locked Pulses

From Eq. (11.6) we find that for a given position z_o inside the laser cavity [Fig. 11.4 (a)], right-traveling pulses occur at times (with $t_o = z_o/c$)

$$t_{\rightarrow} = t_o,\ t_o + \tau_R,\ t_o + 2\tau_R, \cdots, \tag{11.11}$$

and left-traveling pulses at times

$$t_{\leftarrow} = -t_o,\ -t_o + \tau_R,\ -t_o + 2\tau_R, \cdots, \tag{11.12}$$

Thus, a right- (or left-) traveling pulse with a peak intensity N^2I_o and a duration of $0.89/\Delta\nu_{\text{osc}}$ passes any given point every roundtrip time τ_R.

How many pulses are circulating inside the cavity at any given time? To answer this question, let us examine the pulse positions at a fixed time t_o^*. The locations of the right- and left-traveling pulses are given by

$$z_{\rightarrow} = z_o^* \quad \text{and} \quad z_{\leftarrow} = -z_o^* + 2L \quad (\text{where } z_o^* = ct_o^*). \tag{11.13}$$

It is easily verified that only one pulse (either right- or left-going) is present within the cavity bounds ($0 < z < L$) at any given instant [Fig. 11.4(b)].

Note that the spatial extent of a mode-locked pulse, $\Delta z = c\tau_p$, is much shorter than the cavity length. For example, a 1-ps pulse has a spatial length $\Delta z = c\tau_p = 0.3$ mm, whereas

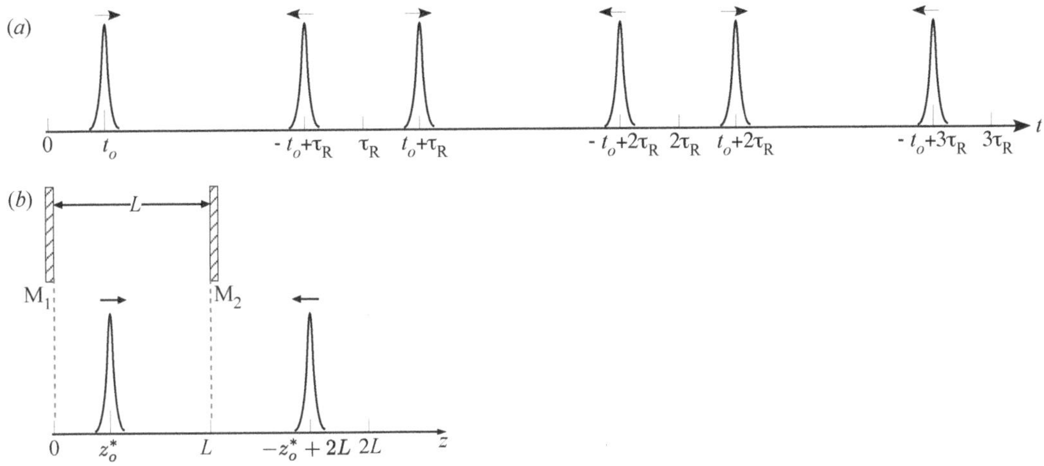

FIGURE 11.4
(a) Pulse sequence observed at a fixed point inside the cavity, where the first right-traveling pulse is observed at time t_o. (b) Spatial distribution of pulses at a fixed instant t_o^*.

a typical laser cavity is several tens of centimeters long. Recall also that in writing the expression for the field, we have suppressed the transverse beam profile. If only the axial modes are oscillating, as is often the case, the fundamental Gaussian mode of the cavity is a good approximation for the transverse spatial profile of these pulses. Combining these concepts, the oscillating pulse inside the laser cavity can be visualized as a single ultrashort pulse of duration τ_p and spot radius w_0 propagating back and forth within the cavity. Thus, a mode-locked laser pulse inside the laser cavity may be visualized as a "light-bullet" bouncing back and forth between the cavity mirrors. Outside the cavity (e.g., to the right of mirror M_2), an output pulse is detected each time a right-traveling pulse hits M_2. The output field from the laser, $A_{\text{out}}(z,t)$, can be expressed as

$$A_{out}(z,t) = \sqrt{\mathcal{T}_2}\, i \frac{A_o}{2} \left(\frac{\sin N\pi\tau/\tau_R}{\sin \pi\tau/\tau_R} \right) e^{-i\omega_c\tau}\,, \tag{11.14}$$

where $\omega_c = 2\pi\nu_c$ is the optical angular frequency and $\tau = t - z/c$. The output thus consists of a sequence of short pulses with period τ_R, duration $0.89/\Delta\nu_{\text{osc}}$, and peak intensity of $N^2 I_o \mathcal{T}_2$, where $\mathcal{T}_2$ is the output coupler transmittance.

The expression for pulse width derived here applies to the equal-amplitude multimode spectrum considered above. The preceding equal-amplitude model illustrates the basic pulse-forming mechanism. In an inhomogeneously broadened laser, a more realistic mode-amplitude distribution follows the gain profile. The mechanism for producing multimode oscillation in a homogeneously broadened system is different and is discussed in the next section.

In an inhomogeneously broadened laser, the Gaussian gain profile dictates that the mode intensities (which are $\propto A_p^2$) follow a Gaussian distribution (assuming the central mode coincides with the line center)

$$A_p^2 = A_o^2 e^{-(p\Delta\nu_{\text{ax}}/\Delta\nu_{\text{osc}})^2 4\ln 2}\,. \tag{11.15}$$

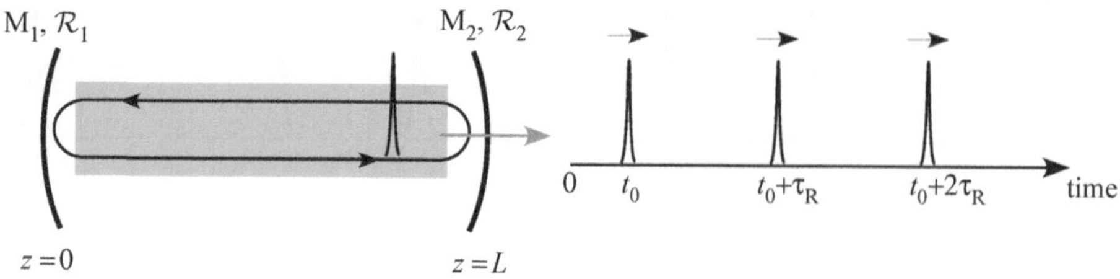

FIGURE 11.5
Output pulse sequence from a mode-locked laser.

With this mode intensity distribution, the sum over the oscillating modes for the right-traveling wave can be evaluated for $N \gg 1$ by replacing the sum with an integral

$$\begin{aligned} A(z,t) &= e^{-i\omega_o \tau} \sum_{p=-(N-1)/2}^{(N-1)/2} \frac{A_p}{2} e^{-i2\pi p\tau/\tau_R}\,, \\ &\approx \frac{A_o}{2} e^{-i2\pi\nu_c\tau} \int_{-\infty}^{\infty} dp \, \exp\left[-\left(p/\tau_R \Delta\nu_{\text{osc}}\right)^2 2\ln 2 - i2\pi p\tau/\tau_R\right]\,, \\ &= \frac{A_o}{2} e^{-i\omega_c\tau} \left(\frac{\tau_R \Delta\nu_{\text{osc}}}{\sqrt{2\ln 2}}\right) \sqrt{\pi}\, e^{-\tau^2\left(\pi\Delta\nu_{\text{osc}}/\sqrt{2\ln 2}\right)^2}\,. \end{aligned} \tag{11.16}$$

Note that this approximation of replacing a discrete sum by an integral is valid only for the duration of a single pulse. While Eq. (11.16) accurately models the individual pulse shape, it washes out the long-term periodic character of the mode-locked pulse train predicted by Eq. (11.5) and observed experimentally. From Eq. (11.16), the pulse intensity has a Gaussian temporal profile

$$I(z,t) = I_{\max} e^{-2\tau^2\left(\pi\Delta\nu_{\text{osc}}/\sqrt{2\ln 2}\right)^2}\,, \tag{11.17}$$

where $I_{\max} = \frac{1}{2}\epsilon_o nc|A(0,0)|^2$ is the maximum intensity. The full width at half maximum (FWHM) of this pulse is

$$\tau_p = \frac{2\ln 2}{\pi\Delta\nu_{\text{osc}}} = \frac{0.44}{\Delta\nu_{\text{osc}}} = \frac{0.44\tau_R}{N}\,. \tag{11.18}$$

The corresponding spatial extent of this pulse is $c\tau_p = 0.44c\tau_R/N$. Such a pulse is said to be transform limited.

Unlike an inhomogeneously broadened laser, a homogeneously broadened laser naturally tends to oscillate in a single mode. However, spatial hole burning and the spatial distribution of gain may permit a small cluster of modes near the center of the gain profile to oscillate. Consequently, $\Delta\nu_{\text{osc}}$ for a homogeneously broadened gain medium is restricted to a narrow region near the center of the gain profile. Nevertheless, if we can feed modes further out by coupling them to modes near the center, we can force the laser into multimode operation. This is precisely the role of an amplitude modulator. As discussed in a later section, the modulator broadens the oscillation spectrum by generating sidebands, pushing the oscillation frequency spectrum outward from the line center. This broadening of the oscillation spectrum is directly opposed by the natural tendency of a homogeneously broadened gain medium to narrow the oscillation spectrum. Mode-locking behavior in a homogeneously broadened laser is therefore a balance between these two opposing tendencies, which is reflected in the expression for the mode-locked pulse width in a homogeneous broadened system

$$\tau_p \approx \frac{0.45}{\sqrt{\nu_m \Delta\nu_{\rm at}}}\,, \tag{11.19}$$

where ν_m is the modulation frequency and $\Delta\nu_{\rm at}$ is the gain bandwidth.

11.2 Methods for Mode-locking

The methods for mode-locking can be classified into two primary categories: active and passive. In active mode-locking, mode coupling is induced by externally modulating a cavity parameter such as the loss, gain, or refractive index. In passive mode-locking, the required modulation arises from nonlinear laser-matter interactions within the cavity.

11.2.1 Active Mode-locking

The two main approaches to active mode-locking are periodic amplitude modulation (AM mode-locking) and phase modulation, often referred to as FM mode-locking of the laser cavity modes. Modulation of the laser pump or gain can also be used for mode-locking. While each method has distinct advantages and limitations, we focus on AM mode-locking, which is widely used and provides a particularly transparent illustration of the mode-locking mechanism.

An amplitude modulator (AM) introduces a periodically varying loss into the laser cavity at an angular frequency ω_m. Both electro-optic (Pockels cell) and acousto-optic amplitude modulators are used for this purpose. Electro-optic modulators are generally preferred in high-gain lasers, whereas acousto-optic modulators are favored in low-gain systems due to their low insertion losses.

We model the periodic modulator loss by taking its amplitude transmission, $t_{_M}(t)$, to be a periodic function of time [see Fig. 11.6(a)] given by

$$t_{_M}(t) = e^{-M(1-\cos\omega_m t)} \approx 1 - M(1-\cos\omega_m t)\,, \tag{11.20}$$

where $M \ll 1$ is the depth of modulation. This transmission function multiplies the amplitude of the right- (or left-) traveling wave component of each axial mode ω_p of the cavity upon each pass through the modulator. The incident amplitude, $A_{p,\rm in}$, and outgoing amplitude, $A_{p,\rm out}$ of mode ω_p are related by

$$\begin{aligned} A_{p,\rm out}(t) &= t_{_M} A_{p,\rm in} e^{-i\omega_p t} = \left[1 - M(1-\cos\omega_m t)\right] A_{p,\rm in} e^{-i\omega_p t}\,, \\ &= \left[(1-M)e^{-i\omega_p t} + \frac{1}{2}Me^{-i(\omega_p-\omega_m)t} + \frac{1}{2}Me^{-i(\omega_p+\omega_m)t}\right] A_{p,\rm in}\,. \end{aligned} \tag{11.21}$$

From Eq. (11.21), it is evident that the cavity mode ω_p develops sidebands at frequencies $\omega_p \pm \omega_m$. If the modulation frequency is close to the axial mode spacing, $\omega_m \approx 2\pi/\tau_{_R} = 2\pi\Delta\nu_{ax}$, these sidebands lie close to the adjacent mode frequencies and act as source terms in their respective equations of motion. Thus, in addition to the usual contributions from cavity gain and loss, the dynamic equation for each cavity mode has contributions arising from the modulation of its nearest-neighbor modes [Fig. 11.6(b)]. This couples the cavity mode equations. While a rigorous analysis of how this coupling stabilizes into a phase-locked state

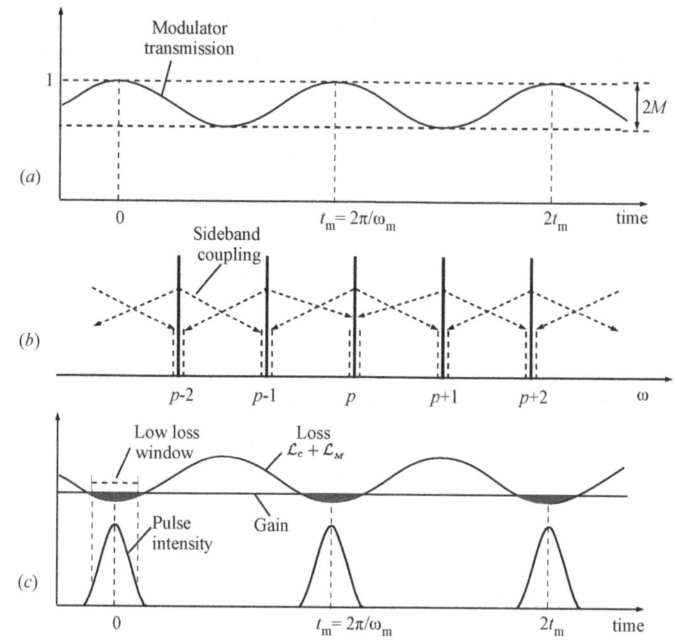

FIGURE 11.6
(a) Amplitude transmission function of an AM modulator. (b) Mode coupling by sideband generation. (c) Steady-state mode-locked behavior of the laser intensity.

is mathematically complex, it is conceptually analogous to the frequency locking observed when an oscillator is injection-locked by an external signal. A simple model of this process was considered in Sec. 10.3.

The basic physics of mode-locking can be grasped with reference to Fig. 11.6(c). The overall roundtrip cavity loss has a modulation given by $\mathcal{L}_M = -\ln(t_M^2) \approx 2M(1 - \cos\omega_m t)$, which establishes a narrow, periodic temporal window of minimum loss in the laser. When the modulation frequency matches the axial mode spacing, $\omega_m = 2\pi\Delta\nu_{ax} = 2\pi/\tau_R$, a short optical pulse passing through the modulator at the instant of minimum loss can circulate in the laser cavity and return exactly when the loss is at a minimum again. If a pulse arrives slightly earlier than the minimum-loss time t_M, its leading edge is selectively attenuated [Fig. 11.7(c)]. This asymmetric clipping effectively delays the peak of the pulse, steering it closer to t_M during successive passes. After many hundreds or thousands of roundtrips, the pulse reaches the steady-state configuration shown in Fig. 11.6(c). While this narrow loss window continuously forces the pulse to shorten, a finite lower limit to the pulse duration exists. This limit arises because shorter pulses require multimode oscillation over a broader bandwidth, but the frequency range over which modes can successfully oscillate is fundamentally limited by the gain bandwidth of the medium [see Fig. 11.7(c)].

The modulator in a mode-locked laser is usually driven at the fundamental axial-mode spacing frequency $\omega_m = 2\pi\Delta\nu_{ax}$. This is known as fundamental mode-locking. It is, however, possible to drive the modulator at a harmonic of this frequency, $\omega_m = 2\pi N\Delta\nu_{ax}$. In this case, the mode ω_p will acquire sidebands at frequencies $\omega_p \pm \omega_m$, coupling cavity modes that are separated by N axial-mode intervals. This technique, called harmonic mode-locking, allows the generation of an output pulse train with repetition rate N/τ_R.

Mode-locking can also be induced by modulating the phase or frequency of cavity modes. This is called FM mode-locking. It also works by generating sidebands of cavity modes.

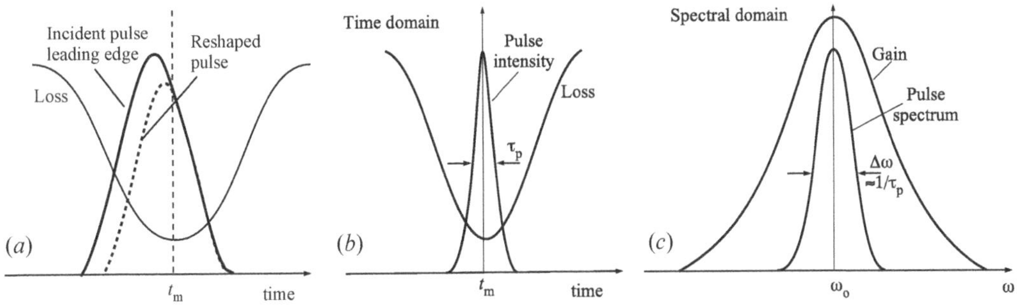

FIGURE 11.7
(a) Pulse reshaping by modulator loss. (b) Steady-state pulse width is determined by a balance between pulse narrowing from the modulator loss and spectral broadening due to finite gain bandwidth. (c) Pulse width and spectral bandwidth limitation dictated by the gain curve.

An FM modulator consists of a transparent material whose refractive index n is sinusoidally modulated at frequency ω_m, subjecting a cavity mode to a time-varying phase shift $(2\pi/\lambda)\ell_m n_o \cos(\omega_m t) \equiv M \cos\omega_m t$, where ℓ_m is the modulator length. The effect of the phase modulator can be expressed mathematically by a complex transmission function of the form

$$t_{_{FM}} = e^{iM\cos(\omega_m t)} = J_o(M) + 2iJ_1(M)\cos\omega_m t - 2J_2(M)\cos 2\omega_m t + \cdots \,, \tag{11.22}$$

where $J_n(M)$ is the Bessel function of the first kind of order n and argument M. For small modulation index M, this series is well-approximated by its first two terms. The transmission function then has the same form as $t_{_M}$ except that the sidebands are $\pi/2$ out of phase relative to the carrier frequency. FM mode-locking is used less frequently because the pulses remain frequency-modulated (chirped), and the mode-locked state tends to be unstable.

11.2.2 Passive Mode-locking

The role of an amplitude modulator in mode-locking can be visualized as a fast shutter inside the cavity that is opened periodically at frequency $\omega_m \approx 1/\tau_{_R}$ for a short fixed duration by an external drive. However, an external signal is not the only way to operate the shutter. If an intracavity medium can transition from absorbing to transparent when exposed to the leading edge of an optical pulse, the pulse itself can act as the shutter switch. An intracavity saturable absorber works precisely in this manner; the rising, high intensity edge of the pulse saturates or bleaches the absorption, allowing the remainder of the pulse to pass unattenuated. Following the passage of the pulse, the absorber relaxes back to its highly lossy state, and the process repeats when the pulse returns after a roundtrip [Fig. 11.8]. Unlike the active modulation configuration, the duration of the low-loss window is not fixed externally; it is determined by the internal dynamics of the laser. Mode-locking induced by a saturable absorber is an example of passive mode-locking.

To gain some insight into how passive mode-locking with an intracavity saturable absorber works, consider a two-level saturable absorber with a population relaxation rate γ_{2a} fast enough to allow it to track rapid variations of light intensity. The absorber loss per pass can be written as [Sec. 10.5.4, Eq. (10.79)]

$$\mathcal{L}_a = \frac{2\alpha_a L_a}{1 + I/I_s} \approx 2\alpha_a L_a \,(1 - I/I_s) \,, \tag{11.23}$$

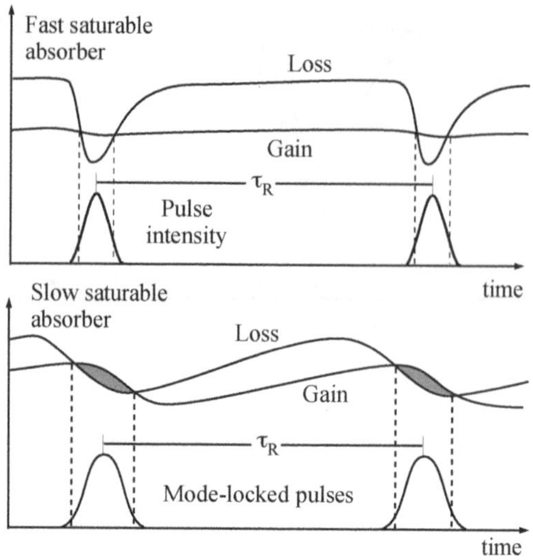

FIGURE 11.8
Passive mode-locking pulse profiles by fast and slow absorbers. For simplicity of visualization, gain and saturable absorption are shown overlapping in time.

where α_a is the small-signal absorption coefficient, L_a is the thickness of the absorber medium, and $I_s = (h\nu/2\sigma_a)\gamma_{2a}$ is the saturation intensity (with σ_a representing the absorption cross section). The total cavity loss, including the cavity loss $\mathcal{L}_c$ without the absorber, becomes

$$\mathcal{L} = \mathcal{L}_c + 2\alpha_a L_a - 2\alpha_a L_a I/I_s \tag{11.24}$$

When the laser is pumped above threshold, initial spontaneous-emission noise creates a multimode oscillation spectrum with random mode phases and amplitudes, resembling the chaotic noise bursts in Fig. 11.2(a). Low-intensity bursts experience balanced roundtrip gain and loss ($2g_o L_g = \mathcal{L}_c + 2\alpha_a L_a$) and fail to grow. However, the most intense noise spike selectively bleaches the absorber, experiences a lower net loss, and grows faster than competing fluctuations during subsequent roundtrips. Over hundreds of roundtrips, this spike suppresses background noise and forms the steady-state pulse train shown in Fig. 11.8. This intensity-dependent loss gating continuously narrows the circulating pulse. Because the lower-intensity leading and trailing edges are clipped while the high-intensity peak passes unattenuated, the pulse shortens on every roundtrip. The ultimate steady-state pulse width is determined by the balance between this saturable-loss narrowing and the spectral filtering imposed by the finite gain bandwidth as in the case of active mode-locking [Fig. 11.7].

Mode-locking or Q-switching (or Both)?

As demonstrated by the physics of Q-switching (Sec. 10.5.4) and mode-locking, a saturable absorber can facilitate both phenomena, though their parameter requirements differ. Mode-locking requires the absorber to respond nearly instantaneously to the cavity field fluctuations. For two noise-driven cavity modes, this means that the interference intensity $I \propto \left|A_1 e^{-i\omega_1 t + i\phi_1} + A_2 e^{-i\omega_2 t + i\phi_2}\right|^2 = \frac{1}{2}(|A_1|^2 + |A_2|^2 + 2|A_1||A_2|\cos[(\omega_2 - \omega_1)t + \phi_1 - \phi_2)]$ exhibits a beat frequency that directly modulates the absorber loss via Eq. (11.23), driving the system into a mode-locked state. Passive Q-switching, on the other hand, requires the absorber's saturation intensity to be significantly lower than that of the gain medium, en-

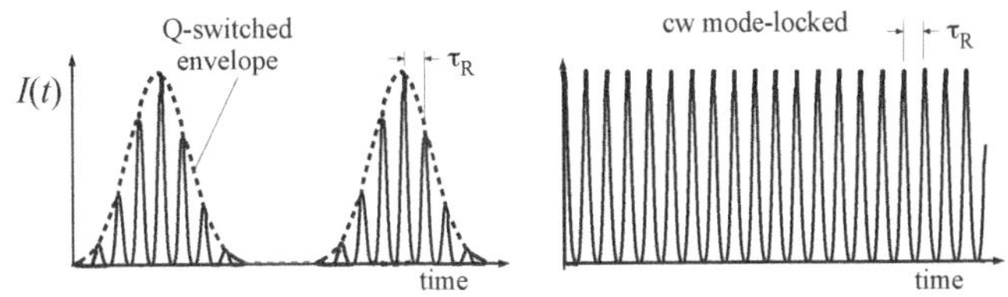

FIGURE 11.9
Laser power as a function of time for (a) a Q-switched mode-locked (QML) laser and (b) a continuous-wave (CW) mode-locked laser.

suring that once the absorber bleaches, a large net gain is created, leading to the formation of a giant single pulse.

In real systems, these regimes often overlap because nonlinear relaxation timescales vary with local field intensity. Consequently, while inserting a saturable absorber can passively mode-lock many solid-state lasers, it can simultaneously introduce Q-switching instabilities. Depending on operating parameters, these instabilities can drive the laser into a hybrid Q-switched mode-locked (QML) regime or completely destabilize the pulse train. The ultrafast pulse repetition rate is governed by the cavity roundtrip time, whereas the slow Q-switching modulation frequency is approximately given by the laser's relaxation oscillation frequency.

11.2.3 Kerr-Lens Mode-locking (KLM)

Kerr lens mode-locking is a fast passive mode-locking technique that relies on the lensing effect induced by the Kerr-type nonlinear response of an optical material (such as quartz or sapphire) when a light beam of sufficiently high intensity traverses it. In a Kerr-type nonlinear medium, the refractive index of the medium is modified by the local field intensity and can be written as

$$n(\boldsymbol{r},t) = n_o + n_2 I(\boldsymbol{r},t)\,, \tag{11.25}$$

where n_2 is the nonlinear refractive-index coefficient, which is directly related to the third-order susceptibility $\chi^{(3)}$ of the medium [see Sec. 12.8.2]. The value of n_2 can be positive or negative. Kerr lens mode-locking requires a medium with a positive n_2, such as fused quartz with $n_2 \approx 4.5 \times 10^{-16}$ cm^2/W or sapphire with $n_2 \approx 3.45 \times 10^{-16}$ cm^2/W. Through the intensity-dependent refractive index, the temporal intensity variation produces self-phase modulation (SPM), whereas the transverse intensity variation produces Kerr lensing. This lens induces self-focusing ($n_2 > 0$) or defocusing ($n_2 < 0$) of the pulse.

Because the pulse is a phase-locked superposition of axial resonator modes, its transverse beam profile is well-approximated by a Gaussian distribution. Consequently, the higher optical intensity along the beam axis creates a larger localized refractive index change in the central part of the beam than in its wings. This modified refractive index profile acts exactly like a positive focusing lens. If we write the transverse beam intensity as $I(\rho) = I_o e^{-2\rho^2/w^2}$, where I_o is the on-axis intensity peak, ρ is the radial distance from the axis, and w is the spot radius, the nonlinear phase shift acquired by the beam in traversing a length ℓ_m of the medium is given by

$$\Delta\phi = \frac{2\pi(\Delta n)\ell_m}{\lambda} = \frac{2\pi n_2 I_o \ell_m}{\lambda} e^{-2\rho^2/w^2} \approx \frac{2\pi n_2 I_o \ell_m}{\lambda}\left(1 - 2\rho^2/w^2\right). \tag{11.26}$$

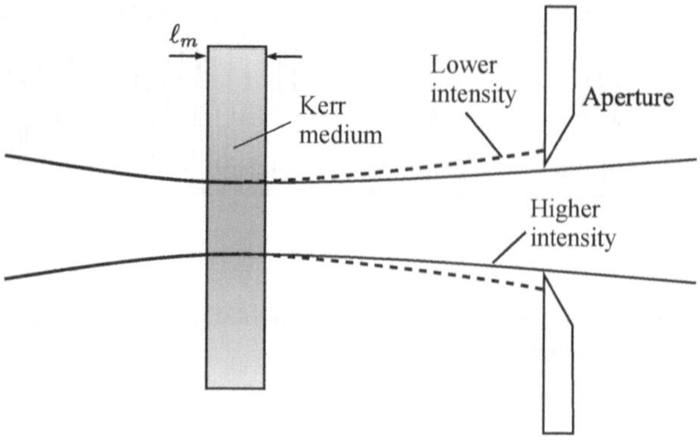

FIGURE 11.10
A Kerr lens in combination with a suitably placed aperture of appropriate size acts as a fast saturable absorber.

To first order in $(\rho/w)^2$, the phase change $\Delta\phi$ exhibits a parabolic transverse profile characteristic of a positive spherical lens [see Sec. 2.8.1]. The dependence of the Kerr-lens optical power on pulse intensity implies that more intense light pulses are focused more tightly. By placing a hard aperture of appropriate size at a suitable location relative to the Kerr-lens medium, a saturable loss element is realized [Fig. 11.10]. A stronger Kerr lens at higher intensities focuses the beam to a smaller spot radius at the aperture plane, thereby reducing the clipping losses experienced by the pulse. The electronic Kerr-lens response is extraordinarily fast and can be treated as instantaneous for all practical purposes. Thus, the effective loss-gating mechanism of Kerr-lens mode-locking is analogous to that of a fast saturable absorber. Using this technique within broadband gain media, such as titanium-doped sapphire (Ti:sapphire) lasers, ultrashort mode-locked pulses down to 6 to 8 fs have been achieved.

The optical power of the Kerr lens increases as the beam radius decreases. As the beam radius decreases due to self-focusing, optical intensities become even higher, further amplifying the self-focusing effect. This positive feedback can eventually lead to very high optical intensities that can trigger catastrophic beam collapse and optical damage to the medium. To avoid this runaway effect, the peak power must remain below a critical threshold given by $P_{crit} \approx 0.148\lambda^2/n_o n_2$, where λ is the vacuum wavelength, and n_o is the background linear refractive index. For fused silica fibers in the 1-μm wavelength region, this critical power is of the order of 4 MW. The wavelength dependence of the critical power underscores the mitigating role of diffraction (neglected here), which becomes more pronounced at longer wavelengths and counteracts beam collapse.

11.2.4 Mode-locked Pulse Profile and Duration

Analytical prediction of the pulse intensity profile and width is a complex mathematical problem involving competing time scales and spectral responses. Finding a simple, closed-form analytic expression to represent pulse shape is even more challenging [see Sec. 11.4]. However, valuable insights can be gleaned directly from physical arguments. From the pulse-shaping mechanism depicted in Fig. 11.7(a), it is clear that each time the pulse passes through the modulator, its leading and trailing edges are attenuated, while the peak passes unaffected by the time-varying loss. Therefore, it is reasonable to expect steady-state pulses to exhibit a symmetric, bell-shaped temporal profile. A Gaussian fulfills this condition, as does the sinc-squared function introduced in Sec. 11.1. A Lorentzian also features a bell-shaped profile; however, the attenuation of its relatively long tails by the modulator loss in each pass constitutes too strong a perturbation to allow a stable steady-state to develop. Consequently, a Lorentzian is not expected to provide a stable steady-state pulse profile in this simple loss-gating picture. Other viable pulse shapes will be encountered later in this chapter.

The manner in which a finite gain bandwidth limits the steady-state mode-locked pulse width differs fundamentally between homogeneous and inhomogeneous gain media. An inhomogeneously broadened laser operating sufficiently far above threshold naturally supports multimode oscillation over a broad spectrum that spans essentially the entire gain bandwidth $\Delta\nu_{\rm osc} \approx \Delta\nu_{\rm at}$. In this case, the primary role of the modulator is merely to lock the phases of these already oscillating modes. Under the synchronization condition $\omega_m \approx \Delta\omega_{ax}$ (fundamental mode-locking) and assuming a Doppler-broadened Gaussian gain profile, we obtain a Gaussian mode-locked pulse with a FWHM pulse width given by Eq. (11.18),

$$\tau_p \approx \frac{0.441}{\Delta\nu_D}, \tag{11.18*}$$

where $\Delta\nu_D$ is the inhomogeneous (Doppler) linewidth.

In contrast, a homogeneously broadened laser naturally tends to oscillate in a single mode experiencing the highest net gain near the line center. Although spatial and spectral hole-burning may permit a small group of modes to oscillate simultaneously, the bandwidth remains restricted to a narrow region around the line center frequency ω_o. Mode-locked oscillations in such a laser will thus build up from an initial unlocked multimode oscillation of relatively narrow bandwidth. The spontaneous noise pulses arising from this background [see Fig. 11.2] are initially much longer than the inverse of the transition bandwidth $1/\Delta\nu_{\rm at}$. In this case, the time-gating mechanism described in Fig. 11.7(a) is highly effective at narrowing the temporal pulse profile and forcing the pulse spectrum to broaden.

However, this spectral broadening is strongly opposed by the finite gain bandwidth. As the pulse spectrum expands to fill the gain curve, its spectral wings experience diminished gain. Thus, passing through the gain medium acts as a spectral filter, inducing spectral narrowing of the pulse. The steady-state balance of this competition yields a pulse whose intensity profile is well-approximated by a Gaussian, but whose FWHM width τ_p [Eq. (11.19)] is determined by the inverse of the geometric mean of modulator frequency and atomic gain bandwidth,

$$\tau_p \approx \frac{0.45}{\sqrt{\nu_m \Delta\nu_{\rm at}}} \tag{11.19*}$$

where $\nu_m = \omega_m/2\pi \approx \Delta\nu_{ax}$ is the fundamental modulation frequency.

For the same gain bandwidth $\Delta\nu_{\rm at}$, and since $\nu_m/\Delta\nu_{\rm at} \ll 1$, Eqs. (11.18) and (11.19) give $\tau_{p,\rm hom} \gg \tau_{p,\rm inhom}$. Thus, an inhomogeneously broadened laser can produce considerably shorter mode-locked pulses than a homogeneously broadened laser of the same gain bandwidth. In the inhomogeneous case, short noise bursts with durations of order $1/\Delta\nu_{\rm at}$ are already present before mode-locking, so the modulator acts primarily as a temporal selector that favors the pulse arriving at the minimum-loss time.

11.3 Laser Pulse Propagation

Our discussion of laser light has taken us from the highly stable monochromatic output of single-mode lasers to the quasi-monochromatic output of free-running single- or few-mode lasers, and finally to short and ultrashort pulses containing thousands to millions of phase-coherent axial modes from mode-locked lasers. Our analytic description and understanding of monochromatic and quasi-monochromatic light from a laser relied on treating light as a sinusoidal wave at a dominant carrier frequency, modulated by a slowly varying amplitude envelope. We use the same approach here to develop analytic methods for describing short pulse phenomena in lasers.

Recall that the intracavity field of a laser is affected by the frequency-dependent response of various elements, including saturable gain and absorption, optical losses, and dispersion in the host medium and other intracavity elements. For narrowband fields, it is sufficient to evaluate the responses of these elements at the dominant frequency. However, due to the ultrashort temporal durations and corresponding ultrabroad spectral bandwidths of modern laser pulses, the frequency dependence of these elements can no longer be ignored. In this section we develop an analytical framework to model and understand short-pulse phenomena in lasers by considering the propagation of a light pulse in a dispersive medium.

11.3.1 Carrier Wave and Pulse Envelope Equation

To maintain continuity with our established notation, we begin by writing the space- and time-dependent complex electric field of a linearly polarized, paraxial light pulse traveling in the $+z$-direction as

$$E(\boldsymbol{r},t) = A(z,t)e^{-i(\omega_o t - k_o z)}u(x,y,z) \equiv E(z,t)u(x,y,z)\,, \tag{11.27}$$

where $u(x,y,z)$ accounts for the transverse spatial variation of field amplitude due to diffraction and satisfies the paraxial wave equation.

For wideband fields, the pulse envelope amplitude $A(z,t)$ can be written as a continuous superposition of monochromatic plane waves of positive frequencies [see Eq. (2.20)]

$$A(z,t) = \int_o^\infty \frac{d\omega}{2\pi}\tilde{A}(\omega)e^{-i[(\omega-\omega_o)t-(k(\omega)-k_o)z]}\,. \tag{11.28}$$

Note that the exponent depends on the frequency detuning $\omega - \omega_o$ and the wave-number mismatch $\Delta k = k(\omega) - k(\omega_o)$, because we have explicitly factored out the carrier plane wave component $e^{-i(\omega_o t - k_o z)}$ from the envelope amplitude. As established in Chapter 5, the propagation constant $k(\omega)$ is frequency-dependent because the refractive index $n(\omega)$ varies with frequency

$$k(\omega) = \frac{n(\omega)\omega}{c}\,. \tag{11.29}$$

Let us write the field frequency as $\omega = \omega_o + \Omega$, where ω_o is a reference frequency located near the center of the pulse spectrum. The detuning frequency $\Omega \equiv \omega - \omega_o$ lies in the range $-\omega_o < \Omega < \infty$. While this mirrors the analytical signal representation introduced in Section 2.1, we must now account for the frequency dependence of the propagation constant $k(\omega)$ in making approximations and evaluating spectral integrals. To do this, we expand $k(\omega)$ as a Taylor series in powers of Ω about the center frequency ω_o,

$$k(\omega_o + \Omega) = k_o + \beta_1\Omega + \frac{1}{2}\beta_2\Omega^2 + \frac{1}{6}\beta_3\Omega^3 + \cdots\,, \tag{11.30a}$$

so that $$\Delta k(\Omega) \equiv k(\omega_o + \Omega) - k_o = \beta_1\Omega + \frac{1}{2}\beta_2\Omega^2 + \frac{1}{6}\beta_3\Omega^3 + \cdots\,. \tag{11.30b}$$

Here we have used the standard shorthand notation for the dispersion coefficients

$$\beta_n = \left.\frac{d^n k(\omega)}{d\omega^n}\right|_{\omega=\omega_o} \equiv \frac{d^n k(\omega)}{d\omega_0^n}\,, \tag{11.30c}$$

with $$k_o = k(\omega_o)\,,\; \beta_1 = \frac{dk}{d\omega_o} \equiv \frac{1}{v_g},\; \beta_2 = \frac{d^2 k}{d\omega_0^2} = \frac{d}{d\omega_o}\left(\frac{1}{v_g}\right). \tag{11.30d}$$

The physical meaning of various terms introduced in Eq. (11.30d) will unfold as we proceed. For now, note that k_o is the carrier wave propagation constant, while the first-order derivative $\beta_1 = \frac{dk}{d\omega_o}$ possesses the dimensions of inverse speed, s/m, and is defined as the reciprocal of the group velocity v_g. The second-order derivative β_2 represents group velocity dispersion (GVD), which measures how rapidly the inverse group velocity varies with frequency. Higher-order terms ($\beta_3, \ldots$) represent higher-order dispersion.

Using these relations, the electric field amplitude can be expressed in terms of detuning Ω as

$$\begin{aligned} A(z,t) &= \int_{-\omega_o}^{\infty} \frac{d\Omega}{2\pi} \tilde{A}(\omega_o + \Omega) e^{-i[\Omega t - \Delta k(\Omega) z]}, \\ &\approx \int_{-\infty}^{\infty} \frac{d\Omega}{2\pi} \tilde{A}(\omega_o + \Omega) e^{i\Delta k(\Omega) z} e^{-i\Omega t}, \end{aligned} \tag{11.31}$$

where we have extended the lower integration limit from $-\omega_o$ to $-\infty$. This approximation remains highly accurate even for ultrashort laser pulses, because the pulse spectrum vanishes well before Ω reaches the lower limit $-\omega_o$. For example, even a 6 fs pulse centered at 750 nm, has a bandwidth covering only a small fraction of the carrier frequency, so its spectrum remains far from the lower bound $\Omega = -\omega_o$.

Under this approximation, the spectral amplitude at the input boundary ($z = 0$) is the direct Fourier transform of the initial envelope profile $A(0, t)$

$$\tilde{A}(0, \Omega) = \int_{-\infty}^{\infty} dt A(0,t) e^{i\Omega t} \equiv \tilde{A}(\omega_o + \Omega)\,, \tag{11.32a}$$

The spectral amplitude at any propagation distance z evolves as

$$\tilde{A}(z, \Omega) = \tilde{A}(0, \Omega) e^{i\Delta k(\Omega) z}\,. \tag{11.32b}$$

The spatial evolution of the pulse envelope can then be evaluated by taking the inverse Fourier transform

$$A(z,t) = \int_{-\infty}^{\infty} \frac{d\Omega}{2\pi} \tilde{A}(0,\Omega) e^{i\Delta k(\Omega) z} e^{-i\Omega t} \equiv \int_{-\infty}^{\infty} \frac{d\Omega}{2\pi} \tilde{A}(z,\Omega) e^{-i\Omega t}\,. \tag{11.33}$$

Two critical insights emerge from this formulation:

1. **Spectral Phase versus Intensity:** Equation (11.32b) dictates that the input and output spectral amplitudes differ exclusively by a phase factor determined by the medium dispersion and propagation distance. Crucially, while dispersion shifts the spectral phase, the spectral intensity (the power spectrum) of the pulse remains strictly invariant during propagation through a linear, lossless dispersive medium.

2. **Mathematical Simplification:** Equation (11.33) shows that the pulse envelope can be modeled as the Fourier superposition of low-frequency (baseband) plane waves. This approximation filters out fast optical oscillations, removing the rapid optical oscillations from the envelope description, allowing us to simplify the resulting equations and obtain analytic solutions that offer useful physical insight into ultrafast pulse dynamics.

To derive the governing (partial differential) equation for the pulse envelope $A(z,t)$, we take the spatial derivative of its Fourier transform $\tilde{A}(z,\Omega)$ in Eq. (11.32b), yielding

$$\frac{\partial \tilde{A}(z,\Omega)}{\partial z} = i\Delta k(\Omega)\tilde{A}(z,\Omega) = i\left[\frac{\Omega}{v_g} + \frac{1}{2}\beta_2\Omega^2 + \frac{1}{6}\beta_3\Omega^3 + \cdots\right]\tilde{A}(z,\Omega)\,. \tag{11.34}$$

We transform this back into the time domain by invoking the operational properties of the Fourier transform–specifically, mapping $\tilde{A}(z,\Omega) \to A(z,t)$ and replacing the frequency product $\Omega\tilde{A}(z,\Omega)$ with the differential operator $i\frac{\partial A(z,t)}{\partial t}$. This leads to the generalized pulse envelope propagation equation

$$\frac{\partial A(z,t)}{\partial z} = -\frac{1}{v_g}\frac{\partial A(z,t)}{\partial t} - i\frac{\beta_2}{2}\frac{\partial^2 A(z,t)}{\partial t^2} + \frac{\beta_3}{6}\frac{\partial^3 A(z,t)}{\partial t^3} + \cdots\,. \tag{11.35}$$

Equations (11.34) and (11.35) offer complementary, mathematically equivalent descriptions of pulse propagation. The former describes the evolution in the frequency domain, while the latter describes it in the time domain. The Fourier transform connects the two descriptions. We utilize whichever domain simplifies the problem at hand; a frequency-domain approach is ideal for managing spectral phases, whereas the time-domain framework provides a highly intuitive physical picture of pulse reshaping.

In summary, the temporal pulse profile at any distance can be obtained either by directly solving Eq. (11.35) or by solving the simpler frequency-domain relation in Eq. (11.34) and taking the inverse Fourier transform via Eq. (11.33). While general closed-form solutions do not exist when complex boundary conditions are present, these equations supply the mathematical framework for all ultrafast laser physics approximations.

11.3.2 Effect of Linear Dispersion

In a medium where only linear dispersion is important, β_2 and higher-order dispersion terms in Eq. (11.34) can be neglected. The equation governing the envelope $\tilde{A}(z,\Omega)$ under pure first-order dispersion simplifies to

$$\frac{\partial \tilde{A}(z,\Omega)}{\partial z} = i\beta_1\Omega\tilde{A}(z,\Omega)\,. \tag{11.36}$$

By substituting $\beta_1 = 1/v_g$, this equation transforms into the time domain as the following partial differential equation for the pulse envelope

$$\left[\frac{\partial}{\partial z} + \frac{1}{v_g}\frac{\partial}{\partial t}\right]A(z,t) = 0\,, \tag{11.37}$$

which also follows directly from Eq. (11.35). This is a first-order wave equation with solutions in which z and t occur exclusively in the traveling-wave combination $\tau = t - z/v_g$ [see Sec. 2.3]. Consequently, the carrier envelope and the total electric field can be written as

$$A(z,t) = F(\tau)\,, \tag{11.38a}$$

$$E(z,t) = F(\tau)e^{-i(\omega_o t - k_o z)} \quad \text{where } \tau = t - \beta_1 z = t - \frac{z}{v_g}\,. \tag{11.38b}$$

This electric field $E(z,t)$ represents a carrier plane wave with phase $\phi(z,t) = k_o z - \omega_o t$ modulated by the envelope $F(\tau)$. As discussed in Sec. 2.3, the phasefronts (surfaces of constant carrier phase) propagate at the phase velocity v_p and the carrier envelope, being a function of only the variable $\tau = t - z/v_g$, moves without change of shape at the group velocity v_g:

$$\text{phase velocity} \qquad v_p \equiv \left.\frac{dz}{dt}\right|_{\phi=\text{const}} = \frac{\omega_o}{k_o} = \frac{c}{n_o}, \tag{11.39}$$

$$\text{group velocity} \qquad v_g \equiv \left.\frac{dz}{dt}\right|_{\tau=\text{const}} = \left[\frac{dk}{d\omega_o}\right]^{-1} = \left[\frac{n_o}{c} + \frac{\omega_o}{c}\frac{dn}{d\omega_o}\right]^{-1},$$
$$= \frac{v_p}{1 + \frac{\omega_o}{n}\frac{dn}{d\omega_o}}. \tag{11.40}$$

Group velocity describes the speed of the carrier envelope formed by the superposition of the entire group of waves in the spectrum. Phase velocity describes the speed of carrier phasefronts. The energy density–proportional to $|A(z,t)|^2 = |F(\tau)|^2$–also propagates entirely at the group velocity v_g. The parameter $T_g(z) = z/v_g$ represents the group delay. This means an envelope function injected at $z = 0$ reaches $z = L$ perfectly intact after a time delay of L/v_g. First-order dispersion does not change the pulse shape, width, spectrum, or amplitude. Because no particular shape was assumed for $F(\tau)$, this holds true for all traveling wave packets.

The Gaussian Pulse

An important example of an envelope function is a Gaussian. Accordingly, consider a pulse with a Gaussian envelope at the input face ($z = 0$) of a linear dispersive medium

$$E(0,t) = E_o e^{-t^2/\tau_o^2} e^{-i\omega_o t}, \tag{11.41}$$

where E_o is a constant and τ_o characterizes the pulse duration. The intensity associated with this pulse is

$$I(t) = I_0 e^{-2t^2/\tau_o^2}. \tag{11.42}$$

The full width at half maximum (FWHM) duration $\Delta\tau$ of the pulse intensity is related to τ_o by

$$\Delta\tau = \sqrt{2\ln 2}\,\tau_o = 1.177\tau_o. \tag{11.43}$$

Taking the Fourier transform of $E(0,t)$ in Eq. (11.41) yields the spectral amplitude $\tilde{A}(0,\Omega)$ and spectral density $S(0,\Omega)$

$$\tilde{A}(0,\Omega) = \int_{-\infty}^{\infty} E_o e^{-t^2/\tau_o^2} e^{i\Omega t} dt = E_o\sqrt{\pi\tau_o^2}\, e^{-\Omega^2\tau_o^2/4}, \tag{11.44}$$

$$S(0,\Omega) \equiv |\tilde{A}(0,\Omega)|^2 = S_o e^{-\Omega^2\tau_o^2/2}, \tag{11.45}$$

where $S_o = \pi^2 E_o^2 \tau_o^2$. Pulse spectral density $S(0,\Omega)$ is also a Gaussian with an FWHM of $\Delta\Omega = 2\sqrt{2\ln 2}/\tau_o$. The time-bandwidth product for the pulse has the physical minimum limit $\Delta\nu\Delta\tau = \Delta\Omega\Delta\tau/2\pi = 2\ln 2/\pi = 0.441$ for a transform-limited pulse.

After traveling a distance z, the electric field of the pulse, according to Eq. (11.38), is given by

$$E(z,t) = E_o e^{-(t-z/v_g)^2/\tau_o^2} e^{-i\omega_o(t-z/v_p)}. \tag{11.46}$$

A plot of the real physical electric field of this pulse is given by

$$\mathcal{E}(z,t) = \mathcal{R}eE(z,t) = E_o e^{-(t-z/v_g)^2/\tau_o^2}\cos\omega_o(t - z/v_p). \tag{11.47}$$

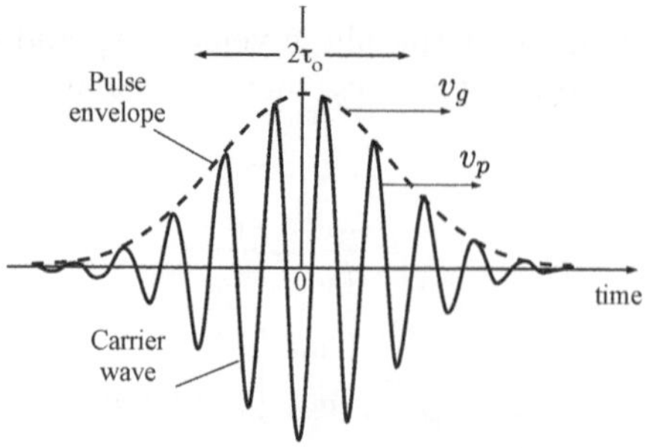

FIGURE 11.11
Carrier oscillations and envelope for a Gaussian pulse.

As shown in Fig. 11.11, the Gaussian pulse envelope acts as a modulation (dashed curve) over sinusoidal carrier wave oscillations. Because phase and group velocities differ in dispersive media, the absolute maxima of the carrier oscillations and the envelope do not stay aligned over distance.

The pulse intensity at a point z,

$$I(z,t) = I_o e^{-2(t-z/v_g)^2/\tau_o^2}, \tag{11.48}$$

remains a Gaussian with exactly the same peak intensity and width as the initial pulse at $z = 0$. The spectral amplitude remains $\tilde{A}(z,\Omega) = e^{i\Omega z/v_g}\tilde{A}(0,\Omega)$, proving the pulse spectrum is unchanged. Thus, linear dispersion simply replaces the free-space envelope delay time z/c with the group delay $\beta_1 z = z/v_g$.

11.3.3 Effect of Quadratic Dispersion

For short optical pulses with broad spectra, second- and higher-order dispersion terms must also be considered. If the spectral bandwidth is sufficiently narrow that only the first- and second-order dispersion terms contribute significantly, the third- and higher-order terms $(\beta_3, \beta_4, \ldots)$ in Eq. (11.34) may be neglected. The resulting equation for $\tilde{A}(z,\Omega)$ can be transformed into the time-domain equation using the Fourier-transform correspondences $\tilde{A}(z,\Omega) \longleftrightarrow A(z,t)$ and $\Omega\tilde{A}(z,\Omega) \longleftrightarrow -i\partial A(z,t)/\partial t$. This yields the following time-domain equation describing the pulse-envelope evolution in the presence of first- and second-order dispersion

$$\left[\frac{\partial}{\partial z} + \frac{1}{v_g}\frac{\partial}{\partial t} + \frac{i\beta_2}{2}\frac{\partial^2}{\partial t^2}\right] A(z,t) = 0\,. \tag{11.49}$$

This equation can also be obtained directly from Eq. (11.35) by retaining only the first- and second-order dispersion terms. With the change of variables

$$\begin{aligned}
&\zeta = z, \quad \tau = t - \frac{z}{v_g}, \\
&\frac{\partial}{\partial z} = \frac{\partial \zeta}{\partial z}\frac{\partial}{\partial \zeta} + \frac{\partial \tau}{\partial z}\frac{\partial}{\partial \tau} = \frac{\partial}{\partial \zeta} - \frac{1}{v_g}\frac{\partial}{\partial \tau}, \\
&\frac{\partial}{\partial t} = \frac{\partial \zeta}{\partial t}\frac{\partial}{\partial \zeta} + \frac{\partial \tau}{\partial t}\frac{\partial}{\partial \tau} = \frac{\partial}{\partial \tau},
\end{aligned} \tag{11.50}$$

the pulse-envelope evolution equation becomes

$$\frac{\partial A}{\partial z} + \frac{i\beta_2}{2}\frac{\partial^2 A}{\partial \tau^2} = 0\,, \tag{11.51}$$

where, for simplicity, we have replaced ζ by z. Equation (11.51) has many solutions, but a particularly important and useful one is relatively easy to find. We note that this equation is mathematically identical to the paraxial wave equation [Eq. (3.10)] in one transverse dimension

$$\frac{\partial \psi_o}{\partial z} - \frac{i}{2k}\frac{\partial^2 \psi_o}{\partial x^2} = 0\,, \tag{11.52}$$

under the correspondence $z \to z$, $k \to -\frac{1}{\beta_2}$, and $x \to \tau$. The only important distinction is that β_2 may be either positive or negative whereas the wave number k is always positive. Nevertheless, this formal correspondence allows many results from Gaussian beam diffraction to be applied directly to pulse propagation in quadratic dispersive media. Using the Gaussian-beam solution of the paraxial wave equation [Eq. (3.13)] together with the above correspondence, the pulse envelope can be written as

$$\psi_o(z,x) = A_o\sqrt{\frac{w_0}{w(z)}}\exp\left[-\frac{x^2}{w^2(z)} + i\frac{2zx^2}{kw_0^2w^2(z)} - i\frac{1}{2}\tan^{-1}\frac{z}{z_R}\right]\,, \tag{11.53}$$

where

$$w(z) = w_0\sqrt{1+\frac{z^2}{z_R^2}}\,,\quad R(z) = z + \frac{z_R^2}{z} = \frac{1}{4}\frac{k^2w_0^2w^2(z)}{z}\,,\quad z_R = \frac{1}{2}kw_0^2\,, \tag{11.54}$$

have their usual meanings defined in Chapter 3. Using the above correspondence between parameters identified earlier, the envelope can be written as

$$A(z,\tau) = A_o\sqrt{\frac{\tau_o}{\tau_p(z)}}\exp\left[-\frac{\tau^2}{\tau_p^2(z)} - i\frac{z\tau^2}{\Lambda\tau_p^2(z)} + i\frac{1}{2}\tan^{-1}\frac{z}{\Lambda}\right]\,, \tag{11.55}$$

where $$\tau_p(z) = \tau_o\sqrt{1+\frac{z^2}{\Lambda^2}}\,,\quad \Lambda = \frac{\tau_o^2}{2\beta_2}\,. \tag{11.56}$$

Here $\tau_p(z)$ characterizes the pulse width and plays the same role as the spot radius of a Gaussian beam. The parameter $|\Lambda|$ is the characteristic dispersion length, analogous to the Rayleigh range of a Gaussian beam. For propagation distances $z \ll |\Lambda|$, dispersive effects are weak and the pulse width remains nearly unchanged. When $z \gtrsim |\Lambda|$, dispersion becomes significant and the pulse broadens appreciably. Note that the input pulse envelope $A(0,\tau)$ of this wave is a transform-limited Gaussian pulse

$$A(0,\tau) = A_o\exp\left[-\frac{\tau^2}{\tau_o^2}\right]\,, \tag{11.57}$$

Equation (11.55) describes the effect of quadratic dispersion on an initially transform-limited Gaussian pulse. Although the temporal profile changes during propagation, the spectral intensity $|\tilde{A}(z,\Omega)|^2$ remains unchanged because linear, lossless dispersion modifies only the spectral phase. Additional nonlinearity or loss, however, can alter the spectral intensity. The temporal broadening and frequency chirp introduced by second-order dispersion play a central role in ultrafast laser systems and are examined in the following sections.

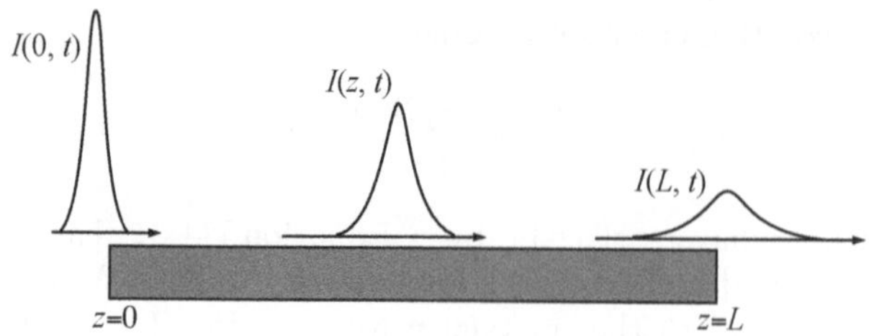

FIGURE 11.12
Pulse spreading in a medium with quadratic dispersion.

11.3.4 Dispersive Pulse Broadening

The optical intensity proportional to $|A(z,\tau)|^2$ associated with the field given by Eq. (11.55) is

$$I(z,\tau) = I_o \frac{\tau_o}{\tau_p(z)} \exp\left[-2\frac{\tau^2}{\tau_p^2(z)}\right]. \tag{11.58}$$

This profile is a Gaussian with a full width at half maximum (FWHM) duration $\Delta\tau = \sqrt{2\ln 2}\,\tau_p(z)$. The pulse width increases monotonically with the propagation distance z into the medium, reaching its minimum value at $z = 0$ (which corresponds to the transform-limited pulse under consideration):

$$\begin{aligned}\Delta\tau &= \sqrt{2\ln 2}\,\tau_o\sqrt{1+z^2/\Lambda^2} \\ &\xrightarrow{z\gg|\Lambda|} \sqrt{2\ln 2}\,\tau_o\frac{z}{|\Lambda|} = 2\sqrt{2\ln 2}\frac{\beta_2|z}{\tau_o} = |\beta_2|z\Delta\Omega\,,\end{aligned} \tag{11.59}$$

where we used the fact that the FWHM spectral width is defined as $\Delta\Omega \approx 2\sqrt{2\ln 2}/\tau_o$. Thus, in the presence of second-order dispersion, the pulse width increases continuously with propagation distance [Fig. 11.12], a phenomenon that occurs irrespective of the mathematical sign of β_2. However, the temporal shape of the pulse remains Gaussian for all values of z. Note that because $\Omega = \omega - \omega_o$ it follows that $\Delta\Omega = \Delta\omega$. Similarly, for a given position z, the local time $\tau = t - z/v_g$ implies that the temporal width measurements are equivalent ($\Delta\tau = \Delta t$).

11.3.5 Frequency Chirp

Another consequence of second-order dispersion is that the wave develops a time-dependent instantaneous frequency. This can be deduced from Eqs. (11.31) and (11.55) by noting that the phase of the field amplitude $A(z,t) \sim e^{i\phi(z,t)}$ is time dependent

$$\phi(z,t) = -\omega_o t - \frac{2\beta_2 z(t-z/v_g)^2}{\tau_o^2\tau_p^2(z)} + k_o z + \frac{1}{2}\tan^{-1}\frac{z}{\Lambda}. \tag{11.60}$$

Differentiating this phase function yields the instantaneous angular frequency, $\omega(t)$

$$\omega(t) \equiv -\frac{\partial\phi}{\partial t} = \omega_o + \frac{4\beta_2 z(t-z/v_g)}{\tau_o^2\tau_p^2(z)} \equiv \omega_o + \delta\omega(t)\,. \tag{11.61}$$

In this case, the instantaneous frequency varies linearly with time. If $\beta_2 > 0$, the field frequency increases continuously across the pulse profile. By analogy with a time-dependent

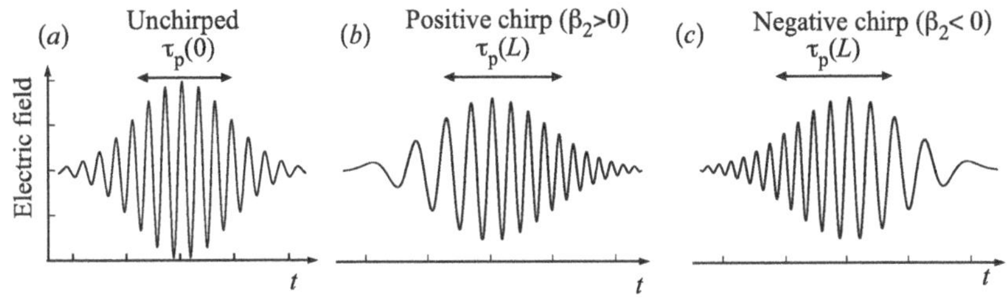

FIGURE 11.13
Frequency chirp in a medium with quadratic dispersion.

acoustic pitch, this phenomenon is referred to as chirp. The case $\beta_2 > 0$ is termed positive chirp ($\delta\omega(t) > 0$ for the trailing edge where $t > z/v_g$) whereas $\beta_2 < 0$ yields a negative chirp ($\delta\omega(t) < 0$ for the trailing edge). Note that although the pulse width changes, there is no change in the pulse spectrum as we already noted following Eq. (11.32b).

11.3.6 Origin of Chirp and Pulse Broadening

Frequency chirp and pulse broadening are directly related to the variation of group velocity with frequency, a phenomenon known as group velocity dispersion (GVD). The parameter β_2 [Eq. (11.30d)], which characterizes this frequency-dependent variation of group velocity, can be expressed as

$$\beta_2 = \left[\frac{d}{d\omega}\left(\frac{1}{v_g}\right)\right]_{\omega_o} = -\frac{1}{v_g^2(\omega_o)}\frac{dv_g(\omega_o)}{d\omega_o}, \tag{11.62}$$

Quadratic variation of the wave number with frequency implies that the variation of group velocity over the pulse spectrum cannot be ignored. The group velocity variation away from the carrier frequency ω_o can be linearly approximated as

$$v_g(\omega) \approx v_g(\omega_o) + \frac{dv_g(\omega_o)}{d\omega_o}\Omega = v_g(\omega_o)\left[1 - \beta_2 v_g(\omega_o)\Omega\right], \tag{11.63}$$

where we have used Eq. (11.62). Consequently, a localized group of frequencies near ω will be delayed by

$$\begin{aligned} t_d(\omega) &= \frac{z}{v_g(\omega)} = \frac{z}{v_g(\omega_o)[1 - \beta_2 v_g(\omega_o)\Omega]} \\ &\approx \frac{z}{v_g(\omega_o)}\left[1 + \beta_2 v_g(\omega_o)\Omega\right] = \frac{z}{v_g(\omega_o)} + z\beta_2\Omega. \end{aligned} \tag{11.64}$$

Thus, over a pulse spectral width of $\Delta\omega(= \Delta\Omega)$, that is, for frequencies ranging from $\omega_o - \Delta\Omega/2$ to $\omega_o + \Delta\Omega/2$, the overall group delay spread is given by

$$\Delta t_d = |\beta_2| z \Delta\Omega, \tag{11.65}$$

which reproduces the result in Eq. (11.59). Note that neither the GVD coefficient (β_2) nor the propagation distance z alone defines how a pulse spreads over a given distance. Instead, it is the group delay dispersion (GDD), defined as $D = |\beta_2|z$, that determines the total temporal expansion of the pulse.

Frequency chirp is easy to understand if we recall that for $\beta_2 > 0$ [where $dv_g/d\omega < 0$ from Eq. (11.30d)], the higher-frequency Fourier components travel at a lower velocity than the lower-frequency components. Hence at a given position z, the higher frequency components arrive later than the lower frequency components, causing the instantaneous frequency across the pulse profile to increase with time.

Pulse spreading considered in the preceding paragraphs assumes positive group velocity dispersion ($\beta_2 > 0$), which is typical for most materials in the visible and near-infrared part of the optical spectrum. Similar conclusions apply to the negative group velocity dispersion regime [$\beta_2 < 0$], where the lower-frequency Fourier components travel slower than the higher-frequency components. The end result is the same – a Gaussian pulse with an initial $1/e^2$ intensity half-width τ_o will experience temporal broadening. Managing group velocity dispersion is an important problem in ultrashort pulse generation and propagation. In addition to bulk material media, optical elements such as prisms and gratings can be arranged to introduce positive or negative group velocity dispersion.

All materials exhibit group velocity dispersion to some extent, and pulse spreading becomes especially significant for short pulses. For example, titanium-doped sapphire (Ti:sapphire) lasers are widely used for ultrashort pulse generation. For short pulse propagation in sapphire, which features a dispersion coefficient of $\beta_2 = 58\ \mathrm{ps}^2/\mathrm{km} = 58 \times 10^{-6}\ \mathrm{ps}^2/\mathrm{mm}$ at 800 nm, the pulse duration scales as

$$\tau_p(z) = \tau_o\sqrt{1 + 4\beta_2^2 z^2/\tau_o^4} = \tau_o\sqrt{1 + 1.35 \times 10^{-8} z^2/\tau_o^4}\,, \tag{11.66}$$

where $\tau_p(z)$ and τ_o are expressed in picoseconds and z is in millimeters. A pulse with carrier wavelength of 800 nm and initial width $\tau_o = 1$ ps experiences a negligible change in its width when passing through a 20 mm sapphire rod. In contrast, a $\tau_o = 10$ fs pulse traversing the same rod will be stretched to 232 fs.

Group velocity dispersion also becomes important in long-distance pulse propagation in fiber-optic telecommunications, where temporal pulse spreading can cause successive signal pulses to overlap. This overlap blurs individual pulse identity, degrading the information integrity and ultimately limiting the maximum information transmission rate. For the common optical telecommunications fiber with $\beta_2 \approx 20\ \mathrm{ps}^2/\mathrm{km}$ at $\lambda = 1.55\ \mu\mathrm{m}$, an initial $\tau_o = 10$ ps pulse will stretch to approximately 61 ps after traveling 15 km. Assuming that the information is transmitted as a series of ones and zeros, the maximum transmission rate before significant pulse overlap occurs is $1/(61\ \mathrm{ps}) \approx 16$ Gb/s.

11.3.7 Controlling Pulse Spread

The effect of second-order dispersion on an initially transform-limited Gaussian pulse is to broaden the pulse in time. This exact mechanism is used to compress a chirped pulse. To see this, consider a chirped pulse incident on a medium with quadratic dispersion. As established in the preceding section, we know that the field of Eq. (11.55) at a propagation distance $z = L$ represents a linearly chirped Gaussian pulse. Its temporal envelope can be written as

$$A(t) = A_0 \exp\left[-(a + ib)t^2\right]\,, \tag{11.67}$$

where all constant phase factors and amplitude prefactors have been absorbed into the front coefficient A_o and the parameters a and b are given by

$$a = \frac{1}{\tau_p^2(L)} \quad \text{and} \quad b = \frac{2\beta_2 L}{\tau_o^2\tau_p^2(L)}\,. \tag{11.68}$$

The linear time delay has been removed by choosing the time origin at the group-delay center $t_o = L/v_g$. This is clearly a Gaussian pulse since its intensity scales as $|A(t)|^2 \propto \exp[-2at^2]$.

By taking the Fourier transform of Eq. (11.67), the corresponding spectral amplitude $\tilde{A}(\Omega)$ and the corresponding spectral intensity, apart from an overall normalization factor, are

$$\tilde{A}(\Omega) = A_0 \int_{-\infty}^{\infty} dt e^{-(a+ib)t^2+i\Omega t} = A_0 \sqrt{\frac{\pi}{a+ib}} \exp\left[-\frac{\Omega^2(a-ib)}{4(a^2+b^2)}\right], \tag{11.69}$$

$$S(\Omega) = \frac{\pi}{\sqrt{a^2+b^2}} \exp\left[-\frac{a\Omega^2}{2(a^2+b^2)}\right]. \tag{11.70}$$

Now, we let the chirped pulse of Eq. (11.67) serve as the input pulse $A_2(0,t)$ for a second medium of length L_2 with the quadratic wavenumber mismatch $\Delta k(\Omega) = \beta_1'\Omega + \frac{1}{2}\beta_2'\Omega^2$. The spectral amplitude of the pulse after traversing the second medium is given by [using Eq. (11.32b)]

$$\tilde{A}_2(L_2,\Omega) = \tilde{A}_2(0,\Omega) e^{i\beta_1'\Omega L_2 + i\beta_2'\Omega^2 L_2/2}. \tag{11.71}$$

Dropping the amplitude prefactors, the temporal envelope of the pulse emerging from the second medium is given by the inverse Fourier transform of $\tilde{A}_2(L_2,\Omega)$

$$A(L_2,t) = \int_{-\infty}^{\infty} \frac{d\Omega}{2\pi} \exp\left[-\frac{\Omega^2}{4}\left(\frac{a-ib}{a^2+b^2} - 2i\beta_2' L_2\right) - i\Omega(t-\beta_1' L_2)\right]. \tag{11.72}$$

The integral can be evaluated by using the identity

$$\int_{-\infty}^{\infty} e^{-Ax^2-2Bx} dx = \sqrt{\frac{\pi}{A}} e^{B^2/A}, \quad \mathcal{R}e[A] > 0. \tag{11.73}$$

Even without evaluating the integral, a direct comparison of the Ω^2 coefficient in Eq. (11.72) to that in Eq. (11.69) reveals how the quadratic spectral phase is modified. The imaginary part of the Ω^2 coefficient (quadratic spectral phase) in the exponent has changed from $\frac{b}{4(a^2+b^2)}$ to $\left[\frac{b}{4(a^2+b^2)} - \frac{1}{2}\beta_2' L_2\right]$. Because this quadratic spectral phase term is directly responsible for the temporal chirp, this result shows that we can control both the chirp parameter and the pulse width. This is accomplished by passing the pulse through an optical system with an appropriately chosen group-delay dispersion (GDD) $\beta_2' L_2$–in both sign and magnitude. To eliminate the linear chirp completely, one must design an optical device whose second-order spectral phase exactly cancels the quadratic spectral phase of the input pulse. The effects of higher-order spectral phases can be similarly controlled and managed.

As an example, consider the standard single-mode silica fiber used in optical telecommunications operating at 1.5 μm. This wavelength regime falls within the anomalous dispersion region, where $\beta_2 \approx -20$ ps^2/km and the third-order dispersion coefficient is $\beta_3 = 0.1$ ps^3/km. The spectral amplitude of a pulse down the fiber in this case evolves as

$$\tilde{A}(z,\Omega) = \tilde{A}(0,\Omega) \exp\left[\frac{i\beta_2 z}{2}\Omega^2 + \frac{i\beta_3 z}{3!}\Omega^3 + \cdots\right] \tag{11.74}$$

As shown earlier, the quadratic dispersion term induces a diffraction-like expansion that spreads the pulse in the time domain. The cubic (β_3) and higher-order dispersion terms introduce complex deviations from this behavior. These effects in optical fiber communication systems have been largely overcome by engineering the refractive index profile of the fiber core to shift the zero-dispersion wavelength to the 1.55 μm region, where attenuation is lowest ($\sim$ 0.22 dB/km). For such dispersion-shifted fibers (DSF) operating near their zero-dispersion point, β_2 is suppressed close to zero.

In addition to dispersion-induced changes, ultrashort pulses are strongly affected by optical nonlinearities because short pulses usually have high peak intensities inducing nonlinear contributions to pulse spectral phase. For example, the electronic Kerr nonlinearity, discussed earlier, gives rise to self-phase modulation (SPM) that induces an intensity-dependent frequency chirp. Under appropriate dispersion conditions, the effects of GVD and self-phase modulation can oppose one another, and under optimized conditions, the temporal broadening caused by GVD can be perfectly balanced by the nonlinear compression induced by SPM. In an anomalous-dispersion medium, the temporal effect of GVD can balance the nonlinear phase modulation produced by the Kerr effect, allowing a stable pulse to propagate without changing shape. Some of these effects will be considered in the next section as well as in Chapter 12.

11.3.8 Dispersion Control

It is clear from the preceding discussion that cavity dispersion will have a deleterious effect on short pulse generation in lasers. To mitigate this effect, dispersion-compensating optical elements are incorporated into the cavity to reduce the net cavity dispersion to an acceptable level.[3] The four-prism pulse compressor shown in Fig. 11.14(a) introduces negative group-delay dispersion (GDD), allowing it to compensate for the typically positive GDD due to propagation through materials within the cavity. It can also physically separate and then recombine the spectral components of a pulse, making it routinely useful for stretching and subsequently compressing ultrashort pulses by factors of 1,000 or more.

The negative GDD of the four-prism configuration can be understood by recalling that a material prism exhibits positive GDD–meaning that the blue components experience a larger group delay than the red components inside the glass. The prism naturally refracts shorter wavelengths (B for high frequencies) more sharply toward its base than longer wavelengths (R for low frequencies). By tracing the paths of the long-wavelength components (dashed lines) and short-wavelength components (solid lines) through the four-prism sequence, we observe that the longer wavelengths traverse geometrically longer paths through the prisms. Consequently, the red components are delayed relative to the blue components in passing through the assembly. The compressor as a whole thus contributes negative GDD to the transmitted pulse.

The net GDD of the four-prism setup can be continuously tuned by adjusting the geometric spacing between the prisms and translating the prisms into or out of the beam paths, thereby shifting the relative balance of path lengths in air versus glass.

The two-prism setup shown in Fig. 11.14(b), effectively one half of the four-prism configuration, operates in a similar fashion and is ideally suited for integration within linear laser cavities.

Another class of dispersion-compensating devices uses pairs of blazed gratings, as illustrated in Figs. 11.15(a) and (b). These configurations also provide negative GDD. Other variations based on four-grating sequences or hybrid combinations of devices in Figs. 11.14 and 11.15(a) and (b) can be found in the literature. Indeed, both a prism compressor and a grating compressor in tandem are sometimes used; because their respective third-order dispersion coefficients possess opposite signs, this hybrid arrangement can be used to achieve independent control of the second- and third-order spectral phases.

A chirped Bragg mirror or multilayer dielectric mirror [Fig. 11.15(c)] is another device used for dispersion compensation. Such a mirror consists of a large number of alternating thin-film layers of low- and high-refractive-index materials. The optical thicknesses of the layers are varied systematically with depth into the coating. The resulting amplitude reflection coefficient is engineered to be of the form $r(\omega) = |r(\omega)|e^{i\phi(\omega)}$ such that $|r(\omega)| = 1$ across the pulse bandwidth, while $\phi(\omega)$ is tailored to cancel out the specific spectral phase of the

[3] E. B. Treacy, "Optical pulse compression with diffraction gratings," IEEE J. Quantum Electron. **QE-5**, 454–458 (1969); R. L. Fork, O. E. Martinez, and J. P. Gordon, "Negative dispersion using pairs of prisms," Opt. Lett. **9**, 150–152 (1984); R. L. Fork, C. H. Brito Cruz, P. C. Becker, and C. V. Shank, "Compression of optical pulses to six femtoseconds by using cubic phase compensation," Opt. Lett. **12**, 483–485 (1987).

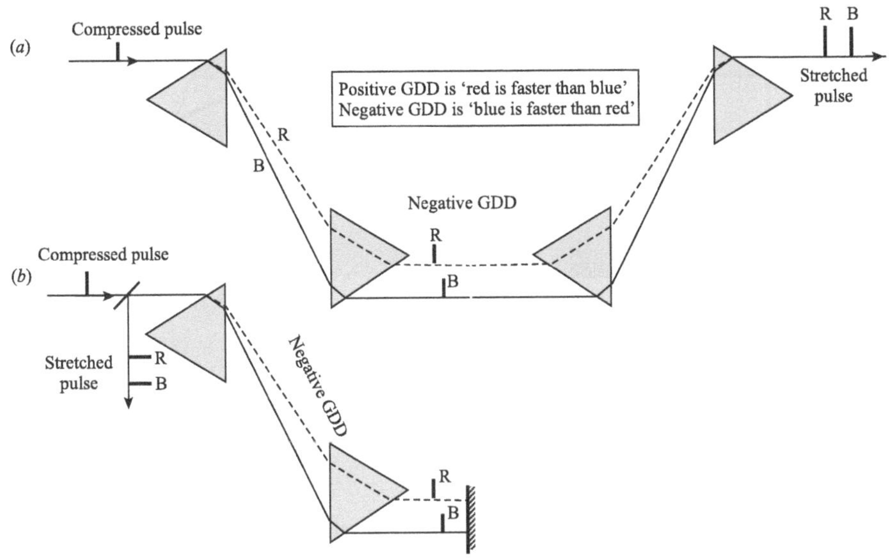

FIGURE 11.14
(a) Four-prism dispersion-control arrangement providing negative GDD; (b) two-prism arrangement providing negative GDD.

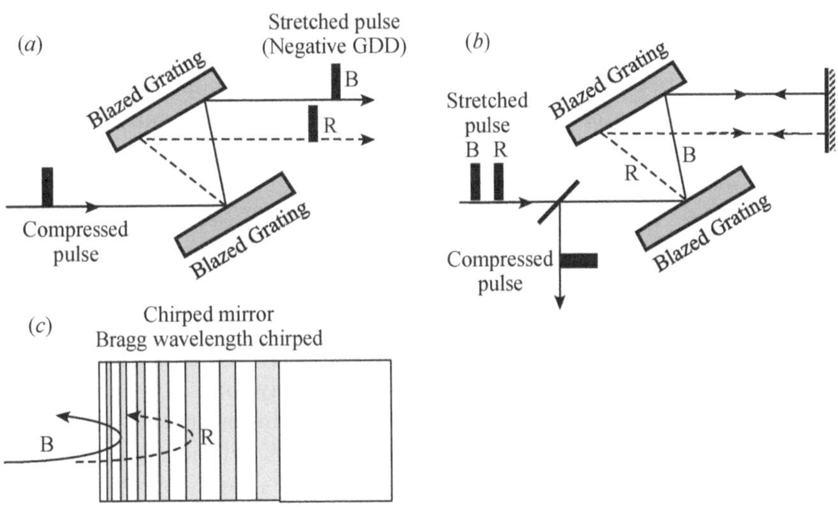

FIGURE 11.15
(a) and (b): Negative dispersion devices based on pairs of blazed gratings. Figure (a) illustrates stretching of a compressed (dispersionless) incident pulse and (b) illustrates compensation of positive dispersion of an incident pulse. (c) Chirped mirrors with an engineered dispersion of the reflection coefficient.

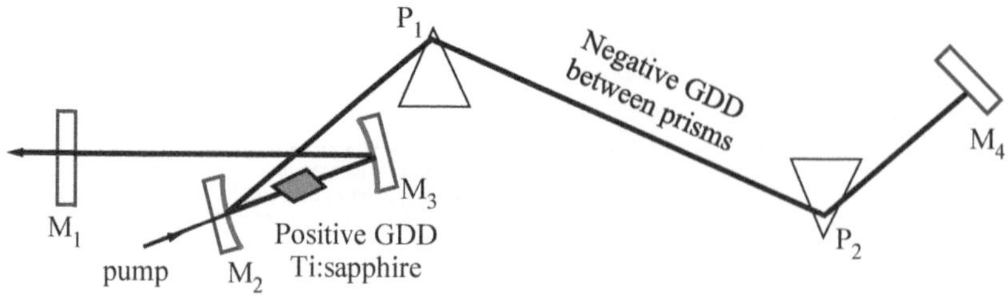

FIGURE 11.16
Compensating positive group velocity dispersion using a prism pair in a Ti:sapphire laser.

incident pulse. This mechanism imposes a wavelength-dependent delay on the individual Fourier components, providing custom compensation for arbitrary material dispersion.

Ultrashort-Pulse Ti:Sapphire Laser

As an example of dispersion control, consider ultrashort pulse generation within a Ti:sapphire laser. Figure 11.16 shows a typical configuration of such a laser in which a 10-mm-long Ti:sapphire rod in a Z-shaped standing-wave laser cavity is pumped by a focused Ar-ion or a doubled YAG laser beam. The optical Kerr effect induced in the sapphire rod, combined with an appropriately sized intracavity aperture, establishes Kerr-lens mode-locking (KML) as described in Sec. 11.2.3. At a central wavelength of $\lambda = 800$ nm, the material dispersion of the crystal adds a positive GDD of $\beta_2 L = 580$ fs^2 to the pulse circulating in the cavity during each pass. This is compensated by incorporating a two-prism sequence (the first half of the four-prism sequence of Fig. 11.14) into the cavity. Under optimized operating conditions where the two-prism sequence exactly cancels the positive GDD of the sapphire rod, pulses as short as 30 fs can be obtained. By reducing the sapphire rod's length (which reduces the positive GDD of the active medium at the expense of a lower single-pass gain), the residual positive GDD of the active medium can be compensated by the negative GDD of specialized chirped mirrors. This reduction simultaneously suppresses the third-order dispersion within the cavity, allowing the generation of ultrashort optical pulses down to approximately 6 fs.

Colliding-Pulse Mode-locked (CPML) Dye Laser

As a second example of dispersion control, consider a bi-directional mode-locked ring laser that supports two counter-propagating pulse trains and uses saturable absorption for passive mode-locking. Figure 11.17(a) shows a representative layout for such a laser. The gain medium consists of a flowing dye jet located at one tightly focused beam waist within the cavity (between mirrors M_1 and M_2). The saturable absorber is another suitably chosen dye jet of thickness $\sim 10\,\mu$m located at the second beam waist between mirrors M_3 and M_4.

The gain and absorber jets are physically separated by exactly one quarter of the cavity perimeter ($L/4$), where L is the ring cavity perimeter. This configuration ensures that the two counter-propagating pulses pass through the gain medium independently, separated in time by half the roundtrip time ($L/2c$). However, the two pulses "collide" within the thin saturable absorber jet. This overlap strongly enhances absorber saturation, generating a significantly higher peak net gain [Fig. 11.17(b)]. The very small thickness of the absorber jet confines their overlap to a short propagation distance and hence to a brief interaction time. This balanced configuration ensures that both counter-propagating pulses experience identical amplification when they pass through the gain jet.

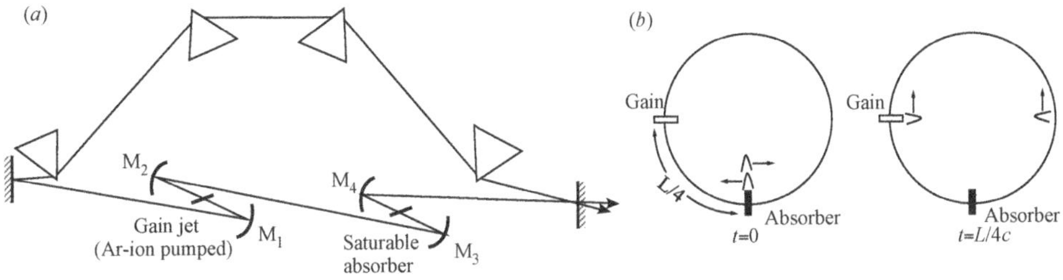

FIGURE 11.17
(a) Schematic layout of a colliding pulse mode-locked (CPML) laser using two-prism pairs for dispersion compensation. (b) Timing diagrams showing two oppositely directed pulses colliding (overlapping) in the absorber jet at $t = 0$ and at $t = L/4c$ while one pulse passes through the gain jet, the other is located half a cavity perimeter away.

The positive GDD of pulses caused by the cavity dispersion is compensated by incorporating a four-prism sequence [Fig. 11.14(a)] into the ring cavity. With the net GDD minimized this way, stable pulses of 50 fs can be generated. Under suitable operating conditions, self-phase modulation (SPM) within the dye jets broadens the pulse spectrum and introduces an intensity-dependent chirp. In combination with appropriately adjusted cavity dispersion, this nonlinear spectral broadening can support stable ultrashort pulses as short as 30 fs.

11.4 Analytical Pulse Formation

We now use the mathematical framework developed in the preceding sections to present a simple analytical treatment aimed at providing physical insight into the complex transient dynamics that lead to ultrafast pulse generation.

Different cavity elements, such as the gain medium (with its finite gain bandwidth), optical losses (whose time dependence may be introduced by an optical modulator or a saturable absorber), dispersion, and optical nonlinearities such as the Kerr effect, modify the laser field amplitude during each cavity roundtrip. Which of these effects must be included depends on the particular laser system under consideration. In this section, we consider the effects of cavity gain, loss, and absorption, and analyze how these elements interact through the feedback provided by the intracavity field to establish the conditions required for ultrafast pulse generation.

To keep the discussion simple, we restrict our treatment of mode-locking dynamics to the case of a slow gain medium, for which the population (or gain) recovery time is much longer than the pulse duration ($\gamma_2^{-1} \gg \tau_{\text{pulse}}$). This regime applies to many important ultrafast lasers. Examples include doped solid-state gain media such as Ti:sapphire, Nd:YAG, and Er-doped glass fibers. Because the gain medium responds slowly compared with the pulse duration, gain saturation during a single pulse (dynamic gain saturation) is negligible and will therefore be ignored. We also neglect complications arising from transverse spatial variations of the intracavity field, the pump distribution, and the cavity losses. Finally, we conclude the chapter with a brief discussion of the influence of other important cavity effects, including optical nonlinearities, dispersion, and diffraction, on ultrafast optical pulses.

11.4.1 Analytical Model of Active Mode-locking

We begin with the circulating-wave model of the laser field in the form of a light pulse circulating inside the laser cavity with spectral amplitude $\tilde{A}(z,\Omega)$ and a slowly varying complex envelope $A(z,t)$, defined by Eqs. (11.32b) and (11.33). The evolution equations for these amplitudes in a dispersive medium were derived in Sec. 11.3. Here we extend that treatment by explicitly incorporating the frequency dependence of the cavity gain and losses. For the present discussion, the effects of background-medium dispersion are neglected.

It is convenient to work with the complex spectral amplitude $\tilde{A}(z,\Omega)$. A key step is to recognize that $\tilde{A}(z,\Omega)$ may be regarded as the amplitude of a narrowband field with absolute frequency $\omega = \omega_o + \Omega$. Gain and loss terms can therefore be incorporated into the evolution equation for $\tilde{A}(z,\Omega)$ following the same procedure used to derive Eqs. (7.102) and (7.107) for narrowband fields.

Expressed in terms of the detuning frequency $\Omega = 2\pi(\nu - \nu_o)$, the small-signal gain coefficient for a homogeneously broadened medium [Eq. (7.66a)], neglecting dynamic saturation, is

$$g(\Omega) = \frac{g_o}{1+(2\Omega/\Delta\omega_H)^2} \approx g_o\left(1 - 4\frac{\Omega^2}{\Delta\omega_H^2}\right), \tag{11.75}$$

where g_o is the line-center small-signal gain coefficient and $\Delta\omega_H$ is the homogeneous linewidth (FWHM) of the gain medium. Similarly, for an inhomogeneously broadened medium, the gain coefficient [Eq. (7.73a)] expressed in terms of Ω is

$$g(\Omega) = g_o \exp\left[-4\ln 2\left(\frac{\Omega^2}{\Delta\omega_D^2}\right)\right] \approx g_o\left[1 - 4\ln 2\left(\frac{\Omega^2}{\Delta\omega_D^2}\right)\right], \tag{11.76}$$

where g_o is the peak small-signal gain coefficient and $\Delta\omega_D$ is the Doppler width of the gain profile. To second order in Ω, both gain profiles are quadratic and may therefore be written in the unified form

$$g(\Omega) = g_o\left(1 - \frac{\Omega^2}{\Omega_g^2}\right), \tag{11.77}$$

where
$$\Omega_g = \begin{cases} \frac{1}{2}\Delta\omega_H & \text{homogeneous gain}, \\ \frac{1}{2}\left(\frac{\Delta\omega_D}{\sqrt{\ln 2}}\right) & \text{inhomogeneous gain}. \end{cases} \tag{11.78}$$

The cavity losses are incorporated by adding half the intensity loss per roundtrip, $\frac{1}{2}\mathcal{L}_c/c\tau_R = \frac{1}{2}\ell_c$, where $c\tau_R$ is the roundtrip cavity length, to the field amplitude equation as described in Secs. 4.3 and 7.9. The evolution equation for the spectral envelope function is then:

$$\begin{aligned} \frac{\partial \tilde{A}_o(z,\Omega)}{\partial z} &= \left[\frac{1}{2}g_o\left(1 - \frac{\Omega^2}{\Omega_g^2}\right) - \frac{1}{2}\ell_c\right]\tilde{A}_o(z,\Omega), \\ &= \left[\frac{1}{2}(g_o - \ell_c) - \frac{g_o}{2\Omega_g^2}\Omega^2\right]\tilde{A}_o(z,\Omega). \end{aligned} \tag{11.79}$$

This frequency-domain equation is transformed into the time-domain using the Fourier-transform correspondences $\tilde{A}_o(z,\Omega) \to A(z,t)$ and $\Omega\tilde{A}_o(z,\Omega) \to -i\partial A(z,t)/\partial t$. This yields the following time domain equation describing the pulse envelope evolution

$$\frac{\partial A(z,t)}{\partial z} = \left[\frac{g_o}{2\Omega_g^2}\frac{\partial^2}{\partial t^2} + \frac{1}{2}(g_o - \ell_c)\right]A(z,t)\,. \tag{11.80}$$

The loss introduced by an amplitude modulator can be included by treating the modulator transmission as a survival factor for the intracavity field. Using the expression for the amplitude transmission t_M from Eq. (11.20), the corresponding loss coefficient is

$$\ell_m = \frac{1}{2}\left(\frac{-\ln t_M^2}{c\tau_R}\right) = \frac{1}{2c\tau_R} 2M(1-\cos\omega_m t) \approx \frac{1}{2}\mu\omega_m^2 t^2\,, \quad \mu = \frac{2M}{c\tau_R}\,. \tag{11.81}$$

The modulator is treated as a localized loss element, similar to the cavity mirrors, since the transit time through the modulator is only a small fraction of the cavity roundtrip time. The quadratic expansion of the cosine is justified because the pulse passes through the modulator near the point of minimum loss (maximum transmission) and experiences the modulation over only a short time interval. If $\omega_m \approx 2\pi/\tau_R$ and the pulse duration is much shorter than the cavity roundtrip time ($|t| \ll \tau_R$), the loss can be accurately approximated by a quadratic function of time. Combining these contributions yields the master equation governing the pulse envelope

$$\frac{\partial A(z,t)}{\partial z} = \left[\frac{g_o}{2\Omega_g^2}\frac{\partial^2}{\partial t^2} - \frac{1}{2}\mu\omega_m^2 t^2 + \frac{1}{2}(g_o - \ell_c)\right] A(z,t)\,. \tag{11.82}$$

Within this averaged propagation model, a steady-state pulse that reproduces itself after each roundtrip satisfies $\frac{\partial A(z,t)}{\partial z} = 0$. The steady-state field equation then simplifies to

$$\left[-\frac{g_o}{2\Omega_g^2}\frac{\partial^2}{\partial t^2} + \frac{1}{2}\mu\omega_m^2 t^2 - \frac{1}{2}(g_o - \ell_c)\right] A(z,t) = 0\,. \tag{11.83}$$

This equation has the same mathematical form as the Schrödinger equation for a one-dimensional quantum harmonic oscillator, with the independent variable mapped from position to time. The scaling used for the harmonic oscillator in quantum mechanics, together with the dimensions of the parameters appearing above, suggests introducing the characteristic time

$$\tau_g = \left[\frac{g_o}{\mu\omega_m^2\Omega_g^2}\right]^{1/4}\,. \tag{11.84}$$

Equation (11.83) can then be written compactly as

$$\left[-\frac{1}{2}\tau_g^2\frac{d^2}{dt^2} + \frac{1}{2}(t/\tau_g)^2 - \frac{1}{2}\frac{(g_o-\ell_c)}{\mu\omega_m^2\tau_g^2}\right] A(z,t) = 0\,. \tag{11.85}$$

This is mathematically identical to the quantum harmonic oscillator equation with mass $m = 1$, natural frequency $\omega = 1$, and coordinate mapping $x \to t/\tau_g$.[4] All of the laser parameters are contained in the final term, which plays the role of the energy eigenvalue.

The ground-state solution of Eq. (11.85) is the Gaussian pulse

$$A(t) = A_o e^{-t^2/\tau_o^2}\,, \quad \tau_o = \sqrt{2}\tau_g\,, \tag{11.86a}$$

which corresponds to the lowest eigenvalue

$$\frac{1}{2}\left(\frac{g_o-\ell_c}{\mu\omega_m^2\tau_g^2}\right) = \frac{1}{2} \implies g_o = \ell_c + \frac{1}{2}\mu\omega_m^2\tau_o^2\,. \tag{11.86b}$$

The full width at half maximum (FWHM) intensity width of this pulse is

$$\tau_p = \sqrt{2\ln 2}\,\tau_o = 0.44\left[\frac{g_o}{\mu}\right]^{1/4}\sqrt{\frac{1}{\nu_m\Delta\nu_g}}\,. \tag{11.87}$$

[4]V. S. Mathur and S. Singh, *Concepts in Quantum Mechanics* (CRC Press, 2008), Chapter 5.

Although this expression for the pulse width is usually derived for homogeneously broadened gain media, the quadratic approximation to the gain profile is equally valid for inhomogeneously broadened media, as shown above. Equation (11.86b) establishes the gain-loss balance condition required for a fundamental Gaussian pulse. For continuous-wave (cw) operation, the roundtrip gain simply equals the cavity loss ($g_o = \ell_c$). In contrast, stable pulse operation requires a slightly larger steady-state gain to compensate for the additional time-dependent loss introduced by the modulator. The increased gain allows oscillation over a broader spectral bandwidth and is particularly important for homogeneously broadened media, where cw multimode oscillation would otherwise be confined to a narrow spectral region near the center of the gain profile. In practice, however, this increase is small because $|\omega_m \tau_o| = 2\pi|\tau_o/\tau_R| \ll 1$.

11.4.2 Analytic Model of Passive Mode-locking

Passive mode-locking relies on an intracavity saturable absorber whose optical loss decreases at a faster rate than the gain during the passage of the pulse. For simplicity, we consider only a fast saturable absorber whose response to the optical field is effectively instantaneous. The absorption coefficient of such an absorber can be written as [cf. Eq. (10.82)]

$$\alpha(|A|^2) = \frac{\alpha_o}{1 + \frac{|A(t)|^2}{|A_s|^2}}, \tag{11.88}$$

where A_s represents the field amplitude corresponding to the saturation intensity of the absorber. Assuming a thin absorber of length L_a, the corresponding intensity transmission coefficient (the roundtrip survival factor) is $e^{-2\alpha L_a}$. Its contribution to field-amplitude loss is one-half of the intensity loss averaged over a cavity roundtrip, τ_R,

$$\begin{aligned}\frac{1}{2}\left(\frac{-\ln(e^{-2\alpha(|A|^2)L_a})}{c\tau_R}\right) &= \frac{\alpha(|A|^2)L_a}{c\tau_R} \approx \frac{\alpha_o L_a}{c\tau_R}\left(1 - \frac{|A(t)|^2}{|A_s|^2}\right) \\ &= \frac{\alpha_o L_a}{c\tau_R} - \gamma|A|^2, \quad \gamma = \frac{\alpha_o L_a}{c\tau_R|A_s|^2}\end{aligned} \tag{11.89}$$

Including this saturable absorption in the field-amplitude equation yields

$$\frac{\partial A(z,t)}{\partial z} = \left[\frac{1}{2}\frac{g_o}{\Omega_g^2}\frac{d^2}{dt^2} + \gamma|A(z,t)|^2 + \frac{1}{2}(g_o - \ell)\right] A(z,t), \tag{11.90}$$

where the constant unsaturated absorption loss, $2\alpha_o L_a/c\tau_R$, has been combined with the intrinsic cavity loss ℓ_c and incorporated into the total cavity loss parameter ℓ.

The steady-state field amplitude then satisfies the nonlinear differential equation

$$\left[-\frac{1}{2}\frac{g_o}{\Omega_g^2}\frac{d^2}{dt^2} - \gamma|A(z,t)|^2 - \frac{1}{2}(g_o - \ell)\right] A(z,t) = 0. \tag{11.91}$$

This equation has the mathematical form of the stationary Schrödinger equation for a reflectionless potential in quantum mechanics. Exploiting this analogy, the steady-state solution can be written as

$$A(t) = A_o \operatorname{sech}\left(\frac{t}{\tau_s}\right), \tag{11.92}$$

where the pulse duration τ_s and peak field amplitude A_o satisfy

$$\gamma|A_o|^2 = \frac{g_o}{\Omega_g^2 \tau_s^2}, \tag{11.93a}$$

$$g_o - \ell + \frac{g_o}{\Omega_g^2 \tau_s^2} = 0. \tag{11.93b}$$

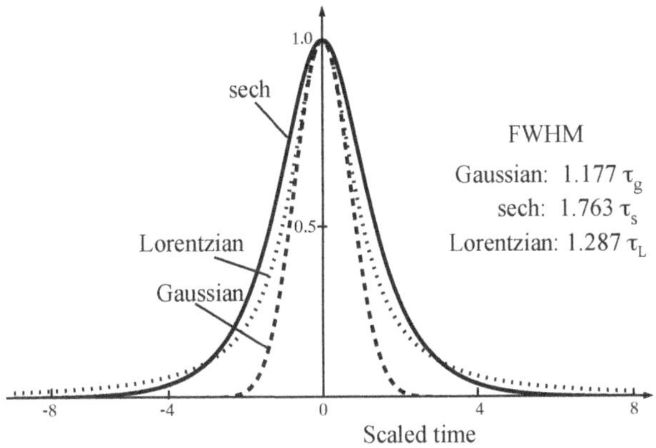

FIGURE 11.18
Normalized intensity profiles for several common pulse shapes, plotted as functions of scaled time (time expressed in units of the characteristic pulse durations, τ_o, τ_s, τ_L). The indicated widths are intensity FWHM values.

Once again, we find that, provided the laser gain, background loss, and saturable-absorption parameters satisfy the conditions given by Eq. (11.93), a localized pulse constitutes the natural steady-state solution of the laser cavity. In this case, however, the pulse has a hyperbolic-secant rather than a Gaussian temporal profile.

Figure 11.18 and Table 11.1 compare the temporal profiles, pulse widths, and spectral bandwidths of Gaussian, hyperbolic-secant, and two other commonly encountered optical pulse envelopes.

Our purpose has been to demonstrate that, with an appropriate balance among the cavity gain, loss, saturable absorption, and other operating parameters, a pulse solution emerges as the natural steady-state operating mode of a laser. One important physical mechanism omitted from the present treatment is cavity dispersion, that is, the frequency dependence of the refractive index of the intracavity components. As discussed in Sec. 11.3, the stability of the pulse shape depends critically on the cavity dispersion profile.

11.4.3 Including Material Dispersion

For ultrashort laser pulses, the effects of group-velocity dispersion within the intracavity materials become significant. Following the discussion in Sec. 11.3, it is straightforward to

TABLE 11.1
Time-bandwidth product (TBP) for some common pulse shapes.

Field Envelope	Intensity profile	τ_p (FWHM)	Spectral profile	$\Delta\omega_p$ (FWHM)	TBP
Gaussian	$e^{-2(t/\tau_o)^2}$	$1.177\tau_o$	$e^{-(\Omega\tau_o)^2/2}$	$2.355/\tau_o$	0.441
sech	$\text{sech}^2(t/\tau_s)$	$1.763\tau_s$	$\text{sech}^2(\pi\Omega\tau_s/2)$	$1.12/\tau_s$	0.315
Lorentzian	$[1+(t/\tau_L)^2]^{-2}$	$1.287\tau_L$	$e^{-2\|\Omega\|\tau_L}$	$0.693/\tau_L$	0.142
rectangular	$\begin{cases} 1 & \|t\| < \tau_r \\ 0 & \text{otherwise} \end{cases}$	τ_r	$\text{sinc}^2(\Omega\tau_r)$	$2.78/\tau_r$	0.443

incorporate material dispersion into the equation of motion for the field envelope. We retain only the linear and quadratic terms in the Taylor expansion of the propagation constant $k(\omega)$ [Eq. (11.30)],

$$k(\omega) \approx k_o + \frac{\Omega}{v_g} + \frac{1}{2}\beta_2\Omega^2, \quad \Omega = \omega - \omega_o. \tag{11.94}$$

Following the steps leading to Eq. (11.49), and including the cavity gain and loss terms, we obtain the following evolution equation for the pulse envelope,

$$\frac{\partial A(z,t)}{\partial z} = \left[-\frac{1}{v_g}\frac{\partial}{\partial t} + \frac{1}{2}\left(\frac{g_o}{\Omega_g^2} - i\beta_2\right)\frac{\partial^2}{\partial t^2} + \frac{1}{2}(g_o - \ell_c)\right] A(z,t). \tag{11.95}$$

Using the change of variables given by Eq. (11.50), and including the saturable-absorber term, this equation becomes

$$\frac{\partial A(z,\tau)}{\partial z} = \left[\frac{1}{2}\left(\frac{g_o}{\Omega_g^2} - i\beta_2\right)\frac{\partial^2}{\partial \tau^2} + \gamma|A(z,\tau)|^2 + \frac{1}{2}(g_o - \ell_c)\right] A(z,\tau). \tag{11.96}$$

11.4.4 Including the Kerr Effect

The Kerr effect, introduced in Sec. 11.2.3, appears as an intensity-dependent contribution, $\Delta n_{NL} = n_2 I(t)$, to the refractive index of a medium when an intense optical pulse propagates through it. This nonlinear change in refractive index adds an intensity-dependent contribution, $\Delta n_{NL}\omega/c = n_2 I\omega/c$, to the propagation constant [Eq. (11.94)], giving

$$k(\omega) = k_o + \frac{\Omega}{v_g} + \frac{\beta_2}{2}\Omega^2 + \frac{\omega}{c}n_2 I(t) \equiv k_o + \frac{\Omega}{v_g} + \frac{\beta_2}{2}\Omega^2 + \delta|A(t)|^2, \tag{11.97}$$

where we have used the fact that the pulse intensity satisfies $I(t) \propto |A(t)|^2$. The proportionality factors are absorbed into δ, whose dimensions depend on the normalization chosen for A so that $\delta|A|^2$ has dimensions of inverse length. The pulse-envelope equation then becomes

$$\frac{\partial A(z,\tau)}{\partial z} = \left[\frac{1}{2}\left(\frac{g_o}{\Omega_g^2} - i\beta_2\right)\frac{\partial^2}{\partial \tau^2} + (\gamma + i\delta)|A|^2 + \frac{1}{2}(g_o - \ell_c)\right] A(z,\tau). \tag{11.98}$$

The inclusion of dispersion and the nonlinear refractive-index change leads to a considerably more general evolution equation. The story does not end here, however. Because laser pulses possess a transverse spatial structure, the nonlinear refractive index also gives rise to Kerr lensing, causing the transverse field profile to evolve during propagation. This introduces diffraction into the problem and adds yet another level of complexity.

We shall not pursue these effects further. Instead, we conclude by observing that the generalized pulse-propagation equation provides a catalog of the physical processes that influence the formation and propagation of ultrashort laser pulses. The key challenge is to identify which terms are important for a particular laser system and under what operating conditions. Appropriate approximations can then be made to obtain analytical solutions. Ultimately, these solutions must be tested against experimental observations and the underlying physical principles to develop physical insight and refine the theoretical model.

11.5 Laser Frequency Comb

The frequency-domain picture of mode-locked pulses presented in the opening sections of this chapter assumed that all longitudinal cavity modes within the gain bandwidth are uniformly spaced in frequency and oscillate in phase. This is clearly an oversimplification. The number of phase-coherent modes can be estimated from the ratio of the cavity roundtrip time to the pulse width. For a meter-long laser cavity producing a train of 100 fs pulses, we find that about 100,000 longitudinal modes contribute to the pulse bandwidth. Because of the dispersion of the intracavity elements, the axial-mode frequencies cannot remain uniformly spaced over the entire bandwidth involved in pulse formation.

11.5.1 Cavity Modes and Modes of a Mode-Locked Laser

Recalling that the cavity mode frequency is determined by the condition that the roundtrip phase for the mode be equal to $2\pi m$, the axial mode condition for a two-mirror cavity, when dispersion of various elements in the cavity is taken into account, can be written as

$$\Delta\phi(\nu_m) = \frac{4\pi\nu_m}{c}\sum_i n_i(\nu_m)L_i \equiv \frac{2\pi\nu_m}{c}n(\nu_m)L = 2\pi m\,, \tag{11.99}$$

where $n_i(\nu_m)$ is the refractive index at frequency ν_m of element i of length L_i in the cavity. The quantity $\sum_i n_i(\nu_m)L_i \equiv n(\nu_m)L$ is the sum of the optical path lengths of all cavity elements, which can be written as $n(\nu_m)L$, where $n(\nu_m)$ is the average effective refractive index and L is the geometrical length of the cavity. Then, the cavity mode frequencies are given by

$$\nu_m = \frac{mc}{2n(\nu_m)L}\,. \tag{11.100}$$

From the phase condition (11.99) for two successive modes, the axial mode separation $\Delta\nu_{ax} \equiv \nu_{m+1} - \nu_m$ is determined by the relation

$$\Delta\phi(\nu_{m+1}) - \Delta\phi(\nu_m) = 2\pi\,. \tag{11.101}$$

For a large number of axial modes with small mode spacing, this phase difference can be approximated as

$$\left.\frac{d\Delta\phi}{d\nu}\right|_{\nu=\nu_m}\Delta\nu_{ax} = 2\frac{dk_m}{d\nu}L\Delta\nu_{ax} = \frac{4\pi L}{v_g(\nu_m)}\Delta\nu_{ax} = 2\pi\,, \tag{11.102}$$

where $v_g(\nu)$ is the group velocity in the vicinity of frequency ν. From this equation, we find that the axial mode spacing $\Delta\nu_{ax}$ is given by

$$\Delta\nu_{ax}(\nu) = \frac{v_g(\nu)}{2L}\,. \tag{11.103}$$

It depends on mode frequency ν and is not uniform. The actual frequency dependence of the mode spacing depends on the cavity dispersion lumped into the effective refractive index $n(\nu)$. The cavity mode spectrum thus is a nonuniform frequency comb. The superposition of such a nonuniform frequency comb will result in a non-periodic, nonuniform pulse train. Unlike the case of equally spaced modes, the field amplitude, intensity, and power will not be periodic functions of time.

In contrast to the picture of laser emission based on the mode structure of a passive laser cavity, a mode-locked laser, despite dispersion in the laser cavity, emits a train of equally spaced pulses with period τ_R in the time domain. Via Fourier series, such a periodic pulse train corresponds to a periodic comb of frequency "spikes" in the spectral domain with constant spacing, $\nu_R = 1/\tau_R$. Indeed, experiments with stabilized mode-locked lasers show that the spikes of the frequency comb are equally spaced throughout the pulse bandwidth to better than 10^{-16}.

These equally spaced phase-coherent frequencies should therefore be distinguished from the resonances of the passive (cold) cavity, whose spacing is frequency dependent because of intracavity dispersion. The theory of the mode-locked laser discussed in Sec. 11.4 shows that this transformation of the unequal mode spacing of the passive cavity into a perfect frequency comb is the result of an interplay between dispersion and nonlinear optical processes inside the laser.

11.5.2 The Comb Equation

Following the preceding discussion, comb mode frequencies can be written as harmonics of the repetition rate ν_R

$$\nu_m = m\nu_R + \nu_o\,, \tag{11.104}$$

where $\nu_o < \nu_R$ represents a nonzero carrier-envelope offset frequency. The meaning of ν_o will become apparent shortly. Equation (11.104) is referred to as the comb equation. It states that the frequency comb underlying mode-locked pulses from a laser, consisting of hundreds of thousands of optical frequencies spanning hundreds of terahertz, is defined by only two parameters: (i) the pulse repetition rate ν_R and (ii) the offset frequency ν_o.

These equidistant phase-locked comb modes characterized by an angular frequency $\omega_m = 2\pi\nu_m$ and a spectral amplitude A_m result in a pulse train with an electric field given by

$$E(t) = \sum_{m=m_i}^{m_f} A_m e^{-i\omega_m t} \equiv A(t)e^{-i(\omega_c t+\phi_o)}\,, \tag{11.105}$$

where the mode index m is a large integer varying from m_i to m_f spanning hundreds of thousands to a million depending on the mode-locked laser. Here, $A(t)$ is the slowly varying pulse envelope, ω_c is the carrier angular frequency, and ϕ_o is the carrier-envelope offset (CEO) phase [Fig. 11.19]. The carrier frequency ω_c may conveniently be chosen as the frequency of a comb tooth near the center of the spectrum, $\omega_c = 2\pi(m_c\nu_R + \nu_o)$.

From Eq. (11.105), it is evident that while the temporal pulse envelope $A(t)$ repeats exactly after one roundtrip time τ_R, the carrier phase ϕ_o evolves and is generally not an integer multiple of 2π. Indeed, the phase accumulated by a mode with index m during one roundtrip changes by

$$2\pi\nu_m\tau_R = 2\pi\nu_o\tau_R + 2\pi m\nu_R\tau_R = 2\pi\nu_o\tau_R + 2\pi m\,. \tag{11.106}$$

Consequently, independent of the specific mode index m, every mode acquires an identical additional phase shift (modulo 2π) relative to the pulse envelope, given by the first term in Eq. (11.106)

$$\Delta\phi_o = 2\pi\nu_o\tau_R\,. \tag{11.107}$$

This extra phase slip manifests as a pulse-to-pulse carrier-envelope phase slip, $\Delta\phi_o$, as shown in Fig. 11.19. Rearranging Eq. (11.107), we can express the offset frequency ν_0 directly in terms of this pulse-to-pulse phase slip as

$$\nu_o = \frac{\Delta\phi_o}{2\pi\tau_R} = \frac{\Delta\phi_o}{2\pi}\nu_R\,. \tag{11.108}$$

Based on this discussion, the pulse train emitted by a mode-locked laser can be visualized as a pulse that propagates back and forth inside the laser cavity. Each time the pulse hits the cavity output coupler, a small fraction of its power is transmitted. This mechanism results in a periodic output pulse train with a repetition period determined by the group velocity $v_g(\nu_c)$ evaluated at the carrier frequency:

$$\tau_R = 2L/v_g(\nu_c)\,, \tag{11.109}$$

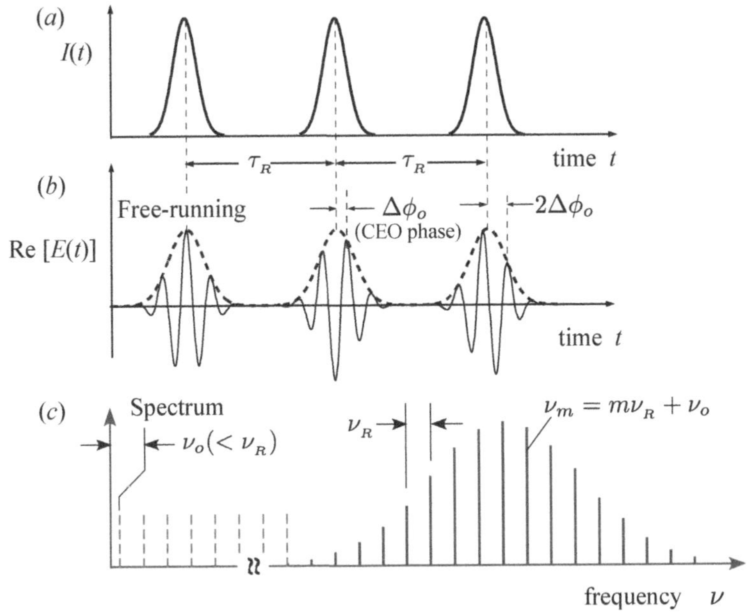

FIGURE 11.19
Time-domain representation of a mode-locked ultrafast laser pulse train and the corresponding frequency-domain representation of frequency comb. Extrapolation of the comb spectrum to zero frequency yields a lowest frequency offset from zero by ν_o.

This roundtrip time corresponds to the fundamental repetition rate $\nu_R = 1/\tau_R$ at which the pulse circulates inside the laser cavity. While the pulse envelope reproduces its shape on each roundtrip, the underlying carrier phase ϕ_o slips from pulse to pulse by the discrete amount $\Delta\phi_0$. Stabilization of carrier phase ϕ_o requires measuring the phase slip $\Delta\phi_o$ and using an active electronic feedback loop to lock it to a fixed value.

The relative phase slip between the carrier wave and its envelope can also be understood as arising from the difference between the phase and group velocities within dispersive media in the laser cavity. In cases where the refractive index can be approximated as a constant over the entire gain bandwidth, the group velocity equals the phase velocity. In this case, $\nu_o = 0$, and the mode frequencies become exact integer multiples of the repetition rate ν_R. For $\nu_o = 0$, the pulse period τ_R is an integer number of optical cycles $\tau_m = 1/\nu_m$ of the oscillating modes [Fig. 11.20].

11.5.3 Comb Stabilization and Applications

The pulse repetition frequency ν_R is a microwave frequency that connects directly to optical frequencies via $\nu_m = m\nu_R$, where $m \sim 10^5$ to 10^6 is a large positive integer. Any changes to ν_R cause an accordion-like expansion and contraction of the frequency comb. The offset frequency ν_0 ($\nu_0 < \nu_R$) for the comb is also a microwave frequency. However, a change in ν_o shifts all comb lines uniformly by the same amount. Precise knowledge of the offset frequency ν_o is, therefore, the key to the determination of the absolute frequencies of the individual comb modes. Without this knowledge, individual comb frequencies are only known to within an uncertainty window of $\pm\nu_R$. This uncertainty represents a relative error of 10^{-5} to 10^{-6} at optical frequencies, depending on the comb spacing.

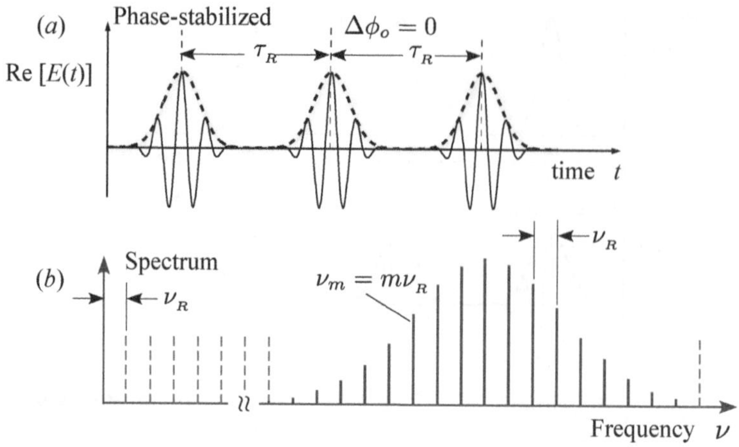

FIGURE 11.20
Time-domain representation of a mode-locked ultrafast pulse train and the corresponding frequency-domain representation of the frequency comb. Extrapolation of the comb spectrum to zero frequency yields a lowest frequency offset from zero ν_o.

A free-running optical frequency comb generated by a reasonably stable mode-locked laser in a typical research laboratory can provide this level of accuracy, which is quite adequate for many applications, such as measuring Doppler-broadened linewidths. For applications requiring higher accuracy or long-term signal averaging, both ν_R and ν_o must be actively stabilized. This is achieved by using negative feedback to control the laser cavity length (using a piezoelectric transducer, for example) to stabilize ν_R, and intracavity dispersion (such as tilting a mirror or modulating pump power) to control ν_o. With good mechanical stability combined with high-bandwidth electronic feedback loops, cavity length can be maintained to within a femtometer.

The laser repetition rate, ν_R, can be measured by detecting the optical pulse train with a fast photodetector. This detection results in an electronic pulse train [Fig. 11.19(a)] whose spectrum consists of coherently related microwave Fourier harmonics, $m\nu_R$. However, the measurement of ν_o is not as straightforward. Several ingenious schemes exist to measure ν_o using nonlinear optical processes. The simplest is based on second-harmonic generation (SHG) but requires a comb spectrum sufficiently broad to cover an optical octave (a factor of 2 in frequency). Although standard mode-locked lasers do not emit an octave-spanning spectrum, this limitation is bypassed by sending the short pulses into a highly nonlinear photonic crystal fiber to generate an ultra-broadband white-light supercontinuum [see Sec. 12.8.4]. Assuming an octave-spanning spectrum is available, the second harmonic of the mth mode from the low-frequency wing of the comb will overlap with the $2m$th mode from the high-frequency wing of the fundamental comb. The heterodyne beat note between the second-harmonic comb and the fundamental comb yields a microwave frequency (equal to the offset ν_o)

$$2\nu_m - \nu_{2m} = 2(m\nu_R + \nu_o) - (2m\nu_R + \nu_o) = \nu_o \,. \tag{11.110}$$

Many modes within the phase-matching range of the SHG [see Chapter 12] will contribute to the second harmonic signal and, therefore, to the heterodyne beat-note spectrum. This collective contribution drastically improves the signal-to-noise ratio of the heterodyne beat-note spectrum, thereby enhancing the signal. The lowest fundamental beat frequency detected is exactly ν_o; other beat frequencies, which will be multiples or mixtures of ν_R and

ν_o, can be electronically filtered out. Once ν_R and ν_o are known and stabilized (locked to high-precision references, such as an atomic clock), the laser frequency comb is transformed into an absolute optical frequency ruler, with its phase-locked modes acting as perfectly spaced tick marks.

The existence of a perfectly regular frequency comb has revolutionized optical metrology. The comb can be used as a ruler to measure the exact frequency spacing between any pair of unknown optical frequencies, ν_1 and ν_2. This can be done by measuring the beat note $\Delta\nu_1$ between ν_1 and the closest comb tooth of index m_1, and similarly measuring the beat note $\Delta\nu_2$ between ν_2 and its nearest comb tooth m_2. The absolute frequency difference between the two unknown sources is then given, algebraically, by $\nu_2 - \nu_1 = \Delta\nu_2 - \Delta\nu_1 + (m_2 - m_1)\nu_R$. The signs are determined by whether the unknown frequencies lie above or below their respective adjacent teeth.

As another application, a frequency comb can serve as a broadband light source to measure the absorption spectrum of a sample by sending the comb into a sample cell.[5] Assuming that an absorber transition is resonant with only a single comb line and all other lines are detuned, the absorption spectrum of the sample can be extracted by spectrally resolving the transmitted comb and comparing it with a reference input spectrum; the transmission is obtained from their ratio. Two-photon excitation with a frequency comb can be quite efficient, as thousands of comb line pairs of the same sum frequency contribute coherently to the excitation.

In summary, an optical frequency comb (OFC) is a mode-locked laser that emits a train of ultrashort pulses at a constant repetition rate ν_R. Dispersion in the laser cavity induces a pulse-to-pulse phase slip $\Delta\phi_o$ of the carrier wave relative to the peak of the pulse envelope. This phase slip $\Delta\phi_o$ shifts every harmonic of ν_R in the pulse spectrum by a constant amount called the carrier-envelope offset frequency, $\nu_o = \nu_R \Delta\phi_o / 2\pi$. The comb line optical frequencies are represented by the comb equation (11.104). Remarkably, every mode in the optical spectrum is fully parametrized and controlled via only two characteristic microwave frequencies, ν_R and ν_o, both of which can be measured and stabilized with great precision. The comb frequency spectrum consists of thousands of phase-coherent laser lines uniformly spaced by ν_R. Rigorous experiments have demonstrated that the fractional difference between the spacing of the optical modes of a Kerr-lens mode-locked Ti:sapphire laser and the measured pulse repetition frequency is less than 10^{-15} and the frequency spacing of the comb modes is uniform to better than 10^{-16}.[6] An optical frequency comb thus serves as a frequency ruler to measure the absolute frequencies of unknown sources, and its broad spectrum enables new approaches to high-resolution spectroscopy over a vast spectral frontier.

Summary

In this chapter, we investigated the physics of multimode laser operation under conditions in which the output consists of a highly stable train of mode-locked ultrashort pulses. We developed a frequency-domain model of mode-locked pulses by analyzing the coherent linear superposition of a large number of uniformly spaced longitudinal modes oscillating with fixed phase relationships. Such a superposition produces periodic constructive interference, resulting in an ultrafast pulse train in the time domain.

[5]Nathalie Picqué and T. W. Hänsch, Nature Photonics **13**, 146-157 (2019).

[6]T. Udem, J. Reichert, R. Holzwarth, T. W. Hänsch, "Accurate measurement of large optical frequency differences with a mode-locked laser," Opt. Lett. **24**, 881–883 (1993).

This model was used to develop a physical description of the intracavity field and to determine how the pulse width and repetition rate depend on the cavity length, the available gain bandwidth, and the lineshape parameters of the active medium. This framework was then used to explain how active and passive mode-locking techniques establish phase coherence across an otherwise uncoupled multimode spectrum. We described the physical mechanisms by which active modulators and passive saturable absorbers modify the cavity loss and thereby favor pulsed operation over continuous-wave operation.

In the second part of the chapter, we developed a simple semiclassical analytical treatment of mode-locked operation. The wave-envelope equations used earlier for narrowband fields were extended in Sec. 11.3 to describe the broadband, multifrequency character of ultrashort pulses. These evolution equations were subsequently used in Sec. 11.4 to derive steady-state pulse solutions, including Gaussian profiles for active mode-locking and hyperbolic-secant profiles for passive mode-locking, illustrating the balance among gain, loss, bandwidth, dispersion, and nonlinear effects required to sustain ultrafast laser pulses.

Finally, in Sec. 11.5 we introduced the laser frequency comb, an important application of highly coherent mode-locked laser operation. Its distinctive properties were illustrated with examples showing its importance in precision optical measurement.

Problems

1. **Mode-locking a He:Ne laser:** A 100-cm-long He:Ne laser has a Doppler-broadened gain linewidth (FWHM) of 1.5 GHz. Brewster windows on the discharge tube ensure oscillation in a single polarization. The laser is pumped so that the unsaturated line-center gain is four times the threshold gain.

   (a) Estimate the frequency range over which longitudinal TEM_{00} modes can oscillate and hence the number of such modes that can oscillate simultaneously.

   (b) If all of these modes are phase-locked, estimate the duration and repetition frequency of the mode-locked pulses.

2. **Mode-locking an Ar-ion gas laser:** A 100 cm long Ar-ion laser has a Doppler-broadened gain linewidth (FWHM) of 3.5 GHz. Brewster windows on the gain tube ensure oscillation in a single polarization. The laser is pumped so that the unsaturated line-center gain is four times the threshold gain.

   (a) Estimate the number of longitudinal TEM_{00} modes that can oscillate simultaneously. If all of these modes are phase-locked, estimate the repetition frequency and duration of the mode-locked pulses.

   (b) The pulse duration obtained in part (a) is relatively long. Explain why liquid-dye lasers and many solid-state lasers can generate much shorter mode-locked pulses.

3. **Active mode-locking and gain broadening:** A standing-wave laser cavity of optical length $L_e = 1.5$ m has an amplitude modulator located near one end of the cavity. The laser has an available oscillation bandwidth of 120 GHz.

   (a) Calculate the longitudinal-mode spacing and estimate the number of longitudinal cavity modes that lie within the oscillation bandwidth.

   (b) Assuming that the modulator phase-locks the oscillating longitudinal modes, determine the pulse repetition frequency. For an inhomogeneously broadened gain medium, estimate the shortest pulse duration that could be obtained if essentially the full oscillation bandwidth participates in mode-locking. Discuss qualitatively why the effective mode-locking bandwidth, and hence the attainable pulse duration, may be different for a homogeneously broadened gain medium.

(c) How would your answers to parts (a) and (b) change for a unidirectional ring laser having the same roundtrip optical length and oscillation bandwidth?

4. **Dye laser:** A dye laser has an oscillation bandwidth of 500 GHz, and the spacing between adjacent longitudinal cavity modes is 180 MHz. Assume that the oscillating modes have equal amplitudes.

   (a) For fundamental active mode-locking, estimate the duration of the output pulses and the pulse repetition time. Use the result for actively mode-locked lasers derived in the text.

   (b) If, instead, the laser had an inhomogeneously broadened gain profile with the same 500-GHz oscillation bandwidth and essentially the full bandwidth participated in mode-locking, estimate the transform-limited pulse duration.

5. **Mode-locking in an Nd:YAG laser:** An Nd:YAG laser operating at $\lambda = 1.06\,\mu$m has a homogeneously broadened gain line with FWHM $\Delta\nu_{_H} = 195$ GHz. The standing-wave cavity has length $L = 1.5$ m. Assume that the oscillation bandwidth is approximately equal to the homogeneous gain linewidth.

   (a) Using the active mode-locking result derived in the text, estimate the pulse duration when the laser is mode-locked with an acousto-optic modulator (AOM).

   (b) What modulation frequency should be applied to the AOM for fundamental mode-locking?

   (c) If the gain profile were instead inhomogeneously broadened with the same oscillation bandwidth, and essentially the full bandwidth could participate in mode-locking, estimate the transform-limited pulse duration.

6. **Ti:sapphire laser:** A Ti:sapphire laser has a 1.0-m-long cavity and operates near a center wavelength $\lambda_o = 800$ nm with an available wavelength bandwidth $\Delta\lambda = 100$ nm. The laser is passively mode-locked by Kerr-lens mode-locking. Assuming that essentially the full bandwidth participates in forming a transform-limited Gaussian pulse, estimate the shortest pulse duration.

7. **Gaussian pulse propagation in a dispersive medium:** Consider a Gaussian pulse propagating in the $+z$-direction. At the input plane $z = 0$ of a dispersive medium, its complex electric field is

$$E(0,t) = A_o e^{-t^2/\tau_o^2} e^{-i\omega_c t}\,.$$

Writing the field in terms of a slowly varying envelope,

$$E(0,t) = A(0,t)e^{-i\omega_c t}\,, \qquad A(0,t) = A_o e^{-t^2/\tau_o^2}\,,$$

the Fourier transform of the envelope is

$$\widetilde{A}(0,\Omega) = \tau_o A_o \sqrt{\pi}\, e^{-\Omega^2\tau_o^2/4}\,,$$

apart from the normalization factor associated with the Fourier-transform convention being used. Here $\Omega = \omega - \omega_c$ is the frequency detuning.

The input envelope can therefore be represented as a superposition of monochromatic components,

$$A(0,t) = \frac{1}{2\pi}\int_{-\infty}^{\infty} d\Omega\, \widetilde{A}(0,\Omega)e^{-i\Omega t}\,.$$

Assume that the propagation constant of the medium can be approximated to second order about ω_c as

$$k(\omega) \approx k_c + \frac{\Omega}{v_g} + \frac{1}{2}\beta_2\Omega^2\,.$$

Show that after propagating a distance z, the field can be written in the form

$$E(z,t) = \frac{e^{-i(\omega_c t - k_c z)}}{2\pi}\int_{-\infty}^{\infty} d\Omega\, \widetilde{A}(0,\Omega)\exp\left[-i\Omega\left(t-\frac{z}{v_g}\right) + i\frac{1}{2}\beta_2 z\Omega^2\right].$$

Evaluate the Gaussian integral by completing the square or by contour integration. Determine the propagated pulse envelope and discuss how its temporal width changes with propagation distance. In particular, identify the distance scale over which group-velocity dispersion produces significant pulse broadening.

8. **RMS time-bandwidth product of a Gaussian pulse:** We have generally used full width at half maximum (FWHM) to characterize the temporal and spectral widths of pulses. Another useful measure of the width of a distribution is its root-mean-square (rms) width, defined as the square root of the variance,

$$(\Delta x)_{\mathrm{rms}} \equiv \sqrt{\langle (x-\langle x\rangle)^2\rangle}\,, \qquad \langle (x-\langle x\rangle)^2\rangle = \int_{-\infty}^{\infty} (x-\langle x\rangle)^2 P(x)\,dx\,, \tag{11.111}$$

where $P(x) \geq 0$ is a normalized distribution.

Use this definition to calculate the rms temporal width and rms angular-frequency width of the transform-limited Gaussian pulse

$$A(z,t) = A_o e^{-\tau^2/\tau_o^2} e^{-i(\omega_o t - k_o z)}\,, \tag{11.112}$$

where τ is the local time relative to the pulse center. Calculate the corresponding rms time-bandwidth product. Also express the time-bandwidth product in terms of the ordinary frequency $\nu = \omega/2\pi$.

9. **Dispersion broadening of an ultrashort pulse:** A transform-limited Gaussian pulse with an intensity FWHM duration of 10 fs and center wavelength 800 nm propagates through a 10-mm-thick fused-quartz plate. At the center wavelength, the dispersion parameters are $\beta_1 = 4.8$ ps/mm and $\beta_2 = 36$ fs^2/mm.

   (a) Calculate the group delay acquired by the pulse in passing through the plate.

   (b) Calculate the intensity FWHM duration of the pulse after it emerges from the plate.

10. **Space-time analogy:** Noting the correspondence $(z,k,x) \longleftrightarrow \left(\eta, -\frac{1}{\beta_2}, \tau\right)$, between propagation distance, wave number, and transverse coordinate in Gaussian-beam diffraction and the corresponding variables in dispersive pulse propagation, we were able to use the Gaussian-beam solution of Chapter 3 to describe the effect of quadratic dispersion on pulse evolution in Sec. 11.3.3. Discuss the physical and mathematical basis of this space-time analogy. In particular, identify the temporal analogs of diffraction and phasefront curvature, and discuss the significance of the sign of β_2.

The space-time analogy was originally recognized in the 1960s.[7] This problem presents another example of the space-time analogy by looking at the temporal analog of the far-field diffraction (Fraunhofer diffraction). In far-field diffraction, the spatial profile of the output field is the Fourier transform of the input field. We can create a highly dispersive path using a long length of a dispersive material such as an optical fiber. Assuming that β_3 and higher-order dispersion terms can be neglected, we expect the pulse shape at the output of such a path to be the Fourier transform of the input field, which is simply the spectrum of the pulse. The pulse at the output of such a dispersive path can be written as

$$A(z,t) = \int_{-\infty}^{\infty} \tilde{A}(0,\Omega) \exp\left[\frac{i\beta_2 z}{2}\Omega^2 - i\Omega t\right] d\Omega\,.$$

Evaluate this integral for long distances and show that the output pulse shape is a scaled version of the input spectrum with the conversion factor $\Omega = t/(\beta_2 z)$.

11. **Time lens:** Another application of the space-time analogy is the concept of a time lens. An ordinary thin lens imposes a quadratic phase variation across the transverse profile of a beam. In one transverse dimension, its effect can be written as

$$u_{\text{out}}(x) = \exp\left(\frac{ikx^2}{2f}\right) u_{\text{in}}(x)\,,$$

where f is the focal length of the lens. The lens does not change the transverse intensity profile at the plane of the lens; rather, the quadratic spatial phase changes the subsequent propagation of the beam and can cause it to focus.

Using the space-time analogy, discuss how an analogous "time lens" might be created. What form of temporal phase modulation would be required, and what physical processes might be used to produce it?

[7]P. Tournois, "Analogie optique de la compression d'impulsion," C.R. Acad. Sci. **258**, 3839–3842 (1964); S. A. Akhmanov, A. P. Sukhorukov, and A. S. Chirkin, "Nonstationary nonlinear optical effects and ultrashort light pulse formation," IEEE J. Quantum Electron. **QE-4**, 598–605 (1968); S. A. Akhmanov, A. P. Sukhorukov, and A. S. Chirkin, "Stationary phenomena and space-time analogy in nonlinear optics," Sov. Phys. JETP **28**, 748–757 (1969).

12

Nonlinear Optics

Due to the high power and large electric fields associated with intense laser beams, the material response to light cannot be fully described by linear polarization alone; contributions from higher-order powers of the electric field become significant. This nonlinear material response leads to many novel phenomena, such as the generation of light at frequencies different from those of the exciting waves, and forms the foundation of the field of nonlinear optics. Since its beginning with the observation of second-harmonic generation in 1961, nonlinear optics has developed into a vast field encompassing a wide variety of phenomena. While gain and absorption saturation in lasers, as well as two-photon excitation, are logically examples of a material medium's nonlinear response to an optical field, these topics are traditionally not studied under the umbrella of nonlinear optics. Yet, even with their exclusion, it is impossible to exhaustively discuss all topics of interest within this vast field.

Our interest here is in nonlinear processes such as optical harmonic generation, frequency mixing, and intensity-dependent refractive index changes leading to self-focusing and self-phase modulation. These phenomena play an important role either in the physics of lasers themselves or in applications of laser light to extend the range of coherent radiation into ultraviolet (UV), mid-infrared, and terahertz (THz) regions. Although these effects are usually small, they can be enhanced significantly under carefully controlled conditions.

To analyze these phenomena, the medium's polarization is expressed as a Taylor series in powers of the electric field. Here, we consider only the nonlinear polarization terms involving the second and third powers of the applied electric field, as they are sufficient to describe the most important nonlinear optical processes relevant to laser physics. Furthermore, these processes are treated within the parametric limit, which assumes that the frequencies of the light waves are detuned far from material resonances. Under these conditions, absorption is negligible and the nonlinear susceptibilities may be treated as approximately real. Their frequency dependence is also weak enough over the spectral range of interest that, where appropriate, we neglect it in developing the basic physics. Nonlinear effects arising from second- and third-order polarizations are discussed separately in the following sections.

12.1 Second-Order Nonlinear Polarization

To keep the notation consistent, real time-domain signals (fields and polarizations) are denoted by script letters, and their analytic representatives by the corresponding Roman letters, as we have done throughout the book [see Sec. 2.1].

The linear (first-order) polarization $\boldsymbol{\mathcal{P}}^{(1)}$ of a medium for a narrowband electric field $b\mathcal{E}$ is written as

$$\boldsymbol{\mathcal{P}}^{(1)}(t) = \epsilon_o \chi \boldsymbol{\mathcal{E}}(t)\,, \tag{12.1}$$

where χ is the dielectric susceptibility. This equation assumes that the medium's response is independent of the direction of the electric field relative to the crystal axes. Nonlinear

DOI: 10.1201/9781003584179-12

optics also involves media with anisotropic responses to an applied electric field. In such cases, the i-th component of linear polarization $\mathcal{P}_i^{(1)}$ is related to the electric components $\mathcal{E}_j$ by

$$\mathcal{P}_i^{(1)}(t) = \epsilon_o \chi_{ij} \mathcal{E}_j(t) \,, \tag{12.2}$$

where χ_{ij} is the second-rank linear-susceptibility tensor. The indices (i, j, k) denote the Cartesian components of a vector with summation implied over repeated Cartesian indices (such as j on the right).[1] For example, the $i = x$-component of linear polarization is given by $\mathcal{P}_x = \epsilon_o \chi_{xj} \mathcal{E}_j = \epsilon_o(\chi_{xx}\mathcal{E}_x + \chi_{xy}\mathcal{E}_y + \chi_{xz}\mathcal{E}_z)$.

Equation (12.2) represents a linear relation between the induced polarization and the electric field. Considering Eq. (12.2) as the first term in a Taylor series expansion of the medium's polarization, the second-order nonlinear polarization is written as

$$\mathcal{P}_i^{(2)}(t) = \epsilon_o \chi_{ijk} \mathcal{E}_j(t) \mathcal{E}_k(t) \equiv 2\epsilon_o d_{ijk} \mathcal{E}_j \mathcal{E}_k \,, \tag{12.3}$$

where χ_{ijk} is the second-order nonlinear susceptibility tensor. For historical reasons, it is written as $2d_{ijk}$, and $\mathcal{E}_j$ and $\mathcal{E}_k$ are Cartesian components of the electric field.

Equations (12.2) and (12.3) assume that the polarization at time t depends on the instantaneous values of the electric field, which implies an instantaneous medium response to the applied field. This is a good approximation in the parametric limit considered here, in which the field frequencies are assumed to be detuned far from medium resonances. Under these conditions, d_{ijk} are taken to be real and only weakly dependent on frequency. In general, however, the nonlinear susceptibilities depend on the frequencies of the applied fields (i.e., the medium is dispersive and lossy), and Eqs. (12.2) and (12.3) must be replaced by frequency-dependent relations between the Fourier components of the polarization and the field. The resulting complication is of little value for our purpose, which is to illustrate the new physics that arises when intense laser fields interact with material media and the special conditions that optimize its observation. The Taylor series expansion of polarization assumes that the applied electric field is small compared to some characteristic field. Using the classical expression for non-resonant atomic polarizability, $\alpha = 4\pi\epsilon_o a^3$, where a is the atomic size (size of the electron cloud), the linear polarization for a collection of atomic oscillators with number density $\mathcal{N}$ can be written as

$$P^{(1)} \equiv \epsilon_o \chi^{(1)} E = \mathcal{N} \alpha E = \mathcal{N} 4\pi\epsilon_o a^3 E = \mathcal{N} e a \left(\frac{E}{E_a} \right) \,, \tag{12.4}$$

where $E_a = e/4\pi\epsilon_o a^2$ is the internal atomic electric field experienced by the electron cloud. From this we conclude that $\chi^{(1)} = \mathcal{N} e a/\epsilon_o E_a$ and since the average atomic spacing in a solid is of order a, the number density $\mathcal{N} \sim 3/4\pi a^3$, so that the linear dielectric susceptibility is of order $\chi^{(1)} = \mathcal{N} a^3 4\pi \sim 3$. This corresponds to a refractive index $n \sim \sqrt{1+3} \approx 2$, which is of the expected order of magnitude for typical optical materials.

We also observe that the small parameter of expansion is the ratio of the applied electric field E to the internal atomic electric field $E_a = e/4\pi\epsilon_o a^2$. The Taylor series expansion of polarization is thus valid for $E/E_a \ll 1$. On the other hand, from Eqs. (12.2) and (12.3) we see that the second-order nonlinear polarization term will become comparable to the linear polarization term when $E \approx E_a$, giving us $\chi^{(2)} E_a^2 \approx \chi^{(1)} E_a$. Thus, the magnitude of the second-order nonlinear susceptibility is expected to be $\chi^{(2)} \approx \chi^{(1)}/E_a$. Taking twice the Bohr radius as an estimate of atomic size, $a = 2a_o = 0.1$ nm, we find $E_a \approx 1.4 \times 10^{11}$ V/m and

$$\chi^{(2)} \approx \chi^{(1)}/E_a \sim 2 \times 10^{-11}\,\mathrm{m/V}. \tag{12.5}$$

[1]Throughout the text, each Cartesian index is assumed to take on the values $(1, 2, 3)$ or (x, y, z).

Typical values of second-order nonlinear optical susceptibility $\chi^{(2)}$ range from 10^{-12} m/V to 10^{-10} m/V. A similar estimate for the third-order susceptibility yields

$$\chi^{(3)} \approx \chi^{(1)}/E_a^2 = 1.5 \times 10^{-22}\,\mathrm{m}^2/\mathrm{V}^2. \tag{12.6}$$

For comparison, $|\chi^{(3)}|$ ranges from approximately $3 \times 10^{-22}\,\mathrm{m}^2/\mathrm{V}^2$ for typical dielectric media like fused silica and BK-7 glasses, up to $2 \times 10^{-18}\,\mathrm{m}^2/\mathrm{V}^2$ for organic nonlinear optical materials. For air, it has a significantly lower value of $2 \times 10^{-25}\,\mathrm{m}^2/\mathrm{V}^2$.

We now assume that the optical field $\boldsymbol{\mathcal{E}}(t)$ incident upon a second-order nonlinear optical medium consists of two distinct frequency components $\boldsymbol{E}(\omega_1)$ and $\boldsymbol{E}(\omega_2)$ (suppressing the spatial dependence for simplicity). We write it as

$$\boldsymbol{\mathcal{E}}(t) = \mathcal{R}e\left(\boldsymbol{E}(\omega_1)e^{-i\omega_1 t} + \boldsymbol{E}(\omega_2)e^{-i\omega_2 t}\right) = \frac{1}{2}\left(\boldsymbol{E}(\omega_1)e^{-i\omega_1 t} + \boldsymbol{E}(\omega_2)e^{-i\omega_2 t} + c.c.\right), \tag{12.7}$$

where *c.c.* stands for the complex conjugate of the expression to its left inside the parentheses. Substituting this in Eq. (12.3) with tensor d_{ijk}, we find

$$\begin{aligned}
\mathcal{P}_i^{(2)}(t) &= \frac{\epsilon_o d_{ijk}}{2}\left(E_j(\omega_1)e^{-i\omega_1 t} + E_j(\omega_2)e^{-i\omega_2 t} + c.c.\right)\left(E_k(\omega_1)e^{-i\omega_1 t} + E_k(\omega_2)e^{-i\omega_2 t} + c.c.\right), \\
&= \frac{\epsilon_o dijk}{2}\Big[E_j(\omega_1)E_k(\omega_1)e^{-i2\omega_1 t} + E_j(\omega_2)E_k(\omega_2)e^{-i2\omega_2 t} + c.c. \\
&\quad + \Big(E_j(\omega_1)E_k(\omega_2) + E_j(\omega_2)E_k(\omega_1)\Big)e^{-i(\omega_1+\omega_2)t} + c.c. \\
&\quad + \Big(E_j(\omega_1)E_k^*(\omega_2) + E_j^*(\omega_2)E_k(\omega_1)\Big)e^{-i(\omega_1-\omega_2)t} + c.c. \\
&\quad + E_j(\omega_1)E_k^*(\omega_1) + E_j(\omega_2)E_k^*(\omega_2) + c.c.\Big].
\end{aligned} \tag{12.8}$$

This equation shows that the induced nonlinear polarization has a response at new frequencies, namely, the second-harmonics $2\omega_1$ and $2\omega_2$, the sum frequency $\omega_1 + \omega_2$, the difference frequency $\omega_1 - \omega_2$, and a zero-frequency (DC) nonlinear polarization associated with optical rectification. These frequencies did not exist in the incident field. Writing the ith component of induced polarization at frequency ω as

$$\mathcal{P}_i^{(2)}(\omega) = \mathcal{R}e[P_i^{(2)}(\omega)e^{-i\omega t}] = \frac{1}{2}\left[P_i(\omega)e^{-i\omega t} + c.c.\right], \tag{12.9}$$

in Eq. (12.8), we obtain the following expressions for the nonlinear polarization at the new frequencies

$$P_i^{(2)}(2\omega_1) = \epsilon_o d_{ijk}E_j(\omega_1)E_k(\omega_1), \tag{12.10a}$$

$$P_i^{(2)}(2\omega_2) = \epsilon_o d_{ijk}E_j(\omega_2)E_k(\omega_2), \tag{12.10b}$$

$$P_i^{(2)}(\omega_1 + \omega_2) = \epsilon_o d_{ijk}\left[E_j(\omega_1)E_k(\omega_2) + E_j(\omega_2)E_k(\omega_1)\right], \tag{12.10c}$$

$$P_i^{(2)}(\omega_1 - \omega_2) = \epsilon_o d_{ijk}\left[E_j(\omega_1)E_k^*(\omega_2) + E_j^*(\omega_2)E_k(\omega_1)\right], \tag{12.10d}$$

$$P_i^{(2)}(0) = \epsilon_o d_{ijk}[E_j(\omega_1)E_k^*(\omega_1) + E_j(\omega_2)E_k^*(\omega_2)]. \tag{12.10e}$$

12.1.1 Contracted Notation for Nonlinear Polarization

The tensor d_{ijk} is symmetric in its last two indices, $d_{ijk} = d_{ikj}$ because, within the instantaneous nondispersive approximation adopted here, interchange of the two input-field factors leaves the nonlinear polarization unchanged. This so-called intrinsic permutation symmetry allows us to replace d_{ijk} by the matrix d_{im} according to the prescription

jk	11	22	33	23, 32	31,13	12, 21
m	1	2	3	4	5	6

where both numerical and Roman indices $1, 2, 3$ and x, y, z have been used to denote Cartesian components. Then the nonlinear polarization for second-harmonic generation ($\omega \to 2\omega$) can be written as

$$\begin{bmatrix} P_x^{(2)}(2\omega) \\ P_y^{(2)}(2\omega) \\ P_z^{(2)}(2\omega) \end{bmatrix} = \epsilon_o \begin{bmatrix} d_{11} & d_{12} & d_{13} & d_{14} & d_{15} & d_{16} \\ d_{21} & d_{22} & d_{23} & d_{24} & d_{25} & d_{26} \\ d_{31} & d_{32} & d_{33} & d_{34} & d_{35} & d_{36} \end{bmatrix} \begin{bmatrix} E_x(\omega)^2 \\ E_y(\omega)^2 \\ E_z(\omega)^2 \\ 2E_y(\omega)E_z(\omega) \\ 2E_x(\omega)E_z(\omega) \\ 2E_x(\omega)E_y(\omega) \end{bmatrix} \tag{12.11}$$

and the nonlinear polarization for sum-frequency generation ($\omega_3 = \omega_1 + \omega_2$) as

$$\begin{bmatrix} P_x^{(2)}(\omega_3) \\ P_y^{(2)}(\omega_3) \\ P_z^{(2)}(\omega_3) \end{bmatrix} = 2\epsilon_o \begin{bmatrix} d_{11} & d_{12} & d_{13} & d_{14} & d_{15} & d_{16} \\ d_{21} & d_{22} & d_{23} & d_{24} & d_{25} & d_{26} \\ d_{31} & d_{32} & d_{33} & d_{34} & d_{35} & d_{36} \end{bmatrix} \begin{bmatrix} E_x(\omega_1)E_x(\omega_2) \\ E_y(\omega_1)E_y(\omega_2) \\ E_z(\omega_1)E_z(\omega_2) \\ E_y(\omega_1)E_z(\omega_2) + E_z(\omega_1)E_y(\omega_2) \\ E_x(\omega_1)E_z(\omega_2) + E_z(\omega_1)E_x(\omega_2) \\ E_x(\omega_1)E_y(\omega_2) + E_y(\omega_1)E_x(\omega_2) \end{bmatrix} . \tag{12.12}$$

Symmetries associated with the spatial structure of the material further reduce the number of independent elements of d_{im}. An extreme example of this is that d_{ijk} and, therefore, d_{im} vanish for a medium with inversion symmetry (centrosymmetric). Consequently, second-harmonic generation and other second-order nonlinear processes can occur only in media lacking inversion symmetry. More generally, the spatial symmetry of the crystal lattice requires some elements of the d_{im} matrix to be equal and others to be zero. When the interacting optical frequencies are far from material resonances and lie within the transparency range of the material, the nonlinear susceptibility may be treated as approximately frequency independent. Under these conditions, the indices of tensor d_{ijk} can be freely permuted. This is known as the Kleinman symmetry. As an example, for the important class of nonlinear crystals of the KDP type, only d_{14}, d_{25}, and d_{36} are non-zero and under Kleinman symmetry these three d-coefficients are equal. For these crystals, only one element d_{36} needs to be specified.

12.1.2 Effective Coupling d_{eff}

The induced nonlinear polarization acts as a source of new frequency components in Maxwell's equations and, therefore, in the wave equation and leads to a coupling of various frequency components via their nonlinear interaction. To optimize the generation of new frequencies, the interacting waves must propagate in certain directions and the electric field of each wave must possess a specific polarization (electric field orientation) relative to crystal axes. Then, for a fixed geometry (fixed propagation direction and field polarizations), the nonlinear polarizations can be expressed by scalar relations. For example, the nonlinear polarization for sum-frequency generation (12.12) is given by the scalar relation

$$P^{(2)}(\omega_3) = 2\epsilon_o d_{\text{eff}} E(\omega_1)E(\omega_2) \,, \tag{12.13}$$

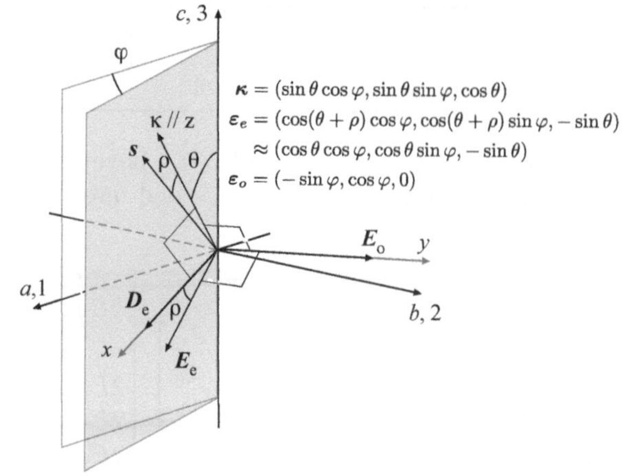

FIGURE 12.1
Geometry of a three-wave interaction in KDP. Axes 1-2-3 (or a, b, c) are the crystallographic axes. Unit vector $\boldsymbol{\kappa}$ ($\parallel$ wavevector $\boldsymbol{k}$), in the direction of the laboratory z-axis, makes an angle θ with the c-axis. $\boldsymbol{\varepsilon}_e$ is a unit vector in the direction of $\boldsymbol{D}_e$. The angle ρ between the vectors $\boldsymbol{D}_e$ and $\boldsymbol{E}_e$ is usually small and we can take $\boldsymbol{\varepsilon}_e$ to be the direction of $\boldsymbol{E}_e$. The ray vector $\boldsymbol{s}$ ($\parallel$ to the Poynting vector), $\boldsymbol{\kappa}$, $\boldsymbol{D}_e$, and $\boldsymbol{E}_e$ vectors are coplanar and lie in a plane making an angle φ with the crystallographic 1-3 plane. The ordinary polarization vector $\boldsymbol{\varepsilon}_o$($\parallel$ $\boldsymbol{E}_o$ and $\perp$ $\boldsymbol{\kappa}$) lies in the crystallographic 1-2 plane at an angle $\varphi + \pi/2$ to the crystallographic 1-3 plane.

where $P^{(2)}(\omega) = |\boldsymbol{P}^{(2)}(\omega)|$ and $E(\omega) = |\boldsymbol{E}(\omega)|$ and d_{eff} is an effective coupling constant, which can be expressed as a combination of d_{im} coefficients multiplied by appropriate trigonometric functions of the angles defining the direction of wave propagation relative to the crystal axes. This can be seen by writing each field as $\boldsymbol{\varepsilon}_m E(\omega_m)$, where $\boldsymbol{\varepsilon}_m$ is the unit polarization vector and $E(\omega_m)$ its scalar amplitude. The scalar amplitude $P(\omega)$ (scalar projection along the interacting polarization) is computed by first calculating the nonlinear polarization $\boldsymbol{P}^{(2)}(\omega)$ according to Eqs. (12.12). The coefficient d_{eff} is then given by[2]

$$d_{\text{eff}} = \varepsilon_i(\omega_3) d_{ijk} \varepsilon_j(\omega_1) \varepsilon_k(\omega_2)\,, \tag{12.14}$$

where the summation over repeated indices is implied. All Cartesian components refer to the crystallographic axes. Figure 12.1 illustrates the geometry for a three-wave interaction in KDP, a crystal belonging to the $\bar{4}2m$ crystal class. Note that the direction of wave propagation, which we usually take to be the (laboratory) z-axis, does not coincide with the crystallographic 3-axis. To avoid confusion the crystallographic axes are labeled as 1, 2, 3 with axis 3 corresponding to the crystallographic c-axis, which is also the optic axis for KDP. Using the geometry of Fig. 12.1 for Type-I phase matching in KDP [second-harmonic as an extraordinary wave (e-wave) and the two fundamental waves as ordinary waves (o-waves)], we can show [Problem 1]

$$d_{\text{eff}} = d_{14} \sin\theta \sin(2\varphi)\,. \tag{12.15}$$

[2]A method for calculating d_{eff}, along with a table of its expressions for crystals of different symmetry classes, can be found in F. Zernike and J. E. Midwinter, *Applied Nonlinear Optics*, Dover, New York, NY, 2006.

12.2 Amplitude Equations for Nonlinear Frequency Generation

In the last section we considered the second-order nonlinear response of a material system and showed that it gives rise to polarization terms at new frequencies not present in the incident light. The nonlinear polarization then acts as a source of new frequency components in Maxwell's equations and, therefore, in the wave equation. We now wish to describe the generation of these new components and how different frequency components become coupled by the nonlinear interaction.

Consider a light wave at frequency ω_m traveling in the z-direction (not necessarily co-incident with the z-axis of the crystal) in a second-order nonlinear medium. For a fixed geometry (implying a fixed propagation direction and fixed field polarizations), the analytic signal representing the scalar field component at frequency ω_m can be written as

$$E_m(z,t) = A_m(z)e^{i(k_m z - \omega_m t)}\,, \tag{12.16}$$

where $k_m = n_m\omega_m/c$ is the wavevector magnitude, n_m denotes the refractive index of the medium at frequency ω_m, and c is the speed of light in vacuum. The z-dependence of field amplitudes is explicitly indicated to account for their evolution as the light waves propagate through the medium. By expressing the total medium polarization at each frequency as the sum of the linear and second-order nonlinear polarizations,

$$P_m(z,t) = P_m^{(1)}(z,t) + P_m^{(2)}(z,t) = \left[P_m^{(1)}(z) + P_m^{(2)}(z)\right] e^{-i\omega_m t}\,. \tag{12.17}$$

Here the spatial dependence, including the appropriate wavevector phase factors, is contained in the polarization amplitudes $P_m^{(1)}(z)$ and $P_m^{(2)}(z)$. Substituting Eq. (12.17) into Maxwell's equations following the procedure established in Sec. 2.2, we find that each field component obeys a wave equation of the form (2.30). In this case, the atomic source term P_{at} is replaced by the nonlinear polarization $P_m^{(2)}$ at frequency ω_m

$$\frac{\partial^2 E_m(z,t)}{\partial z^2} - \mu_o\epsilon_m\frac{\partial^2 E_m}{\partial t^2} - \mu_o\sigma_m\frac{\partial E_m}{\partial t} = \mu_o\frac{\partial^2 P_m^{(2)}}{\partial t^2}\,. \tag{12.18}$$

The linear polarization is taken into account, in the usual way, by introducing the dielectric constant ϵ_m (or the refractive index $n_m = \sqrt{\epsilon_m/\epsilon_o}$) at frequency ω_m. The field amplitude A_m is allowed to vary slowly with z as a result of the nonlinear interaction among the waves. Substituting $E_m(z,t)$ from Eq. (12.16) into the wave equation (12.18), invoking the slowly varying envelope approximation $\left|d^2A_m/dz^2\right| \ll k_m\left|dA_m/dz\right|$ on the left-hand side, and using $\partial^2 P_m^{(2)}(z,t)/\partial t^2 = -\omega_m^2 P_m^{(2)}(z,t)$ on the right-hand side, we get

$$\frac{dA_m}{dz} + \frac{\alpha_m}{2}A_m = i\frac{\omega_m}{2\epsilon_o n_m c}P_m^{(2)}\,, \tag{12.19a}$$

where
$$\alpha_m = \frac{\sigma_m}{\epsilon_o n_m c}, \tag{12.19b}$$

is the intensity absorption coefficient, and σ_m, $n_m = \sqrt{\epsilon_m/\epsilon_o}$, and ϵ_m are, respectively, the conductivity, refractive index, and the dielectric constant of the medium at frequency ω_m. We have also used the relation between wavevector magnitude and frequency $k_m = n_m\omega_m/c$. Equation (12.19a) is the basic equation that will be used to describe coherent frequency generation through the nonlinear interaction of intense laser light with material media.

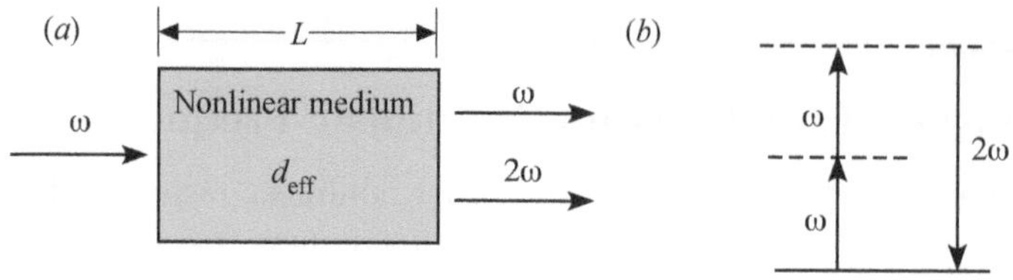

FIGURE 12.2
second-harmonic generation: (a) Interaction geometry and (b) energy-level diagram. The dashed lines represent virtual levels.

12.3 Second-Harmonic Generation

For second-harmonic generation (SHG), we have two interacting light waves with frequencies $\omega_1 = \omega$ and $\omega_2 = 2\omega$ [Fig. 12.2]. According to Eq. (12.11), the nonlinear polarization amplitudes at these frequencies are given by[3]

$$P_2^{(2)}(z) = \epsilon_o d_{\text{eff}} A_1^2 e^{2ik_1 z}\,. \tag{12.20a}$$

$$P_1^{(2)}(z) = 2\epsilon_o d_{\text{eff}} A_2 A_1^* e^{i(k_2-k_1)z}. \tag{12.20b}$$

Substituting these expressions into Eq. (12.19a), and dividing both sides of the resulting equation by $e^{ik_m z}$, we get the coupled-amplitude equations for the two frequency components

$$\frac{dA_2}{dz} = \frac{i\omega_2 d_{\text{eff}}}{2n_2 c} A_1^2 e^{i\Delta kz}, \tag{12.21a}$$

and

$$\frac{dA_1}{dz} = \frac{i\omega_1 d_{\text{eff}}}{n_1 c} A_2 A_1^* e^{-i\Delta kz}, \tag{12.21b}$$

where

$$\Delta k = 2k_1 - k_2. \tag{12.21c}$$

Here we have neglected medium losses ($\sigma_m = 0$) for simplicity. While these coupled equations can be solved exactly in terms of elliptic integrals, we instead focus on two special cases that highlight the importance of SHG for extending the frequency range of coherent laser light and identify the parameters governing conversion efficiency.

First, consider the spatial variation of the wave intensities. Recalling the standard relation $I_m = \frac{1}{2}\epsilon_o n_m c|A_m|^2$ and using the coupled amplitude equations, the evolution of field intensities is given by

$$\frac{dI_2}{dz} = \frac{1}{2}\epsilon_o n_2 c\left(A_2^*\frac{dA_2}{dz} + A_2\frac{dA_2^*}{dz}\right) = -\frac{1}{2}\epsilon_o\omega_2 d_{\text{eff}}\,\mathcal{I}m\left[A_2^* A_1^2 e^{i\Delta kz}\right], \tag{12.22}$$

$$\frac{dI_1}{dz} = \frac{1}{2}\epsilon_o n_1 c\left(A_1^*\frac{dA_1}{dz} + A_1\frac{dA_1^*}{dz}\right) = -\epsilon_o\omega_1 d_{\text{eff}}\,\mathcal{I}m\left[A_2 A_1^{*2} e^{-i\Delta kz}\right]. \tag{12.23}$$

Noting that $\omega_2 = 2\omega_1$ and a complex number and its conjugate have imaginary parts of opposite signs, the spatial derivatives of wave intensities satisfy

$$\frac{dI_2}{dz} = -\frac{dI_1}{dz} = -\frac{1}{2}\epsilon_o\omega_2 d_{\text{eff}}\,\mathcal{I}m\left[A_2^* A_1^2 e^{i\Delta kz}\right], \tag{12.24}$$

[3]For SHG, the two input fields in the sum-frequency process have the same frequency, $\omega + \omega = 2\omega$. We therefore relabel the generated frequency ω_3 of Eq. (12.11) as $\omega_2 = 2\omega$.

Equation (12.24) shows that the increase in the intensity of one wave exactly equals the decrease in the intensity of the other. Consequently, the total intensity $I_1 + I_2$ remains constant, as expected for a lossless medium. If we recall that the intensity ($\mathrm{W/m^2}$) of narrowband fields is related to photon flux (photons $\mathrm{m^{-2}\ s^{-1}}$) via $\Phi_m = I_m/(\hbar\omega_m)$, Eq. (12.24) implies

$$\frac{d}{dz}\frac{I_2}{\omega_2} = -\frac{1}{2}\frac{d}{dz}\frac{I_1}{\omega_1}, \tag{12.25}$$

which states that the rate at which second-harmonic (ω_2) photons are created equals half the rate at which fundamental (ω_1) photons are destroyed. This is the Manley-Rowe relation for second-harmonic generation. Physically, this means that generation of one second-harmonic photon requires the simultaneous annihilation of two fundamental photons or, conversely, the destruction of one second-harmonic photon produces two fundamental photons.

Equations (12.21) and (12.25) reveal that the direction of power flow (growth or attenuation) depends on the relative phases of the field amplitudes and the accumulated phase mismatch Δkz. Because dI_2/dz and dI_1/dz always have opposite signs, the direction of power flow between the waves changes periodically during propagation.

Now, let us consider second-harmonic generation when only the fundamental wave is present at the input $z = 0$ of a nonlinear medium of length L, that is, $A_1(0) \neq 0$ and $A_2(0) = 0$ [Fig. 12.2]. When the accumulated mismatch $|\Delta k|L$ is large, the conversion remains small. We can therefore treat the fundamental field as approximately constant $A_1(z) \approx A_1(0)$, corresponding to the undepleted-pump approximation in Eq. (12.21a). The resulting equation is readily integrated over the crystal length to yield

$$A_2(L) = \frac{i\omega_1 d_{\text{eff}} L}{n_2 c} A_1^2(0) e^{i\Delta kL/2} \left(\frac{\sin(\Delta kL/2)}{\Delta kL/2}\right). \tag{12.26}$$

The resulting second-harmonic intensity grows according to

$$I_2 = \frac{2\omega_1^2 d_{\text{eff}}^2 L^2}{\epsilon_o n_2 n_1^2 c^3} I_1^2 \left(\frac{\sin(\Delta kL/2)}{\Delta kL/2}\right)^2. \tag{12.27}$$

By introducing a characteristic second-harmonic-generation length

$$L_{\text{SHG}} = \left(\frac{\epsilon_o n_1^2 n_2 c^3}{2\omega_1^2 d_{\text{eff}}^2 I_1}\right)^{1/2} = \frac{\lambda_1\sqrt{n_1 n_2}}{2\pi d_{\text{eff}}|A_1(0)|}, \tag{12.28}$$

the second-harmonic-generation efficiency η_{SHG} can be expressed as

$$\eta_{\text{SHG}} \equiv \frac{I_2}{I_1(0)} = \frac{L^2}{L_{\text{SHG}}^2}\left(\frac{\sin(\Delta kL/2)}{\Delta kL/2}\right)^2 = \frac{4}{(\Delta kL_{\text{SHG}})^2}\sin^2(\Delta kL/2). \tag{12.29}$$

The maximum efficiency of SHG is thus $\frac{L^2}{L_{\text{SHG}}^2}$, proportional to L^2, which is strictly attained under perfect phase-matching $\Delta k = 0$. For nonzero wavevector mismatch, the efficiency oscillates with crystal length; its maximum value $4/(\Delta kL_{\text{SHG}})^2$ is typically much smaller than its phase-matched value $\frac{L^2}{L_{\text{SHG}}^2}$.

As conversion efficiency increases, the undepleted pump approximation breaks down, and Eq. (12.29) no longer accurately describes the phase-mismatch dependence. Nevertheless, even in the regime of strong pump depletion, efficient conversion over long interaction lengths requires phase matching, with $\Delta k = 0$ representing the ideal condition. For the perfectly phase-matched case ($\Delta k = 0$) with only the fundamental wave incident on the crystal, it can be shown [Problem 3] that the depletion dynamics follow

$$\frac{I_2}{I_1(0)} = \tanh^2(z/L_{\text{SHG}}), \qquad \frac{I_1}{I_1(0)} = \text{sech}^2(z/L_{\text{SHG}}). \tag{12.30}$$

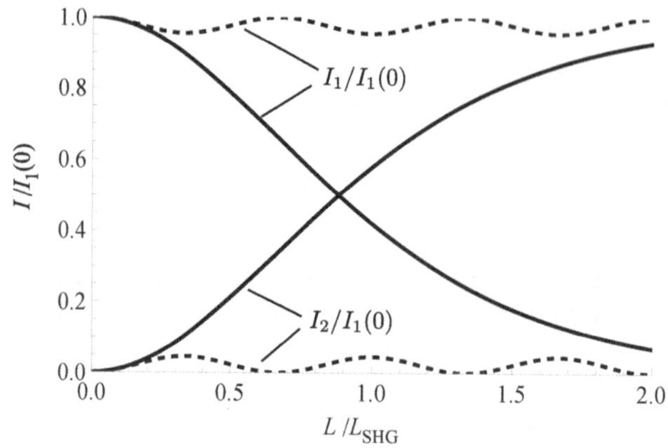

FIGURE 12.3
Normalized second-harmonic intensity $I_2/I_1(0)$ (the same as the efficiency, η_{SHG}) and normalized fundamental intensity $I_1/I_1(0)$ as functions of the normalized crystal length L/L_{SHG} for perfect phase-matching ($\Delta k L_{SHG} = 0$, continuous curves) and for a finite phase mismatch ($\Delta k L_{SHG} = 3\pi/2$, dashed curves).

12.4 Phase Matching

The plot of η_{SHG} in Figure 12.3 for finite Δk shows that the SHG efficiency oscillates as L increases. The oscillatory behavior arises from the phase accumulated because of the wavevector mismatch. When $\Delta k L = \pi$ (modulo 2π), the second-harmonic wave is out of phase with its driving nonlinear polarization [Eq. (12.21a)], causing power to flow back into the fundamental. The distance $L_{coh} = \pi/\ |\Delta k|$ is defined as the coherent buildup length of the second-harmonic. For a nonzero wavevector mismatch, the first maximum of $I_2/I_1(0)$ occurs at $L = L_{coh}$. For typical wavevector mismatch in real crystals, this conversion efficiency remains severely limited. Conversely, the behavior of conversion efficiency under perfect phase-matching ($\Delta k = 0$), described by Eq. (12.30) and represented by the solid lines in Fig. 12.3, demonstrates that appreciable conversion of fundamental-beam power into second-harmonic power is possible.

The phase-matching condition $\Delta k = 0$ is often difficult to achieve because the refractive index of lossless (transparent) materials in the optical range (ω to 2ω) is an increasing function of frequency (normal dispersion). Consequently, perfect phase matching of co-propagating beams requires $2k_1 - k_2 = [2\omega_1 n(\omega_1) - \omega_2 n(\omega_2)]/c = 0$. For second-harmonic generation, with $\omega_2 = 2\omega_1$, this condition requires

$$n(\omega_1) = n(2\omega_1)\,, \tag{12.31}$$

which clearly cannot be achieved when $n(\omega)$ is a monotonically increasing function of ω. In principle, phase-matching can be achieved if the two frequencies fall in the region of anomalous dispersion, where the refractive index decreases with increasing frequency.

12.4.1 Birefringence Phase Matching (BPM)

The most common method for phase matching relies on birefringence displayed by many crystals. Birefringence is an optical property for which, for a given propagation direction, the refractive index experienced by a light wave depends on its polarization relative to the crystallographic axes. To achieve phase matching through birefringence, the polarization of the higher frequency wave is chosen such that it experiences the lower of the two possible refractive indices. For a negative uniaxial crystal, this corresponds to the extraordinary polarization (e). In the following we restrict our considerations to negative uniaxial media. There are two choices for the polarizations of the lower-frequency waves, which are termed Type I and Type II phase matching.

In Type I phase matching, the two lower-frequency waves share the same polarization, whereas in Type II phase matching, they have orthogonal polarizations. For many commonly used nonlinear crystals and interactions, Type I phase matching is experimentally simpler than Type II. Birefringent phase matching (BPM) requires precise angular orientation of the crystal with respect to the propagation vector $\boldsymbol{k}$ of the incident light. It is most simply described for a uniaxial crystal. We restrict our discussion to this case.

Uniaxial crystals are characterized by a unique direction known as the optic axis (or c-axis corresponding to the crystallographic z-axis). Light wave polarization perpendicular to the plane containing the wavevector $\boldsymbol{k}$ and the optic axis is called ordinary (o) polarization. A light wave with o-polarization experiences the ordinary refractive index n_o, independent of the direction of $\boldsymbol{k}$. Polarization within the plane containing $\boldsymbol{k}$ and the optic axis is called the extraordinary (e) polarization. A light wave with e-polarization experiences an extraordinary refractive index $n_e(\theta)$ that depends on the angle θ between the optic axis and the wavevector $\boldsymbol{k}$, given by

$$\frac{1}{n_e^2(\theta)} = \frac{\sin^2\theta}{n_e^2} + \frac{\cos^2\theta}{n_o^2}. \tag{12.32}$$

where n_e is the principal value of the extraordinary refractive index, which occurs when the wave propagates perpendicular to the c-axis ($\theta = 90^o$). Note that $n_e(\theta)$ reduces to the principal value n_e for $\theta = 90^o$, and to n_o for $\theta = 0$. Phase matching is achieved by tuning the propagation angle to a specific value $\theta = \theta_m$, such that the refractive index $n_e(\theta_m)$ satisfies the condition $\Delta k = 0$. For this reason, birefringent phase matching is also referred to as angle phase matching.

Uniaxial crystals with $n_e < n_o$ are classified as negative uniaxial and those with $n_e > n_o$ as positive uniaxial. Phase matching types are also described by explicitly stating the wave polarizations in order of increasing frequency. For example, Type I phase matching in a negative uniaxial crystal is referred to *ooe* phase matching, and Type II as *eoe* or *oee* phase matching, where the final entry always denotes the highest-frequency wave.

Beam Walk-off in BPM and SHG with Gaussian Beams

If the angle θ between the propagation direction and the optic axis is neither 0 nor 90 degrees, the Poynting vector $\boldsymbol{S}$ (energy flow vector) and the vector $\boldsymbol{k}$ are not parallel for an extraordinary wave. Consequently, the ordinary and extraordinary waves with collinear wavevectors spatially diverge from one another as they propagate through the crystal. The walk-off angle ρ is given by $\tan(\theta + \rho) = \frac{n_o^2}{n_e^2}\tan\theta$. For small birefringence Δn, the walk-off angle can be approximated by $\rho \approx \frac{n_e - n_o}{n_o}\sin(2\theta)$, which gives a signed walk-off angle.

For finite-sized beams, such as Gaussian laser beams, this walk-off effect limits the spatial overlap of the interacting waves and decreases the efficiency of the nonlinear mixing process. Under these conditions, increasing the crystal length will yield diminishing returns. Weak focusing will reduce conversion efficiency due to a lower beam intensity. Excessively tight focusing will degrade conversion efficiency due to angular phase-mismatch and pronounced beam walk-off. The optimum configuration balances these factors by matching the confocal parameter of the fundamental beam, $b = 2z_R = 2(\pi n_1 w_{o1}^2/\lambda)$, to the crystal length L.

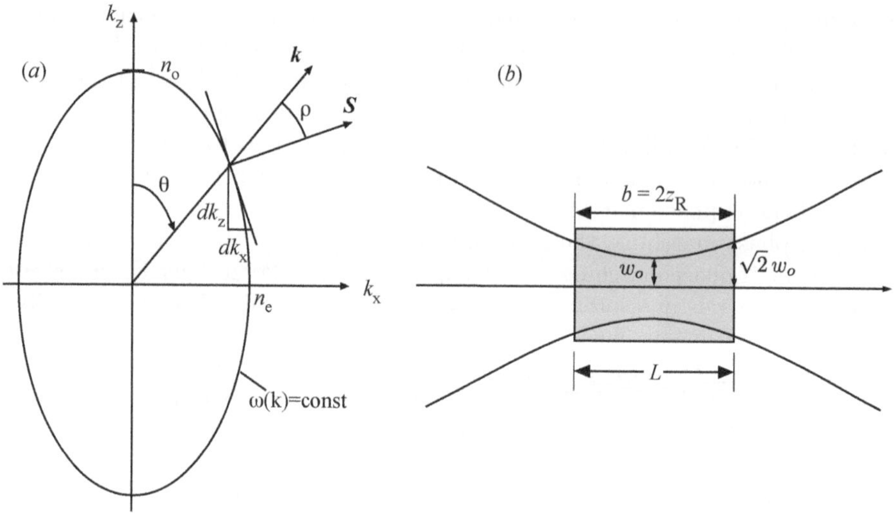

FIGURE 12.4
(a) Beam walk-off angle and (b) Gaussian beam focusing in SHG.

As illustrated in Fig. 12.4, for $L > b$, the beam cross-section increases and the conversion efficiency drops. In the near-zone regime, $(L \leq b)$, we can approximate the fundamental intensity as $I_1 = 2P_1/\pi w_{o1}^2$, where w_{o1} is the beam waist radius of the fundamental beam inside the crystal. The generated second-harmonic inherits the same confocal parameter as the fundamental. Analogous to the plane-wave case, the phase-matched $(n_1 = n_2)$ efficiency for SHG can be written in terms of optical powers using Eq. (12.29) as

$$\frac{P_2}{P_1} = \frac{1}{2}\frac{2\omega_1^2 d_{\text{eff}}^2}{\epsilon_o n_1^3 c^3} L^2 \frac{2P_1}{\pi w_{o1}^2} = \frac{2\omega_1^3 d_{\text{eff}}^2}{\epsilon_o \pi n_1^2 c^4} P_1 L\,, \tag{12.33}$$

where we have set $L = b = \frac{2\pi n_1 w_{o1}^2}{\lambda_1}$. A more careful optimization yields $L = 2.84b$ for maximum conversion efficiency. Using these optimized focusing conditions, maximum power efficiency is given by

$$\eta_{\text{opt}}^{\text{power}} = \frac{P_2}{P_1} = \frac{5.68\omega_1^3 d_{\text{eff}}^2}{\epsilon_o \pi c^4 n_1^2} P_1 L = \frac{5.68\omega_1^3 d_{\text{eff}}^2}{\epsilon_o \pi c^4 n_1^2} P_1 L\,. \tag{12.34}$$

Example 12.1 *Consider SHG with 1 W of fundamental light at $\lambda_1 = 1.06\,\mu m$ in a 10-mm-long KDP crystal $(n_1 = 1.5)$ using Type-I phase matching with $d_{\text{eff}} = 0.24$ pm/V.*

Using these parameters and assuming phase matching $(\theta_m \approx 41^0)$ and optimum Gaussian beam focusing, we obtain

$$\begin{aligned}\eta_{\text{opt}}^{\text{power}} &= \left(\frac{4}{4\pi\epsilon_o}\right)\frac{5.68(8\pi^3)d_{\text{eff}}^2}{\lambda_1^3 c n_1^2} P_1 L = 36\times 10^9 \frac{5.68(8\pi^3)(0.24\times 10^{-12})^2}{(1.06\times 10^{-6})^3(3\times 10^8)(1.5)^2}\cdot 1\cdot 10^{-2} \\ &= 4\times 10^{-5}\end{aligned}$$

This corresponds to approximately 40 μW of second-harmonic light at 0.53 μm.

12.4.2 Quasi-Phase-Matching (QPM)

Phase matching using birefringence is not always possible either because the material lacks sufficient birefringence, as in gallium arsenide, or because its natural birefringence is

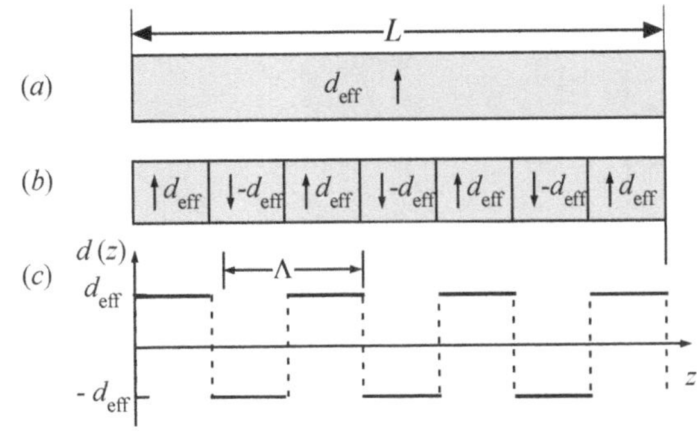

FIGURE 12.5
Second-order nonlinear optical material as (a) a homogeneous single crystal, (b) a periodically poled material in which the positive c-axis alternates in orientation with period Λ, (c) spatial variation of d_{eff}.

insufficient to compensate for the dispersion of the linear refractive indices over the wavelength range of interest. In such cases another technique known as quasi-phase-matching (QPM) can be used. QPM is often implemented in a Type-0 configuration, in which all interacting waves have the same polarization. The principle of quasi-phase-matching can be grasped from Fig. 12.5, which compares a single crystal of nonlinear optical material [Fig. 12.5(a)] with a periodically poled material structure [Fig. 12.5(b)]. In the PPM structure, the orientation of the c-axis flips by 180° from one layer to the next. Inverting the direction of the c-axis changes the sign of the effective nonlinear coefficient d_{eff}. This periodic spatial modulation of the sign of d_{eff} can compensate for the wavevector mismatch Δk. Figure 12.6 compares the growth of SH intensity under quasi-phase-matched conditions with its growth under perfectly and imperfectly phase-matched conditions.

In a perfectly phase-matched ($\Delta k = 0$) interaction, the second-harmonic intensity grows quadratically with propagation distance. For nonzero wavevector mismatch ($\Delta k \neq 0$), the SH intensity oscillates with propagation distance with very low conversion efficiency. With quasi-phase-matching in a periodically poled medium, the inversion period of d_{eff} is chosen to be equal to twice the coherent buildup length, $\Lambda = 2L_{\text{coh}}$. Consequently, the intensity grows monotonically as the propagation distance increases, albeit less rapidly than in a perfectly phase-matched interaction. In this case, each time the SH wave amplitude is about to decrease due to the destructive interference caused by the wavevector mismatch, the reversal of the sign of d_{eff} shifts the relative phase by π. This reset allows the field amplitude to continue growing monotonically, accompanied by minor spatial ripples.

For an analytic description of quasi-phase-matching, the spatial variation of the nonlinear coupling coefficient shown in Fig. 12.5(b) can be approximated by a square-wave function $d(z)$ [Fig. 12.5(c)], which has the Fourier series representation

$$d(z) = d_{\text{eff}} \sum_{m=-\infty}^{\infty} c_m e^{i2\pi m z/\Lambda} . \tag{12.35}$$

In a periodically poled material, this position-dependent $d(z)$ replaces d_{eff} in the coupled-amplitude Eqs. (12.21). To achieve quasi-phase-matching, we can isolate and focus on a single spatial harmonic from the Fourier series. Because the Fourier amplitude c_m decreases with increasing harmonic order m, we consider the fundamental component $m = 1$, which has an amplitude given by $c_1 = 2/\pi$. The effective phase-matching condition then becomes

$$2k_1 - k_2 - 2\pi/\Lambda = 0 . \tag{12.36}$$

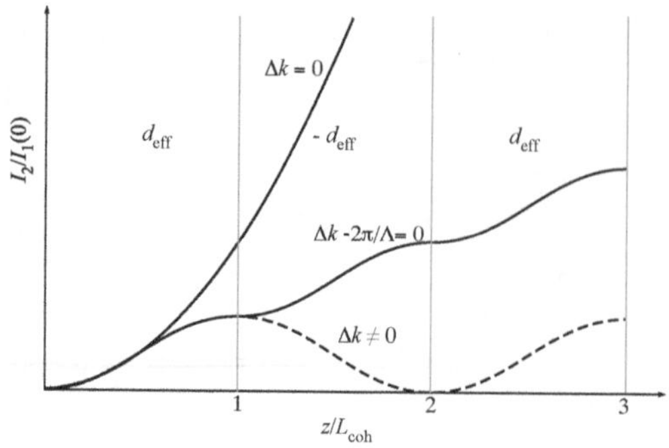

FIGURE 12.6
Growth of SH field intensity for three different phase matching conditions as labeled: (i) $\Delta k = 0$, (ii) $\Delta k - 2\pi/\Lambda = 0$, and (iii) $\Delta k \neq 0$.

This condition is satisfied if the spatial period Λ of the PPM structure is chosen such that

$$\Lambda = 2L_{\text{coh}} = 2\pi/|2k_1 - k_2| = \frac{\lambda_1}{2|n(\omega_1) - n(\omega_2)|}. \tag{12.37}$$

The required PPM period Λ typically ranges from 5 μm to 100 μm. For example, for second-harmonic generation of 1.06 μm in lithium niobate, the index mismatch $\Delta n = |n_2 - n_1| \approx 0.09$ gives a coherent buildup length of $L_{\text{coh}} = \lambda_1/4\Delta n \approx 2.9\,\mu$m and a poling period of $\Lambda = 5.8\,\mu\text{m} \approx 6\,\mu$m. Although the effective nonlinear coefficient is reduced from d_{eff} to $2d_{\text{eff}}/\pi$, this reduction can be readily made up by using longer interaction lengths.

QPM was proposed as a phase-matching technique before BPM. The first QPM structures were fabricated by slicing a nonlinear optical medium into micron-thick, coherence-length segments and rotating alternating segments by 180°. Handling these ultrathin slices was difficult and the increased number of surfaces added significant loss that degraded conversion efficiency. The major breakthrough came with the discovery that the orientation of ferroelectric domains in a thin lithium niobate crystal can be permanently flipped by applying a sufficiently large voltage across patterned electrodes, thereby producing a static electric field that reverses the ferroelectric domains. Lithium niobate (LN) and lithium tantalate (LT) are among the most widely used periodically poled ferroelectric materials. The d_{33} coefficient in LN (28 pm/V) and LT (16 pm/V) are among the largest in commonly used ferroelectrics. Both materials offer transparency from the near-UV to the infrared (300 nm–4 μm).

In summary, QPM offers several advantages over traditional birefringent phase-matching: it can utilize a large nonlinear coefficient (d_{33} in LN and LT); it is highly effective in isotropic crystals that exhibit weak or no birefringence, (such as GaAs); and it can eliminate birefringent spatial walk-off when implemented in a collinear Type-0 configuration, in which all interacting waves have the same polarization.

Example 12.2 *Let us revisit SHG with 1W of 1.06 μm fundamental light in a 1-cm-long periodically poled lithium niobate (PPLN) crystal for which the material coefficient is $d_{\text{eff}} = d_{33} = 28$ pm/V, $n = 2.14$. Use of QPM in this case allows all beams to have the same polarization (Type-0) and thus access the largest nonlinear coefficient d_{33}.*

For QPM in PPLN, $d_{\text{eff}} \to 2d_{33}/\pi = 18\,\text{pm/V}$. Using this in Eq. (12.33), we find power conversion efficiency

$$\begin{aligned}\eta_{\text{opt}}^{\text{power}} &= \left(\frac{4}{4\pi\epsilon_o}\right)\frac{5.68(8\pi^3)(d_{\text{eff}})^2}{\lambda_1^3 c n_1^2}P_1 L \\ &= 36\times10^9\frac{5.68(8\pi^3)(18\times10^{-12})^2}{(1.06\times10^{-6})^3(3\times10^8)(2.14)^2}10^{-2} = 0.1\end{aligned}$$

This means about 100 mW of second-harmonic power with 1 W of fundamental.

12.5 Sum- and Difference-Frequency Generation

12.5.1 Sum-Frequency Generation

In sum frequency generation (SFG), as shown in Fig. 12.7, two low-frequency waves at ω_1 and ω_2 interact to generate a high-frequency wave ω_3 such that $\omega_3 = \omega_1 + \omega_2$. The nonlinear polarization $P_m^{(2)}$ at each frequency ω_m $(m = 1, 2, 3)$ in scalar form can be written as

$$P_3^{(2)}(z) = 2\epsilon_o d_{\text{eff}} A_1(z) A_2(z) e^{i(k_1+k_2)z}, \tag{12.38a}$$

$$P_1^{(2)}(z) = 2\epsilon_o d_{\text{eff}} A_2^*(z) A_3(z) e^{i(k_3-k_2)z}, \tag{12.38b}$$

$$P_2^{(2)}(z) = 2\epsilon_o d_{\text{eff}} A_1^*(z) A_3(z) e^{i(k_3-k_1)z}, \tag{12.38c}$$

Using these in Eq. (12.19a) and ignoring medium losses, we obtain the following coupled-amplitude equations for the three interacting waves:

$$\frac{dA_3}{dz} = i\frac{\omega_3}{n_3 c} d_{\text{eff}} A_1 A_2 e^{i\Delta kz}, \tag{12.39a}$$

$$\frac{dA_2}{dz} = i\frac{\omega_2}{n_2 c} d_{\text{eff}} A_3 A_1^* e^{-i\Delta kz}, \tag{12.39b}$$

$$\frac{dA_1}{dz} = i\frac{\omega_1}{n_1 c} d_{\text{eff}} A_3 A_2^* e^{-i\Delta kz}, \tag{12.39c}$$

with

$$\Delta k = k_1 + k_2 - k_3\,. \tag{12.39d}$$

An important application of these equations is the generation of short-wavelength radiation at ω_3 by mixing two lower-frequency laser beams inside a suitable nonlinear crystal. For

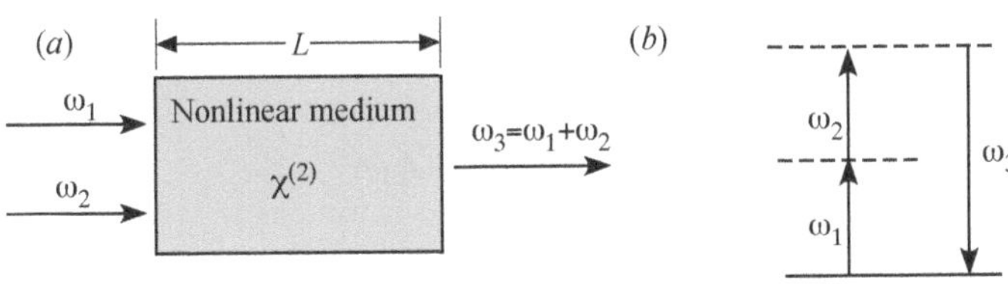

FIGURE 12.7
Sum frequency generation: (a) Interaction geometry; (b) energy-level diagram. The dashed lines represent virtual levels.

example, exciting the two-photon transition ($1s \rightarrow 2s$) in atomic hydrogen requires 243 nm light. This can be synthesized via SFG by combining 351 nm radiation from an argon-ion laser (ω_1) with 789-nm tunable radiation from a Ti:sapphire laser (ω_2) in KDP. Similar to the case of SHG with low conversion efficiency, we can assume the input amplitudes $A_1(0)$ and $A_2(0)$ to remain approximately constant over a crystal of length L. Solving Eq. (12.39a) under this approximation yields the sum-frequency output intensity $I_3 = \frac{1}{2}\epsilon_o n_3 c|A_3(L)|^2$):

$$I_3(L) = \frac{2\omega_3^2 d_{\text{eff}}^2}{\epsilon_o n_1 n_2 n_3 c^2} L^2 I_1 I_2 \left(\frac{\sin \Delta kL/2}{\Delta kL/2} \right)^2 . \tag{12.40}$$

Phase matching considerations here are similar to the SHG case, requiring $\omega_3 n(\omega_3) = \omega_1 n(\omega_1) + \omega_2 n(\omega_2)$ to be satisfied for suitable polarizations of the three waves. For phase-matched operation, the expression simplifies to

$$I_3 = \frac{2\omega_3^2 d_{\text{eff}}^2}{\epsilon_o n_1 n_2 n_3 c^2} L^2 I_1 I_2 . \tag{12.41}$$

To generate 243 nm light described above, phase matching is achieved using Type I (*ooe*) polarization in KDP. Recalling that $\omega = 2\pi c/\lambda$, the energy conservation requirement $\omega_3 = \omega_1 + \omega_2$ dictates that the wavelengths must satisfy $\frac{1}{\lambda_3} = \frac{1}{\lambda_1} + \frac{1}{\lambda_2}$. Given $\lambda_1 = 351$ nm and $\lambda_2 = 789$ nm, the exact output wavelength is $\lambda_3 = 242.92$ nm. For perfect phase matching, the three refractive indices must satisfy

$$\frac{n_o(351)}{351} + \frac{n_o(789)}{789} = \frac{n_e(\theta, 243)}{242.92} . \tag{12.42}$$

Using the refractive index values $n_o(351) = 1.53236, n_o(790) = 1.50226, n_e(243) = 1.522638$, the left-hand side evaluates to $\frac{n_o}{351} + \frac{n_o}{789} = \frac{1}{159.497}$ (where the tabulated index values are used as approximations). Exact phase matching is achieved by finding the angle for which $n_e(\theta, 243) = \frac{242.92}{159.497} = 1.52304$. Substituting the principal values $n_e(243) = 1.522638$ and $n_o(243) = 1.57274$ into Eq. (12.32) yields a phase-matching angle of $\theta_m \approx 85^o$.

12.5.2 Frequency Upconversion

Frequency upconversion is a special case of sum frequency generation where a weak, low-frequency signal at ω_1 is converted to a higher frequency at ω_3 by mixing it with a strong laser field at ω_2. This technique is useful for converting weak IR signals into the visible for better detection or analysis. Under the assumption of a strong undepleted pump field at ω_2 and perfect phase matching $\Delta k = 0$, the system reduces to two coupled differential equations

$$\frac{dA_3}{dz} = i\frac{\omega_3 d_{\text{eff}}}{n_3 c} A_1 A_2 , \tag{12.43a}$$

$$\frac{dA_1}{dz} = i\frac{\omega_1 d_{\text{eff}}}{n_1 c} A_3 A_2^* , \tag{12.43b}$$

subject to the boundary conditions $A_3(0) = 0, A_1(0) = A_{10}$, and $A_2(0) = A_2 = |A_2|e^{i\phi_2}$. Differentiating the equation for A_1 and using the equation for A_3 yields an uncoupled second-order differential equation for A_1. Its solutions are sine and cosine functions.

Applying the boundary conditions then gives

$$A_1(z) = A_{10}\cos\kappa z\,, \tag{12.44a}$$

$$A_3(z) = \left(\frac{dA_1(z)}{dz}\right)\left(\frac{n_1 c}{i\omega_1 d_{\text{eff}} A_2^*}\right) = i\sqrt{\frac{n_1\omega_3}{n_3\omega_1}}A_{01}e^{i\phi_2}\sin\kappa z \tag{12.44b}$$

$$\kappa = \sqrt{\frac{\omega_1\omega_3 d_{\text{eff}}^2|A_2|^2}{n_1 n_3 c^2}}\,. \tag{12.44c}$$

This solution shows that the up-converted signal wave inherits the phase ϕ_2 of the pump field A_2. From these amplitudes, we obtain wave intensities as

$$I_1(z) = I_{10}\cos^2\kappa z\,, \tag{12.45}$$

$$I_3(z) = \frac{\omega_3}{\omega_1}I_{10}\sin^2\kappa z\,. \tag{12.46}$$

These equations show that complete conversion occurs for $\kappa z = \pi/2$, where the maximum intensities satisfy $I_{3\text{max}}/\omega_3 = I_{10}/\omega_1$. This result directly confirms the Manley-Rowe relations, which state that every generated up-converted photon corresponds to the destruction of one fundamental signal photon. When the interaction length exceeds $\kappa z = \pi/2$, energy flows back into ω_1.

12.5.3 Difference-Frequency Generation and Parametric Amplification

Difference-frequency generation (DFG) represents an interesting case of three-wave interaction in which two incident optical waves at frequencies ω_3 (pump) and ω_2 (signal) interact to generate a lower frequency $\omega_1 = \omega_3 - \omega_2$ (idler). It is used to generate long-wavelength infrared radiation with readily available near-infrared pump and signal laser sources. For example, by mixing the outputs of a 1.562-μm diode laser and a 1.083-μm ytterbium-fiber laser in a nonlinear crystal, one can produce a mid-infrared output at 3.5 μm. Similarly, terahertz radiation can be produced by mixing two closely spaced visible or near-infrared signals. Two frequency components of a broadband ultrashort pulse interacting in a nonlinear crystal can also result in difference-frequency generation at terahertz frequencies. The DFG output inherits the coherence properties of the pump lasers.

If the input pump A_3 and signal A_2 are sufficiently strong that their depletion can be neglected over a crystal of length L, and if perfect phase matching $\Delta k = 0$ is assumed, the generated idler intensity is easily found to be

$$I_1 = \frac{2\omega_1^2 d_{\text{eff}}^2}{\epsilon_o n_1 n_2 n_3 c^3}L^2 I_3 I_2\,. \tag{12.47}$$

This expression is similar to the expression for sum frequency generation.

We now consider the case where a strong pump at ω_3 (so that A_3 can be treated as a constant) converts light from ω_2 to ω_1. The coupled wave equations in this case are given by

$$\frac{dA_1}{dz} = i\frac{\omega_1 d_{\text{eff}}}{n_1 c}A_3 A_2^* e^{-i\Delta k z}\,, \tag{12.48a}$$

$$\frac{dA_2}{dz} = i\frac{\omega_2 d_{\text{eff}}}{n_2 c}A_3 A_1^* e^{-i\Delta k z}\,, \tag{12.48b}$$

subject to the boundary conditions $A_3(0) = A_3 = |A_3|e^{i\phi_3}$, $A_1(0) = 0$, and $A_2(0) = A_{20} = |A_{20}|e^{i\phi_2}$. For $\Delta k = 0$, these can be solved following the procedure used in solving the

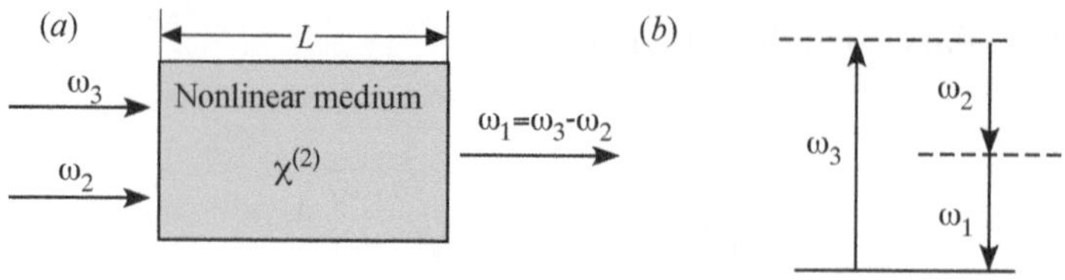

FIGURE 12.8
Difference-frequency generation: (a) Interaction geometry; (b) energy-level description. The dashed lines represent virtual levels.

coupled-amplitude equations for upconversion. Note that on the right-hand side we have the complex conjugates of signal and idler fields. Differentiating Eq. (12.48a) with respect to z and using Eq. (12.48b) to eliminate dA_2^*/dz from the right-hand side, we find that A_1 obeys a second-order linear differential equation whose fundamental solutions are hyperbolic sine and cosine functions. Armed with this knowledge, and using the boundary conditions, we find that the signal and idler amplitudes evolve according to

$$A_1(z) = i\sqrt{\frac{n_2\omega_1}{n_1\omega_2}}|A_{20}|e^{i(\phi_3-\phi_2)}\sinh\kappa z\,, \tag{12.49a}$$

$$A_2(z) = A_{20}\cosh\kappa z \tag{12.49b}$$

$$\kappa = \sqrt{\frac{\omega_1\omega_2 d_{\text{eff}}^2|A_3|^2}{n_1 n_2 c^2}} = \sqrt{\frac{2\omega_1\omega_2 d_{\text{eff}}^2 I_3}{\epsilon_o n_1 n_2 n_3 c^3}}\,. \tag{12.49c}$$

From the signal and idler wave amplitudes, we find the intensities of the two waves evolve according to

$$I_1(z) = \frac{\omega_1}{\omega_2}I_{20}\sinh^2\kappa z, \tag{12.50a}$$

$$I_2(z) = I_{20}\cosh^2\kappa z\,. \tag{12.50b}$$

Comparing this solution with the one for frequency upconversion, we find a qualitatively different behavior; namely, both the signal (ω_2) and idler (ω_1) waves grow exponentially for $\kappa z \gg 1$. We also observe that while the amplified signal field ω_2 retains its initial phase ($A_{20} = |A_{20}|e^{i\phi_2}$), the phase of the generated idler wave ω_1 depends on the phase difference between the pump and signal waves. The energy required for the growth of signal and idler waves is supplied by the pump wave, though it is masked here by the small-signal

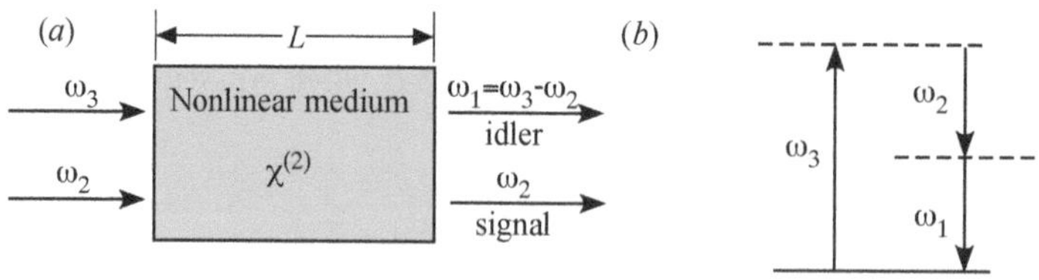

FIGURE 12.9
Optical parametric amplification: (a) Interaction geometry (b) Energy-level description. The dashed lines denote virtual levels.

assumption that A_3 = const. In the high conversion efficiency regime ($\kappa z \gg 1$), the ratio of the signal and idler intensities satisfies the relation

$$\frac{I_1}{\omega_1} = \frac{I_2}{\omega_2}, \tag{12.51}$$

which states that the photon fluxes of both fields are equal.

The amplification of the fields may be viewed heuristically as stimulated parametric generation: the incident signal at ω_2 stimulates the conversion of pump photons into signal-idler photon pairs. Consequently, the difference-frequency generation (DFG) process leads to both an amplification of the input signal wave and the growth of an idler wave from the background. For this reason, difference-frequency generation under these operating conditions is also referred to as optical parametric amplification (OPA).

Even in the absence of injected signal and idler fields, pump photons can spontaneously convert into correlated signal-idler photon pairs. This process is known as spontaneous parametric downconversion (SPDC), or parametric fluorescence. In this case, the phase-matching condition is the vector equation $\boldsymbol{k}_1 + \boldsymbol{k}_2 = \boldsymbol{k}_3$, which can be satisfied by non-collinear (traveling in different directions) signal and idler photon pairs. Light generated in SPDC exhibits certain properties (correlations). The correlations exhibited by SPDC light require a quantum description of the electromagnetic field. While this falls outside the scope of this book, we present a heuristic description of these photon correlations in the next section.

12.5.4 Spontaneous Parametric Downconversion (SPDC)

The properties of SPDC differ qualitatively for Type-I and Type-II phase-matching configurations. For simplicity, we assume a degenerate ($\omega_1 = \omega_2$) downconversion process in a negative uniaxial crystal and consider the two cases separately. The downconversion geometry for a negative uniaxial crystal under Type-I phase-matching conditions is shown in Fig. 12.10. At the exact collinear phase-matching angle (θ) the downconverted light is observed

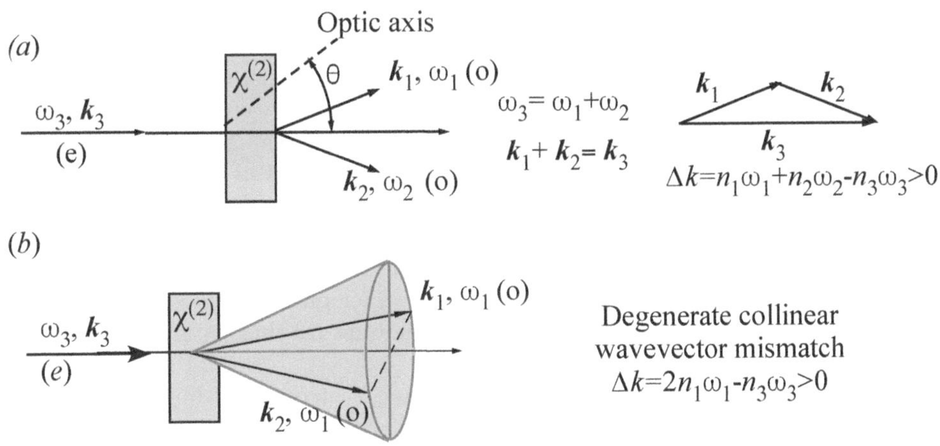

FIGURE 12.10
(a) Type-I spontaneous parametric downconversion (fluorescence) in a negative uniaxial nonlinear crystal. Both energy and momentum conservation must be obeyed at the single photon level. Many photon pairs $(\boldsymbol{k}_1, \omega_1)$ and $(\boldsymbol{k}_2, \omega_2)$ satisfy these conditions. (b) Degenerate ($\omega_1 = \omega_2$) photon pairs in Type-I SPDC with $\Delta k > 0$ are emitted in a cone centered on the pump beam.

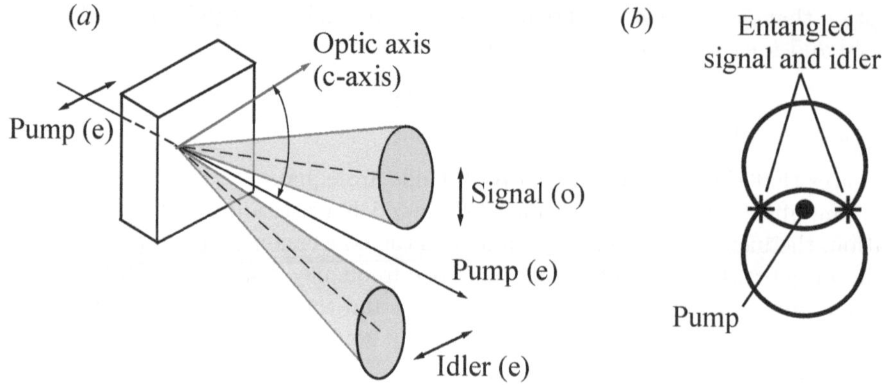

FIGURE 12.11
Geometry of Type-II SPDC in a negative uniaxial crystal.

on the axis defined by the pump beam. As the pump angle θ is increased, phase-matching is satisfied under noncollinear conditions. The generated idler and signal photons are then emitted along a cone centered on the transmitted pump beam. Momentum conservation requires that if a photon is observed in a certain direction along the cone, another photon is simultaneously observed along a diametrically opposite direction on the emission cone. Because we are considering the degenerate case, the signal and idler photons are identical in frequency and polarization, making them fundamentally indistinguishable – the labels 'signal' and 'idler' are therefore interchangeable. In this case, it really does not matter; the most striking aspect of this process is that photon numbers detected in diametrically opposite correlated modes exhibit fluctuations in their difference below the shot-noise level expected for independent classical beams.

Unlike the Type-I downconversion, Type-II SPDC in a negative uniaxial crystal driven by an e-polarized pump generates signal and idler photons with orthogonal (o and e) polarizations. The geometry of the Type-II SPDC process is shown in Fig. 12.11. Each downconverted signal and idler photon is again emitted in the form of a ring-like spatial pattern. However, because the photons possess orthogonal polarizations, they experience different refractive indices. Consequently, the ordinary and extraordinary emission rings are centered on different propagation directions. Under certain phase-matching conditions, the two distinct rings intersect at two specific locations as shown in Fig. 12.11. At these intersections, the two photons are found to be polarization-entangled. Entanglement in the present context means that while a photon detected at either intersection has a 50/50 chance of being measured as o or e, it is found that whenever the photon in one intersection region is found to be o-polarized, its twin photon from the other intersection region is found to be e-polarized, and vice versa. The downconverted photons are thus entangled in their states of polarization.

If we define the horizontal polarization state as $|H\rangle$ (corresponding to o-polarization) and the vertical state as $|V\rangle$ (corresponding to e-polarization), the polarization state vector $|\psi\rangle$ of the emitted photon pair at these intersections is represented by the state

$$|\psi\rangle = \frac{1}{\sqrt{2}}(|H\rangle_s |V\rangle_i + |V\rangle_s |H\rangle_i). \tag{12.52}$$

Polarization entanglement of this type has many applications in quantum information science, quantum cryptography, and computing. These quantum correlations complement and

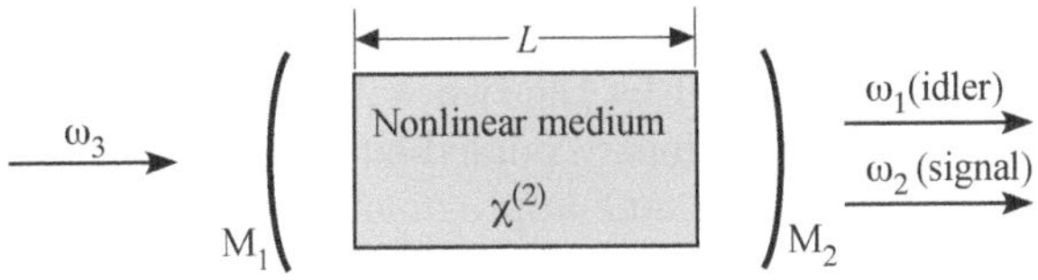

FIGURE 12.12
The optical parametric oscillator. The cavity is resonant at either one downconverted frequency (SRO) or both downconverted frequencies (DRO).

go beyond those implied solely by energy conservation, $\omega_1 + \omega_2 = \omega_3$, and momentum conservation, $\boldsymbol{k}_1 + \boldsymbol{k}_2 = \boldsymbol{k}_3$.

12.6 Optical Parametric Oscillation

We have already seen that difference-frequency generation can produce gain at both signal and idler frequencies. This gain is referred to as optical parametric gain. As in the case of laser gain, parametric gain can be combined with suitable optical feedback to construct an optical parametric oscillator (OPO). For this purpose, the nonlinear crystal is placed in an optical resonator shown in Fig. 12.12 and is pumped by a suitably focused pump beam. The cavity can be resonant either at ω_1 only (singly resonant oscillator, SRO) or at both ω_1 and ω_2 (doubly resonant oscillator, DRO). We assume that the cavity mirrors are transparent to the pump beam. Oscillation will occur when parametric gain exceeds the losses of the optical resonator. Oscillation occurs at both ω_1 and ω_2 with the particular pair of values of ω_1 and ω_2 determined by the energy conservation and phase-matching conditions. The OPO has proven to be a versatile source of frequency-tunable coherent radiation throughout the infrared, visible, and ultraviolet spectral regions. It can be operated as a source of continuous-wave or pulsed radiation, and even as a periodic train of nanosecond, picosecond, or femtosecond pulses.

We now consider how the equations for parametric amplification, Eqs. (12.48), can be applied to an optical parametric oscillator as shown in Fig. 12.12. We first assume that the cavity is resonant at both signal and idler frequencies (doubly resonant cavity). Both waves will then travel back and forth within the cavity, but the parametric gain will occur only when they are traveling in the same direction as the pump wave ω_3, because only then can the phase-matching condition be satisfied. This means that while loss occurs on every pass through the cavity, parametric gain occurs only every other pass. To calculate gain and arrive at the threshold condition for parametric oscillations, the amplitude equations must include distributed losses represented by the conductivity term in Eq. (12.19b), which we have so far neglected. To keep the calculations simple, assume that the nonlinear medium fills the entire cavity and consider only phase-matched operation. Also neglect pump-wave depletion by the parametric process. We can then set $\Delta k = 0$ and $A_3(z) \approx A_3$, where A_3 is the field amplitude of the pump wave. The equations describing the OPO are then obtained from Eqs. (12.48) by adding the loss terms:

$$\frac{dA_2}{dz} = -\frac{\alpha_2}{2}A_2 + i\frac{\omega_2}{n_2 c}d_{\text{eff}}A_3A_1^*\,, \tag{12.53a}$$

$$\frac{dA_1}{dz} = -\frac{\alpha_1}{2}A_1 + i\frac{\omega_1}{n_1 c}d_{\text{eff}}A_3A_2^*\,, \tag{12.53b}$$

where the loss coefficients α_1 and α_2 [Eq. (12.19b)] take into account distributed losses (scattering, absorption, etc.) suffered by the two waves. We have seen that without losses, the fundamental solutions of these linear coupled equations are hyperbolic sine and cosine functions. For simplicity, we now assume approximately equal distributed losses, $\alpha_1 \approx \alpha_2$. The resulting solutions are [Problem 3]

$$A_1(L) = e^{-\alpha L}\left[A_{10}\cosh gL + i\frac{\kappa_1 A_3 A_{20}^*}{g}\sinh gL\right] \tag{12.54a}$$

$$A_2(L) = e^{-\alpha L}\left[A_{20}\cosh gL + i\frac{\kappa_2 A_3 A_{10}^*}{g}\sinh gL\right], \tag{12.54b}$$

where $$\alpha = \frac{1}{4}(\alpha_1+\alpha_2), \quad g = \sqrt{\kappa^2 + (\alpha_1-\alpha_2)^2/16} \tag{12.54c}$$

and $$\kappa_i = \frac{\omega_i d_{\text{eff}}}{n_i c}, \quad \kappa = \sqrt{\kappa_1\kappa_2|A_3|^2} = \sqrt{\frac{2\omega_1\omega_2 d_{\text{eff}}^2}{\epsilon_o n_1 n_2 n_3} I_3}. \tag{12.54d}$$

Here A_{10} and A_{20} denote the idler and signal wave amplitudes at the input of the nonlinear medium. We can now use these equations to derive the threshold condition for parametric oscillations to occur.

For the two-mirror cavity shown in Fig. 12.12, assuming the mirror reflectances $\mathcal{R}_1$ and $\mathcal{R}_2$ are the same for both waves, and recalling that parametric gain is available only when the signal and idler waves are co-propagating with the pump, the threshold condition is that the wave amplitudes replicate themselves (both in amplitude and phase) after each cavity roundtrip. This leads to the condition

$$A_{10}^* = e^{-\mathcal{L}_1/2}\left[A_{10}^*\cosh gL - i\frac{\kappa_1 A_3^* A_{20}}{g}\sinh gL\right], \tag{12.55a}$$

$$A_{20} = e^{-\mathcal{L}_2/2}\left[A_{20}\cosh gL + i\frac{\kappa_2 A_3 A_{10}^*}{g}\sinh gL\right]. \tag{12.55b}$$

Note that we have written the replication condition in terms of the amplitude of one wave and the complex conjugate amplitude of the other wave to simplify the calculations. Here $\mathcal{L}_1$ and $\mathcal{L}_2$ denote fractional intensity losses per roundtrip for the idler and signal fields, respectively,

$$\mathcal{L}_1 = \alpha_1 L + 2\alpha L - \ln \mathcal{R}_1\mathcal{R}_2, \quad \mathcal{L}_2 = \alpha_2 L + 2\alpha L - \ln \mathcal{R}_1\mathcal{R}_2. \tag{12.56}$$

The additional term $\alpha_i L$ accounts for the passive loss experienced by field i on the return pass, when it propagates opposite to the pump and therefore receives no parametric gain. For nonzero-amplitude solution of the system of equations in (12.55), the determinant of the coefficient matrix must vanish. This leads to the condition that must be satisfied by cavity loss and gain,

$$1 + e^{-\frac{1}{2}(\mathcal{L}_1+\mathcal{L}_2)}\left(\cosh^2 gL - \frac{\kappa^2}{g^2}\sinh^2 gL\right) = \left(e^{-\frac{1}{2}\mathcal{L}_1} + e^{-\frac{1}{2}\mathcal{L}_2}\right)\cosh gL. \tag{12.57}$$

For approximately equal distributed losses for the signal and idler fields, $\alpha_1 \approx \alpha_2$, which is often the case, the factor $\kappa^2/g^2 \approx 1$ and this expression simplifies to

$$\cosh gL = 1 + \frac{(1-e^{-\mathcal{L}_1/2})(1-e^{-\mathcal{L}_2/2})}{e^{-\mathcal{L}_1/2} + e^{-\mathcal{L}_2/2}}. \tag{12.58}$$

Finally, for a resonator with small losses and operation near threshold, we can use the approximations,

$$e^{-\mathcal{L}_1/2} \approx 1 - \tfrac{1}{2}\mathcal{L}_1, \quad e^{-\mathcal{L}_2/2} \approx 1 - \tfrac{1}{2}\mathcal{L}_2, \quad \text{and} \quad \cosh gL \approx 1 + \tfrac{1}{2}g^2L^2, \tag{12.59}$$

which lead to the threshold condition for a doubly resonant OPO (DRO)

$$g^2L^2 = \frac{1}{4}\mathcal{L}_1\mathcal{L}_2\,. \tag{12.60}$$

If only the signal field is resonant and no idler feedback to the crystal is assumed ($e^{-\mathcal{L}_1/2} \to 0$), then for small signal loss $\mathcal{L}_2$, we obtain the threshold condition for singly resonant OPO (SRO)

$$g^2L^2 = \mathcal{L}_2\,. \tag{12.61}$$

The ratio of threshold pump power for the SRO to the threshold pump power for the DRO is $4/\mathcal{L}_1$. For example, if $\mathcal{L}_1 = \mathcal{L}_2 = 0.02$, the SRO threshold is approximately $4/0.02 = 200$ times the DRO threshold. Recalling the definition of g from Eq. (12.54) with $\alpha_1 \approx \alpha_2$, we see that $g^2 \propto |A_3|^2$. Thus the threshold conditions mean that a certain threshold pump intensity is required for parametric oscillation to occur. Assuming a Gaussian pump, we can express the pump intensity in terms of pump power as $I_3 = 2P/\pi w_3^2$. The threshold pump power is then

$$P_{3\text{th}} = \frac{\mathcal{L}_1\mathcal{L}_2}{16} \cdot \frac{\epsilon_o n_1 n_2 n_3 c^3}{\omega_1\omega_2 d_{\text{eff}}^2 L^2}\pi w_3^2\,. \tag{12.62}$$

Despite the advantage of low threshold in DRO, both signal and idler waves must belong to the cavity modes simultaneously, making the resonance condition more restrictive. A DRO is also highly sensitive to small cavity-length fluctuations and is therefore less commonly used in practice.

12.7 Electro-optic Modulators

The electro-optic effect is the change in the refractive index experienced by a light wave in the presence of a DC (or low-frequency) electric field in a nonlinear optical material. This effect is known as the Pockels effect. It can be seen as a frequency-mixing interaction between the optical field and an externally applied DC or low-frequency electric field. For example, in sum-frequency generation, $\omega_3 = \omega_1 + \omega_2$, if we write the total (real) electric field as $\mathcal{E}(t) = \mathcal{E}_o + \mathcal{E}_\omega(t)$, then the induced second-order polarization can be written as

$$\mathcal{P}^{(2)}(t) = \epsilon_o 2d_{\text{eff}}\left(\mathcal{E}_o^2 + 2\mathcal{E}_o\mathcal{E}_\omega + \mathcal{E}_\omega^2(t)\right). \tag{12.63}$$

This gives complex nonlinear polarization at frequency ω

$$P^{(2)}(\omega) = 4\epsilon_o d_{\text{eff}}\mathcal{E}_o E_\omega \equiv \epsilon_o\chi^{(2)}E_\omega\,. \tag{12.64}$$

Combining this DC-field-induced contribution to the susceptibility $\Delta\chi = 4d_{\text{eff}}\mathcal{E}_o$ with the first-order susceptibility $\chi^{(1)}$ changes the refractive index from n to $\sqrt{n^2 + \Delta\chi}$:

$$\sqrt{1 + \chi^{(1)} + \Delta\chi} = \sqrt{n^2 + 4d_{\text{eff}}\mathcal{E}_o} \approx n + \frac{2d_{\text{eff}}\mathcal{E}_o}{n} \equiv n + \Delta n. \tag{12.65}$$

The second term, $\Delta n = 2d_{\text{eff}}\mathcal{E}_o/n$, is the change in the refractive index due to the applied DC field $\mathcal{E}_o$. For historical reasons, this change in n due to the applied DC field $\mathcal{E}_o$ is computed as a change in $1/n^2$ in the form $\Delta(1/n^2) = r_{\text{EO}}\mathcal{E}_o$, where r_{EO} is the electro-optic coefficient. Since the change in n is usually small, we can write $\Delta(1/n^2) = -2\Delta n/n^3$ and get an expression for Δn in terms of electro-optic coefficient $\Delta n = -n^3 r_{EO}\mathcal{E}_o/2$. The two expressions for Δn in terms of $d_{\rm eff}$ and $r_{\rm EO}$ must, of course, be equal because they describe the same physical effect. Equating the two expressions gives the electro-optic coefficient in terms of the nonlinear coefficient

$$r_{\rm EO} = -\frac{4d_{\rm eff}}{n^4}. \tag{12.66}$$

The electro-optic coefficient r, like the nonlinear coefficient d, is a third-rank tensor with similar symmetry properties and one uses the same contracted notation. Within the present approximation, and suppressing the refractive-index factors associated with different principal axes, the corresponding contracted coefficients have the schematic relation

$$r_{ij} = -\frac{4d_{ji}}{n^4}. \tag{12.67}$$

As an example, the electro-optic coefficient r_{63} appearing in the description of electro-optic modulators is related to d_{36} by $r_{63} = -4d_{36}/n^4$.

A typical arrangement of an electro-optic modulator is shown in Fig. 12.13 below. It consists of an electro-optic crystal placed between two crossed polarizers, which are oriented at 45° with respect to the electrically induced dielectric axes x', y' when an electric field is applied along the z-axis. The arrangement shown can be achieved using a potassium dihydrogen phosphate (KDP) crystal. Also included in the path is a naturally birefringent crystal plate that introduces a fixed retardation so that the overall retardation when the beam hits the output polarizer (analyzer) is the sum of the retardation due to this plate and the electrically induced retardation.

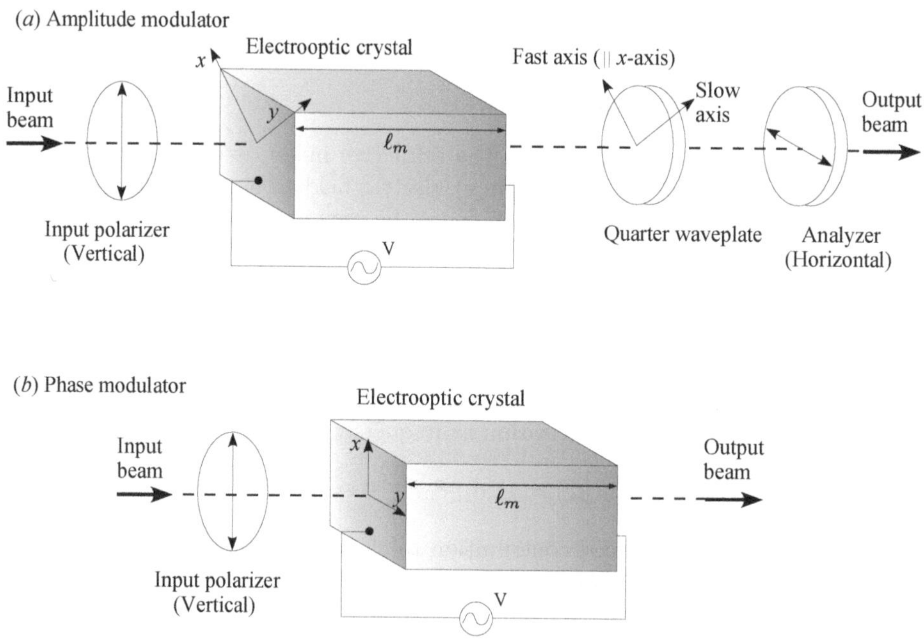

FIGURE 12.13
Electro-optic (a) amplitude and (b) phase modulators based on different orientations of a Pockels cell. A voltage difference V across the crystal sets up a DC electric field of magnitude $\mathcal{E}_o = V/\ell_m$.

12.7.1 Amplitude Modulator

The components of the vertically polarized incident field along the x and y axes of the crystal see refractive indices n_x and n_y inside the crystal. For a field of amplitude A_o incident normally on the input face $z = 0$ of the crystal, the x and y components of the field emerging from the output face $z = \ell_m$ [see Fig. 12.13] will be

$$A_x(\ell_m, t) = \frac{A_o}{\sqrt{2}} e^{-i\omega(t - n_x \ell_m/c)}, \tag{12.68a}$$

$$A_y(\ell_m, t) = \frac{A_o}{\sqrt{2}} e^{-i\omega(t - n_y \ell_m/c)} = A_x(\ell_m, t) e^{i\omega(n_y - n_x)\ell_m/c}, \tag{12.68b}$$

with $$n_x = n_o - \frac{n_o^3}{2} r_{63} \left(\frac{V}{\ell_m}\right), \quad n_y = n_o + \frac{n_o^3}{2} r_{63} \left(\frac{V}{\ell_m}\right), \tag{12.68c}$$

and $$\delta = \frac{\omega}{c}(n_y - n_x)\ell_m = \frac{\omega}{c} n_o^3 r_{63} V. \tag{12.68d}$$

Here r_{63} is the electrooptic coefficient of the crystal [KDP, in this example] and n_o is the ordinary refractive index of the crystal. The output field from the analyzer is horizontally polarized and given by

$$A_{\text{out}}(\ell_m, t) = -A_x(\ell_m, t)\cos(\pi/4) + A_y(\ell_m, t)\cos(\pi/4) = -\frac{A_x(\ell_m, t)}{\sqrt{2}} \left(1 - e^{i\delta}\right). \tag{12.69}$$

The intensity of the beam emerging from the modulator is then

$$I_{\text{out}} = \frac{1}{2}\epsilon_o c \left|A_{\text{out}}(\ell_m, t)\right|^2 = \frac{1}{2}\epsilon_o c \left|A_o\right|^2 \sin^2(\delta/2) = I_{\text{in}} \sin^2(\delta/2). \tag{12.70}$$

For amplitude modulation, the modulator is biased with a fixed voltage $V_{\pi/2}$ to produce $\delta_o = \pi/2$ and 50% transmission. A small sinusoidal voltage modulation is superimposed on $V_{\pi/2}$ giving a retardation

$$\delta = \frac{\omega}{c} n_o^3 r_{63} \left[V_{\pi/2} + V_m \cos\omega_m t\right] = \frac{\pi}{2} + \delta_m \cos\omega_m t, \quad \delta_m = \frac{\omega}{c} n_o^3 r_{63} V_m, \tag{12.71a}$$

and $$\frac{I_{\text{out}}}{I_{\text{in}}} = \sin^2\left[\pi/4 + (\delta_m/2)\cos\omega_m t\right] = \frac{1}{2}\left[1 + \sin(\delta_m \cos\omega_m t)\right]$$

$$\approx \frac{1}{2}(1 + \delta_m \cos\omega_m t), \quad \text{for } \delta_m \ll 1. \tag{12.71b}$$

Thus, a small modulation voltage (small δ_m) will cause a nearly identical modulation of the transmitted intensity. If δ_m is not small, the intensity variation will be distorted and contain higher (odd) harmonics of ω_m. Because $\delta_m \ll 1$ in Eq. (12.71b), another parametrization $\delta_m = 2M/(1 - 2M)$ allows us to write

$$\frac{I_{\text{out}}}{I_{\text{in}}} = \frac{[1 - 2M(1 - \cos\omega_m t)]}{2(1 - 2M)} \approx \frac{e^{-2M(1-\cos\omega_m t)}}{2(1 - 2M)} \equiv \frac{t_{\text{M}}^2}{2(1 - 2M)}, \tag{12.72a}$$

where $$t_{\text{M}} = e^{-M(1-\cos\omega_m t)}. \tag{12.72b}$$

This analytic parametrization of power loss in terms of t_{M}^2 (for a fixed M) is convenient and useful for specifying the time-dependence of field amplitude modulation in lasers. The constant prefactor $1/[2(1 - 2M)]$ can be absorbed into the average cavity loss.

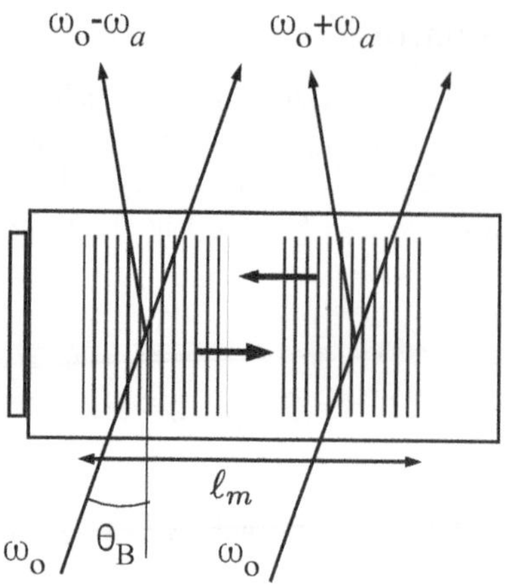

FIGURE 12.14
Acousto-optic amplitude modulator for mode-locking based on standing acoustic waves.

12.7.2 Phase Modulator

In this case, the electro-optic crystal is oriented such that one of its induced birefringent axes x or y coincides with the polarization of the incident light. The application of the electric field along the z-direction in this case merely changes the phase of the output beam. An incident field polarized along the crystal's x-axis, for example, at the input face $z = 0$, will emerge as

$$A_{\text{out}}(\ell_m, t) = A_o e^{-i\omega(t - n_x \ell_m/c)} = A_o e^{-i\omega(t - n_o \ell_m/c) - i\delta}, \tag{12.73a}$$

with $$\delta = \frac{\omega n_o^3 r_{63} \ell_m}{2c} \frac{V}{\ell_m} = \frac{\omega n_o^3 r_{63} V}{2c}. \tag{12.73b}$$

If the applied voltage is $V_m \cos \omega_m t$, the output field can be written as

$$A_{\text{out}}(\ell_m, t) = A_o e^{-i\omega(t - n_o \ell_m/c)} e^{iM \cos \omega_m t} \equiv A_{\text{in}} t_{\text{FM}}, \tag{12.74a}$$

where $$t_{\text{FM}} = e^{-iM \cos \omega_m t}, \quad M = \frac{\omega n_o^3 r_{63} V_m}{2c}. \tag{12.74b}$$

In this case, the transmitted field is phase-modulated with a modulation index M.

12.7.3 Acousto-optic Amplitude Modulator

Acousto-optic amplitude modulators use the strain induced by a standing acoustic wave to generate a periodic refractive index grating through the photoelastic effect. This periodic index grating diffracts the incoming light, leading to a periodic loss modulation for the non-diffracted beam. The depth of loss modulation can be controlled by the intensity of the acoustic wave. The acousto-optic modulator uses standing acoustic waves, rather than the traveling acoustic wave commonly used for Q-switching [see Fig. 10.13].

A standing acoustic wave can be considered as a superposition of two plane waves propagating in opposite directions. The diffracted optical field can be viewed as the coherent sum of components shifted by $\pm\omega_a$ through scattering from the two counterpropagating acoustic waves

$$\propto \left| A_o e^{-i(\omega+\omega_a)t} + A_o e^{-i(\omega-\omega_a)t} \right|^2 = 2A_o^2 (1 + \cos 2\omega_a t) \tag{12.75}$$

This means that the oppositely traveling acoustic waves introduce a periodic loss modulation at twice the acoustic frequency. For fundamental mode-locking, the modulator is placed close to one cavity mirror and the modulation frequency $2\omega_a \approx 2\pi(c/2L)$; that is, the transducer is driven at a frequency equal to $c/4L$. For example, $\nu_a = 50$ MHz for a 1.5-m-long standing-wave cavity.

12.8 Third-order Nonlinear Optical Processes

In centrosymmetric media, the lowest-order nonlinearity is third-order in the applied electric field. The most general form of third-order nonlinear polarization is given by

$$\mathcal{P}_i^{(3)}(t) = \epsilon_o \chi_{ijk\ell} \mathcal{E}_j(t) \mathcal{E}_k(t) \mathcal{E}_\ell(t) , \tag{12.76}$$

where $\chi_{ijk\ell}$ is the third-order susceptibility of the medium, which is a tensor of fourth rank. The equation has been written under the assumption that the interacting field frequencies are detuned from medium resonances, allowing $\chi_{ijk\ell}$ to be treated as approximately real and weakly frequency dependent over the spectral range of interest. This nonlinearity leads to a rich set of novel phenomena whose description can be quite complex and is beyond the scope of this book. We restrict our discussion to enumerating the primary effects by considering the scalar form of Eq. (12.76)

$$\mathcal{P}^{(3)}(t) = \epsilon_o \chi^{(3)} \mathcal{E}^3(t) , \tag{12.77}$$

where $\chi^{(3)}$ is the effective third-order susceptibility. Even when restricted to the scalar form, the full expression for $\mathcal{P}^{(3)}$ remains exceedingly complicated if the applied electric field $\mathcal{E}$ has several different frequency components. Therefore, we consider the third-order nonlinear polarization when the applied field is a monochromatic field of frequency ω with analytic signal representation

$$\mathcal{E}(t) = \mathcal{R}e E(\omega) e^{-i\omega t} = \frac{1}{2} E(\omega) e^{-i\omega t} + c.c = \frac{1}{2} A(\omega) e^{i[k(\omega) z - \omega t]} + c.c. \tag{12.78}$$

The corresponding third-order polarization is written as

$$\mathcal{P}^{(3)}(t) = \mathcal{R}e P(\omega) e^{-i\omega t} = \frac{1}{2} P(\omega) e^{-i\omega t} + c.c. , \tag{12.79}$$

By substituting the field profile into Eq. (12.77), we can express the nonlinear polarization (12.77) as

$$\mathcal{P}^{(3)}(t) = \frac{1}{4} \epsilon_o \chi^{(3)} \left[E^3(\omega) e^{-i3\omega t} + 3|E(\omega)|^2 E(\omega) e^{-i\omega t} + c.c. \right] . \tag{12.80}$$

The first term in this equation describes a material response at frequency 3ω, which corresponds to the third-harmonic of the applied driving frequency ω. This term acts as a source that generates a new field component at frequency 3ω, which was not present in the incident field. In the photon picture, this process corresponds to the simultaneous absorption of three photons of frequency ω and the coherent emission of a single photon of frequency 3ω.

The second term describes a nonlinear response at the fundamental frequency of the applied field, which gives rise to an intensity-dependent refractive index. As noted in Sec. 12.1, although the magnitude of $|\chi^{(3)}|$ is typically quite small, both of these effects play vital roles in the generation of coherent radiation as well as in the design and operation of lasers. We consider each process in further detail.

12.8.1 Third-Harmonic Generation

The first term in Eq. (12.80) describes a material response at frequency 3ω induced by an applied field at frequency ω. This term leads to the third-harmonic generation (THG) of the driving field. Unlike second-order processes, THG is possible in both centrosymmetric and noncentrosymmetric media. Consequently, it can be observed in solids, liquids, and gases alike. Arguably, the most interesting cases involve the generation of ultraviolet (UV) and extreme ultraviolet (EUV) radiation in gases.

The third-order polarization term for THG can be written as

$$P^{(3)}(3\omega) = \frac{1}{4}\epsilon_o \chi^{(3)} E^3(\omega)\,, \tag{12.81}$$

Substituting this source polarization into the slowly varying envelope equation (12.19a)–in place of the second-order polarization terms–we obtain the spatial evolution equation for the slowly varying third-harmonic field amplitude,

$$\frac{dA(3\omega, z)}{dz} = i\frac{3\omega\chi^{(3)}}{8n_3 c} A^3(\omega) e^{i\Delta k\, z}, \quad \Delta k = 3k(\omega) - k(3\omega)\,. \tag{12.82}$$

Assuming a low conversion efficiency, which is almost always the case, we can treat the fundamental amplitude $A(\omega)$ as remaining essentially unaffected by third-harmonic generation. Integrating this differential equation over an interaction medium of length L yields the third-harmonic output intensity

$$I(3\omega, L) = \left(\frac{3\omega\chi^{(3)}}{4\epsilon_o c^2}\right)^2 \frac{L^2 I^3(\omega)}{n_3 n_1^3} \left(\frac{\sin(\Delta k L/2)}{\Delta k L/2}\right)^2 . \tag{12.83}$$

For Gaussian laser beams, the peak intensity and total power within the near-zone waist region are related by $I \approx 2P/\pi w_{o1}^2$. Equation (12.83) for THG intensity can be re-expressed as a power conversion efficiency

$$\frac{P(3\omega)}{P(\omega)} = \frac{1}{3}\left(\frac{3\omega\chi^{(3)}}{4\epsilon_o c^2}\right)^2 \left(\frac{2L}{\pi w_{o1}^2}\right)^2 \frac{P^2(\omega)}{n_3 n_1^3} \left(\frac{\sin(\Delta k L/2)}{\Delta k L/2}\right)^2 . \tag{12.84}$$

The numerical factor $\frac{1}{3}$ arises from the fact that the generated third-harmonic Gaussian beam has a tighter spot radius of $w_{o3} = w_{o1}/\sqrt{3}$, where w_{01} is the fundamental spot radius. This geometric reduction is a direct consequence of the cubic dependence of the driving polarization on the fundamental field, $P^{(3)}(3\omega) \propto E^3(\omega)$. Under perfect phase-matching condition $\Delta k = 0$, and assuming optimum focusing where the medium length equals the fundamental confocal parameter, $L = b = 2\pi n_1 w_{o1}^2/\lambda^2$, THG power conversion efficiency is given by

$$\eta_{THG}^{(\text{power})} \equiv \frac{P(3\omega, L)}{P(\omega)} = \frac{16}{3}\left(\frac{3\omega\chi^{(3)}}{4\epsilon_o c^2}\right)^2 \frac{P^2(\omega)}{n_3 n_1^3 \lambda^2}\,. \tag{12.85}$$

More careful treatments with focused Gaussian beams show that the optimum focusing is $L = 1.7b$. [G. C. Bjorklund, IEEE J. Quantum Electron. QE-11, 287 (1975).] In practice, perfect phase matching for direct THG is difficult to achieve in solids. For this reason, third-harmonic light is commonly generated via a cascaded second-order process: the fundamental wave is first frequency-doubled ($\omega + \omega \to 2\omega$), and the resulting second-harmonic is mixed with the remaining fundamental light via sum frequency generation ($2\omega + \omega \to 3\omega$). In contrast, direct THG within gaseous media can be phase-matched effectively by utilizing a binary mixture of gases to artificially cancel out normal dispersion. Through these optimized techniques, conversion efficiencies of up to 10% have been achieved.

12.8.2 Intensity-Dependent Refractive Index

The second term in Eq. (12.80) represents a nonlinear contribution to the material polarization at the incident field frequency. This term can be naturally combined with the linear polarization to construct an intensity-dependent effective permittivity

$$\epsilon = \epsilon_o \left(1 + \chi^{(1)} + \frac{3}{4}\chi^{(3)}|E(\omega)|^2\right) = \epsilon_o \left(n^2 + \frac{3}{4}\chi^{(3)}|E(\omega)|^2\right), \tag{12.86}$$

where $n = \sqrt{1+\chi^{(1)}}$ is the conventional refractive index defined by the medium's linear response. This intensity-dependent contribution to the permittivity yields an intensity-dependent contribution to the refractive index experienced by the wave

$$\sqrt{\epsilon/\epsilon_o} = \sqrt{n^2 + \frac{3}{4}\chi^{(3)}|E(\omega)|^2} \approx n + \frac{3}{8n}\chi^{(3)}|E(\omega)|^2 . \tag{12.87}$$

This effect is known as the optical Kerr effect. The second term in Eq. (12.87) represents the net index change Δn induced by the intense light wave itself in a medium with a third-order nonlinearity. Expressed directly in terms of the optical intensity $I = \frac{1}{2}\epsilon_o nc|E(\omega)|^2$, the change Δn can be written as

$$\Delta n = \frac{3}{8n}\chi^{(3)}|E(\omega)|^2 = n_2 I , \tag{12.88a}$$

where

$$n_2 = \frac{3\chi^{(3)}}{4n^2\epsilon_o c} . \tag{12.88b}$$

represents the optical nonlinear refractive index, expressed in SI units of m^2/W. While many other processes contribute to an intensity-dependent refractive index change, we focus exclusively on the non-resonant contribution originating from electronic polarization. Typical values of n_2 for bound electronic responses are $\sim 10^{-20}$ m^2/W (or $\sim 10^{-16}$ cm^2/W). This means that a laser beam of intensity 1 GW/cm^2 will only produce a refractive index change of 10^{-7}. The fractional change in refractive index is thus very small, yet refractive index changes of this order accumulate significant phase delays over long propagation path lengths and can lead to dramatic self-action effects. We have already considered two self-action effects resulting from the intensity-dependent refractive index change in Chapter 11 in the context of laser mode-locking. We briefly revisit them here to point out their relevance beyond mode-locking in lasers.

One of the processes that can occur as a result of the intensity-dependent refractive index is self-focusing considered in Sec. 11.2.3. Self-focusing occurs when a beam of light having a nonuniform transverse intensity distribution propagates through a material for which n_2 is positive. Under these conditions, the nonlinear material response effectively turns the medium into a positive lens, causing the incident beam to focus. This is the mechanism responsible for Kerr-lens mode-locking in a Ti:sapphire laser. This effect occurs for cw as well as pulsed laser beams. It is of great practical importance for high peak-power laser pulses because the intensity at the focal spot of the self-focused beam can be large enough to cause optical damage of the material. Another effect of intensity-dependent refractive index is time-dependent modulation of wave phase leading to frequency chirp and pulse spectrum broadening in short pulse propagation when pulse intensities reach sufficiently high values. This effect is of special interest for the generation of coherent radiation via the processes known variously as supercontinuum generation and white light generation. These effects are easier to describe for linearly polarized light waves in an isotropic medium. This is the case considered below.

12.8.3 Self-phase Modulation (SPM)

For a linearly polarized beam, the third-order nonlinear polarization relevant to refractive index is given by

$$P^{(3)}(z,t) = \epsilon_0 \frac{3}{4}\chi^{(3)}|A|^2 A(z,t)\,. \tag{12.89}$$

Note that, because of time-dependent phase modulation, we have retained both the slowly varying spatial and temporal dependence of the nonlinear polarization. The equation for the evolution of the slowly varying wave amplitude due to nonlinear polarization (12.89) is then obtained by neglecting the $\partial^2 A/\partial\tau^2$ term as well as the gain, loss, and absorption terms on the right-hand side of Eq. (11.98),

$$\frac{dA(z,\tau)}{dz} = i\left(\frac{3\omega_o\chi^{(3)}}{8nc}\right)|A(z,\tau)|^2 A(z,\tau)\,. \tag{12.90}$$

By writing the field amplitude as $A(z,\tau) = |A(z,\tau)|e^{i\Phi(z,\tau)}$ in this equation and equating the real and imaginary parts from the two sides, we get

$$\frac{d|A(z,\tau)|}{dz} = 0, \tag{12.91a}$$

$$\frac{d\Phi(z,\tau)}{dz} = \left(\frac{3\omega_o\chi^{(3)}}{8nc}\right)|A(z,\tau)|^2\,. \tag{12.91b}$$

The first of these equations shows that SPM does not affect the pulse shape, that is, $|A(z,\tau)| = |A(0,\tau)|$. Using this in Eq. (12.91b), we find that SPM gives rise to an intensity-dependent phase, which grows linearly with the distance of propagation into the nonlinear medium,

$$\Phi(z,\tau) = \Phi(0,\tau) + \left(\frac{3\omega_o\chi^{(3)}}{8nc}\right)|A(0,\tau)|^2 z\,. \tag{12.92}$$

Thus, for an input pulse $A(0,\tau)$, the pulse inside the nonlinear medium is

$$A(z,\tau) = A(0,\tau)e^{i\phi(\tau)}, \tag{12.93a}$$

where

$$\phi(\tau) = \left(\frac{3\omega_o\chi^{(3)}}{8nc}\right)z|A(0,\tau)|^2 = \frac{\omega_o n_2 z}{c}I(0,\tau)\,. \tag{12.93b}$$

In the last step we have used the definition of n_2 [Eq. (12.88b)] and recalled the definition of intensity in terms of field amplitude. The time dependence of the pulse intensity $I(0,\tau)$ leads to a time-dependent modulation of the pulse phase $\phi(\tau)$. Recalling that a time-varying phase implies frequency chirp, that is, the pulse has local instantaneous frequency given by $\omega_o + \delta\omega$, where

$$\delta\omega(\tau) = -\frac{\partial\phi}{\partial\tau}\,. \tag{12.94}$$

The SPM-induced chirp introduces new frequency components, which broaden the pulse spectrum. The spectral density is obtained by taking the modulus squared of the Fourier transform of Eq. (12.93),

$$\tilde{A}(z,\Omega) = \int_{-\infty}^{\infty} d\tau A(z,\tau)e^{i\Omega\tau}. \tag{12.95}$$

The integral, in general, must be evaluated numerically. However, many important insights can be obtained without evaluating it. For common symmetric bell-shaped Gaussian or hyperbolic secant pulses, the chirp is temporally non-monotonic; it reaches a negative

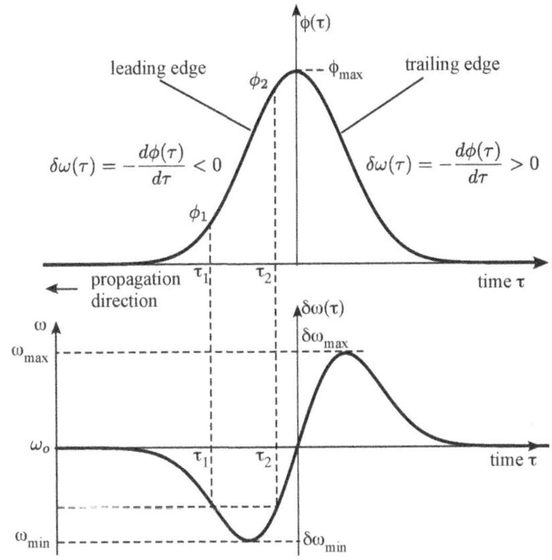

FIGURE 12.15
(a) Variation of SPM phase for an initially unchirped symmetric pulse, (b) Variation of SPM frequency shift for the pulse in (a).

minimum and a positive maximum and approaches zero as $|\tau| \to \infty$. The non-monotonic time dependence of chirp for a bell-shaped pulse implies that the pulse has the same instantaneous frequency at two distinct points, but since these points are shifted in time, the two waves of the same frequency will have different phases and will interfere giving the spectrum an oscillatory character.

The structure of the spectrum can be understood by referring to Fig. 12.15, which shows the phase $\phi(\tau)$ and frequency shift $\delta\omega(\tau)$ for a Gaussian pulse. The phase curve [Fig. 12.15(a)] mirrors the intensity envelope of the laser pulse while the frequency shift [Fig. 12.15(b)] is the negative derivative of $\phi(\tau)$. The leading half is shifted toward lower frequencies (red-shifted), whereas the trailing half is shifted toward higher frequencies (blue-shifted). The inflection points of the ϕ-curve possess the largest slope on the curve and naturally lead to the two extreme frequencies $\omega_{\min} = \omega_o + \delta\omega_{\min}$ and $\omega_{\max} = \omega_o + \delta\omega_{\max}$. The net spread, $\delta\omega_{\max} - \delta\omega_{\min} = \Delta\omega_{\max}$ is a reasonable measure of the magnitude of self-phase modulation spectral broadening.

Crucially, the non-monotonic nature of the chirp function [Fig. 12.15(b)] implies that the pulse exhibits identical instantaneous frequencies at two distinct points, say τ_1 and τ_2. Because these points are physically separated in time, these two waves with the same instantaneous frequency carry different accumulated phases, ϕ_1 and ϕ_2. Consequently, they will interfere with one another with a phase difference $\Delta\phi_{12} = \phi_2 - \phi_1$. This interference is constructive if the phase difference $\Delta\phi_{12}$ is an integer multiple of 2π and destructive if $\Delta\phi_{12}$ is an odd integer multiple of π. This phase interference imparts a highly oscillatory character to the pulse's power spectrum [Fig. 12.16]. If the incident temporal envelope is symmetric, the resulting SPM spectrum remains symmetric about the central laser frequency ω_o. The total number of distinct spectral peaks can be closely approximated by the relation: $N \approx \frac{\phi_{\max}}{\pi} + \frac{1}{2}$.

For a standard Gaussian laser pulse whose intensity distribution

$$I(\tau) = I_o e^{-2\tau^2/\tau_o^2}, \tag{12.96}$$

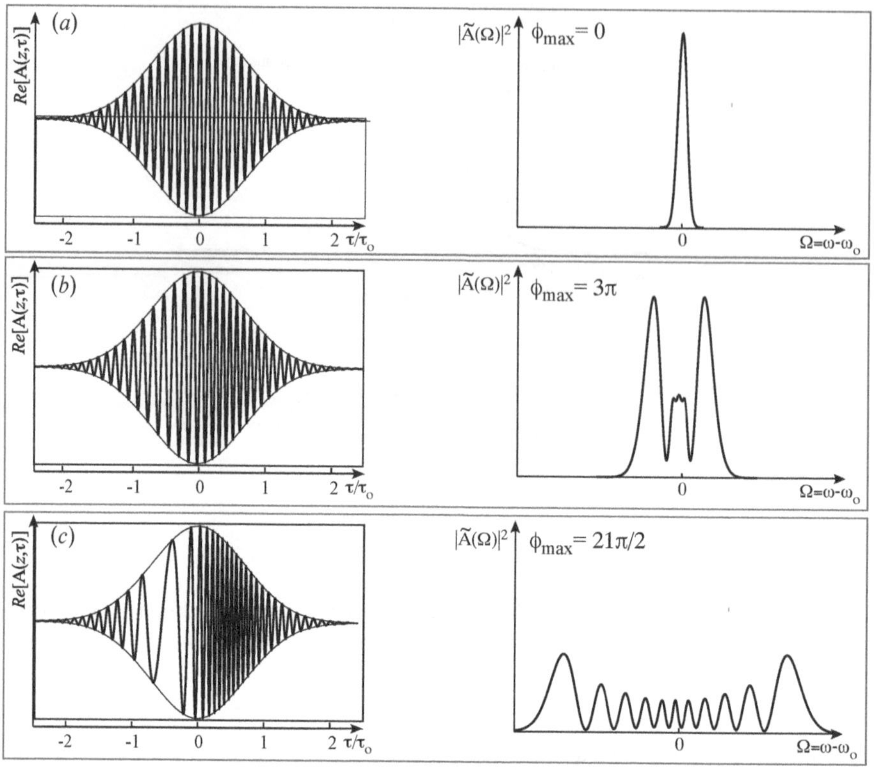

FIGURE 12.16
Spectral evolution of an initially unchirped Gaussian pulse driven by self-phase modulation (SPM): (a) input spectrum; (b) intermediate SPM ($\phi_{\max} = 3\pi$); and (c) strong SPM ($\phi_{\max} = 21\pi/2$).

is temporally symmetric, the maximum phase shift $\phi_{\max} = \frac{\omega_o n_2 z}{c} I_o$ occurs at the pulse center $\tau = 0$. The total SPM spectral broadening $\Delta\omega_{SPM} \equiv \delta\omega_{\max} - \delta\omega_{\min}$ is determined by maximizing the derivative of the phase

$$\Delta\omega_{SPM} = 2\left|\frac{d\phi}{d\tau}\right|_{\max} = 8\phi_{\max}\left|\frac{\tau}{\tau_o^2}e^{-2\tau^2/\tau_o^2}\right|_{\max} = \frac{4}{\sqrt{e}}\frac{\phi_{\max}}{\tau_o} = 2.426\frac{\omega_o n_2 z}{c\tau_o}I_o\,, \tag{12.97}$$

As illustrated by the numerical profiles in Figure 12.16, increasing the peak phase shift ϕ_{max} expands the outer boundaries of the emission spectrum while generating a progressively denser network of interference fringes. SPM spectral broadening is inversely proportional to the incoming width τ_o – the shorter the incoming pulse, the greater the broadening. The spectral broadening is also linearly proportional to I_o and z. The dependence of SPM spectral broadening on propagation length z means new frequency components are generated continuously as the pulse propagates in the medium. Using the relation for FWHM spectral width $\Delta\omega = 2.355/\tau_o$ for a Gaussian pulse [Table 11.1], the SPM spectral broadening can be expressed as

$$\frac{\Delta\Omega_{SPM}}{\Delta\omega} = 1.032\phi_{\max} = 1.032\frac{\omega_o n_2 z}{c}I_o\,. \tag{12.98}$$

The broadening factor is thus approximately equal to the maximum phase shift $\phi_{\max}$. Since $\phi_{\max}$ as large as 100 can be realized, this means SPM can broaden the spectrum dramatically. In the case of intense ultrashort pulses, the broadened spectrum can extend over 100 THz or more, especially if SPM is accompanied by other nonlinear processes such as stimulated Raman scattering and four-wave mixing.

The experimental situation is not as ideal as the simple theory presented here describes. In real media, delayed nonlinear responses and other propagation effects can break the symmetry predicted by the simple instantaneous-Kerr model. The laser beam has a finite cross section and will undergo diffraction. The transverse intensity variation of the laser pulse also causes a refractive index $\delta n(r)$ that varies in the transverse directions leading to self-focusing of the beam and complicating the simultaneously occurring SPM of the beam. In a more rigorous theory, one should expect a distortion of the pulse shape as it propagates. However, the qualitative discussion presented here applies even when the $\phi(\tau)$ curve is an asymmetric bell-shaped profile.

12.8.4 White-Light Supercontinuum Generation

The spectral broadening we have just described is referred to as continuum generation (CG) or supercontinuum generation (SCG). White-light supercontinuum (WLCG) emphasizes generation that covers the visible spectrum (white light). Supercontinuum generation was first observed in the late sixties and early seventies by focusing picosecond laser pulses into bulk crystals and glasses and later even in liquids, gases, and optical fibers. In SCG, laser pulses of narrow spectral bandwidth are converted to pulses with very broad spectral bandwidth. The pulse emerging from the material appears as a white light flash, regardless of whether the input pulse is in the near IR or near UV range. Self-phase modulation is one of the principal mechanisms responsible for the initial spectral broadening, often accompanied by other nonlinear and dispersive effects.

The materials used for SCG are sapphire (Al_2O_3) and fused silica. Sapphire is a transparent centrosymmetric dielectric with high damage threshold. The nonlinear refractive index of both fused silica and sapphire is $n_2 \approx 3.2 \times 10^{-20}\,\mathrm{m^2/W}$. To get a π phase shift, $\phi_{\max} = \omega_o n_2 L I_o/c = 2\pi n_2 L I_o/\lambda = \pi$, in a fused silica fiber of length 1.0 m, we need peak intensity at $\lambda = 1.06\,\mu\mathrm{m}$ of $I_o = \frac{\lambda}{2Ln_2} = 5 \times 10^{-7}/3.2 \times 10^{-20}\,\mathrm{W/m^2} = 1.56\mathrm{GW/cm^2}$. For $(10\,\mu\mathrm{m})^2$ core this corresponds to 1.56 kW power, which is possible with pulsed lasers. For anAs_2S_3 glass fiber with a $(20\,\mu\mathrm{m})^2$ core but with much larger nonlinear index, $n_2 = 2.0 \times 10^{-17}\,\mathrm{m^2/W}$, only 10 W will suffice.

The observation of a white light supercontinuum covering an octave (largest frequencies are twice the smallest frequencies) using a new type of photonic crystal fiber in the late 1990s revolutionized the field. Extreme light confinement and specifically engineered dispersion of these fibers enabled supercontinuum generation using low-energy pulses from widespread tabletop femtosecond lasers. Most useful supercontinuum light today is generated in optical fibers, spanning the visible and near-infrared spectral regions. Because the supercontinuum is generated within a guided optical mode, it retains high spatial coherence while providing a broad optical bandwidth. These properties enable applications such as multiwavelength microscopy and wavelength-division multiplexing.

Summary

In this chapter, we discussed second- and third-order nonlinear optical processes. These processes arise when a high-intensity optical field induces polarization components that depend quadratically or cubically on the electric field, in addition to the usual linear polarization.

We limited our discussion to nonlinear processes used either to generate coherent radiation at new wavelengths or to support important laser functions. Most of the important nonlinear frequency-generation processes considered here are based on second-order nonlinear polarization. In the electric-dipole approximation, a bulk second-order nonlinear polarization requires a medium without inversion symmetry. We briefly introduced the notation used to express the induced polarization in terms of field components and the elements of the nonlinear-susceptibility tensor for different crystal symmetries. Using the explicit form of the nonlinear polarization in Maxwell's equations, we derived in Sec. 12.2 the coupled equations for the amplitudes of the interacting fields. These equations were used to discuss second-harmonic generation (SHG) and sum- and difference-frequency generation in Secs. 12.3–12.5. The efficiencies of these nonlinear processes and their optimization through phase matching and focusing were discussed in Sec. 12.4.

Of particular importance is parametric amplification, in which a lower-frequency signal is amplified through a second-order nonlinear interaction driven by a higher-frequency pump wave. This process is a special case of difference-frequency generation. When combined with suitable optical feedback, parametric amplification leads to optical parametric oscillation. Optical parametric oscillators, which are important sources of tunable coherent radiation, were briefly discussed in Sec. 12.6.

A special case of sum- or difference-frequency mixing occurs when an optical wave propagates through a second-order nonlinear medium in the presence of an applied DC electric field. Under these conditions, the nonlinear polarization produces a change in the refractive index experienced by the optical wave. This phenomenon, known as the linear electro-optic or Pockels effect, is conventionally described using its own notation. In Sec. 12.7, we related it to second-order nonlinear polarization and used it to explain electro-optic amplitude and phase modulators.

Of the many third-order nonlinear optical effects, we considered only third-harmonic generation and the intensity-dependent refractive-index change in Sec. 12.8. The latter plays an important role in Kerr-lens mode-locking of Ti:sapphire and dye lasers. For short optical pulses, it also produces a time-dependent nonlinear phase through self-phase modulation (SPM). Under suitable conditions, SPM can produce substantial spectral broadening and contribute to continuum, supercontinuum, and white-light generation. A simple analytical model was presented to show how self-phase modulation broadens the pulse spectrum.

Problems

1. **Second-harmonic generation in KDP:** Consider second-harmonic generation in KDP, for which the only nonzero elements of the d_{im} matrix are $d_{14} = d_{25} = d_{36}$.

   (a) For Type-I phase matching, the second-harmonic propagates as an extraordinary wave and the fundamental as an ordinary wave. Using Fig. 12.1 and Eq. (12.11), show that
   $$d_{\text{eff}}^{(I)} = d_{14} \sin\theta \sin(2\varphi)\,.$$

   (b) For Type-II phase matching, the second-harmonic propagates as an extraordinary wave, while one fundamental wave is ordinary and the other is extraordinary. Using Fig. 12.1 and Eq. (12.11), show that
   $$d_{\text{eff}}^{(II)} = -d_{14} \sin(2\theta) \cos(2\varphi)\,.$$

2. **Second-harmonic generation in lithium iodate:** For lithium iodate (Class 6), the nonzero elements of the d_{im} matrix are $d_{15} = d_{24} = d_{31} = d_{32}$ and d_{33}. Show that for Type-I phase matching

$$d_{\text{eff}}^{(I)} = -d_{15}\sin\theta\,,$$

whereas for Type-II phase matching

$$d_{\text{eff}}^{(II)} = 0\,.$$

3. **Exact solution for phase-matched second-harmonic generation:** Consider Eqs. (12.21) for second-harmonic generation with perfect wave-vector matching, $\Delta k = 0$. Assume that only the fundamental wave is present at the input. Recalling that the total irradiance

$$I = I_1 + I_2 = I_1(0)$$

remains constant during propagation, introduce dimensionless real amplitudes u_1 and u_2 and phases ϕ_1 and ϕ_2 through

$$u_1 e^{i\phi_1} = \sqrt{\frac{\epsilon_o n_1 c}{2I}}E_1\,, \qquad u_2 e^{i\phi_2} = \sqrt{\frac{\epsilon_o n_2 c}{2I}}E_2\,.$$

(a) Show that u_1, u_2, and the relative phase $\phi = \Delta kz + \phi_2 - 2\phi_1$ satisfy

$$\begin{aligned}\frac{du_1}{dz} &= -\frac{1}{L_{\text{SHG}}}u_1u_2\sin\phi\,,\\ \frac{du_2}{dz} &= \frac{1}{L_{\text{SHG}}}u_1^2\sin\phi\,,\\ \frac{d\phi}{dz} &= \Delta k + \frac{\cos\phi}{\sin\phi}\frac{d}{dz}\ln\left(u_2u_1^2\right)\,.\end{aligned}$$

(b) For $\Delta k = 0$, show that $u_2u_1^2\cos\phi$ is a constant of motion.

(c) Using the result of part (b), together with

$$u_1^2 + u_2^2 = 1 \quad \text{and the initial conditions} \quad u_1(0) = 1\,, \quad u_2(0) = 0\,,$$

show that the solutions corresponding to efficient energy transfer from the fundamental to the second-harmonic are

$$u_1(z) = \text{sech}(\kappa z)\,, \qquad u_2(z) = \tanh(\kappa z)\,, \qquad \kappa \equiv \frac{1}{L_{\text{SHG}}}\,.$$

4. **Energy conservation in self-phase modulation:** We have seen that self-phase modulation (SPM) generates new frequencies and broadens the pulse spectrum. Is the total pulse energy conserved in the SPM process? Explain.

5. **Optical parametric amplification with phase mismatch:** Consider an optical parametric process with phase mismatch Δk and equal linear loss coefficients α for the signal and idler. In the undepleted-pump approximation, the coupled-amplitude equations are

$$\begin{aligned}\frac{dA_1}{dz} &= -\frac{\alpha}{2}A_1 + i\kappa_1A_3A_2^*e^{i\Delta kz}\,,\\ \frac{dA_2}{dz} &= -\frac{\alpha}{2}A_2 + i\kappa_2A_3A_1^*e^{i\Delta kz}\,,\end{aligned}$$

where $\quad \kappa_i = \dfrac{\omega_i d_{\text{eff}}}{n_i c}\,.$

Assume that the pump amplitude A_3 remains constant.

(a) Show that the solutions are

$$A_1(z)e^{\alpha z/2 - i\Delta kz/2} = A_1(0)\left[\cosh gz - i\frac{\Delta k}{2g}\sinh gz\right] + \frac{i\kappa_1 A_3}{g}A_2^*(0)\sinh gz\,,$$
$$A_2(z)e^{\alpha z/2 - i\Delta kz/2} = A_2(0)\left[\cosh gz - i\frac{\Delta k}{2g}\sinh gz\right] + \frac{i\kappa_2 A_3}{g}A_1^*(0)\sinh gz\,,$$

where

$$g = \sqrt{\kappa^2 - \frac{(\Delta k)^2}{4}}\,, \qquad \kappa^2 = \frac{2\omega_1\omega_2 d_{\text{eff}}^2}{\epsilon_o n_1 n_2 n_3 c^3} I_3\,.$$

(b) Now consider the case in which only A_2 is present at the input:

$$A_2(0) \neq 0\,, \qquad A_1(0) = 0\,.$$

Show that the intensity gain G, defined by $\frac{|A_2(L)|^2}{|A_2(0)|^2} = e^G$, is

$$G = -\alpha L + \ln\left[1 + \kappa^2 L^2 \left(\frac{\sinh gL}{gL}\right)^2\right]\,.$$

Bibliography

General References for Chaps. 1-4

[1] M. Sargent III, M. O. Scully, and W. E. Lamb, Jr., *Laser Physics* (Addison-Wesley, 1974).

[2] A. E. Siegman, *Lasers* (University Science Books, 1986).

[3] A. Yariv, *Quantum Electronics* (Wiley, 1989).

[4] G. P. Agrawal and N. K. Dutta, *Semiconductor Lasers* (Van Nostrand, 1993).

[5] J. T. Verdeyen, *Laser Electronics*, 3rd edition, (Prentice-Hall, 1995).

[6] L. Mandel and E. Wolf, *Optical Coherence and Quantum Optics* (Cambridge, 1995).

[7] M. O. Scully and M. S. Zubairy, *Quantum Optics* (Cambridge, 1997).

[8] B. E. A. Saleh and M. C. Teich, *Fundamentals of Photonics* (Wiley, 2007).

[9] P. W. Milonni and J. H. Eberly, *Lasers* (Wiley, 2010).

[10] O. Svelto, *Principles of Lasers* (Plenum, 2010).

Chapter 3

[11] H. Kogelnik and T. Li "Laser Beams and Resonators," Applied Optics **5**, 1550–1567 (1966).

[12] U. T. Schwarz, M. A. Bandres, and J. C. Gutierrez-Vega, "Observation of Ince-Gaussian modes in stable resonators," Opt. Lett. **29**, 1870–1872 (2004).

[13] J. C. Gutierrez-Vega, M. D. Iturbe-Castillo, G. A. Ramirez, E. Tepichin, R. M. Rodriguez-Dagnino, S. Chavez-Cerda, and G. H. C. New, "Experimental demonstration of optical Mathieu beams," Opt. Commun. **195**, 35–40 (2001).

[14] F. Gori, G. Guattari, and C. Padovani, "Bessel-Gauss beams," Opt. Commun. **64**, 491–495 (1987).

[15] A. Forbes, "Structured Light from Lasers," Laser Photonics Rev. **13**, 1900140 (2019).

[16] A. Zdagkas, C. McDonnell, J. Deng, Y. Shen, G. Li, T. Ellenbogen, N. Papasimakis, N. I. Zheludev, "Observation of Toroidal Pulses of Light," Nat. Photonics **16**, 523–528 (2022).

[17] A. W. Snyder and J. D. Love, *Optical Waveguide Theory* (Chapman & Hall, 1996).

[18] G. P. Agrawal, *Fiber-optic Communication Systems* (Wiley, 1997).

Chapter 6

[19] R. Loudon, *The Quantum Theory of Light*, (Oxford U. Press, 2000), Chap. 2.

[20] M. Sargent III, M. O. Scully, and W. E. Lamb, *Laser Physics* (Addison-Wesley, 1974), Chaps. 1-3.

[21] R. Loudon, *The Quantum Theory of Light*, (Oxford U. Press, 2000).

[22] G. Grynberg, A. Aspect, and C. Fabre, *Introduction to Quantum Optics: From the Semi-classical Approach to Quantized Light* (Cambridge, 2010).

Chapter 7

[23] F. P. Schäfer, ed., *Dye Laser* (Springer-Verlag, 1990).

[24] W. Koechner, *Solid-State Laser Engineering* (Springer, 2006).

[25] F. Trager, ed., *Handbook of Lasers and Optics* (Springer, 2007).

Chapter 8

[26] A. Siegman, *Lasers* (University Science Books, 1986), Chaps. 12& 13.

[27] J.T. Verdeyen, *Laser Electronics* (Prentice Hall, 1995), Chaps. 8 & 9.

[28] P. W. Milonni and J. H. Eberly, *Laser Physics* (Wiley, 2010), Chaps. 4 & 5.

[29] O. Svelto, *Principles of Lasers* (Springer, 1998), Chap. 7.

Chapter 9

[30] M. Sargent III, M. O. Scully, and W. E. Lamb, Jr., *Laser Physics* (Addison-Wesley, 1974)

[31] M. Scully and S. Zubairy, *Quantum Optics* (Cambridge U. Press, 2012)

[32] L. Mandel and E. Wolf, *Optical Coherence and Quantum Optics* (Cambridge U. Press, 1995)

Chapter 10

[33] W. E. Lamb, Jr., "Theory of an optical maser," Phys. Rev. **134**, A1429–A1450 (1964).

[34] C. L. O'Bryan, III and M. Sargent, III, "Theory of Multimode Laser Operation," Phys. Rev. A **8**, 3071–3092 (1973).

[35] M. Sargent III, M. O. Scully, and W. E. Lamb, Jr., *Laser Physics* (Addison-Wesley, 1974), Chaps. 9-12.

[36] A. Siegman, *Lasers* (University Science Books, 1986), Chaps. 25 & 26

[37] A. Yariv, *Quantum Electronics* (Wiley, 1989).

[38] O. Svelto, *Principles of Lasers* (Plenum, 1998), Chap. 8.

Chapter 11

[39] H. A. Haus, "Mode-locking of lasers," J. Sel. Top. Quantum Electron. **6**, 1173–1185 (2000).

[40] H.R. Telle, G. Steinmeyer, A.E. Dunlop, J. Stenger, D.H. Sutter, and U. Keller, "Carrier-Envelope Offset Phase Control: A Novel Concept for Absolute Optical Frequency Measurement and Ultrashort Pulse Generation," Appl. Phys. **B 69**, 327–332 (1999).

[41] T. Udem, J. Reichert, R. Holzwarth, and T. W. Hänsch, "Accurate measurement of large optical frequency differences with a mode-locked laser," Opt. Lett. **24**, 881–883 (1999).

[42] S. A. Diddams, D. J. Jones, J. Ye, S. T. Cundiff, J. L. Hall, "Direct Link between Microwave and Optical Frequencies with a 300 THz Femtosecond Laser Comb," Phys. Rev. Lett. **84**, 5102–5105 (2000).

[43] D. J. Jones, S. A. Diddams, J. K. Ranka, A. J. Stentz, R. S. Windeler, J. L. Hall, S. T. Cundiff, "Carrier-envelope phase control of femtosecond mode-locked lasers and direct optical frequency synthesis," Science **288**, 635–640 (2000).

[44] R. Holzwarth, Th. Udem, T. W. Hänsch, J. C. Knight, W. J. Wadsworth, P. St. J. Russell, "Optical frequency synthesizer for precision spectroscopy," Phys. Rev. Lett. **85**, 2264–2267 (2000).

[45] J. Ye and S. Cundiff, *Femtosecond Optical Frequency Comb: Principle, Operation and Applications* (Springer, 2005).

[46] J. Diels and W. Rdolph, *Ultrashort Laser Pulse Phenomena* (Elsevier, 2006).

[47] A. M. Weiner, *Ultrafast Optics* (Wiley, 2009).

[48] U. Keller, *Ultrafast lasers* (Springer-Verlag, 2022).

Chapter 12

[49] J. A. Armstrong, N. Bloembergen, J. Ducuing, P. S. Pershan, "Interactions between Light Waves in a Nonlinear Dielectric," Phys. Rev. **127**, 1918–1939 (1962).

[50] G. D. Boyd and D. A. Kleinman, "Parametric interaction of focused Gaussian beams," J. Appl. Phys. **39**, 3597–3639 (1968).

[51] H. Cruz-Ramirez, R. Ramirez-Alarcon, M. Corona, K. Garay-Palmett and A. B. U'Ren, "Spontaneous parametric processes in modern optics," Optics & Photonics News **37**, 36–41 (2011).

[52] C. Lin and R. H. Stolen, "New nanosecond continuum for excited-state spectroscopy," Appl. Phys. Lett. **28**, 216–218 (1976).

[53] J. K. Ranka, R. S. Windeler, A. J. Stentz, "Visible continuum generation in air–silica microstructure optical fibers with anomalous dispersion at 800 nm", Opt. Lett. **25**, 25–27 (2000).

[54] F. Zernike, J. E. Midwinter, *Applied Nonlinear Optics* (Wiley, 1973).

[55] R. R. Alfano, ed. *The Supercontinuum Laser Source: Fundamentals with Updated References* (Springer-Verlag, 2006).

[56] J. M. Dudley and J. R. Taylor, eds. *Supercontinuum Generation in Optical Fibres* (Cambridge University Press, 2010).

[57] G. I. Stegemann and R. A. Stegemann, *Nonlinear optics* (Wiley, 2012) Chaps. 2-4.

[58] R. Boyd, *Nonlinear Optics* (Academic Press, 2020), Chaps. 1 and 2.

Index

For Product Safety Concerns and Information please contact our
EU representative GPSR@taylorandfrancis.com Taylor & Francis
Verlag GmbH, Kaufingerstraße 24, 80331 München, Germany